AF252173

NANOCOMPOSITE THIN FILMS AND COATINGS

Processing, Properties and Performance

NANOCOMPOSITE THIN FILMS AND COATINGS

Processing, Properties and Performance

editors

Sam Zhang
Nanyang Technological University, Singapore

Nasar Ali
University of Aveiro, Portugal

Imperial College Press

Published by

Imperial College Press
57 Shelton Street
Covent Garden
London WC2H 9HE

Distributed by

World Scientific Publishing Co. Pte. Ltd.

5 Toh Tuck Link, Singapore 596224

USA office: 27 Warren Street, Suite 401-402, Hackensack, NJ 07601

UK office: 57 Shelton Street, Covent Garden, London WC2H 9HE

British Library Cataloguing-in-Publication Data
A catalogue record for this book is available from the British Library.

Cover photo: HRTEM image (by Y. T. Pei) of magnetron sputtered nc-TiC/a-C (by X. L. Bui)

NANOCOMPOSITE THIN FILMS AND COATINGS

ISBN-13 978-1-86094-784-1
ISBN-10 1-86094-784-0

Typeset by Stallion Press
Email: enquiries@stallionpress.com

Printed by FuIsland Offset Printing (S) Pte Ltd, Singapore

CONTENTS

Chapter 1

Magnetron Sputtered Hard and Yet Tough Nanocomposite Coatings with Case Studies: Nanocrystalline TiN Embedded in Amorphous SiN$_x$

Sam Zhang, Deen Sun and Xuan Lam Bui

Chapter 2

Magnetron Sputtered Hard and Yet Tough Nanocomposite Coatings with Case Studies: Nanocrystalline TiC Embedded in Amorphous Carbon

Sam Zhang, Xuan Lam Bui and Deen Sun

Chapter 3

Properties of Chemical Vapor Deposited Nanocrystalline Diamond and Nanodiamond/Amorphous Carbon Composite Films   167

S. C. Tjong

Chapter 4

Synthesis, Characterization and Applications of Nanocrystalline Diamond Films   207

Zhenqing Xu and Ashok Kumar

Chapter 5

Properties of Hard Nanocomposite Thin Films   **281**

J. Musil

Chapter 6

Nanostructured, Multifunctional Tribological Coatings   **329**

*John J. Moore, In-Wook Park, Jianliang Lin, Brajendra Mishra
and Kwang Ho Kim*

Chapter 7

Nanocomposite Thin Films for Solar Energy Conversion   **381**

Yongbai Yin

Chapter 8

Application of Silicon Nanocrystal in Non-Volatile Memory Devices

419

T. P. Chen

Chapter 9

Nanocrystalline Silicon Films for Thin Film Transistor and Optoelectronic Applications

473

Youngjin Choi, Yong Qing Fu and Andrew J. Flewitt

Chapter 10

Amorphous and Nanocomposite Diamond-Like Carbon Coatings for Biomedical Applications   513

T. I. T. Okpalugo, N. Ali, A. A. Ogwu, Y. Kousar and W. Ahmed

Chapter 11

Nanocoatings for Orthopaedic and Dental Application   573

Weiqi Yan

CHAPTER 1

MAGNETRON SPUTTERED HARD AND YET TOUGH NANOCOMPOSITE COATINGS WITH CASE STUDIES: NANOCRYSTALLINE TiN EMBEDDED IN AMORPHOUS SiN$_x$

Sam Zhang*, Deen Sun and Xuan Lam Bui

School of Mechanical and Aerospace Engineering
Nanyang Technological University
50 Nanyang Avenue, Singapore 639798
**msyzhang@ntu.edu.sg*

1. Introduction

Nanocomposite thin films comprise at least two phases, a nanocrystalline phase and a matrix phase, where the matrix can be either nanocrystalline or amorphous phase. The general characteristics of nanocomposite coating are a host material with another material homogenously embedded in it, with one (or both) of these materials having a characteristic length scale of 1–100 nm as schematically illustrated in Fig. 1.1.

An example is given in Fig. 1.2 where 10~20 nm (TiCr) $C_x N_y$ crystals were embedded into a diamond-like carbon (DLC) matrix to reach a hardness of 40 GPa [1].

Nanocomposite thin films represent a new class of materials which exhibit special mechanical [2–4], electronic [5, 6] magnetic [7] and optical properties due to their size-dependent phenomena [8–10]. Recently it attracted increasing interest due to the endless possibilities of the synthesizing materials of unique properties. By convention, hard materials usually refer to materials with Vickers hardness less than 40 GPa, and superhard materials with Vickers hardness exceeding 40 GPa. Hard and superhard two- and multiphase thin films exhibit high hardness significantly exceeding that given by the rule of mixture, i.e.

$$H(A_a B_b) = \frac{aH(A) + bH(B)}{a + b},\qquad(1.1)$$

1

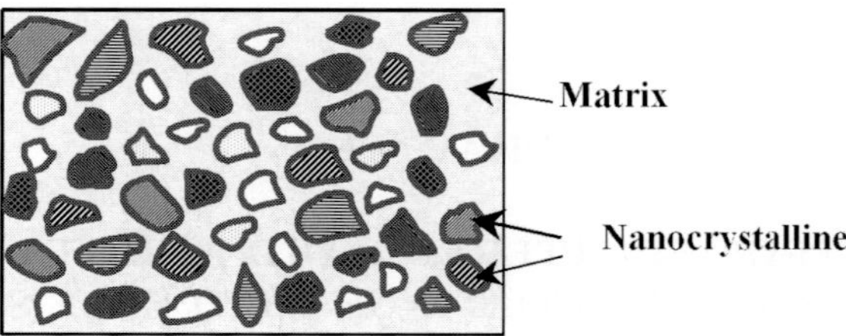

Fig. 1.1. Schematic diagram of nanocomposite coating microstructure, showing nanocrystalline phase embedded in matrix.

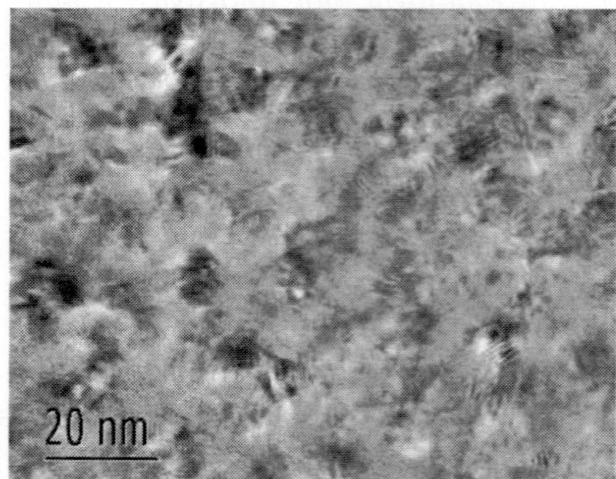

Fig. 1.2. Transmission electron microscopy (TEM) photo of TiCrCN nanocomposite film showing nanosize TiCrCN crystals embedded into the amorphous DLC matrix [1].

where $H(A)$ is the hardness of pure A, $H(B)$ the hardness of pure B. a is the composition of A in the mixture and b is the composition of B in the mixture. $H(A_a B_b)$ is the hardness of the mixture.

The use of coated materials in engines, machines, tools and other wear-resistant components is steadily increasing and has achieved a high level of commercial success, compared to the common non-coated materials such as steel [11]. Wear-resistant, superhard thin films for high speed dry machining would allow the industry to increase the productivity of expensive automated machines and to save on the high costs presently needed for environmentally hazardous coolants. Depending on the kind of machining, the recycling costs of these coolants amount to 10–40% of the total machining costs. For example, in Germany alone, these costs can be close to one billion US dollars per year [11, 12].

2. Deposition

2.1. *Design of Microstructure*

In conventional bulk materials, refining grain size is one of the possibilities for hardness increase. The same is true for films or coatings. With a decrease in grain size, the multiplication and mobility of the dislocations are hindered, and the hardness of materials increases according to the Hall–Petch relationship [13, 14]:

$$H(d) = H_0 + Kd^{-1/2}, \qquad (2.1)$$

where H is hardness, K is a constant and d the grain size. This effect is especially prominent for grain size down to tens of nanometers. However, dislocation movement, which determines the hardness in conventional materials, has little effect when the grain size is less than approximately 10 nm. At this size scale, further reduction in grain size brings about a decrease in hardness (Fig. 2.1) because of grain boundary sliding [15]. Softening caused by grain boundary sliding is mainly attributed to large amount of defects in grain boundaries, which allow fast diffusion of atoms and vacancies under

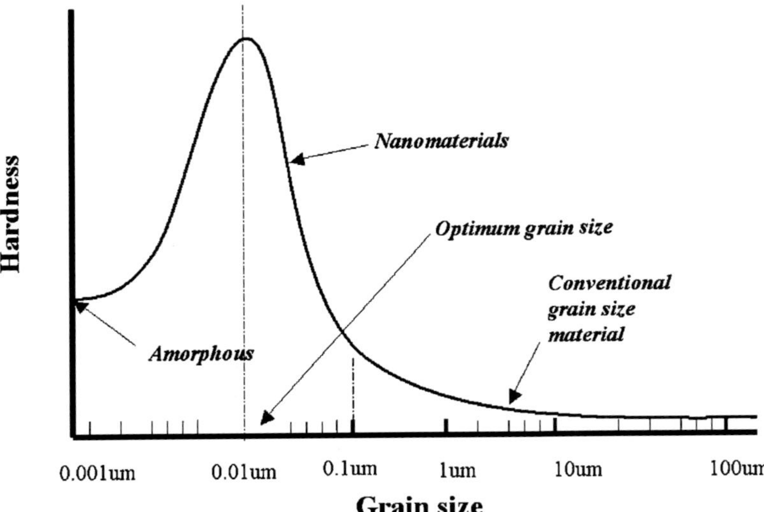

Fig. 2.1. Hardness of material as a function of grain size. When grain size is greater than the optimum value, with a decrease in grain size, hardness increases (Hall–Petch relationship). When grain size is less than the optimum value, with a decrease in grain size, hardness decreases (anti-Hall–Petch relationship) [16].

stress [16, 17]. As such, further increase in hardness requires hindering of grain boundary sliding. This can be realized through proper microstructural design, i.e. by increasing the complexity and strength of grain boundaries [18]. Multiphase structures are expected to have interfaces with high cohesive strength, since different crystalline phases often exhibit different sliding systems and provide complex boundary to accommodate a coherent strain, thus preventing the formation of voids or flaws [19]. A variety of hard materials can be used in nanocomposite coating microstructural design. Figure 2.2 shows potential hard materials.

Apart from hardness, good mechanical properties also include high toughness. High toughness can be obtained in nanocomposite thin films through the nanosize grain structure as well as deflection, meandering and termination of nanocracks [21]. Veprek proposed a design concept for novel superhard ceramic/ceramic nanocomposite coatings with high toughness [2, 12, 22]. In this design, multiphase structure is used to maximize the interface complexity and ternary or quaternary systems with strong tendency of segregating into binary compounds used to form sharp and strong interface in order to avoid grain boundary sliding [15]. The crystallite size is controlled to approximately 3–4 nm and the separation distance between crystallites maintained at less than 1 nm. Based on this design concept, Veprek and co-workers prepared nc-TiN/a-Si_3N_4/a- & nc-$TiSi_2$ [23],

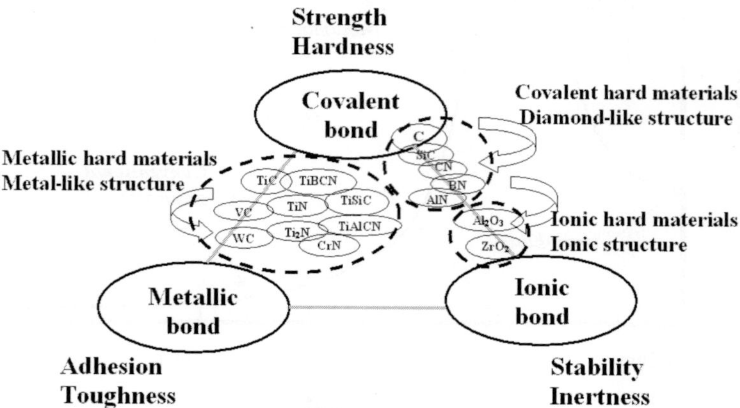

Fig. 2.2.   Hard materials for nanocomposite coatings in the bond triangle and changes in properties with the change in chemical bonding. They include: (1) covalent bonding materials with high hardness and high temperature strength, (2) metallic bonding materials with good adhesion and toughness, and (3) ionic bonding materials with good stability and inertness [20].

nc-TiN/a-Si$_3$N$_4$ [24], nc-W$_2$N/a-Si$_3$N$_4$ [25], nc-VN/a-Si$_3$N$_4$ [26], nc-TiN/a-BN [27] and nc-TiN/a-BN/a-TiB$_2$ [28] superhard nanocomposite coatings by means of plasma CVD. In nc-TiN/a-Si$_3$N$_4$/a- & nc-TiSi$_2$ nanocomposite coating system, TiN nanocrystals were embedded in amorphous Si$_3$N$_4$ (existing in grain boundary). Also existing in grain boundaries were amorphous TiSi$_2$ and crystallite TiSi$_2$ (Fig. 2.3). An ultrahigh hardness (H_v exceeding 100 GPa [23]) was obtained for this system. In addition, the indentation test did not induce microcracks, indicating good toughness. The authors attributed the toughness enhancement to the lack of large stress concentration at the tip of the crack due to the nanoscale of the crystals. In fact, the stress concentration factor at the tip can be estimated through [29]

$$\frac{\sigma_{\text{tip}}}{\sigma_{\text{applied}}} = 1 + 2\sqrt{\frac{c/2}{r}}, \qquad (2.2)$$

where c is crack length and r crack tip radius. For $c/2 = 1$–2 nm and $r = 0.2$–0.3 nm (one atomic bond length), the stress concentration factor is only 4–6 (according to Eq. (2.2)), a very small number compared to 30–100 in conventional microstructure [23]. The crack propagation is

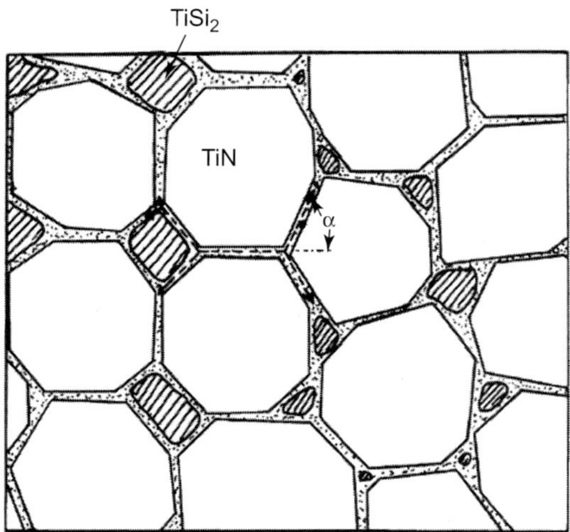

Fig. 2.3. Schematic diagram of the nanostructure of nc-TiN/a-Si$_3$N$_4$/a- & nc-TiSi$_2$ nanocomposite. TiN nanocrystals were embedded in amorphous Si$_3$N$_4$ (existing in grain boundary). Also existing in grain boundaries were amorphous TiSi$_2$ and crystallite TiSi$_2$. TiN crystallite size is approximately 3–4 nm and the separation distance between crystallites is less than 1 nm [23].

also strongly hindered in the nanocomposite structure due to the absence of dislocation activity in crystallite of 3–6 nm in size. In this case, the crack can propagate only along grain boundaries. When the crack size approaches that of the crystallite, the crack has to undergo bending or branching. After that, only the component of the applied stress normal to the plane of the bent or branched crack may cause further propagation. Thus the stress that drives crack propagation decreases, resulting in a deceleration of crack propagation. In Veprek's design, two immiscible nitrides (nc-TiN and a-Si_3N_4) were used [30, 31] to achieve thermal stability. This, however, may degrade the cohesive strength of the interface between the crystal and boundary [3]. When local tensile stress at the crack tip is high enough, unstable crack propagation eventually results [32].

In order to obtain superhardness, usually plastic deformation is strongly prohibited, and dislocation movement and grain boundary sliding are prevented, thus probably causing a loss in ductility. Today, more and more researchers realize that a certain degree of grain boundary sliding is necessary in order to improve toughness of nanocomposite coatings. Usually, to overcome the brittleness of ceramic bulk materials, a second ductile phase is incorporated to improve the toughness of the composite [33–35]. In recent years, the microstructure of these composites has been further refined by the addition of nanometer-sized metal particles [36, 37]. This should also apply to thin films and coatings. Musil and co-workers embedded crystalline nitrides in metallic Cu [38, 39], Ni [40–42] and Y [43], etc. In these coatings, the crystallite size of the ceramic phase is normally controlled less than 10 nm, and the volume of the boundary is greater than that of hard phase [44]. The hardness of these coating systems varies from 35 GPa to approximately 60 GPa. The existence of metal matrix is expected to improve toughness by crack tip blunting and/or the increase of the work of plastic deformation. In such a design, however, the metallic grain boundary thickness cannot be too thin: a very thin grain boundary renders the toughening mechanism ineffective because in the case of nanoscale grains, dislocation movement is restricted [45]. Should this happen, the mechanical response of the ceramic/metal nanocomposite will be effectively similar to that of ceramic/ceramic nanocomposite (which will defeat the purpose of incorporating the soft metallic phase). Another way of enhancing toughness is to allow grain boundary sliding to take place (rather than cracking) to release the accumulated strain. Voevodin *et al.* [45, 46] embedded hard nanocrystalline carbide of 10–20 nm in amorphous carbon (a-C) matrix. Crystallite size of this magnitude can restrict initial crack size and create a large volume of grain boundaries [1]. The thickness of amorphous boundary should be maintained above 2 nm to prevent interaction of atomic planes in the adjacent grains and to facilitate grain boundary sliding, but less than 10 nm to restrict path of a straight crack. As a result, nc-TiC/a-C and nc-WC/a-C nanocomposite thin films achieved a "scratch toughness" [47] 4–5 folds that of the nanocrystalline carbide alone, at expense of some hardness.

2.2. *Synthesis of Thin Films*

Nanocomposite thin films can be prepared by chemical vapor deposition (CVD) or physical vapor deposition (PVD) techniques (Table 2.1). In both CVD and PVD methods, one of the most critical deposition factors is the kinetic energy of the vapor phase particle, which can generally be divided according to the range of typically reported energies [48], into three regimes as shown in Fig. 2.4:

1. *Thermal regime* $(0 \sim 0.3\,\text{eV})$, in which particles have low energy. Techniques within this range include chemical vapor deposition and thermal evaporation;
2. *Mediate regime* $(1 \sim 100\,\text{eV})$, in which particles have energies ranging from the bonding strength to the lattice penetration threshold. Techniques within this range include sputtering deposition, and arc vapor deposition;
3. *Implantation regime* $(>100\,\text{eV})$ in which particles energies are well able to cause surface penetration and implantation. Technique within this range includes ion implantation.

Table 2.1. Main preparation methods for nanocomposite thin films.

Group	Sub-group	Methods
Physical vapor deposition (PVD)	Thermal evaporation (TE)	Pulsed laser deposition (PLD) Electron beam deposition (EB-PVD)
	Sputter deposition	Magnetron sputtering Ion beam sputtering
	Arc vapor deposition	Vacuum arc deposition Filtered arc deposition
	Ion implantation	Ion beam deposition (IBD)
Chemical vapor deposition (CVD)		Plasma enhanced CVD (PECVD) Plasma assistant CVD (PACVD) Electron cyclotron resonance CVD (ECR-CVD)

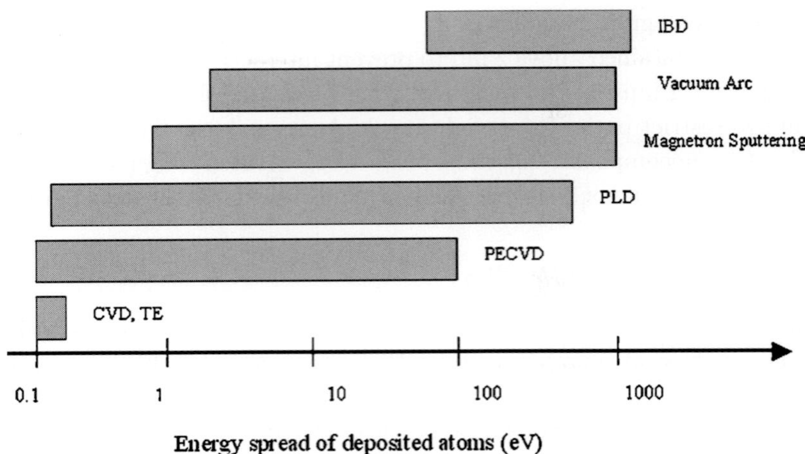

Fig. 2.4. Approximate energy range (in electron volts) of deposited particles produced by selected deposition methods.

Different techniques are now available for the preparation of nano-composite thin films. The most promising methods are chemical vapor deposition (CVD) [49, 50] and magnetron sputtering [51], although other methods such as laser ablation, thermal evaporation, ion beam deposition and ion implantation are also used by various researchers [52]. High deposition rate and uniform deposition for complicated geometries are the advantages of the CVD method compared to magnetron sputtering. However, the main concern for the CVD method is that the precursor gases, $TiCl_4$, $SiCl_4$ or SiH_4, may pose problems in production because they are corrosive in nature and are fire hazards. Moreover, the incorporation of chloride in protective films may induce interface corrosion problems during exposure to elevated temperatures under working condition. For most applications, a low deposition temperature is required to prevent substrate distortion and loss of mechanical properties. This is difficult to realize in the CVD process.

At present, significant effort has been devoted to the preparation of nanocomposite thin films using magnetron sputtering since this technology is a low temperature and far less dangerous method compared to CVD [53]. Also, it is easily scalable for industrial applications. In magnetron sputtering, energetic ion bombardment is used to vaporize the source material, often referred to as the target, as illustrated in Fig. 2.5. The deposition system is filled with a noble gas, often argon, to a total pressure of 0.01 to 0.1 mbar. A negative potential of some kV is applied to the target. Positive

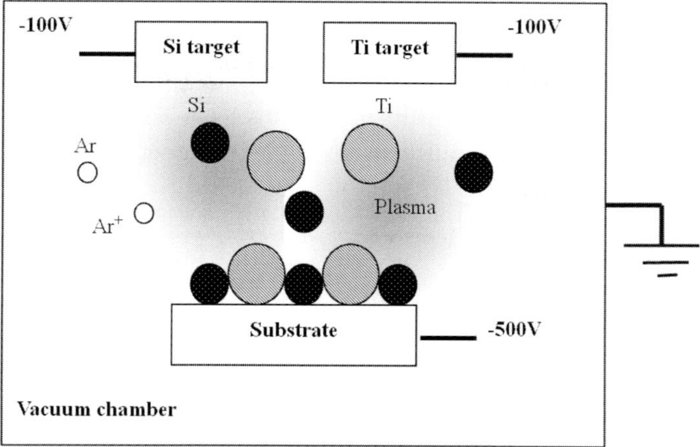

Fig. 2.5.   Schematic diagram of the sputtering process. Argon gas is ionized by electrons within the plasma. The argon ions are attracted and accelerate towards the target surface where they result in sputtering.

ions naturally occurring in the gas will therefore be accelerated towards the target. When they impinge on the target, they transfer their momentum to surface atoms of the target, and if the value of the momentum in both directions is higher than the surface binding energy, a target atom will be ejected, i.e. sputtered. This ejected flux of target atoms, which has a main direction, is then transported through the space towards the substrate. Depending on the gas pressure and the distance between substrate and target, the flux will be more or less scattered by the gas. The average distance an atom can travel before a collision is called the mean free path. The mean free path l_m can be estimated through [54]

$$l_m = \frac{kT_g}{\sqrt{2}\pi P_g d_g^{\,2}}, \qquad (2.3)$$

where k is the Boltzmann constant, T_g and P_g the gas temperature and pressure respectively, and d_g the diameter of the gas molecule ($d_{Ar} = 0.364\,\mathrm{nm}$). During sputtering process, the film surface is ion bombarded, which can densify the growing film by enhancing the surface atom mobility. In addition, ion bombardment of the growing film can restrict the grain growth and permit the formation of nanocrystalline. The size and crystallographic orientation of grains can be controlled by the energy of bombarding ions. Kinetic energy of ionized particles can be estimated by [55–57]:

$$U_k \propto \frac{D_w V_s}{P_g^{\,1/2}}, \qquad (2.4)$$

where U_k is the kinetic energy, D_w the target power density, V_s the substrate bias and P_g the gas pressure.

E303A magnetron sputtering system as shown in Fig. 2.6 was conducted to deposit nanocomposite thin films. The system includes four 4-inch planar high performance water-cooled magnetrons and a heated

(a)

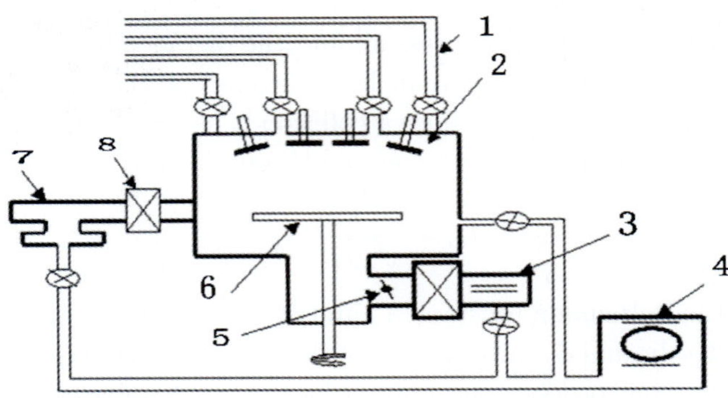

(b)

Fig. 2.6.   Sputtering system used in this project: (a) the outlook, and (b) the schematic structure of the main chamber. There are four individual targets in the chamber and co-sputtering can be conducted.

rotatable substrate holder. The substrate-to-target distance was 100 mm to fully reactive between the working gas and sputtered target atoms. The substrate can be heated to a maximum temperature of 500°C. Two 600 W RF generators and two 1 kW DC power supply can be automatically selected to individual targets. Also, the substrate holder can be earthed, floated or powered by a dedicated 600 W RF generator. This facilitates pre-sputter wafer cleaning and reverse bias sputter etching. The vacuum system is of high specification, with 800 l/s cryogenic, ultrahigh vacuum and high performance rotary backing pump. Gas flow control is via mass flow control valves on three gas lines, with pressure control set point achieved by automatic close loop throttle valve control. Reactive sputtering is assisted by inert argon gas to avoid target poising and to allow higher sputtering rate. The system is designed to coat multiple 2–4 inch wafers, and single 6–8 inch wafer at one time. A load lock is fitted with an automatic wafer transfer arm. The dominant parameters during deposition are target power density, deposition temperature, substrate bias an gas ratio.

3. Characterization

Film characterization is an inevitable and vital step in ensuring high quality film for the intended application. Different characterization techniques can be used to identify nanocomposite thin films; X-ray photoelectron spectroscopy (XPS), Auger electron spectroscopy (AES), Rutherford back scattering spectroscopy (RBS) and energy dispersive X-ray analysis (EDX) are powerful tools for characterizing the chemical composition [58], each with different penetrations and accuracies, as illustrated in Table 3.1.

3.1. *Composition*

For composition characterization, the most useful tool is X-ray photoelectron spectroscopy (XPS). In XPS, if we measure the energy of the ejected

Table 3.1.   Composition analysis methods [58].

Analysis methods	Elemental range	Detection limits (at.%)	Spatial resolution	Penetration
XPS	Li–U	0.1–1	100 μm	1.5 nm
AES	Li–U	0.5	10 nm	0.5–7.5 nm
RBS	Li–U	1.0	1–4 mm	2–30 nm
EDX	Be–U	0.1	0.5–2.0 μm	1–3 μm

photoelectrons we can calculate the binding energy, which is the energy required to remove the electron from its atom:

$$E_{\mathrm{B}} = h\nu - E_{\mathrm{K}} - W, \qquad (3.1)$$

where E_{B} is binding energy, hv the photon energy, E_{K} the kinetic energy of the electron, and W the spectrometer work function. From the binding energy we can learn some important facts about the sample under investigation [59]:

- the relative quantity of each element;
- the elements from which it is made;
- the chemical state of the elements present;
- depth distribution (profile).

In the calculation of the relative quantity of each element, the principle is as follows: the complete XPS spectrum of a material contains peaks that can be associated with the various elements (except H and He) present in the outer 10 nm of that material. The area under these peaks is related to the amount of each element present. Therefore, by measuring the peak areas and correcting them for the appropriate instrumental factors, the percentage of each element detected can be determined. The equation that is commonly used for the calculation is [60]:

$$I_{ij} = K^* T(E_{\mathrm{K}})^* L_{ij}(\gamma)^* \sigma_{ij}{}^* n_{ij}{}^* \lambda(E_{\mathrm{K}}) \cos\theta, \qquad (3.2)$$

where I_{ij} is the area of peak j from element i, K the instrumental constant, $T(E_{\mathrm{K}})$ the transmission function of analyzer, L_{ij} the angular asymmetry factor of obital j of element i, and σ_{ij} the photoionization cross-section of peak j from element i. E_{K} is the kinetic energy of the emitted electron, θ the take-off angle of the photoelectrons measured with respect to surface normal, and n_i the concentration of element i.

There are several factors which affect the binding energy, therefore affecting the composition determination. For conducting samples, it is the work function of the spectrometer (W) that is important. This can be calibrated by placing a clean Au standard in the spectrometer and adjusting the instrumental settings such that the known E_{B} values for Au are obtained. However, some materials do not have sufficient electrical conductivity or cannot be mounted in electrical contact with the spectrometer. These samples require an additional source of electrons to compensate for the positive charge built up by the emission of photoelectrons. Therefore, the measured E_{B} of an insulated sample depends on its work function and the energy of the compensative flooding electrons. Under these conditions it is best to use an internal reference. Generally speaking, all the factors that affect the E_B can affect the composition determination.

Chemical composition of as-prepared thin films was determined by XPS analysis using a Kratos-Axis spectrometer with monochromatic Al K$_\alpha$ (1486.71 eV) X-ray radiation (15 kV and 10 mA) and hemispherical electron energy analyzer. The base vacuum of the chamber was 2.67×10^{-7} Pa. The survey spectra in the range of 0–1100 eV were recorded in steps of 1 eV for each sample, followed by high resolution spectra over different element peaks in steps of 0.1 eV, from which the detailed composition was calculated. The spectra were referenced to the C 1s line of 284.6 eV [61]. Curve fitting was performed after a Shirley background [62] subtraction by nonlinear least square fitting using a mixed Gauss/Lorentz function. In the least square fitting analysis of Ti 2p spectra, the area ratio of the $2p_{3/2}$ to $2p_{1/2}$ envelope was kept constant at two with a constant energy difference of 5.8 eV. The parameters used in fitting the Ti 2p, Si 2p and Ni 2p spectra are listed in Table 3.2. Also listed are binding energies available in the literature. Sputter depth profiles of the films were obtained by recording the XPS spectra after sputtering with an accelerating voltage of 4 keV Ar ion beam. The bombardment was performed at an angle of incidence of 45° with respect to the surface normal. The sputter rate determined on a 30 nm thick SiO_2 sample was 3.0 nm/min. Half of the intensity of the oxygen plateau was taken as the measure of the oxide layer thickness. (It is useful to note that the thickness of the specimen studied can only be roughly estimated from the thickness and sputtering rate of the silicon dioxide standard, because the sputtering yield of the elements in the specimen differs substantially from that in the standard.)

3.2. *Topography*

Topography or surface morphology of the as-deposited film was characterized using atomic force microscopy (AFM) (Shimadzu SPM-9500J2). The measurement was conducted in ambient atmosphere in contact mode with a Si_3N_4 tip. The scan resolution is 256 pixels × 256 pixels, set point 2.000 V and scan rate 1.000 Hz. In order to quantitatively describe the surface morphology, scaling theory [75, 76] was used to analyze the AFM roughness data. The height–height correlation function $G(r)$ was defined as [77]

$$G(r) \equiv \langle [h(x, y) - h(0, 0)]^2 \rangle. \tag{3.3}$$

Table 3.2.   Reported binding energy (eV) values for Ti 2p, Si 2p and Ni 2p photoelectron spectra.

| Element | Ti $2p_{3/2}$ | | | | Si 2p | | | | Ni 2p | | | |
| | | | | | | | | | Ni $2p_{1/2}$ | Satel. peak | Ni $2p_{3/2}$ | |
State	TiO_2	TiN_xO_y	TiN	Ref	SiO_2	Si_3N_4	Free-Si	Ref		Metallic Ni		Ref
Position(eV)	458.0	456.5	455.0	[63]	103.4	101.9	99.6	[64]	870.7	859.3	852.8	[65]
	459.1 ± 0.2	~ 457.0	455.2 ± 0.2	[66]	103.6	101.6		[67]	870.7	859.6	853.0	*
	459.2	457.5	455.2	[68]	$103 \sim 104$		99.6	[69]				
	458.8	457.1	455.6	[70]	103.3	101.8	99.3	[71]				
	459.0	457.5 ± 0.1	455.2	[72]	103.4	101.8	99.6	*				
	458.0	456.4	455.0	[73]								
	458.7	456.9	454.8	[74]								
	459.0	457.6	455.0	*								

*Present work.

Thus the dynamic scaling hypothesis [78] suggests that:

$$G(r) = \begin{cases} r^{2\alpha} & \text{as } r \ll \xi \\ 2\omega^2 & \text{as } r \gg \xi \end{cases}, \tag{3.4}$$

where the interface width ω describes the vertical growth and the lateral correlation length ξ characterizes the lateral growth.

3.3. *Microstructure*

The microstructure of the nanocomposite films was characterized using TEM (JEOL JEM 2010). The acceleration potential was 200 kV and the camera constant changed from 1.5 to 2.5 nm × mm. An image analyzer was used to measure the crystalline size and amount from TEM micrographs. The size of the crystallite was measured from the TEM micrographs using the image analyzer by measuring two perpendicular dimensions of a crystallite. The average of these two-dimensional measurements was taken as the crystallite size. The amount of the crystalline phase, termed crystalline fraction, was estimated as the area ratio of the nanocrystalline phase to the total image area. For each sample, a large number (at least 10) of TEM micrographs were taken and the images analyzed for the average crystalline fraction. Though this method results in approximately 10% uncertainties due to a contrast problem, it does give a first-degree approximation with a visual advantage. The sample preparation procedures for TEM/HRTEM study are as follows.

1. Prepare potassium bromide (KBr) tablets with a diameter of 10 mm and thickness of 3 mm. The roughness (R_a) of the tablet is measured as 80 μm.
2. Deposit nc-TiN/a-SiN$_x$ and Ni-toughened nc-TiN/a-SiN$_x$ nanocomposite thin films on different potassium bromide (KBr) tablets for 20 min for a thin film of about 100 nm in thickness. Other deposition parameters are kept the same as its counterpart which deposited on silicon wafer for 120 min.
3. Float the as-prepared nanocomposite thin films off from the KBr tablets by dissolving the coated tablets in de-ionized water followed by scooping the films out onto a copper grid.
4. Dry the films for two hours for the TEM/HRTEM study.

To supplement the TEM results, XRD was used for microstructure and phase identification [Philips PW1830 with Cu tube anode ($\lambda = 0.15418$ nm) at 30 kV and 20 mA]. The step size was 0.01° and step time 0.5 seconds. In order to reduce the interference from the substrate in case of thin films, GIXRD (Rigaku MAX 2000 with Cu K$_\alpha$ radiation) was conducted at scan rate of 2°/min, step size 0.032° and incident angle from 0.1° through 0.5° and 1.0° to 1.5°. Based on the XRD/GIXRD data, grain size, preferential orientation and the lattice parameter were obtained.

Grain Size

To estimate the grain size d, the well-known Scherrer formula [79] was used by measuring the full width at half maximum (FWHM) of the XRD/GIXRD peak at the angle of interest:

$$d = \frac{C\lambda}{\beta \cos\theta},$$ (3.5)

where C is a constant ($C = 0.91$), d the mean crystallite dimension normal to diffracting planes, and λ the X-ray wavelength. β in radians is the peak width at half maximum peak height, and θ is the Bragg angle. It should be noted that the Scherrer formula does not take into account the peak broadening induced by microstraining.

Preferential Orientation

The degree of preferential orientation was quantitatively represented through a coefficient of texture T_{hkl}, defined as [80]:

$$T_{hkl} = \frac{I_m(hkl)/I_0(hkl)}{\frac{1}{n} \sum_1^n \left[I_m(hkl)/I_0(hkl)\right]},$$ (3.6)

where $I_m(hkl)$ is the measured relative intensity of the reflection from the (hkl) plane, $I_0(hkl)$ is that from the same plane in a standard reference sample and is listed in Table 3.3 for TiN. n is the total number of reflection peaks from the coating. In the present study, $n = 3$, since only three major peaks are selected (i.e. (111), (200) and (220) diffraction plane). For the extremely preferential orientation, $T_{hkl} = 3$, while for the random one, $T_{hkl} = 1$.

Table 3.3. Relative intensity of XRD peaks in a standard reference TiN sample (JCPDS 38-1420).

Peaks	(111)	(200)	(220)
Relative intensity	72	100	45

Lattice Parameter

Lattice parameter of the solid solution for cubic structure was calculated according to peak positions or the distance between the crystal planes through [81]:

$$a_{\exp} = d_{hkl}\sqrt{h^2 + k^2 + l^2} = f(\theta), \tag{3.7}$$

$$a_{\exp} = a_{\text{true}} + \Delta a = a_{\text{true}} + C\Phi(\theta), \tag{3.8}$$

$$\Phi(\theta) = \frac{\cos^2\theta}{2}\left(\frac{1}{\sin\theta} + \frac{1}{\theta}\right), \tag{3.9}$$

where $a_{\exp}$ is the experimentally calculated lattice parameter, a_{true} the true value, Δa the experiment error, d_{hkl} the interplanar spacing, and (hkl) the crystal plane indices. C is a constant, θ is the Bragg angle, and Δa is 0 when $\Phi(\theta)$ is 0. Therefore plotting $a_{\exp}$ against $\Phi(\theta)$ and extrapolating to $\Phi(\theta) = 0$ gives rise to the true lattice parameter.

3.4. Mechanical Properties

3.4.1. Hardness

The materials handbook defines hardness as the resistance of materials to plastic deformation, usually by indentation. Hardness can be calculated through [19]:

$$H = \frac{P}{A}, \tag{3.10}$$

where P is the test indentation load and A the contact area. Microindentation and nanoindentation can be carried out to measure thin film hardness. The effect of the substrate on determining the mechanical properties of films using indentation has been discussed in detail by Saha and Nix [82, 83]. It is widely accepted that when the indentation depth is less than one tenth of the thickness of the thin film, the effect of substrate on the film hardness can be neglected [84]. In microindentation, the tester applies the selected test loads from several hundred mN to about 10 N using deadweights. The indentations are typically made using a Vickers indenter which is a regular pyramid made of diamond with an angle of 136° between the opposite faces. The hardness value of thin film as thin as a few microns can be calculated using the test load and the contact area. Whereas in the nanoindentation test, features less than 100 nm across, and thickness as thin as less than 5 nm, can be evaluated [58, 85]. In contrast to microindentation where the indentation force ranges from several hundred mN to about $10\,\mathrm{N}$, nanoindentation uses only $100\,\mu\mathrm{N}$ to a few mN. Nanoindentation can be carried out with a sharp indenter. The Berkovich indenter with an angle of $115°$ is useful when one wishes to probe properties at the smallest possible scale [86]. During the indentation, the indenter is forced into the surface at a selected rate and to a selected maximum force or a selected maximum indentation depth. A load-displacement curve is obtained and the depth of the indentation is measured to evaluate the actual contact area during indentation which yields the hardness and elastic properties.

Today, the most widely used hardness measurement is that of Oliver–Pharr method [84]. This method assumes that there is always some downward elastic deflection or "sink-in" situation in the material surface upon indentation (Fig. 3.1).

In this method, a small indentation is made with a Berkovich indenter, and displacement (or penetration depth) h, is continuously recorded during a complete cycle of loading and unloading and therefore a load-displacement curve is obtained as illustrated in Fig. 3.2. The contact depth h_c is estimated from the load-displacement curve through

$$h_c = h_{\mathrm{max}} - \varepsilon \frac{P_{\mathrm{max}}}{S},\tag{3.11}$$

where h_{max} is the maximum indentation depth, P_{max} the peak indentation load, S the contact stiffness, and ε a constant which depends on the indenter geometry. Empirical studies have shown that $\varepsilon = 0.75$ for a Berkovich indenter. The contact area A is estimated by evaluating an indenter shape

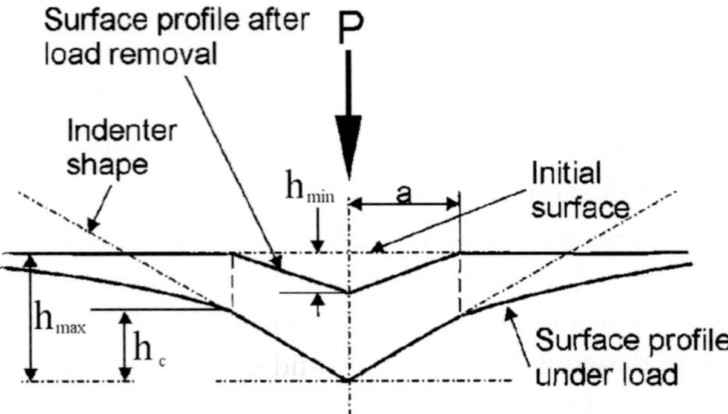

Fig. 3.1. Schematic representation from a cross-section through an indentation, showing the maximum indentation depth h_{max}, contact depth h_c, minimum indentation depth h_{min}, and sink-in situation.

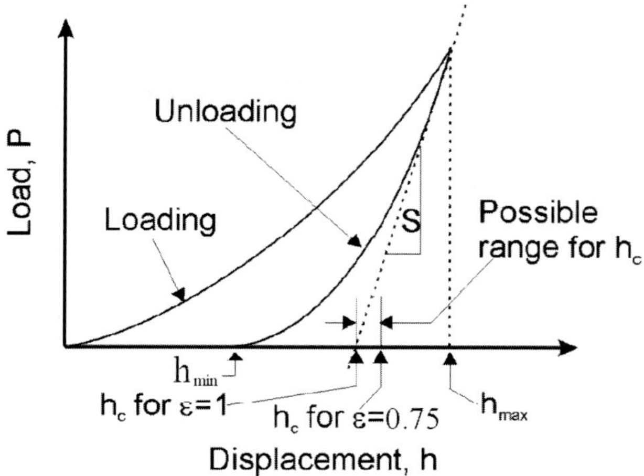

Fig. 3.2. Load-displacement curves for Oliver–Pharr method, showing the maximum indentation depth h_{max}, contact depth h_c, and minimum indentation depth h_{min}.

function at the contact depth h_c:

$$A = f(h_c). \qquad (3.12)$$

The shape function, $f(h_c)$, relates the cross-sectional area of the indenter to the distance h_c from its tip. For the Berkovich indenter, the shape function is given by $f(h_c) = 24.56\,h_c{}^2$. Once the contact area is determined from the load-displacement curve, the hardness H, and effective elastic modulus E_{eff} follow from:

$$H = \frac{P_{\mathrm{max}}}{A}, \qquad (3.13)$$

and

$$E_{\mathrm{eff}} = \frac{1}{\delta_1}\frac{\sqrt{\pi}}{2}\frac{S}{\sqrt{A}}, \qquad (3.14)$$

where δ_1 is a constant which depends on the geometry of the indenter, and $\delta_1 = 1.034$ for the Berkovich one [84, 87].

The Oliver–Pharr method of calculating contact area does not account for the "extra" area due to pile-up, and hence underestimates the contact area, thus overestimating the hardness and effective elastic modulus [83]. However, if the thin film is harder than the substrate material, this method can be conveniently used to determine hardness with high precision, because for a hard film on soft substrate, there is usually the "sink-in" situation on the coating surface during indentation and very minimum pile-up.

Hardness was evaluated using a Nano II^{TM} with a Berkovich indenter. The indentation depth was set less than one tenth of the film thickness to avoid substrate effect. At least five indentations were made for each sample. The results were an average of these readings from Oliver–Pharr method.

3.4.2. *Toughness*

Indentation is perhaps the most widely used tool in assessing thin film toughness. Plastic deformation leads to stress relaxation in materials. The easier the stress relaxation proceeds, the larger plasticity is inherent in the material. Thus, comparing the plastic strain with the total strain in an indention test directly gives a simple and rough, but quick indication of how "tough" the material is. Plasticity is defined as the ratio of the plastic displacement over the total displacement in the load-displacement curve [88] (Fig. 3.3):

$$\text{Plasticity} = \frac{OA}{OB}, \tag{3.15}$$

where OA is plastic deformation, and OB total deformation. A superhard DLC film with hardness of 60 GPa has only 10% plasticity [48], whereas a "tough" nc-TiC/a-C film with a hardness of 32 GPa has 40% plasticity [46,47]. Hydrogen-free amorphous carbon films with hardness of 30 GPa has

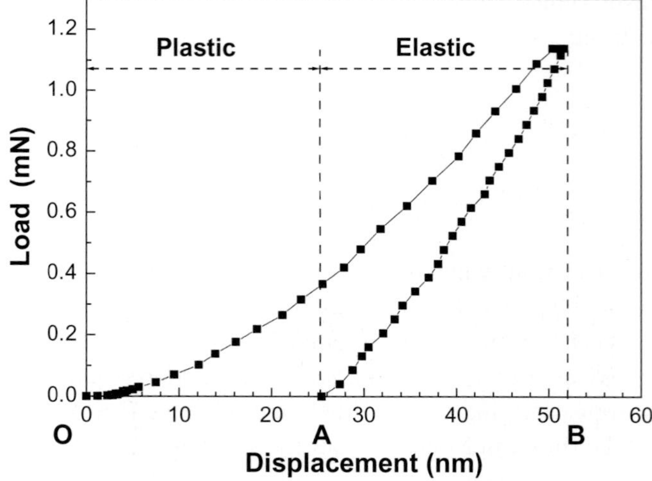

Fig. 3.3.   Schematic diagram of load-displacement curve obtained from nanoindentation. Plasticity is calculated by the ratio of OA/OB.

a toughness of 50–60% in plasticity [89] depending on bias voltage during sputtering. Magnetron sputtered $1\,\mu$m thick $\text{Ti}_{1-x}\text{Al}_x\text{N}$ films with hardness 31 GPa obtained a plasticity of 32% [90].

However, "plasticity" is not fracture toughness. To measure a film's proper fracture toughness, and to avoid the difficulties in making the pre-crack, many researchers directly indent the films without a pre-crack. When the stress exceeds a critical value, a crack or spallation will be generated. Failure of the film is manifested by the formation of a kink or plateau in the load-displacement curve or crack formation in the indent impression [91–93]. As a qualitative, crude and relative assessment, Holleck and Schulz [94] compare the crack length under the same load, and Kustas *et al.* [95] measures the "spall diameter" — the damage zone around the indenter. More quantitatively, the length c of radial crack (Fig. 3.4) is related to the toughness K_{IC} through [96]:

$$K_{\text{IC}} = \delta \left(\frac{E}{H} \right)^{1/2} \left(\frac{P}{c^{3/2}} \right), \qquad (3.16)$$

where P is applied indentation load, and E and H are elastic modulus and hardness of the film, respectively. δ is an empirical constant which depends on the geometry of the indenter. For the standard Vickers diamond pyramid indenter, the value of δ is taken as 0.016 ± 0.004 [97] and 0.0319 [98, 99], respectively. The criterion for a well-defined crack is taken as $c \geq 2a$ [97], where a is the half of the diagonal length of the indent (Fig. 3.4). Both E and H can be determined from an indentation test at a much smaller

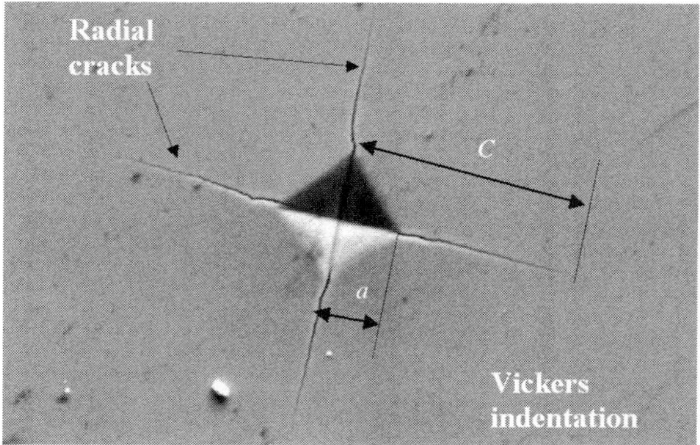

Fig. 3.4.   Scanning electron microscopy (SEM) observation of radial cracks at Vickers indentation. a is half of the diagonal length of the impression and c is the crack length. A well-defined crack is taken as $c \geq 2a$.

load and analyzing of indentation load-displacement data [84]; crack length c can be obtained using SEM, thus implementation of the method seems straightforward [87].

However, the difficulty lies in the existence of cracking threshold, locating crack position and determining crack length. Although indentation can be realized with a Vickers indenter, Berkovich indenter, or cube corner indenter [100, 101], there exists a cracking threshold below which indentation cracks cannot form. Existence of the cracking threshold causes severe restrictions on achievable spatial resolution. The occurrence of the indentation cracking depends on the condition of the indenter tip [102]. Harding *et al.* [103] found that indentation-cracking threshold could be significantly reduced by employing a sharper indenter (cube corner indenter compared to the Berkovich and Vickers indenters). The cube corner indenter induces more than three times the indentation volume as compared to that by the Berkovich indenter at the same load. Consequently, the crack formation is easier with the cube corner indenter thereby reducing the cracking threshold. For the cube corner indenter, the angle between the axis of symmetry and a face is 35.3° (compared to 65.3° for the Berkovich indenter), and there are three cracks lying in directions parallel to the indentation diagonal (Fig. 3.5). Cracks that are well defined and symmetrical around the cube corner indentation are used to calculate the toughness. Different researchers used different δ values: 0.0319 [98, 99], 0.040 [103], and 0.0535 [104]. Despite the inherited problems, due to its simplicity, the indentation method is

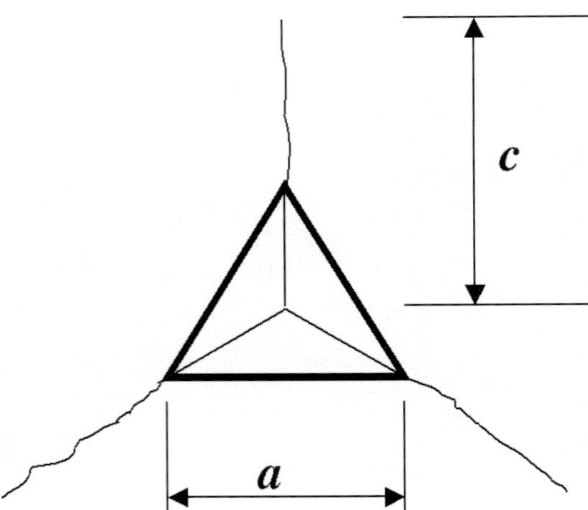

Fig. 3.5.   Schematic diagram of median-radial crack systems for cube corner indentation. *a* is the length of the edge of impression; *c* is the length of crack parallel to the indentation diagonal direction.

widely used in toughness evaluation of thin films. To cite a few: sputter deposited DLC film (1.92 μm thick, 1.57 MPa m$^{1/2}$) [105], plasma sprayed Al$_2$O$_3$ (200–300 μm thick, containing 13% TiO$_2$, 4.5 MPa m$^{1/2}$) [106], atmospheric pressure CVD SiC (3 μm, 0.78 MPa m$^{1/2}$) [107], plasma enhanced CVD nc-TiN/SiN$_x$ ($\sim$1.5 μm, 1.3–2.4 MPa m$^{1/2}$) [108] and TiC$_x$N$_y$/SiCN (2.7–3.3 μm, $\sim$1 MPa m$^{1/2}$) [109].

Micro hardness tester (DMH-1) was used to obtain well-defined radial cracks for coatings on silicon wafer substrate. To ensure measurable crack length, the indentation was conducted at load of 10.00, 5.00, 3.00, 2.00, 1.00, 0.50 and 0.25 N, respectively. For each load, at least five readings were obtained. In order to reduce substrate effect on thin film toughness, Nano II^{TM} with Berkovich indenter was also used to characterize these samples. The indentation was performed at the depths of 1300, 1000, 700, 400 and 200 nm, respectively. For each depth, at least five readings were obtained. Only samples with well-defined radial cracks were used to calculate thin film toughness.

3.4.3. *Residual Stress*

Residual stress was measured using a Tencor FLX-2908 laser system by testing the curvature changes of silicon substrate before and after film deposition. At least five measurements were performed for each sample at different orientation of silicon wafer. The residual stress σ can be calculated through Stony equation:

$$\sigma = \frac{Et_s^2}{(1 - v)6Rt_f}, \tag{3.17}$$

where $E/(1 - \nu)$ is the biaxial elastic modulus of the substrate, and t_s and t_f are thickness of the substrate and film, respectively. R is the substrate radius of curvature, which is calculated through:

$$R = \frac{1}{\left(\frac{1}{R_2} - \frac{1}{R_1}\right)} = \frac{R_1 R_2}{(R_1 - R_2)}, \tag{3.18}$$

where R_1 and R_2 are the radius of silicon substrate before and after film deposition.

3.4.4. *Adhesion*

Scratch test was conducted to testing films adhesion to substrate using Shimadzu SST-101 scan scratch tester: the X–Y table of the scratch tester was moved in the X-direction, and also moved sideways (in the Y-direction)

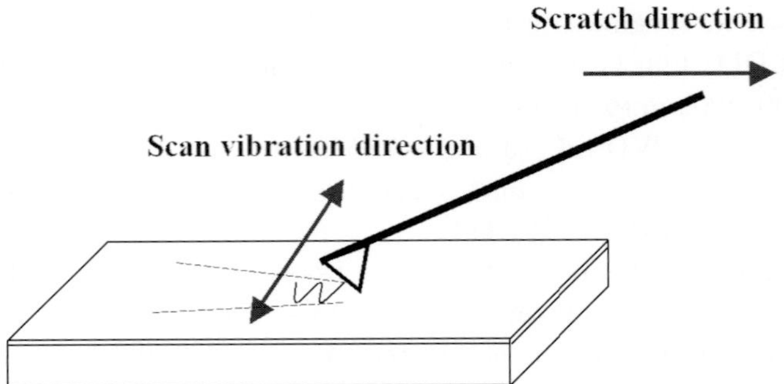

Fig. 3.6. Schematic diagram of scan scratch test used in this project. The X–Y table of the scratch tester moves in the X-direction, and also moves sideways (in the Y-direction) to generate a "scanning scratch" effect, thus greatly increasing the coverage of each scratch.

to generate a "scanning scratch" effect and thus greatly increase the coverage of each scratch (Fig. 3.6). At the same time, an increasing normal load was applied continuously to the surface of the film through a diamond indenter of $15\,\mu$m in radius until total failure of the film. A scratch speed of $2\,\mu$m/s and scanning amplitude of $50\,\mu$m were used. At least three tests were performed on each sample.

3.5. *Oxidation Resistance*

To study films oxidation resistance, oxidation was performed using Elite Furnace (BRF14/5-2416) at 450°C, 550°C, 625°C, 700°C, 750°C, 800°C, 850°C and 900°C to 1000°C in static hot air. After a 10 min heating ramp, the temperature was kept constant for 15 min. After the specimen was oxidized for the pre-determined time, the specimen was cooled down to a temperature around 300°C within 20 min, then drawn from the furnace and left until room temperature thereafter for oxidation resistance study using XPS, AFM and GIXRD.

4. Case Studies: Silicon Nitride Nanocomposite Coating

4.1. *Nanocrystalline TiN Embedded in Amorphous SiN_x or nc-TiN/a-SiN$_x$*

DC magnetron power was applied to Ti (99.99%) target, while RF power was applied to Si_3N_4 (99.999%) target. Four-inch silicon (100) wafers were used as substrates. The substrates were cleaned using an ultrasonic bath (ethanol absolute 99%) for 15 min to remove the contamination, such as spots, oils, grease, dust, fingerprints, etc. The substrates were heated to 450°C for 30 min prior to deposition. This process mainly removed water molecules, which were adsorbed on surface. The substrate holder rotated with 15 rpm during the deposition for uniformity. Deposition was performed at a substrate temperature of 450°C for 2 hours to obtain a film thickness of 0.6 μm. Base pressure for deposition was 1.33×10^{-5} Pa. During deposition, the gas (purity of N_2 are Ar are 99.9995%) pressure was 0.67 Pa and gas flow rate was 30 sccm (standard cubic centimetre per minute). Film deposition parameters are listed in Table 4.1 according to which hard and superhard nc-TiN/a-SiN$_x$ nanocomposite thin films were prepared.

The as-prepared nc-TiN/a-SiN$_x$ nanocomposite thin films were studied using different characterization techniques, such as XPS, AFM, XRD/ GIXRD, TEM/HRTEM, scratch, microindentation, and nanoindentation tests. The results and discussions are presented in this chapter.

4.1.1. *Composition*

The nc-TiN/a-SiN$_x$ nanocomposite thin films were etched with Ar ion beam for 15 min before composition measurement. The chemical compositions obtained from XPS are listed in Table 4.2, in which the atomic concentrations of Si, Ti and N are used to describe the samples. The results show that N content for all samples are about 50 ± 5 at.%, while Si and Ti contents vary greatly with the experimental conditions, mainly Si_3N_4 target power density.

4.1.1.1. Quantitative Compositional Analysis

Figure 4.1 shows a detailed XPS survey scan spectrum with indexed peaks. XPS survey scan spectrum with binding energy from 0 to 1100 eV is

Table 4.1. Experimental conditions for deposition of hard and superhard nc-TiN/a-SiN$_x$ nanocomposite thin films.

	Sample code	P1	P2	P3	P4	P5	P6	P7	P8
Deposition condition	Si_3N_4 power density (W/cm^2)	1.1	2.2	3.3	4.4	5.5	6.6	6.6	7.7
	Power ratio of Ti to Si_3N_4	5.00	2.50	1.67	1.25	1.00	0.67	0.33	0
	Gas ratio of N_2 to Ar	1.0	1.0	1.0	1.0	1.0	0.5	0.5	0

Table 4.2.   Chemical composition characteristics of hard and superhard nc-TiN/a-SiN$_x$ nanocomposite thin films.

Sample code	XPS chemical composition (at.%)		
	Si	Ti	N
P1	1.7	50.4	47.9
P2	5.2	43.1	51.7
P3	7.2	38.1	54.7
P4	11.3	35.1	53.6
P5	14.9	29.4	55.7
P6	23.6	21.4	55.0
P7	32.5	15.7	51.8
P8	52.7	0	47.3

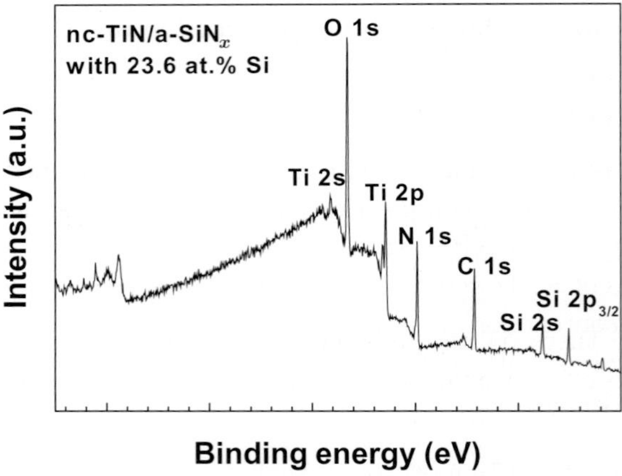

Fig. 4.1.   XPS survey scan of nc-TiN/a-SiN$_x$ nanocomposite thin film with 23.6 at.% Si (sample P6). The dominant signals are from C, O, N, Si, and Ti.

recorded. The dominant signals are from C, O, N, Si and Ti. The oxygen and carbon contamination exist because the film is exposed to air (ambient laboratory) and the XPS spectrum is obtained before the film surface is etched.

Figure 4.2 shows one of the typical XPS depth profiles of the nc-TiN/a-SiN$_x$ nanocomposite thin film. There is an inevitable oxygen contamination in the topmost layer of the film. It is possible that the reaction with residual gases in the spectrometer chamber or the ionic transport via impurities in the Ar gas causes this residual oxygen content in films. Though oxygen

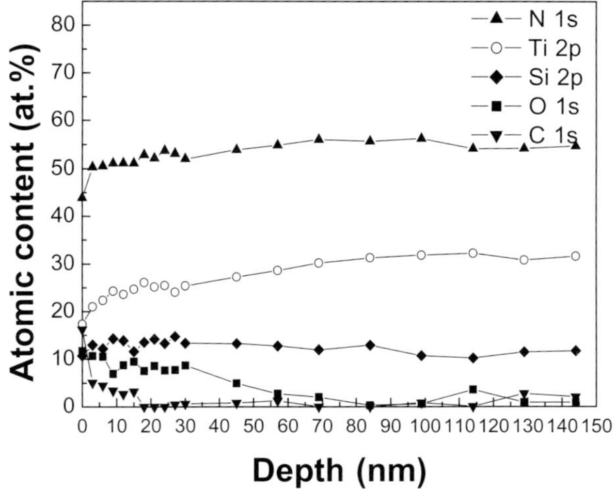

Fig. 4.2. XPS depth profile of nc-TiN/a-SiN$_x$ nanocomposite thin film with 14.9 at.% Si (sample P5). Silicon content remains almost constant, whereas nitrogen and titanium content slightly increases in the first 10 nm and then remains at constant values.

exists as deep as 75 nm, and carbon at approximately 25 nm, their concentrations are very low, which are expected to render a very insignificant effect on the film's mechanical properties, such as nanoindentation hardness. (Should there be any influence on hardness owing to the existence of oxides, the effect would be such that the measurements would underestimate the hardness because both Ti and Si oxides have low hardness: 16 GPa for DC magnetron sputtered TiO$_x$ [110], approximately 10 GPa in the case of reactive cathodic vacuum arc deposited TiO$_2$ [111], and 8 GPa in the case of pulsed magnetron sputtered SiO$_2$ [112], etc.) Silicon content remains almost constant, whereas nitrogen and titanium content slightly increases in the first 10 nm and then remains at constant values. From the depth profile (Fig. 4.2), the oxygen and carbon concentrations are too low in comparison to Ti, Si and N, and are thus ignored in composition computation.

Figure 4.3 shows the XPS spectra of Ti 2p and Si 2p as a function of silicon content. It has been documented that the shoulders observed on the high binding energy side of the Ti 2p$_{3/2}$ and Ti 2p$_{1/2}$ component peaks are inherent characteristic features of stoichiometric TiN [113, 114]. This stoichiometric TiN characteristic feature has been observed in our nc-TiN/a-SiN$_x$ nanocomposite thin films [Fig. 4.3 (a)]. Figure 4.3 (b) shows

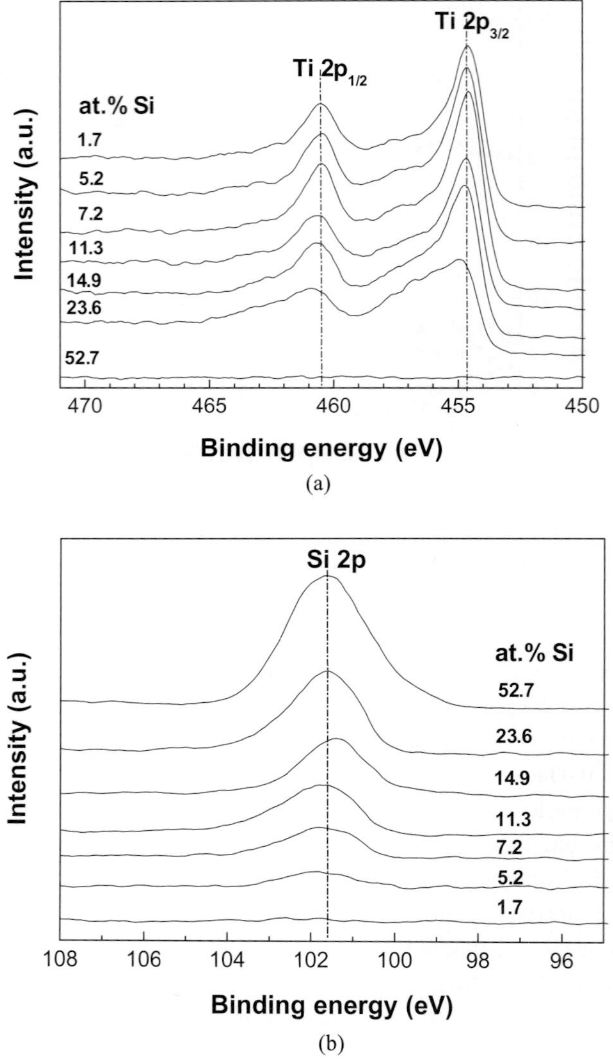

Fig. 4.3. Stacks of (a) Ti 2p, and (b) Si 2p XPS core level spectra for nc-TiN/a-SiN$_x$ nanocomposite thin films with different Si content after surface layer removal.

the Si 2p peak intensity increases significantly with the increase in silicon content.

In order to obtain more detailed information on the chemical composition, the acquired XPS core level spectrum of Ti 2p and Si 2p for the nc-TiN/a-SiN$_x$ nanocomposite thin film with 23.6 at.%Si (sample P6) after

15 min Ar ion beam etching were deconvoluted. Curve fitting was performed after a Shirley background subtraction by nonlinear least square fitting using a mixed Gauss/Lorentz function. The parameters used in the fitting procedure are listed in Table 3.2. Figure 4.4 shows the XPS core line fit for Ti 2p spectra. There are three main components of Ti–N, Ti–O and Ti–X (combination of TiN_x, TiN_xO_y and satellite peak); no unbound Ti metallic bonds can be detected. The peak at 455.0 eV is attributed to Ti–N bond or TiN. The peak at 459.0 eV is related to Ti–O or TiO_2. The peak at 457.6 eV is contributed from Ti–X. From Fig. 4.4, it can be seen that the main contribution comes from TiN and Ti–X, and nearly no TiO_2. This is due to the high base vacuum ($\times\ 10^{-5}$ Pa) before film deposition. Another reason is that before XPS narrow scan and core line fit, a 15 min Ar ion beam etching was conducted to remove the surface oxide due to air exposure (ambient laboratory).

Figure 4.5 shows the quantitative deconvolution results of Ti 2p spectra of nc-TiN/a-SiN_x nanocomposite thin films with different silicon content. The relative atomic ratios of Ti in TiN, TiO_2 and TiX are calculated. Ti–O remains low in concentration for all the films. Ti–X peak component increases and TiN component decreases significantly with increases

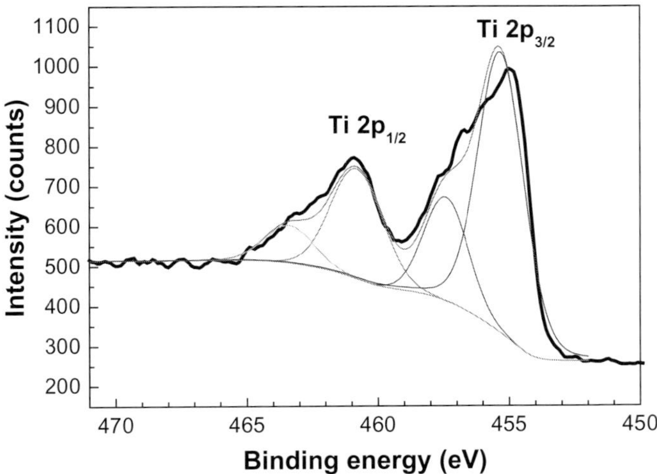

Fig. 4.4. Ti 2p deconvolution of nc-TiN/a-SiN_x nanocomposite thin film with 23.6 at.%Si (sample P6). There are three main components of Ti–N, Ti–O and Ti–X (combination of TiN_x, TiN_xO_y and satellite peak). The peak at 455.0 eV is attributed to Ti–N bond. The peak at 459.0 eV is related to Ti–O bond. The peak at 457.6 eV is contributed from Ti–X.

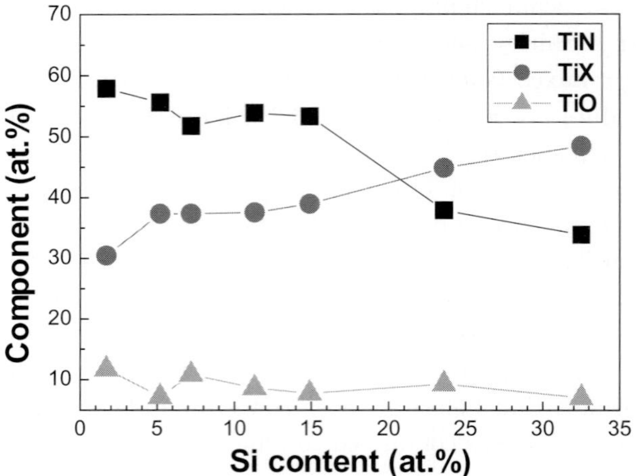

Fig. 4.5.   Changes of the different Ti 2p components of nc-TiN/a-SiN$_x$ nanocomposite thin films with different Si content after deconvolution. Ti–O remains low in concentration for all the films. Ti–X peak component increases and TiN component decreases significantly with an increase in Si content.

in silicon content. There is a gradual shift of the Ti 2p peak to a higher binding energy and the Ti–X component increases significantly with the increases in silicon content. This can be partly attributed to the increase in the non-stoichiometric component; another reason could be due to silicon involvement into the titanium structures.

Figure 4.6 shows the Si 2p core level spectrum of nc-TiN/a-SiN$_x$ nanocomposite thin film with 23.6 at.%Si (sample P6). Three chemically distinct components are found in the Si 2p core level photoelectron spectrum. The peak corresponding to 101.8 eV is attributed to Si–N bond of stoichiometric Si$_3$N$_4$. Another weak component at approximately 103.4 corresponds to the Si–O bond. Traces of Si peak (at 99.6 eV) can be detected, belonging to free silicon. The existence of free Si is often observed to exist in nc-TiN/a-SiN$_x$ thin films synthesized by magnetron sputtering [115]. The Ti–Si bonding energy is at 98.8 eV — too close to that of elemental Si, and thus difficult to eliminate the possibility of existence of TiSi$_x$ in the system. However, one thing is for sure: its amount is minute even if it exists. Take TiSi as an example, from thermodynamics, the formation of Si$_3$N$_4$ is much more favored compared to the formation of TiSi, since the formation energy for Si$_3$N$_4$ (-665.4 kJ/mol) is much more negative than that for TiSi (-132.2 kJ/mol).

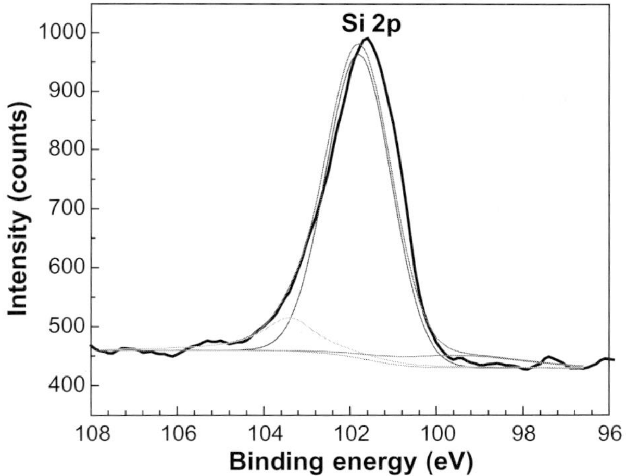

Fig. 4.6. Si 2p deconvolution of nc-TiN/a-SiN$_x$ nanocomposite thin film with 23.6 at.% Si (sample P6). The peak corresponding to 101.8 eV is attributed to the Si–N bond of stoichiometric Si$_3$N$_4$. Another weak component at approximately 103.4 eV corresponds to the Si–O bond. Traces of Si peak at 99.6 eV can be detected, belonging to free silicon.

Figure 4.7 shows the quantitative deconvolution results of Si 2p spectra of nc-TiN/a-SiN$_x$ nanocomposite thin films with different silicon content. The silicon element exists mostly as the Si–N bond, although Si–O as well as traces of Si–Si and Si–Ti bonds can be occasionally observed.

4.1.1.2. Effect of Deposition Conditions

Figure 4.8 shows the relationship between silicon content and Si$_3$N$_4$ target power density (from 1.1 to 5.5 W/cm^2) for the nc-TiN/a-SiN$_x$ nanocomposite thin films (samples P1 to P5) deposited at 450°C with a constant Ti target power density of 5.5 W/cm^2. It is obvious that as Si$_3$N$_4$ target power density increases from 1.1 to 5.5 W/cm^2, the silicon content in the as-prepared nanocomposite thin films increases linearly from 1.7 at.% to 14.9 at.%.

Figure 4.9 shows the relationship between silicon content and deposition target power ratio of Ti to Si$_3$N$_4$ for the nc-TiN/a-SiN$_x$ nanocomposite thin films (samples P1 to P8) deposited at 450°C. With the increase in target power ratio of Ti to Si$_3$N$_4$ from 0 to 1.5, the silicon content in the as-deposited nc-TiN/a-SiN$_x$ nanocomposite thin films decreases significantly from 52.7 at.% to 7.2 at.%, and then tails off.

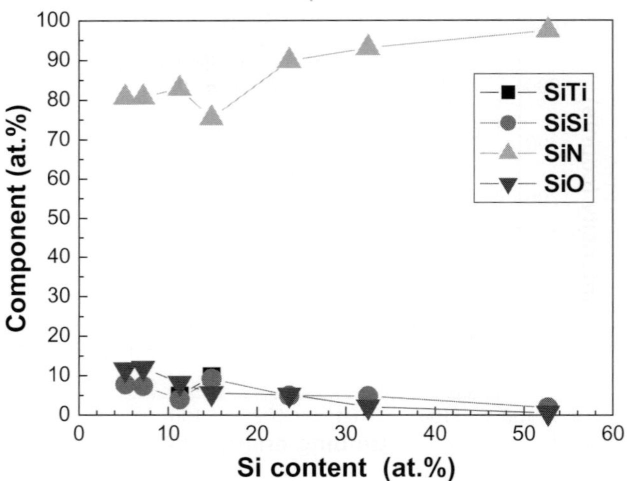

Fig. 4.7. Changes of the different Si 2p components of nc-TiN/a-SiN$_x$ nanocomposite thin films with different Si content after deconvolution. The Si element exists primarily in the Si–N bond.

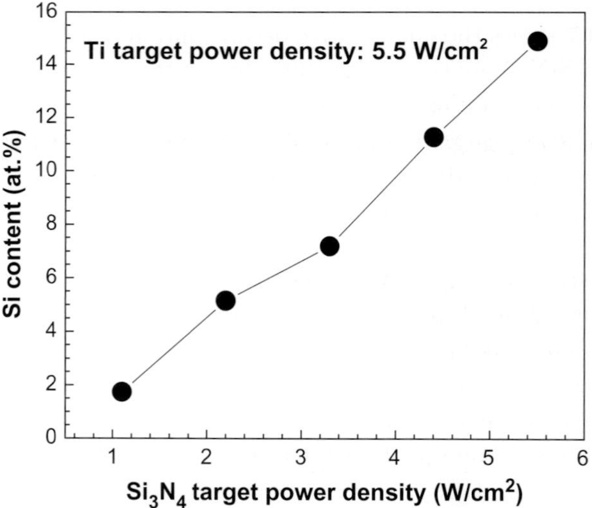

Fig. 4.8. Si content changes linearly with Si$_3$N$_4$ target power density. With the increase of Si$_3$N$_4$ target power density from 1.1 to 5.5 W/cm^2, the Si content in the as-prepared thin films increases linearly from 1.7 at.% to 14.9 at.%.

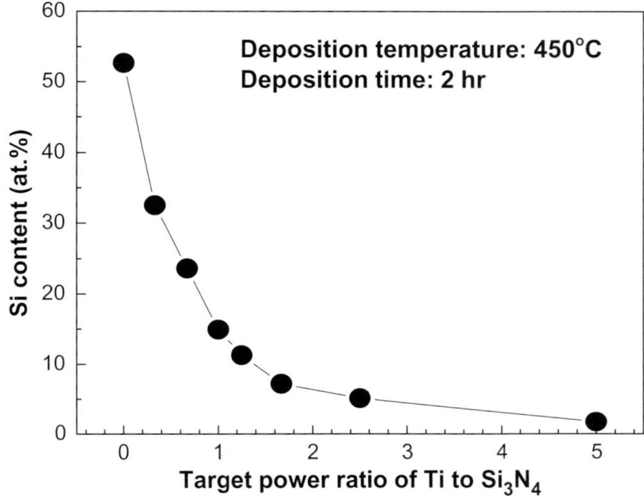

Fig. 4.9. Si content changes with target power ratio of Ti to Si_3N_4 target. With an increase in target power ratio from 0 to 1.5, the Si content in the as-deposited nc-TiN/a-SiN_x thin films decreases significantly from 52.7 at.% to 7.2 at.%; with further increase in power ratio, the trend no longer appears scientific.

4.1.2. *Topography*

To evaluate the effect of target power density of Si_3N_4 on surface morphology, the Si_3N_4 target power density is changed from 1.1 to 5.5 W/cm^2 while Ti target power density is kept constant at 5.5 W/cm^2. Topography characteristics of the as-prepared nanocomposite thin films are tabulated in Table 4.3.

Figure 4.10 shows the AFM images (300 nm $\times$ 300 nm) of the nc-TiN/a-SiN_x nanocomposite thin films with different Si_3N_4 target power densities. With increasing Si_3N_4 target power density, the roughness gradually

Table 4.3. Topography characteristics of hard and superhard nc-TiN/a-SiN_x nanocomposite thin films.

Sample code	Roughness R_a(nm)	Interface width ω (nm)	Lateral correlation length ξ (nm)
P1	5.8	7.3	28.0
P2	2.5	3.4	13.8
P3	1.5	1.8	12.8
P4	0.9	1.0	9.6
P5	0.7	1.0	9.4

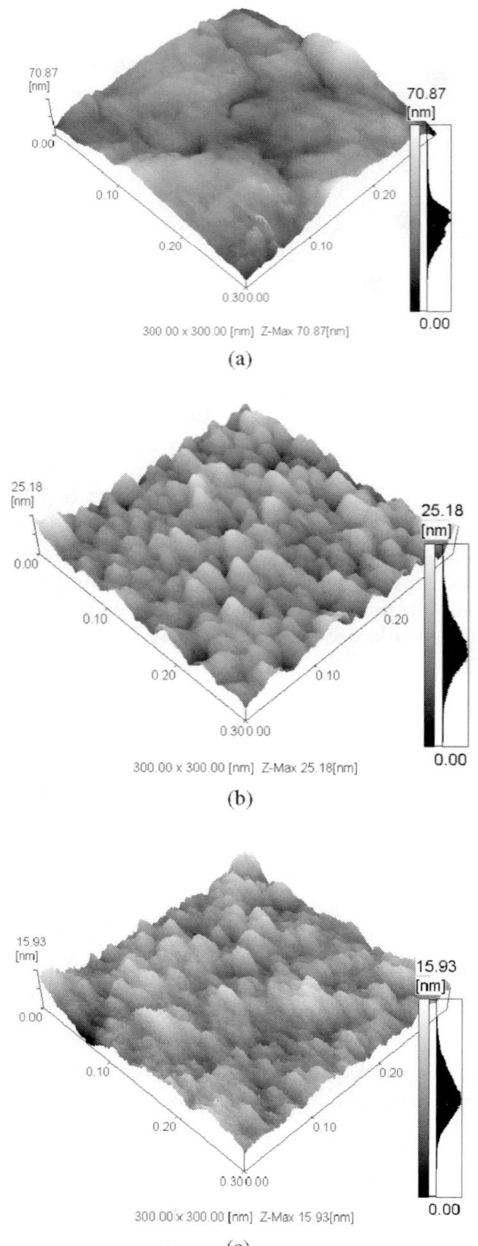

Fig. 4.10. AFM topography of the nc-TiN/a-SiN$_x$ nanocomposite thin films with increasing Si$_3$N$_4$ target power density (in W/cm^2): (a) 1.1, (b) 2.2, (c) 3.3, (d) 4.4, and (e) 5.5, with roughness (R_a, nm) 5.8, 2.5, 1.5, 0.9 and 0.7, respectively.

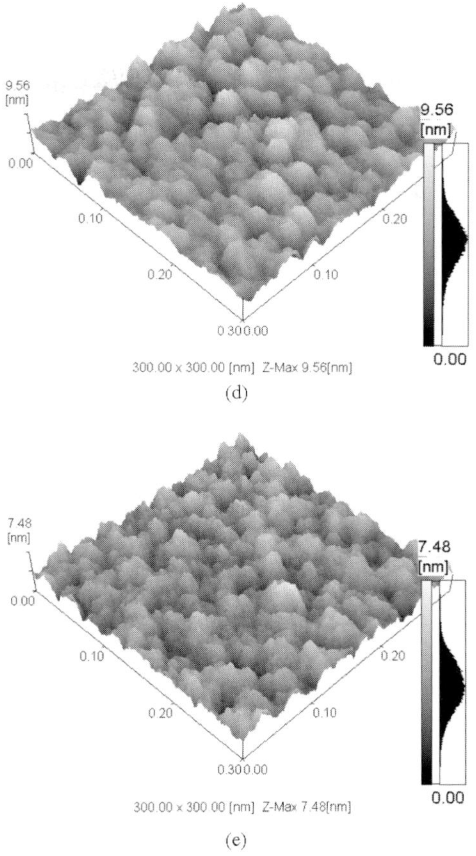

(d)

(e)

Fig. 4.10.　(*Continued*)

decreases. Numeric treatment of the images gives rise to the height–height correlation function $G(r)$. Figure 4.11 plots $G(r)$ as a function of r for sample represented by Fig. 4.10 (e), i.e. 5.5 W/cm^2 Si$_3$N$_4$ target power density. As predicted by Eq. (3.4), when r is small, $G(r)$ has a power law dependency on distance r. At "distant" locations (as r is large), $G(r)$ is nearly constant. Fitting the curve to Eq. (3.4) gives the lateral correlation length $\xi = 9.4$ nm, the interface width $\omega = 1.0$ nm and the smoothness exponent $\alpha = 0.854$. The oscillation is due to the insufficient sampling size [77]. Similar treatment of other samples listed in Table 4.3 results in different parameters of ω, ξ, and α as a function of Si$_3$N$_4$ target power density. Plotting these values gives rise to Figs. 4.12 and 4.13.

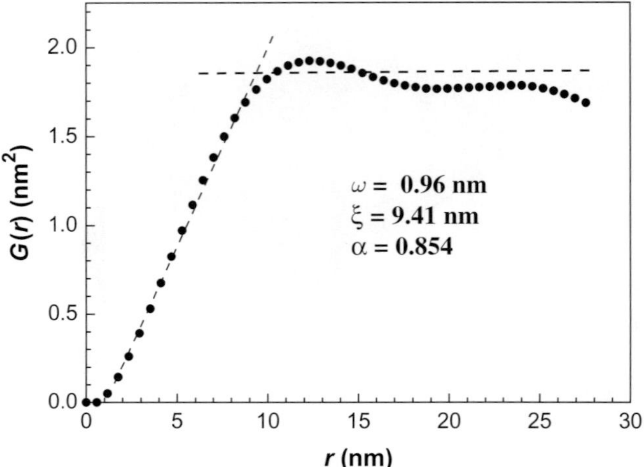

Fig. 4.11. Height–height correlation function $G(r)$ for the nc-TiN/a-SiN$_x$ nanocomposite thin film with Si$_3$N$_4$ target power density of 5.5 W/cm^2 (Fig. 4.10 (e)). The oscillation is due to the insufficient sampling size.

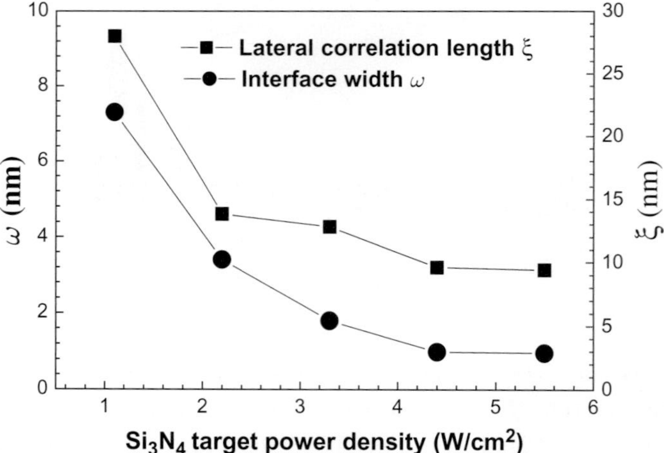

Fig. 4.12. Interface width ω and lateral correlation length ξ vary with Si$_3$N$_4$ target power density. With increase in Si$_3$N$_4$ target power density, both interface width and lateral correlation length decrease.

The morphology of the growing surfaces is determined by the competition of vertical build-up and lateral diffusion (Fig. 4.14). The vertical build-up is caused by the random angle incident of the arriving atoms (due to the uniform rotation of the substrate), and the growth produces

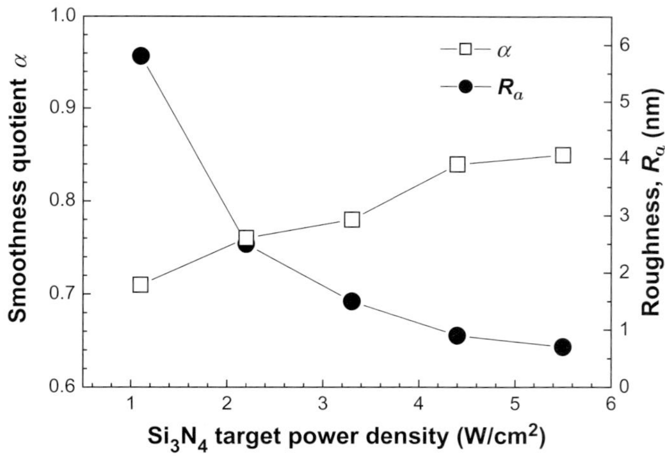

Fig. 4.13. Surface smoothness quotient α and roughness R_a vary with target power density. With increase in Si_3N_4 target power density, surface roughness R_a decreases, while the smoothness quotient α increases.

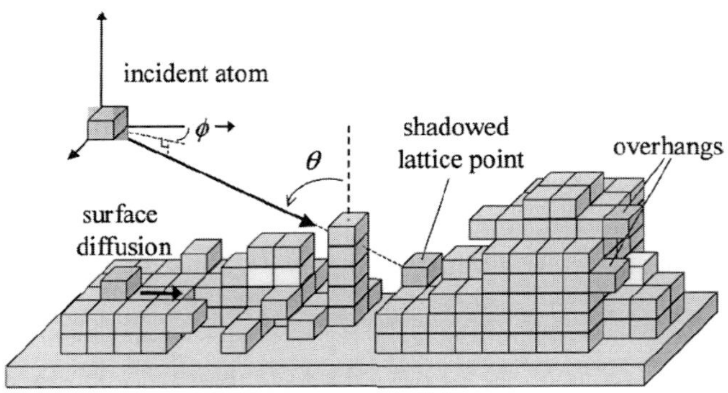

Fig. 4.14. Schematic diagram of the competition of vertical build-up. Lateral diffusion determines the morphology of the growing surfaces [118].

columnar film structure. The lateral growth depends on surface diffusion which is largely determined by kinetic energy of the arriving ions. This is clearly seen from the trends of ω and ξ in Fig. 4.12. As Si_3N_4 target power density increases from 1.1 to 5.5 W/cm^2, the interface width ω decreases from $\sim$7 to $\sim$1 nm (Fig. 4.12), indicating that the film becomes smoother (recall that ω is the root mean square of the vertical fluctuation), which is more directly observable in terms of increasing smoothness quotient α or decreasing roughness R_a in Fig. 4.13. As target power density increases, ξ decreases from $\sim$28 to $\sim$10 nm. Since ξ depicts the distance within which "height" values (i.e. roughness) are correlated, a larger ξ means that surface topography is correlated in a wider area. This is seen as larger "humps", as in Fig. 4.10(a), for film deposited at a lower power density $(1.1\,\mathrm{W/cm^2})$. The growth kinetic is controlled by the mobility of the impinging atoms on the surface before they condense and become entrapped in the film. This mobility can be enhanced by inputting energy to the system, such as by increasing deposition temperature or supplying impact energy through ion bombardment. At low target power densities, ions have low mobility and thus would be more likely to "stick" at where it arrives: the surface diffusion is slow and the chances of erasing the peaks and filling up the "valleys" are small, thus resulting in a much rougher surface. In the same token, small ξ indicates that the surface topography is correlated only in a small area, as seen in Fig. 4.10 progressively from (b) through (e) as target power density increases. As the Si_3N_4 target power density increases, the kinetic energy obtained by each Si or SiN_x (as well as amount) increases, which transforms into faster lateral diffusion and smoothens out the roughness at locations further away, making the "hump" more localized and smaller, giving rise to a smaller value of ξ and greater values of α [116, 117].

4.1.3. *Microstructure*

The effect of sputtering power density of Si_3N_4 target on microstructure includes changes in crystal phase, grain size and distribution, preferential orientation and lattice parameter. Microstructure characteristics of the as-prepared nanocomposite thin films are tabulated in Table 4.4.

Table 4.4. Microstructure characteristics of hard and superhard nc-TiN/a-SiN$_x$ nanocomposite thin films.

Sample Code	P1	P2	P3	P4	P5	P6	P7	P8
Crystallite								
Fraction (%)	3.6	6.7	7.3	7.1	10.8	1.1	0.0	0.0
Size (nm)	7.2	8.2	7.5	5.8	6.0	3.2	0	0
Coefficient of								
Texture T_{hkl}								
(111)	2.84	0.01	0.62	—	0.00	—	—	—
(200)	0.02	1.34	1.79	—	0.13	—	—	—
(220)	0.14	1.64	0.60	—	2.87	—	—	—
Lattice parameter								
a_{TiN} (nm)	0.42350	0.43788	0.42994	—	0.44408	—	—	—

4.1.3.1. Crystal Phase and Amorphous Matrix

Figure 4.15 shows the bright-field HRTEM morphological appearance of nc-TiN/a-SiN$_x$ nanocomposite thin film with 11.3 at.%Si (sample P4). Crystallites are embedded in matrix. The grain size is about 6 nm. Analysis of the selected area diffraction (SAD) pattern shows that these crystallites are polycrystalline TiN. No crystalline TiSi$_x$ and Si$_3$N$_4$ are found. In general, TiN crystallographic planes of (111), (200) and (220) exhibit more distinct rings than other diffraction rings (Fig. 4.16). Proof of the crystallites being TiN also comes from XRD analysis (Fig. 4.17). The substrate on which film was deposited is silicon wafer (100), and the peak at about 69° in Fig. 4.17 belongs to silicon (400) diffraction.

Analysis of the SAD pattern (where there is no crystallite) gives rise, on the other hand, to a diffuse pattern typical of an amorphous phase (Fig. 4.18). Together with XPS analysis (Sec. 4.1.1), where silicon is mostly in Si–N bond, the results confirm that the amorphous phase is amorphous silicon nitride (a-SiN$_x$). Usually, the TiN deposited at such conditions is in crystalline phase. Silicon nitride, however, stays amorphous even at 1100°C [119]. These results are in agreement with the idea of spontaneous formation of such nanostructures due to spinodal decomposition which occurs during deposition [31].

Figure 4.19 shows an empiric model for the phase formation in Ti–Si–N coating system, including the formation of nanocrystalline titanium nitride (nc-TiN) and amorphous silicon nitride matrix (a-SiN$_x$). The deposition conditions for the Ti–Si–N coating system are: deposition temperature

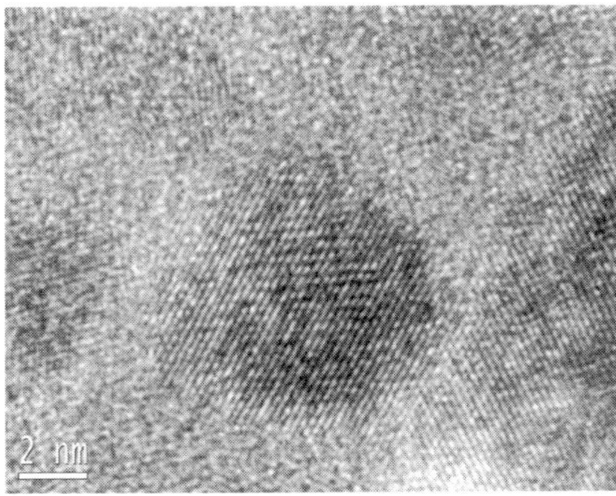

Fig. 4.15.   HRTEM bright-field micrograph of nc-TiN/a-SiN$_x$ nanocomposite thin film with 11.3 at.%Si (sample P4) showing the crystalline in size of 6 nm embedded in matrix.

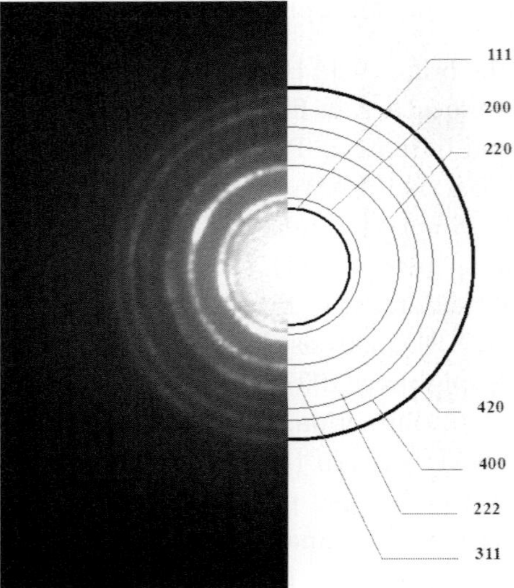

Fig. 4.16. Selected area diffraction (SAD) pattern of nc-TiN/a-SiN$_x$ nanocomposite thin film with 11.3 at.%Si (sample P4) showing the crystalline TiN (111), (200) and (220) crystallographic planes exhibit more distinct rings than other diffraction rings.

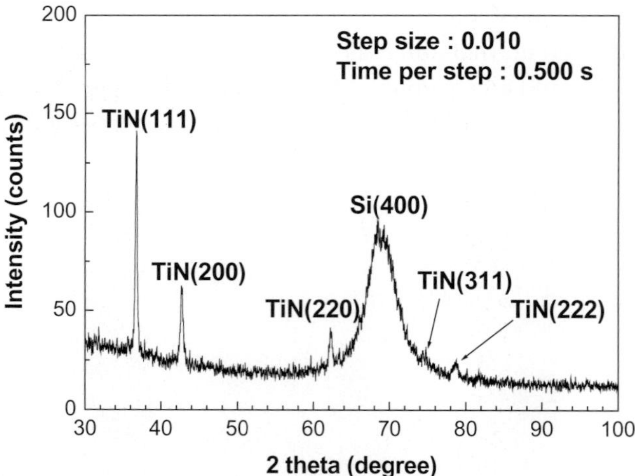

Fig. 4.17. XRD pattern of nc-TiN/a-SiN$_x$ nanocomposite thin film with 11.3 at.%Si (sample P4). Formation of crystallite TiN is confirmed.

Fig. 4.18.   SAD pattern (where there is no crystallite) of nc-TiN/a-SiN$_x$ nanocomposite thin film with 11.3 at.%Si (sample P4) showing a diffuse pattern typical of an amorphous phase.

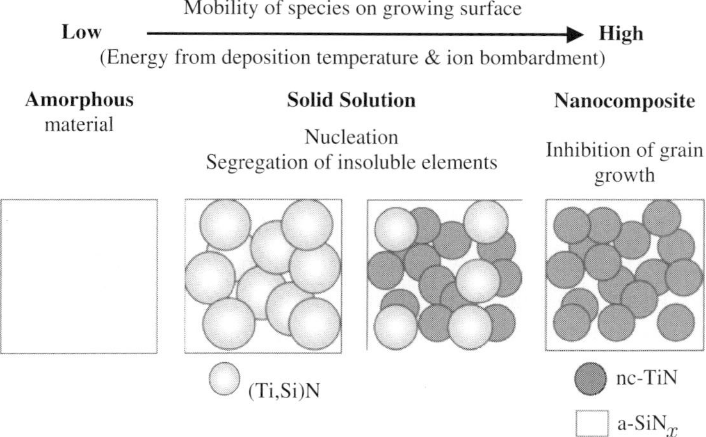

Fig. 4.19.   Empiric model for phase formation in Ti–Si–N coating system: formation of nanocrystalline titanium nitride (nc-TiN) and amorphous silicon nitride matrix (a-SiN$_x$) [120].

at 450°C, and total deposition power densities of (Ti + Si$_3$N$_4$) targets greater than 6.6 W/cm^2. In such deposition conditions, the mobility of the species at the growing Ti–Si–N surface can be enhanced by the high deposition temperature and the ion bombardment effect which is due to the high deposition target power density. This enhancement in mobility is high enough to assure nanocrystals/amorphous phase segregation, forming the nc-TiN/a-SiN$_x$ nanocomposite.

4.1.3.2. Grain Size and Distribution

Usually, XRD results are used to calculate grain size with Scherrer formula [79] or Williamson–Hall plot analysis [121] owing to convenient sample preparation compared to that of the TEM sample. However, grain size calculated based on XRD results can be affected by some factors, such as residual stress and texture. Thus, the bright-field TEM morphological appearance of the nc-TiN/a-SiN$_x$ nanocomposite thin film is used to determine grain size and crystallite fraction. Typical bright-field TEM microphotos and the corresponding SAD patterns of nc-TiN/a-SiN$_x$ thin films with different silicon content are presented in Fig. 4.20 (with high magnification) and Fig. 4.21 (with low magnification). Grain size and crystallite fraction obtained based on the high magnification photos are consistent with those based on the low magnification ones.

Figure 4.22 shows that nc-TiN grain size changes as a function of silicon content. The grain size of nc-TiN basically decreases with an increase in silicon content. During the deposition, the Ti target power density does not vary (5.5 W/cm^2) while Si$_3$N$_4$ target power density varies from 1.1 to 5.5 W/cm^2 (samples P1 through P5) with N$_2$ to Ar gas ratio as unity. Though the presence of a large amount of N$_2$ may result in Ti target poisoning to a certain degree (by forming TiN on Ti target surface) that would contribute to some lessening of the Ti ion partial pressure, this effect would be the same for samples P1 to P5 since they have the same Ti target power density and the same N$_2$ flow rate. However, the increase in of Si$_3$N$_4$ target power inevitably increases silicon ion mobility and partial pressure while reducing the titanium ion partial pressure, giving rise to a higher probability of SiN$_x$ formation than TiN crystals. By the same token, the growth of TiN nanocrystals is hindered, resulting in a decrease in the crystallite size. At Si$_3$N$_4$ target power density of 6.6 and 7.7 W/cm^2 (samples P6 through P8) and Ti target power density of 4.4 W/cm^2 down to zero, the titanium ions do not have enough mobility or energy to form nc-TiN, thus the films deposited are basically amorphous.

Figure 4.23 shows the relationship between crystallite fraction and silicon content. The increase in silicon composition comes from the increase in silicon nitride target power density. At a higher sputtering power density of Si$_3$N$_4$, more ionic nitrogen should be available for the formation of TiN crystallites from more complete ionization of reactive N$_2$ gas and a certain degree of dissociation from the Si$_3$N$_4$ compound. Also, higher target power density effectively exerts stronger ion bombardment on the growing film,

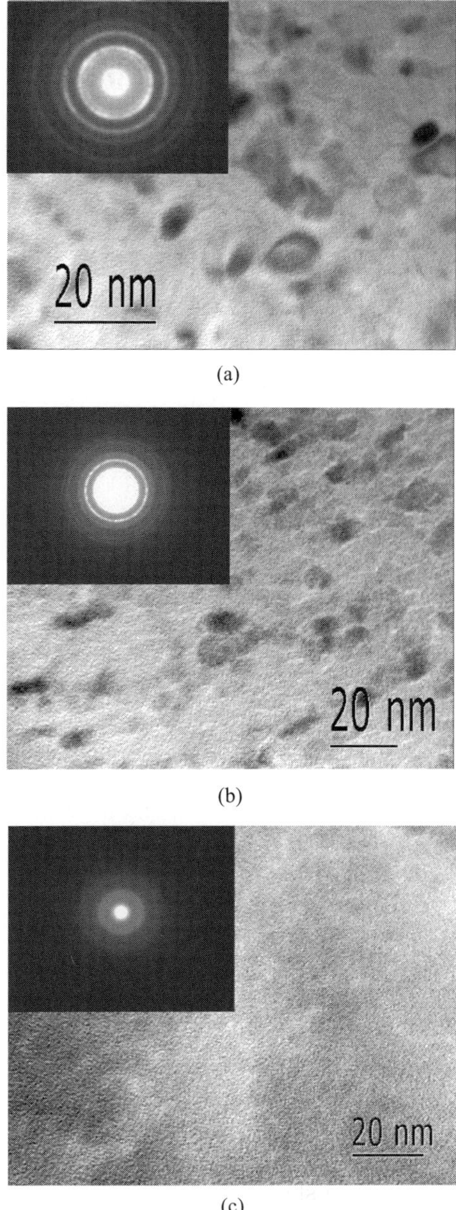

Fig. 4.20. Bright-field TEM images (high magnification) and corresponding SAD patterns of nc-TiN/a-SiN$_x$ nanocomposite thin films with different Si content: (a) 5.2 at.% (b) 14.9 at.%, and (c) 52.7 at.%. Grain size decreases with an increase in Si content.

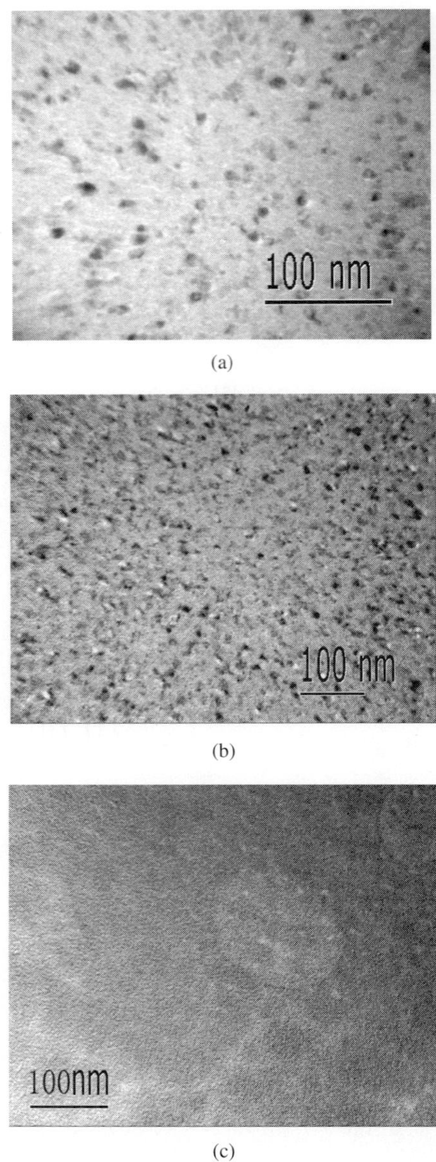

Fig. 4.21. Bright-field TEM images (low magnification) of nc-TiN/a-SiN$_x$ nanocomposite thin films with different Si content: (a) 5.2 at.% (b) 14.9 at.%, and (c) 52.7 at.%. Grain size decreases with an increase in Si content.

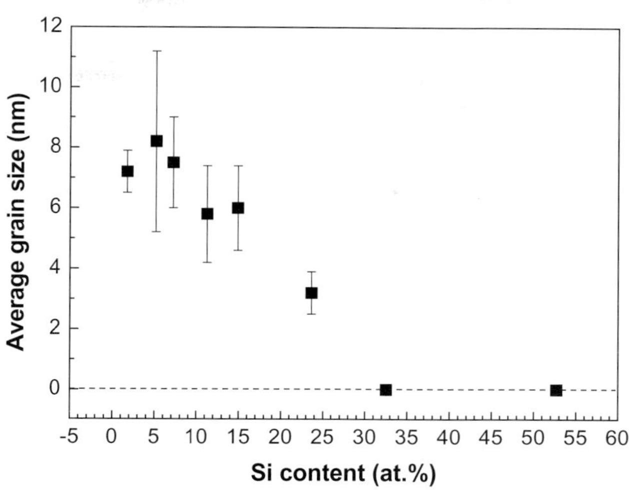

Fig. 4.22.   TiN crystallite size vs. Si content based on the TEM image observation. As Si content increases from 1.7 at.% to 52.7 at.%, TiN crystallite size decreases from about 8 nm to zero.

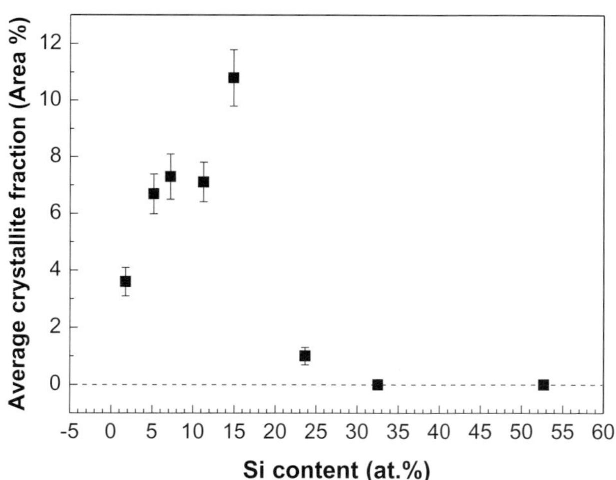

Fig. 4.23.   TiN crystallite fraction vs. Si content based on the TEM image observation. As Si content increases from 1.7 at.% to 14.9 at.%, TiN crystallite fraction increases from 4% to 11%. Further increase in Si content to 52.7 at.% results in nearly no TiN crystallite phase.

thus supplying more nucleus for the formation of crystalline TiN. In addition, more kinetic energy can enhance segregation and formation of TiN. Although this effect applies equally to the SiN_x matrix, it does not show up since it is amorphous. As a result, the nc-TiN crystallite fraction increases with silicon content. The drastic dip in nc-TiN crystallite fraction from 23.6 at.%Si actually comes from the reduction in Ti target power density from $4.4\,W/cm^2$ down to zero, and very little or no nc-TiN is formed.

It should be pointed out that deposition time is only 20 min for the TEM samples, compared to 120 min for those deposited on silicon wafers. Therefore, the average grain size measured based on TEM microphotos might be slightly smaller than the 120 min ones due to grain growth.

4.1.3.3. Preferential Orientation

Figure 4.24 reveals that Si_3N_4 target power density has a significant effect on the orientation of the TiN crystallites. The degree of preferential orientation can be quantitatively represented using coefficient of texture, T_{hkl}, through Eq. (3.6). The calculated coefficient of texture is tabulated in Table 4.4 and plotted in Fig. 4.25. At low Si_3N_4 target power density level, the nc-TiN is mainly oriented in the [111] direction (open squares in Fig. 4.25). Increasing the Si_3N_4 sputtering target power density level, the preferential orientation changes from [111] to [200] (solid circles) and [220] (open circles). Further increase of the target power density aligns

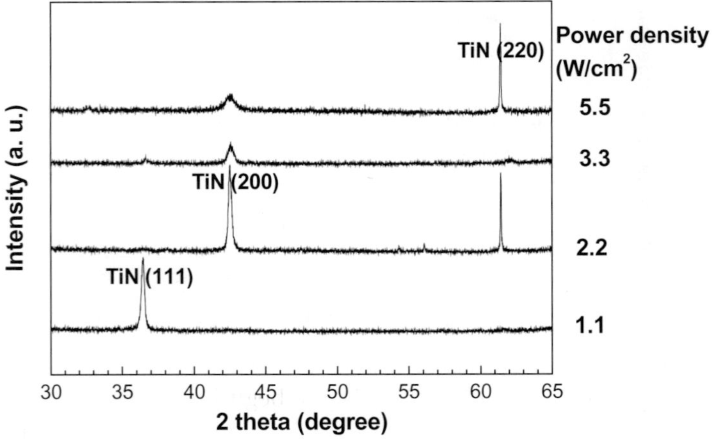

Fig. 4.24. XRD patterns of the nc-TiN/a-SiN$_x$ nanocomposite thin films with different Si$_3$N$_4$ target power densities, revealing a significant effect of Si$_3$N$_4$ target power density on the orientation of the nc-TiN crystallites.

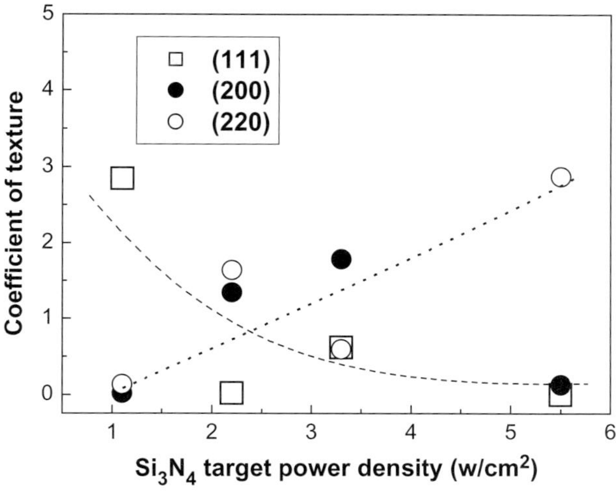

Fig. 4.25.   Coefficient of texture of nc-TiN crystallites changing with Si_3N_4 target power density. As Si_3N_4 power density increases, the preferential orientation changes from [111] to [200] (solid circles) and [220] (open circles). Further increase of the target power density aligns the nc-TiN crystallites in mainly the [220] direction. At very high Si_3N_4 target power densities, the crystallites predominantly orient along the [220] direction.

the nc-TiN crystallites mainly in the [220] direction. At very high Si_3N_4 target power density, the crystallites predominantly orient along [220] direction.

The orientation and the evolution of the growing film depend basically on the effective total energy: the rule of the lowest energy. This is also true in the case of preferential growth orientation of TiN thin films [122]. The most prominent competition energies are surface energy [123] and strain energy [124]. The surface energy of a plane can be calculated using the sublimation energy and the number of free bonds per surface area. For TiN, the surface energies for (111), (200) and (220) are 400, 230 and $260\,J/m^2$, respectively. Sublimation energy is 6.5×10^{-19} J/atom [123, 125]. The (200) plane is the plane of the lowest surface energy in the TiN crystal (which has the NaCl-type structure [126, 127]). Thus, as the surface energy is the main driving force, the TiN crystallites will orient on (200) plane (or [200] growth direction). At the very beginning of the deposition, it should be a surface energy controlled process because the strain energy has not kicked in yet, thus (200) crystallographic plane should be predominantly parallel to the substrate surface. Since the total strain energy is proportional to the layer thickness and depends on the mean elastic moduli acting in the (*hkl*) plane parallel to the interface, competition from strain energy kicks in as the film grows. For pure TiN, the strain energy is minimized when the (111) plane is parallel to the interface [125]. In our experiment, even though the crystallite is not pure TiN but a (Ti, Si)N solid solution, at low Si_3N_4 target power densities, the amount of Si atoms in the TiN lattice is very limited (less than 2 at.% in the whole film at $1.1\,W/cm^2$, as revealed by XPS), thus the strain energy of the (111) plane is dominant, resulting in the (111) texture. As deposition power density increases, the ions obtain greater kinetic energy which is even high enough to force itself into the empty space in the TiN lattice and form interstitial solid solution, as manifested in the lattice parameter increase (Table 4.4). Meanwhile, simple calculation reveals that for the fcc structure (Fig. 4.26), the (111) plane has the highest planar density (PD) [19] (defined as number of atoms per unit area in the plane): $0.29/r^2$, where r is the radius of the atom forming fcc lattice. Thus squeezing into (111) plane is more difficult (this would cause further increase in total system energy). For (200), PD $= 0.25/r^2$, and for (220), PD $= 0.18/r^2$. Therefore, it is more probable for arriving ions to insert into (200) or (220) planes, resulting in the least increase in the total system energy is (less strain energy is generated because the atoms or ions do not have to squeeze as hard). Consequently, since the (220) plane has the lowest planar density, at high deposition power density, the preferential orientation is the [220] direction.

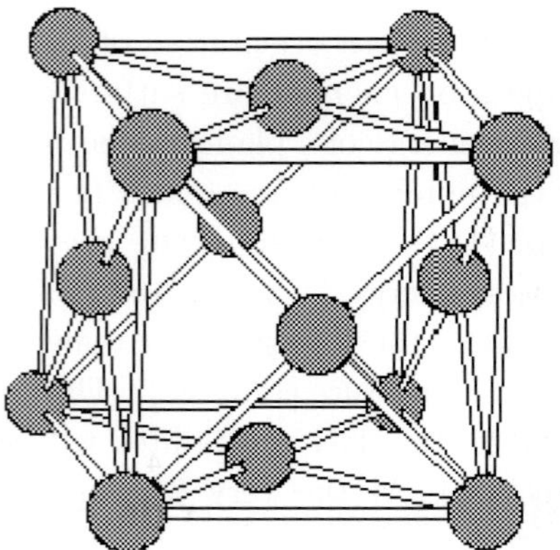

Fig. 4.26. Schematic diagram of fcc structure. The (111) plane has the highest planar density: $0.29/r^2$, where r is the radius of the atom forming fcc lattice. For (200), PD $= 0.25/r^2$, and for (220), PD $= 0.18/r^2$.

4.1.3.4. Lattice Parameter

During co-sputtering of Ti and Si_3N_4 targets in Ar and N_2 atmosphere, aside from forming amorphous SiN_x in the matrix, silicon also enters into TiN crystallites as solid solute which can affect the nc-TiN lattice parameters. The calculated lattice parameters are listed in Table 4.4 and plotted in Fig. 4.27 as a function of Si_3N_4 target power density. With an increase in Si_3N_4 target power density, the lattice parameter increases from 0.42350 to 0.44408 nm. It is interesting to note that the measured TiN lattice parameter of 0.42350 nm at the lowest Si_3N_4 target power density (1.1 W/cm^2) is smaller than that of the database value of 0.42417 nm; all other values are larger than this value. At the 1.1 W/cm^2 target power density level, there is reason to believe that the silicon ions for this deposition has very low kinetic energy and thus very low mobility. In such a case, silicon forms substitutional solid solution with TiN [128], i.e. (Ti, Si)N. Since the ionic radius of Si $^{4+}$ (0.041 nm) is only about half of that of Ti $^{3+}$ (0.075 nm) [129], substitution of titanium with silicon results in the reduction in lattice parameter. Thereafter, as the Si_3N_4 target power density is increased to 2.2, 3.3 and 5.5 W/cm^2, the silicon ion would have obtained much more kinetic

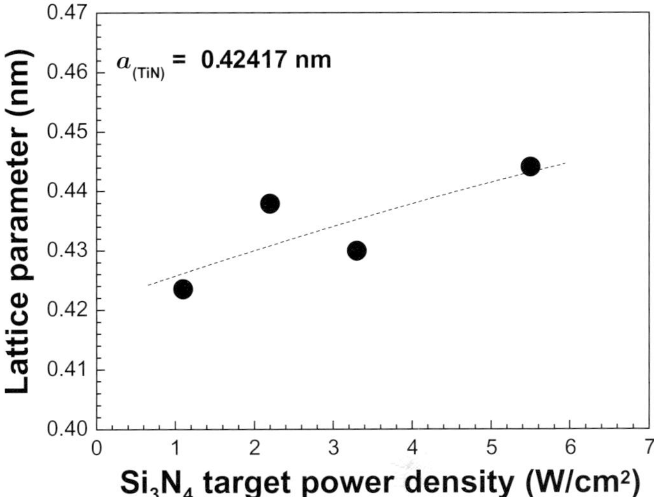

Fig. 4.27.   Calculated lattice parameter as a function of Si_3N_4 target power density. With an increase in Si_3N_4 target power density from 1.1 to 5.5 W/cm^2, lattice parameter increases from 0.42350 to 0.44408 nm.

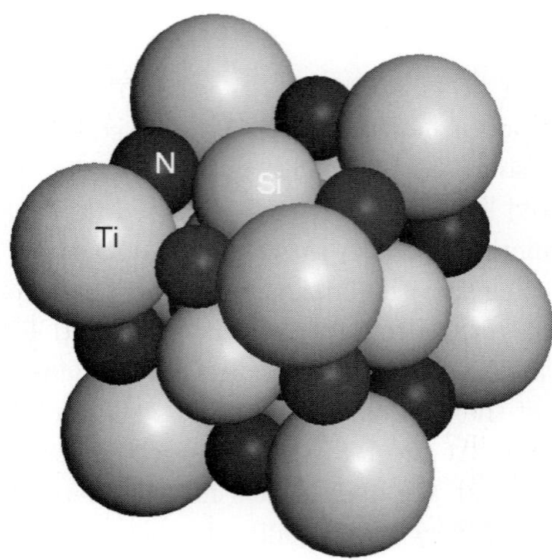

Fig. 4.28. Schematic diagram of microstructure of (Ti, Si)N solid solution, where Si is in the interstitial position of TiN lattice.

energy, thus forming interstitial solid solution (Fig. 4.28) [129]. Therefore, an increase in lattice parameter is observed.

4.1.4. *Mechanical Properties*

The effects of the sputtering power density of Si_3N_4 target on mechanical properties, such as residual stress, Young's modulus, hardness and adhesion strength are summarized in Table 4.5.

Table 4.5. Mechanical characteristics of hard and superhard nc-TiN/a-SiN$_x$ nanocomposite thin films.

Sample code	P1	P2	P3	P4	P5	P6	P7	P8
Residual stress σ (MPa)	-4.2	-5.6	-40.2	-82.0	-150.0	—	-67.5	-52.5
Hardness H (GPa)	7.6	16.6	20.7	25.8	36.8	16.4	15.5	15.0
Young's modulus E (GPa)	196	209	222	276	324	211	185	184
Critical load (mN) L_{C1}	200	300	250	225	175	50	50	50
L_{C2}	350	375	450	375	325	200	225	250

4.1.4.1. Residual Stress

The residual stress data obtained from Eq. (3.17) is plotted in Fig. 4.29 as a function of Si_3N_4 target power density. With an increase of Si_3N_4 target power density, the residual stress increases from near-zero to about $-0.15\,GPa$. The minus sign indicates that the stress is in the compressive state. This is a very small residual stress for magnetron sputtering deposited thin films.

Assuming that the stress is zero, which results purely from a change of temperature (as a result of slow cooling from deposition temperature to room temperature), the residual stress σ in magnetron sputtered films is structure related and results from the three aspects are as follows (Eq. (4.1)): thermal stress σ_T, growth-induced stress σ_g and structural mismatch-induced stress σ_m:

$$\sigma = \sigma_T + \sigma_g + \sigma_m. \tag{4.1}$$

The thermal stress σ_T is an extrinsic stress resulting from the difference in the coefficient of thermal expansion (CTE). The latter two stresses (growth-induced stress σ_g and structure mismatch-induced stress σ_m) constitute the intrinsic stress [130, 131]. In our nc-TiN/a-SiN_x nanocomposite

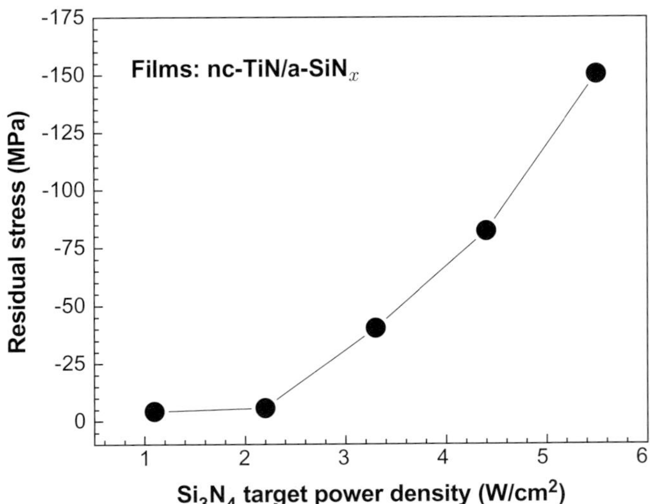

Fig. 4.29. Relationship of residual stress with Si_3N_4 target power densities. With increasing Si_3N_4 target power density, the residual stress increases from near-zero to about $-0.15\,GPa$.

thin films, since the matrix is amorphous, basically no effort is needed in structure "matching" between nc-TiN and the a-SiN$_x$ matrix, therefore $\sigma_m \sim 0$. Thus in our system, the residual stress σ mainly comes from thermal stress σ_T and growth-induced stress σ_g, which are analyzed in detail below.

1. Thermal Stress (σ_T)

The thermal stresses originate from the fact that the CTE is different between the thin film and the substrate. Hence the shrinkage of the film differs from that of the substrate during post deposition cooling. The stress in the film measured at room temperature T_{rm} (usually 25°C) can be calculated using the following equation [132, 133]:

$$\sigma_T = \frac{E_f}{1 - \nu_f} \int_{T_{\mathrm{dep}}}^{T_{\mathrm{rm}}} (\alpha_s(T) - \alpha_f(T)) \cdot dT, \tag{4.2}$$

where T_{dep} and T_{rm} represent the deposition temperature and room temperature, respectively. E_f and ν_f are the elastic modulus and the Poisson ratio of the film. $\alpha_f(T)$ and $\alpha_s(T)$ are CTE of the thin film and substrate, respectively. Though CTE normally varies with temperature, for a rough estimation, if we do not take the temperature effect into account, we may use the CTE data from the literature (listed in Table 4.6) to estimate the thermal stresses. Our XPS analysis shows there is no trace of Ti–Si; even if there is some, the amount is quite small. So we do not consider the TiSi phase effect during the residual stress analysis. Using XPS composition results (Table 4.2) and assuming only TiN and Si$_3$N$_4$ (for the simplicity of estimating the fraction of a-SiN$_x$) exist in the film, TiN fraction is estimated. Based on the rule of mixtures between TiN and Si$_3$N$_4$, CTE values of the nc-TiN/a-SiN$_x$ nanocomposite thin films are calculated and listed in Table 4.7. Also listed in Table 4.7 are the thermal stresses calculated using Eq. (4.2). Note that these values are positive, i.e. a tensile stress is generated in the film due to thermal mismatch between the film and the

Table 4.6.   Material constants used for calculation of thermal stress σ_T.

Constant	Notation	Value	Ref.
CTE (Si substrate)	α_{Si}	2.6×10^{-6}/K	[134]
CTE (TiN)	α_{TiN}	9.4×10^{-6}/K	[135]
CTE (Si$_3$N$_4$)	α_{Si3N4}	2.5×10^{-6}/K	[135]
Poisson's ratio (for all films)	ν_{f}	0.25	[136]

Table 4.7. Calculated CTE and thermal stress σ_T.

Sample code	Si_3N_4 power density (W/cm^2)	TiN (%)	CTE ($\times 10^{-6}$/K)	E (GPa)	σ_T (GPa)
P1	1.1	98.86	9.3	196	0.75
P2	2.2	96.17	9.1	209	0.78
P3	3.3	94.07	9.0	222	0.80
P4	4.4	90.33	8.7	276	0.95
P5	5.5	85.56	8.4	324	1.05

substrate. This is easily understood since the CTE of the film is greater than that of the substrate. Upon cooling, the shrinkage in the film would be more than that in the substrate. Since the film is not delaminated, the film must then be "stretched" (thus tensile stress) to satisfy the continuity between the film and the substrate.

2. Growth-Induced Stress (σ_g)

The stress resulting from the growth of sputtered thin film depends on substrate temperature, gas pressure and bias voltage, i.e. the kinetic energy of the ions. In magnetron sputtering, the kinetic energy of the ions is of the order of 100 eV [130, 137], sufficient to create defects in the growing layer. Increasing the bombardment leads to an increase in stress [131]. In general, as sputtering power density increases, the kinetic energy of the impinging ions increases, which results in an increase in the strain of the film [138]. Within a certain sputtering power range, this can be best understood by relating the deposition ion energy U_k with the target power density D_w together with substrate bias V_s and gas pressure P_g through Eq. 4.3:

$$U_k \propto \frac{D_w V_s}{P_g^{1/2}}. \tag{4.3}$$

Therefore, this energetic ion generates stress in both the amorphous matrix and the nanocrystalline phase. In the amorphous matrix, the energetic ionic bombardment induces atomic distortion or atom displacement from their equilibrium (stress-free) locations. In the crystalline phase, this energetic ion generates lattice distortion in terms of substitutional or interstitial solid solution (Table 4.4), resulting in the increase in the lattice parameter of nc-TiN from 0.42350 nm to 0.44408 nm. The total growth-induced stress σ_g can be seen from the difference of the total residual stress σ and the thermal stress σ_T. From Eq. (4.1),

$$\sigma_g = \sigma - \sigma_T. \tag{4.4}$$

Since the total residual stress σ (neglecting structural mismatch-induced stress σ_m) is from near-zero to -0.15 GPa, and the thermal stresses σ_T are about 1 GPa (Table 4.5), the growth-induced stress σ_g is therefore about -1 GPa. These two components of residual stresses are opposite in sign, thus they mostly cancel each other out, resulting in an almost stress-free thin film.

4.1.4.2. Hardness

4.1.4.2.1. *Nanoindentation*

Figure 4.30 shows one typical nanoindentation load-displacement profile of nc-TiN/a-SiN$_x$ nanocomposite thin film with 14.9 at.%Si (sample P5). The film thickness is 600 nm. The film roughness (R_a) in 300 nm $\times$ 300 nm areas is about 0.8 nm (less than 1% of the film thickness), which is identified using AFM. The maximum load during indentation is 1.14 mN and the maximum indentation depth is 50 nm (less than one tenth of the film thickness). The hardness and Young's modulus analyzed using Oliver–Pharr method are 36.8 GPa and 324 GPa, respectively.

4.1.4.2.2. *Effect of Indentation Depth*

Nanoindentation with different indentation depth were performed on nc-TiN/a-SiN$_x$ nanocomposite thin film with 11.3 at.%Si (sample P4, with

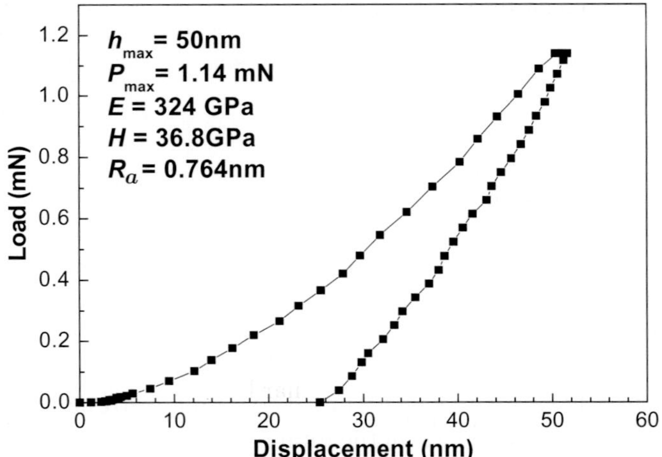

Fig. 4.30. Nanoindentation load-displacement profile of nc-TiN/a-SiN$_x$ nanocomposite thin film with 14.9 at.%Si (sample P5). Nanoindentation depth is less than one tenth of the films thickness so as to avoid the effect of substrate on film hardness.

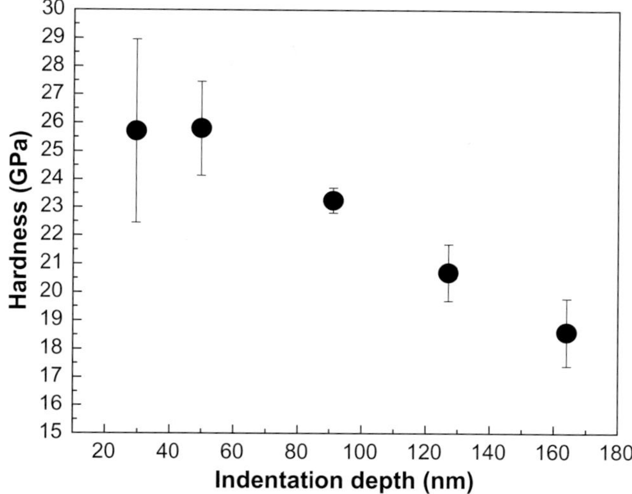

Fig. 4.31. Hardness as a function of indentation depth. When indentation depth is less than one tenth of the coating thickness, the hardness near keeps a constant value of about 25.8 GPa. With increase in indentation depth, the measured hardness decreases due to the effect of Si substrate (hardness 9.8 GPa).

600 nm in thickness) to study the effect of indentation depth on film hardness. Figure 4.31 shows the measured hardness as a function of indentation depth. When indentation depth is within one tenth of film thickness, the average measured hardness is nearly constant, which represents the intrinsic hardness. The measured intrinsic hardness of nc-TiN/a-SiN$_x$ nanocomposite thin film with 11.3 at.%Si is 25.8 GPa. Under lower indentation depth, such as 25 nm, the deviation from the average value is much larger than that under high indentation depth, for example 50 nm. This is due to the fact that small indentation depth tends to be significantly affected by surface roughness compared to high indentation depth. When indentation depth exceeds one tenth of film thickness, with further increase in depth, the measured hardness decreases linearly. This is because the elastic–plastic deformation field in the film just below the indenter is not confined to the film itself, but to a long range field that extends into the silicon substrate. The effect of silicon substrate on film hardness is inevitable. The silicon (100) wafer substrate has much lower hardness (9.8 GPa) than that of nc-TiN/a-SiN$_x$ nanocomposite thin film (25.8 GPa) with 11.3 at.%Si (sample P4). Therefore, with an increase in indentation depth, the measured hardness decreases.

 S. Zhang et al.

4.1.4.2.3. *Effect of Residual Stress*

Figure 4.32 shows the relationship between hardness, residual stresses and silicon content for nc-TiN/a-SiN$_x$ nanocomposite thin films. Veprek [12] pointed out that the intrinsic hardness values for film are those with residual stress less than 1 GPa. Under that condition, the residual stress will not contribute to the measured hardness value. Since all the as-prepared nc-TiN/a-SiN$_x$ nanocomposite thin films have residual stresses far less than 1 GPa, the enhancement of the hardness in these nanocomposite films is not due to residual stress.

4.1.4.2.4. *Effect of Grain Size and Crystallite Fraction*

Plotting nanoindentation hardness against nc-TiN crystallite size gives rise to Fig. 4.33. This clearly demonstrates: (a) as crystallite size decreases (going from right to left on X-axis), film hardness increases drastically (Hall–Petch relationship); (b) the maximum (37 GPa) is approximately 6 nm crystallite size; (c) further decrease in crystallite size brings about a decrease in hardness (the anti-Hall–Petch effect); and (d) the hardness tails off to the hardness of an amorphous phase as crystallite size diminishes. This result matches extremely well with Schiotz's theoretical prediction [16].

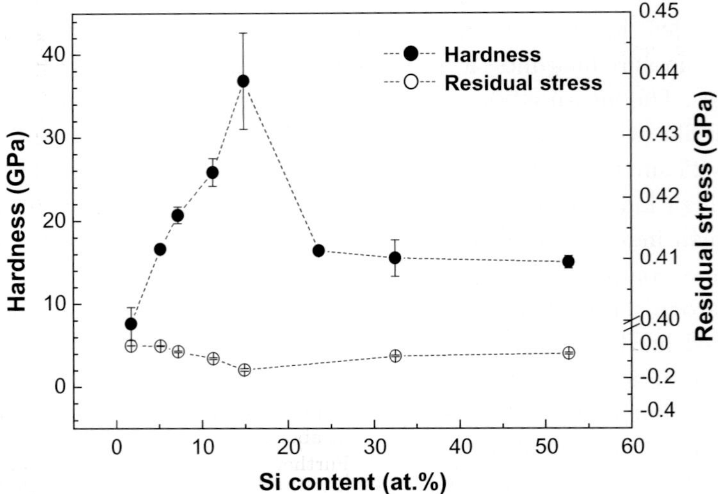

Fig. 4.32.   Relationship between hardness, residual stress and Si content for nc-TiN/a-SiN$_x$ nanocomposite thin films. There is no impact on the film hardness from the residual stress.

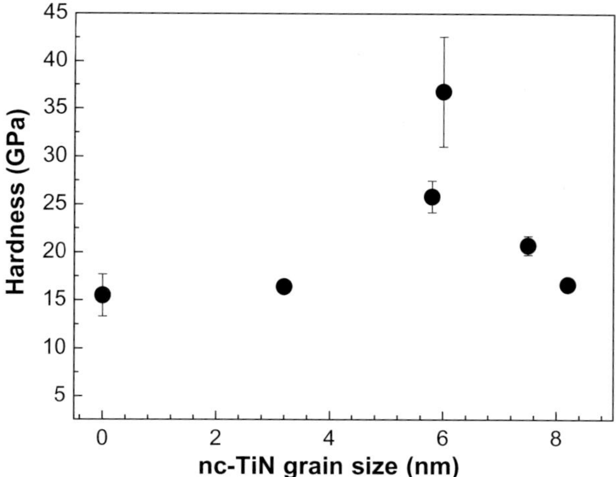

Fig. 4.33. Nanoindentation hardness vs. nc-TiN crystallite size in the nc-TiN/a-SiN$_x$ nanocomposite thin films: both Hall–Petch and anti-Hall–Petch phenomena are observed.

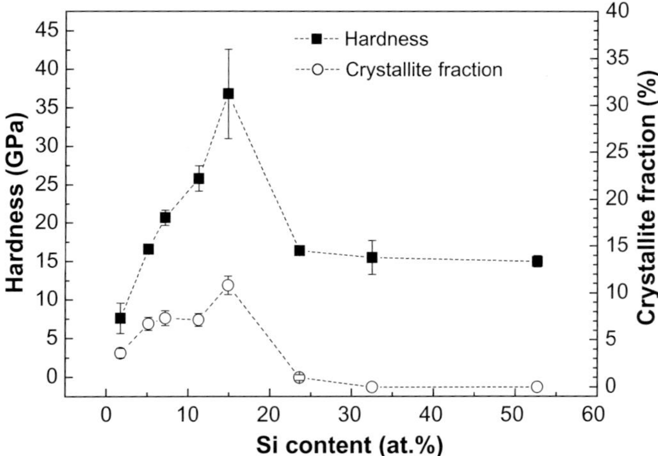

Fig. 4.34. Hardness varies with nc-TiN crystallite fraction and Si content. Film hardness increases significantly to about 37 GPa at about 14.9 at.%Si, corresponding to the maximum crystallite fraction of about 11%. Further increase in Si brings about drastic decrease in hardness to about 15 GPa, which is a common value reported for Si$_3$N$_4$ film; this again, corresponds to the reduction in crystallite fraction.

Figure 4.34 displays the relationship between film hardness together with crystallite fraction as a function of silicon content. The film hardness increases significantly to about $37\,\mathrm{GPa}$ at about 14.9 at.%Si, corresponding to the maximum crystallite fraction about 11%. Further increase in silicon brings about drastic decrease in hardness to about $15\,\mathrm{GPa}$, which is a common value reported for Si_3N_4 thin film. This again, corresponds to the reduction in crystallite fraction. In order to simplify the calculation, assuming crystallite is a ball-shape with diameter of d and the grain boundary is the near out-layer (thickness Δd) of the crystallite, then the volume of grain boundary V_{boundary} can be roughly estimated using

$$V_{\mathrm{boundary}} = \frac{V_{\mathrm{crystal}}}{\frac{4}{3}\pi(\frac{d}{2})^3} \cdot \left[\frac{4}{3}\pi \left(\frac{d}{2} + \Delta d \right)^3 - \frac{4}{3}\pi \left(\frac{d}{2} \right)^3 \right]. \qquad (4.5)$$

Taking the first two terms of Taylor expansion, Eq. (4.5) is simplified to

$$V_{\mathrm{boundary}} \approx \frac{V_{\mathrm{crystal}}}{\frac{4}{3}\pi(\frac{d}{2})^3} \cdot 4\pi \left(\frac{d}{2} \right)^2 (\Delta d) = 6\frac{V_{\mathrm{crystal}}}{d} \cdot \Delta d, \qquad (4.6)$$

where V_{crystal} is the volume of crystalline. From Eq. (4.6) it can be seen that the grain boundary volume is proportional to crystalline volume and reversely to grain size. With increase in silicon content to about 14.9 at.%, the grain size decreases to about $5\,\mathrm{nm}$ (Fig. 4.22) while crystalline fraction (represented in area ratio in this case) increases to 11% (Fig. 4.23). Therefore, the grain boundary volume should increase significantly. Thus, there is more grain boundary to effectively hinder dislocation propagation, which contributes to the increase in hardness. With further increase in silicon content, the crystallite fraction (represented in area ratio in this case) decreases to less than 1%, and at the same time, grain size decrease to $3\,\mathrm{nm}$. Grain boundary strengthening is weakened. The grain boundary sliding is more active, which contributes to the decreases in hardness.

4.1.4.2.5. *Relationship Between Hardness and Young's Modulus*

Figure 4.35 shows the relationship between hardness and Young's modulus for a number of nc-TiN/a-SiN$_x$ nanocomposite thin films with different silicon content. The experiment data show an approximately linear increase in the Young's modulus with the hardness. The Young's modulus varies by a factor of 5.45, showing a proportionality $E = [122 \pm 8]+[5.45 \pm 0.37] \times H$. It is generally accepted that hardness of bulk material scales with the energy

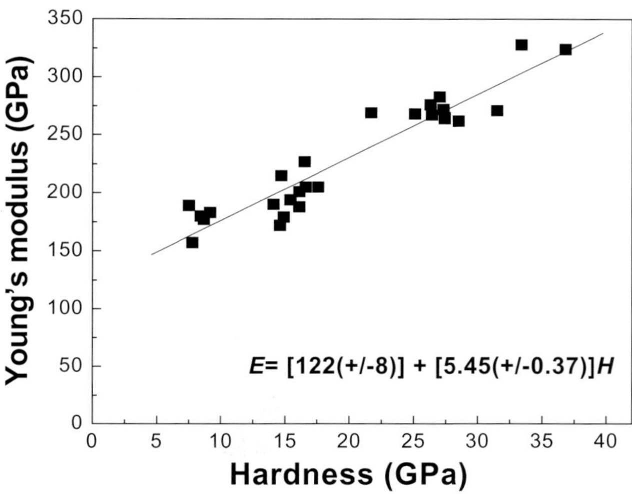

Fig. 4.35. Correlation between indentation hardness and Young's modulus for nc-TiN/a-SiN$_x$ nanocomposite thin films.

of dislocation because the plastic deformation of crystalline materials is due to dislocation activity, and the energy of a dislocation is proportional to the shear modulus through [139]:

$$U_{\mathrm{screw}} = G\bar{b}^2, \tag{4.7}$$

$$U_{\mathrm{edge}} = G\frac{\bar{b}^2}{1-v}, \tag{4.8}$$

where U_{screw} and U_{edge} are energies of screw dislocation and edge dislocation, respectively. $\bar{b}$ is Burgers' vector. G is the shear modulus, which is related to Young's modulus E. Therefore, it is not surprising that such correlation between the indentation hardness H and the Young's modulus E can be found for the hard and superhard nc-TiN/a-SiN$_x$ nanocomposite thin films.

Figure 4.36 shows H^3/E^2 as a function of thin film hardness. The ratio of H^3/E^2 is regarded by Tsui *et al.* [140] as an important criterion of the resistance against plastic deformation for hard materials and is emphasized by Musil [141] as an important criterion of mechanical properties for hard nanocomposite thin films. Measured values of hardness H and Young's modulus E enable calculation of the ratio H^3/E^2, which gives information on the resistance of materials to plastic deformation. The higher the ratio of H^3/E^2, the higher the resistance to plastic deformation. From Fig. 4.36,

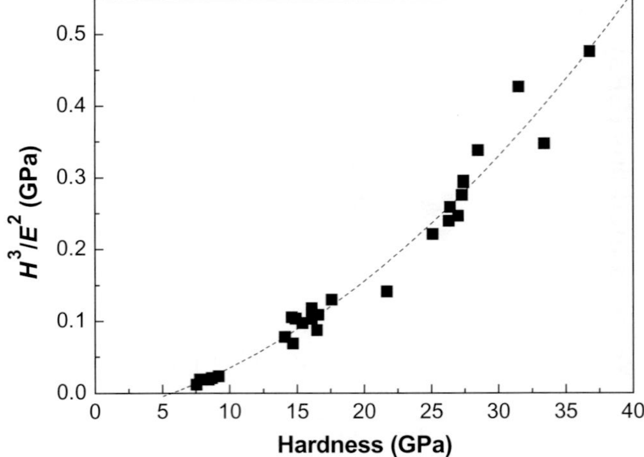

Fig. 4.36. H^3/E^2 as a function of hardness. The resistance to plastic deformation increases with the increase in film hardness.

it can be clearly seen that the resistance to plastic deformation increases with the increase in film hardness.

4.1.4.3 Toughness

To investigate the repeatability of microindentation method in determining thin film toughness, microindentation under different indentation load were conducted on 600 nm thick nc-TiN/a-SiN$_x$ nanocomposite coatings with different silicon content. The well-defined ($c \geq 2a$) cracks generated by microindentation were used to calculate the thin films toughness through Eq. (3.16). In order to obtain well-defined radial crack ($c \geq 2a$) and to focus and observe conveniently under SEM, indentation loads of 10.00, 5.00, 2.00 and 1.00 N were used. The sample information and calculated toughness values are tabulated in Table 4.8 and plotted in Fig. 4.37. The extrapolated toughness values in Table 4.8 were obtained by plotting the calculated toughness value and then extrapolating the curve to depth of about one tenth of the film thickness.

Figure 4.37 shows the calculated toughness of nc-TiN/a-SiN$_x$ thin films with different silicon content under different indentation loads. All the curves obtained under different loads, such as 10.00, 5.00, 2.00 and 1.00 N, confirm that the nc-TiN/a-SiN$_x$ nanocomposite thin film with 7.2 at.%Si (sample P3) possesses the highest toughness. In addition, all the curves plotted show a consistent trend, which indicates that the microindentation

Table 4.8.   Toughness of nc-TiN/a-SiN$_x$ nanocomposite thin films with different Si content under different indentation loads.

| Sample code | Indentation load (N) | | | | Extrapolated toughness value (MPa m$^{1/2}$) |
| | 10.00 | 5.00 | 2.00 | 1.00 | |
	Calculated toughness value (MPa m$^{1/2}$)				
P1	1.13	1.40	1.18	1.53	1.31
P2	0.75	0.96	0.82	0.92	0.86
P3	1.20	1.40	1.89	2.00	1.62
P4	0.81	0.89	1.11	1.46	1.07
P5	0.75	0.73	0.72	0.94	0.78
P6	0.86	0.83	1.13	1.16	0.99
P7	0.95	0.80	0.99	1.15	0.97
P8	0.88	1.04	0.87	1.23	1.00

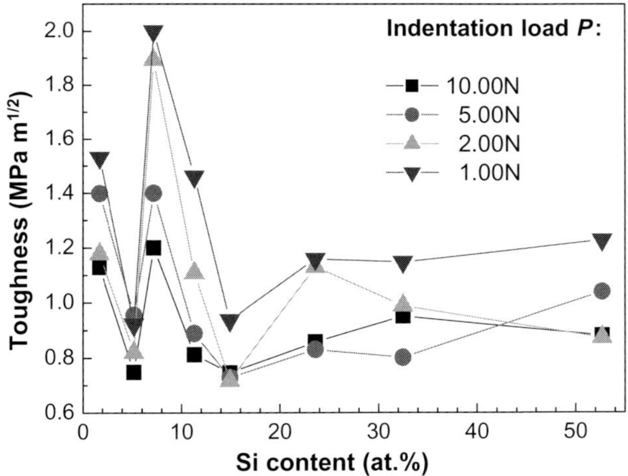

Fig. 4.37.   Toughness (under different test loads: 10.00, 5.00, 2.00 and 1.00 N) of nc-TiN/a-SiN$_x$ nanocomposite thin films with different Si content. From the four curves plotted, it can be concluded that microindentation is a reasonable method for thin film toughness measurement with high consistency and repeatability.

method has high repeatability and can be used as a fairly reliable representation of toughness for thin films. However, it should be pointed out that the measured toughness is affected by the silicon substrate under such high indentation load due to the small thickness (0.6 μm) of the coatings.

It is worth pointing out that, in this section, we assess the toughness measurement methodology. The drop of toughness for nc-TiN/a-SiN$_x$ with 5.2 at.%Si is not due to the test method. This trend has also been observed for other testing approaches, such as scratch and nanoindentation (Fig. 4.42). The mechanism of toughness of nc-TiN/a-SiN$_x$ nanocomposite thin film varying with silicon content needs further study.

4.1.4.4 Adhesion

In a scratch test, the critical loads L_{c1} and L_{c2} can be determined through a combination of optical observation (Fig. 4.38) and a sudden kink in friction (expressed in terms of voltage ratio in Fig. 4.39) [12, 142]. The complete peeling off of film from substrate is represented by a sudden large change in friction.

Figure 4.38 shows the optical observations of scan scratch track on nc-TiN/a-SiN$_x$ nanocomposite thin film surface at both low and high magnifications. Below the lower critical load L_{c1}, the thin film is scratched by the scratch tip, associated with the plastic flow of materials. At and after

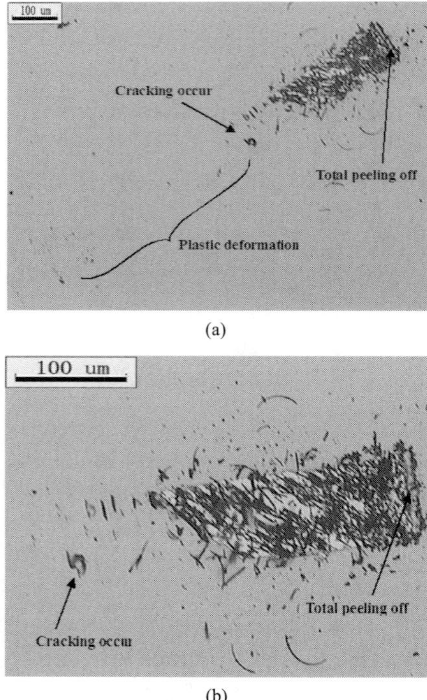

Fig. 4.38. Optical observation with (a) low magnification, and (b) high magnification of the film surface after scanning scratch, showing that cracking occurs and there is a complete peeling off of coating from substrate.

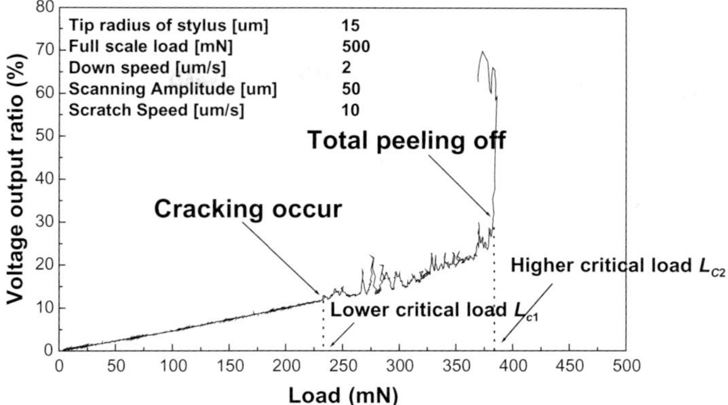

Fig. 4.39.   One example of scratch test showing lower critical load L_{c1}, under which the cracking occurs in coating, and higher critical load L_{c2}, under which there is a complete peeling off of coating from substrate.

the lower critical load L_{c1}, cracks were observed. The schematic diagram of scratch damage mechanism for the nc-TiN/a-SiN$_x$ nanocomposite thin film is shown in Fig. 4.40. Before the lower critical load L_{c1}, if a film has good toughness, scratch associated with the plastic flow of materials is responsible for the damage of film [Fig. 4.40(a)]. If a film has a poor toughness, cracking could occur during the scan scratch process, associated with cracks and debris [Fig. 4.40(b)]. When the load increases up to the higher critical load L_{c2}, delamination or buckling will occur, which induces complete peel off of films from substrate [Fig. 4.40(c)].

Figure 4.41 shows the scratch toughness, which is represented by lower critical load L_{c1} varying with silicon content in nc-TiN/a-SiN$_x$ nanocomposite thin films. The scratch toughness is characterized using lower critical load L_{c1} obtained from the scratch test. The highest scratch toughness can achieve approximately 300 mN for nc-TiN/a-SiN$_x$ nanocomposite films with 5.2 at.%Si. However, the critical load is not "fracture toughness" (and, of course, the unit is wrong for fracture toughness). What the lower critical load represents is a load bearing capacity, or crack initiation load. Perhaps it can be treated as some sort of "crack initiation resistance": the higher the L_{c1}, the more difficult it is to initiate a crack in the film. However, initiation of a crack does not necessarily result in fracture in the film; what important is how long the film can hold and withstand further loading before a catastrophic fracture occurs. Thus, the film toughness should be proportional to both the lower critical load L_{c1} and the load difference $(L_{c2} - L_{c1})$ between

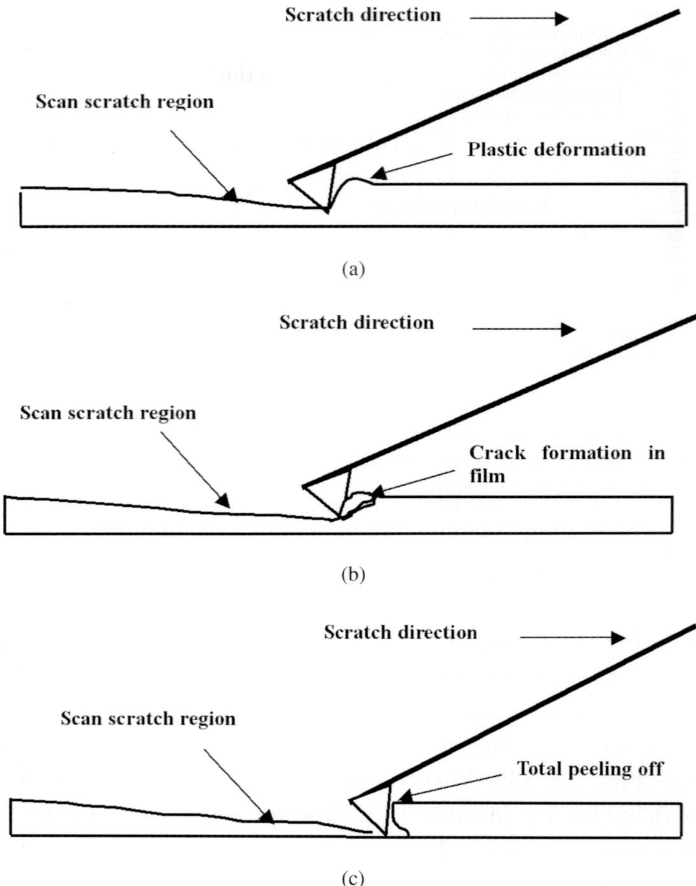

Fig. 4.40. Schematic diagram of scan scratch damage mechanism of the nc-TiN/a-SiN$_x$ nanocomposite thin film: (a) scan scratch associated with the plastic flow of materials, (b) crack formation in film at the lower critical load L_{c1}, and (c) total peeling off of coating from substrate at the higher critical load L_{c2}.

the higher and the lower critical load. The product of these two terms is termed "scratch crack propagation resistance", or CPR$_\text{s}$:

$$\text{CPR}_\text{s} = L_{c1}(L_{c2} - L_{c1}) \qquad (4.9)$$

The parameter CPR$_\text{s}$ can be used as a quick qualitative indication of the film toughness or used in a quality control process for tough film.

As a first degree approximation, the parameter "scratch crack propagation resistance" (or CPR$_\text{s}$) is used to qualitative estimate the film

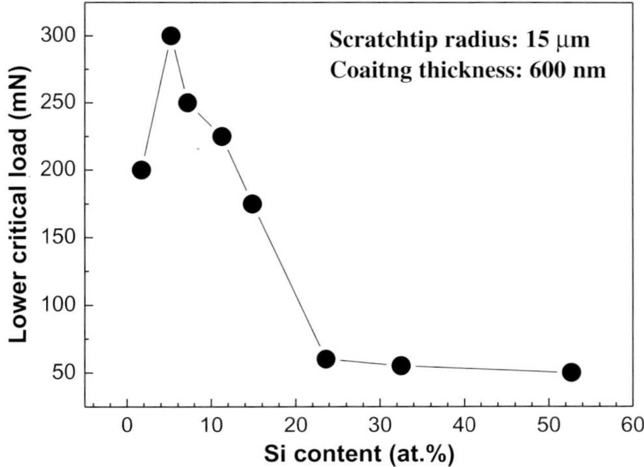

Fig. 4.41.   Scratch toughness represented by lower critical load L_{c1} for nc-TiN/a-SiN$_x$ nanocomposite thin films with different Si content. Sample P2 shows the best toughness property.

toughness. The film toughness represented by CPR$_s$ for nc-TiN/a-SiN$_x$ nanocomposite thin films with different silicon content is shown in Fig. 4.42. A high L_{c1} value means that the film has a high ability to resist cracking, whereas a high L_{c2} value means that the film is more durable, i.e. even if cracking occurs, the film is not totally damaged. From a machining application viewpoint, the nc-TiN/a-SiN$_x$ nanocomposite thin film prepared with silicon content of 7.2 at.% is preferred for its good comprehensive mechanical properties. This film has high crack initiation resistance (with $L_{c1} = 250\,\mathrm{mN}$) and high adhesion strength (with $L_{c2} = 450\,\mathrm{mN}$). At the same time, the film has a high hardness of about 20 GPa. While the nc-TiN/a-SiN$_x$ nanocomposite thin film with silicon content of 5.2 at.% shows poor mechanical properties, the crack propagation resistance (CPR$_s$) is the lowest (Fig. 4.42). The ability to resist crack for this film (with 5.2 at.%Si) is not too low ($L_{c1} = 300\,\mathrm{mN}$). However, if cracking occurs, the film will be totally damaged in a short time due to its lower adhesion strength ($L_{c2} = 375\,\mathrm{mN}$).

4.1.5. *Summary*

The effect of Si$_3$N$_4$ target power density (and/or target power ratio of Ti to Si$_3$N$_4$) on mechanical properties, such as residual stress, hardness and

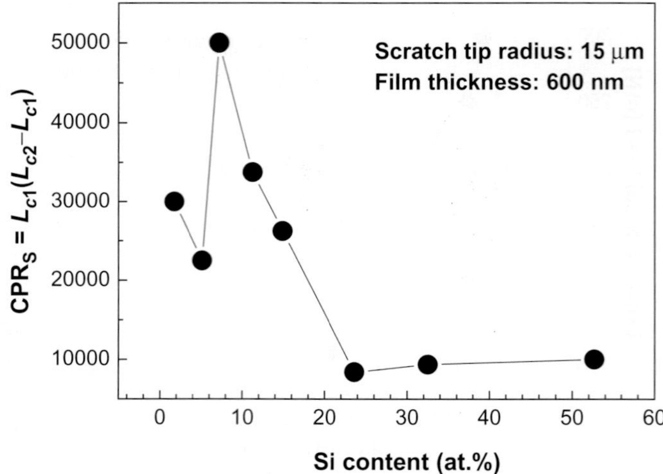

Fig. 4.42. Toughness (denoted by crack propagation resistance) of nc-TiN/a-SiN$_x$ nanocomposite thin films with different Si content. Sample P3 shows the best toughness property.

toughness, of the nc-TiN/a-SiN$_x$ nanocomposite thin films can be summarized as follows:

1. The residual stress resulting from sputtering deposition of nc-TiN/a-SiN$_x$ nanocomposite thin films comprise mainly of growth-induced stress and thermal stress. Growth-induced stress increases with sputtering target power density, but only reaches about -1 GPa, even when the highest target power density is used. Growth-induced stress is opposite in sign and roughly equal in quantity to the thermal stress induced by the difference in the coefficient of thermal expansion between the film and the silicon wafer substrate. As a result, the magnetron sputtered nc-TiN/a-SiN$_x$ nanocomposite thin films become almost stress-free.

2. Hardness was measured using nanoindentation with a depth less than one tenth of film thickness. The indentation data was analyzed using the Oliver–Pharr method. The enhancement in hardness is due not to low residual stress (less than -1 GPa), but to microstructure. The relationship between the nc-TiN/a-SiN$_x$ nanocomposite thin film hardness and the crystallite size of nc-TiN matches the Hall–Petch relationship at larger crystallite size and anti-Hall–Petch relationship as the crystallite size becomes very small (below 6 nm in this study). The relationship

between hardness and Young's modulus is agreeable with that obtained from the other nanocomposite thin film system.

3. Toughness measurement and assessment evaluation results confirm that microindentation is a convenient and repeatable technique used to characterize toughness of hard thin film deposited on brittle substrate. At high indentation load, well-defined $(c \geq 2a)$ radial crack lengths generated are clear and can be easily used to calculate toughness. However, substrate effect may come into play and affect the actual values of thin film toughness. At lower indentation load, substrate effect on thin film toughness measurement can be reduced. However, there come other difficulties, including generating a well-defined $(c \geq 2a)$ radial crack, locating the crack, focusing the crack under electron microscopy and measuring the crack length. A nanoindentation test can be conducted to characterize thin film toughness both quantitatively and qualitatively. In quantitative determination of thin film toughness, the substrate effect can be significantly reduced, however, difficulties come from locating the indentation impression and measuring crack length. In qualitative characterization, plasticity can present consistent trends for series samples. However, plasticity is not "toughness". Scratch test data can be used indirectly to qualitatively characterize thin film toughness. Crack propagation resistance parameter has been proposed. However, these parameters are not the termed "toughness". Two-step tensile test can be conducted to measure the toughness of the thin films in the case of satisfying the assumptions.

4. The toughness of as-prepared nc-TiN/a-SiN$_x$ nanocomposite thin films (samples P1 to P8) has been measured using the microindentation method due to its convenience. To deduce the film toughness from the substrate-effected data, K_{IC} values are plotted and the curve is then extrapolated to a depth of about one tenth of the film thickness to obtain the film toughness. The results show that the nc-TiN/a-SiN$_x$ nanocomposite thin film with 7.2 at.%Si has the highest toughness value of 1.62 MPa m$^{1/2}$, while the thin film with 14.9 at.%Si has the poorest toughness (0.78 MPa m$^{1/2}$).

4.2. *Ni-Toughened nc-TiN/a-SiN$_x$*

Irie *et al.* [143] proposed doping metallic nickel into hard and superhard TiN coatings by cathodic arc ion plating to improve toughness. TiN and Ni were selected for the high hardness phase and high toughness phase, respectively. In this study, in order to obtain thin films with high hardness in combination with high toughness, Ni-toughened nc-TiN/a-SiN$_x$ films were deposited by doping Ni into nc-TiN/a-SiN$_x$ with high hardness. TiNi (atomic ratio of 50/50, 99.99%) target was used as the source of Ni. DC magnetron powers were applied to Ti and TiNi targets, while RF power was used on Si$_3$N$_4$ target. Four-inch silicon (100) wafers were used as substrate. The substrate treatment was the same as that for deposition of nc-TiN/a-SiN$_x$ samples. The base pressure for deposition was 1.33×10^{-5} Pa. During deposition, gas pressure was 0.67 Pa, and gas flow rates for N_2 and Ar were both 15 sccm. Deposition was performed at a substrate temperature of 450°C for two hours to obtain films with thickness of about 0.7 μm. Film deposition parameters are listed in Table 4.9 according to the Ni-toughened nc-TiN/a-SiN$_x$ nanocomposite thin films prepared.

The as-prepared Ni-toughened nc-TiN/a-SiN$_x$ nanocomposite thin films were studied using different characterization techniques, such as XPS, AFM, XRD/GIXRD, TEM/HRTEM, scratch, microindentation and nanoindentation. The results and discussions are presented in this section.

4.2.1. *Composition*

All Ni-toughened nc-TiN/a-SiN$_x$ nanocomposite thin films are etched with an Ar ion beam for 15 min before composition measurement. The chemical compositions obtained from XPS are listed in Table 4.10, in which the atomic concentration of Ni, Ti, Si and N are used to describe the samples. The results show that titanium content for all samples is about 30 ± 5 at.%. Ni content for all samples varies greatly from zero to 38.8 at.% with different experimental conditions, mainly target power ratio of TiNi/(TiNi + Ti).

Table 4.9. Experimental conditions for deposition of Ni-toughened nc-TiN/a-SiN$_x$ nanocomposite thin films.

	Sample code	S1	S2	S3	S4	S5	S6	S7	S8
Deposition condition	Si$_3$N$_4$ target power density (W/cm^2)	6.6	6.6	6.6	6.6	6.6	6.6	6.6	4.4
	TiNi target power density (W/cm^2)	0	0.2	0.7	1.1	1.4	1.6	2.2	4.4
	Target power ratio of TiNi/(TiNi+Ti)	0	0.04	0.12	0.20	0.25	0.30	0.40	1.00

Table 4.10. Chemical composition characteristics of Ni-toughened nc-TiN/a-SiN$_x$ nanocomposite thin films.

	Sample code	S1	S2	S3	S4	S5	S6	S7	S8
Deposition condition	Si$_3$N$_4$ power density (W/cm^2)	6.6	6.6	6.6	6.6	6.6	6.6	6.6	4.4
	TiNi power density (W/cm^2)	0	0.2	0.7	1.1	1.4	1.6	2.2	4.4
	Power ratio of TiNi/(TiNi + Ti)	0	0.04	0.12	0.20	0.25	0.30	0.40	1.00
XPS Chem. Com.	Ni (at.%)	0	2.1	4.3	6.3	13.0	16.4	19.0	38.8
	Ti	35.7	37.2	33.2	29.3	26.2	26.6	27.4	26.8
	Si	12.5	10.6	14.0	15.3	17.0	14.1	11.5	4.7
	N	51.8	50.1	48.5	49.1	43.8	42.9	42.1	29.7

4.2.1.1 Quantitative Compositional Analysis

Figure 4.43 shows one detailed XPS survey scan spectrum with indexed peaks. XPS survey scan spectra with binding energy from 0 to 1100 eV is recorded. The dominant signals are from C, O, Ni, Ti, Si and N. Oxygen and carbon contaminations exist because of film exposure to air (ambient laboratory). The XPS spectrum is obtained without prior surface bombardment by Ar ion beam.

Figure 4.44 shows one typical XPS depth profile of the Ni-toughened nc-TiN/a-SiN$_x$ nanocomposite thin film with 4.3 at.%Ni (sample S3). There is an inevitable oxygen and carbon contamination layer on the film surface, which is expected to render an insignificant effect on the film's mechanical properties such as nanoindentation hardness. Ni and Si contents increase at the beginning and then remain almost constant, whereas Ti and N contents increase dramatically at the beginning and then increase slightly afterwards.

Figure 4.45 shows the Ni 2p core level spectrum. The peaks at the binding energy value of 853.0 eV and 870.7 eV are confirmed to be 2p$_{3/2}$ and 2p$_{1/2}$ of metallic nickel, respectively. The peak at 859.6 eV is the satellite peak, which is probably a consequence of sputter-damaged crystallites [69]. No nickel nitride exists, as has been reported by [143, 144]: nickel will not react with N$_2$ because the formation of nickel nitride is much more thermodynamically unfavorable than the formation of TiN.

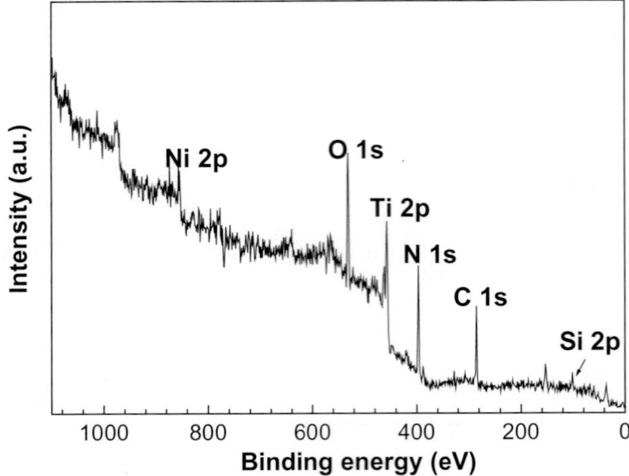

Fig. 4.43. XPS survey scan of Ni-toughened nc-TiN/a-SiN$_x$ nanocomposite thin film. The main signals come from C, O, N, Si, Ni and Ti.

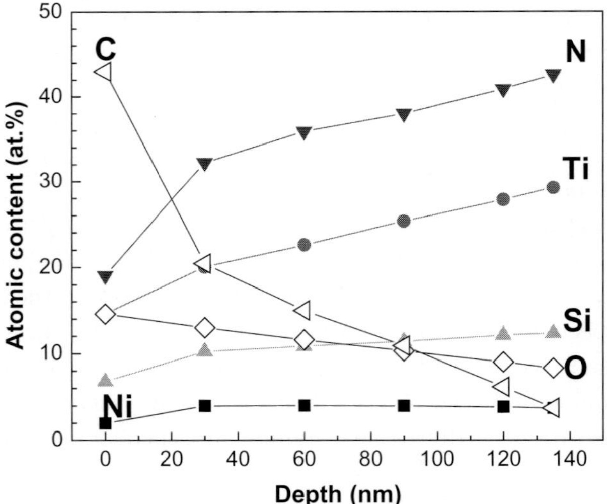

Fig. 4.44. XPS depth profile of the Ni-toughened nc-TiN/a-SiN$_x$ nanocomposite thin film with 4.3 at.%Ni (sample S3).

4.2.1.2 Effect of Deposition Conditions

Figure 4.46 shows the relationship between chemical composition and target power ratio of TiNi/(TiNi + Ti) for the Ni-toughened nc-TiN/a-SiN$_x$ nanocomposite thin films (samples S1 to S8) deposited at 450°C. It is

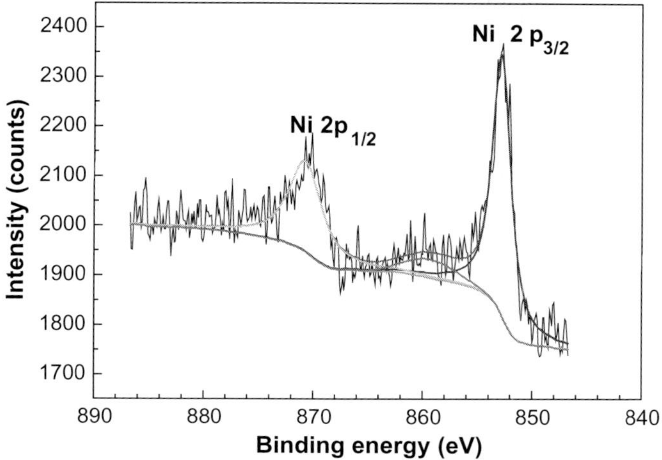

Fig. 4.45.   Ni 2p core level spectrum. The peaks at the binding energy value of 853.0 and 870.7 eV are confirmed to be 2p$_{3/2}$ and 2p$_{1/2}$ of metallic Ni, respectively. The peak at 859.6 eV is the satellite peak, which is probably a consequence of sputter-damaged crystallites.

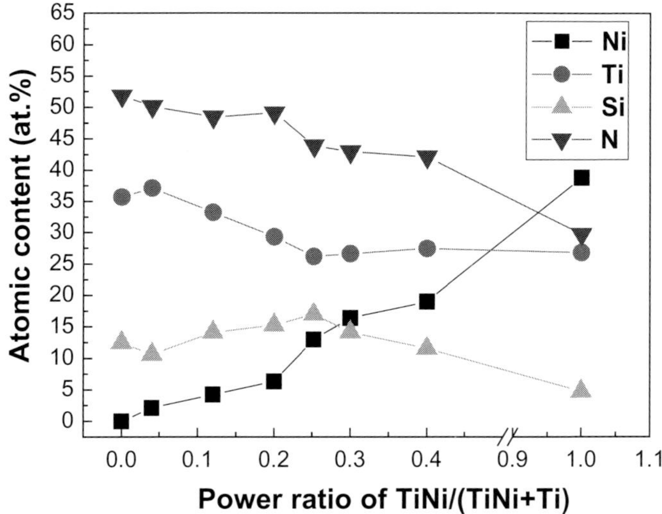

Fig. 4.46.   Content changing with target power ratio of TiNi/(TiNi + Ti). With an increase of TiNi/(TiNi + Ti) from zero to unity, Ni content increases linearly from zero to about 39 at.%.

obvious that Ni content in the as-prepared thin films increases linearly from 0 to 39 at.% with an increase in TiNi/(TiNi + Ti) from zero to unity, while N content decreases linearly from 52 at.% to 30 at.%. Ti content decreases from about 35 at.% to 26 at.% with an increase of TiNi/(TiNi + Ti) from zero to 0.25; with further increase in target power ratio, Ti content keeps constant at about 26 at.%. For Si content, with an increase in target power ratio from zero to 2.5, silicon content increases from 12 at.% to 17 at.%; with further increase of TiNi/(TiNi + Ti) to unity, silicon content decreases to 5 at.%.

4.2.2. *Topography*

The effect of sputtering target power ratio of TiNi/(TiNi + Ti) on surface topography is evaluated using AFM. Film topography characteristics are listed in Table 4.11.

Figure 4.47 shows AFM images ($1\,\mu\text{m} \times 1\,\mu\text{m}$) of Ni-toughened nc-TiN/a-SiN$_x$ nanocomposite thin films prepared with different target power ratios of TiNi/(TiNi + Ti). With an increase in target power ratio, the roughness gradually decreases. Numeric treatment of the images gives rise to the height–height correlation function $G(r)$ as described through Eq. (3.1).

Figure 4.48 plots $G(r)$ as a function of r for the Ni-toughened nc-TiN/a-SiN$_x$ nanocomposite thin films (samples S1 to S7) represented by Fig. 4.47. The oscillation is due to the insufficient sampling size [77]. As predicted by Eq. (3.4), when r is small, $G(r)$ has a power law dependence on distance r. At "distant" locations (i.e. r is quite large), $G(r)$ is nearly a constant. Fitting the curves using Eq. (3.4) gives the interface width ω and lateral correlation length ξ. Plotting these values gives rise to Fig. 4.49. The competition of vertical build-up and lateral diffusion determines the morphology of the growing surfaces. The vertical build-up is caused by the random

Table 4.11. Topography characteristics of Ni-toughened nc-TiN/a-SiN$_x$ nanocomposite thin films.

	Sample code		S1	S2	S3	S4	S5	S6	S7	S8
Topography	Roughness (nm)	R_a	6.3	6.1	5.4	4.2	3.8	2.5	1.6	—
	Interface width (nm)	ω	8.0	7.4	6.6	6.4	3.9	3.2	1.3	—
	Lateral correlation length (nm)	ξ	14.6	13.9	11.3	10.5	9.8	9.4	—	—

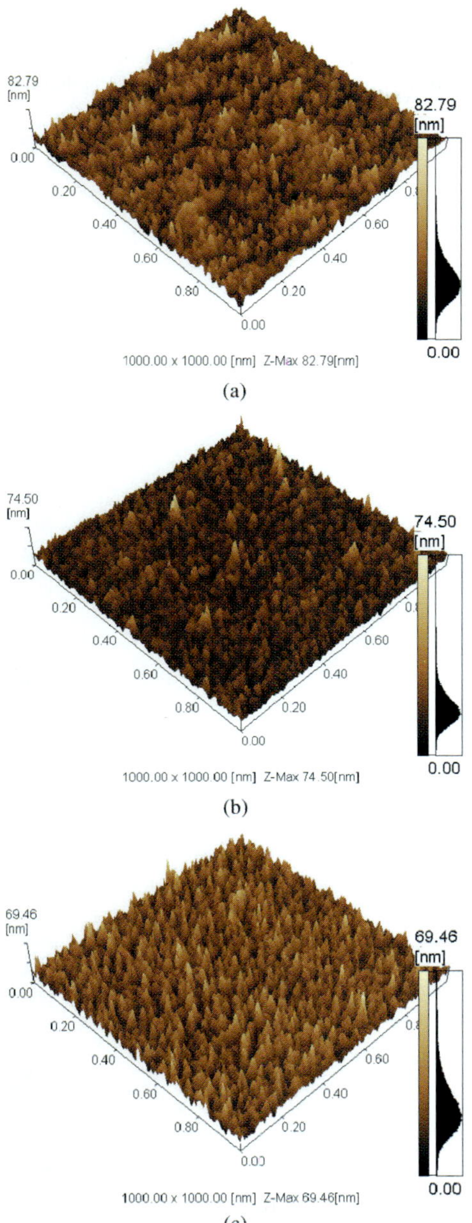

Fig. 4.47.   AFM topography of the Ni-toughened nc-TiN/a-SiN$_x$ nanocomposite thin films with increasing power ratio of TiNi/(TiNi + Ti): (a) 0, (b) 0.04, (c) 0.12, (d) 0.20, (e) 0.25, (f) 0.30, and (g) 0.40, with roughness (R_a, nm) 6.3, 6.1, 5.4, 4.2, 3.8, 2.5, and 1.6, respectively.

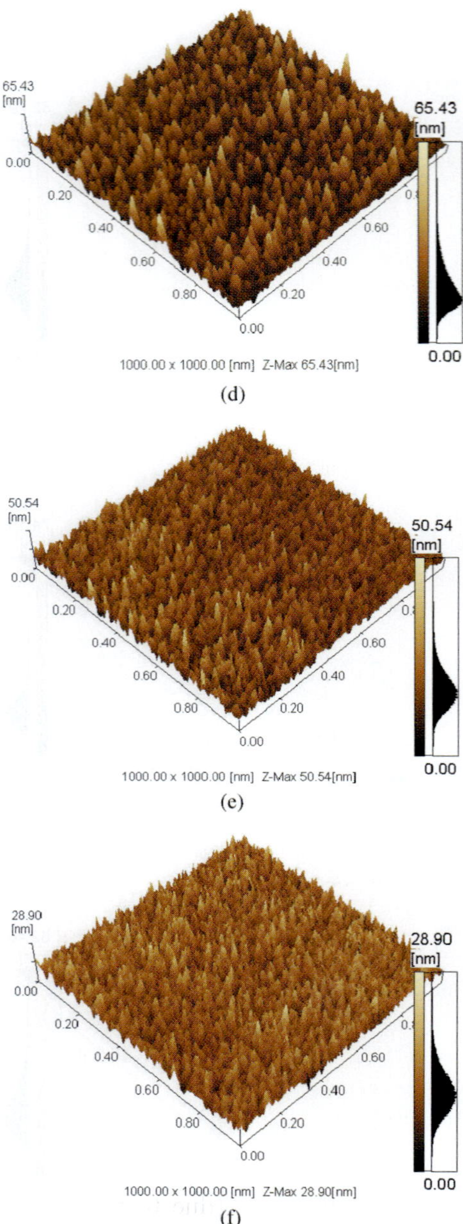

Fig. 4.47.   (*Continued*)

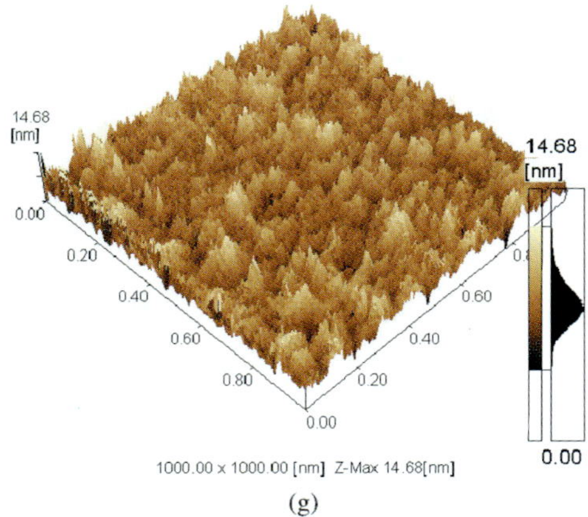

Fig. 4.47.   (*Continued*)

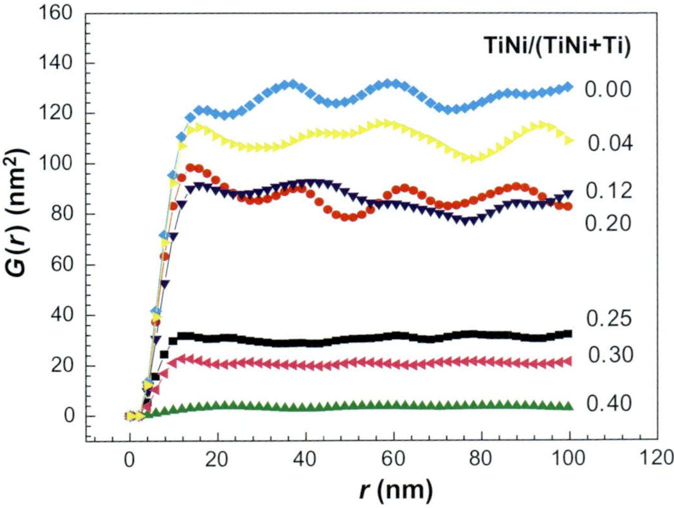

Fig. 4.48. Height–height correlation $G(r)$ for the thin films with different TiNi/(TiNi + Ti) power ratio. The oscillation is due to the insufficient sampling size.

angle incident of the arriving atoms (due to the uniform rotation of the substrate), and the growth produces columnar film structure. The lateral growth depends on surface diffusion which is largely determined by kinetic energy of the arriving ions. This is clearly seen from the trends of ω and ξ in Fig. 4.49. As target power ratio of TiNi/(TiNi + Ti) increases from 0 to 0.3,

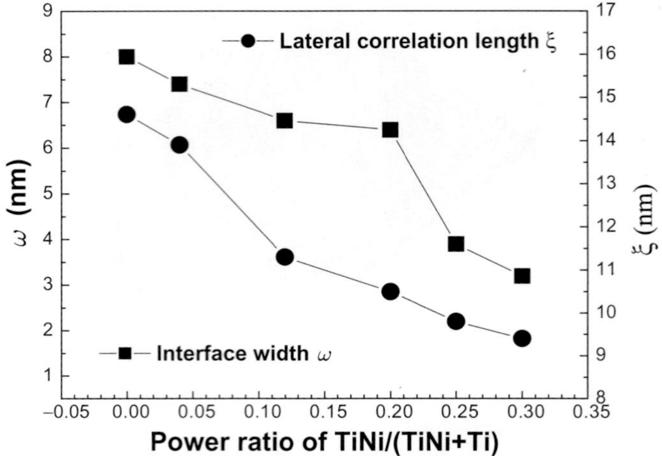

Fig. 4.49. Interface width ω and lateral correlation length ξ vary with target power ratio of TiNi/(TiNi + Ti). As target power ratio of TiNi/(TiNi + Ti) increases from 0 to 0.3, ω decreases from $\sim$8 to $\sim$3 nm, and ξ decreases from $\sim$15 to $\sim$9 nm.

the interface width ω decreases from $\sim$8 to $\sim$3 nm (Fig. 4.49), indicating that the film becomes smoother (recall that ω is the root mean square of the vertical fluctuation). As target power ratio of TiNi/(TiNi + Ti) increasing, ξ decreases from $\sim$15 to $\sim$9 nm. Since ξ depicts the distance within which the "height" values (i.e. roughness) are correlated, smaller ξ means the surface topography is correlated in a narrow area.

The growth kinetics is controlled by the mobility of the impinging atoms on the surface before they condense and become entrapped in the film. This mobility can be enhanced by inputting energy to the system, such as increasing deposition temperature or supplying impact energy through ion bombardment. The preferential sputtering of a compound target by ion bombardment is well known [145]. Since titanium and nickel have different melting points (1660 and 1453°C [146]), Ni is apt to be sputtered. In addition, the presence of a large amount of N_2 can result in a certain degree of Ti target poisoning (by forming TiN on Ti target) that would contribute to the lessening of Ti ion partial pressure. Preferential sputtering and target poisoning are the reasons for ion bombardment effect strengthening with increase of TiNi target power even though the (TiNi + Ti) total power is kept constant for the Ni-toughened nc-TiN/a-SiN$_x$ thin films (samples S1 to S7). Thus, at low target power ratio of TiNi/(TiNi + Ti), ions have low mobility and would thus be more likely to "stick" at where it arrives: the surface diffusion is slow, and the chances of erasing the peaks and filling up the "valleys" are small, thus increasing a much rougher surface. In the same token, small ξ indicates that the surface topography is correlated only in a small area. As target power ratio of TiNi/(TiNi + Ti) increases, the kinetic energy obtained by each Ni and/or TiN (as well as amount) increases, which transforms into faster lateral diffusion and smoothens out the roughness at locations further away, making the "hump" more localized and smaller, giving rise to smaller values of ξ and ω [116, 117].

4.2.3. *Microstructure*

The effect of sputtering target power ratio of TiNi/(TiNi + Ti) on microstructure, such as crystalline phase, grain size, preferential orientation and lattice parameter are tabulated in Table 4.12.

4.2.3.1 Crystal Phase and Amorphous Matrix

Figure 4.50 shows the bright-field HRTEM morphological appearance of Ni-toughened nc-TiN/a-SiN$_x$ nanocomposite thin film with 2.1 at.%Ni (sample S2). Crystallites are embedded in matrix. The grain size is about 8 nm. Analysis of the SAD patterns shows that these crystallites are polycrystalline TiN. No crystalline Ni, TiSi and Si$_3$N$_4$ are found (Fig. 4.51). In general, (111), (200) and (220) TiN crystallographic planes exhibit more distinct rings than other diffraction rings. Proof of the crystallites being TiN also comes from the analysis of the GIXRD pattern (Fig. 4.52).

Figure 4.52 confirms the existence of crystalline TiN. In addition, no peaks from crystalline metallic Ni and Si$_3$N$_4$ can be identified in the GIXRD patterns. SAD analysis of the matrix (where there is no crystallite) gives rise to a diffuse pattern typical of an amorphous phase. Together with XPS analysis in Sec. 4.1.1, where nickel is in the Ni–Ni bond and silicon is mostly in the Si–N bond, the results verify that the amorphous matrix comprises amorphous metallic Ni and amorphous silicon nitride (a-SiN$_x$).

4.2.3.2 Grain Size and Distribution

Grain size calculated using Scherrer formula is listed in Table 4.12 and plotted in Fig. 4.53. In small amounts of Ni, for example, less than 4.3 at.%Ni, the nc-TiN crystallite size increases as Ni content increases. This may be due to the reason that small quantities of metallic nickel can decrease

Table 4.12. Microstructure characteristics of Ni-toughened nc-TiN/a-SiN$_x$ nanocomposite thin films.

	Sample code		S1	S2	S3	S4	S5	S6	S7	S8
Microstructure	Grain size (nm)	d	3.9	8.8	11.8	7.8	4.6	4.4	3.2	3.2
	Latt. Param. (nm)	a_{TiN}	0.41889	0.41959	0.41678	0.41701	0.41603	0.41135	0.4225	0.43237
	Coefficient of Texture	T_{111}	1.20	0.61	0.36	0.54	1.06	0.58	1.01	0.90
		T_{200}	1.05	1.54	1.67	1.62	0.97	1.62	1.22	1.25
		T_{220}	0.76	0.84	0.97	0.86	0.97	0.80	0.77	0.85

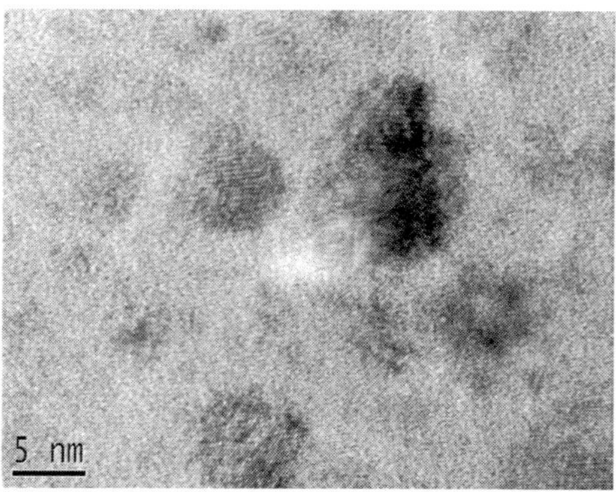

Fig. 4.50.   HRTEM bright-field micrograph of Ni-toughened nc-TiN/a-SiN$_x$ nanocomposite thin film with 2.1 at.%Ni (sample S2) showing the crystallite of size 8 nm.

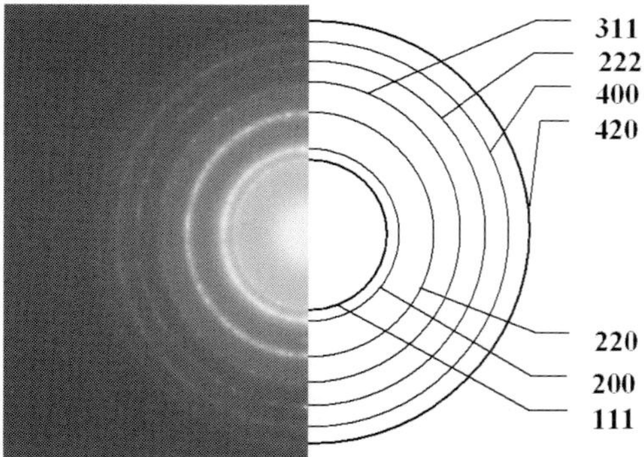

Fig. 4.51.   SAD pattern of the crystallite with pattern index, showing the crystalline TiN phase.

the formation energy of TiN and excite crystallite growth. When Ni content is greater than 4.3 at.%, with an increase in target power ratio TiNi/(TiNi + Ti), because surface mobility is sufficient, the segregated Si and Ni are sufficient to nucleate and develop the SiN$_x$ phase and metallic Ni phase, respectively, which form a layer on the growth surface, covering the TiN nanocrystallites and limiting their growth. From the analysis of peak widths in the GIXRD patterns, a decrease in grain size from $\sim$12 nm to $\sim$3 nm (Fig. 4.53) is also evident, which is consistent with Si and Ni segregation.

Grain size changes with the target power ratio of TiNi/(TiNi + Ti), which is confirmed from TEM micrograph observation (Fig. 4.54). It

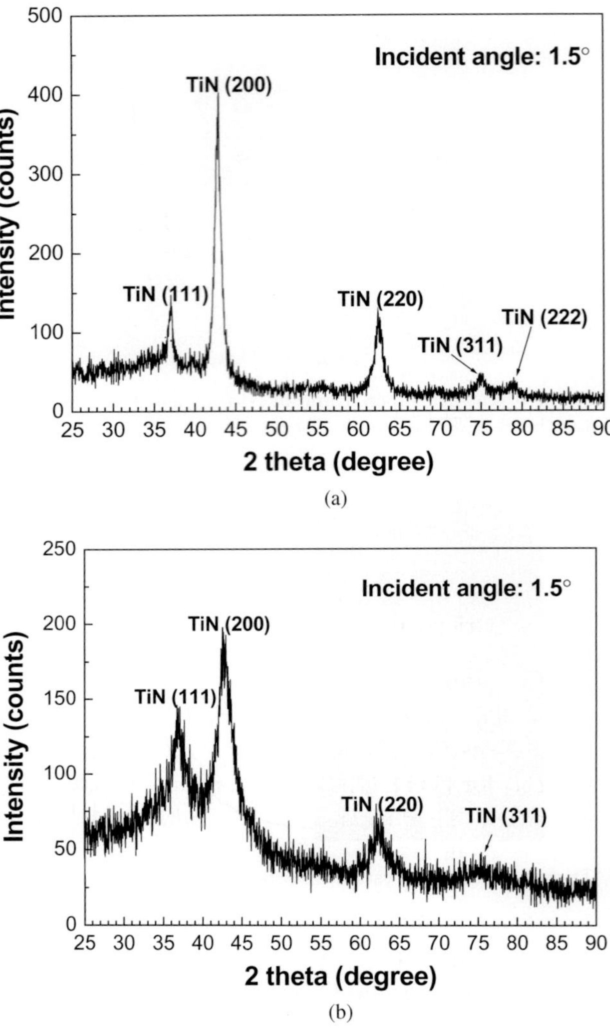

Fig. 4.52. GIXRD patterns of Ni-toughened nc-TiN/a-SiN$_x$ nanocomposite thin films with (a) 2.1 at.%Ni (sample S2), and (b) 19.0 at.%Ni (sample S7), showing the existence of crystalline TiN.

should be noted that the grain sizes in the TEM samples are smaller than that from XRD analysis, because the deposition times for the TEM samples are about 20 min, while the XRD ones are about 2 hours (120 min).

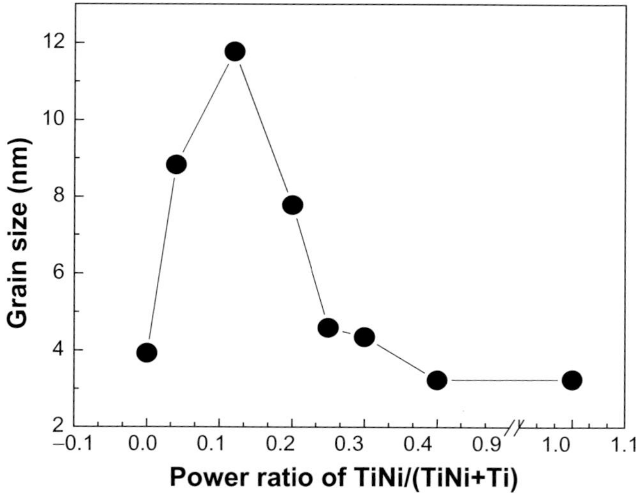

Fig. 4.53.   Change of grain size with target power ratio of TiNi/(TiNi + Ti).

4.2.3.3. Preferential Orientation

The peak orientation can be observed directly from Fig. 4.55. The degrees of
the preferential orientation denoted by the coefficient of texture calculated
through Eq. (3.6) are listed in Table 4.12 and plotted in Fig. 4.56. The coefficient of texture T_{hkl} for (111), (200) and (200) are close to unity, indicating
a random orientation. That means that the addition of Ni results in the
deterioration of preferential orientation. In the synthesis of nc-TiN/a-SiN$_x$
nanocomposite thin films, a competitive growth has been put forward to
explain the development of TiN (220) preferential orientation (Sec. 4.1.3.3).
Taking into consideration the effect of metallic Ni addition, it is reasonable
to deduce that the segregated additives Ni inhibit the growth of TiN crystals
to stimulate spores of other TiN nuclei. Consequently, competitive growth
is suppressed, resulting in weak texture. This is in agreement with the effect
of soft metal additives (Cu, Ag, etc.) on the weakening of texture reported
in [147, 148].

4.2.3.4. Lattice Parameter

Table 4.12 lists the lattice parameter measurements of the nc-TiN crystallites calculated using the GIXRD patterns shown in Fig. 4.55. The result is
also plotted in Fig. 4.57 together with the standard lattice parameter data

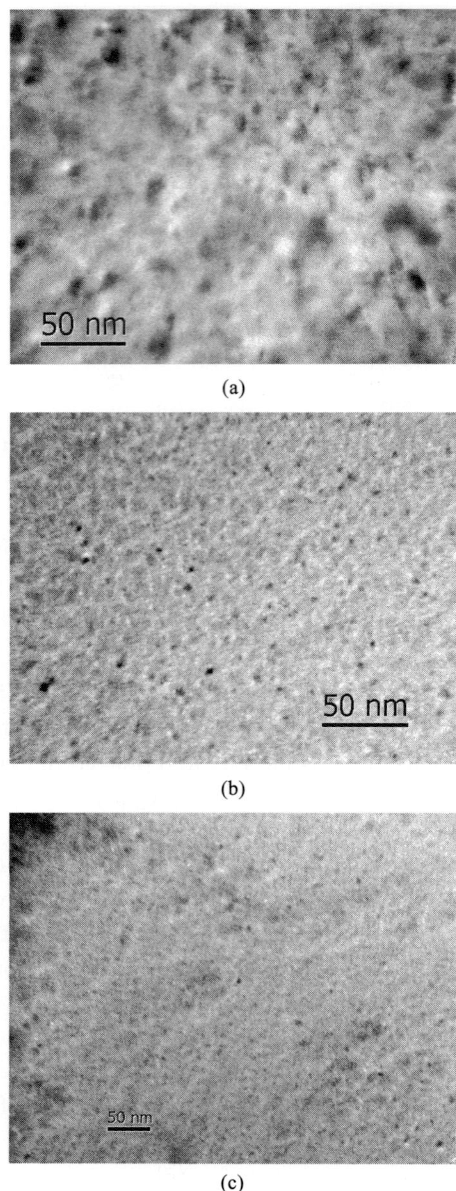

Fig. 4.54. TEM micrographs of Ni-toughened nc-TiN/a-SiN$_x$ nanocomposite thin films with different power ratios of TiNi/(TiNi + Ti): (a) 0.04, (b) 0.20, and (c) 0.40, showing a reduction in grain size.

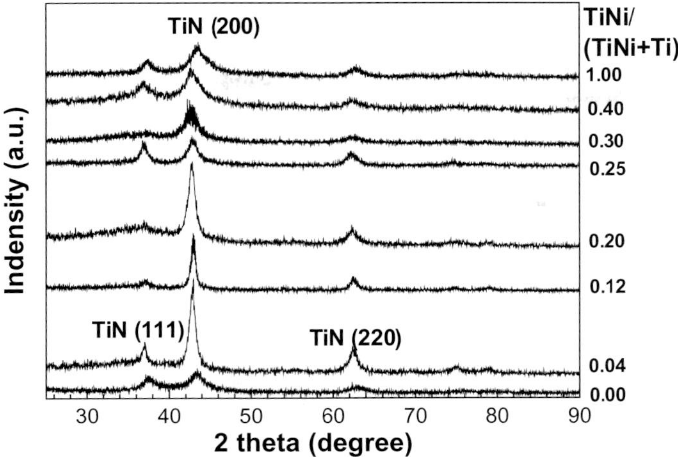

Fig. 4.55.   GIXRD patterns of Ni-toughened nc-TiN/a-SiN$_x$ nanocomposite thin films (samples S1 to S8) with different target power ratios of TiNi/(TiNi + Ti), showing random orientations.

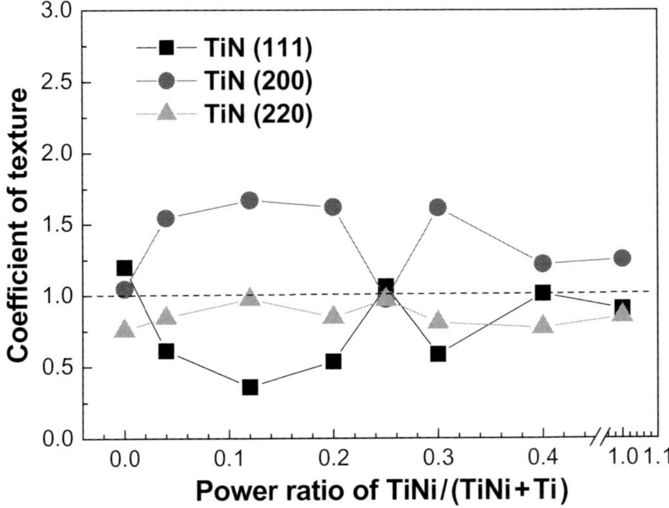

Fig. 4.56.   Calculated coefficient of texture of Ni-toughened nc-TiN/a-SiN$_x$ nanocomposite thin films (samples S1 to S8) with different target power ratios of TiNi/(TiNi + Ti). The coefficient of texture T_{hkl} for (111), (200) and (200) are close to 1, which indicates a random orientation.

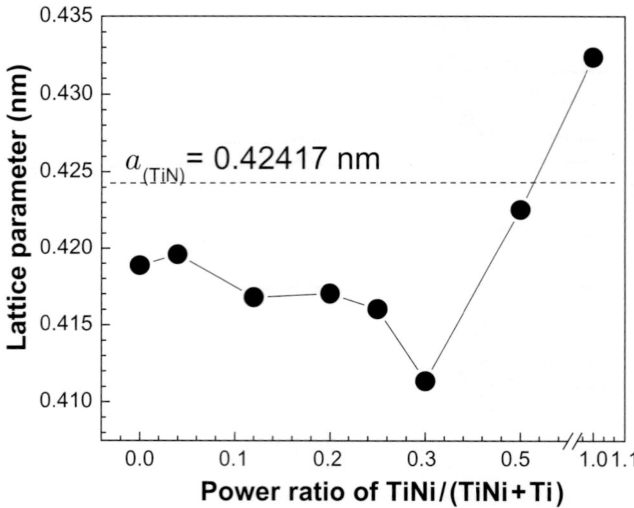

Fig. 4.57. Calculated lattice parameter as a function of target power ratio of TiNi/(TiNi + Ti). With an increase of TiNi/(TiNi + Ti) from zero to 0.3, the lattice parameter decreases from 0.41887 nm to 0.41135 nm. With further increase in TiNi/(TiNi + Ti), the lattice parameter increases to 0.43237 nm.

for TiN crystals. At Ni = 0 (without doping of Ni), the lattice parameter of nc-TiN is 0.41889 nm. However, pure TiN crystals should have a lattice constant of 0.42417 nm (JCPDS 38-1420). This is because the nc-TiN embedded in a-SiN$_x$ already forms solid solution with Si to become nc-(Ti, Si)N. Since Si^{4+} has a radius of 0.041 nm, which is only about half of that of Ti^{3+} (0.075 nm) [129], substitution of Ti with Si results in a reduction in lattice parameter [128]. With the increase of TiNi/(TiNi + Ti) from zero to 0.3, the lattice parameter decreases from 0.41889 to 0.41135 nm. With further increase of TiNi/(TiNi + Ti) to unity, the lattice parameter increases to 0.43237 nm. When target power ratio of TiNi/(TiNi + Ti) is less than 0.3, the measured TiN lattice parameter is smaller than that of the database value of 0.42417 nm (JCPDS 38-1420). The ionic radius of Ni^{3+} (0.056 nm [149]) is less than that of Ti^{3+} (0.075 nm), thus the reason for further reduction of the (Ti, Si)N lattice parameter is clear: Ni enters nanocrystalline (Ti, Si)N by substituting Ti, thus forming nc-(Ti, Si, Ni)N. When target power ratio of TiNi/(TiNi + Ti) becomes greater than 0.3, nc-(Ti, Si, Ni)N becomes saturated with Ni, thus further increases in Ni forces it to enter in the interstitial position. This results in an abrupt increase in lattice parameter.

4.2.4. *Mechanical Properties*

The effect of sputtering target power ratio of TiNi/(TiNi + Ti) on mechanical properties, such as residual stress, Young's modulus, hardness and toughness are tabulated in Table 4.13.

4.2.4.1. Residual Stress

Figure 4.58 shows the measured residual stress of the Ni-toughened nc-TiN/a-SiN$_x$ thin films as a function of target power ratio of TiNi/(TiNi + Ti). With an increase of TiNi/(TiNi + Ti) target power ratio from zero to 0.12, the residual stress increases from -400 to -1300 MPa. With further increase of TiNi/(TiNi + Ti) target power ratio to 0.30, the residual stress decreases to near-zero. The minus sign indicates that stresses are in compressive state. With an increase of TiNi/(TiNi + Ti) target power ratio from 0.30 to 1.00, the residual stress changes from compressive to tensile state with a value of 300 MPa. The increase in compressive residual stress from -407 to -1355 MPa is possibly due to the ion bombardment. However, the decrease with further increase of Ni needs to be studied further.

4.2.4.2. Hardness

Figure 4.59 displays the relationship between hardness and Ni content, including a typical nanoindentation load-depth profile. The measured hardness of Ni-toughened nc-TiN/a-SiN$_x$ nanocomposite thin films increases from 28 to 33 GPa with Ni content increased from 0 to about 2.1 at.%. A further increase in Ni content brings a decrease in hardness to 14 GPa. When Ni content is less than 2.1 at.%, the growth of amorphous metal-

Table 4.13. Mechanical characteristics of Ni-toughened nc-TiN/a-SiN$_x$ nanocomposite thin films.

	Sample code		S1	S2	S3	S4	S5	S6	S7	S8
Mechanical properties	Residual stress (MPa)	σ	-407	-566	-1355	-1122	-661	-33	230	320
	Hardness (GPa)	H	28.4	32.6	28.3	28.5	20.2	19.5	18.8	14.4
	Young's modulus (GPa)	E	295	296	298	278	265	267	226	250
	Toughness (MPa m$^{1/2}$)	K_{IC}	1.15	1.21	1.36	1.23	1.73	1.95	2.25	2.60
	Adhesion (mN)	L_{c2}	587	618	628	627	821	882	785	717

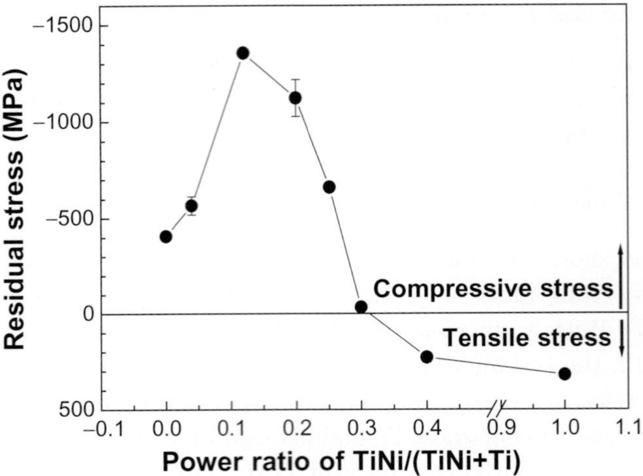

Fig. 4.58.   Relationship of residual stress with target power ratio of $\mathrm{TiNi}/(\mathrm{TiNi}+\mathrm{Ti})$.

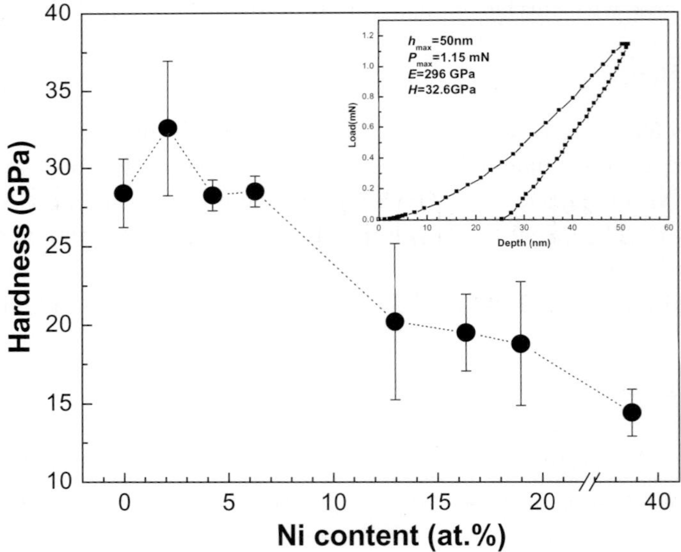

Fig. 4.59.   Hardness varying with Ni content in the Ni-toughened nc-TiN/a-SiN$_x$ nanocomposite thin films. With increase of Ni content, hardness decreases due to grain boundary sliding.

lic Ni is inhibited by the shortage of Ni atoms. The nickel atoms disperse in the TiN lattice and result in lattice distortion (Fig. 4.57). Since lattice distortion can stop dislocation propagation, an increase in measured hardness is expected, due mainly to solution hardening. Further increase in Ni results in more lattice distortion, thus the solution hardening effect should be enhanced. However, the increase in nickel composition comes from the increase in target power ratio of TiNi/(TiNi + Ti), which can improve the mobility of sputtered Ni adatoms. Thus Ni adatoms conglomerate to form network phase, together with a-SiN$_x$ phase to block the growth of TiN crystals. When the TiN grain size decreases below a certain limit, the fraction of grain boundary (Ni + Si$_3$N$_4$) increases rapidly and the hardness would decrease due to grain boundary sliding. This should account for the decrease in hardness with further increase in Ni content.

4.2.4.3. Toughness

The toughness of Ni-toughened nc-TiN/a-SiN$_x$ thin films was studied using microindentation method, since the microindentation method can achieve highly consistent results. In order to deduce the film toughness from the substrate effected data, K_{IC} values are plotted and the curve is then extrapolated to a depth one tenth of the film thickness. Figure 4.60 shows

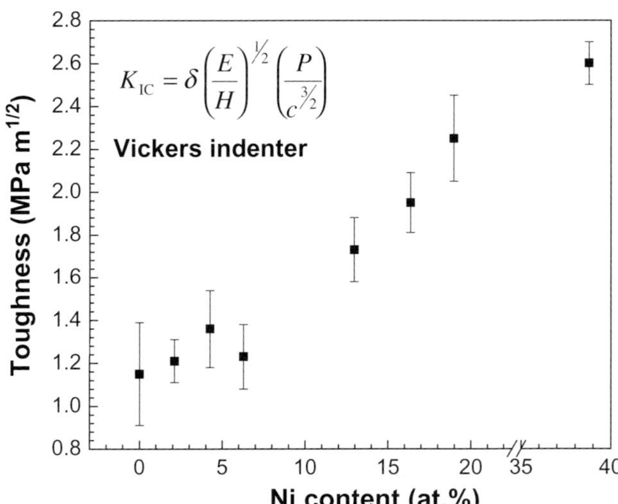

Fig. 4.60.   Toughness as a function of Ni content. With an increase in Ni content, the film toughness increases, indicating a significant effect of ductile metallic Ni on film toughness enhancement.

the toughness of Ni-toughened nc-TiN/a-SiN$_x$ nanocomposite thin films as a function of Ni content. Compared to the high hardness nc-TiN/a-SiN$_x$ nanocomposite thin film (sample S1) where there is no Ni, the Ni-toughened nc-TiN/a-SiN$_x$ nanocomposite thin films (samples S2 to S8) show an increased toughness. With increase of Ni content, the film toughness increases. With increase in nickel content in the as-prepared thin films, the main mechanisms responsible for toughness enhancement are (Fig. 4.61): (1) Relaxation of the strain field around the crack tip through ductile phase (metallic nickel) deformation or crack blunting, whereby the work for plastic deformation is increased. (2) Ni adatoms can form network phase surrounding the TiN crystals. Bridging of cracks by ligaments of the ductile metallic nickel phase behind the advancing crack tip, whereby the work for plastic deformation is also increased [150, 151].

4.2.5. *Oxidation Resistance*

4.2.5.1. Oxidation Variation with Depth

Chemical state of Ti, Si and Ni

Figure 4.62 shows XPS depth profiles of the Ni-toughened nc-TiN/a-SiN$_x$ nanocomposite thin film with 2.1 at.% Ni (sample S2) oxidized at 850°C. Roughly judging from the oxygen and nitrogen concentration, the profiles in Fig. 4.62 can be divided into five regions. Detailed analysis of the chemical state of Ti (Fig. 4.63) gives rise to composition evolution from TiO$_2$ to TiN$_x$O$_y$, and then to TiN while passing through the oxidation layer.

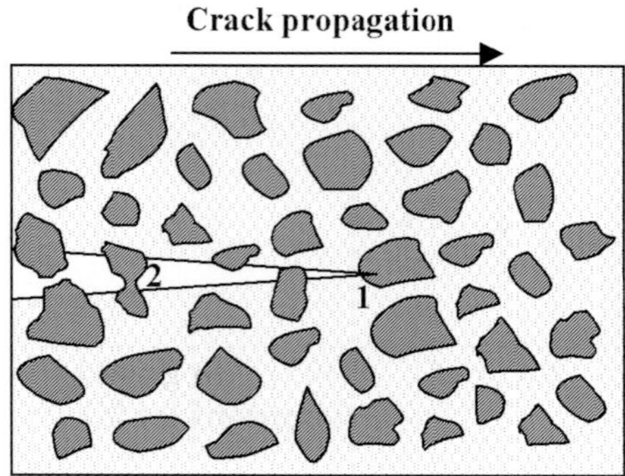

Fig. 4.61. Schematic diagram of ductile phase toughening mechanism through (1) ductile phase deformation or crack blunting, and (2) crack bridging.

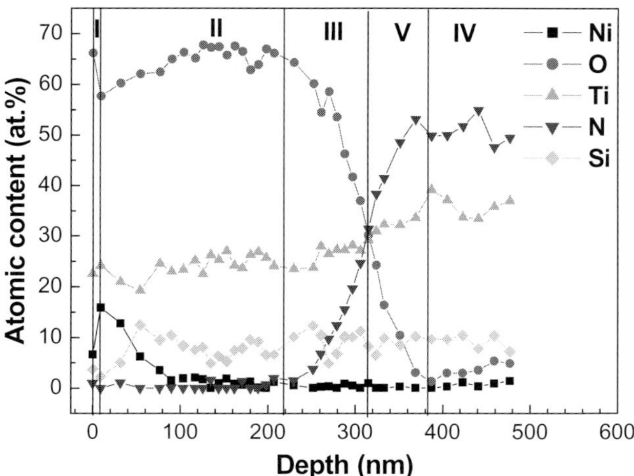

Fig. 4.62.   XPS depth profile of the Ni-toughened nc-TiN/a-SiN$_x$ nanocomposite thin film with 2.1 at.%Ni (sample S2) oxidized at 850°C in hot air for 15 min. A sharp oxide/nitride interface exists and a nickel on the top layer of the oxidized coating is observed. Five regions can be distinguished in this area from the surface to the non-oxidized material core according to the chemical state of Ti and Si.

Figure 4.63(a) plots Ti 2p core level spectra in the binding energy range from 452 to 468 eV for all the five regions. Sampling for Region I is at the surface; for Region II–V, in the middle of each region. Figure 4.63(b) is the quantitative deconvolution result of relative concentration of Ti in TiO$_2$, TiN$_x$O$_y$ and TiN.

In Fig. 4.63(a), Ti 2p peaks of the oxidized film consist of three doubles: Ti 2p$_{3/2}$ at 459.0, 457.6, 455.0 eV, and Ti 2p$_{1/2}$ at $(459.0 + 5.8)$, $(457.6 + 5.8)$, $(455.0 + 5.8)$ eV. The pair at 459.0 and $(459.0 + 5.8)$ eV is assigned to TiO$_2$; the pair at 457.6 and $(457.6 + 5.8)$ eV is ascribed to oxynitride TiN$_x$O$_y$; the pair at 455.0 and $(455.0 + 5.8)$ eV is for TiN (Table 3.2). From Fig. 4.63(a), it is obvious that deep inside the coating (Region V), the main composition is still TiN (the nanocrystalline TiN in the nanocomposite thin film); moving more towards the surface (Fig. 4.62), regions IV and III, the amount of TiN decreases while TiN$_x$O$_y$ increases. At the same time, TiO$_2$ appears. As the oxidation degree becomes even more severe (Regions II and I), both TiN$_x$O$_y$ and TiN decrease to give way to the formation of more and more TiO$_2$. At the surface, TiN and TiN$_x$O$_y$ completely disappear while TiO$_2$ prevails [Fig. 4.63(b)].

As seen also from Fig. 4.62, nitrogen content drops drastically from its bulk composition of about 50 at.% in Region V to less than 2 at.% in the oxidation layer (Regions II and I). This is in agreement with earlier analysis of

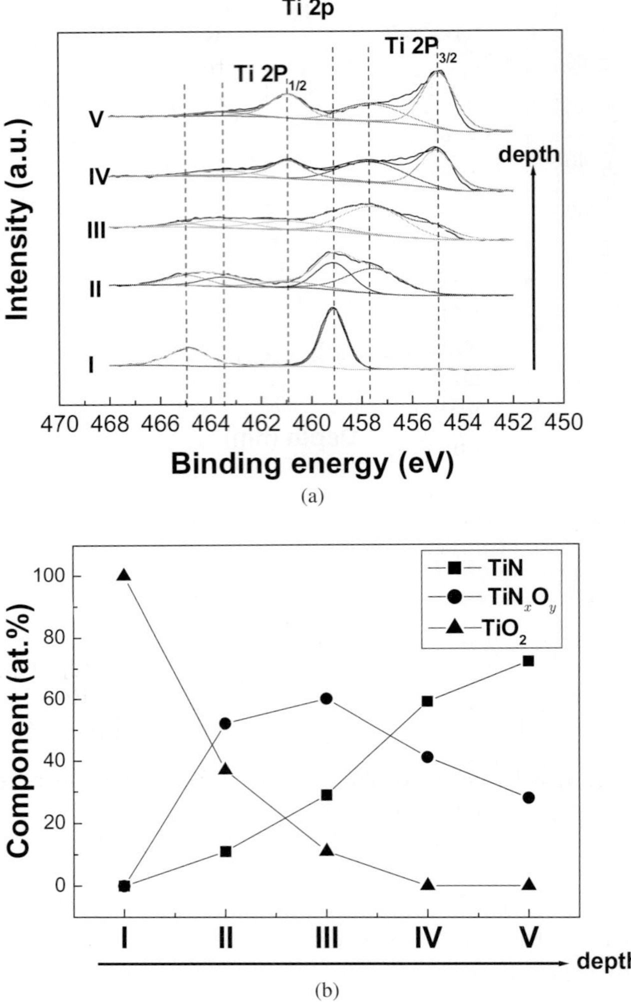

Fig. 4.63. (a) Ti 2p core level XPS spectra evolution, a shift of the Ti 2p$_{3/2}$ signal towards higher binding energies indicates the formation of a TiO$_2$ layer on top of the oxidized coating. (b) Change of the different Ti 2p components in different regions (from Regions I through II, III and IV to V).

the evolution of the compounds as depth varies in the oxide layer. Since the total amount of N is low and it is responsible for TiN, TiN$_x$O$_y$, and Si$_3$N$_4$ (to be discussed later), it is reasonable to assume that x in TiN$_x$O$_y$ is very small while y is large (oxygen content is greater than 65 at.% in Region II).

From the shape of the oxygen and nitrogen profiles, it is obvious that during the oxidation process, oxygen diffuses inward and nitrogen diffuses outward. It is interesting to note that nickel diffuses towards the surface from the depth and builds up close to the surface (the drop of the relative amount at the surface comes from the calculation involving surface adsorbed C). It is believed that the surface enrichment of Ni will benefit the oxidation resistance of the film.

Figure 4.64(a) plots Si 2p core level spectra in different regions of the oxide layer. Si 2p spectrum has three possible states (Table 3.2): $99.6\,\mathrm{eV}$ for atomic silicon (Si^0); $101.8\,\mathrm{eV}$ corresponds to the Si–N bond (stoichiometric Si_3N_4); $103.4\,\mathrm{eV}$ for the Si–O bond (stoichiometric SiO_2). Some authors have reported Si–Ti bonding at $98.8\,\mathrm{eV}$ or the existence of titanium silicide [67]. This experiment does not observe any Si–Ti bonding. Figure 4.64(b) is the quantitative deconvolution result of relative concentration of Si in SiO_2, Si_3N_4, and Si^0.

Going from deep in the film outward to the surface of the oxide layer [from Region V down to Region I in Fig. 4.64(a)], it is worth noting:

1. The amorphous silicon nitride matrix is actually prominently a-Si_3N_4 with a very small amount of free silicon [<10 at.% of all silicon in the film; see Fig. 4.64(b)]; the existence of free Si is due to a deficit in nitrogen source compared to Si source during deposition process.
2. Moving towards the surface, the amount of free silicon decreases to zero, the amount of Si in the state of Si_3N_4 drops from about 80 at.% to 15 at.% while Si in SiO_2 increases from about 10 at.% to 85 at.%. In other words, the free silicon and some of the Si_3N_4 become oxidized into SiO_2.

TiN films start to oxidize at a temperature level of $550°C$ [152], while Si_3N_4 is more oxidation resistant than TiN, as has been reported by Gogotsi and Porz [153, 154]. In fact, this is also noticed by comparing the surface oxidation state of Ti and Si from this experiment: at the surface, TiN is completely oxidized into TiO_2 [Region I in Fig. 4.63(b)]; while also at the surface of the same sample, there is still about 15 at.%Si in the form of Si_3N_4 free from oxidation [Region I in Fig. 4.64(b)].

Figure 4.65 shows Ni 2p core level spectra. Metallic nickel (Ni^0) has a binding energy value of $853.0\,\mathrm{eV}$ ($2p_{3/2}$) and $870.7\,\mathrm{eV}$ ($2p_{1/2}$) (Table 3.2). The peak at $859.6\,\mathrm{eV}$ is the satellite peak probably due to the consequence

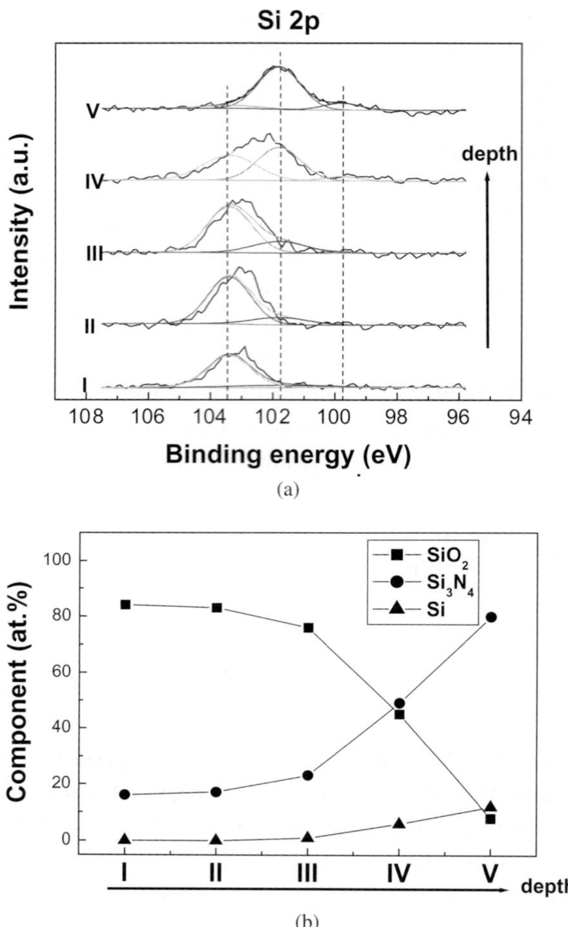

Fig. 4.64.   (a) Si 2p core level XPS spectra, a shift of Si 2p signal towards higher binding energy indicates the formation of a SiO_2 layer on top of the oxidized coating. (b) Change of different Si 2p components in different regions (from Regions I through II, III and IV to V).

of sputter-damaged crystallites [69]. A few points worth noticing in Fig. 4.65:

1. There is no peak shift for Ni 2p, which indicates that the Ni does not react with oxygen at 850°C.
2. From deep within the film towards the surface, Ni peak intensities increase from Region V to Region I, indicating Ni diffusion towards the sample surface, as also illustrated in Fig. 4.62.

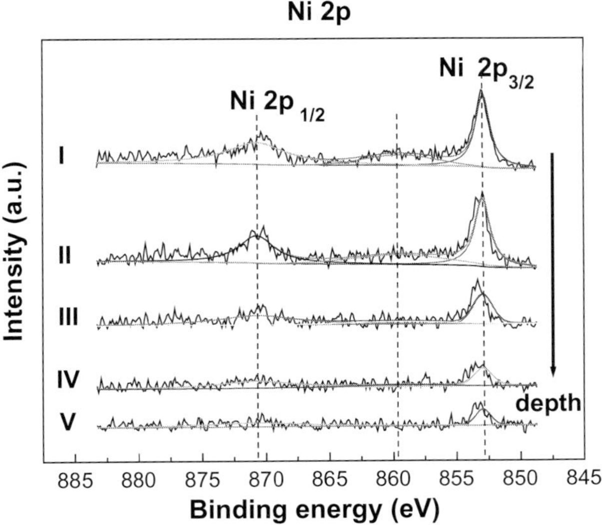

Fig. 4.65. Ni 2p core level XPS spectra eVolution in different regions (from Regions I through II, III and IV to V). Metallic nickel is confirmed and the decreasing of peaks intensity from outer layer to inner layer indicates the existence of Ni enrichment in the outer layer.

Figure 4.66 shows XPS depth profile of nc-TiN/a-SiN$_x$ nanocomposite thin film oxidized at 850°C for 15 min. The oxidation treatment for this sample is the same as that for the Ni-toughened nc-TiN/a-SiN$_x$ nanocomposite thin film. The thickness of the oxide layer is about 420 nm, compared to that ($\sim$315 nm, Fig. 4.62) for Ni-toughened nc-TiN/a-SiN$_x$ nanocomposite thin films; it can be concluded that Ni addition can improve the oxidation resistance of the nc-TiN/a-SiN$_x$ nanocomposite thin films.

Phase Identification

Figure 4.67 shows the GIXRD patterns with different incident angles for the Ni-toughened nc-TiN/a-SiN$_x$ nanocomposite thin film with 2.1 at.%Ni (sample S2) oxidized at 550°C for 15 min. No XRD peaks are observed for crystalline Si$_3$N$_4$, SiO$_2$ and Ni. Since XPS analysis indicates the existence of Si$_3$N$_4$ and SiO$_2$ [Fig. 4.64(b)], thus these phases must be in amorphous state. XPS results indicate the existence of metallic Ni (Fig. 4.65) from surface to the in-depth place of the film; this is also evident from Fig. 4.67: as incident angle increases to 1.5°, no crystalline peaks of Ni are observed.

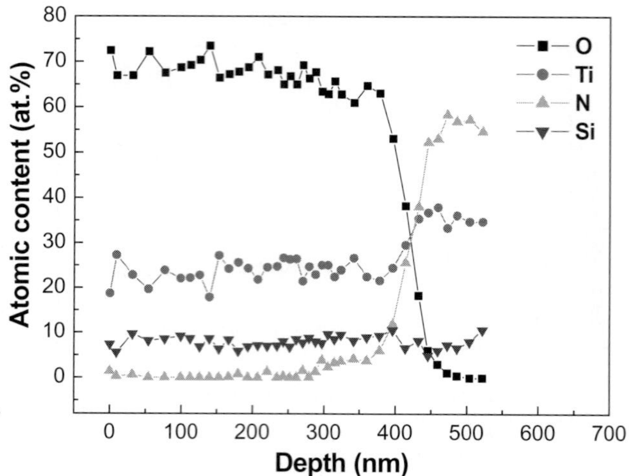

Fig. 4.66.   XPS depth profile of the nc-TiN/a-SiN$_x$ nanocomposite thin film (without Ni) oxidized at 850°C in hot air for 15 min.

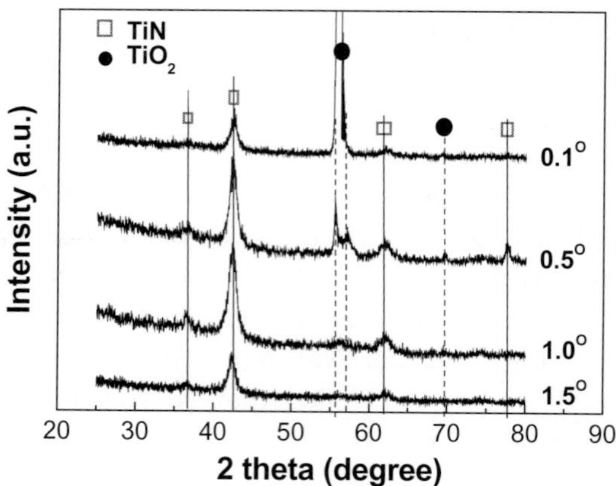

Fig. 4.67.   GIXRD patterns with different incident angles (from 0.1° through 0.5° and 1.0° to 1.5°) for the Ni-toughened nc-TiN/a-SiN$_x$ nanocomposite thin film with 2.1 at.%Ni (sample S2) oxidized at 550°C in hot air for 15 min. On the surface (under low incident angle 0.1°) most are crystallite TiO$_2$. In the inner coating (under high incident angle 1.5°), most are crystalline TiN.

At the low incident angle (such as 0.1°), the presence of well-crystallized TiO$_2$ (rutile) and TiN are observed. With an increase in the incident angle, the intensity of TiO$_2$ peaks decreases while the intensity of TiN peaks increases, indicating reduction in oxidation with depth. (X-ray penetration depth increases as incidence increases.) This agrees well with the XPS analysis [Fig. 4.63(b)].

4.2.5.2. Oxidation Variation with Temperature

Chemical Composition

Figure 4.68 shows the surface composition of the thin films oxidized at 450–1000°C. With an increase in oxidation temperature, oxygen content increases slightly, silicon decreases slightly, and titanium remains constant, while nitrogen decreases sharply. At 450°C nitrogen is about 25 at.%. A hundred-degree increase to 550°C brings about a significantly decrease to 7 at.% with further increase of temperature to 625°C, N decreases to <4 at.%. The significant decrease in N comes from the oxidation of Ti from TiN and the oxidation of Si from Si_3N_4. These reactions deplete N through the formation of N_2. The fact that nitrogen still persists in the surface layer indicates that Si_3N_4 still exists even at 1000°C. Also note from Fig. 4.68: as the oxidation temperature increases from 450 to 900°C, surface nickel increases from 1 at.% to 9 at.%; i.e. higher temperature promotes outward

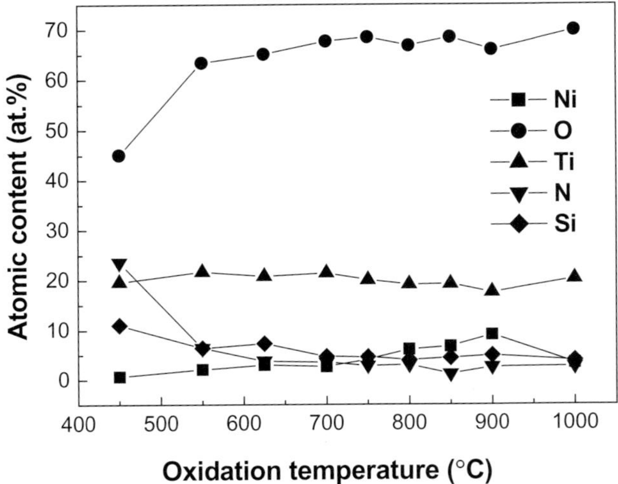

Fig. 4.68.   Change of Ti, Ni, Si, N and O species in the thin film surface with elevated oxidation temperature. With an increase in oxidation temperature, oxygen content in surface area increases slightly and titanium remains constant, while a decrease in nitrogen amount is observed. A gradual enrichment of nickel and deficiency in silicon with elevated oxidation temperature to 900°C is observed.

diffusion of nickel. With further increase to 1000°C, however, surface nickel drastically decreases to 3 at.% (the reason for this is still unclear).

Phase Identification

Figure 4.69 shows the GIXRD patterns of the Ni-toughened nc-TiN/a-SiN$_x$ nanocomposite thin film with 2.1 at.%Ni (sample S2) oxidized at elevated temperatures from 550°C through 625°C, 700°C, 750°C and 850°C to 950°C. Since an incident angle of 0.5° is low enough to observe both TiN and TiO$_2$ peaks (Fig. 4.67), 0.5° is chosen for all the samples in Fig. 4.69. As temperature increases, the intensity of the TiO$_2$ peaks slowly increases and that of TiN peaks decreases. At 850°C and above, the number and intensity of TiO$_2$ peaks increase significantly, signaling the total collapse of the film's oxidation resistance.

Based on the GIXRD analysis for the Ni-toughened nc-TiN/a-SiN$_x$ nanocomposite thin film with 2.1 at.%Ni (sample S2) oxidized at 950°C, taking $2\theta = 42.528°, 74.176°$ and $77.824°$, respectively, the calculated lattice parameter of TiN is 0.4247 nm, which agrees with the standard value of TiN ($a_{JCPDF} = 0.42417$ nm) very well. (As stipulated by [155], an error of 0.0005 nm can be considered as perfect match.) This indicates that there is no interstitial or substitutional solid solution existing after oxidation.

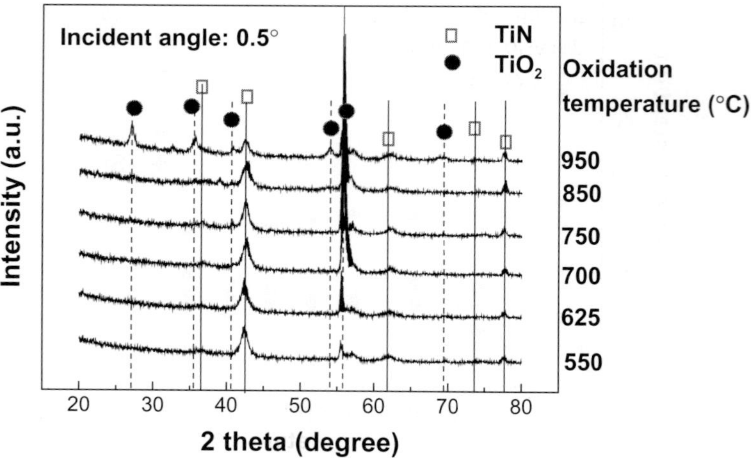

Fig. 4.69. GIXRD patterns of Ni-toughened nc-TiN/a-SiN$_x$ nanocomposite thin film with 2.1 at.%Ni (sample S2) oxidized at temperatures from 550°C through 625°C, 700°C, 750°C and 850°C to 950°C in static hot air for 15 min. Formation of crystallized TiO$_2$ on the surface is observed.

In addition, the Ni-toughened nc-TiN/a-SiN$_x$ is stress-free after oxidation even though transformation of Si$_3$N$_4$ into SiO$_2$ leads to a 96% molar volume increase. It is understandable that the high temperature involved facilitated effective atomic diffusion that neutralized possible generation of internal stresses.

Topography

Figure 4.70 shows the roughness change of Ni-toughened nc-TiN/a-SiN$_x$ nanocomposite thin film with 2.1 at.% Ni (sample S2) oxidized at 450°C, 550°C, 625°C, 750°C, 850°C and 950°C to 1000°C. With increases of oxidation temperature from 450°C to 850°C, the roughness increases slightly from $\sim$1 to $\sim$4 nm. However, with further increases to 1000°C, the roughness increases drastically to $\sim$41 nm. The metallic nickel is known for its anti-oxidation properties and its presence as a passive layer to maintain the stability of the Ni-toughened nc-TiN/a-SiN$_x$ nanocomposite thin film at low oxidation temperature by hindering migration of oxygen atoms. Above a threshold temperature, in this case, 850°C (depends on nickel content), the barrier effect of metallic nickel can no longer prevent oxygen diffusion. Consequently, rutile nucleation and growth occur, which induces

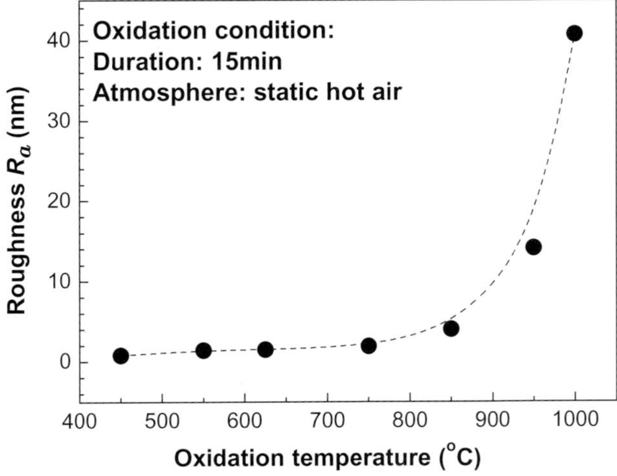

Fig. 4.70. Surface roughness change with elevated oxidation temperature. With increases in oxidation temperature from 450°C to 850°C, the roughness increases slightly from $\sim$1 to $\sim$4 nm. With further increases to 1000°C, the roughness increases dramatically to $\sim$41 nm.

a strong roughening of the surface that, in turn, allows penetration of oxygen atoms further into the film to fuel an acceleration of the oxidation process.

4.2.5.3. Discussions

The experiment suggests that the oxidation of the Ni-toughened nitride is mainly a diffusional process; nickel atoms diffuse outward and oxygen ions diffuse inward (Fig. 4.62). Oxidation is proceeded by a progressive replacement of nitrogen with diffused oxygen. The oxidation of titanium nitride to rutile or titanium dioxide (TiO_2) starts at 450°C according to:

$$TiN(s) + O_2(g) \leftrightarrow TiO_2(s) + \frac{1}{2}N_2(g). \qquad (4.10)$$

This mechanism is in agreement with the general theory for the oxidation of nitride, which considers that the process is controlled by anionic diffusion [156–160]. The oxidation of silicon nitride to stoichiometric SiO_2 starts at $\sim$450°C according to this reaction:

$$Si_3N_4(s) + 3O_2(g) \leftrightarrow 3SiO_2(s) + 2N_2(g). \qquad (4.11)$$

This oxidation is the result of the inward diffusion of oxygen through the oxide layer [161,162]. Five regions can be distinguished in the oxidized layer (Fig. 4.62) consisting of the following six components: TiO_2, TiN_xO_y, TiN, SiO_2, Si_3N_4 and Ni (Figs. 4.63–4.65). The TiO_2 and TiN are crystalline phases while Si_3N_4, SiO_2 are amorphous phases (Fig. 4.67). Following XPS depth profile (Fig. 4.62) and GIXRD result at different incident angles (Fig. 4.67) of the oxidized film, a schematic representation of the phase distribution in the oxidized layer is proposed in Fig. 4.71, where white circles represent crystalline TiO_2, gray circles crystalline TiN_xO_y, black circles crystalline TiN, small solid circles metallic Ni, and the background is amorphous Si_3N_4 with amorphous SiO_2.

From top to the core of the film, or from left to the right in Fig. 4.71, Region I is composed of mainly crystalline TiO_2, amorphous SiO_2 and metallic Ni in the amorphous matrix of Si_3N_4; in Region II the extent of oxidation is lessened (less amount TiO_2 and lots of TiN_xO_y); in Region III, more TiN presents while TiN_xO_y reduces; in Region IV, even less TiN crystalline is oxidized (reduced amount of TiO_2 and TiN_xO_y); in Region V, although there are still TiN_xO_y, basically no TiO_2 exists. The presence of a Ni-rich zone at the top of the oxide layer effectively blocked the inward

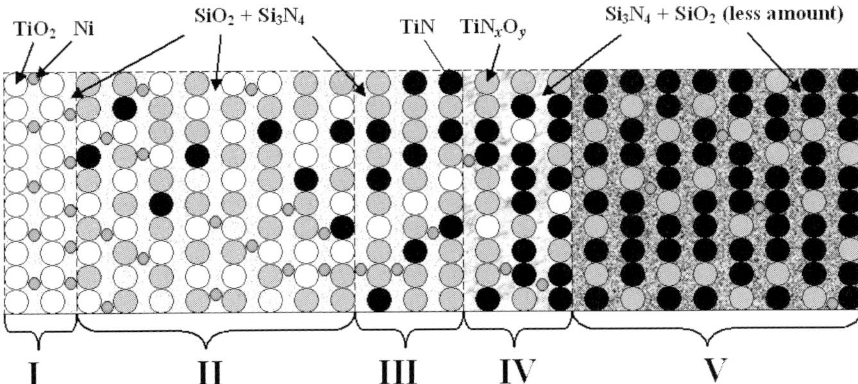

Fig. 4.71. Proposed microstructural model for the Ni-toughened nc-TiN/a-SiN$_x$ nanocomposite thin film after oxidation. Five regions can be distinguished from surface to core material.

diffusion of O. This is similar to the oxidation resistance behavior of TiAlN, where a strong aluminum migration to the surface forms a passive layer — in this case, alumina — to protect the TiAlN coating [163]. As oxidation temperature increases, the oxidation process becomes faster, presumably by the increase in the diffusivity of O$_2$ and N$_2$. Combining the components information at the oxide area from XPS (Fig. 4.68), GIXRD (Fig. 4.69) and the topographical information from AFM observations (Fig. 4.70) allows us to assume there is a threshold temperature (depending on the nickel content), in this case 850°C, below which the Ni-toughened nc-TiN/a-SiN$_x$ forms a stable nickel rich layer, the number of diffusion paths (grain boundaries, defects) would considerably decrease leading to the passivation regime, limiting oxygen diffusion and thus increasing the oxidation resistance. Above the threshold temperature, the barrier effect of metallic nickel can no longer prevent oxygen diffusion. Consequently, oxide thickness increases significantly, rutile nucleation and growth along random directions occur, and roughness of the surface increases drastically.

4.2.6. *Summary*

The effect of target power ratio of TiNi/(TiNi + Ti) on chemical composition, topography, microstructure, mechanical properties, and oxidation resistance of the Ni-toughened nc-TiN/a-SiN$_x$ nanocomposite thin films can be summarized as follows:

1. The Ni-toughened nc-TiN/a-SiN$_x$ nanocomposite thin film can be prepared by co-sputtering of Ti, TiNi and Si$_3$N$_4$ targets in Ar/N$_2$ atmosphere. XPS analysis shows that Ni is in the metallic state without reaction with N$_2$. Together with TEM and GIXRD results, it can be concluded that metallic nickel is in an amorphous state.

2. As target power ratio of TiNi/(TiNi + Ti) increases from 0 to 0.3, interface width decreases from $\sim$8 nm to $\sim$3 nm and lateral correlation length decreases from $\sim$15 nm to $\sim$9 nm, which confirms that increasing the target power ratio of TiNi/(TiNi + Ti) results in smoother coating surface.

3. The coefficient of texture for (111), (200) and (220) crystal plane of nc-TiN in the Ni-toughened nc-TiN/a-SiN$_x$ nanocomposite are close to unity, which indicates that the addition of Ni results in a random orientation of nc-TiN.

4. Results from nc-TiN lattice calculation show that at low target power ratio, the solid solution is substitutional with nickel taking the place of titanium. With further increase of target power ratio the interstitial components increase due to high energy ion bombardment.

5. The residual stress of Ni-toughened nc-TiN/a-SiN$_x$ thin films increases from -400 to -1300 MPa with increase in TiNi/(TiNi + Ti) target power ratio from zero to 0.12. With further increase of TiNi/(TiNi + Ti) target power ratio to unity, the residual stress decreases and changes from compressive to tensile state with value of 300 MPa.

6. The hardness of Ni-toughened nc-TiN/a-SiN$_x$ nanocomposite thin films keeps constant with little increase in Ni content, which is due to the solid solution hardening. With further increase in Ni content, hardness decreases, which is due to the grain boundary sliding.

7. The toughness of the Ni-toughened nc-TiN/a-SiN$_x$ nanocomposite thin films increases with an increase in Ni content. Doping from 0 to 40 at.%Ni in nc-TiN/a-SiN$_x$ brings about an increase in toughness from 1.15 to 2.60 MPa m$^{1/2}$ at some expense of hardness (dropping from 30 to 14 GPa). Ductile phase toughening through crack blunting and crack bridging is responsible for the toughness increase.

8. The oxidation of Ni-toughened nc-TiN/a-SiN$_x$ thin film is mainly a diffusional process; nickel atoms diffuse outward and oxygen ions diffuse inward. Oxidation is proceeded by a progressive replacement of nitrogen with diffused oxygen.

Symbols

A	Contact area
$A(I)$	Area between loading and unloading curves
$A(II)$	Area under loading curve
D	Diagonal of Vickers indentation impression
D_w	Target power density
E	Elastic modulus
E_{eff}	Effective elastic modulus
E_f	Elastic modulus of film
E_i	Elastic modulus of indenter
E_r	Reduced elastic modulus
G	Shear modulus
G_c	Critical strain energy release rate
$G(r)$	Height–height correlation function
H	Hardness
H_p	Plastic hardness
H_U	Universal hardness
H_v	Vickers hardness
$I_m(hkl)$	Measured relative intensity of the reflection from (hkl) plane
$I_0(hkl)$	Standard relative intensity of the reflection from (hkl) plane
K	Stress intensity factor
K_{IC}	Fracture toughness
L	Original length of beam (or substrate)
L_{c1}	Lower critical load in scratch
L_{c2}	Higher critical load in scratch
P	Test indentation load
P_c	Load at fracture
P_g	Gas pressure
P_{max}	Peak indentation load
P_{N2}	Partial pressure of nitrogen
R	Radius of curvature
R_a	Roughness
R_1	Radius of silicon substrate before being coated
R_2	Radius of silicon substrate after being coated
S	Contact stiffness
T_{dep}	Deposition temperature
T_g	Gas temperature

T_{hkl}	Coefficient of texture
T_m	Melting point
T_{rm}	Room temperature
T_s	Substrate temperature
ΔU	Strain energy difference before and after cracking
U_{fr}^c	Energy dissipated during indentation chipping
U_{edge}	Energy of edge dislocation
U_k	Kinetic energy of ionized particles
U_{screw}	Energy of screw dislocation
V_{boundary}	Volume of grain boundary
V_{crystal}	Volume of crystalline
V_s	Substrate bias
Δa	Experiment error for lattice parameter
a	Half length of the diagonal of Vickers indentation impression
	Or length of the edge of cube corner indentation impression
a_{exp}	Experimental calculated lattice parameter
a_{true}	Lattice parameter true value
$\bar{b}$	Burgers' vector
b	Crack spacing
c	Crack length
d	Grain size
d_g	Diameter of gas molecule
d_{hkl}	Interplanar spacing
f_{gb}	Volume fraction of grain boundary
h	Indentation depth
h_c	Contact depth
h_{corr}	Corrected depth
h_f	Thickness of film on tensile substrate beam
h_{max}	Maximum indentation depth
h_{min}	Minimum indentation depth
h_s	Thickness of tensile substrate beam
(hkl)	Crystal plane indice
k	Boltzmann constant
l_m	Mean free path
r	Crack tip radius
s	Span between two supporting positions
t'	Effective thickness
t_f	Thickness of film

t_s	Thickness of substrate
w	Width of the specimen
α	Smoothness quotient
α_f	Coefficient of thermal expansion of film
α_s	Coefficient of thermal expansion of substrate
β	Peak width at half maximum peak height
δ	Indenter geometry constant for fracture toughness calculation
δ_1	Indenter geometry constant for elastic calculation
δ_2	Indenter geometry constant for hardness calculation
δ_s	Geometrical parameter of specimen
ε	Indenter geometry constant for contact depth calculation
ε_f	Elastic strain of film
ε_s	Elastic strain of substrate
λ	X-ray wavelength
ν	Poisson's ratio
ν_f	Poisson's ratio of film
ν_i	Poisson's ratio of indenter
θ	Bragg angle
ρ	Crack density
σ	Residual stress
σ_{applied}	Applied stress
σ_c^f	Effective critical cracking stress
σ_g	Growth-induced stress
σ_m	Mismatch-induced stress
σ_t	Tensile stress
σ_{tip}	Stress at the tip of the crack
σ_{T}	Thermal stress
σ_y	Yield stress
σ_y^s	Yield stress of the substrate
ω	Interface width
ξ	Lateral correlation length

Abbreviations

AES	Auger electron spectroscopy
AFM	Atomic force microscopy
CPR	Crack propagation resistance
CVD	hemical vapor deposition
CTE	Coefficient of thermal expansion
DC	Direct current

DLC	Diamond-like carbon
ECR-CVD	Electron cyclotron resonance CVD
EDX	Energy dispersive X-ray analysis
FWHM	Full width at half maximum
GIXRD	Grazing incidence X-ray diffraction
HRTEM	High resolution transmission electron microscopy
IBD	Ion beam deposition
PACVD	Plasma assistant CVD
PD	Planar density
TE	Thermal evaporation
PECVD	Plasma enhanced CVD
PLD	Pulsed laser deposition
PVD	Physical vapor deposition
RBS	Rutherford back scattering spectroscopy
RF	Radio frequency
SAD	Selected area diffraction
SEM	Scanning electron microscopy
TEM	Transmission electron microscopy
XPS	X-ray photoelectron spectroscopy
XRD	X-ray diffraction

References

1. S. Zhang, Y.Q. Fu, H.J. Du, X.T. Zeng and Y.C. Liu, *Surf. Coat. Technol.* **162** (2002) 42–48.
2. S. Veprek and S. Reiprich, *Thin Solid Films* **268** (1995) 64–71.
3. V. Provenzano and R.L. Holtz, *Mater. Sci. Eng.* **A204** (1995) 125–134.
4. S. Veprek and A.S. Argon, *Surf. Coat. Technol.* **146–147** (2001) 175–182.
5. L. Maya, W.R. Allen, A.L. Glover and J.C. Mabon, *J. Vac. Sci. Technol.* **B13**(2) (1995) 361–365.
6. T. Onishi, E. Iwamura, K. Takagi and K. Yashikawa, *J. Vac. Sci. Technol.* **A14**(5) (1995) 2728–2735.
7. F. Mazaleyrat and L.K. Varga, *J. Magnetism and Magnetic Mater.* **215–216** (2000) 253–259.
8. R.A. Andrievski and A.M. Glezer, *Scripta Mater.* **44** (2001) 1621–1624.
9. B. Cantor, C.M. Allen, R. Dunin-Burkowski, M.H. Green, J.L. Hutchinson, K.A.Q.Q'. Reilly and A.K. Petford-Long, *Scripta Mater.* **44** (2001) 2055–2059.
10. A.I. Gusev, *Physics-Uspekhi* **41**(1) (1998) 49–76.
11. T. Cselle and A. Barimani, *Surf. Coat. Technol.* **76–77** (1995) 712–718.
12. S. Veprek, *J. Vac. Sci. Technol.* **A17**(5) (1999) 2401–2420.
13. E.O. Hall, *Proc. Phys. Soc. Sec.* **B64**(9) (1951) 747–753.

14. N.J. Petch, *J. Iron and Steel Institute* **174** (1953) 25–28.
15. J. Schiotz, F.D.D. Tolla and K.W. Jacobsen, *Nature* **391** (1998) 561–563.
16. J. Schiotz, *Proc. 22nd Riso Int. Symp. Mat. Sci.*, eds. A.R. Dinesen, M. Eldrup, D.J. Jensen, S. Linderoth, T.B. Pedersen, N.H. Pryds, A.S. Pedersen and J.A. Wert, (Roskilde, Denmark, 2001) 127–139.
17. J. Schiotz, T. Vegge, F.D.Di. Tolla and K.W. Jacobsen, *Phys. Rev.* **B60**(17) (1999) 11971–11983.
18. Henk, H. Westphal, *Euro PM99 Conf. Adv. Hard Mater. Production* (Turin, Italy, 1999).
19. W.D. Callister Jr., *Materials Science and Engineering: An Introduction 3rd Edition* (John Wiley & Sons, New York, 1994) 67–88.
20. H. Holleck, in *Surface Engineering: Science and Technology I*, eds. A. Kumar, Y.W. Chung, J.J. Moore and J.E. Smugeresky (The Minerals, Metals and Materials Society, 1999) 207–218.
21. S. Veprek, P. Nesladek, A. Niederhofer, F. Glatz, M. Jilek and M. Sima, *Surf. Coat. Technol.* **108–109** (1998) 138–147.
22. S. Veprek, *Surf. Coat. Technol.* **97** (1997) 15–22.
23. S. Veprek, A. Niederhofer, K. Moto, T. Bolom, H.-D. Mannling, P. Nesladek, G. Dollinger and A. Bergmaier, *Surf. Coat. Technol.* **133–134** (2000) 152–159.
24. S. Veprek, S. Reiprich and S. Li, *Appl. Phys. Lett.* **66**(20) (1995) 2640–2642.
25. S. Veprek, M. Haussmann and S. Reiprich, *J. Vac. Sci. Technol.* **A14**(1) (1996) 46–51.
26. S. Veprek, M. Haussmann and S. Li, *13th Int. Conf. on CVD, Los Angeles, May 1996, The Electrochem. Soc. Proc.* **96–5** (1996) 619.
27. P. Karvankova, M.G.J. Veprek-Heijman, O. Zindulka, A. Bergmaier and S. Veprek, *Surf. Coat. Technol.* **163–164** (2003) 149–156.
28. P. Karvankova, M.G.J. Veprek-Heijman, M.F. Zawrah and S. Veprek, *Thin Solid Films* **467** (2004) 133–139.
29. T.L. Anderson, *Fracture Mechanics: Fundamental and Applications* (CRC press, Boca Raton, 2005) 27–29.
30. P. Nesladek and S. Veprek, *Physica Status Solidi* **A177** (2000) 53–62.
31. H.-D. Mannling, D.S. Patil, K. Moto, M. Jilek and S. Veprek, *Surf. Coat. Technol.* **146–147** (2001) 263–267.
32. C. Mitterer, P.H. Mayrhofer, M. Bechliesser, P. Losbichler, P. Warbichler, F. Hofer, P.N. Gibson, W. Gissler, H. Hruby, J. Musil and J. Vicek, *Surf. Coat. Technol.* **120–121** (1999) 405–411.
33. M. Ruhle and A.G. Evans, *Prog. Mater. Sci.* **33** (1989) 85–167.
34. O. Raddatz, G.A. Schneider, W. Mackens, H. Vob and N. Claussen, *J. Eur. Ceram. Soc.* **20** (2000) 2261–2273.
35. P.A. Mataga, *Acta Metal.* **37**(12) (1989) 3349–3359.
36. D.G. Morris, *Proc. 22nd Riso Int. Symp. Mat. Sci.*, eds. A.R. Dinesen, M. Eldrup, D.J. Jensen, S. Linderoth, T.B. Pedersen, N.H. Pryds, A.S. Pedersen and J.A. Wert (Roskilde, Denmark, 2001) 89–104.
37. R.S. Mishra and A.K. Mukherjee, *Mater. Sci. Eng.* **A301** (2001) 97–101.
38. J. Musil and P. Zeman, *Vacuum* **52** (1999) 269–275.

39. J. Musil, P. Zeman, H. Hruby and P.H. Mayrhofer, *Surf. Coat. Technol.* **120–121** (1999) 179–183.

40. J. Musil, P. Karvankova and J. Kasl, *Surf. Coat. Technol.* **139** (2001) 101–109.

41. J. Musil and F. Regent, *J. Vac. Sci. Technol.* **A16**(6) (1998) 3301–3304.

42. M. Misina, J. Musil and S. Kadlec, *Surf. Coat. Technol.* **110** (1998) 168–172.

43. J. Musil and H. Polakova, *Surf. Coat. Technol.* **127** (2000) 99–106.

44. J. Musil and J. Vlcek, *Mater. Chem. Phys.* **54** (1998) 116–122.

45. A.A. Voevodin and J.S. Zabinski, *Thin Solid Films* **370** (2000) 223–231.

46. A.A. Voevodin, S.V. Prasad and J.S. Zabinski, *J. Appl. Phys.* **82**(2) (1997) 855–858.

47. A.A. Voevodin and J.S. Zabinski, *J. Mater. Sci.* **33** (1998) 319–327.

48. A.A. Voevodin and M.S. Donley, *Surf. Coat. Technol.* **82** (1996) 199–213.

49. J.R. Creighton and P. Ho, Introduction to Chemical Vapor Deposition (CVD) *Chemical Vapor Deposition*, eds. J.-H. Park and T.S. Sudarshan (ASM International, Materials Park, 2001) 1–22.

50. K.L. Choy, *Prog. Mater. Sci.* **48** (2003) 57–170.

51. P.J. Kelly and R.D. Arnell, *Vacuum* **56** (2000) 159–172.

52. R.F. Bunshah (ed.), *Handbook of Deposition Technologies for Films and Coatings 2nd Edition* (Noyes Publications, New Jersey, 1994) 27–54.

53. I. Safi, *Surf. Coat. Technol.* **127** (2000) 203–218.

54. M. Berger, *Development and Tribological Characterization of Magnetron Sputtered TiB2 and Cr/CrN Coatings* (Uppsala University, Uppsala Sweden, 2001) 11–12.

55. R.M. Bradley, J.M.E. Harper and D.A. Smith, *J. Appl. Phys.* **60**(12) (1986) 4160–4164.

56. B. Rauschenbach and K. Helming, *Nul. Instrum. Methods. Phys. Res.* **B42** (1989) 216–223.

57. D.L. Smith, *Thin-film Deposition: Principles and Practice* (McGraw-Hill, New York, 1995) 402–445.

58. Handbook of Analytical Methods for Materials. Materials Evaluation and Engineering, Inc. Plymouth, Mn55441–5447, http://www.mee-inc.com/hamm72d.pdf.

59. J.F. Watts and J. Wolstenholme, *An Introduction to Surface Analysis by XPS and AES* (John Wiley, New York, 2003) 59–78.

60. J.C. Vicherman, *Surface Analysis-The Principal Techniques* (John Wiley, New York, 1997) 61–67.

61. T.L. Barr and S. Seal, *J. Vac. Sci. Technol.* **A13**(3) (1995) 1239–1246.

62. D.A. Shirley, *Phys. Rev.* **B5**(12) (1972) 4709–4714.

63. F.-H. Hu and H.-Y. Chen, *Thin Solid Films* **355–356** (1999) 374–379.

64. J.A. Taylor, *Appl. Surf. Sci.* **7** (1981) 168–184.

65. S.C. Kuiry, S. Wannaparhun, N.B. Dahotre and S. Seal, *Scripta Mater.* **50** (2004) 1237–1240.

66. N.C. Saha and H.G. Tompkins, *J. Appl. Phys.* **72**(7) (1992) 3072–3079.

67. M. Jilek, P. Holubar, M.G.J. Veprek-Heijman and S. Veprek, *Mater. Res. Soc. Symp. Proc.* **750** (2002) 393–396.

68. I. Milosev, B. Navinsek and H.-H. Strehblow, *Surf. Coat. Technol.* **74–75** (1995) 897–902.

69. Y.Q. Fu, H.J. Du and S. Zhang, *Thin Solid Films* **444** (2003) 85–90.

70. F. Esaka, K. Furuya, H. Shimada, M. Imamura, N. Matsubayashi, H. Sato, A. Nishijima, A. Kawana, H. Ichimura and T. Kikuchi, *J. Vac. Sci. Technol.* **A15**(5) (1997) 2521–2528.

71. J.F. Moulder, W.F. Stickle, P.E. Sobol and K.D. Bomben, in *Handbook of X-ray Photoelectron Spectroscopy: A Reference Book of Standard Spectra for Identification and Interpretation of XPS Data*, ed. J. Chastian (Physical Electronics, Eden Prairie, Minnesota, 1995) 238–238.

72. I. Milosev, H.-H. Strehblow and B. Navinsek, *Surf. Interface Analys.* **26** (1998) 242–248.

73. C. Ernsberger, J. Nickerson, A.E. Miller and J. Moulder, *J. Vac. Sci. Technol.* **A3**(6) (1985) 2415–2418.

74. H.C. Man, Z.D. Cu and X.J. Yang, *Appl. Surf. Sci.* **199** (2002) 293–302.

75. F. Family and T. Vicsek, *Dynamics of Fractal Surfaces* (World Scientific, Singapore, 1991) 5–11.

76. A.L. Barabasi and H.E. Stanley, *Fractal Concepts in Surface Growth* (Cambridge University Press, Cambridge, 1995) 19–37.

77. H.-N. Yang, Y.-P. Zhao, A. Chan, T.-M. Lu and G.-C. Wang, *Phys. Rev.* **B56**(7) (1997) 4224–4232.

78. H.-N. Yang, Y.-P. Zhang, G.-C. Wang and T.-M. Lu, *Phys. Rev. Lett.* **76**(20) (1996) 3774–3777.

79. P. Scherrer, *Gottinger Nachrichten* **2** (1918) 98–103.

80. D.M. Lee, *J. Mater. Sci.* **24** (1989) 4375–4378.

81. H.P. Klug and L.E. Alexander, *X-Ray Diffraction Procedures for Ploycrystalline and Amorphous Materials* (Wiley-Interscience, New York, 1974) 422–422.

82. R. Saha and W.D. Nix, *Mater. Sci. Eng.* **A319–321** (2001) 898–901.

83. R. Saha and W.D. Nix, *Acta Mater.* **50**(1) (2002) 23–38.

84. W.C. Oliver and G.M. Pharr, *J. Mater. Res.* **7** (1992) 1564–1583.

85. W.D. Nix, *Mater. Sci. Eng.* **A234–236** (1997) 37–44.

86. N. Kikuchi, M. Kitagawa, A. Sato, E. Kusano, H. Nanto and A. Kinbara, *Surf. Coat. Technol.* **126** (2000) 131–135.

87. G.M. Pharr, *Mater. Sci. Eng.* **A253** (1998) 151–159.

88. Y.V. Milman, B.A. Galanov and S.I. Chugunova, *Acta Metall. Mater.* **41**(9) (1993) 2523–2532.

89. S. Zhang, X.L. Bui and Y.Q. Fu, *Surf. Coat. Technol.* **167** (2003) 137–142.

90. P.W. Shum, K.Y. Li, Z.F. Zhou and Y.G. Shen, *Surf. Coat. Technol.* **185** (2004) 245–253.

91. J. Malzbender and G. de With, *Surf. Coat. Technol.* **137** (2001) 72–76.

92. J. Malzbender and G. de With, *Surf. Coat. Technol.* **127** (2000) 265–272.

93. R. McGurk and T.F. Page, *J. Mater. Res.* **14** (1999) 2283–2295.

94. H. Holleck and H. Schulz, *Surf. Coat. Technol.* **36** (1988) 707–714.

95. F. Kustas, B. Mishra and J. Zhou, *Surf. Coat. Technol.* **120–121** (1999) 489–494.

96. B.R. Lawn, A.G. Evans and D.B. Marshall, *J. Am. Ceram. Soc.* **63** (1980) 574–581.

97. G.R. Anstis, P. Chantikul, B.R. Lawn and D.B. Marshall, *J. Am. Ceram. Soc.* **64** (1981) 533–538.

98. G.M. Pharr, D.S. Harding and W.C. Oliver, in *Mechanical Properties and Deformation Behavior of Materials Having Ultra-Fine Microstructure*, eds. M. Nastasi, M. Parkin Don and H. Gleiter (Klumer Academic Press, Netherlands, 1993) 449–461.

99. A.A. Volinsky, J.B. Vella and W.W. Gerberich, *Thin Solid Films* **429** (2003) 201–210.

100. A.E. Giannakopoulos, P.-L. Larsson and R. Vestergaard, *Int. J. Solids Structures* **31**(19) (1994) 2679–2708.

101. W.C. Oliver and G.M. Pharr, *J. Mater. Res.* **19** (2004) 3–20.

102. T. Lube and T. Fett, *Eng. Fract. Mech.* **71** (2004) 2263–2269.

103. D.S. Harding, W.C. Oliver and G.M. Pharr, *Mater. Res. Soc. Symp. Proc.* **356** (1995) 663–668.

104. T.W. Scharf, H. Deng and J.A. Barnard, *J. Vac. Sci. Technol.* **A15**(3) (1997) 963–967.

105. P. Kodali, K.C. Walter and M. Nastasi, *Trib. Int.* **30**(8) (1997) 591–598.

106. Y. Xie and H.M. Hawthorne, *Wear* **233–235** (1999) 293–305.

107. X. Li and B. Bhushan, *Thin Solid Films* **340** (1999) 210–217.

108. P. Jedrzejowski, J.E. Klemberg-Sapieha and L. Martinu, *Thin Solid Films* **426** (2003) 150–159.

109. P. Jedrzejowski, J.E. Klemberg-Sapieha and L. Martinu, *Thin Solid Films* **466** (2004) 189–196.

110. T. Sonoda, A. Watazu, J. Zhu, A. Kamiya, T. Nonami, T. Kameyama, K. Naganuma and M. Kato, *Thin Solid Films* **386** (2001) 227–232.

111. H. Takikawa, T. Matsui, T. Sakakibara, A. Bendavid and P.J. Martin, *Thin Solid Films* **348** (1999) 145–151.

112. O. Zywitzki, H. Sahm, M. Krug, H. Morgner and M. Neumann, *Surf. Coat. Technol.* **133–134** (2000) 555–560.

113. L. Porte, L. Roux and J. Hanus, *Phys. Rev.* **B28**(6) (1983) 3214–3224.

114. C.W. Low, I.L. Strydom and K.H.J.V. Staden, *Surf. Coat. Technol.* **49** (1991) 348–352.

115. E.A. Lee and K.H. Kim, *Thin Solid Films* **420–421** (2002) 371–376.

116. W.M. Tong, R.S. Williams, A. Yanase, Y. Segawa and M.S. Anderson, *Phys. Rev. Lett.* **72** (1994) 3374–3377.

117. J.H. Jeffries, J.-K. Zuo and M.M. Craig, *Phys. Rev. Lett.* **76** (1996) 4931–4934.

118. T. Karabacak, J.P. Singh, Y.-P. Zhao, G.-C. Wang and T.-M. Lu, *Phys. Rev.* **B68**(12) (2003) 125–408.

119. M. Diserens, J. Patscheider and F. Levy, *Surf. Coat. Technol.* **108–109** (1998) 241–246.

120. F. Vaz, L. Rebouta, Ph. Goudeau, T. Girardeau, J. Pacaud, J.P. Riviere and A. Traverse, *Surf. Coat. Technol.* **146–147** (2001) 274–279.

121. G.K. Williamson and W.H. Hall, *Acta. Metall.* **1**(1) (1953) 22–31.

122. J. Pelleg, L.Z. Zevin and S. Lungo, *Thin Solid Films* **197** (1991) 117–128.

123. J. Pelleg, L.Z. Zevin, S. Lungo and N. Croitoru, *Thin Solid Films* **197** (1989) 117–128.

124. U.C. Oh and J.H. Je, *J. Appl. Phys.* **74**(3) (1993) 1692–1696.

125. E. Zoestbergen, *X-Ray Analysis of Protective Coatings* (Groningen University Press, Groningen, 2000) 38–44.

126. J.H. Je, D.Y. Noh, H.K. Kim and K.S. Liang, *J. Appl. Phys.* **81**(9) (1997) 6126–6133.

127. L. Hultmann, J.-E. Sundgren and J.E. Greene, *J. Appl. Phys.* **66**(2) (1989) 536–544.

128. F. Vaz, L. Rebouta, B. Almeida, P. Goudeau, J. Pacaud, J.P. Riviere and J.B. Sousa, *Surf. Coat. Technol.* **120–121** (1999) 166–172.

129. B.H. Park, Y.I. Kim and K.H. Kim, *Thin Solid Films* **348** (1999) 210–214.

130. H. Oettel and R. Wiedemann, *Surf. Coat. Technol.* **76–77** (1995) 265–273.

131. S. Zhang, H. Xie, X. Zeng and P. Hing, *Surf. Coat. Technol.* **122** (1999) 219–224.

132. Y.Q. Fu, H.J. Du and C.Q. Sun, *Thin Solid Films* **424** (2003) 107–114.

133. G.A. Slack and S.F. Bartram, *J. Appl. Phys.* **46**(1) (1975) 89–98.

134. http://ssd-rd.web.cern.ch/ssd-rd/Data/Si-General.html

135. H. Holleck, *J. Vac. Sci. Technol.* **A4**(6) (1986) 2661–2669.

136. http://www.brycoat.com/physprop.html

137. W. Ensinger, *Nucl. Instrum. Methods Phys. Res.* **B127–128** (1997) 796–808.

138. H. Windischmann, *J. Vac. Sci. Technol.* **A9**(4) (1991) 2431–2436.

139. D. Roylance, *The Dislocation Basis of Yield and Creep* (Massachusetts Institute of Technology, Cambridge, 2001) 1–11.

140. T.Y. Tsui, G.M. Pharr and W.C. Oliver, *Mater. Res. Soc. Symp. Proc.* **383** (1995) 447–452.

141. J. Musil, *Surf. Coat. Technol.* **125** (2000) 322–330.

142. J. Ligot, S. Benayoun and J. Hantzpergue, *J. Wear* **243** (2000) 85–91.

143. M. Irie, H. Ohara, A. Nakayama, N. Kitagawa and T. Nomura, *Nucl. Instrum. Methods Phys. Res* **B121** (1997) 133–136.

144. M.F. Zhu, I. Suni, M.-A. Nicolet and T. Sands, *J. Appl. Phys.* **56**(10) (1984) 2740–2745.

145. W.O. Hofer, in *Topics in Applied Physics: Sputtering by Particle Bombardment iii Characteristics of Sputtered Particles, Technical Applications*, eds. R. Behrisch and K. Wittmaack (Springer-Verlag, Berlin, 1991) 65–67.

146. E.M. Trent, *Wear* **128** (1988) 65–81.

147. Z.G. Li, J.L. He, T. Matsumoto, T. Mori, S. Miyake and Y. Muramatsu, *Surf. Coat. Technol.* **174–175** (2003) 1140–1144.

148. J.G. Han, H.S. Myung, H.M. Lee and L.R. Shaginyan, *Surf. Coat. Technol.* **174–175** (2003) 738–743.

149. A. Mehta and P.J. Heaney, *Phys. Rev.* **B49**(1) (1994) 563–571.

150. S. Hogmark, S. Jacobson and M. Larsson, *Wear* **246** (2000) 20–33.

151. L.S. Sigl, P.A. Mataga, B.J. Dalgleish, R.M. McMeeking and A.G. Evans, *Acta. Metall.* **36**(4) (1988) 945–953.

152. W.-D. Munz, *J. Vac. Sci. Technol.* **A4**(6) (1986) 2717–2725.

153. Y.G. Gogotsi, F. Porz and G. Dransfield, *Oxid. Met.* **39**(1–2) (1993) 69–91.

154. Y.G. Gogotsi and F. Porz, *Corros. Sci.* **33**(4) (1992) 627–640.

155. M. Diserens, J. Patscheider and F. Levy, *Surf. Coat. Technol.* **120–121** (1999) 158–165.

156. J. Desmaison, P. Lefort and M. Billy, *Oxid. Met.* **13**(6) (1979) 505–517.

157. A. Bellosi, A. Tampieri and Y.Z. Liu, *Mater. Sci. Eng.* **A127** (1990) 115–122.

158. A. Tampieri, E. Landi and A. Bellosi, *Br. Ceram. Trans. J.* **90** (1991) 194–196.

159. H. Ichimura and A. Kawana, *J. Mater. Res.* **8** (1993) 1093–1100.

160. F. Deschaux-Beaume, T. Cutard, N. Frety and C. Levaillant, *J. Am. Ceram. Soc.* **85**(7) (2002) 1860–1866.

161. L.U.J.T. Ogbuji and S.R. Bryan, *J. Am. Ceram. Soc.* **78**(5) (1995) 1272–1278.

162. L.U.J.T. Ogbuji, *J. Am. Ceram. Soc.* **78**(5) (1995) 1279–1284.

163. F. Vaz, L. Rebouta, M. Andritschky, M.F. Da Silva and J.C. Soares, *J. Eur. Ceram. Soc.* **17** (1997) 1971–1977.

CHAPTER 2

MAGNETRON SPUTTERED HARD AND YET TOUGH NANOCOMPOSITE COATINGS WITH CASE STUDIES: NANOCRYSTALLINE TiC EMBEDDED IN AMORPHOUS CARBON

Sam Zhang[*], Xuan Lam Bui and Deen Sun

School of Mechanical and Aerospace Engineering
Nanyang Technological University
50 Nanyang Avenue, Singapore 639798
[*]*msyzhang@ntu.edu.sg*

1. Al-doped Amorphous Carbon: a-C(Al)

Amorphous carbon (a-C) or diamond-like carbon (DLC) is a preferred material for wear protective coatings in a variety of applications because of its high hardness, high wear resistance and very low friction when sliding against most engineering materials, compared to conventional hard ceramics such as TiN, TiAlN, CrN, etc. However, the following drawbacks limit its applications in engineering fields.

Deposition-inherited high residual stress: high residual stress limits the coating thickness to less than 2 μm since residual stress weakens the adhesion strength of coatings on substrates [1]. A coating thickness of less than 2 μm is not suitable for severe tribological applications where long working life is crucial. Pure a-C also exhibits brittle behavior at applied high load [2], thus its load-bearing capability is limited. Also, the thermal stability and oxidation resistance of pure a-C is very poor, which restricts its application at temperatures less than 400°C [3]. Modification of the a-C structure to overcome these drawbacks becomes imperative for effective utilization of a-C in engineering applications.

Doping of aluminum is an effective way of reducing the deposition-inherited residual stress but hardness of the coating sacrifies. To bring back the coating hardness, nanocrystalline TiC grains are embedded in the amorphous matrix to form a nanocomposite. This chapter starts with doping of Al into amorphous carbon to form Al-doped a-C (denoted as "a-C(Al)")

111

and moves on to the incorporation of nanocrystalline TiC (denoted as "nc-TiC") into pure a-C matrix and into a-C(Al) matrix to form nc-TiC/a-C and nc-TiC/a-C(Al) nanocomposite coatings. In the end, the nc-TiC/a-C(Al) nanocomposite coating of more than $20\,\mu$m thick is deposited on a piston ring and tested in engine operation. Results show that the coating is superior to the best commercial coatings available in the market.

1.1. *Composition and Microstructure*

Generally, at the same power density, the sputtering yield of Al is 6–7 times higher than that of C [4], therefore the power density of the Al target in this study was selected at a much lower level (from 0 to $1.8\,\mathrm{W/cm^2}$) than that of the graphite target ($10.5\,\mathrm{W/cm^2}$). The composition of the C and Al in the coating was determined from X-ray photoelectron spectroscopy (XPS) by calculating the area under the peaks of C 1s at 284.6 eV, Al 2p at 72.6 eV, and O 1s at 532.4 eV [5–8]. Since the process pressure was low enough (0.6 Pa), the Ar inclusion in the coating was negligible and was not taken into account during composition calculation. Figure 1.1 shows a typical XPS spectrum of Al-doped a-C (without ion etching of the coating surface). Except for the difference in the intensity and width of the peaks, the XPS profile is almost the same for all the different power densities applied to Al target. It is easy to recognize a large amount of oxygen contaminating the coating. Al 2p spectra (with fitting curves) of the Al-doped

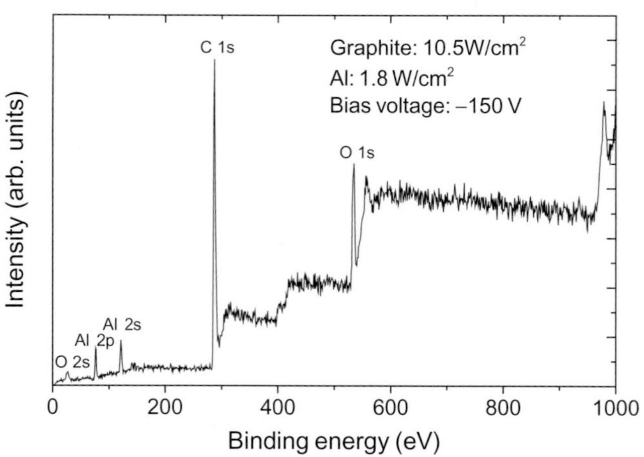

Fig. 1.1.   XPS spectrum of Al-doped a-C coating.

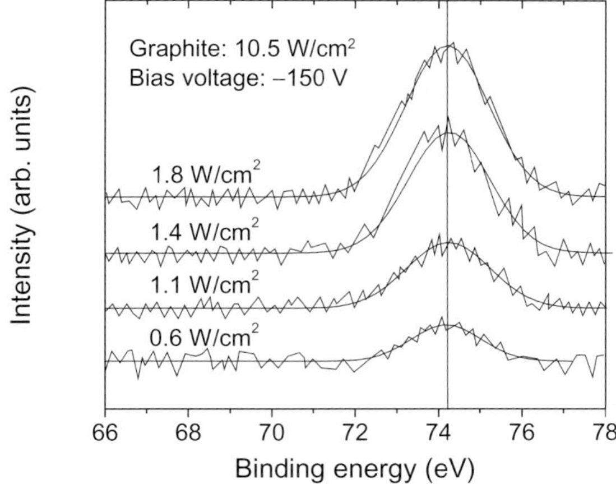

Fig. 1.2.   XPS spectra of the Al 2p for Al-doped a-C deposited at different power densities of the Al target under constant substrate bias and graphite target power density.

a-C coatings deposited under $-150\,\text{V}$ bias with different power densities of Al target are shown in Fig. 1.2.

The symmetrical single peaks at $74.2\,\text{eV}$ indicate that almost all the Al on the surface were bonded with oxygen to form aluminum oxide [9]. Al is extremely reactive with oxygen. Therefore immediately after unloading from the deposition chamber, Al on the coating surface was oxidized. If the oxidation layer is removed, the Al 2p XPS spectra should shows shift of binding energy and increase in intensity, as is shown in Fig. 1.3 before and after 15 min ion etching. After etching, the peak was shifted to $72.6\,\text{eV}$ with a higher intensity. It indicates that Al exists in the coating as elemental aluminum. The higher intensity of the Al 2p peak after etching is a result of the removal of carbon and oxygen contamination on the coating surface. Also, a small amount of Al–O bonds was detected from the deconvoluted peak at $74.2\,\text{eV}$. The oxygen is believed to have contaminated the coating during the deposition process. Although the chamber was pumped to a base pressure of 1.33×10^{-5} Pa, there was always a small amount of oxygen in the chamber available to contaminate the coating. No bond between Al and C at $73.6\,\text{eV}$ [6] was seen in the coating even when the Al target was at $1.8\,\text{W/cm}^2$.

Figure 1.4 shows the XPS spectrum (after etching) of the C 1s where the peaks of the C–C bonds at $284.6\,\text{eV}$ and C–O bonds at $286.3\,\text{eV}$ were seen. It again confirms that no chemical bonding exists between C and Al since no peak at $281.5\,\text{eV}$ was observed. The composition of coatings

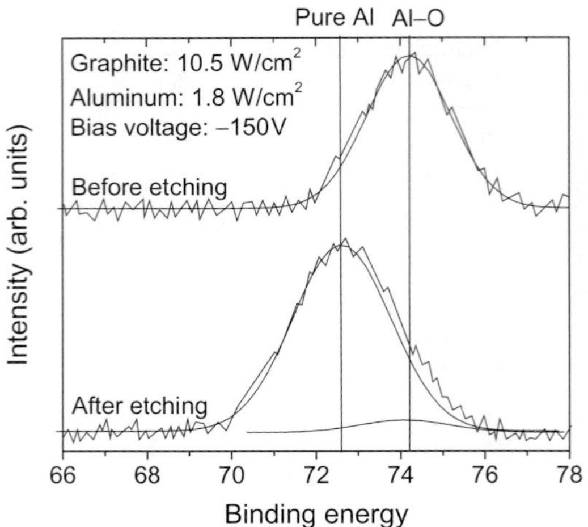

Fig. 1.3.   XPS spectra of Al 2p for Al-doped a-C before and after ion etching for 15 min.

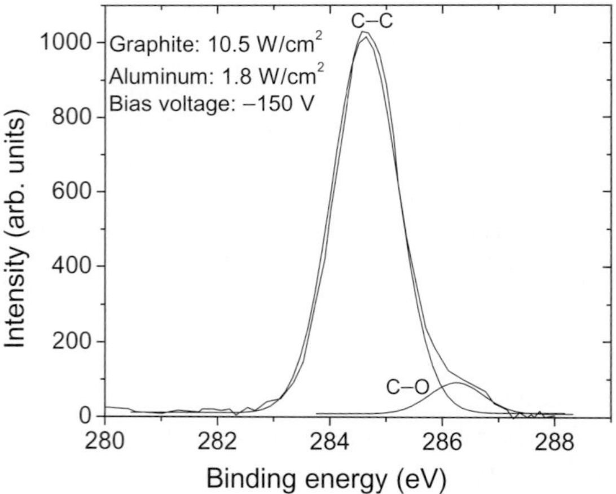

Fig. 1.4.   XPS spectrum of C 1s for Al-doped a-C after etching for 15 minutes.

as calculated from XPS spectra (after etching for 15 min) is tabulated in Table 1.1. The Al content in the coating increased from 5.1 to 19.6 at.% as the power density of the Al target increased from 0.6 to 1.8 W/cm^2. The oxygen content in the coating was 2–4 at.%, a small amount.

At the same target power density, the change of Al concentration at different bias voltages (in the range from -20 to -150 V) was not significant, as seen from Fig. 1.5 where the Al concentration in the Al-doped a-C is

Table 1.1. Composition of the Al-doped a-C coating deposited under $-150\,\mathrm{V}$ bias and power density of the graphite target of $10.5\,\mathrm{W/cm^2}$.

Power density of Al target ($\mathrm{W/cm^2}$)	C (at.%)	Al (at.%)	O (at.%)
0.6	93	5	2
1.1	87	9	4
1.4	83	14	3
1.8	77	19	4

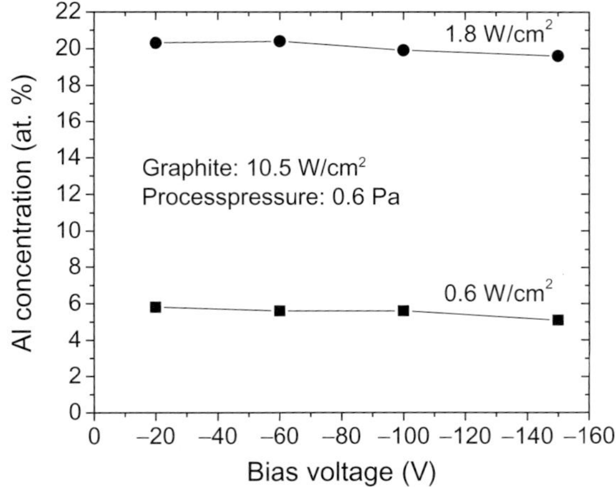

Fig. 1.5.   Relationship between bias voltage and Al concentration in Al-doped a-C coatings deposited under different power densities of the Al target.

plotted as a function of bias voltage. Therefore, the parameter that can be used to effectively control the composition of metal in the coating is the power density on the metallic target.

Al-doped a-C coatings are amorphous. Figure 1.6 shows the XRD spectra of an Al-doped a-C coating (19 at.%Al) on Si(100) wafer together with that of the Si wafer without coating. The XRD profile of the Al-doped a-C did not have any noticeable peak except the peak from the Si substrate at around $69°$ 2θ. This indicates that the Al in the coating is not crystalline. This is supported by the TEM image (Fig. 1.7) with selected area diffraction pattern where a broad halo was seen.

Figure 1.8 shows Raman spectra of Al-doped a-C for different Al compositions in the wavenumber range from 850 to $1900\,\mathrm{cm^{-1}}$. Outside this range, no feature was observed. The $I_\mathrm{D}/I_\mathrm{G}$ ratios calculated from the intensity of the G and D peaks have also been put in the diagram. The $I_\mathrm{D}/I_\mathrm{G}$ ratio

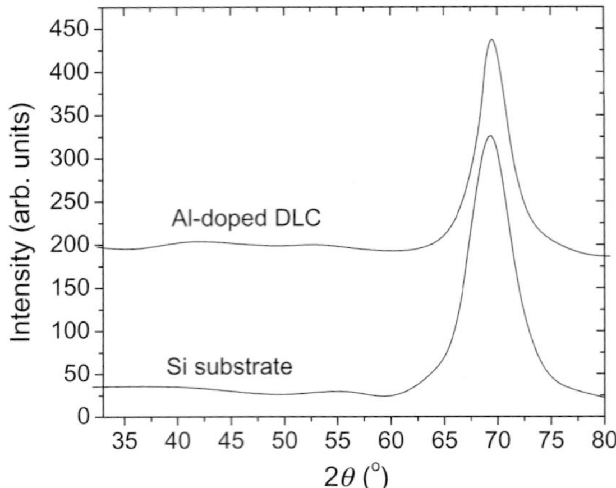

Fig. 1.6. XRD spectra of Al-doped a-C coating (19 at.%Al) and a Si(100) substrate. No noticeable peak is seen except the one of Si(100) wafer at 69° 2θ.

Fig. 1.7. TEM micrograph with diffraction pattern of Al-doped a-C coating (19 at.% Al). The coating is amorphous: a broad halo is seen from the diffraction pattern.

increased with increasing power density of the Al target. Without Al, the I_D/I_G ratio was 1.1 and it increased to 2.7 as the Al content increased to 19 at.%. Since the sp^3/sp^2 ratio is inversely proportional to the I_D/I_G ratio, an increasing I_D/I_G ratio indicates that incorporation of Al hinders the formation of sp^3 sites leading to an increase in sp^2 fraction in the coating. As the I_D/I_G ratio becomes high, more sp^2 sites become available such that they start organizing into small graphitic clusters [10, 11]. This is directly attributed to the decreased energies of carbon atoms/ions coming to the

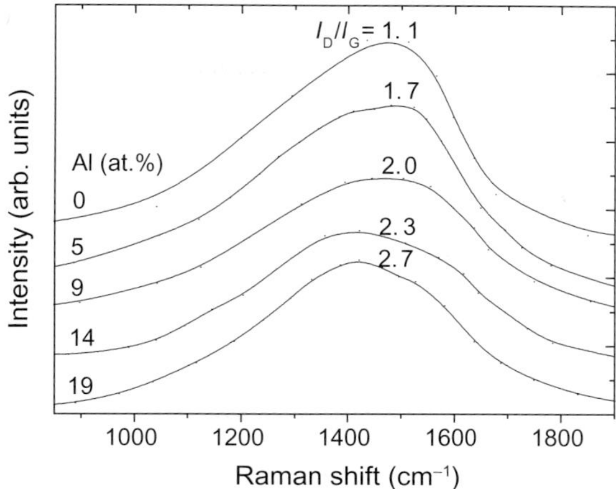

Fig. 1.8.　Raman spectra of Al "doped" a-C coatings with different Al concentrations.

substrate when Al is co-sputtered: a higher power density applied on the Al target (leading to a higher Al content in the coating) results in more collisions. At low bombarding energies, surface diffusion takes place. It should be noted that surface diffusion is easier to take place than bulk diffusion, since the activation energy is lower. Diffusion in the surface layers of the coating tends to generate ordered clusters with high sp^2 content, i.e. with structures closer to the thermodynamically stable graphite phase than a typical DLC arrangement.

1.2. *Mechanical Properties*

The hardness and Young's modulus of Al-doped a-C coatings decreased with increasing Al. Without Al, the hardness of a pure a-C coating was 32. 5 GPa and it decreased to a low level of 7.8 GPa when the coating was "doped" with 19 at.%Al as seen in Fig. 1.9. This is the consequence of the increase in sp^2 bonding structure as Al is added. In addition, Al is a soft material; therefore the hardness of the coating is further decreased. However, the increase of the sp^2 bonding in the coating contributes to the relaxation of the residual stress. It has been proven by Sullivan *et al.* [12] that when the atomic volume of the sp^2 sites exceeds that of the sp^3 sites, the in-plane size is less due to its shorter bond length. Thus the formation of sp^2 sites with their σ plane aligned in the plane of compression greatly relieves biaxial compressive stress.

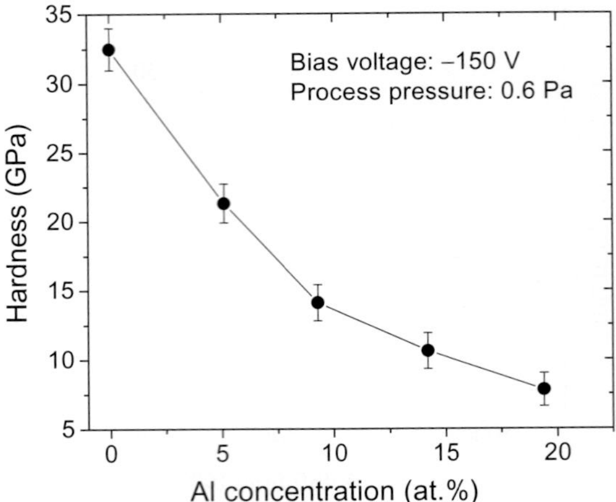

Fig. 1.9. Hardness of a-C coating as a function of Al concentration.

Figure 1.10 plots the relationship between residual stress of Al-doped a-C coatings (thickness of about $1.2\,\mu$m) and Al content. Without Al, the residual stress was as high as $4.1\,$GPa. When 5 at.%Al was added, the stress dropped to $0.9\,$GPa (a 24.5 times reduction) and the $I_{\mathrm{D}}/I_{\mathrm{G}}$ ratio increased from 1.1 to 1.7 (a 21.5 times increase). For a-C coatings with high sp^3 content, the stress decreases considerably with very little variation in strains in the coating [13]. Therefore, continued addition of Al only resulted in a gradual decrease in stress. At 19 at.%Al, the residual stress reduced to $0.2\,$GPa.

Adhesion strength is related to the interfacial properties between coating and substrate. Among the factors which affect the interfacial properties, residual stress is an important issue. Addition of Al results in a considerable increase in adhesion strength of the coating to the substrate. This is due to two reasons: (1) the low residual stresses in the coating, and (2) the high toughness achieved because Al is a tough material and the coating contains a more graphite-like structure. Both effects can be seen from the results of scratch tests: high critical load combined with plastic behavior.

Figure 1.11 shows the optical micrographs of scratches after scratch tests on a-C "doped" with 5 and 19 at.%Al. Coatings with different Al content exhibited different behaviors in the scratch tests.

The damage of coatings doped with 5 at.%Al occurred at an applied load of about $367\,$mN. However, at such a high load, the coating exhibited only partial damage. This is different from the results obtained from pure

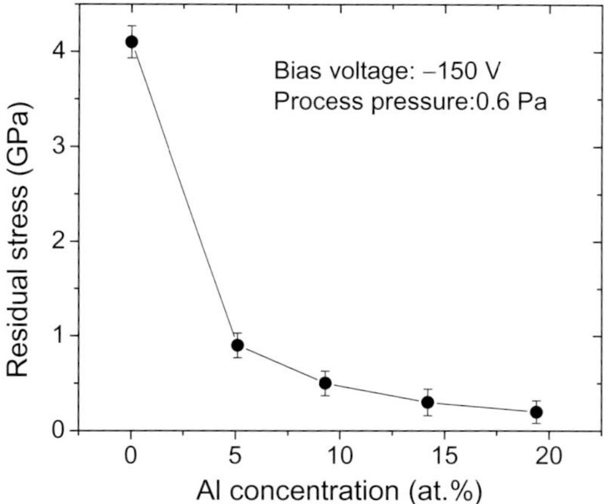

Fig. 1.10.   Residual stress as a function of Al content in Al-doped a-C coatings.

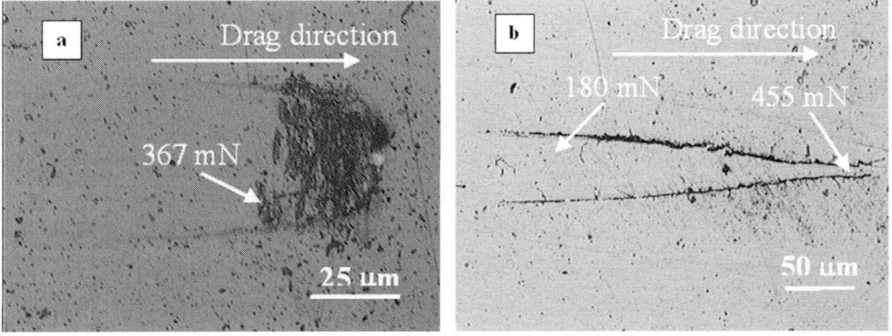

Fig. 1.11.   Optical micrographs of scratches on the coatings after scratch tests of
Al-doped a-C coatings: (a) 5 at.%Al and (b) 19 at.%Al.

a-C coatings where at low applied loads (less than 240 mN), the coatings
were totally damaged. As 19 at.%Al was added, no fracture was seen. Since
this coating had low hardness (7.8 GPa), as the load reached about 180 mN,
the diamond tip had already ploughed into the coating causing a gradual
decrease in the scanning amplitude. Even until the end of the test, when
the load reached 455 mN, no fracture or interfacial failure was observed, an
indication of superior toughness and adhesion strength. It is clear that the
more the Al the more the adhesion improvement. However, this superiority
of toughness comes with the high price of hardness drop, therefore limitation
in applications.

In summary, co-sputtering of Al and graphite targets results in a very low stress and tough Al-doped a-C (or a-C(Al)) coating at the expense of its hardness. Al exists as clusters of atoms in the coating.

2. Nanocrystalline TiC Embedded in Amorphous Carbon: nc-TiC/a-C

2.1. *Composition*

Ti was added into a-C by co-sputtering of graphite and Ti targets. The power density on the graphite target was $10.5\,\mathrm{W/cm^2}$ whereas that on the Ti target was changed for different compositions. The process pressure was maintained at $0.6\,\mathrm{Pa}$ and the substrate temperature was kept constant at $150°\mathrm{C}$ during deposition.

After unloading from the deposition chamber, the coatings were contaminated with large amounts of oxygen and carbon from the ambient atmosphere. Therefore ion etching was carried out for 15 min before XPS analysis. The oxygen incorporated into the coatings during deposition was less than 5 at.% and was excluded from the calculations. Fitting the C 1s and the Ti $2p_{3/2}$ peaks allows discrimination among the three concurrent phases in the coating: a-C, TiC and metallic Ti. The C 1s positions of the C–C in a-C and C–Ti in TiC are 284.6 and 281.8 eV, respectively [5]. The Ti $2p_{3/2}$ peak of Ti in TiC is at 454.9 eV and that of metallic Ti is at 453.8 eV [14]. The Ti 2p also has another peak (Ti $2p_{1/2}$) at 461 eV for Ti in TiC and 459.9 for metallic Ti (neutral or zero valence state). The composition of the coatings deposited at different power densities on the Ti target and under $-150\,\mathrm{V}$ bias is tabulated in Table 2.1. Those values were calculated from the C 1s and Ti 2p XPS spectra, which are shown in Fig. 2.1 and Fig. 2.2. At the same power density on the Ti target, the Ti content in the coating remained almost unchanged when different bias voltages (in the range from -20 to $-150\,\mathrm{V}$) were applied. This is similar to the case of adding Al to a-C (Fig. 1.5).

From Fig. 2.1, at 0 at.%Ti (i.e. 100% a-C), only the C–C bond was observed at 284.6 eV. As the Ti content increased, the carbide (TiC) peak at 281.8 eV appeared and grew in intensity while the a-C peak decreased. At 3 at.%Ti, almost no TiC was seen — Ti existed in the coating as elemental Ti, which gave a peak at 453.8 eV of Ti 2p in Fig. 2.2. The limitation of the formation of TiC at low Ti content can be attributed to the low power density applied to the Ti target (e.g. $0.8\,\mathrm{W/cm^2}$ for 3 at.%Ti). At low power densities, the energy of the sputtered Ti ions may not be high enough to

Table 2.1.   Composition of Ti doped a-C coatings deposited at different power densities of Ti target under $-150\,\text{V}$ bias and graphite target power density of $10.5\,\text{W/cm}^2$.

Power density of Ti target (W/cm^2)	Ti (at.%)	C (at.%)	Coating structure
0	0	100	a-C
0.8	3	97	a-C(Ti)
1.4	8	92	nc-TiC/a-C
1.8	16	84	nc-TiC/a-C
2.1	25	75	nc-TiC/a-C
2.4	30	70	nc-TiC/a-C
2.7	36	64	nc-TiC/a-C
3.0	45	55	nc-TiC/a-C
3.3	48	52	TiC
4.0	53	47	TiC

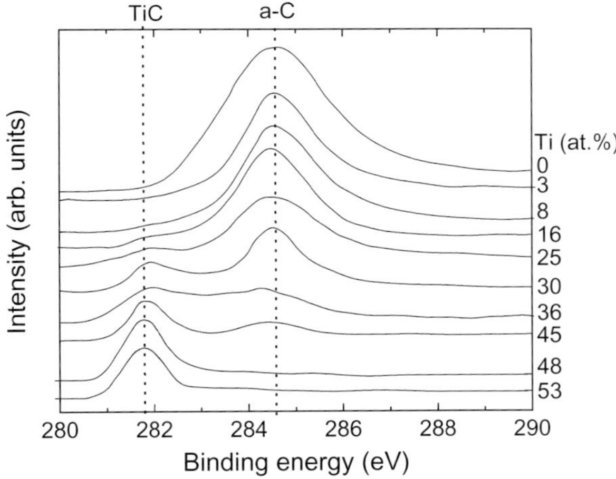

Fig. 2.1.   XPS spectra of C 1s of Ti doped a-C coatings deposited under $-150\,\text{V}$ bias [15].

support the formation of Ti–C bonds. The formation of TiC was seen as 8 at. %Ti was added (the power density of the Ti target was $1.4\,\text{W/cm}^2$) and from the Ti 2p spectrum only part of Ti bonded with C to form TiC. At 16 at.%Ti and more, most Ti in the coating bonded with C to form TiC (as seen from the Ti 2p$_{3/2}$ XPS spectra in Fig. 2.2; a single peak, which can be attributed to TiC, was seen at 454.9 eV). As the Ti content exceeded 48 at.%, the peak associated with a-C was almost undetectable from the XPS spectra of the C 1s (Fig. 2.1) indicating that there was virtually no a-C in

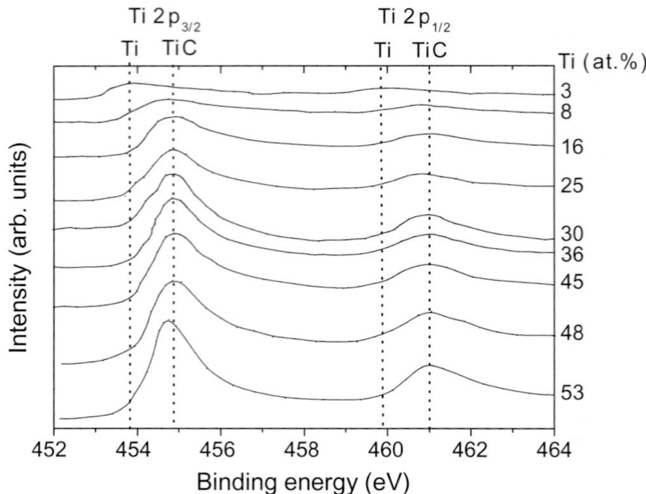

Fig. 2.2.　XPS spectra of Ti 2p in Ti doped a-C coatings deposited under -150 V bias.

these coatings (all the carbon atoms bonded with Ti to form TiC). Also, from Fig. 2.2, a considerable amount of metallic Ti was seen in the coating "doped" with 53 at.%Ti (the Ti $2p_{3/2}$ peak was considerably shifted towards 453.8 eV). Figure 2.3 presents the composition of a-C, TiC, and Ti (calculated from XPS) in the coatings at different atomic concentrations of Ti. The amount of metallic Ti was calculated by subtracting the carbide contribution to the Ti 2p obtained by peak fitting from the total Ti content. From the figure, the concentration of a-C and TiC was in the range 50–50 at.% at about 27 at.%Ti incorporation. Therefore, at Ti less than 27 at.%, a-C is dominant and TiC becomes dominant if Ti content exceeds 27 at.%.

2.2. *Topography*

Figure 2.4 plots AFM surface roughness of nc-TiC/a-C coatings (1.2 μm thickness, deposited on a Si wafer) as a function of Ti content. With increasing Ti, surface roughness of the coatings increased. At 3 at.%Ti, the surface roughness of the coating was 1.9 nm R_a, a little higher than that of pure a-C (1.1 nm R_a). After that, R_a increased considerably to about 5 nm at 8 at.%Ti. From 8 to 36 at.%Ti, R_a very slowly increased from 5 to about 8 nm. From 36 at.%Ti, R_a increased drastically and reached a very high

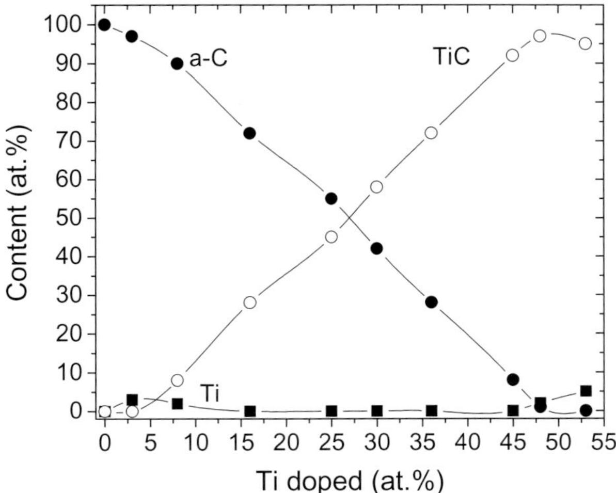

Fig. 2.3.   Content of a-C, TiC and metallic Ti as a function of Ti addition (calculated from the XPS data presented in Figs. 2.1 and 2.2).

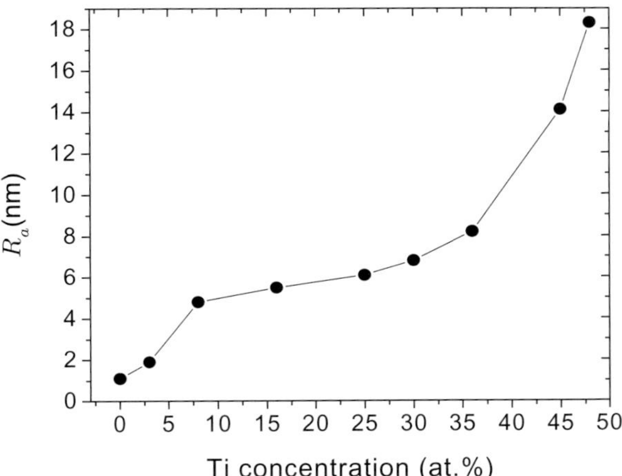

Fig. 2.4.   Surface roughness (R_a) as a function of Ti concentration.

value of 18.3 nm at 48 at.%Ti. From XPS, XRD, TEM and Raman results, it was clear that more Ti incorporation resulted in a large fraction of crystalline phase (thus less amorphous carbon) with larger grain size, which led to rougher surface morphologies.

2.3. *Microstructure*

Figure 2.5 shows the XRD spectra of the coatings at different Ti. At 3 at.%Ti, the coating is amorphous. Also, as mentioned above, Ti exists in this coating as elemental Ti. Therefore this coating was denoted as a-C(Ti) in Table 2.1. Above 8 at.%Ti incorporation, the formation of the TiC crystalline phase was observed. Peaks at 35.9, 41.7, and 60.4° (2θ) were attributed to the (111), (200), and (220) diffraction planes of TiC. The intensity of those peaks increased with increasing Ti. No dominant texturing was observed. From 8 to 45 at.%Ti, as seen from XPS and XRD results (Figs. 2.3 and 2.5), the crystalline TiC and a-C co-existed. Therefore the coatings were denoted as nc-TiC/a-C (the prefix nc- stands for nanocrystalline since the size of the TiC grains is in the nanometer range, as will be shown later) in Table 2.1. At higher content of Ti, there was almost no a-C in the coating but only crystalline TiC and a small amount of metallic Ti. The coatings, in this case, were denoted as nc-TiC in Table 2.1. It can be seen that, at $-150\,\mathrm{V}$ bias, nanocomposite can be obtained if the Ti content is from 8 to about 45 at.%Ti. Outside this range, the resultant coating is a-C(Ti) or crystalline TiC. From 25 at.%Ti onwards, a small shift of the TiC peaks to the smaller Bragg angle was seen. That was believed to have come from the compressive stress generated in the coating during the deposition process. It was mainly intrinsic stress because the coatings were deposited under low deposition temperature thus the thermal

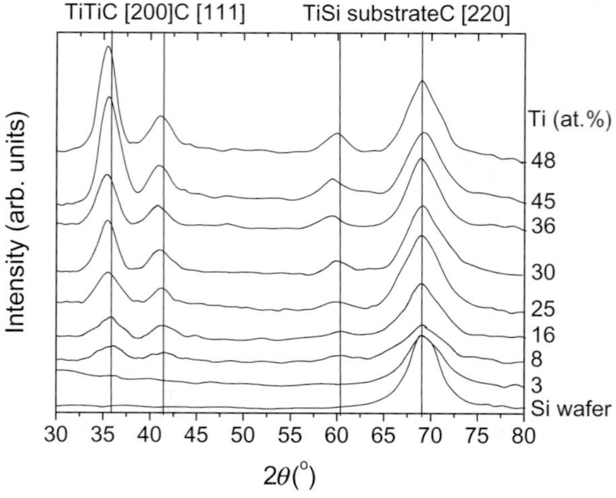

Fig. 2.5.   XRD spectra of Ti-doped a-C coatings deposited under $-150\,\mathrm{V}$ bias.

stress was negligible. At lower Ti content, the residual stress was low and no shift of the peaks was seen.

Ignoring the effect of microstraining due to the compressive stress, the average TiC crystalline size can be estimated using the Debye–Scherrer formula [16]:

$$D = \frac{K\lambda}{\beta\cos(\theta)}\,[\text{nm}],$$

where D is the mean crystalline dimension normal to diffracting planes, K is a constant ($K = 0.91$), λ is the X-ray wavelength ($\lambda = 0.15406\,\text{nm}$), β in radians is the peak width at half maximum peak height and θ is the Bragg angle. The calculated grain size is plotted in Fig. 2.6.

The grain size increased with increasing Ti concentration. On average, TiC [111] crystallites were largest, followed by that of TiC [200]; the TiC [220] crystallites were smallest. For TiC [111], the grain size increased from 3 to 17 nm as Ti increased from 8 to 48 at.%. With increasing Ti, there are more Ti^{4+} ions readily available for the growth of TiC crystallites. At the same time, as Ti increases, the relative amount of a-C is reduced (Fig. 2.3), thus the constraints exerted on the growth of the crystallites are alleviated. All these combine to result in an increase in the grain size of TiC with increasing Ti incorporation.

Figure 2.7 shows the TEM micrographs of coatings with different Ti content. At 3 at.%Ti, no TiC grains were seen; the coating was amorphous.

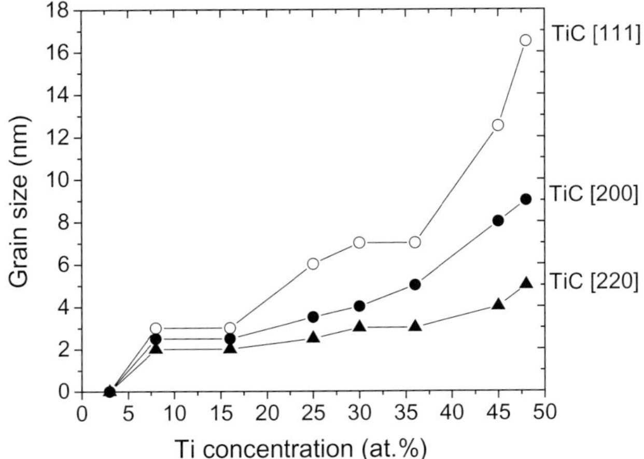

Fig. 2.6.   Grain size of TiC as a function of Ti concentration (calculated from XRD results in Fig. 2.5).

     S. Zhang et al.

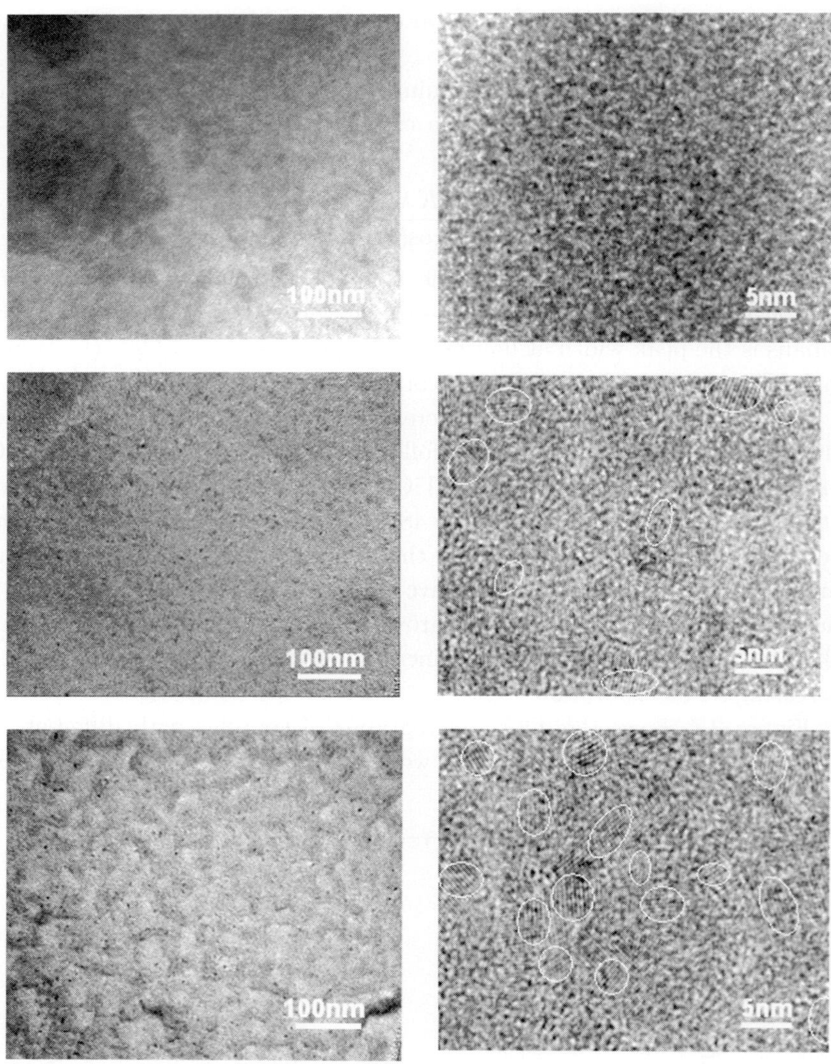

Fig. 2.7. TEM micrographs of coatings with different Ti content. At 3 at.%Ti, the coating is amorphous. TiC nanograins are observed from 8 at.%Ti onwards. The size and fraction of crystallites increase with increasing Ti content. At 45 at.%Ti, the coating contains almost TiC grains [bright-field (BF TEM) is added for an easier recognition of grains].

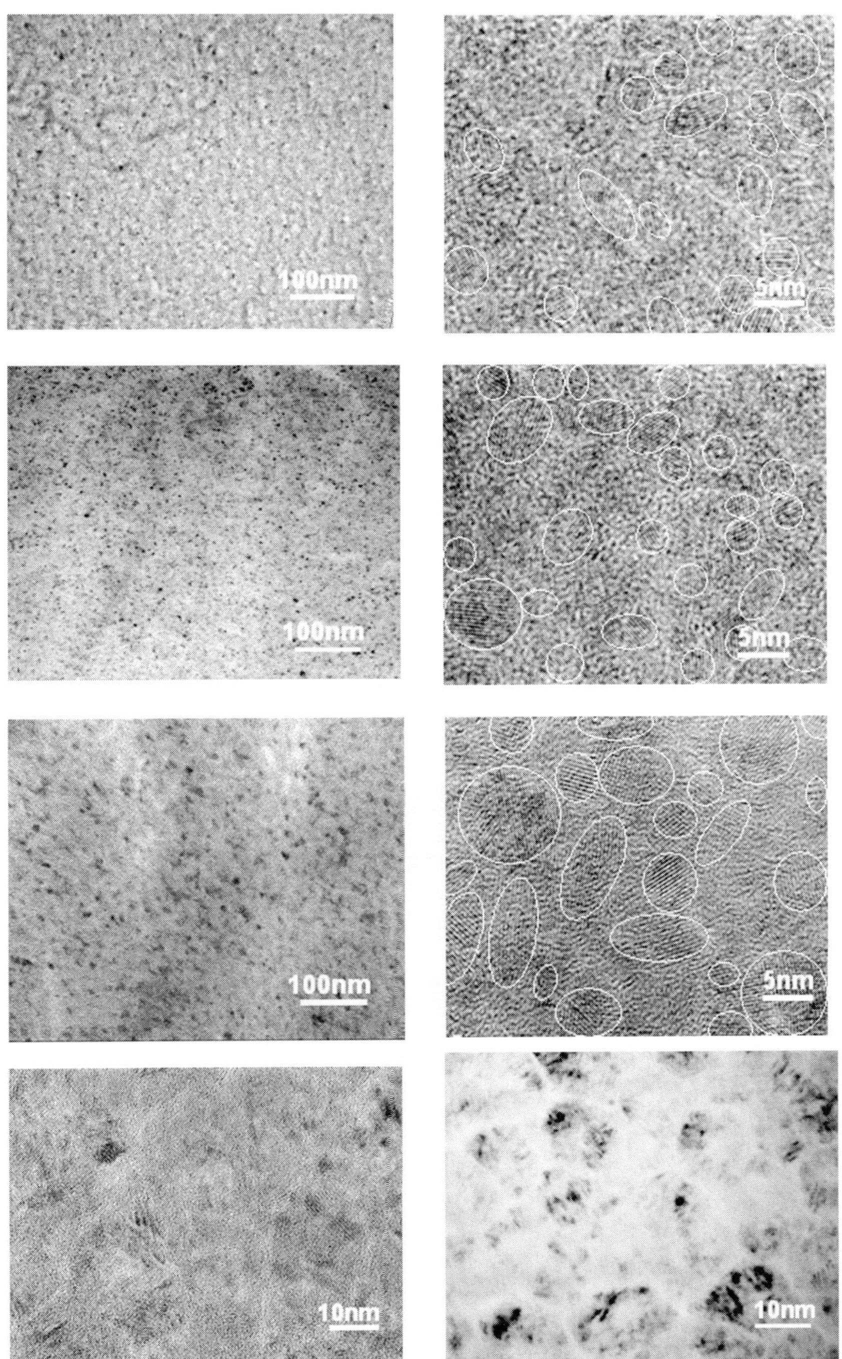

Fig. 2.7.   (*Continued*)

At 8 at.%Ti, the TiC grains in the a-C matrix were observable: a few grains scattered within an a-C matrix. The density of grains increased with increasing Ti content in the coating. At 45 at.%Ti, very little a-C was seen. The coating contains almost only large TiC grains. Also, it is difficult to recognize the grains in the matrix. Therefore, a bright field image is included for easier recognition of the grains and estimation of their size. Above 8 at.%Ti, the diffraction patterns of coatings were almost similar: sharp rings with almost uniform intensity for the [111], [200] and [220] directions (Fig. 2.8). This indicates a random orientation ([111], [200] or [220]) of the TiC grains in the a-C matrix. The observation from TEM agrees well with results obtained from XRD. The intensity of the TiC [311] and TiC [222] is very weak, indicating that very few TiC crystallites are oriented these directions. Therefore, the Braggs peaks of TiC [311] and TiC [222] (at 72.3° and 76.2° 2θ, respectively) were not observable in the XRD spectra. Grain sizes observed from the TEM micrographs as compared to that calculated from XRD spectra are plotted in Fig. 2.9. The trend of grain sizes observed from TEM and XRD was the same: grain sizes increased as Ti content in the coating increased. However, the grain size calculated from XRD was smaller than that observed from TEM. In the calculation of grain size (using the Scherrer formula) the effect of microstrains (due to the residual stress) in broadening the Braggs peak was ignored.

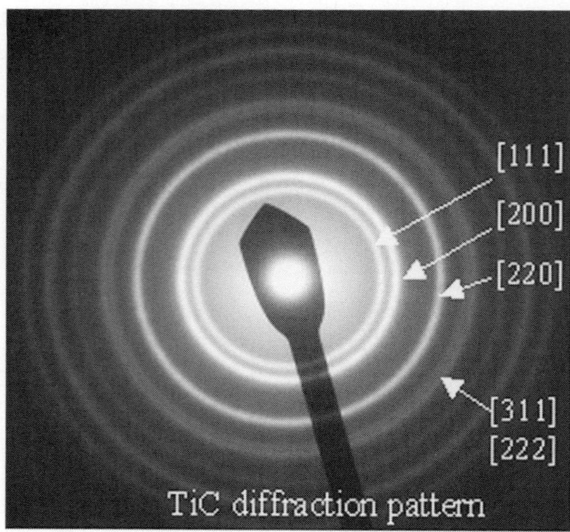

Fig. 2.8.   Diffraction pattern of nc-TiC/a-C coatings indicating the random orientation of the TiC crystallites.

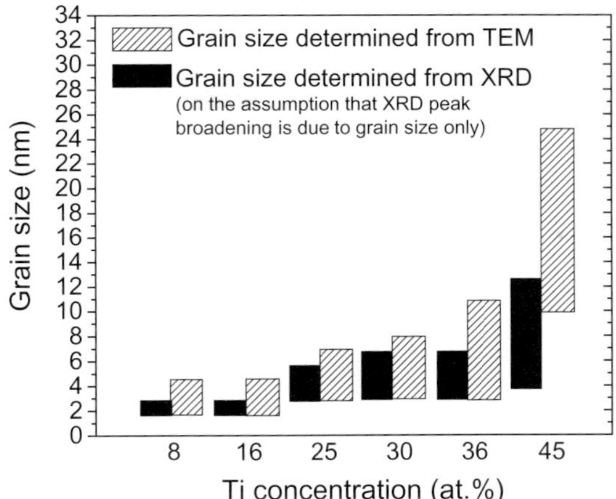

Fig. 2.9.   Grain size of TiC determined from TEM and XRD. The grain size increases as Ti content increases.

Taking microstraining into account, the grain size can be obtained through Williamson–Hall Plot [17], thus,

$$D = \frac{K\lambda}{\beta\cos(\theta) - 4\varepsilon\sin(\theta)}.$$

Ignoring microstrain ε thus enlarges the denominator, resulting in smaller calculated grain size.

Raman spectra and the I_D/I_G ratio for nc-TiC/a-C coatings are shown in Fig. 2.10. From the figure, the intensity of the carbon peak decreased as more Ti was added into the coating and no peak was seen for the coating "doped" with 45 at.%Ti. This is due to the decrease of a-C when more Ti was added (Ti bonds with C to form TiC). As seen from Fig. 2.3, when 45 at.%Ti was added, the coating contained only 10 at.%C. Such a small amount was not enough to produce a peak in the Raman spectrum.

The increase in the I_D/I_G ratio indicates that the sp^3 fraction decreases when Ti is added. That is the same as the case of addition of Al. However, the addition of Al resulted in a greater decrease in sp^3 bonding (increase in I_D/I_G ratio). From Figs. 1.8 and 2.10, addition of 16 at.%Ti resulted in an I_D/I_G ratio of 1.9, whereas the addition of only 9 at.%Al resulted in an I_D/I_G ratio of 2. The Al added goes into the a-C matrix and disturbs the carbon structure in comparison to the addition of Ti where the Ti forms nanograin TiC with C.

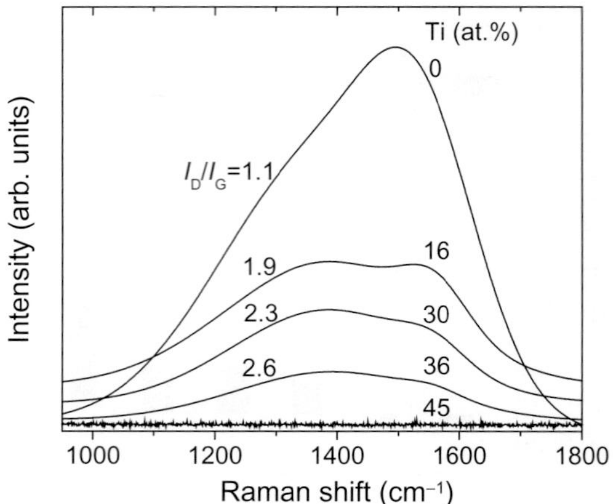

Fig. 2.10. Raman spectra of nc-TiC/a-C nanocomposite coatings for compositions of Ti of 0, 16, 30, 36, and 45 at.% [15].

2.4. *Mechanical Properties*

2.4.1. *Hardness and Residual Stress*

The trend of hardness and Young's modulus when "doping" Ti is different from the case of "doping" Al. Al exists in the coating as elemental Al and this leads to a decrease of hardness as more Al is added due to the decrease of sp^3 bonding in a-C, plus Al itself is a soft metal. In the case of Ti, however, as Ti is added above a certain level, nanograins of strong phase TiC form. In this case, the hardness of the coatings is not only dependent on the sp^3 fraction of a-C but also on the nanocrystalline phase of TiC (the size of grains, their volume fraction and distribution in the a-C matrix). Figure 2.11 plots the coating hardness and Young's modulus as a function of Ti concentration.

Pure a-C coating has hardness and Young's modulus of 32.5 and 342.6 GPa, respectively. On increasing Ti content, the coating hardness decreases and then increases owing to different mechanisms: as discussed before, at low Ti content (less than 16 at.%), the addition of Ti in the coating only serves as "doping" (virtually no TiC formation or very few TiC crystallites are found scattered in the a-C matrix, Fig. 2.7 and Table 2.2). Addition of Ti decreases the amount of sp^3 bonding in the a-C matrix (Fig. 2.10), which results in a decrease in hardness. Starting from 16 at.%Ti, a considerable amount of TiC nanograins form and the amount increases with increasing Ti (Table 2.2) resulting in a recovery of hardness to 32 GPa

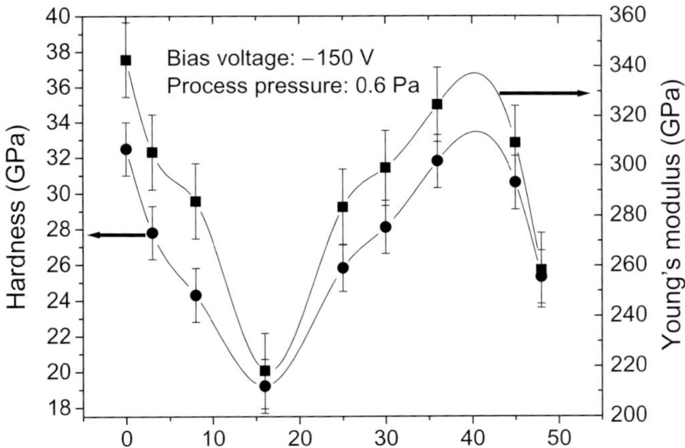

Fig. 2.11. Hardness and Young's modulus of nc-TiC/a-C as a function of Ti concentration.

Table 2.2. Summary of the structure of nc-TiC/a-C deposited under −150 V bias with different compositions of Ti (from XPS, XRD and TEM results).

Ti content (at.%)	Composite description	Tic grain size (nm)	Tic phase volume (%)	a-C interface thickness (nm)
0	Amorphous	0	0	N/A
3	Amorphous	0	0	N/A
8	Random inclusion of TiC grains	2–5	5	> 10
16	encapsulated into a-C matrix	2–5	18	< 10
25	Considerable amount of TiC	3–7	28	< 10
30	grains encapsulated in the a-C	3–8	39	< 8
36	matrix	3–11	69	< 5
45	Almost TiC grains with a minor amount of a-C phase	10–25	93	< 2

at 36 at.%Ti after offsetting the effect of the high sp^2 fraction in the matrix. With further increase of Ti, the coating hardness decreases again as a result of grain coarsening. A fitting curve connects the experimental data in Fig. 2.11 indicating that a maximum hardness of about 32–33 GPa, which is comparable to that of pure a-C, can be obtained at Ti content of 38–42 at.%. In this range of Ti content, the grain sizes of the TiC are about 3–25 nm, the a-C interface thickness is less than 5 nm and the volume fraction of TiC nanograins is about 70–90% (Table 2.2). It is clear that, superhardness (higher than 40 GPa) is not achievable in this case because the two simultaneous conditions required for superhardness are not satisfied. Firstly, the grain size of TiC is not small enough (<10 nm) to totally suppress the operation of dislocations. Secondly, the a-C interface is not thin enough (<1 nm) to support the coherence strain-induced enhancement of hardness [18].

The relationship between residual stress and Ti concentration is shown in Fig. 2.12. When Ti was co-sputtered, the residual stress in the coating decreased from 4.1 GPa (a-C) to 0.9 GPa (16 at.%Ti) then increased gradually to 2.1 GPa at 45 at.%Ti. At higher Ti content, it slightly decreased. The decrease of residual stress as Ti is added is understandable from the increase in sp^2 fraction, which helps to relax the compressive stress accumulated in the coating during the deposition process. Also, it should be noted that the energy of the sputtered species coming to the substrate during co-sputtering of two targets (graphite and Ti) is considerably lower owing to higher probability of collisions between species in the plasma compared to the case in which only the graphite target is sputtered. When a considerable amount of TiC is formed (from 16 at.%Ti), associated with less a-C inclusion, the hard phase of TiC hinders the relaxation of stress. The more TiC nanograins form, the less stress is relaxed, leading to an increase in residual stress as more Ti is added. However, the highest residual stress in nc-TiC/a-C is about 2 GPa, which is only half of that of the pure

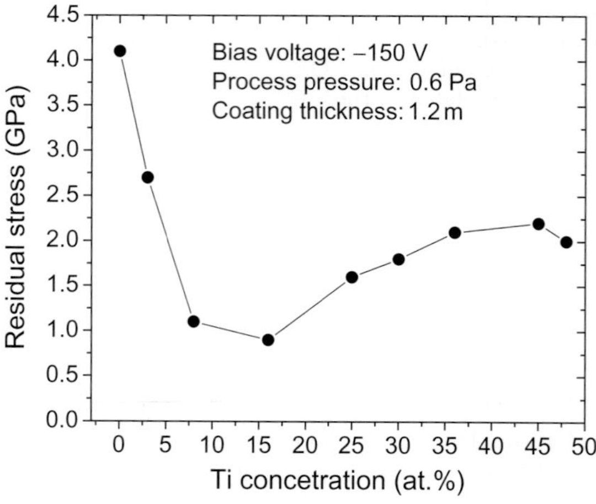

Fig. 2.12.   Residual stress of nc-TiC/a-C coatings as a function of Ti concentration.

a-C (about 4 GPa). However, a high hardness (about 32 GPa) is obtained. From Figs. 1.10 and 2.12, it is clear that Al is better than Ti in reduction of residual stress.

2.4.2. *Tribology*

Figure 2.13 shows the coefficient of friction of the coatings containing various amounts of Ti sliding against a 100Cr6 steel ball in ambient air (22°C, 75% humidity). From the figure, the coefficient of friction increases from 0.17 to 0.24 as the Ti content increases from 3 to 36 at.%. The a-C matrix influences significantly the friction of nc-TiC/a-C coatings through formation of a graphite-rich layer which acts as solid lubricant in humid air. When more Ti is added, more TiC crystallites form, thus there is less a-C matrix in direct contact with the wearing ball. This results in less solid lubricant between the two sliding surfaces and a higher coefficient of friction is thus seen. Another reason causing the increase in friction is the increase in surface roughness as more Ti is added as mentioned above. The development of the coefficient of friction exhibits an abrupt change at about 45 at.%Ti. At that Ti content, the concentration of a-C is approximately 10 at.% (Fig. 2.3). Such a low amount of a-C can no longer transform into graphitic lubricant, leading to a drastic increase in coefficient of friction. The coatings, in this case, exhibit the same friction behavior as polycrystalline TiC.

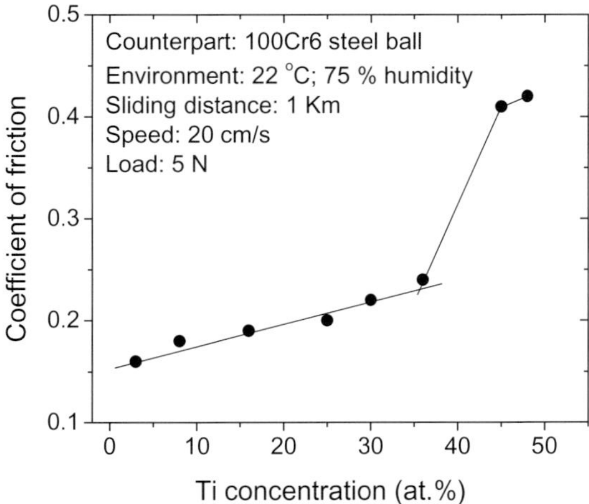

Fig. 2.13. Coefficient of friction of nc-TiC/a-C coatings with different Ti content. The coefficient of friction increases as more Ti is added to the a-C.

2.5. *Summary*

Co-sputtering of graphite and Ti forms a nanocomposite with TiC nanograins embedded in a matrix of amorphous carbon (nc-TiC/a-C). The size of the nanograins is strongly dependent on the Ti content (thus the power density of Ti target) and varies from a few to about 30 nm. Higher Ti content (i.e. higher power density on the Ti target) results in a larger grain size and rougher surface morphology. The hardness of the nanocomposite coatings is not only dependent on the sp^3 fraction of the a-C matrix but also (mainly) on the TiC nanograins (the size, the volume fraction and the a-C interface thickness). At low Ti content (less than 8 at.%) the formation of nanograins is very limited with very few nanograins scattered in the a-C matrix. At high Ti content (more than 45 at.%) the coatings mostly consist of nanocrystalline TiC and they exhibit polycrystalline TiC-like properties (rough surface and high friction). A high hardness of 32 GPa can be obtained with 36 at.%Ti (the grain size is 3–11 nm and the a-C interface thickness is less than 5 nm) while the residual stress of this coating is only 2 GPa (half of that of pure a-C).

3. Al-Toughened nc-TiC/a-C

As a result of the studies of a-C, a-C(Al) and nc-TiC/a-C, it is evident that a combination of hard crystalline TiC and a tough, relatively hard amorphous matrix of a-C(Al) would yield a coating of high toughness, low residual stress and adequate hardness, and therefore good wear resistance, thermal stability and oxidation resistance. The coating of interest is nc-TiC/a-C(Al), or nanocrystalline TiC embedded in amorphous carbon matrix doped with Al.

The power density on graphite target has been set at $10.5\,\mathrm{W/cm^2}$ for all samples. From analysis before, a power density of $2.7\,\mathrm{W/cm^2}$ on the Ti target yields an amount of TiC crystalline phase high enough to maintain a high hardness while the a-C content is adequate for self-lubrication to take place. The power density on the Al target is varied from 0.6 to $1.8\,\mathrm{W/cm^2}$ for different Al compositions. The substrate is biased at $-150\,\mathrm{V}$ and the process pressure kept constant at 0.6 Pa during deposition. The composition of the coatings is calculated from the areas under the C 1s, Ti 2p and Al 2p peaks in XPS spectra. The results are tabulated in Table 3.1 (the coatings are etched using an Ar ion beam for 15 min before analysis; less than 4 at.% of oxygen contaminated is ignored in the calculations).

Table 3.1. Composition of the nc-TiC/a-C(Al) nanocrystalline coating.

Power density of Al target (W/cm^2)	Composition (at.%)			Notation
	C	Ti	Al	
0.6	62	35	3	$C_{62}Ti_{35}Al_3$
1.1	60	34	6	$C_{60}Ti_{34}Al_6$
1.4	57	32	11	$C_{57}Ti_{32}Al_{11}$
1.8	56	31	13	$C_{56}Ti_{31}Al_{13}$

3.1. *Composition*

Figure. 3.1 shows the C 1s, Ti 2p and Al 2p XPS profiles of $C_{56}Ti_{31}Al_{13}$ coating. It should be noted that except for the intensity of the peaks, there was no difference between the XPS spectra of all four coatings. The results revealed from XPS spectra of the four coatings meet the expectations of the coating design:

- The Al existed in the coating as elemental Al. A symmetrical single peak at 72.6 eV was seen in the Al 2p spectra.
- Except for the bonds of the oxides, Ti–C was the only chemical bond found in the coating. The peaks at 281.8 eV in the C 1s profile and at 454.9 eV (461 eV for Ti $2p_{1/2}$) in the Ti 2p profile are attributed to TiC. This indicates that under the deposition conditions, other carbides such as aluminum carbide or aluminum–titanium carbide did not form.
- All Ti bonded with C to form TiC. No shift of the Ti 2p peaks to lower binding energies was seen from the spectra. Therefore there was no elemental Ti in the coatings.

3.2. *Microstructure*

The XRD spectra of the nc-TiC/a-C(Al) nanocomposite coatings are shown in Fig. 3.2. TiC was the only crystalline phase in these coatings and the TiC nanograins were randomly oriented in [111], [200], and [220] directions. The more Al added, the lower the intensity of the peaks, since the amount of TiC was lower as more Al was added into the coating. However, as more Al was added, the peak intensities and the widths in all directions became more consistent, indicating uniform grain size and random orientation. With Al, no shift of the peaks was seen, which indicated that the coating had low residual stress. The grain size of all

 S. Zhang et al.

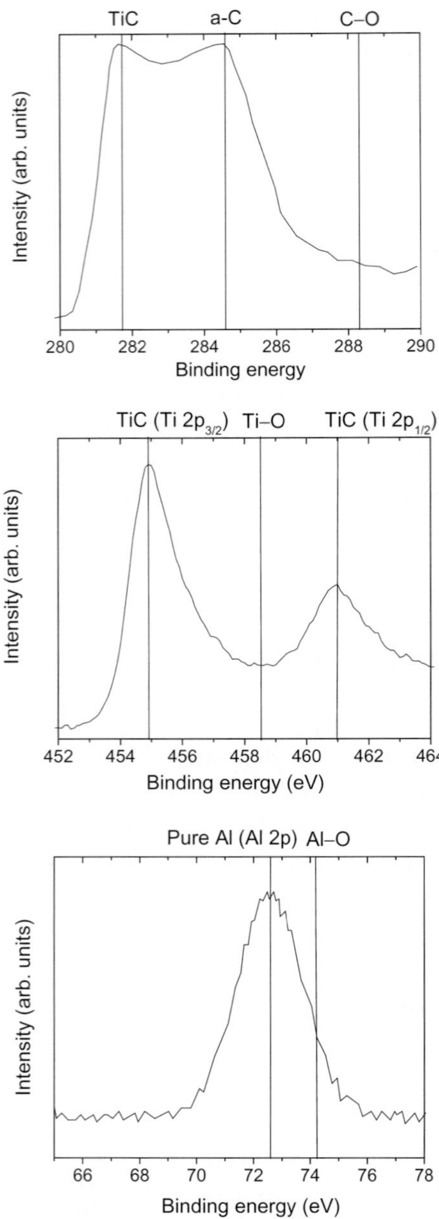

Fig. 3.1.   C 1s, Ti 2p and Al 2p XPS spectra of $C_{56}Ti_{31}Al_{13}$ coating. (Power density of graphite: $10.5\,W/cm^2$, Ti: $2.7\,W/cm^2$, and Al: $1.8\,W/cm^2$. 15 min etching was applied before analysis.).

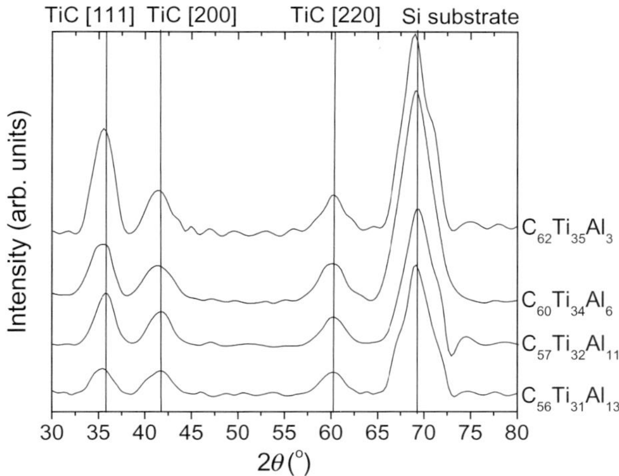

Fig. 3.2.   XRD spectra of the nc-TiC/a-C(Al) nanocomposite coatings.

nc-TiC/a-C(Al) coatings was 2–5 nm as calculated from the XRD spectra using Scherrer formula. The calculated values should be very near the real dimension since nc-TiC/a-C(Al) had low residual stress (not much shift of TiC peaks was seen compared to the case of nc-TiC/a-C). Therefore the effect of residual stress on Braggs peaks' broadening would be negligible.

Figure 3.3 shows the bright field (BF) TEM image with (a) diffraction patterns and (b) HRTEM micrographs of the $C_{56}Ti_{31}Al_{13}$ coating. The diffraction pattern confirmed the existence of randomly oriented TiC nanograins. Also the BF TEM image shows a high volume fraction of TiC nanograins in the matrix of Al-doped a-C. The grain size was determined to be 2–6 nm, which was very consistent with the results calculated from XRD (2–5 nm). From Tables 2.1 and 2.2, co-sputtering of graphite (at $10.5\,W/cm^2$) and Ti (at $2.7\ W/cm^2$) resulted in a nanocomposite nc-TiC/a-C coating (36 at.%Ti) with grain size from 3 to 11 nm. When Al was incorporated by co-sputtering of graphite (at $10.5\ W/cm^2$), Ti (at $2.7\ W/cm^2$) and Al (at 0.6–$1.8\,W/cm^2$), the grain size of the TiC was less than 6 nm. It is clear that incorporation of Al reduces the grain size of TiC. The interface thickness of the amorphous phase was determined to be less than 5 nm, comparable to that of the nc-TiC/a-C coating.

                     S. Zhang et al.

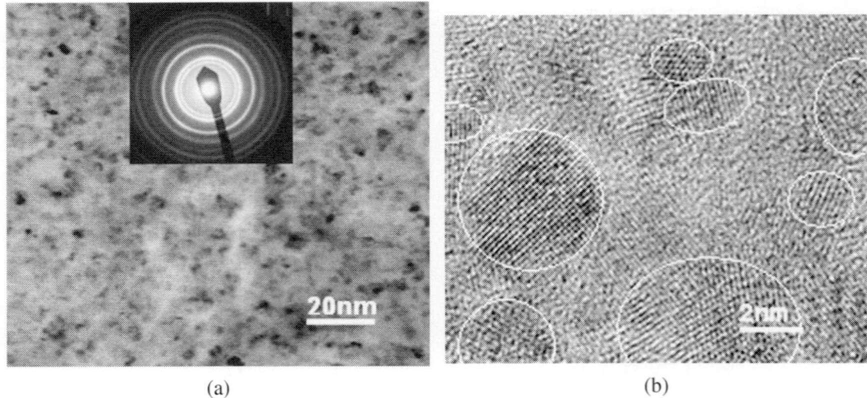

(a)   (b)

Fig. 3.3. (a) BF TEM (with diffraction pattern), and (b) HRTEM of a nc-TiC/a-C(Al) ($C_{56}Ti_{31}Al_{13}$) nanocomposite coating. In BF TEM, the dark spots are TiC nanograins.

Figure 3.4 shows the surface morphology of the $C_{62}Ti_{35}Al_3$ and $C_{56}Ti_{31}Al_{13}$ coatings. The surface roughness of $C_{62}Ti_{35}Al_3$ and $C_{56}Ti_{31}Al_{13}$ was 6.9 and 5.4 nm R_a, respectively. The nc-TiC/a-C(Al) coating has a smoother morphology than nc-TiC/a-C and more Al added resulted in an even smoother surface. The surface morphoplogy is closely related with thin films growth. The growth kinetic is controlled by the mobility of the imping-ing atoms on the surface before they condense and become entrapped in the film. This mobility can be enhanced by inputting energy to the sytem, such as increasing deposition temperature or supplying impact energy through ion bombardment. In the present study, the C and Ti target power densities are kept constant for all the samples while the Al target power density increased from 0.6 to 1.8 W/cm^2. The Al atoms obtain more energy at higher Al target power density therefore the mobility of Al atom is higher, which results in a smoother surface.

3.3. Mechanical Properties

Mechanical properties of a-C, nc-TiC/a-C and nc-TiC/a-C(Al) coatings investigated in this study are tabulated in Table 3.2.

3.3.1. Hardness, Toughness and Adhesion

The hardness and Young's modulus of nc-TiC/a-C(Al) were less than that of a-C and nc-TiC/a-C. However the big benefit here was the low resid-ual stress of the nc-TiC/a-C(Al) coatings. Addition of Al or Ti into a-C

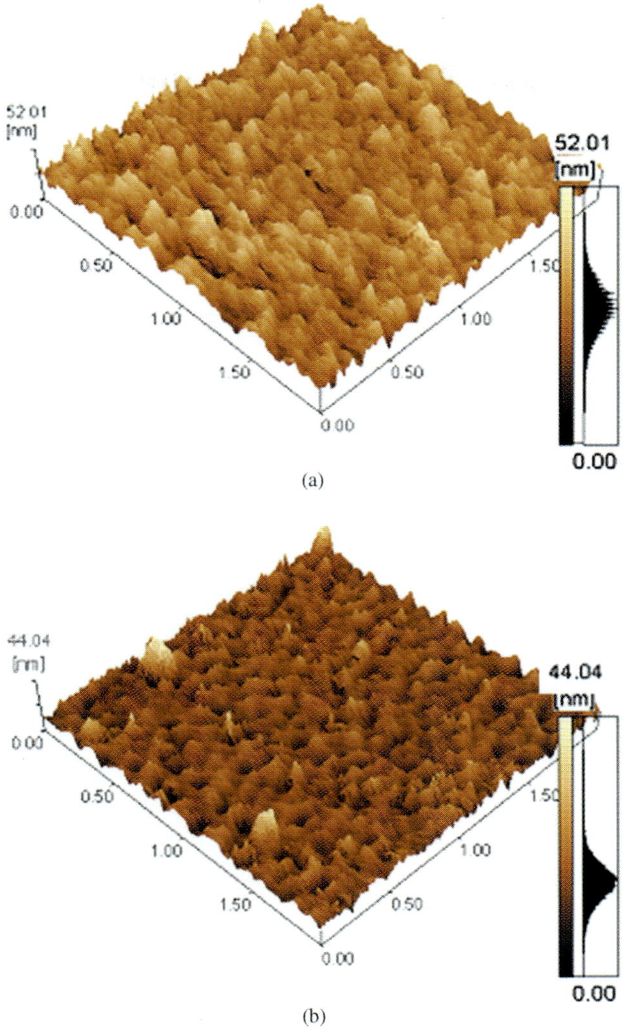

Fig. 3.4. AFM images of (a) $C_{62}Ti_{35}Al_3$, and (b) $C_{56}Ti_{31}Al_{13}$ nanocomposite coatings.

considerably reduced the residual stress of the coating. When Al and Ti were added into a-C at the same time a low residual stress was obtained. Note that $C_{56}Ti_{31}Al_{13}$ had very low residual stress of 0.5 GPa while its hardness was at a relatively high level of about 20 GPa. More importantly, the low residual stress allowed thick adherent coatings to be deposited, leading to a long working life.

Table 3.2. Mechanical properties of a-C, nc-TiC/a-C and nc-TiC/a-C(Al) coatings.

Coatings	Deposition [Power density $(W/cm)^2$]				Stoichiometry	Hardness (GPa)	Young's modulus (GPa)	Stress (GPa)
	C	Ti	Al	Bias (V)				
a-C	10.5	0	0	−150	C	32.5	342.6	4.1
nc-TiC/a-C	10.5	2.7	0	−150	$C_{64}Ti_{36}$	31.8	324.5	2.1
nc-TiC/a-C(Al)	10.5	2.7	0.6	−150	$C_{62}Ti_{35}Al_3$	27.3	297.5	1.3
nc-TiC/a-C(Al)	10.5	2.7	1.8	−150	$C_{56}Ti_{31}Al_{13}$	19.5	220.4	0.5

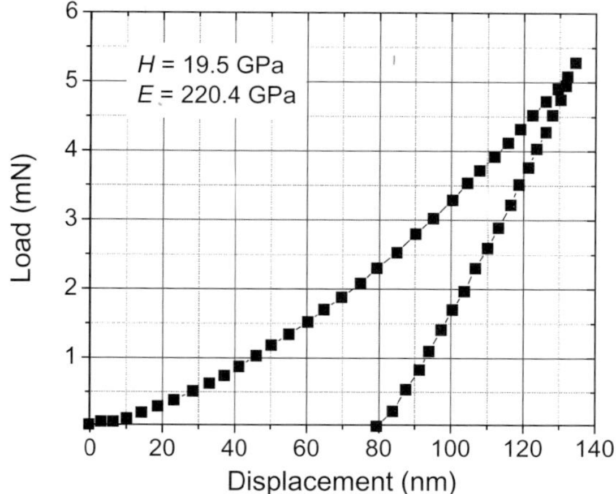

Fig. 3.5. Load and unload curves from the nanoindentation of a nc-TiC/a-C(Al) coating. The plasticity of the coating was 58%.

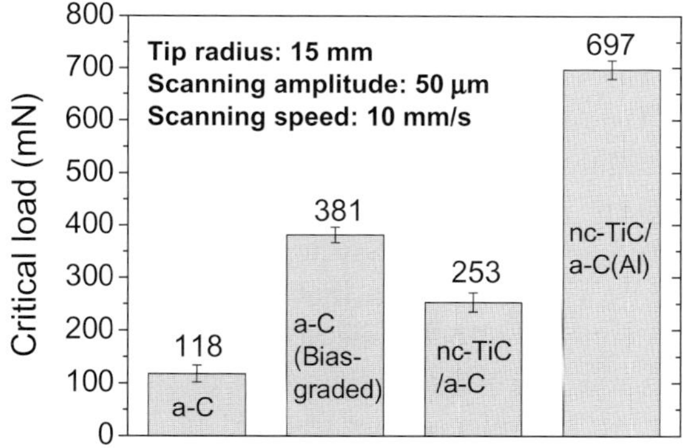

Fig. 3.6. Lower critical load obtained from scratch tests of a-C (-140 V bias), a-C (bias-graded), nc-TiC/a-C, and nc-TiC/a-C(Al) coatings.

Figure 3.5 plots the load and unload curve of the $C_{56}Ti_{31}Al_{13}$ coating. The plasticity (ratio of the unloaded displacement over total displacement) during indentation deformation was estimated to be 58%. Such a high plasticity indicates a very high toughness, as can be visually verified from scratch tests.

Figure 3.6 plots the adhesion strength (in terms of the lower critical load from scratch tests) of a-C (deposited at a constant bias of -140 V), bias-graded a-C (bias voltage is increased from 0 to -140 V at a rate of

$-2\,$V/min, see details in [19]), nc-TiC/a-C ($C_{64}Ti_{36}$) and nc-TiC/a-C(Al) ($C_{56}Ti_{31}Al_{13}$) coatings deposited on 440C steel substrates. As can be seen from the plot, a-C deposited under a high bias voltage of $-140\,$V exhibited the lowest adhesion strength (118 mN). For bias-graded a-C, the adhesion strength was more than three times higher (381 mN). Since the surface roughness was low for both a-C deposited under a constant bias of $-140\,$V and bias-graded a-C (1.1 and 1.4 nm R_a on the Si wafer), the considerable increase in adhesion was attributed to the combination of low residual stress and high toughness of the bias-graded a-C, especially at the interface between the coating and substrate. The adhesion strength of the nc-TiC/ a-C coating was 253 mN, more than twice that of a-C deposited under $-140\,$V bias owing to a much lower residual stress and the toughness enhancement of the nanocomposite configuration (the size of the cracks was limited and their propagation was hindered). However, this critical load was still lower than that of the bias-graded a-C coating (381 mN) because of two reasons. Firstly, the residual stress of nc-TiC/a-C was higher than that of bias-graded a-C (2.1 GPa versus 1.5 GPa). Furthermore, it should be noted that the residual stress was obtained from the change in curvature of a whole Si(100) wafer, i.e. an average of the whole coating regardless of possible variation due to structural grading in case of bias-graded deposition. In bias-graded a-C coatings, the sp^3 fraction increases from the substrate-coating interface towards the outer surface of the coating. The local residual stress at the interface should be a lot lower than that close to the surface (where sp^3 is the highest). Secondly, the roughness of nc-TiC/a-C was 8.2 nm R_a (on a Si wafer), which was much higher than that of the bias-graded a-C coatings (1.4 nm R_a on Si wafer). This resulted in higher friction and thus higher shear stress at the contact area. These reasons gave rise to a higher critical load (better adhesion) for the bias-graded a-C compared to that of nc-TiC/a-C. In the case of the nc-TiC/a-C(Al), a very high critical load of 697 mN was obtained. The extremely low residual stress of 0.5 GPa played an important role in this coating. Another important contribution is attributed to the extremely high toughness as a consequence of adding Al to form an a-C(Al) matrix in which nanosized TiC grains were embedded. In this case, the propagation of the microcracks generated in the scratch process was hindered at the boundaries between the matrix and the grains. Meanwhile, the crack propagation energy was relaxed in the tough matrix. The smoothness of the surface (5.4 nm R_a on Si) also contributed towards the high critical load.

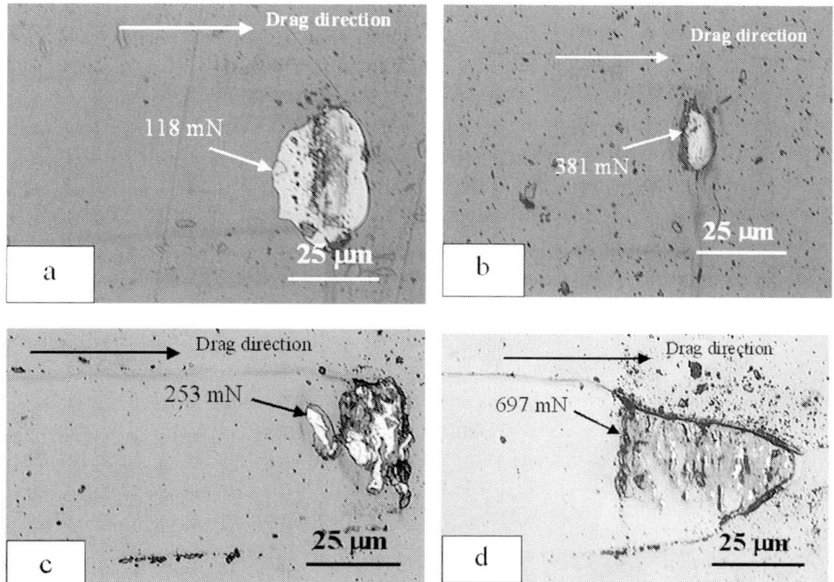

Fig. 3.7.   Optical micrographs of scratch tracks on the coating of (a) a-C deposited under constant bias of -140 V, (b) bias-graded a-C, (c) nc-TiC/a-C, and (d) nc-TiC/a-C(Al).

As can be clearly seen from the optical micrographs of the scratch tracks in Fig. 3.7, a-C coating deposited under constant -140 V bias delaminated in a brittle manner as the load reached 118 mN, and the lower critical load and the higher critical load were not distinguishable. For bias-graded a-C and nanocomposite coatings, however, the scratch damages inflicted on the coating was not continuous (sporadic) as the load continued to increase [Figs. 3.7(b), (c) and (d)]. The fractured surface of the nc-TiC/a-C(Al) coating appeared very "plastic" [Fig. 3.7(d)]: the cracks formed but did not produce spallation, instead, the tip was seen to plough into the coating. As the tip ploughed deeper, the scanning amplitude decreased because the force exerted on the tip to vibrate in the transverse direction was not enough to overcome the resistance created by the material pile-up.

3.3.2. *Tribology*

Tribological properties of a $C_{56}Ti_{31}Al_{13}$ nanocomposite coating were investigated in comparison with that of the pure a-C (deposited under a constant -140 V bias) and nc-TiC/a-C ($C_{64}Ti_{36}$).

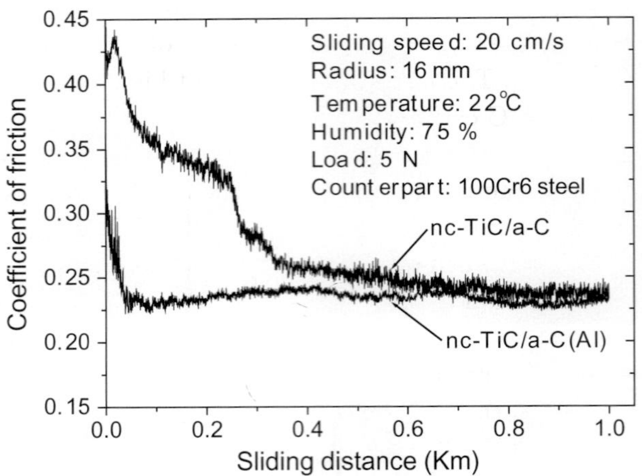

Fig. 3.8. Coefficient of friction versus sliding distance of a nc-TiC/a-C(Al) coating ($C_{56}Ti_{31}Al_{13}$) and a nc-TiC/a-C coating ($C_{64}Ti_{36}$).

3.3.2.1. Dry Tribology

Figure 3.8 plots the coefficient of friction as a function of sliding distance of a nc-TiC/a-C(Al) coating ($C_{56}Ti_{31}Al_{13}$) and a nc-TiC/a-C coating ($C_{64}Ti_{36}$) when sliding against a 100Cr6 steel counterpart in ambient air. It should be noted that the amount of a-C in these nanocomposites was 23 and 30 at.% in the nc-TiC/a-C(Al) coating and in the nc-TiC/a-C coating, respectively. For these two nanocomposite coatings, the a-C in the matrix was mostly graphite-like with a high I_D/I_G ratio (3.1 for nc-TiC/a-C(Al) and 2.6 for nc-TiC/a-C, respectively). Therefore, the graphite-rich phase was available for lubrication. This is different from pure a-C, where the graphite-rich lubricant layer can only form after graphitization takes place.

As seen from the figure, the running-in stage of nc-TiC/a-C(Al) was shorter (about 0.1 Km) with much lower coefficient of friction compared to that of the nc-TiC/a-C coating. The hard nc-TiC/a-C coating (hardness 32 GPa) has a rougher surface (8.2 nm R_a on a Si wafer, Fig. 2.4) than that of the nc-TiC/a-C(Al) coating (hardness 19.5 GPa and 5.4 nm R_a on a Si wafer, Fig. 3.4). Such a rough surface with hard and large asperities (the size of TiC grains in nc-TiC/a-C is 5–11 nm compared to less than 6 nm in nc-TiC/a-C(Al)) in the nc-TiC/a-C resulted in higher friction and more vibrations. It required a considerably longer duration to reach the steady state. At steady state, the coefficient of friction of the nc-TiC/a-C(Al) coating was about 0.23, a bit lower than that of the nc-TiC/a-C coating (0.24) even though the amount of a-C was lower (23 at.% compared to 30 at.%). The main reason was that high hardness and roughness of the nc-TiC/a-C caused more wear of the steel counterpart as seen from the optical micrographs of the wear scars after the test (Fig. 3.9). The higher wear rate of the counterpart contributed more iron oxide into the tribolayer formed between the two sliding surfaces. The wear of the steel counterpart was lower when sliding against nc-TiC/a-C(Al) and the wearing coating contributed not only a-C but also Al (in the form of aluminum hydroxide since the tribotest was carried out in a high humidity condition) into the contact. As mentioned before, aluminum hydroxide has lower shear strength compared to that of iron oxide, giving rise to a slightly lower friction observed in nc-TiC/a-C(Al). It is well known that the coefficient of friction of pure a-C when sliding against a steel counterpart was in the range of 0.11–0.15 depending on the deposition conditions (bias voltage). Compared with these values, the coefficient of friction of the nanocomposite coating was higher. This is due to the fact that the nanocomposite coating has rougher surface and less graphite in the tribolayer (pure a-C contains 100%C). Even so, the coefficients of friction of the a-C nanocomposite coatings are much lower than that of popular ceramic coatings currently used in the industry, such as TiN, TiC, CrN, etc. The coefficients of friction of these generally are from 0.4 to 0.9 [20–30].

The wear track profiles on the coatings and wear scars on the balls after the tribotests for nc-TiC/a-C(Al) and nc-TiC/a-C are shown in Fig. 3.9. For comparison, those of the a-C coating deposited under $-140\,\text{V}$ bias (32.5 GPa hardness) are also added. From the wear track profiles, it can be seen that the wear of nc-TiC/a-C and a-C coatings is very low and the wear track is not detectable. These coatings have much higher hardness (about 32 GPa) compared to that of the steel counterpart (8 GPa). Wear of the counterparts, therefore, was very high, as is seen in the big wear scars. However, the wear scar on the ball sliding against a-C is smaller than that on the ball sliding against the nc-TiC/a-C (the wear scar diameter is 498 and 575 μm, respectively). Understandably, pure a-C has a much smoother morphology compared to that of nc-TiC/a-C and it contains 100% a-C, which produces a larger amount of graphite-like lubricant in the contact area, leading to a lower wear of the counterpart and lower coefficient of friction. Softer than nc-TiC/a-C and a-C, the nc-TiC/a-C(Al) coating exhibits a lower wear resistance. The wear track on nc-TiC/a-C(Al) is observable. Meanwhile, the wear of the counterpart, with a wear scar diameter of 387 μm, is much

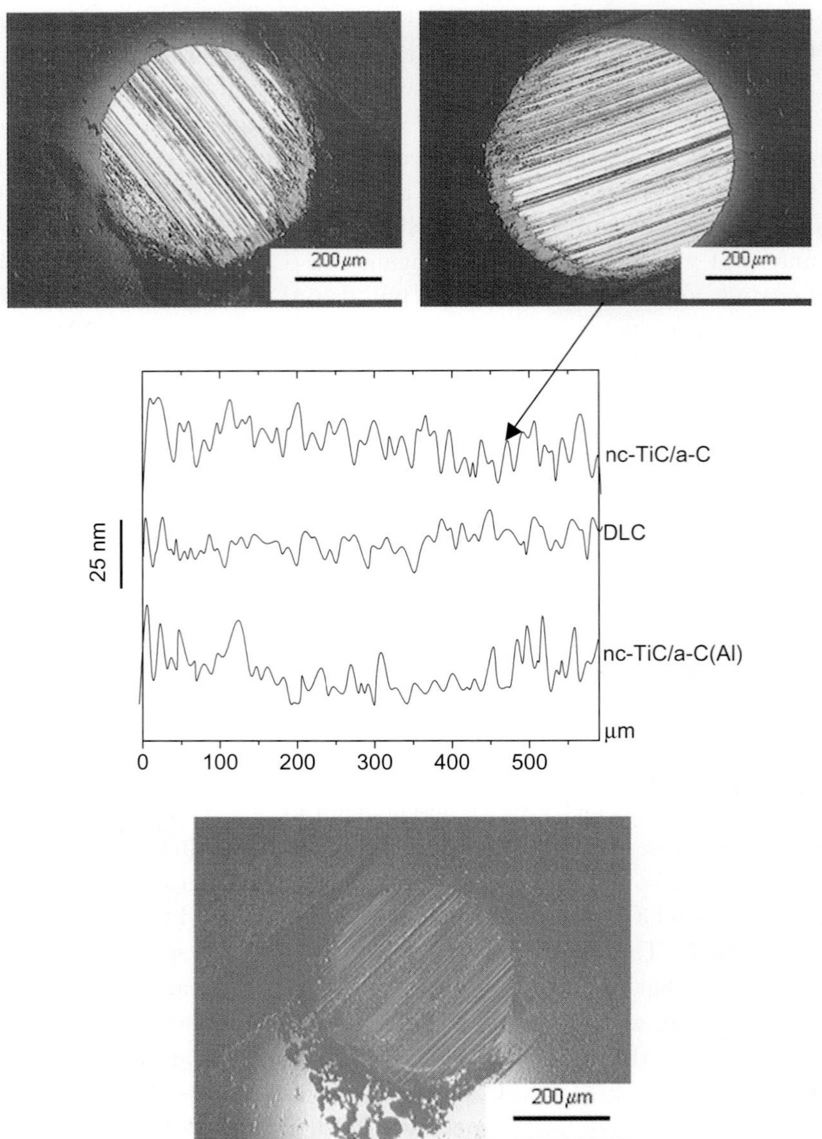

Fig. 3.9. Wear track profiles on the coatings and wear scars on steel balls after 1 Km ball-on-disc tribotests in ambient air (22°C, 75% humidity).

lower than that formed by sliding against nc-TiC/a-C and a-C (where the wear scar diameters are 498 and 575 μm, respectively).

3.3.2.2. Oil-Lubricated Tribology

Under oil-lubricated conditions, the lubrication mechanism is totally different from that under dry conditions. With oil lubrication, oil prevents the formation of the tribologically beneficial layer, changes the friction mechanism at the contact and governs the friction behavior of the oil-lubricated contact. Therefore, the thickness and stability of the oil film between the coating and counterpart surface plays an important role in friction and wear of coatings and counterparts. In oil-lubricated tribotests, because the oil is not pressurized into the contact, the oil film between the two contacting surfaces will not be thick enough to separate the coating surface and the counterpart. The lubrication regime, therefore, is boundary but not hydrodynamic. As such, the asperities of the two surfaces are in contact. Under this lubrication condition, the interaction between the coating surface, counterpart surface with lubricating oil and the additive is vital in order to maintain an oil film (even very thin) in the interface. Generally, the mechanism of boundary lubrication includes [24]:

- Formation of layers of molecules by Van der Waals forces: To form such layers a molecule must have a polar end, which attaches to the metal, and a non-polar end, which associates with the oil solution. Therefore, to assure good adhesion of the oil to the tribosurfaces, the first condition is that an appropriate metal(s) must be added into lubrication oil (Ca, Mg, and Zn are added to Shell Helix oil).
- Formation of high-viscosity layers by reaction of the oil component in the presence of rubbed surfaces: these films may result in a hydrodynamic effect and may be linked to the surface by Van der Waals forces.
- Formation of inorganic layers due to reactions between active oil components and the tribosurface materials: These layers (sulphides, phosphides) have low shear strength and help to prevent scuffing (for this, in Shell Helix oil S and P are added).

Good boundary lubrication occurs when tribosurfaces are metallic and for that purpose, active metals produce better results than inert metals [25]. Figure 3.10 plots the coefficient of frictions of nc-TiC/a-C(Al), nc-TiC/a-C and a-C coatings as a function of sliding distance. Under the oil-lubricated condition, very low vibration allows the accurate estimation

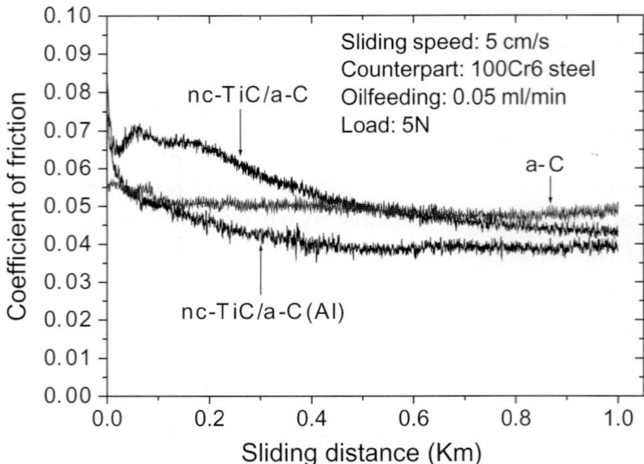

Fig. 3.10.   Coefficients of friction of the nc-TiC/a-C(Al) coating, the pure a-C coating, and the nc-TiC/a-C coating, as functions of sliding distance under oil lubrication.

of the coefficients of friction up to two decimal places. From the figure, after the running-in stage, all three coatings exhibited very low coefficients of friction (0.04–0.05) compared to that under dry conditions (0.15, 0.23 and 0.24 for a-C, nc-TiC/a-C(Al) and nc-TiC/a-C, respectively). The nc-TiC/a-C(Al) coating exhibits the lowest friction followed by the nc-TiC/a-C coating. The pure a-C exhibits the highest coefficient of friction. Experimental results are expected from the nature of these three coatings: pure a-C is a chemically inert material, thus the interaction between the oil and a-C coating is very limited. The a-C surface is therefore a "passive member" in the contact; the chemical interactions occur only at the mating surface (steel ball counterpart). In the case of nanocomposite coatings, there is not only an interaction between steel counterpart and oil but also between the oil and the coating surface because of the metals in the nanocomposite coatings. This results in a thicker and more stable oil film at the interface leading to the observed lower coefficient of friction. It can be easily seen that the nc-TiC/a-C(Al) coating has the best interaction with oil since Al is an active metal and it exists in the coating as elemental Al. Therefore, it is not surprising that nc-TiC/a-C(Al) exhibits the lowest friction under oil lubrication. The wear tracks on all the three coatings are not detectable after the tests, indicating excellent wear resistance.

The wear scars on the ball are shown in Fig. 3.11. The smallest wear scar was the one sliding against nc-TiC/a-C(Al) (with a diameter of

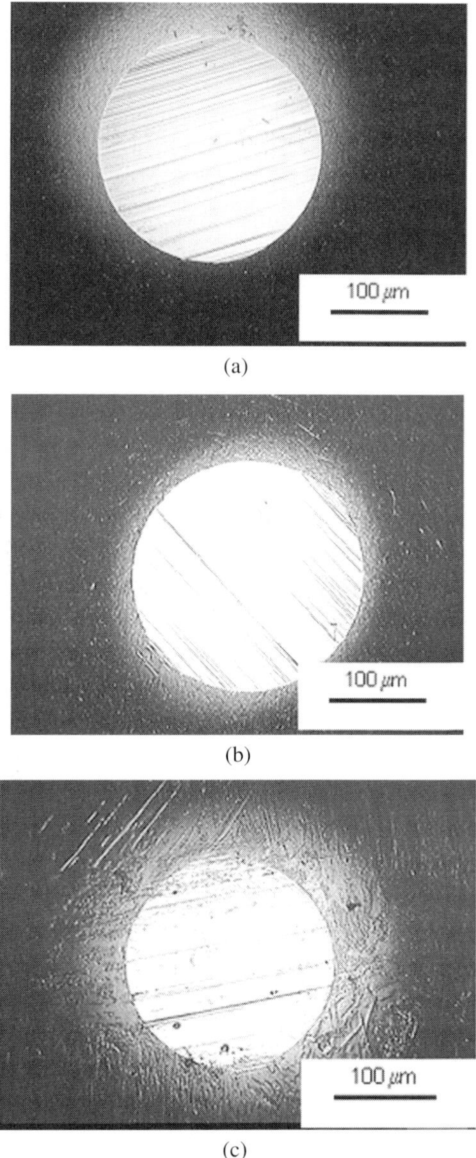

Fig. 3.11. Wear scars on steel counterparts after 1 Km sliding against (a) a-C, (b) nc-TiC/a-C, and (c) nc-TiC/a-C(Al) sliding under oil-lubricated condition.

220 μm). The effective boundary lubrication, as mentioned above, resulted in the low wear of the counterpart. The wear of the ball sliding against the nc-TiC/a-C coating was the highest (the diameter of the scar was 246 μm); a little bit smaller (240 μm scar) against pure a-C coating. The high wear on the ball sliding against nc-TiC/a-C was probably the consequence of the rough surface morphology of the nc-TiC/a-C coating. It is clear that the nc-TiC/a-C(Al) coating is the winner: low coefficient of friction (0.04) and excellent wear resistance (the wear was not detectable after 1 Km sliding against a 100Cr6 steel ball), plus less wear on the counterpart.

3.4. *Thermal Stability and Oxidation Resistance*

A nc-TiC/a-C(Al) coating ($C_{56}Ti_{31}Al_{13}$, hardness 19.5 GPa) was used for this study. Pure thermal stability can be evaluated through annealing in inert gas (thus only temperature is allowed to take effect, not oxidation). Figure 3.12 shows the Raman spectra of the coating after 60 min annealing in an Ar environment at different temperatures. The I_D/I_G ratio of a-C in the matrix increased with increasing temperature. This behavior is the same for pure a-C where a more graphite-like structure was formed at high temperature. The change of the a-C structure happened as temperature exceeded 300°C. At high temperatures (> 500°C) carbon in the matrix was almost graphite-like with large cluster size. Further increase in temperature did not cause much increase in I_D/I_G ratio (I_D/I_G ratio increased from 4.5

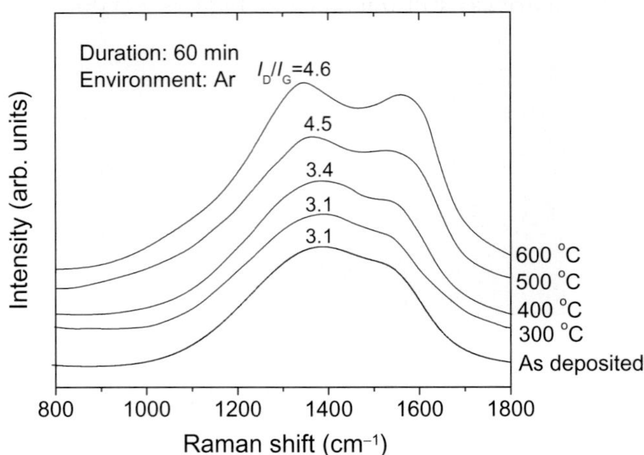

Fig. 3.12. Raman spectra of nc-TiC/a-C(Al) nanocomposite annealed at different temperatures for 60 min in an Ar environment [26].

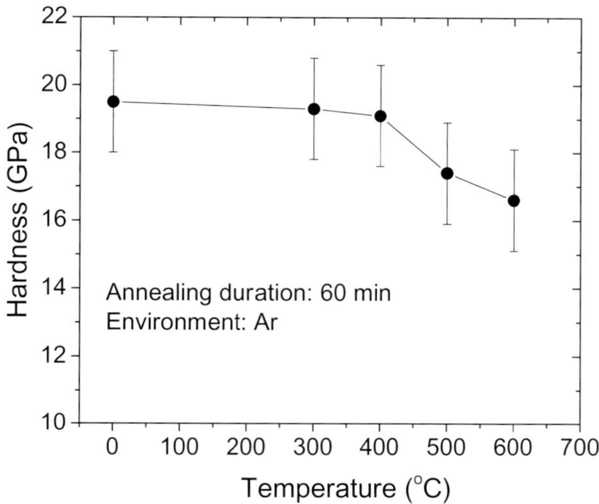

Fig. 3.13. Hardness of nc-TiC/a-C(Al) nanocomposite as a function of annealing temperature [26].

to 4.6 as the temperature increased from 500 to 600°C). Although the amount of graphite-like structure increased with temperature the coating hardness did not change much and remained as high as the original value of about 19 GPa up to 400°C (Fig. 3.13). Even after annealing at 600°C for an hour, the hardness still remained as high as 17 GPa, i.e. about 90% remained. This indicates a very good thermal stability. It should be noted that after annealing for an hour at 400°C, the hardness of the pure a-C coating dropped to only half of its original 32.5 GPa, and at 500°C, only about 8 GPa or about 25% remained. For pure a-C, the major contribution to the coating hardness is the sp^3 bonding structure, which is sensitive to temperature. The hardness of nanocomposite coatings, however, mainly comes from the hard nanocrystalline phase of TiC, which is not influenced as the temperature increases up to 600°C. As can be seen from Fig. 3.14, the XRD pattern of nc-TiC/a-C(Al) before and after annealing in Ar was not noticeably different. Therefore, the hardness of the nanocomposite coating was less sensitive to temperature than pure a-C.

The oxidation resistance can be evaluated from annealing in air. Figure 3.15 shows XPS spectra (Al 2p) of the nanocomposite coating after annealing at different temperatures in air. After unloading from the deposition chamber into atmosphere, all Al on the surface reacted with oxygen

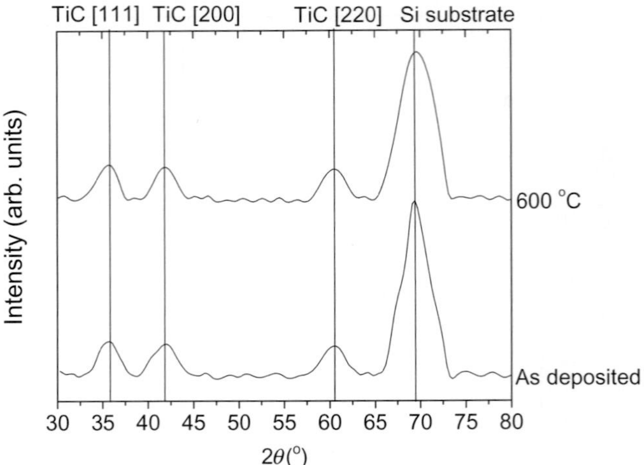

Fig. 3.14. XRD pattern of nc-TiC/a-C(Al) coating before and after 60 min annealing in an Ar environment at 600°C [26].

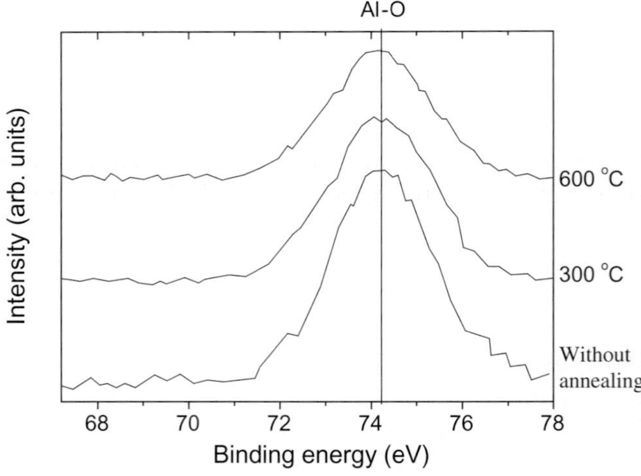

Fig. 3.15. XPS spectra (Al 2p) of nc-TiC/a-C(Al) nanocomposite coating at different annealing temperatures in air for 60 min [26].

to form aluminum oxide (the Al–O peak at 74.2 eV). As the annealing temperature was increased to 600°C, Al–O was the only chemical bond of Al detected. The Ti 2p spectra are shown in Fig. 3.16. The peaks at 454.9 and 461 eV were attributed to TiC ($2p_{3/2}$ and $2p_{1/2}$, respectively), at 458.6 and 464.3 eV to TiO_2, and at 456.2 and 462 eV to TiC_xO_y. As investigated

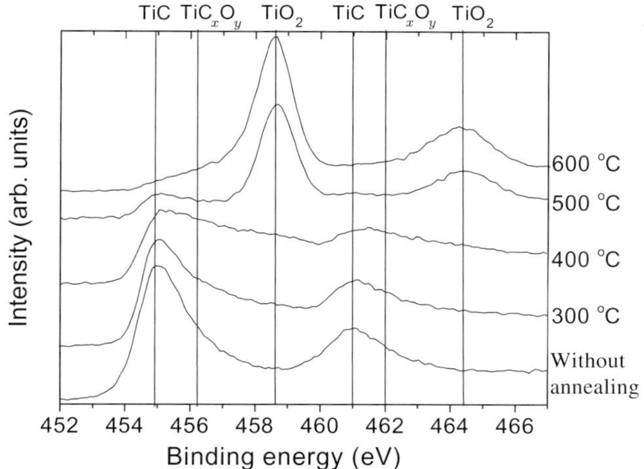

Fig. 3.16.   XPS spectra (Ti 2p) of nc-TiC/a-C(Al) nanocomposite coating at different annealing temperatures in air for 60 min [26].

before, Ti in the coating mostly bonded with C in the TiC nanocrystalline phase. From the spectra, when the temperature was lower than 300°C, the oxidation of TiC was very limited. The oxygen existed in TiC$_x$O$_y$. The formation of TiO$_2$ was seen at an annealing temperature of 400°C. From the results, it can be seen that TiC was considerably oxidized at about 400°C. The oxidation, which leads to the formation of TiO$_2$ is expressed by the following reaction:

$$TiC + 2O_2 \rightarrow TiO_2 + CO_2. \tag{3.1}$$

At 500°C, a large amount of TiO$_2$ was formed (51 at.%Ti bonded with oxygen) and when the annealing temperature was increased to 600°C, almost all the TiC was oxidized to form TiO$_2$.

As mentioned before, the reaction between carbon and oxygen at high temperature (400°C) resulted in the loss of coating thickness:

$$C + O_2 \rightarrow CO_2. \tag{3.2}$$

This also occurred on the surface of the nanocomposite coating where the carbon dioxide was formed from the reactions between TiC and oxygen [Eq. (3.1)] and between C (in the matrix) and oxygen [Eq. (3.2)]. Aside carbon dioxide, aluminum oxide (formed at room temperature) and titanium oxide (formed at high temperature) are also products of the oxidation. These oxides remain on the coating and act as a barrier layer (especially

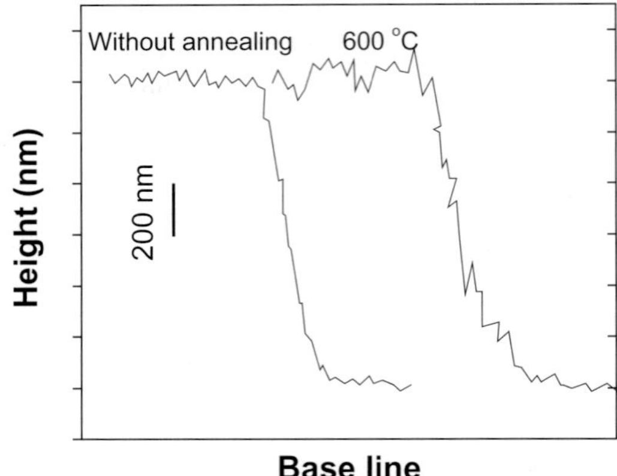

Fig. 3.17.   Coating thickness of nc-TiC/a-(Al) nanocomposite coating before and after annealing in air at 600°C obtained from profilometer [26].

aluminum oxide [27]) to prevent the diffusion of oxygen into the coating. Therefore, the loss of coating thickness was not seen when annealing the nc-TiC/a-C(Al) nanocomposite in ambient air, even at 600°C for 60 min (Fig. 3.17). Also, it can be easily recognized that the formation of the oxide layer resulted in a very rough morphology.

The thickness of the oxide layer was determined from an XPS depth profile based on the oxygen percentage detected. It is determined as the etching depth at which the oxygen content was less than 4 at.% (this oxygen contaminated the coating during the deposition process) and does not change when more etching is continued. Figure 3.18 shows the XPS depth profiles of the coating without annealing, after 60 min annealing in air at 300°C, and after 60 min annealing in air at 600°C.

As seen from the figures, without annealing, the thickness of oxide layer was about 4.2 nm. At 300°C, the increase in thickness of the oxide layer was not significant (the thickness of oxide layer was about 6.3 nm). It should be noted that without annealing the thickness of oxide layer on the nanocomposite (Ti, Cr)CN/a-C was already about 20 nm [28]. At high temperature of 600°C, the thickness of the oxide layer drastically increased to about 55 nm. Under the same annealing condition, the oxide layer on the TiN coating was reported to be 200 nm and that of TiAlN was 35–180 nm depending on the Al content in the coating [27, 29]. These data illustrate the importance of Al in the oxidation resistance of the nanocomposite coating. The addition of Al, even at low amounts considerably enhances the oxidation resistance. As mentioned before, aluminum oxide is a good barrier to prevent oxygen diffusing into and oxidizing the coating. From the experimental results, nc-TiC/a-C(Al) exhibited much better oxidation resistance compared to that of pure a-C, a-C-based nanocomposites which do not contain Al, such as

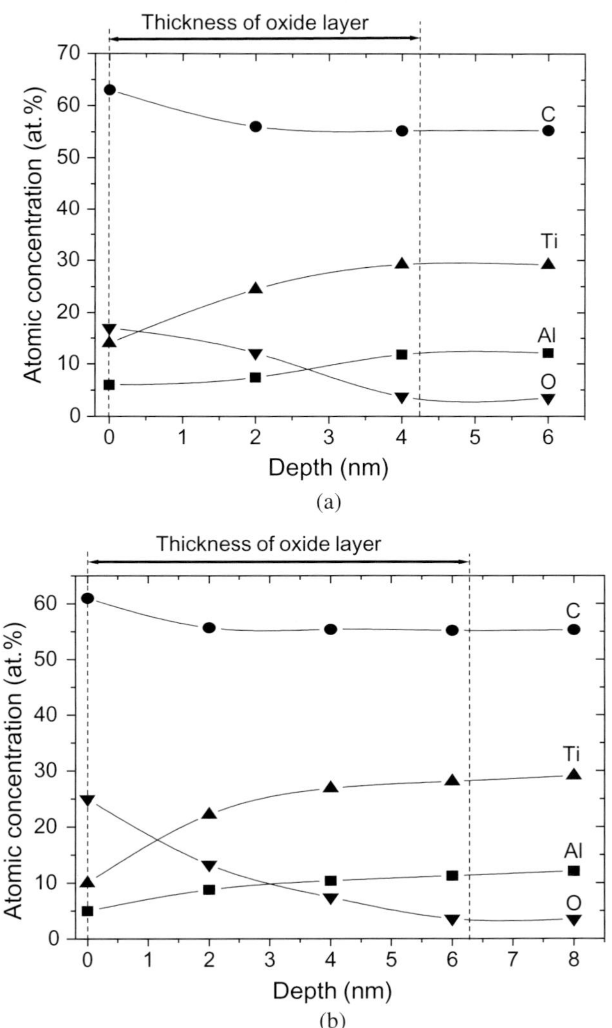

Fig. 3.18.   XPS depth profile of nc-TiC/a-C(Al) nanocomposite coating (a) without annealing (b) after annealing at 300°C for 60 min, and (c) after annealing at 600°C for 60 min in air [26].

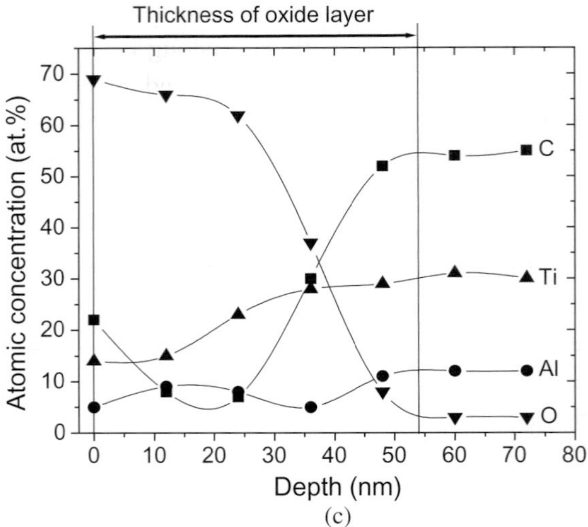

Fig. 3.18. (*Continued*)

(Ti, Cr)CN/a-C, and TiN. The oxidation resistance of nc-TiC/a-C(Al) was comparable to that of TiAlN under testing temperatures up to 600°C.

3.5. *Application in Piston Ring*

Till now, there is little systematic investigation concentrating on the application of a-C or a-C-based nanocomposites as protective coatings for piston rings in internal combustion engines. Pure a-C cannot be used as a protective coating for piston ring since the high residual stress does not allow thick coatings to be deposited. Thickness of less than $1.5\,\mu$m or even a few μm (if bias-graded deposition is applied) cannot meet the requirement of an acceptable mileage within one overhaul life of the engine. Furthermore, pure a-C coatings exhibit brittle behavior, poor thermal stability and oxidation resistance. The characteristics of nc-TiC/a-C(Al) indicate that this material possesses excellent properties for tribological applications. The low friction in dry and boundary-lubricated conditions (much lower than that of conventional nitride and carbide coatings) will contribute much to the reduction of friction losses in the engine, leading to a considerable decrease in fuel consumption. A hardness of $20\,$GPa (comparable to TiN and higher than CrN) combined with good thermal stability and oxidation resistance (much better than TiN) gives the coating good wear resistance under harsh working conditions of piston rings. In addition, the low residual stress allows thick coatings to be deposited for a long working life.

In this study, the engine tests were carried out with a nc-TiC/ a-C(Al) nanocomposite coating ($C_{56}Ti_{31}Al_{13}$). The deposition conditions and mechanical properties of this coating were stated in Table 3.2. As a control, an 18 GPa hard TiN coating was also deposited by reactive magnetron sputtering on a piston ring at Ti target power density of 4 W/cm^2, bias voltage of -80 V and process pressure of 0.6 Pa (40 sccm Ar + 10 sccm N_2).

3.5.1. *Engine*

The engine used in this test is a two-stroke gasoline engine (Fig. 3.19) with cylinder capacity of 41 cc, cylinder bore of 39.8 mm and stroke of 32.5 mm. The cylinder bore is plated with Cr.

A propeller with size of 18–10 (diameter: 18 inches, pitch: 10 inches) was installed to apply the load to the engine. An output of 2 horse power (HP) is required for a revolution speed of 5000 r.p.m for this propeller. The maximum combustion pressure is 5.2 MPa. The maximum pressure on the working surface of the first piston ring (thus, on the coating) is estimated to be about 4 MPa (0.76×5.2). During the tests, the r.p.m of propeller was maintained at 5000 ± 100 by controlling the fuel feeding. The fuel used for the engine was M92 gasoline mixed with 4% Shell Helix Plus engine oil.

3.5.2. *Piston Ring*

The piston ring (Fig. 3.20) was made from cast iron with alloying elements. The chemical composition is shown in Table 3.3.

The geometry of the ring. Inside diameter: 36.32 mm. Outside diameter: 39.37 mm. Thickness: 1.65 mm.

The ring gap area (see Fig. 3.20) was investigated since the wear of the ring gap area is more serious compared to other areas on the circumference of piston ring [30].

3.5.3. *Testing Procedure*

There were two types of samples. One type comprised the whole piston rings, which were utilized for engine tests. The other comprised pieces cut from the rings; after deposition, these pieces were polished to measure the coating thickness.

(a)

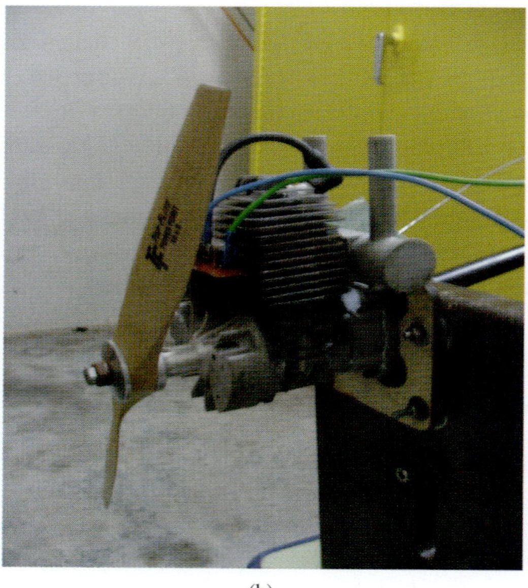

(b)

Fig. 3.19.   The engine: (a) without propeller; (b) installed with propeller for testing.

The engine tests were carried out for 610 hours. After the first 30 hours of running the piston rings were removed from the engine, then ultrasonically cleaned for 20 min to estimate the running-in wear by SEM. After that, the rings were reinstalled and another 580 hours of testing was carried out. Finally, the piston rings were removed then polished for SEM investigation.

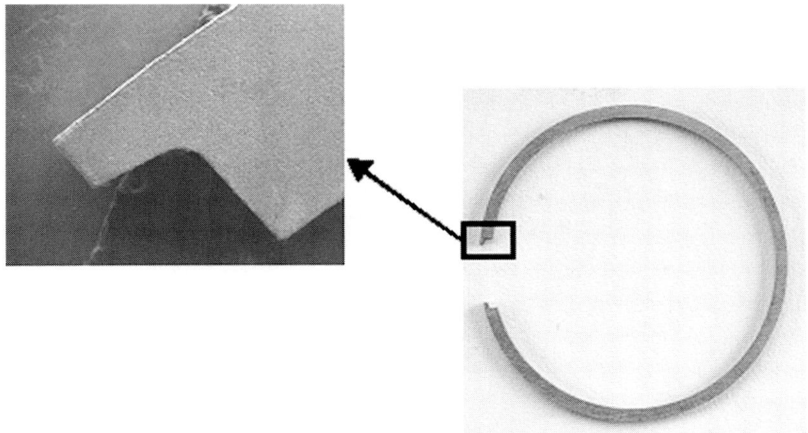

Fig. 3.20.   Piston ring and the ring gap area.

Table 3.3.   Chemical composition of cast ion piston ring.

Element	Fe	C	Si	Mn	Cr	W	V	Ti	P	S
Content (at.%)	92.1	3.9	2.4	0.6	0.2	0.15	0.15	0.05	0.4	0.05

3.5.4. *Results*

The cross-sections of TiN and nc-TiC/a-C(Al) coatings deposited on piston rings are shown in Fig. 3.21. From SEM images, 20 points were randomly chosen for measuring the coating thickness and the average was considered the coating thickness. Coating thickness of 23.2 ± 0.3 and $21.3 \pm 0.3\,\mu$m were estimated for TiN and nc-TiC/a-C(Al) coatings, respectively. The thickness of both coatings were consistent along the circumference of the piston ring. It can be recognized that the nc-TiC/a-C(Al) coating exhibits a denser structure compared to the TiN coating.

Figure 3.22 shows the fractography of the TiN and the nc-TiC/a-C(Al) coatings on piston rings after the first 30 hours testing. The remaining thickness was averaged from the values measured at 20 random points. The remaining thickness was estimated to be 19.9 ± 0.8 and $18.4 \pm 0.5\,\mu$m for TiN and nc-TiC/a-C(Al), respectively.

Figure 3.23 shows the SEM images of the remaining coatings after a "field test" of 610 hours. The TiN coating wore out completely whereas

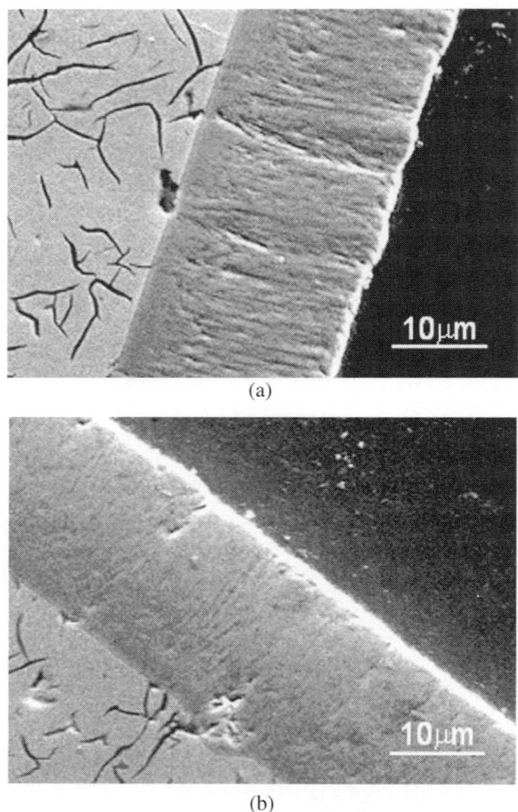

Fig. 3.21. SEM cross-section images of (a) TiN, and (b) nc-TiC/a-C(Al) coatings, deposited on piston ring. The thickness was estimated to be 23.2 ± 0.3 and $21.3 \pm 0.3\,\mu$m for TiN and nc-TiC/a-C(Al), respectively.

the nc-TiC/a-C(Al) coating, which still had a thickness of $2.1 \pm 0.2\,\mu$m remaining, continued to adhere very well to the piston ring with no sign of spallation or peeling. The results from the engine tests are summarized in Table 3.4.

Assuming that the TiN coating wore out right before completion of the test, the wear rate of TiN would be $3.4\,\mu$m/100 h. In reality, the wear rate of the TiN coating may be higher or even a lot higher (depending on when the coating was actually completely worn out). The wear of the nc-TiC/a-C(Al) coating was considerably lower than that of TiN (13% lower in the first 30 hours running-in time and more than 17% lower in the next

(a)

(b)

Fig. 3.22. SEM images of TiN (a) and nc-TiC/a-C(Al) after 30 hours engine test. The thickness remaining was estimated to be 19.9 ± 0.8 and $18.4 \pm 0.5\,\mu$m for TiN and nc-TiC/a-C(Al), respectively.

580 hours). With the nc-TiC/a-C(Al) coated ring, about 3% fuel was saved for the first 30 hours and about 2% for the next 580 hours.

3.6. *Summary*

Bias voltage does not influence the composition of sputtered coatings. The target power density does. Co-sputtering of graphite, Ti and Al deposits a nanocomposite coating where nanosized TiC crystalline phase is embedded in an amorphous carbon matrix doped with Al, or nc-TiC/a-C(Al).

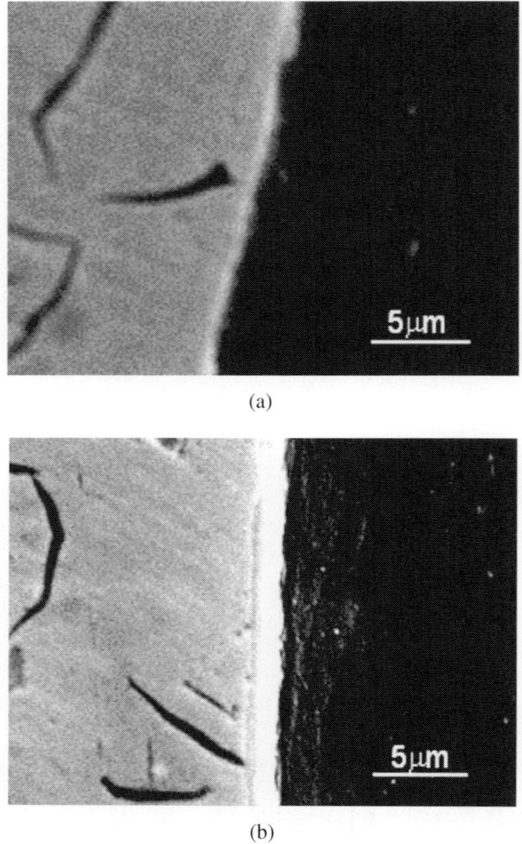

Fig. 3.23. Cross-section after engine test for 610 hours. (a) The TiN coating was completely worn out, but (b) the nc-TiC/a-C(Al) coating still had a thickness of $2.1 \pm 0.2\,\mu$m.

The average grain size of the nanocrystalline TiC grains is less than 6 nm. There is no preferred orientation. The nc-TiC/a-C(Al) coating has a relatively high hardness of about 20 GPa, a very low residual stress of 0.5 GPa, a very smooth surface (5 nm in R_a for a $1.2\,\mu$m thick coating deposited on a Si wafer) and exhibits very high adhesion strength (lower critical load of almost 700 mN) and superior toughness (as evidenced by indenter plaughing into the coating rather than peeling the coating off). The plasticity during indentation is as high as 58%. The nc-TiC/a-C(Al) coating is thermally stable up to 400°C. Even at 600°C for one hour in Ar, 90% or 17 GPa of the hardness remains. The oxidation resistance of the coating is comparable to that of TiAlN, much better than that of pure

Table 3.4. Results from engine tests.

Coatings	Thickness after (μm)			Wear rate (μm/100h)		Specific fuel consumption (g/HP.h)	
	0	30 h	(30 + 580) h	First 30 h	Next 580 h	First 30 h	Next 580 h
TiN	23.2	19.9	0	11	> 3.4	236	225
nc-TiC/a-C(Al)	21.3	18.4	2.1	9.6	2.8	227	221

a-C or TiN. The nc-TiC/a-C(Al) coating exhibits a low coefficient of friction in dry air (0.23) and an extremely low coefficient of friction under oil lubrication (0.04), thus resulting in good wear resistance. Engine test of the coating deposited on piston rings illustrates a reduction of more than 17% in the wear rate and 2% in fuel consumption as compared to TiN coating.

References

1. M. Ham and A. Lou, *J. Vac. Sci. Technol.* **A8**(3) (1990) 2143–2149.
2. A. Matthew and S. S. Eskildsen, *Diamond Relat. Mater.* **3** (1994) 902–911.
3. B.K. Tay, D. Sheeja, S.P. Lau, X. Shi, B.C. Seet and Y.C Yeo, *Surf. Coat. Technol.* **130** (2000) 248–251.
4. R.J. Hill, Physical Vapor Deposition (The BOC Group Inc., CA, 1986).
5. A.A. Voevodin and J.S. Zabinski, *J. Mater. Sci.* **33** (1998) 319–327.
6. M. Maruyama, F.S. Ohuchi and L. Rabenberg, *J. Mater. Sci. Lett.* **9** (1990) 864–866.
7. S. Meskinis, M. Andrulevicius, S. Tamulevicius, V. Kopustinskas, K. Slapikas, J. Jankauskas and B. Ciziute, *Vacuum* **80** (2006) 1007–1011.
8. C. Chang, J. Jao, T. Chang, W. Ho and D. Wang, *Diamond Relat. Mater.* **14** (2005) 2127–2132.
9. H. Randhawa, *J. Vac. Sci. Technol.* **A7(3)** (1989) 2346–2349.
10. Y. Lifshitz, G.D. Lempert, E. Grossman, I. Avigal, C. Uzan-Saguy, R. Kalish, J. Kulik, D. Marton and J.W. Rabalais, *Diamond Relat. Mater.* **4** (1995) 318–323.
11. R.O. Dillon, J. Woollam and V. Katkanant, *Phys. Rev.* **B29** (1984) 3482–3489.
12. J.P. Sullivan, T.A. Friedmann and A.G. Baca, *J. Electronic Mater.* **26** (1997) 1021–1029.
13. A.C. Ferrari, B. Kleinsorge, N.A. Morrison, A. Hart, V. Stolojan and J. Robertson, *J. Appl. Phys.* **85(10)** (1999) 7191–7197.
14. T. Zehnder and J. Patscheider, *Surf. Coat. Technol.* **133–134** (2000) 138–144.
15. S. Zhang, X.L. Bui, J. Jiang and X. Li, *Surf. Coat. Technol.* **198** (2005) 206–211.
16. M.V. Zdujic, O.B. Milosevic and L.C. Karanovic, *Mater. Let.* **13** (1992) 125–128.
17. G. K. Williamson and W.H. Hall, *Act. Metall*, Vol. 1 (1953) 22–31.
18. S. Veprek and S. Reiprich, *Thin Solid Films* **268** (1995) 64–71.
19. S. Zhang, X.L. Bui, Y.Q. Fu, D.L. Butler and H. Du, *Diamond Relat. Mater.* **13(4–8)** (2004) 867–871.
20. K. Hormberg, and A. Matthews, *Coatings Tribology, Tribology Series* **28** (1994).
21. I.L.S. Singer, S. Fayeulle and P.D. Ehni, *Wear* **149** (1991) 375–394.
22. S. Myake and R. Kaneko, *Thin Solid Films* **212** (1992) 256–261.

23. T. Jamal, R. Nimmaggada and R.F. Bunshah, *Thin Solid Films* **73** (1980) 245–254.

24. C.H. Bovington, Friction, wear and the role of additives in their control, in *Chemistry and Technology of Lubricants*, eds. R. M. Mortier and S.T. Orszulik (Blackie Academic and Professional, UK, 1997), p. 320.

25. F. Bowden and D. Tabor, *The Friction and Lubrication of Solid* (Clarendon Press, Oxford, 1986).

26. S. Zhang, X.L Bui and X. Li, *Diamond Relat. Mater.* **15** (2006) 972–976

27. P. Panjan, B. Navisek, M. Cekada and A. Zalar, *Vacuum* **53** (1999) 127–131.

28. S. Zhang, Y.Q. Fu and H. J. Du, *Surf. Coat. Technol.* **162** (2002) 42–48.

29. S. Inoue, H. Uchida, Y. Yoshinaga and K. Koterazawa, *Thin Solids Films* **300** (1997) 171–176.

30. X.L. Bui, Friction and wear of piston rings in internal combustion engines, M. Eng. thesis (Vietnam Maritime University, 1997).

CHAPTER 3

PROPERTIES OF CHEMICAL VAPOR DEPOSITED NANOCRYSTALLINE DIAMOND AND NANODIAMOND/AMORPHOUS CARBON COMPOSITE FILMS

S. C. Tjong

Department of Physics and Materials Science
City University of Hong Kong, Tat Chee Avenue, Kowloon, Hong Kong
aptjong@cityu.edu.hk

1. Introduction

The use of hard coatings with combined functional and structural properties for the protection of materials is widely recognized. Hard coatings improve the durability of substrate materials in hostile environments against wear and oxidation, thus prolonging the tool life. Transition nitride coatings like TiN, TiCN and TiBN have been used frequently in manufacturing sectors to increase the performance of cutting tools and drills. There is an increasing need in industrial sectors for the development of high performance coatings with better oxidation resistance, higher hardness and longer lifetime than conventional TiN coatings. To meet industrial demands for improved coatings, significant effort has been devoted to the design and synthesis of superhard coatings. Diamond is a very attractive material in this respect due to its outstanding and unique properties, such as large bandgap, high chemical inertness, low friction coefficient, high hardness, high refractive index, high thermal conductivity, high optical transparency and good biocompatibility. Potential applications include transparent protective coatings for optical components, tribological coatings for microelectromechanical systems (MEMS), heat sinks, field emission displays and biomedical implant materials, etc. [1–9]. Chemical vapor deposition (CVD) method is one of the most promising techniques to producing low cost, large area and high quality polycrystalline diamond films. Typical CVD deposition involves the placement of silicon or diamond substrates under a flowing gas mixture of hydrogen and hydrocarbon, activated with a hot filament, or plasma at

167

microwave frequency. Diamond crystallites then nucleate on the substrates under appropriate plasma chemistry conditions. The nucleation of diamond is an important step in the growth of diamond thin films, because it strongly affects the growth rate, film morphology and quality [10]. CVD diamond nucleation and growth mechanism on diamond substrates is generally well known [11, 12]. High quality homoepitaxial diamond (100) film can be synthesized from high-power microwave plasma-enhanced CVD [12]. However, the diamond nucleation mechanism on other non-diamond substrates is relatively less well understood due to the difficulty of locating and identifying the nucleation sites. Silicon is commonly employed as a substrate for the diamond deposition owing to its widespread application in the microelectronics industry. The high surface energy, small lattice parameter and small coefficient of thermal expansion prevent diamond from growing heteroepitaxially on Si and other non-diamond substrates. Very few materials such as cubic boron nitride (c-BN) and iridium are reported to be ideal substrates for the hetero-epitaxial diamond nucleation [13–15]. Cubic boron nitride is particularly suitable because its lattice constant (3.615 Å) matches with the diamond lattice parameter (3.567 Å) closely, i.e. with a lattice mismatch of less than 2%. Although the lattice misfit between diamond and iridium is 7.6%, the iridium thermal expansion coefficient is closer to that of diamond [15].

Diamond coatings with submicrometer and micrometer grain sizes fabricated from the CVD method are rather rough and non-uniform over large areas. Low nucleation densities in CVD process require longer deposition times for forming continuous film, leading to large surface roughness due to the large grain size. The high surface roughness poses a major problem for optical coating applications because it causes attenuation and scattering of the transmitted lights [16]. Moreover, CVD coatings generally exhibit a rough faceted surface, which is undesirable for many machining and sliding wear applications [17, 18]. To obtain a smoother surface, mechanical polishing of diamond films is needed. This process is rather tedious and difficult due to the extremely high hardness of diamond. Mechanical polishing can also lead to an increase in the sp^2 content of the diamond films [16]. Roughness of the diamond films can be effectively reduced when their grain size approaches nanometer level. In this respect, efforts have been directed towards the preparation of nanocrystalline diamond (NCD) and ultrananocrystalline diamond (UNCD) coatings having high nucleation rates with desired morphologies and characteristics.

In recent years, nanocrystalline (several nanometers and up to ~100 nm) materials have attracted increasing attention from material scientists, chemists and physicists. When the grain size is below a critical value (~10–20 nm), more than 50 vol.% of atoms are associated with grain boundaries or interfacial boundaries [19]. Bulk ultrafine diamonds with grain sizes of 5–10 nm can be synthesized from detonation of explosives [20, 21] and from fullerene (C_{60}) at a high pressure of 20 GPa [22]. Nanocrystalline diamond films with extremely high grain boundary atoms are emerging as new novel materials with unique electrical, optical and mechanical characteristics [23].

Most NCD films have been deposited on non-diamond substrates via microwave or hot filament CVD plasma consisting of methane–hydrogen mixture (1% CH_4, 99% H_2) under bias-enhanced nucleation (BEN) mode [24–29]. The process involves the surface hydrogen atom abstraction reaction and formation of methyl radical by gas-phase hydrogen abstraction. In this respect, the methyl, CH_3, is considered the growth precursor for the film deposited in a hydrogen-rich plasma. The substrate temperatures for diamond deposition is typically in the range of ~ 600–950°C. Lower substrate temperatures below 600°C would make the hydrogen abstraction reaction become sluggish, thereby causing lower diamond nucleation density and larger grain sizes. Several factors are known to affect the diamond nucleation under BEN controlled mode. These include active carbonaceous species concentration, process pressure, substrate temperature and electrical bias field [24–29]. Recently, Zhou *et al.* [26] reported that composite film consisting of nanocrystalites embedded in an amorphous carbon (a-C) matrix can be prepared by a prolonged bias assisted hot filament CVD. The average crystalline size and volume fraction of nanodiamond can be controlled by changing the CH_4 concentration of the CH_4/H_2 feeding gas.

On the other hand, UNCD films can be synthesized from the microwave (2.45 GHz) assisted hydrogen poor plasmas (e.g. $CH_4/H_2/Ar$ or $C_{60}/H_2/Ar$), with Ar concentrations up to 97% or even without the addition of molecular hydrogen [30–37]. Thus, the plasma chemistry is completely different from that of the hydrogen-rich methane–hydrogen mixture mentioned above. Dense and continuous UNCD films can be grown at substrate temperatures as low as 400°C at reasonably high deposition rates under optimized ultrasonic seeding process [1]. Growing films at lower temperatures permits better control over the final films' thickness and grain size. Gruen and coworkers demonstrated that the carbon dimmer, C_2, is the growth precursor for the UNCD films. Such films exhibit smooth surfaces with grain sizes ranging ~3–10 nm. Excellent mechanical and field emission properties of the UNCD films have been found [36, 37]. The fracture strength of UNCD films is much higher than that of conventional MEMS materials such as polysilicon and SiC [36]. The excellent mechanical and tribological properties render the UNCD coatings an ideal material for MEMS applications. Furthermore, Gruen and coworkers indicated that the microstructure of diamond films can be tailored by changing the gasphase chemistry of the plasma-enhanced CVD process. The sizes of diamond crystallites ranging from microcrystalline to nanocrystalline can be varied continuously by monitoring the Ar/H_2 ratio of the $Ar/H_2/CH_4$ gas mixtures [31]. Therefore, understanding the plasma chemistry, identification of the primary growth precursors and the nucleation sites for diamond crystallites are essential for preparation of the CVD nanodiamond and nanocomposite films with desired mechanical and physical properties.

2. Chemical Vapor Deposition

Two versatile routes are commonly adopted to prepare thin films from vapor phase: physical vapor deposition (PVD) and CVD. The former route includes evaporation, sputtering, ion implantation and laser ablation that involves no chemical reactions. CVD is a process involving chemical reactions between gaseous adsorption species on a hot substrate surface to form thin film or coating with desired properties. The main steps that occur in the CVD process can be summarized as:

- a. Transport of reacting gaseous species to the surface.
- b. Adsorption of the species on the surface.
- c. Heterogeneous surface reaction catalyzed by the surface.
- d. Surface diffusion of the species to growth sites.
- e. Nucleation and growth of the film.
- f. Desorption of gaseous reaction products and transport of reaction products away from the surface [38, 39].

CVD films have better step or surface coverage compared to the films grown by PVD. Moreover, CVD is capable of coating complex-shaped components uniformly. In this aspect, CVD is widely used in the research laboratories and industries to prepare metallic, ceramic and semiconducting thin films. Depending on the activation sources for the chemical reactions, the deposition process can be categorized into thermally activated, laser-assisted and plasma-assisted CVD.

In conventional thermally activated CVD, resistive heating of hot wall reactors provides sufficient high temperatures for the dissociation of gaseous species. This leads to the entire heating of the substrate to a high temperature before the desired reaction is achieved. It precludes the use of substrates having melting points much lower than the reaction temperature. Alternately, one could heat the reacting gases in the vicinity of the substrate by placing a hot tungsten filament inside the reactor. Plasma-enhanced CVD is known to exhibit a distinct advantage over thermal CVD owing to its lower deposition temperature. Various types of energy resources e.g. DC, RF, microwave and electron cyclotron resonance microwave (ECR-MW) radiation are currently used for plasma generation in CVD. In a DC plasma, the reacting gases are ionized and dissociated by an electrical discharge, generating a plasma consisting of electrons and ions. Microwave plasmas are attractive because the excitation microwave frequency (2.45 GHz) can oscillate electrons. Thus high ionization fractions are generated as electrons collide with gas atoms and molecules. These CVD techniques have been successfully used to grow diamond films on various substrates.

Hot filament CVD (HFCVD) presents advantages like low cost, simplicity, ease of scaling and the ability to form uniform diamond films in large areas. In the process, the source gases are ionized by energetic electrons produced by a hot filament. In most cases, the grids have been added in order to accelerate the ions impinging on to the substrate. The major shortcomings are low nucleation density and contamination from the filament materials. Such contamination can affect the electronic and optical properties of the diamond films. Therefore, optical coatings and windows, thermal management and semiconductor applications are commonly prepared by microwave plasma-enhanced CVD [40]. MWCVD system in most research laboratories employed a microwave frequency of 2.45 GHz. To scale up for commercial applications, MWCVD systems equipped with a 915 MHz and 60 kW for generating a large size plasma have been developed [41]. HFCVD is however very effective in producing tribological coatings, as non-planar parts with complex shapes can be homogeneously covered [42]. The first diamond films were synthesized by HFCVD technique from a gas mixture of methane (0.5–2 vol.%) and hydrogen by Matsumoto *et al.* [43]. The gas mixture flows through a refractory metal such as W, Ta or Mo heated at a temperature above 2000°C. Under this condition, H_2 is dissociated into atomic hydrogen, and CH_4 undergoes pyrolysis reaction leading to the formation of radicals such as CH_3, C_2H_2 and other stable hydrocarbon species. The deposition rate is about 1–3 μm h^{-1} over a variety of substrate materials such as titanium, chromium and silicon [44]. The typical chemical reactions occur under thermal activation can be described as follows [40]:

$$H_2 \rightarrow 2H, \tag{2.1}$$

$$CH_4 + H \rightarrow CH_3 + H_2. \tag{2.2}$$

Atomic hydrogen is considered to play a crucial role in CVD deposition. These include adding and abstracting bonded hydrogen, formation of methyl radical by gas-phase hydrogen abstraction and preferential etching of the graphitic phase at the growing diamond surface in CH_4/H_2 plasma [45, 46]. Removal of graphitic species stabilizes the sp^3 structure, thereby producing a good crystalline quality.

Nucleation of diamond, the early stage of crystal formation, is an important step in the growth of diamond films. Despite rapid progress in the CVD diamond synthesis, the mechanism of diamond nucleation on non-diamond substrates still remains much less understood. As mentioned above, the atomic hydrogen has been considered essential for etching the graphitic phase. However, some workers reported that diamond could be synthesized in a hydrogen-free environment [47, 48]. It is therefore necessary to understand the basic chemical processes involved in CVD diamond nucleation.

Nucleation of a new phase involves a balance between the bulk energy per atom and the interfacial or surface energy of the nucleus. The high surface energy, small lattice parameter and small coefficient of thermal expansion result in low nucleation densities on non-diamond substrates. Single crystal silicon is widely used as a substrate material for the deposition of diamond films. However, there exist large lattice mismatch and surface energy differences between silicon and diamond. The lattice constant for silicon is 0.543 nm and surface energy of $\{111\}$ silicon plane is 1.5 JM^{-2}, while the lattice constant of diamond is 0.357 nm and the surface energy of $\{111\}$ diamond plane is 6 JM^{-2} [49, 50]. Diamond nucleation on pristine silicon substrates is usually associated with an incubation period, high localization and low nuclei density of $\sim 10^4$ cm^{-2} [51]. Assuming diamond crystallites are cubic, the minimum nucleation density, N_d, needed to form a continuous layer of thickness d can be estimated from the relation: $N_d = d^{-2}$ [40]. In order to achieve a coalesced polycrystalline film of 1 μm thick, a minimum density of 10^8 cm^{-2} is required. The growth of diamond crystallites on silicon generally follows three-dimensional Volmer–Weber mode. This results in a columnar growth in which the grain size tends to increase with increasing film thickness. In the past, some methods such as seeding or abrading non-diamond substrates with diamond powders, ultrasonically vibrating the substrate in the suspension containing diamond powders were commonly used to enhance the diamond nucleation. Reported nucleation densities after substrate abrasion vary from 10^8–10^{10} cm^{-2} [52–54]. However, the abrasion method would introduce mechanical damage and contamination, producing poor quality diamond–substrate interface. Such contamination is rather difficult to remove through proper degreasing and water rinsing procedures. The effective method for diamond nucleation density enhancement is *in situ* biasing of substrates during the initial stage of deposition [55–57]. This method is referred to as bias-enhanced nucleation (BEN). According to the literature, the density of diamond ranges from 10^8–10^{11} cm^{-2} can be achieved under BEN treatment [28, 50, 56–61]. Proper control of the nucleation and subsequent growth parameters often leads to local heteroepitaxy [62, 63].

Stoner *et al.* have studied in-depth the nucleation and growth of highly oriented diamond (HOD) films on silicon via MPCVD with BEN step using high resolution transmission electron microscopy (HRTEM), Raman, Auger electron spectroscopy and X-ray photoelectron spectroscopy [56]. They reported that an amorphous SiC interfacial layer was developed on silicon before significant diamond nucleation occurred. When the SiC layer had reached a critical thickness ($\sim$90 Å), carbon on the surface tended to form clusters that were eventually favorable for diamond nucleation. It was also found that the biasing pretreatment removed surface oxide and suppressed oxide formation on the surface [57]. They further indicated that the carbide forming nature of the substrate plays a decisive role when performing BEN treatment. Nucleation densities of 10^{10} cm^{-2} can be achieved in certain carbide forming refractory metals such as hafnium and titanium [58]. It is worth noting that diamond nucleation is rather poor on these carbide forming substrates without biasing. This means that biasing is needed for the nucleation of diamond film on these substrates.

The morphology of oriented diamond films can be controlled by taking advantage of the growth competition between different orientations of diamond grains. This behavior can be described by the growth parameter (α) defined as [64]:

$$\alpha = \sqrt{3(V_{100}/V_{111})} \tag{2.3}$$

where V_{100} and V_{111} are the growth rates along the $\langle 100 \rangle$ and $\langle 111 \rangle$ directions, respectively. The growth parameter depends upon the processing conditions such as gas composition, substrate temperature and pressure [65]. When α is close to 1, cubic shape crystals are produced. When α is close to 3, the fastest growth direction is $\langle 100 \rangle$, leading to the formation of octahedral pyramid shape. The morphology of diamond crystals vary from cubic, cubooctahedral to octahedral by increasing the growth parameter from 1 to 3. Fig. 2.1 shows typical cubooctahedral diamond crystals grown on Si(100) substrate from a CH_4/H_2 gas mixture with $\alpha = 1.8$ using MPCVD [63].

In the case of non-diamond substrate forming no carbide, e.g. copper with a cubic structure (lattice constant of 3.61 Å), the nucleation density is relatively low even after BEN treatment. Chuang *et al.* synthesized diamond films on copper via MPCVD with and without BEN-treatment in 2.5–25% CH_4/H_2 gas mixtures [66]. They reported that the diamond density increases from $\sim 10^5$ to $10^7 \, \mathrm{cm}^{-2}$ with increasing methane concentration and saturated at 10% CH_4 under BEN treatment. These nucleation density values are considerably lower than those of BEN-controlled Si substrates, i.e. 10^8–$10^{11} \, \mathrm{cm}^{-2}$ [35, 56–61]. Figures 2.2(a)–(f) are the SEM micrographs showing the effect of bias on diamond nucleation. The nucleation density reached the highest value of $10^6 \, \mathrm{cm}^{-2}$ when the bias voltage was set at $-250 \, \mathrm{V}$ with 5% CH_4 [Fig. 2.2(d)]. The nucleation density tended

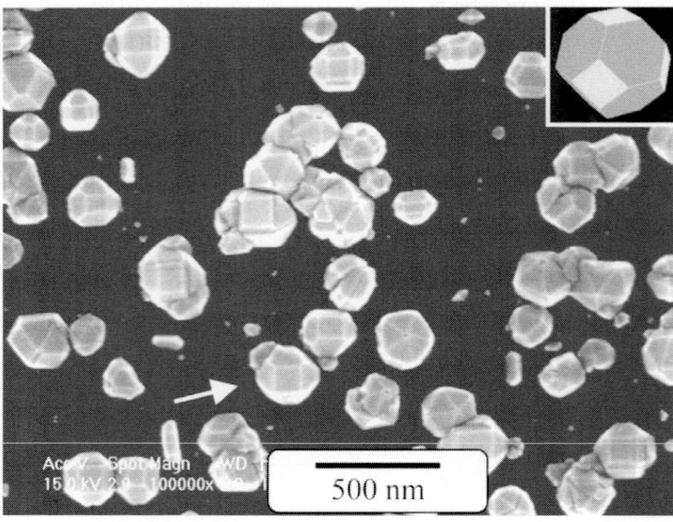

Fig. 2.1. SEM micrograph showing cubooctahedral diamond crystals on Si(100) substrate from a CH_4/H_2 gas mixture with $\alpha = 1.8$ using MPCVD [63].
Reprinted with permission of Elsevier.

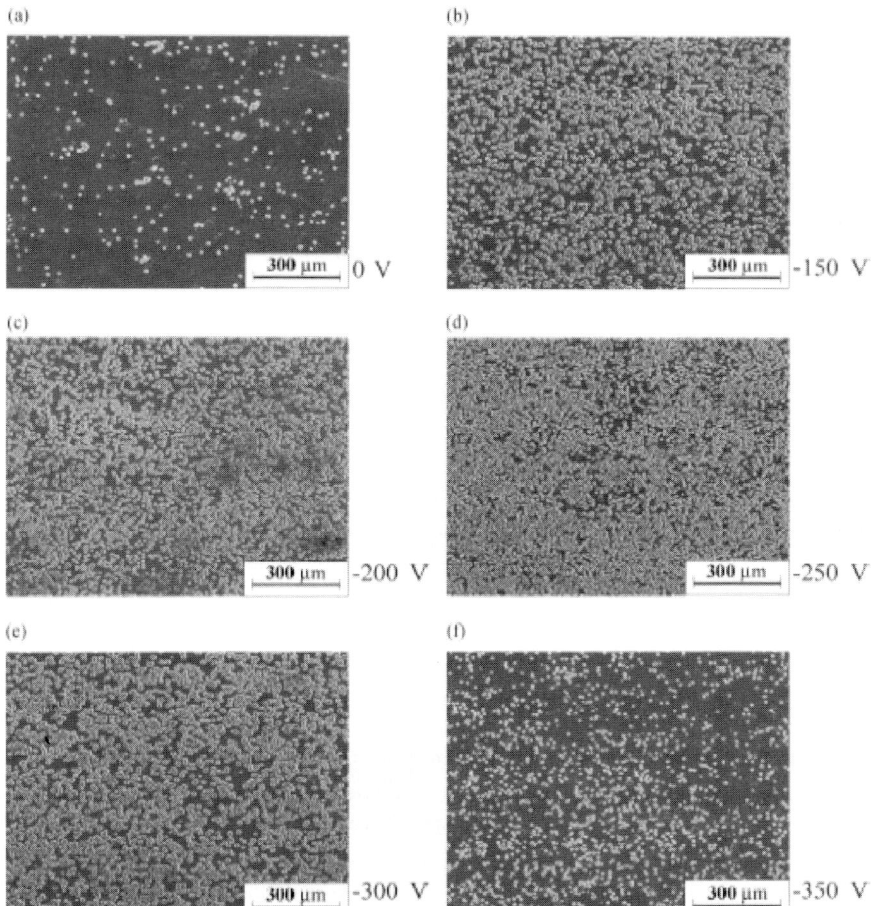

Fig. 2.2.   SEM micrographs showing the effect of bias voltage on diamond nucleation on copper substrate with 5% CH_4: (a) 0 V, (b) -150 V, (c) -200 V, (d) -250 V, (e) -300 V and (f) -350 V [66].
Reprinted with permission of Elsevier.

to decrease with further increase of the bias voltage beyond -250 V due to the etching effect of ions.

For CVD diamond synthesis, methane is commonly used as the source gas of carbon. Wu and coworkers reported that chloromethane (CCl_4) can be used as an alternative replacement for methane during HFCVD diamond on silicon using CCl_4/H_2 gas mixture. Chloromethane facilitates the low temperature formation of diamond films mainly through the enhancement

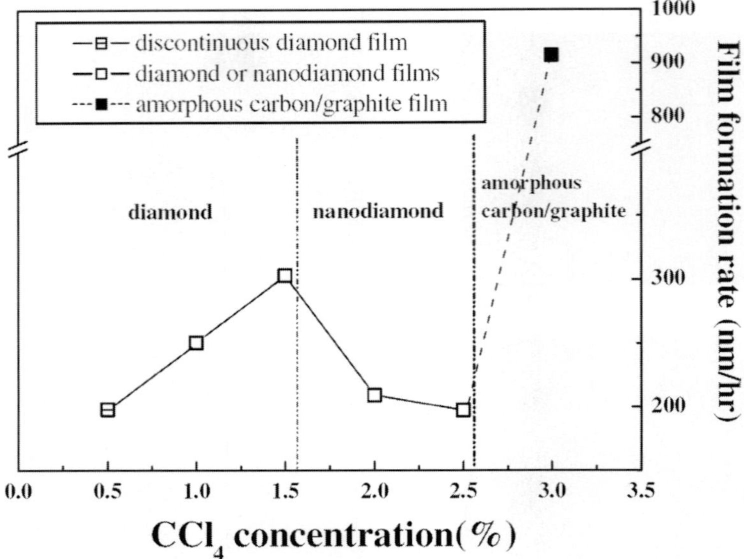

$$CCl_4 \text{ concentration}(\%)$$

Fig. 2.3. Effect of CCl_4 concentration on the formation of NCD and microcrystalline diamond films on silicon substrates [69].
Reprinted with permission of Elsevier.

of nucleation [67, 68]. Because of the weaker C–Cl bond compared to C–H bond of methane, CCl_4 can dissolve readily near the filament, producing a high concentration of gas phase chlorohydrocarbon free radicals ($\cdot CH_2Cl_2$, $\cdot H$, etc.). Moreover, atomic Cl can be formed either through the dissociation of CCl_4 or through the Cl and H exchange reaction, i.e. $H + HCl = Cl + H_2$. Atomic Cl produced enhances the growth rate by increasing both the concentration of methyl radicals $[CH_3]$ in the gas and the concentration of carbon radical $[C_d]$ on the surface. The relatively fast reaction of Cl atoms with surface-bonded C–H compared to that of hydrogen atoms with surface-bonded C–Cl increase the rate of diamond growth [68]. More recently, Ku and Wu reported that NCD film can be formed on silicon via HFCVD using 2 and 2.5% CCl_4/H_4 at a substrate temperature of 610°C (Fig. 2.3). The HRTEM image and selected area diffraction pattern of NCD film deposited from 2.5% CCl_4 are shown in Fig. 2.4. The lattice spacing of 0.205 nm in the HRTEM image corresponds to the d-spacing of {111} crystal planes of diamond [69].

3. NCD Film Formation from Hydrogen-Deficient Plasma

Gruen *et al.* have carried out pioneering research in the synthesis of UNCD films on Si substrates via MPCVD in hydrogen deficient plasmas,

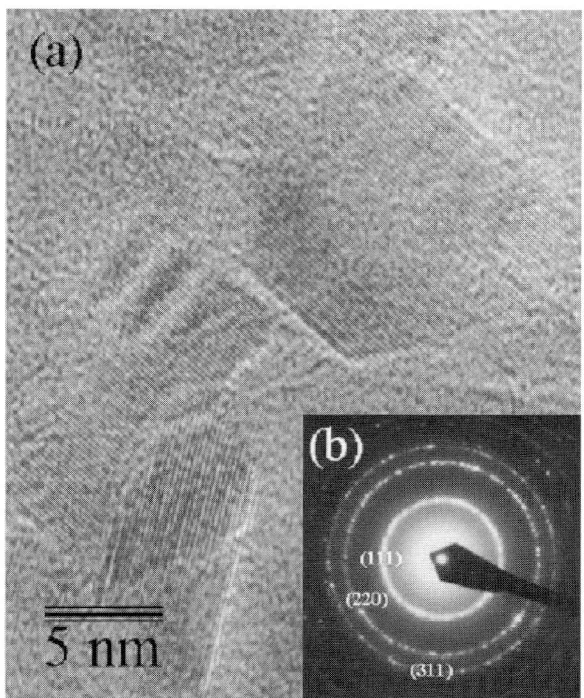

Fig. 2.4.   (a) HRTEM image and (b) selected area diffraction pattern of the diamond film deposited from 2.5% CCl$_4$ [69].
Reprinted with permission of Elsevier.

e.g. CH$_4$(1%)/Ar, or C$_{60}$/Ar and C$_{60}$/Ar/H$_2$ [31, 34, 35]. In the process, Si substrates were either pretreated with mechanical polishing using fine diamond powders or ultrasonic seeding process for enhancing diamond nucleation. The carbon dimmer, C$_2$, is the growth species for UNCD films and directly inserted to the diamond surface. Strong Swan band emission associated with C$_2$ radicals was observed [31]. Argon gas is known to readily discharge into Ar$^+$ and metastable Ar* due to its low ionization potential. The addition of Ar also increases the electron density in the plasma. For the CH$_4$(1%)/Ar plasma at 1600 K, CH$_4$ in the feed gas is thermally converted into C$_2$H$_2$ that is further charge exchanged with Ar$^+$ and Ar* to produce C$_2$H$_2^+$. The subsequent dissociative recombination of the acetylene ion with electron yields C$_2$ dimers [9]. These reactions can be summarized as follows:

$$\mathrm{CH_4} \xrightarrow{1600\mathrm{K}} \mathrm{C_2H_2} + 3\mathrm{H_2} \xrightarrow{\mathrm{Ar^+,Ar^*}} \mathrm{C_2H_2^+} + \mathrm{Ar} \xrightarrow{e^-} \mathrm{C_2} + \mathrm{H_2}. \tag{3.1}$$

In the later case, fragmentation of the fullerenes in the plasma results in strong Swan band emission due to C_2 radicals. C_{60} is employed because it is a hydrogen-free, pure carbon species having high vapor pressure. In the process, a C_{60} sublimator source is attached to a MPCVD chamber. Fullerene vapor is introduced into the plasma chamber by heating the sublimator to $550°C$ while flowing Ar gas through it [34]. The possible reactions involve both charge exchange collisions with Ar^+ and Ar^* to yield C_{60}^+:

$$Ar^+ + C_{60} \rightarrow Ar + C_{60}^+, \tag{3.2}$$

$$Ar^* + C_{60} \rightarrow Ar + C_{60}^+ + e^-. \tag{3.3}$$

Subsequent dissociative recombination of C_{60}^+ with electrons produces C_2 and C_{58} fragment cluster. This reaction can be written as follows:

$$C_{60}^+ + e^- \rightarrow C_{58} + C_2. \tag{3.4}$$

The low level of atomic hydrogen in the plasma allows continuous renucleation of the diamond phase on existing diamond crystallites. Owing to the high renucleation rates of the C_2 dimer, UNCD films have extremely small grain sizes of 3–10 nm. The grains are purely crystalline diamond on the basis of TEM observation and electron energy loss spectroscopy. The grain boundaries are near atomically abrupt as observed with TEM [32]. The grain boundary of UNCD films constitutes about 10% carbon atoms that are π-bonded [9]. The UNCD films produced exhibit exceptionally high surface smoothness. The surface topography and roughness of UNCD films can be observed with the atomic force microscopy (AFM). Figure 3.1 shows the AFM images of microcrystalline and UNCD films grown in a microwave plasma [70]. It can be seen that the UNCD films produced from the $C_{60}/Ar/H_2$ plasma have very smooth surfaces compared to the microcrystalline diamond film counterpart.

For the Ar/CH_4 gas mixture, the plasma may contain less than 1% H_2 due to the decomposition of methane. For the UNCD film grown using an $Ar-1\%$ CH_4 plasma, the diamond grains are likely to nucleate from an amorphous C layer on the Si substrate. However, addition of hydrogen to the Ar/CH_4 gas mixture would suppress formation of the amorphous C layer due to the etching effect, leading to direct nucleation of diamond on the Si surface. Gruen and coworkers studied the effect of hydrogen on the nucleation and morphology of diamond crystallites [31, 71]. They reported that the microstructure of diamond films can be controlled by varying the Ar/H_2 ratio of the $Ar/H_2/CH_4$ gas mixtures [31]. A transition from microcrystalline to nanocrystalline occurs at an Ar/H_2 ratio of 9. They attributed

Conventional CVD diamond film grown from H_2 (99%) - CH_4 (1%)

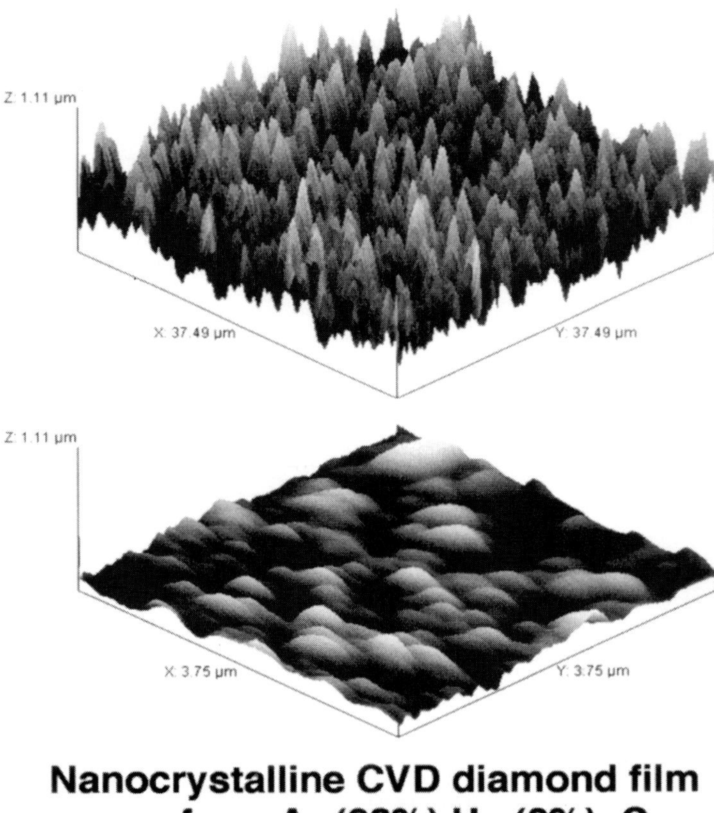

Nanocrystalline CVD diamond film grown from Ar (98%),H_2 (2%), C_{60}

Fig. 3.1.　Atomic force microscopy images showing the surfaces of microcrystalline diamond and UNCD thin films [70].
Reprinted with permission of Elsevier.

this to a change in growth mechanism from $\cdot CH_r$ radical in hydrogen rich content to C_2 species in low hydrogen content plasma. Figures 3.2(a)–(e) are SEM micrographs showing the changes of NCD to microcrystalline diamond morphology by increasing the hydrogen content in the $Ar/H_2/CH_4$ plasma [72]. In the absence of hydrogen, the UNCD film exhibits typical cluster or the so-called ballast type morphology [Fig. 3.2(a)]. As hydrogen is

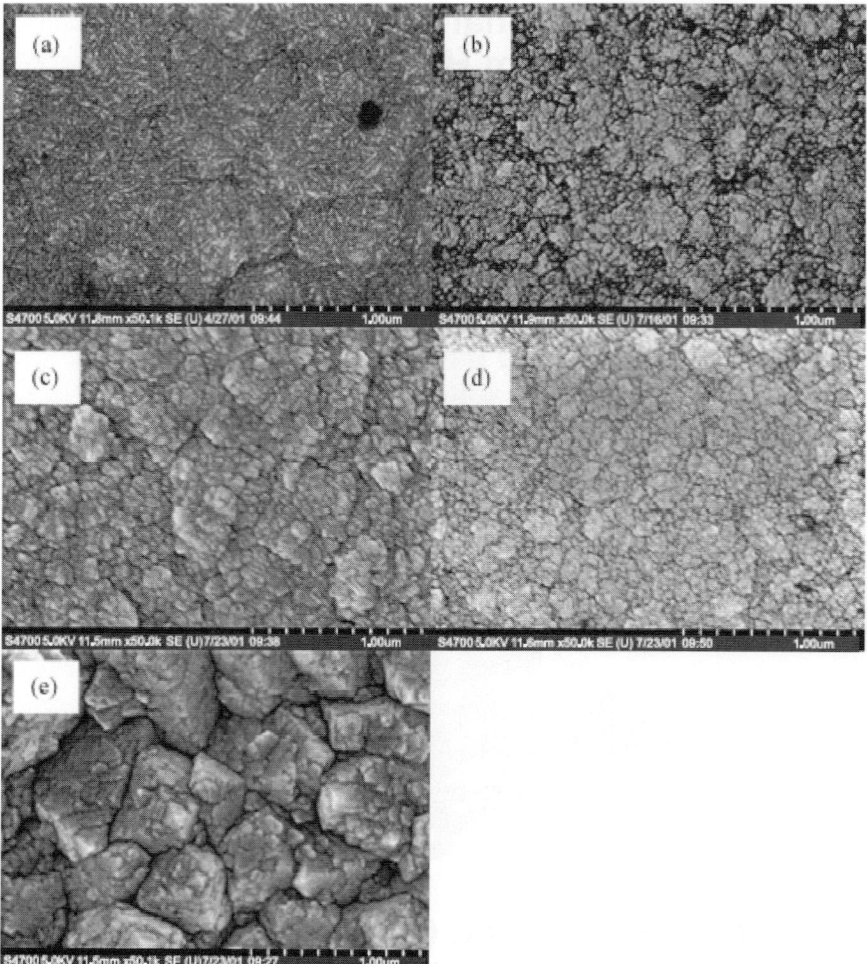

Fig. 3.2. SEM micrographs showing a transition from nanocrystalline to microcrystalline diamond morphology by increasing hydrogen content in the $Ar/H_2/CH_4$ plasma: (a) 0%, (b) 1%, (c) 5%, (d) 10% and (e) 20% H_2 [72].
Reprinted with permission of Elsevier.

introduced into the plasma, the C_2 density in the plasma decreases sixfold for 20% H_2 UNCD [73]. Two obvious changes in the morphology of UNCD films can be observed. The average size of nanograins increases from 5 to 10 nm for 5, 10 and 20% H_2. Moreover, the volume fraction of microcrystalline diamond in the composite film increases. The microcrystalline diamond inclusions over a micron size are major components of the 10%

and 20% H_2 UNCD composite film. As mentioned above, the formation of microcrystalline diamond depends greatly on the presence of methyl radicals and the atomic hydrogen species in the plasma. Increasing the hydrogen content in the $Ar/H_2/CH_4$ plasma inevitably increases the concentration of these two species.

Raman spectroscopy is a versatile tool to determine the vibrational modes of diamond and sp^3 and sp^2 configurations in amorphous carbon or diamond-like coatings (DLC). Diamond has a single Raman active mode at 1332 cm^{-1}, which is a zone center mode of T_{2g} symmetry [74]. In sp^3 hybridization, a carbon atom forms four sp^3 orbitals associated with the combination of one 2s atomic orbital and three 2p atomic orbitals, producing a strong σ covalent bond to the adjacent carbon atom. In the sp^2 configuration, a carbon atom forms three sp^2 orbitals resulting in three σ bonds and the remaining p orbital forms a π bond. The three σ bonds and π bond constitute a ring plane in sp^2 clusters. Both the sp^3 and sp^2-bonded carbon phases constitute the DLC. Due to the resonance effects, the Raman cross-section for sp^2 clusters is much larger than that for sp^3-bonded carbon in conventional, visible Raman excited at 514 or 488 nm. Thus the Raman scattering signal from the sp^2 clusters often predominate [75]. The visible Raman spectrum of DLC generally shows the presence of main G peak at $\sim$1550 cm^{-1} and a D peak shoulder at $\sim$1350 cm^{-1}. Both are attributed to graphitic sp^2 bonding. The G peak arises from the bond stretching modes of the sp^2-bonded carbon in both rings and chains. The D peak is associated with the breathing mode of A_{1g} symmetry involving phonons near the K zone boundary. This peak is forbidden in perfect graphite and only becomes active in the presence of disorder [76]. Recently, UV Raman spectroscopy has been used to characterize the hybridization configurations in the DLC and nanocrystalline diamond thin films. The advantage of UV Raman is its higher excitation energy than visible Raman. The Raman intensity from sp^3-bonded carbon can be enhanced while the dominant resonance Raman scattering from sp^2 cluster is suppressed [77].

It is noted that the sp^2- and sp^3-bonding configurations in the DLC and NCD films can also be analyzed using the electron energy loss spectroscopy (EELS). It is well established that carbon atoms with different structures have very distinct K-shell absorption edge features. The energy absorption caused by the transitions between π and antibonding π^*, and between σ and antibonding σ^* orbitals show different energy levels [78]. The onset of the K absorption edge of the diamond phase with sp^3-type covalent σ bonding occurs at $\sim$289.1 eV due to its excitation into the unoccupied σ^*

electronic states. In contrast, graphite with the sp^2-configuration having π bonding exhibits an additional absorption edge at 284 eV owing to its lower lying antibonding π^* states.

Figures 3.3(a) and (b) show the visible and UV Raman spectra of UNCD thin films grown with successive amounts of hydrogen added to the Ar/H$_2$/CH$_4$ plasma [72]. The 1332 cm^{-1} peak in both the visible and UV Raman spectra is assigned to the microcrystalline diamond phase. The intensity of this peak tends to increase with increasing hydrogen content in the plasma. From Fig. 3.3(a), the diamond peak disappears in UNCD films grown with less than 10% H$_2$ in the plasma. Several broad peaks and shoulders located at 1140, 1330, 1450 and 1560 cm^{-1} can be observed. The peaks at 1140 and 1450 cm^{-1} are assigned to carbon–hydrogen bonds in the grain boundaries. The peaks at 1330 and 1560 cm^{-1} are the D- and G-bands of disordered sp^2-bonded carbon. The peak at 1150 cm^{-1} is often used as a simple criterion for nanocrystalline phase in CVD sample. Ferrari and Robertson indicated that this peak cannot originate from sp^3-bonded C phase, but rather is associated with trans-polyacetylene segments at grain boundaries of the nanocrystalline diamond sample [78]. Recently, Pfeiffer *et al.* demonstrated that the 1150 cm^{-1} in visible Raman is due to the

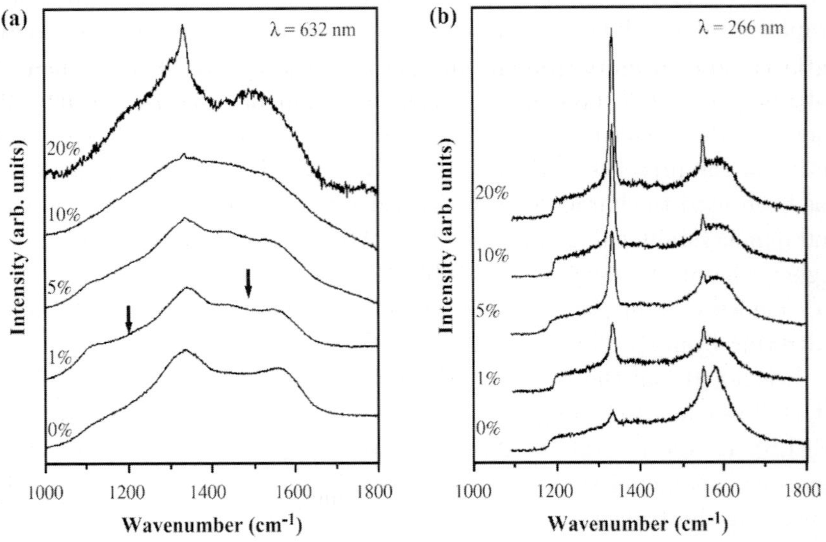

Fig. 3.3. (a) Visible and (b) UV Raman spectra of UNCD films grown with successive amounts of hydrogen added to the Ar/H$_2$/CH$_4$ plasma [72]. Reprinted with permission of Elsevier.

trans-polyacetylene segments at the grain boundaries. It shows a dispersion of $\sim 25\text{cm}^{-1}\,\text{eV}^{-1}$ when it is excited by laser of different wavelength. The $1150\,\text{cm}^{-1}$ mode is always accompanied with the $1480\,\text{cm}^{-1}$, which also shows a similar dispersion [79]. It is worth noting that the transition from microcrystalline to nanocrystalline morphology can also be observed in the film grown from the $Ar/H_2/CH_4$ plasma via HFCVD [80].

As mentioned above, the grain boundaries of UNCD films have a large fraction of sp^2-bonded carbon atoms. Gruen and coworkers demonstrated that the morphology of UNCD films and conductivity are greatly affected by the introduction of nitrogen in a Ar/CH_4 gas mixture, i.e. the presence of CN species in the plasma [81]. TEM observation reveals that the width of grain boundaries of the UNCD films increases dramatically to $\sim 1.5\text{--}2\,\text{nm}$ by doping with nitrogen. As the carbon atoms at grain boundaries are π-bonded, the electrical conductivity of the nitrogen-doped UNCD films is enhanced accordingly. More recently, Popov and coworkers [82–84] and Wu *et al.* [85] reported that the NCD/amorphous carbon nanocomposie coating can be synthesized via MWCVD in hydrogen deficient CH_4/N_2 plasmas with varying methane content. The morphology of synthesized coating consists of diamond nanocrystallites of $3\text{--}5\,\text{nm}$ dispersed in an amorphous matrix [Fig. 3.4(a)]. The bright-field TEM micrograph shows the diamond nanocrcrystallites are separated by a fine network of grain boundaries with a width of $1\text{--}1.5\,\text{nm}$ [Fig. 3.4(b)]. The crystallite phase/amorphous matrix ratio is close to unity. Raman and electron energy loss spectra show the existence of sp^2 carbon in the matrix or grain boundaries [82, 83]. The methane content in the precursor gas mixture has a large influence on the morphology of the coating. The coating prepared at 9% methane exhibit discontinuous and independent nodule-like structure. In contrast, smooth and uniform coating can be achieved using 17% methane [82, 84].

4. NCD Films Formation from Hydrogen-Rich Plasma

Methane and hydrogen mixtures are the most commonly used working gases for the preparation of diamond films in hydrogen-rich plasma. BEN treatment is an effective method for increasing the diamond nucleation density during microwave deposition from a methane/hydrogen gas mixture on silicon substrate [46, 54–57]. High renucleation rates can be achieved by increasing the relative methane content in the methane/hydrogen plasma. Gu and Jiang reported that NCD films with $40\,\text{nm}$ in grain size can be synthesized via MPCVD in the methane/hydrogen plasma under appropriate

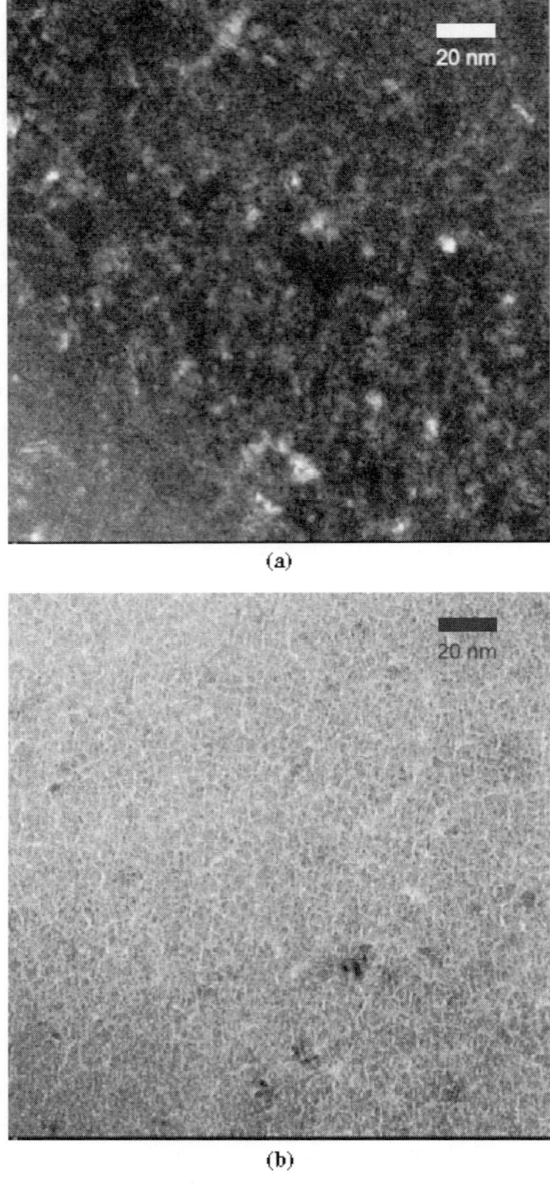

Fig. 3.4. (a) Dark-field and (b) bright-field TEM images of an NCD/a-C amorphous composite coating prepared with $9\%CH_4/91\%N_2$ plasma [83]. Reprinted with permission of Elsevier.

conditions such as high bias voltage, substrate temperature and methane concentration [55]. The proposed mechanisms responsible for diamond nucleation under ion bombardment include preferential sputtering of non-diamond component, etching of substrate surface to form an active step surface and hydrocarbon subplantation. The first model relates preferential sputtering of sp^2 phase compared to sp^3-bonded carbon under ion bombardment [86]. This model does not explanation the actual nucleation of sp^3-bonded C phase that is necessary for diamond growth. The more appropriate model for explaining the mechanism underlying the BEN step is subplantation [87, 88]. This model involves a shallow implantation of hydrocarbon species with energy of $\sim$1–1000 eV near subsurface region. After occurrence of the glow discharge, CH_x^+ ($x = 1, 2, 3$) ions are produced. The incident CH_x^+ ions of sufficient energy penetrate into the surface atomic layers and enter a subsurface interstitial position. The increase in the concentration of trapped ions in the host lattice results in the formation of an inclusion of new phase. It is likely that the preferential displacement of sp^2 atoms by the impact of incoming ions promote sp^3 bonding [88]. McKenzie *et al.* reported that compressive stress can be induced by the shallow implantation of carbon ions during the deposition of tetrahedral amorphous carbon film. Such compressive stress tends to stabilize the sp^3-bonded carbon [89].

Two types of nucleation sites associated with BEN diamond nucleation have been identified, they are the aligned graphitic layer [29, 90, 91], and the SiC interlayer [56, 92]. Robertson and coworkers reported that the ion flux is a critical factor for enhancing diamond nucleation on Si substrate under a negative bias of 200–250 V. They proposed that subplantation of hydrocarbon ions causes deposition of nanocrystalline graphite carbon. Diamond crystallites tend to nucleate on the graphitic planes that are locally oriented perpendicular to the surface. In the process, bias pretreatment causes ion beam deposition of an sp^2-bonded carbon [85, 86]. Typical sp^2 carbon tends to be quite disordered and would have its bonding planes parallel to the surface, leaving a passive surface with few dangling bonds. Moreover, ion deposition also creates the compressive stress which acts to orient some sp^2 planes perpendicular to the surface, and diamond nucleates on dangling bonds of this surface [90, 91]. Garcia *et al.* studied the diamond nucleation on silicon substrate immersed in a hydrogen and methane plasma via MWCVD under BEN treatment. They reported that the carbonaceous species react at the surface forming a graphitic carbon layer. Upon application of a bias voltage, the carbonaceous ions bombard the surface building up a compressive stress in the carbon layer. This leads to a preferential orientation of the graphitic planes normal to the surface. The surface termination of the oriented planes is considered as an active site for the diamond nucleation [29]. Lee and coworkers also confirmed the formation of a graphitic film with its basal planes perpendicular to the substrate on the basis of HRTEM observation. Diamond crystallites then nucleate on graphitic edges [92]. Another possible nucleation site for diamond crystallites is SiC interlayer. Stoner *et al.* reported that an amorphous SiC interfacial layer was developed on silicon prior to diamond nucleation on silicon via bias-enhanced MPCVD [56]. When the SiC layer had reached a critical thickness ($\sim$90 Å), the carbon on the surface tended to form clusters that were eventually favorable for diamond nucleation. Stockel and coworkers also reported that an epitaxial SiC layer of $\sim$100 Å thick was formed on silicon substrate acting as a diffusion barrier for C and Si and as a template for diamond nuclei [93].

5. Nanocomposite Film

As mentioned previously, uniform deposition of the diamond film is rather difficult to achieve via MPCVD under BEN-controlled mode. Microwave reactor creates a plasma ball that is hotter at the center than at the edge. Moreover, the glow discharge associated with the BEN treatment also yields non-uniformity along the radial direction of the substrate [94]. In this case, HFCVD demonstrates distinct advantage of growing uniform diamond films over MPCVD [95]. Recently, Lee and coworkers have studied systematically the diamond nucleation on silicon in a double biased-assisted HFCVD system. The NCD/a-C composite film can be deposited on silicon under prolonged negative bias voltage treatment [24–26, 50, 96]. In the process, a negative bias is applied to the Si substrate and a positive bias voltage is applied to a steel grid placed on top of the hot filament. Electrons can be readily emitted from the hot filament and then accelerated towards the grid. A stable plasma can be generated between the grid and hot filament accordingly. Ions in the plasma are then drawn to the substrate by a negative substrate bias (Fig. 5.1). They reported that diamond crystallites could nucleate either on silicon steps in which the diamond nucleate epitaxially or on an amorphous matrix in which the diamond nucleate randomly. Furthermore, the SiC interlayer cannot be detected [24]. Figures 5.2(A) and (B) are HRTEM images of diamond crystallites with sizes of about 2 and 6 nm that form heteroepitaxially with respect to Si substrate surface, i.e. the step

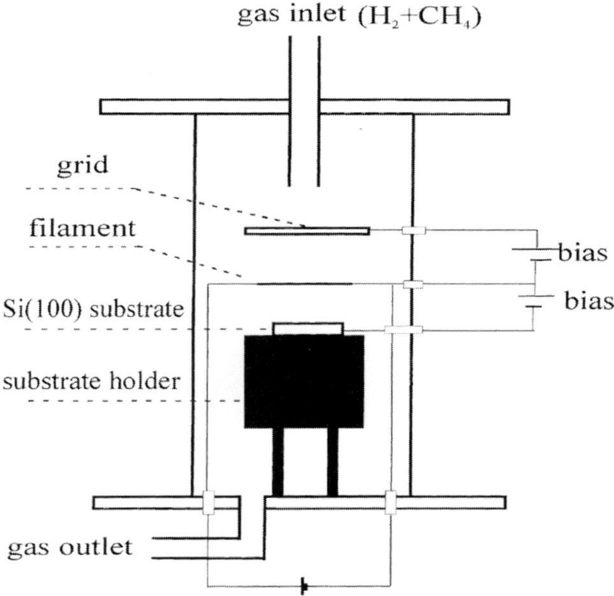

Fig. 5.1.　Schematic diagram showing double bias-assisted HFCVD system [96]. Reprinted with permission of Elsevier.

surface sites of Si. The diamond crystallites were identified by measuring the spacings of the lattice fringes and the angles of intersecting lattice planes. From Fig. 5.2(A), the intersecting angle between the (111) planes of the diamond crystallite and Si substrate is 109.5°. The (111) plane of the diamond crystallite deviates about 1 to 2° from the (111) plane of the Si substrate. This deviation is due to large lattice mismatch between diamond and Si [Fig. 5.2(C)]. However, diamond crystal that nucleates on the {111}–{001} intersecting steps of the Si substrate exhibits perfect epitaxial orientation. No misorientation between the diamond crystal and Si substrate is detected [Fig. 5.2(D)]. They further demonstrated that diamond nucleation and growth from energetic hydrogen and carbon species proceed as follows: (a) formation of dense amorphous hydrogenated carbon (a-C:H) phase via subplantation, (b) spontaneous precipitation of pure sp^3 carbon clusters in the a-C:H phase, a few of which are perfect diamond clusters, (c) annealing of defects in the diamond cluster by incorporation of carbon interstitials and by hydrogen termination, and (d) growth of the diamond cluster by preferential displacement of amorphous carbons at the diamond/amorphous carbon interface [97]. In other words, diamond crystallites initially nucle-

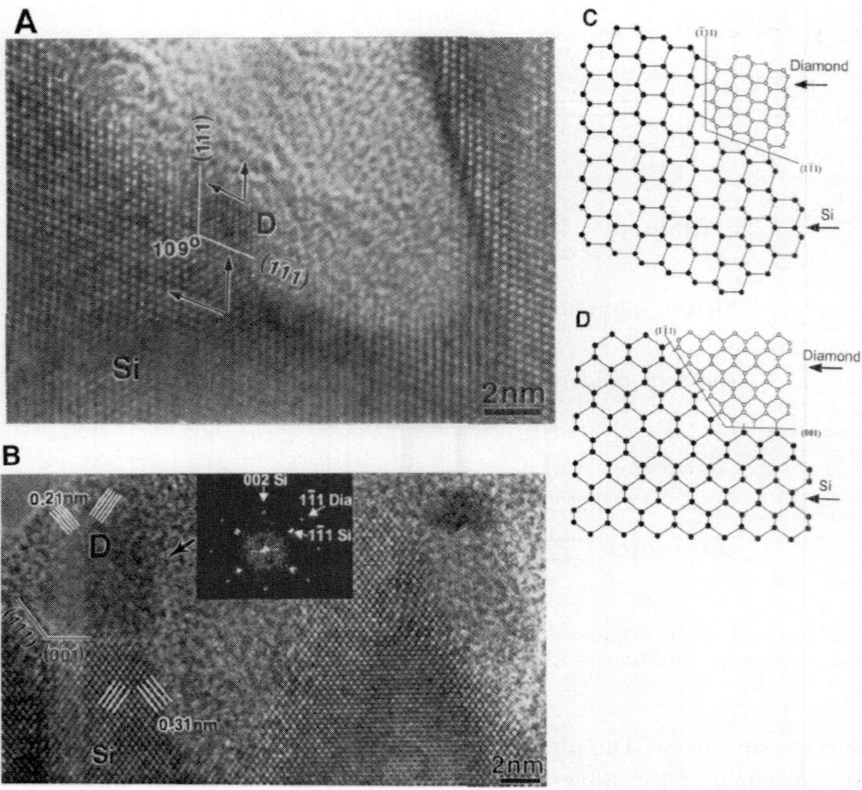

Fig. 5.2. HRTEM images of diamond crystals of (A) 2 nm and (B) 6 nm nucleate directly on a Si step with an epitaxial orientation. (C) and (D) are schematic diagrams showing the interfaces between the diamond crystallites and Si substrate of (A) and (B), respectively [24].
Reprinted with permission of The American Association for the Advancement of Science.

ate on either silicon steps or amorphous carbon. They grow concurrently with their surrounding amorphous carbon matrix, which is not sufficiently etched by the hydrogen-containing plasma. The surrounding amorphous carbon further suppresses the growth of the diamond crystallites. Further exposure to BEN-controlled CVD plasma initiates diamond renucleation on amorphous carbon. Such repeating sequence of diamond encapsulation by surrounding growing amorphous carbon boundaries and diamond renucleation on top of the new amorphous carbon boundaries lead to the formation of nanodiamond/a-C composite [97]. In this respect, composite films consisting of nanodiamond crystallites embedded in an amorphous carbon (a-C) matrix can be achieved by prolonged bias treatment. The structure of the films is determined by the equilibrium between diamond nucleation in the a-C matrix, diamond growth and diamond encapsulation by the a-C matrix. The equilibrium can be manipulated by the CH_4 concentration [25, 26, 96].

Figures 5.3(a)–(c) are the plan view TEM micrographs of the composite films fabricated from 1, 2 and 5% methane under prolonged bias treatment showing nanodiamond crystallites embedded in an amorphous carbon matrix. The inset in Fig. 5.3(a) is the selected area diffraction pattern showing formation of randomly oriented diamond cystallites in amorphous carbon lattice. The EELS spectrum also reveals the presence of amorphous carbon and diamond phases [Fig. 5.3(b)]. The dispersion of diamond crystallites in amorphous carbon matrix can also be readily seen in the cross-sectional TEM micrograph (Fig. 5.4). The sizes of diamond crystallites increase from 6.3 to 11.3 nm with increasing methane content from 1 to 5% on the basis of XRD analysis of (111) peak broadening. The dimensions of diamond crystallites determined from XRD analysis agrees reasonably with those from high resolution TEM images as shown in Figs. 5.5(A)–(C).

6. Mechanical Behavior of NCD Films

In the past decade, extensive efforts have been directed towards the search of nanocomposite coatings of superhardness for wear resistant applications. Typical example is the $M_nN/\alpha\text{-}Si_3N_4$ nanocomposite coatings prepared via plasma induced CVD method [98–101]. M_n represents transition metals like Ti, W, V and Zr. Such coatings consist of the transition metal nitride nanocrystallites with grain sizes in the nanometer range ($\sim$4–6 nm) embedded into <1 nm thin matrix of amorphous Si_3N_4 matrix. In nanocomposite coatings, the transition metal nitride phase is sufficiently hard to bear the load while the amorphous nitride provides structural flexibility. Veprek *et al.* reported that the hardness of these plasma CVD nanocomposite coatings could reach the diamond hardness (70–80 GPa) when the crystallite size approaches about 2 nm [98]. However, the superhardness of the coatings is offset by their low toughness, leading to severe delamination upon application of an external load. Thus, the toughness of the coatings is another critical factor that must be considered during the design of nanocomposite coatings. Apart from high hardness, Voevodin *et al.* demonstrated that toughness of the surface coatings play an important role in enhancing their tribological performances [102].

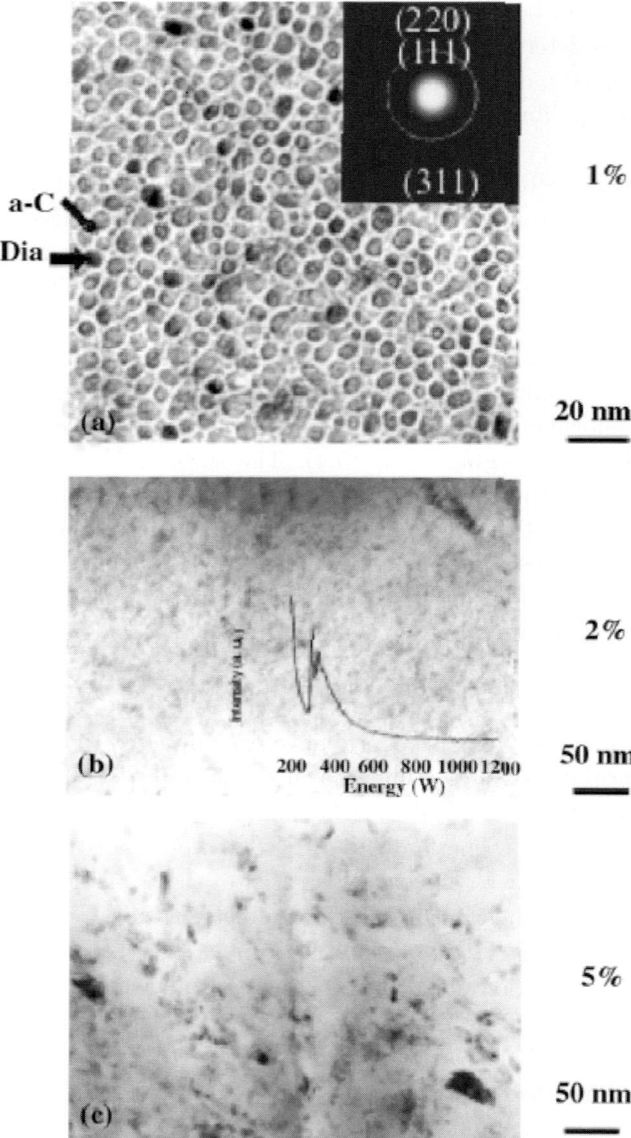

Fig. 5.3.   Plan view TEM micrographs of the composite films prepared using (a) 1%, (b) 2% and (c) 5% methane content under prolonged bias-enhanced HFCVD condition. Insets show the corresponding electron diffraction pattern and electron energy loss spectrum [26].

Reprinted with permission of Elsevier.

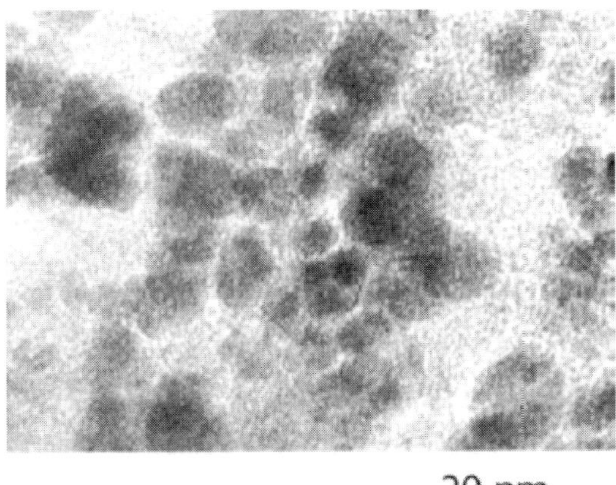

Fig. 5.4.   Cross-section TEM micrograph of the composite film prepared from 1%CH$_4$:
99%H$_2$ [26].
Reprinted with permission of Elsevier.

Because of its chemical inertness, high hardness and fracture strength
and low coefficient of friction, diamond is particularly suited for tribological
applications as biomedical implants, MEMS and microelectronic devices.
In this respect, the wear resistance of diamond coatings is crucial to the
integrity and reliability of MEMS and microelectronic systems. However,
CVD diamond coatings generally exhibit rough surface and large grain size,
resulting in poor tribological performances. This issue can be resolved by
reducing the grain size of diamond coatings from micro- to nanometer scale.

It is well recognized that the mechanical properties of nanocrystalline
materials differ substantially from their microcrystalline counterparts. This
is due to the presence of large volume fraction of atoms at the grain boundaries. Moreover, impurities and other defects derived from the synthesis and
processing also influence the mechanical properties of nanomaterials significantly. For NCD coatings, several parameters such as processing defects,
density and bonding nature of carbon (sp^2 or sp^3) could affect the mechanical properties of diamond coatings as their grain sizes approach the nanometer regime.

The hardness and Young's modulus of CVD diamond are known to vary
considerably with sp^3/sp^2 bonding ratio. Diamond films with higher sp^3-
bonded C yield higher values of stiffness and hardness. Recently, Philip
et al. reported that the NCD films grown on silicon substrate by MWCVD
with high nucleation density in excess of 10^{12} cm^{-2} exhibit Young's modu-

 S. C. Tjong

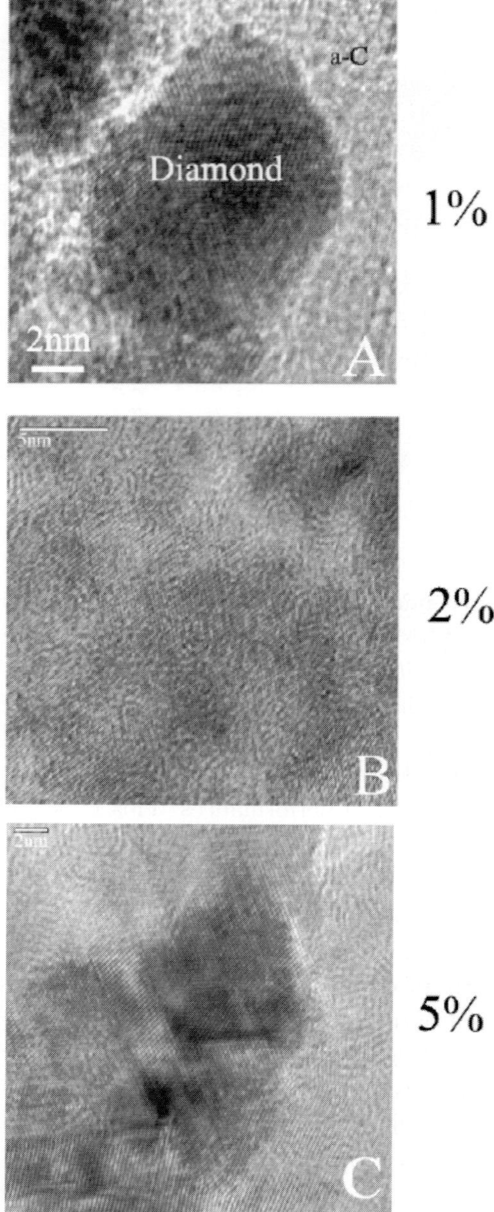

Fig. 5.5. Plan view HRTEM images of the composite films prepared using (A) 1%, (B) 2% and (C) 5% methane content [26].
Reprinted with permission of Elsevier.

lus of 1120 GPa. This value is comparable to that estimated theoretically for ideal polycrystalline diamond. However, the Young's modulus of NCD films grown with lower nucleation densities ($\leq 10^{10}\,cm^{-2}$) is just about half of that with high nucleation density. A higher nucleation density during growth can result in faster film coalescence, lower volume of voids and sp^2-bonded carbon in the films [103]. In this respect, NCD/amorphous carbon nanocomposite films grown by the CVD CH_4/N_2 plasma exhibits much lower hardness ($\sim 40\,GPa$), Young's modulus ($\sim 387\,GPa$) and density ($2.75\,g\,cm^{-3}$) due to the presence of the amorphous carbon matrix. The amorphous carbon matrix is considered to be beneficial to prevent fast brittle failure by improving the toughness of the coating. Moreover, the NCD/a-C nanocomposite film is found to exhibit a strong adhesion on the basis of scratch test [83].

Gruen and coworkers indicated that UNCD films exhibit a fracture strength of $\sim 4.13\,GPa$, which is much higher than those of polysilicon ($\sim 1.5\,GPa$) and SiC ($\sim 1.2\,GPa$) [35]. Therefore, UNCD films with high hardness and fracture strength, low coefficient of friction, smooth surface as well as ease of rapid growth to a thickness of 2 μm show potential applications for MEMS devices [104]. The wear rates of UNCD coating against a Si_3N_4 ball in air and dry nitrogen are two orders of magnitude lower than that of microcrystalline diamond coating rubbing against a Si_3N_4 ball as shown in Fig. 6.1. The smooth UNCN coating renders it to exhibit low coefficient of friction [71, 104].

7. Field Emission Characteristics

Many carbon-based materials such as carbon nanotubes (CNTs) and diamond films show superior electron emission properties at a low threshold field and a high current density owing to the very low or negative electron affinity characteristics [105–107]. They are particularly suitable for electron field emission and display device applications. The electron field emission depends on the work function and the morphology of emitter materials, e.g. sharp edges or tips of surfaces. A good field emitter should exhibit low threshold electric field strength, high electron emission current and good emission stability. A threshold field for 1 μA cm^{-2} is commonly used to rank the emission efficiency of emitter materials. CNTs are generally known to exhibit better electron field emission than diamond films. However, the processability of CNTs on the substrates for fabricating electronic devices is rather poor because of the complication in the synthesis process such as

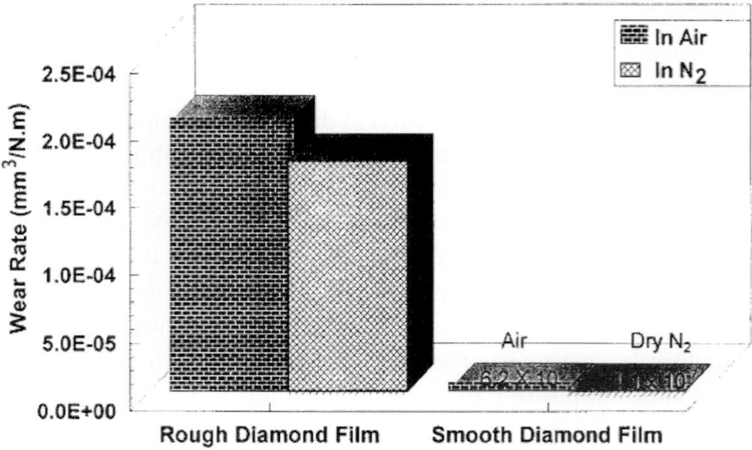

Fig. 6.1.   Wear rates of Si_3N_4 ball slid against smooth UNCD and rough microcrystalline diamond coatings in air and dry nitrogen. Ball-on-disk test conditions: load, 2 N; velocity, 0.05 ms^{-1}; sliding distance, 40 m; ball diameter, 9.55 mm; relative humidity, 37% [104]. Reprinted with permission of Elsevier.

the use of metallic catalysts. In this regard, there has been wide interest in enhancing the electron field emission properties of diamond films by doping with nitrogen or boron [108, 109] or by reducing their grain sizes to nanometer level. The high electron emission capabilities of the B-doped diamond films are derived from the increase in volume fraction of the conductive regions in the film and from high density of emission sites on the film surface [109]. For NCD films, large volume fractions of grain boundaries act as the conducting path for electrons. Therefore, the grain boundaries containing sp^2-bonded C provide effective sites for electron emission. Gruen and coworkers reported that the n-type electrical conductivity of the UNCD films increases by five orders of magnitude (up to 143 Ω^{-1}cm^{-1}) by doping with nitrogen. They proposed that grain boundary conduction involving π-bonded carbon atoms in the grain boundaries is responsible for the high electrical conductivity in these films [81]. Similarly, Ma *et al.* reported that the introduction of nitrogen into the CH_4/H_2 gas mixture leads to a drastic reduction of the resistivity by six orders of magnitude from 10^{11} to 10^5 Ωcm. They attributed this to an increase of the sp^3/sp^2 ratio of carbon bonds and to a reduction in grain sizes associated with the nitrogen doping [110].

Figure 7.1 shows typical field emission characteristics of the MPCVD microcrystalline diamond and NCD films prepared from CH_4/H_2 gas mix-

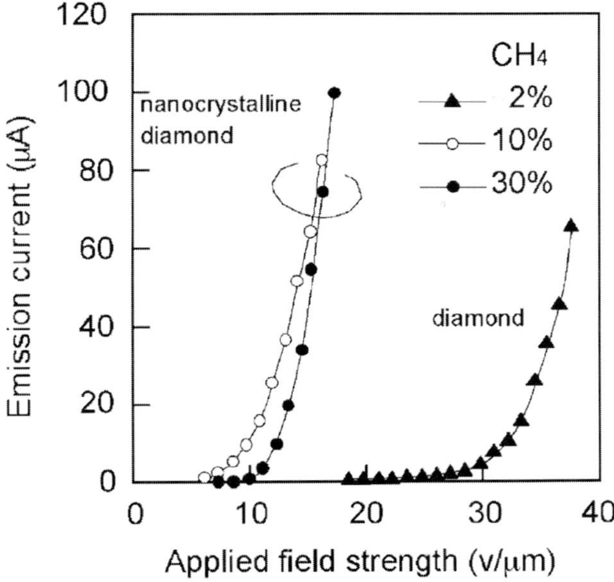

Fig. 7.1.　Field emission characteristics of MPCVD diamond and NCD films [111]. Reprinted with permission of Elsevier.

tures. Different CH_4 concentrations were used to prepare the films. The average grain size was ~ 50 nm for the film grown at 10% CH_4 [111]. Apparently, microcrystalline diamond films exhibit a large emission field strength, i.e. >35 V μm^{-1}. In contrast, the emission of the NCD films is shifted to low field strength region. The threshold field for the emission current of 1 μA is 6 V μm^{-1}. Generally, the electron field emission characteristics of NCD composite films depend mainly on the grain size of the diamond and amorphous carbon content in the films. Low emission threshold of 1 V μm^{-1} and an emission current density of 10 mA cm^{-2} can be attained in the NCD/a-C composite films prepared from the CH_4/N_2 plasma under appropriate growth conditions [85]. NCD films with finer grain sizes and higher amorphous carbon content can supply more electron emission tunnels, thereby enhancing the electron field emission [112]. From these, it appears that the threshold electric field of NCD/a-C composite film is lower than that of boron-doped [109] and undoped microcrystalline diamond films [111], in which electron emission requires a field of ~ 8–40 V μm^{-1}.

The electron field emission behavior of the materials under tunneling mechanism is generally analyzed by the Fowler–Nordheim (F–N) equation.

S. C. Tjong

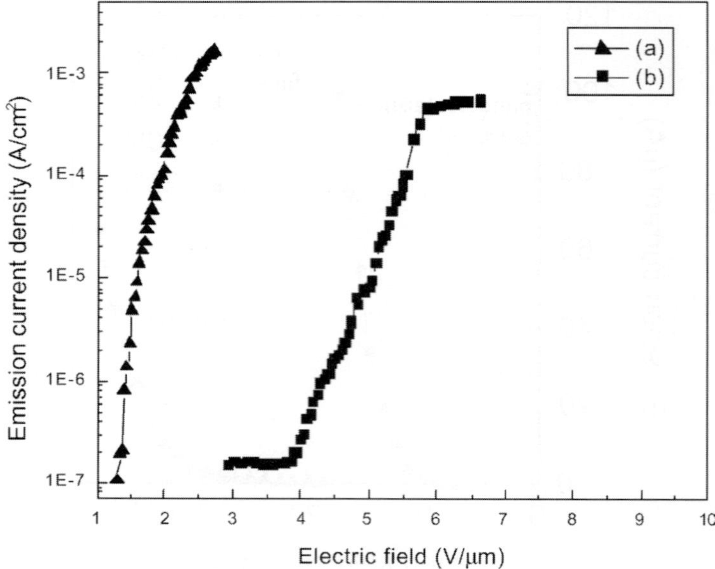

Fig. 7.2. Field emission characteristics of (a) CNT and (b) NCD samples [113]. Reprinted with permission of Elsevier.

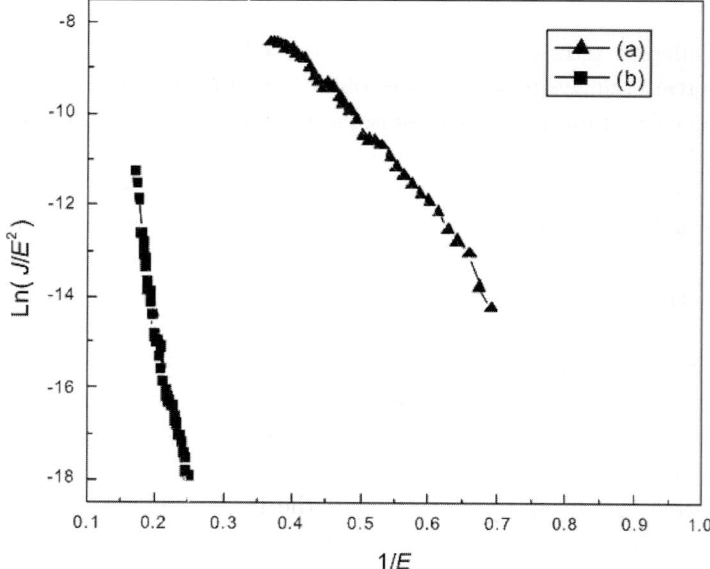

Fig. 7.3. F–N plots for (a) CNT and (b) NCD samples [113]. Reprinted with permission of Elsevier.

In this equation, the field-emission current density, J, can be expressed as a function of the applied electric field, E, the local work function of the emitter tip, ϕ, and a field-enhancing factor, β. Mathematically, F–N equation can be expressed as:

$$J = \frac{A\beta^2 E^2}{\phi} \exp\left(-\frac{B\phi^{3/2}}{\beta E}\right), \qquad (7.1)$$

where A and B are constants. For a field-emission phenomenon, the plot of $\ln(J/E^2)$ versus (E^{-1}) should yield a straight line. The slope $(-B\phi^{3/2}/\beta)$ of F–N plot is related to the work function of the emitter. Figures 7.2 and 7.3 show the field emission behavior of NCD film with grain sizes of 10–15 nm prepared from the CH_4/Ar plasma and its corresponding F–N plot, respectively [113]. It is apparent that the field emission in NCD film can be well described by the F–N plot. The threshold electric field, at which the current density is 1 μA cm^{-2}, is 3.8 V μm^{-1}. The emission current density approaches 540 μA cm^{-2} at an applied field strength of 6.65 V μm^{-1}. For the purpose of comparison, the field emission characteristics of CNT are also plotted in Figs. 7.2 and 7.3. The CNT exhibits a lower threshold electric field strength of 1.3 V μm^{-1} and a much higher emission current density of 1620 μA cm^{-2} at 2.72 V μm^{-1}. However, Wang *et al.* indicated that the NCD film exhibits higher emission stability than the CNT [113].

As discussed above, the addition of nitrogen into CH_4/H_2 plasma is beneficial in reducing the grain sizes and resistivity of UNCD films [81]. Similar beneficial effect in electron field emission was observed by adding nitrogen to CH_4/Ar plasma during the deposition of UNCD film [37]. Figure 7.4(a) shows the plot of onset field for 1×10^{-7} μA versus percent nitrogen in the plasma. A representative field emission current plot against applied field for 2% nitrogen doped UNCD film is shown in Fig. 7.4(b). Apparently, the nitrogen incorporation reduces the onset field from 23 V μm^{-1} for the nitrogen-free films to 5 V μm^{-1} or less for the nitrogen-containing films. This is due to the preferential entering of nitrogen into the grain boundaries, thereby promoting sp^2 bonding in the neighboring atoms. HRTEM images reveal that the width of grain boundaries increases considerably with the addition of 20% N_2 in the plasma [81]. The increase in the sp^2 content is beneficial to enhance the field emission of the UNCD films [37].

We now consider the field emission behavior of boron-doped and undoped NCD films prepared from BEN-controlled, MPCVD-produced CH_4/H_2 plasma [114]. Figure 7.5 shows the effect of negative bias volt-

S. C. Tjong

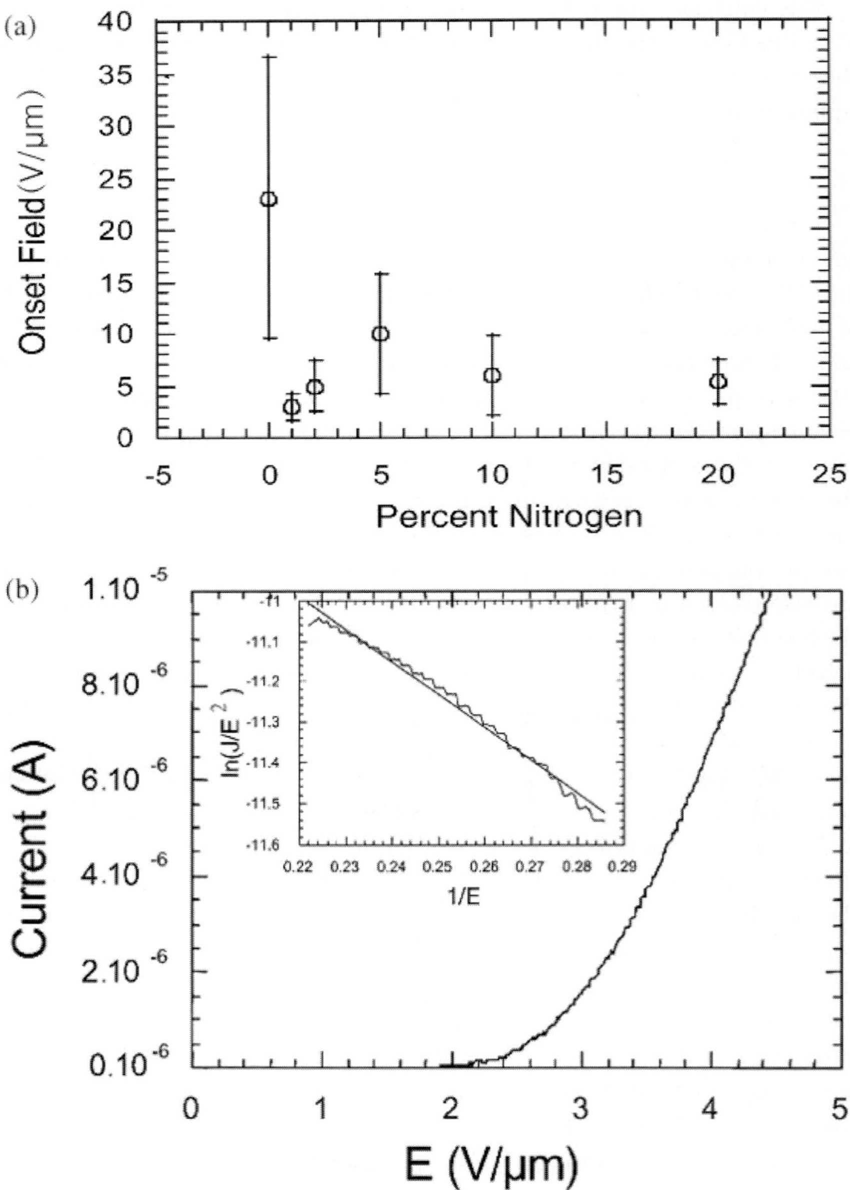

Fig. 7.4.   (a) Onset field versus percent nitrogen in the CH_4/Ar plasma, and (b) current–voltage and F–N plots of a 2% nitrogen containing UNCD film [37].
Reprinted with permission of Elsevier.

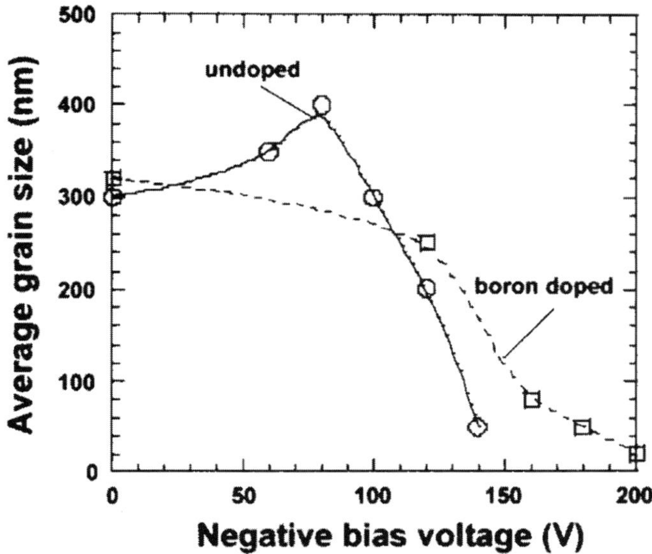

Fig. 7.5. Effect of applied negative bias voltage on the grain size of diamond films synthesized from MPCVD in CH_4/H_2 plasma [114].
Reprinted with permission of The American Institute of Physics.

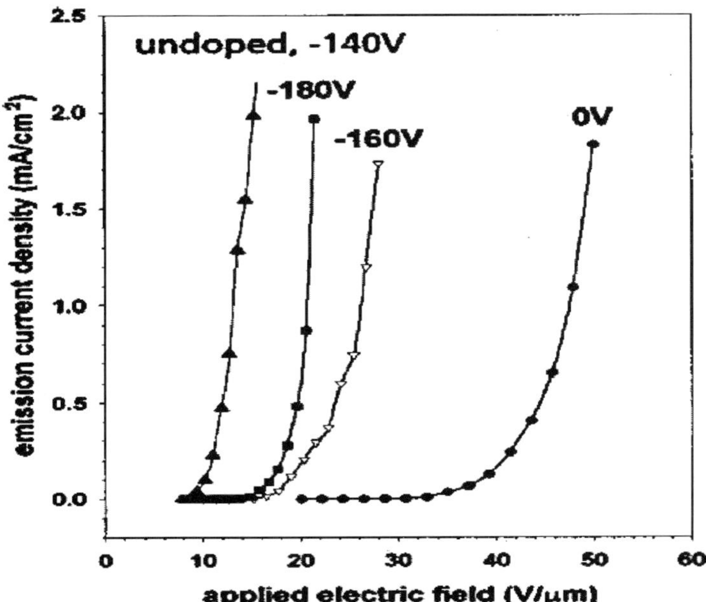

Fig. 7.6. Effect of applied negative bias voltage on the field emission characteristics of boron-doped and undoped diamond films [114].
Reprinted with permission of The American Institute of Physics.

　　　　　　　　　　　S. C. Tjong

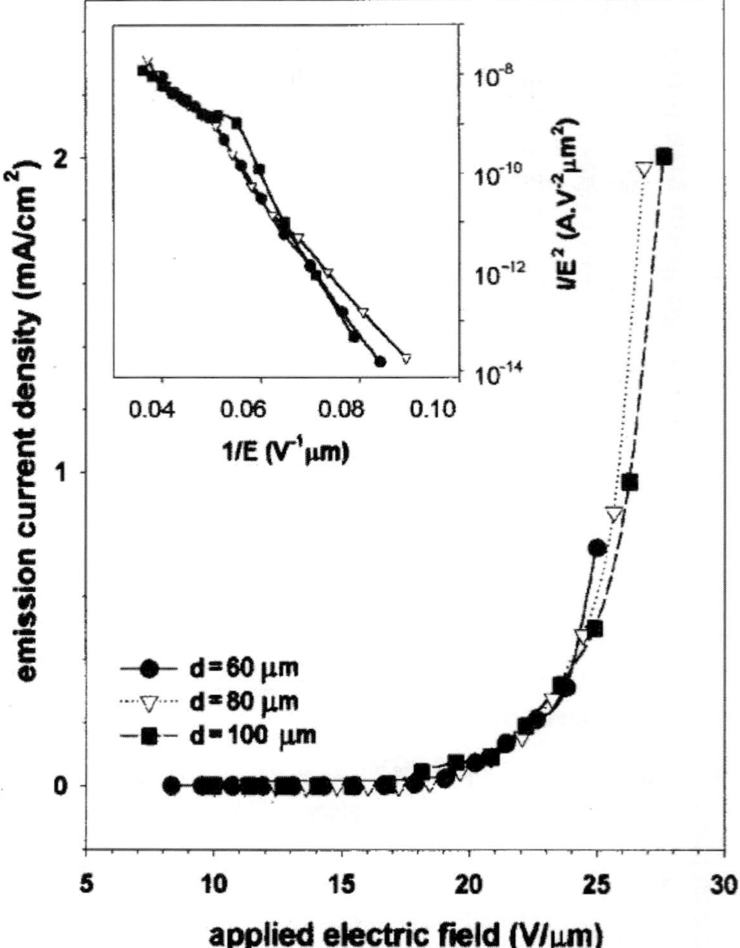

Fig. 7.7.　Field emission characteristics of boron-doped NCD film samples (bias voltage −180 V) under different anode–sample distance (d) values [114].
Reprinted with permission of The American Institute of Physics.

age on the grain sizes of synthetic diamond films. Apparently, a grain size of ∼50 nm can be attained for both the boron-doped and undoped diamond films under bias voltages of −180 and −140 V, respectively. The field emission characteristics of these thin film specimens are depicted in Fig. 7.6. It can be seen that the boron doped (bias voltage −180 V) and undoped (bias voltage −140 V) NCD films exhibit much lower turn-on-field, which is defined as the field required to emit a current of 0.01 $\mathrm{mA\,cm^{-2}}$ among all the thin films investigated. The undoped film (bias voltage $-140\,$V) exhibits lower turn-on voltage ($8\,$V$/\mu$m) than that of boron doped (bias voltage $-180\,$V) NCD film, i.e. $14\,$V$/\mu$m. Although boron doping enhances the p-type conductivity of NCD films, it shows no positive influence on the electron emission. This is in sharp contrast to the microcrystalline diamond films in which boron doping is beneficial for enhancing the conductivity and electron emission of the films [109]. Figure 7.7 shows the emission characteristics of boron doped (bias voltage $-180\,$V) NCD films at different anode–sample distances. The corresponding F–N plots of these samples are shown in the inset of this figure. The linearity observed in the plots of the $\log\left(1/E^2\right)$ versus $1/E$ indicates that these film samples follow the F–N field emission theory.

8. Conclusions

This chapter presents a comprehensive review on the synthesis and properties of nanocrystalline diamond and nanodiamond/a-C composite films prepared via MPCVD technique in different plasma chemistry environments, i.e. hydrogen-deficient and hydrogen-rich plasmas. UNCD films grown from a hydrogen-poor plasma e.g. $CH_4(1\%)$/Ar, or C_{60}/Ar exhibit very fine grain diamond clusters of 3–5 nm. The grain boundaries of UNCD films are near atomically abrupt and constitute about 10% carbon atoms that are π-bonded. The C_2 dimer is considered to be the growth species in hydrogen deficient plasma. The grain boundary width of UNCD films can be increased substantially by doping with nitrogen as the nitrogen atoms enter grain boundary regions. Therefore, the nitrogen-doped UNCD films possess excellent electron conductivity and field emission characteristics. The unique properties of UNCD films, such as high hardness, strength, electron field emission and surface smoothness renders them potential candidate materials for applications in the MEMS and display devices. Moreover, amorphous carbon content in the films can be increased dramatically via MPCVD-produced CH_4/N_2 plasma, leading to the dispersion of nanocrystalline diamonds in an amorphous carbon matrix. The crystallite phase/amorphous carbon matrix ratio is close to unity. The NCD/amorphous carbon nanocomposite films grown from the CH_4/N_2 plasma exhibit much lower hardness ($\sim 40\,$GPa), Young's modulus ($\sim 387\,$GPa) and density ($2.75\,$g cm^{-3}) due to the presence of the amorphous carbon matrix. The amorphous carbon matrix is considered to be beneficial in preventing fast brittle failure by improving the toughness of the films.

The synthesis of the NCD films prepared via the MWCVD technique under BEN-controlled mode in the hydrogen-rich plasma is well recognized. The CH_3 species is responsible for the CVD diamond growth in the CH_4/H_2 plasma. The atomic hydrogen plays some important roles in the nucleation process, such as abstracting bonded hydrogen and etching off the graphite phase. Continuous ion bombardment results in enhancement of the nucleation density and promotion of secondary nucleation, leading to the formation of nanocrystalline diamond films. The sizes of diamond nanocrystallites ($\sim$40–50 nm) in the NCD films prepared in hydrogen-rich plasma appear to be larger compared to those prepared from hydrogen-deficient plasma ($\sim$3–10 nm). Finally, the NCD/a-C composite films can be synthesized in the CH_4/H_2 plasma under prolonged bias-enhanced HFCVD conditions. By using a double bias-assisted HFCVD system, high nucleation density, heteroepitaxial nucleation and growth of diamond crystallites can be achieved. It appears that the diamond crystallites tend to nucleate preferentially on the surface step sites of the silicon substrates.

References

1. X. Xiao, J. Birrell, J.E. Gerbi, O. Auciello and J.A. Carlisle, *J. Appl. Phys.* **96** (2004) 2232–2239.
2. S.C. Tjong, N.B. Wong, G. Li and S.T. Lee, *Mater. Chem. Phys.* **62** (2000) 241–246.
3. S.C. Tjong, H.P. Ho and S.T. Lee, *Diamond Relat. Mater.* **10** (2001) 1578–1583.
4. H.P. Ho, K.C. Lu, S.C. Tjong and S.T. Lee, *Diamond Relat. Mater.* **9** (2000) 1312–1319.
5. T. Sharda, M.M. Rahaman, Y. Nukuya, T. Soga, T. Jimo and M. Umeno, *Diamond Relat. Mater.* **10** (2001) 561–7.
6. Y. Tzeng, C. Liu and A. Hirata, *Diamond Relat. Mater.* **12** (2003) 456–463.
7. M. Amaral, E. Salgueiredo, F.J. Oliveira, A.J. Fernandes, F.M. Osta and R.F. Silva, *Surf. Coat. Technol.* **200** (2006) 6409–6413.
8. M.D. Fries and Y.K. Vohra, *Diamond Relat. Mater.* **13** (2004) 1740–1743.
9. D.M. Gruen, *Annu. Rev. Mater. Sci.* **29** (1999) 211–259.
10. G.F. Zhang and V. Buck, *Surf. Coat Technol.* **132** (2000) 256–61.
11. Y. Mokuno, A. Chayahara, Y. Soda, Y. Horino and N. Fujimori, *Diamond Relat. Mater.* **14** (2005) 1743–1746.
12. T. Teraji, M. Hamada, H. Wada, M. Yamamoto and T. Ito, *Diamond Relat. Mater.* **14** (2005) 1747–1752.
13. M. Yoshikawa, H. Ishida, A. Ishitani, S. Koizumi and T. Inuzuka, *Appl. Phys. Lett.* **58** (1991) 1387–1388.
14. S. Koizumi, T. Murakami, T. Inuzuka and K. Suzuki, *Appl. Phys. Lett.* **57** (1990) 563–565.

15. J.C. Arnault, F. Vonau, M. Mermoux, F. Wyczisk and P. Legagneux, *Diamond Relat. Mater.* **13** (2004) 401–403.

16. K.H. Chen, D.M. Bhusari, J.R. Yang, S.T. Lin, T.Y. Wang and L.C. Chen, *Thin Solid Films* **332** (1998) 34–39.

17. T. Grogler, A. Franz, D. Klaffke, S.M. Rosiwall and R.F. Singer, *Diamond Relat. Mater.* **7** (1998) 1342–1347.

18. M.I. Barros, L. Vandenbulcke and J.J. Blechet, *Wear* **249** (2001) 67–78.

19. S.C. Tjong and H. Chen, *Mater. Sci. Eng. R.* **45** (2004) 1–88.

20. K. Xu and H. Tan, in *High-Pressure Shock Compression of Solids, Vol. 5*, eds. L. Davidson, Y. Horie and T. Sekine (Springer, New York, 2002), p. 139.

21. G.N. Yushin, S. Osswald, V.I. Padalko, G.P. Bogatyreva and Y. Gogotsi, *Diamond Relat. Mater.* **14** (2005) 1721–1729.

22. N. Dubrovinskaia, L. Dubrovinsky, F. Langenhorst, S. Jacobsen and C. Liebske, *Diamond Relat. Mater.* **14** (2005) 16–22.

23. D.M. Gruen, *M.R.S. Bull* **23** (1998) 32–35.

24. S.T. Lee, H.Y. Pang, X.T. Zhou, N. Wang, C.S. Lee, I. Bello and Y. Lifshitz, *Science* **287** (2000) 104–106.

25. X.T. Zhou, Q. Li, F.Y. Meng, I. Bello, C.S. Lee, S.T. Lee and Y. Lifshitz, *Appl. Phys. Lett.* **80** (2002) 3307–3309.

26. X.T. Zhou, X.M. Meng, F.Y. Meng, Q. Li, I. Bello, W.J. Zhang, C.S. Lee, S.T. Lee and Y. Lifshitz, *Diamond Relat. Mater.* **12** (2003) 1640–1646.

27. H.G. Chen and L. Chang, *Diamond Relat. Mater.* **14** (2005) 183–191.

28. H. Liu and D.S. Dandy, *Diamond Relat. Mater.* **4** (1995) 1173–1188.

29. M.M. Garcia, I. Jimenez, O. Sanchez, C. Gomez-Alexandre and L. Vazquez, *Phys. Rev. B* **61** (2000) 10383–10387.

30. L.C. Qin, D. Zhou, A.R. Kraus and D.M. Gruen, *Nanostruct. Mater.* **10** (1998) 649–660.

31. D. Zhou, T.G. McGauley, L.C. Qin, A.R. Krauss and D.M. Gruen, *J. Appl. Phys.* **83** (1998) 540–543.

32. J. Birrell, J.A. Carlisle, O. Auciello, D.M. Gruen and J.M. Gibson, *Appl. Phys. Lett.* **81** (2002) 2235–2237.

33. T.G. McCauley, D.M. Gruen and A.R. Krauss, *Appl. Phys. Lett.* **73** (1998) 1646–1648.

34. D.M. Gruen, S. Liu, A.R. Krauss and X. Pan, *J. Appl. Phys.* **75** (1994) 1758–1763.

35. D.M. Gruen, S. Liu, A.R. Krauss, J. Luo and X. Pan, *Appl. Phys. Lett.* **64** (1994) 1502–1504.

36. H.D. Espinosa, B. Peng, B.C. Prorok, N. Moldovan, O. Auciello, J.A. Carlisle, D.M. Gruen and D.C. Mancini, *J. Appl. Phys.* **94** (2003) 6076–6084.

37. T.D. Corrigan, D.M. Gruen, A.R. Krauss, P. Zapol and R.P.H. Chang, *Diamond Relat. Mater.* **11** (2002) 43–48.

38. M.L. Hitchman and K.F. Jensen, *Chemical Vapor Deposition-Principles and Applications* (Academic Press, London, 1993).

39. A. Sherman, *Chemical Vapor Deposition For Microelectronics-Principles, Technology and Applications* (Noyes Publications, New Jersey, 1987).

40. L.S. Pan and D.R. Kania, *Diamond: Electronic Properties and Applications* (Kluwer Academic Publishers, London, 1995).

41. T. Tachibana, Y. Ando, A. Watanabe, Y. Nishibayashi, K. Kobashi, T. Hirao and K. Oura, *Diamond Relat. Mater.* **10** (2001) 1569–1572.

42. W. Ahmed, H. Sein, N. Ali, J. Gracio and R. Woodwards, *Diamond Relat. Mater.* **12** (2003) 1300–1306.

43. S. Matsumoto, Y. Sato, M. Kamo and N. Setaka, *Jpn. J. Appl. Phys.* **21** (1982) L183–L185.

44. N. Ali, W. Ahmed, Q.H. Fan, I.U. Hassan and C.A. Rego, *Thin Solid Films* **377–378** (2000) 193–197.

45. S. Yugo, T. Kimura and T. Kania, *Diamond Relat. Mater.* **2** (1993) 328–332.

46. S. Yugo, T. Kanai, T. Kimura and T. Muto, *Appl. Phys. Lett.* **58** (1991) 1036–1038.

47. M. Yoshimoto, K. Yoshida, H. Maruta, Y. Hishitani and H. Koinum, *Nature* **399** (1999) 340–343.

48. A.V. Palnichenko, A.M. Jonas, J.C. Charlier, A.S. Aronin and J.P. Issi, *Nature* **402** (1999) 162–165.

49. J.E. Field, *The Properties of Diamond* (Academic Press, London, 1979).

50. K. Ishibashi and S. Furukawa, *Jpn J. Appl. Phys.* **24** (1985) 912–917.

51. S.T. Lee, Y.W. Lam, Z. Lin, Y. Chen and Q. Chen, *Phys. Rev. B* **55** (1997) 15937–15941.

52. S. Iijima, Y. Aikawa and K. Baba, *Appl. Phys. Lett.* **57** (1990) 2646–2648.

53. J.S. Ma, H. Kawarada, T. Yonehara, J.I. Suzuki, J. Wei, Y. Tokota and A. Hiraki, *Appl. Phys. Lett.* **55** (1989) 1071–1073.

54. A. Sawabe and T. Inuzaka, *Thin Solid Films* **137** (1986) 89–99.

55. C.Z. Gu and X. Jiang, *J. Appl. Phys.* **88** (2000) 1788–1793.

56. B.R. Stoner, G.H.M. Ma, S.D. Wolter and J.T. Glass, *Phys. Rev. B* **45** (1992) 11067–11084.

57. B.R. Stoner and J.T. Glass, *Appl. Phys. Lett.* **60** (1992) 698–700.

58. S.D. Wolter, J.T. Glass and B.R. Stoner, *J. Appl. Phys.* **77** (1995) 5119–5124.

59. S.D. Wolter, B.R. Stoner, J.T. Glass, P.J. Ellis, D.S. Buhaenko, G.E. Jenkins and P. Southworth, *Appl. Phys. Lett.* **62** (1993) 1215–1217.

60. S.D. Wolter, D. Borca-Tasciuc, G. Chen, J.T. Prater and Z. Sitar, *Thin Solid Films* **469–470** (2004) 105–111.

61. Q. Chen and J. Lin, *J. Appl. Phys.* **80** (1996) 797–802.

62. Y. Kouzuma, K. Teii, S. Mizobe, K. Uchino and K. Muraoka, *Diamond Relat. Mater.* **13** (2004) 656–660.

63. I.H. Choi, P. Weisbecker, S. Barrat and E. Bauer-Grosse, *Diamond Relat. Mater.* **13** (2004) 574–80.

64. C. Wild, N. Herres and P. Koidl, *J. Appl. Phys.* **68** (1990) 973–978.

65. C. Wild, P. Koidl, W. Muller-Seber, H. Walcher, R. Kohl, N. Herres, R. Locher, R. Samleski and R. Brenn, *Diamond Relat. Mater.* **2** (1993) 158–168.

66. K.L. Chuang, L. Chang and C.A. Lu, *Mater. Chem. Phys.* **72** (2001) 176–80.

67. J.J. Wu, S.H. Yeh, C.T. Su and F.C.N. Hong, *Appl. Phys. Lett.* **68** (1996) 3254–3256.

68. J.J. Wu and F.N.C. Hong, *J. Mater. Res.* **13** (1998) 2498–2504.

69. C.H. Ku and J.J. Wu, *Carbon* **42** (2004) 2201–2205.

70. A.R. Krauss, O. Auciello, D.M. Gruen, A. Jayatissa, A. Sumant, J. Tucek, D.C. Mancini, N. Moldovan, A. Erdemir, D. Ersoy, M.N. Gardos, H.G. Busman, E.M. Meyer and M.Q. Ding, *Diamond Relat. Mater.* **10** (2001) 1952–1961.

71. S. Jiao, A. Sumant, M.A. Kirk, D.M. Gruen, A.R. Krauss and O. Auciello, *J. Appl. Phys.* **90** (2001) 118–22.

72. J. Birrell, J.E. Gerbi, O. Auciello, J.M. Gibson, J. Johnson and J.A. Carlisle, *Diamond Relat. Mater.* **14** (2005) 86–92.

73. A.N. Goyette, J.E. Lawler, L.W. Anderson, D.M. Gruen, T.M. McCauley, D. Zhou and A.R. Krauss, *J. Phys. D: Appl. Phys.* **31** (1998) 1975–1986.

74. J. Robertson, *Mater. Sci. Eng. R* **37** (2002) 129–281.

75. N. Wada, P.J. Gaczi and S.A. Solin, *J. Non-Cryst. Solids* **35–36** (1980) 543–548.

76. A.C. Ferrari and J. Robertson, *Phys. Rev.* **B61** (2000) 14095–14107.

77. A.C. Ferrari and J. Robertson, *Phys. Rev.* **B63** (2002) 121405-1–121405-4.

78. J. Fink, T. Muller-Heinzerling, J. Pfluger, A. Bubenzer, P. Koidl and G. Crecelius, *Solid State Commun.* **47** (1983) 687–691.

79. P. Pfeiffer, H. Kuzmany, P. Knoll, S. Bokova, N. Salk and B. Gunther, *Diamond Relat. Mater.* **12** (2003) 268–271.

80. T. Lin, Y. Yu, A.T. Wee, Z.X. Shen and K.P. Loh, *Appl. Phys. Lett.* **77** (2000) 2692–2694.

81. S. Bhattacharyya, O. Auciello, J. Birrell, J.A. Calisle, L.A. Curtiss, A.N. Goyette, D.M. Gruen, A.R. Krauss, J. Schlueter, A. Sumant and P. Zapol, *Appl. Phys. Lett.* **79** (2001) 1441–1443.

82. C. Popov, W. Kulisch, P.N. Gibson, G. Ceccone and M. Jelinek, *Diamond Relat. Mater.* **13** (2004) 1371–1376.

83. C. Popov, W. Kulisch, S. Boycheva, K. Yamamoto, G. Ceccone and Y. Koga, *Diamond Relat. Mater.* **13** (2004) 2071–2075.

84. W. Kulisch, C. Popov, S. Boycheva, L. Buforn, G. Favaro and N. Conte, *Diamond Relat. Mater.* **13** (2004) 1997–2002.

85. K. Wu, E.G. Wang, Z.X. Cao, Z.L. Wang and X. Jiang, *J. Appl. Phys.* **88** (2000) 2967–2974.

86. J. Angus, P. Koidle and S. Domitz, in *Plasma Deposited Film*, eds. J. Mort and F. Jansen (CRC Press, Boca Raton, Florida, 1986).

87. Y. Lifshitz, S.R. Kasai and J.W. Rabalais, *Phys. Rev. Lett.* **62** (1989) 1290–1293.

88. Y. Lifshitz, S.R. Kasai and J.W. Rabalais, *Phys. Rev.* **B41** (1990) 10468–10480.

89. D.R. McKenzie, D. Muller and B.A. Pailthorpe, *Phys. Rev. Lett.* **67** (1991) 773–776.

90. J. Robertson, J. Geber, S. Sattel, M. Weiler, K. Jung and H. Ehrhardt, *Appl. Phys. Lett.* **66** (1995) 3287–3289.

91. J. Geber, J. Robertson, S. Sattel and H. Ehrhardt, *Diamond Relat. Mater.* **5** (1996) 261–265.

92. Y. Yao, M.Y. Liao, Z.G. Wang and S.T. Lee, *Appl. Phys. Lett.* **87** (2005) 0631031–063103-3.

93. M. Stammler, R. Stockel, M. Albrecht and H.P. Strunk, *Diamond Relat. Mater.* **6** (1997) 747–751.

94. R. Stockel, K. Janischowsky, S. Rohmfeld, J. Ristein, M. Hundhausen and L. Ley, *Diamond Relat. Mater.* **5** (1996) 321–325.

95. P. Bachmann and W. von Enckevort, *Diamond Relat. Mater.* **1** (1992) 1021–1034.

96. X.T. Zhou, H.L. Lai, H.Y. Peng, C. Sun, W.J. Zhang, N. Wang, I. Bello, C.S. Lee and S.T. Lee, *Diamond Relat. Mater.* **9** (2000) 134–139.

97. Y. Lifshitz, Th. Kohler, Th. Frauenheim, I. Guzmann, A. Hoffman, R.Q. Zhang, X.T. Zhou and S.T. Lee, *Science* **297** (2002) 1531–1533.

98. S. Veprek, P. Nesladek, A. Niederhofer, F. Glatz, M. Jilek and M. Sima, *Surf. Coat. Technol.* **108-109** (1998) 138–147.

99. S. Veprek, Nesladek, P., Niederhofer, A. and F. Glatz, *Nanostruct. Mater.* **10** (1998) 679–689.

100. S. Veprek and A.S. Argon, *Surf. Coat. Technol.* **146–147** (2001) 175–182.

101. J. Prochazka, P. Karvankova, G.J. Maritza and S. Veprek, *Mater. Sci. Eng.* **A384** (2004) 102–116.

102. A.A. Voevodin, S.V. Prasad and J.S. Zabinski, *J. Appl. Phys.* **82** (1997) 855–858.

103. J. Philip, P. Hess, T. Feygelson, J.E. Butler, S. Chattopadhyay, K.H. Chen and L.C. Chen, *J. Appl. Phys.* **93** (2003) 2164–2171.

104. A. Erdemir, G.R. Fenske, A.R. Krauss, D.M. Gruen, T. McCauley and R.T. Csencsits, *Surf. Coat Technol.* **120–121** (1999) 565–572.

105. F.J. Himpsel, J.A. Knapp, J.A. Van Vechten and D.E. Eastman, *Phys. Rev.* **B20** (1979) 624–627.

106. V.D. Frolov, A.V. Karabuto, S.M. Pimenov and V.I. Konov, *Diamond Relat. Mater.* **9** (2000) 1196–1200.

107. E. Maillard-Schaller, O.M. Kuettel, L. Diederich, L. Schlapbach and V.V. Zhirnov, *Diamond Relat. Mater.* **8** (1999) 805–808.

108. K. Okano, S. Koizumi, S.R.P. Silva and G.A. Amaratunga, *Nature* **381** (1996) 140–141.

109. M. Itahashi, Y. Umehara, Y. Koide and M. Murakami, *Diamond Relat. Mater.* **10** (2001) 2118–2124.

110. K.L. Ma, W.J. Zhang, Y.S. Zhou, Y.M. Chong, K.M. Leung, I. Bello and S.T. Lee, *Diamond Relat. Mater.* **15** (2006) 626–630.

111. M. Hiramatsu, K. Kato, C.H. Lau, J.S. Foord and M. Hori, *Diamond Relat. Mater.* **12** (2003) 365–368.

112. W. Liu, and C. Gu, *Thin Solid Films* **467** (2004) 4–9.

113. S.G. Wang, Q. Zhang, S.F. Yoon, J. Ahn, D.J. Yang, Q. Wang, Q. Zhou and J.Q. Li, *Diamond Relat. Mater.* **12** (2003) 8–14.

114. X. Jiang, F.C.K. Au and S.T. Lee, *J. Appl. Phys.* **92** (2002) 2880–2883.

SYNTHESIS, CHARACTERIZATION AND APPLICATIONS OF NANOCRYSTALLINE DIAMOND FILMS

Zhenqing Xu and Ashok Kumar*

*Department of Mechanical Engineering
Nanomaterials and Nanomanufacturing Research Center
University of South Florida, Tampa, FL, 33620, USA
akumar@eng.usf.edu

1. Synthesis of Diamond

1.1. *History of Diamond*

Diamond is in an extremely important position for both scientists and the public at large. For ordinary people, the gemstones mean wealth and myth. To the scientists, because of its wide range of extreme properties, diamond is so impressive that research groups all over the world are performing different studies on diamond.

Sir Isaac Newton was the first to propose diamond as an organic material. Later Smithson Tennant discovered that diamond, graphite and coal were the same elements. The structures of these materials were then characterized following the invention of the X-ray and formulation of Bragg's law. It was discovered then that carbon has three different structures: cubic, hexagonal and amorphous [1].

Although initial work in the synthesis of diamond at temperature and pressure was carried out in 1960s [2, 3], worldwide interest in diamond was only triggered years later by a Japanese group that published several methods for diamond deposition on a non-diamond substrate at high rates [4]. Almost 40 years ago, industrial diamond was synthesized commercially using high pressure, high temperature (HPHT) techniques, whereby diamond is crystallized from metal-solvated carbon at pressure $\sim$50–100 kbar and temperature $\sim$1800–2300 K. World interest in diamond has been further increased by the much more recent discovery that it is possible to produce polycrystalline diamond films or coatings by a wide variety of chemical vapor deposition (CVD) techniques using, as process gases, nothing but hydrocarbon gas (typically methane) in an excess of hydrogen. CVD diamond can show mechanical, tribological, and even electronic properties comparable with that of natural diamond. There is currently much optimism in the possibility to scale CVD methods to the extent that they will be able to provide an alternative way to traditional HPHT methods for producing diamond abrasives and heat sinks. Moreover, the possibility of coating large surface areas with a continuous film of diamond will open up a whole new range of potential applications for CVD methods. Thus so with the development of CVD technology, diamond films can be successfully synthesized and used in a variety of fields ranging from electronic to optical and mechanical applications.

1.2. *Structure of Diamond*

Diamonds are exclusively of face-centered diamond cubic structure, as shown in Fig. 1.1. The structure of diamond unit contains eight corner atoms (shown in black), six face-center atoms (shown in light grey), and

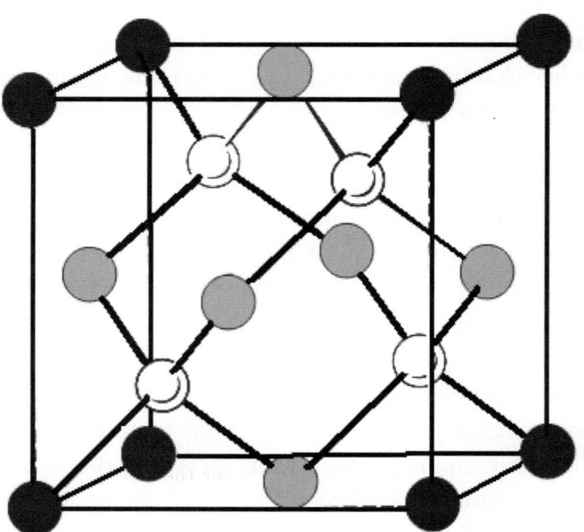

Fig. 1.1.   Face-centered cubic structure of diamond crystal.

four other atoms (shown as empty circles). These four other atoms are from adjacent interpenetrating lattices and are spaced on quarter of a cube diagonal from the former. Every carbon atom is tetrahedrally coordinated to four other carbon atoms via σ bonds emanating from sp^3 hybrid atomic orbital. The four (111) directions in cubic diamond are in the bond directions. In any (111) direction the lattice constant is 3.56 Å and the bond length is 1.54 Å [1].

1.3. *Properties of Diamond*

As Table 1.1 shows, the diamond has many outstanding properties: it is the hardest known material, has the lowest coefficient of thermal expansion (CTE), is chemically inert and wear resistant, offers low friction, has high thermal conductivity, is electrically insulating and optically transparent from the ultraviolet (UV) to the far infrared (IR) range. Given these unique properties, it should be unsurprising that diamond is already used in many diverse applications that include not only electrical but also mechanical and thermal applications. For example, it can be used as a heat sink, as an abrasive, as inserts and/or wear-resistant coatings for cutting tools. Given its many unique properties it is possible to envisage many other potential applications for diamond as an engineering material. However, progress in implementing these ideas has been hampered by the scarcity of natural diamond. Moreover, to synthesize large area diamond thin films with smoothness and uniformity, it becomes the most challenging part for worldwide researchers and scientists to use diamond in industrial applications.

Table 1.1. Some outstanding properties of diamond.

Extreme mechanical hardness (5,700 $\sim$ 10,400 kg/mm^2).
Highest known value of thermal conductivity at room temperature (2 $\times$ 10^3 W/m/K).
High bulk modulus (1.2 $\times$ 10^{12} N/m^2), low compressibility (8.3 $\times$ 10^{-13} m^2/N).
Low thermal expansion coefficient at room temperature (0.8 $\times$ 10^{-6} K).
Broad optical transparency from the deep UV to the far IR region.
Excellent electrical insulator (room temperature resistivity is $\sim$ 10^{16} Ω cm).
Diamond can be doped to change its resistivity over the range 10–10^6 Ω cm.
Wide band gap material (5.5 eV).
Biologically compatible and bioinert.
Very resistant to chemical corrosion and radiation.

1.4. *Chemical Vapor Deposition (CVD)*

As diamond is purely carbon, diamond growth can conceptually be effected by adding one carbon atom at a time to an initial template, so that a tetrahedrally-bonded carbon network (diamond) can be formed. The low cost of the equipment and energy is the main advantage of this method since it can be accomplished at lower pressure. In essence, this is the concept behind the experiments of Derryagin *et al.* [3] and Eversole [5]. In their experiments, thermal depositions of carbon-containing gases at less than 1 atmosphere pressure were used to grow diamond on natural diamond crystals heated to 900°C. The rate of growth in such early CVD experiments was very low. Graphite was also co-deposited with the diamond.

An improvement was made by Angus who demonstrated that atomic hydrogen could etch graphite rather than diamond [2]. Angus was also able to incorporate boron into diamond during growth, giving it semi-conduction properties. Subsequent Russian work extended the possibilities of vapor phase diamond growth by showing that diamond could be grown on non-diamond surfaces [3]. Japanese researchers at the National Institute for Research in Inorganic Materials (NIRIM) were able to bring all these findings together in 1981 by building a "hot filament" reactor in which good quality films of diamond could be grown on non-diamond substrates at significant rates ($\sim$1 μm h^{-1}) [6]. The system operated using a few percent CH_4 in H_2 at 20 Torr (0.026 atm) pressure. Another method was also reported for growing diamond films in a "microwave plasma" reactor in the following years [7]. These discoveries led to worldwide research interest in CVD diamond from the mid 1980s, both in industry and academia, which has continued to present day. Numerous methods for diamond film growth have been developed since, such as DC Plasma, radio frequency (RF) plasma, microwave plasma jet, electron cyclotron resonance (ECR), microwave plasma, and also combustion flame synthesis [1].

A tentative ternary carbon allotropy diagram based on carbon valence bond hybridization is shown in Fig. 1.2 [8]. We consider diamond, graphite, fullerenes, and carbyne as four basic carbon forms. The classification is based on the types of chemical bonds in carbon, including three bonding states corresponding to sp^3 (diamond), sp^2 (graphite), and sp (carbine) hybridization of the atomic orbitals. All other carbon forms are so-called transitional forms, such as diamond-like carbon, fullerenes and nanotubes.

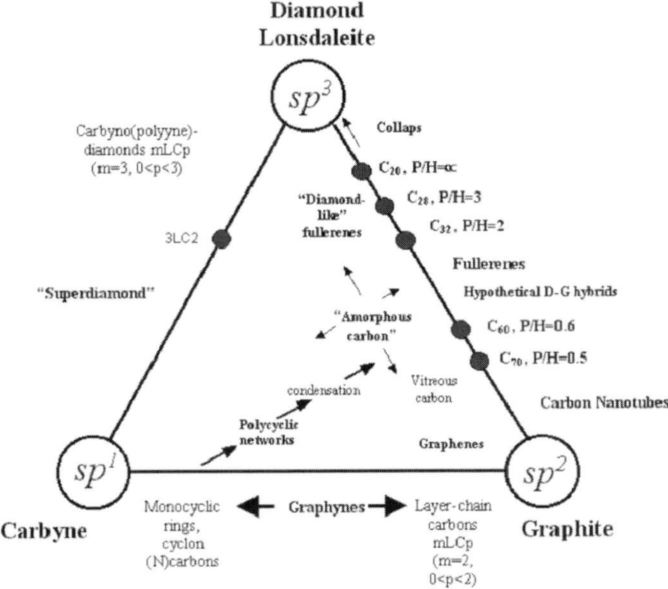

Fig. 1.2. Ternary "phase" diagram of carbon allotropes. P/H corresponds to the ratio of pentagonal/hexagonal rings [8].

The inorganic carbon family consisting of these four members has also been classified by Inagaki [9]. The scheme suggested by Inagaki demonstrates interrelations between organic/inorganic carbon substances at the scale of molecules.

It is widely known that diamond is metastable while graphite is the most stable carbon form at the macroscale. A very interesting transformation between carbon forms at the nanoscale was discovered in the mid-1990s. It was found that nanodiamond particles could be transformed into carbon onions after annealing at around 1300 to 1800 K [10]. Moreover, carbon onions could then be transferred to nanocrystalline diamond under electron irradiation [11]. The phase diagram of carbon has been reconsidered several times and the most recent version is shown in Fig. 1.3. It includes some recently observed phase transitions, such as rapid solid phase graphite to diamond conversion, fast transformation of diamond to graphite, etc [12].

Most of the methods for CVD diamond share the same characteristics [13]:

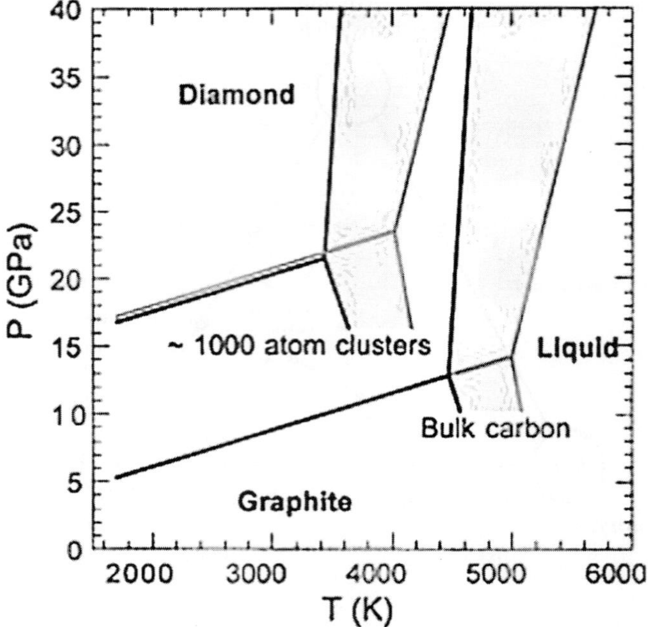

Fig. 1.3. Approximate phase diagram for 1,000 atom carbon clusters. The shadowed region corresponds to estimated uncertainties in location of equilibrium lines derived from available experimental data [12].

- A gas phase must be activated, either by high temperature or by plasma excitation.
- The gas phase must contain carbon-containing species such as hydrocarbon, alcohol, carbon dioxide or carbon monoxide.
- A sufficiently high concentration of a species that etches graphite and suppresses gaseous graphite precursors (e.g. polycyclic aromatic hydrocarbons) must be provided. Atomic hydrogen is most commonly used. However, additional "anti-graphite" species have been proposed, including H_2, OH, O_2, atomic oxygen and F_2.
- The substrate surface must be pretreated to support the nucleation and growth of diamond from the vapor phase. This means that any catalysts that promote formation of graphite should not exit on the substrates surface and the surface should be at or near the solubility limit for carbon so that it can support the precipitation of diamond on the surface, instead of the diffusion of carbon into the substrate.

- A driving force must exist to transport the carbon-containing species form the gas phase to the surface of the substrate. In most CVD methods, the temperature gradient is the driving force for the motion of diamond-producing species via diffusion because the activated gas phase is typically much hotter than the substrate surface.

- The ability to produce a large variety of diamond films and coatings on different materials in either crystalline or vitreous form is the main reason why CVD is the most popular diamond growth method. Several different CVD methods, such as plasma-assisted CVD (MPCVD), hot filament CVD (HFCVD), radio frequency plasma assisted CVD (RFPACVD), wll be introduced and compared in the following text.

Microwave plasmas were first used for diamond synthesis in 1982 at the National Institute for Research in Inorganic Materials (NIRIM), Japan. Because of its simplicity, flexibility and low cost, the microwave plasma-enhanced CVD (MPECVD) system is the most widely used reactor among all diamond deposition approaches. In order to explain the operation of such a reactor a brief explanation of microwave propagation in waveguides will be given as basic knowledge of how to operate the microwave plasma reactor. The whole system, which includes the microwave plasma generator, the waveguide and tuning, the vacuum and the pump, the mass flow system, the substrate heater, the temperature readout system and the exhaust system, will be introduced in the following.

Microwaves are nothing but electromagnetic radiation. So we can use the wave equations and reflective boundary (waveguide) conditions to form an analogue of the particle (wave) in a box problem. The solution to this problem is to provide an integer number of half wavelengths into the box. As a result, a spatially repeating electromagnetic field pattern can produce the waveguide. Hence the electric field strength E and the magnetic field strength B vary periodically along the waveguide. A microwave frequency of 2.45 GHz (used by microwave ovens) is the most commonly used in CVD reactors. However 915 MHz reactors are now becoming more attractive because the lower frequency allows reactor size to be increased.

Unfortunately, in the NIRIM-type reactor, the substrate temperature cannot be independently controlled by changing the plasma parameters because the substrate is immersed in the plasma. To solve this problem, an ASTeX-type reactor was designed by a commercial manufacturer of plasma systems based in the US. Using this reactor, plasma inside the chamber can be generated above the substrate. Hence, the substrate temperature can be controlled independently without the effect of plasma heating on the coating surface.

The ASTeX-type reactor uses a microwave generator and rectangular waveguide, with a mode converter to convert the transverse electric mode in the rectangular waveguide to a transverse magnetic mode in a cylindrical waveguide (the reactor vessel). A moveable antenna maximizes energy coupling between the two waveguide sections. Generation of the plasma occurs by the same mechanisms as discussed above for the NIRIM-type reactor, in the region(s) where the local electric field strength is highest. The ASTeX-type reactor is the most common type of microwave plasma CVD reactor. Of course, in reality, reactors are much more complex than simple cylinders due to the presence of service ports, welds, diagnostic probes, etc. Figure 1.4 shows a schematic drawing of the ASTeX-type MPECVD reactor.

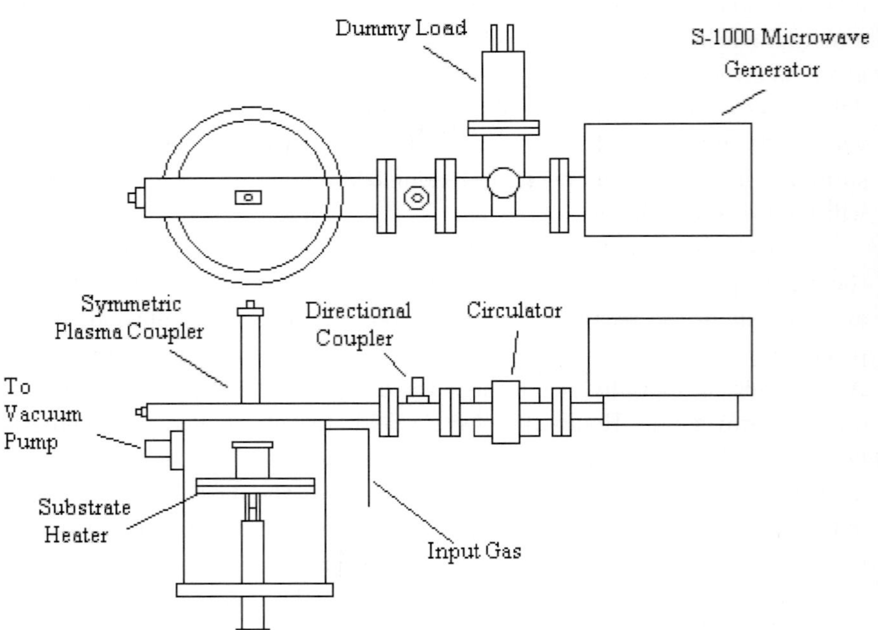

Fig. 1.4.   Schematic drawing of ASTeX-type microwave plasma-enhanced chemical vapor deposition system.

1.5. *Growth Mechanisms of Microcrystalline Diamond (MCD) Films*

The conventional diamond deposition process utilizes temperature and pressure conditions under which graphite is clearly the stable form of carbon, but kinetic factors allow crystalline diamond to be produced by the typical net reaction of

$$CH_4(g) \rightarrow C(\text{diamond}) + 2H_2. \tag{1.1}$$

The reaction mechanism is shown in Fig. 1.5

The chemistry of diamond deposition is complex in comparison to most CVD systems because of competition for deposition among sp^2 and sp^3 types of carbon, and because of the many chemical reactions that can be involved as a result of the complexity of organic systems in comparison to the typical inorganic CVD systems. Instead of only one reaction

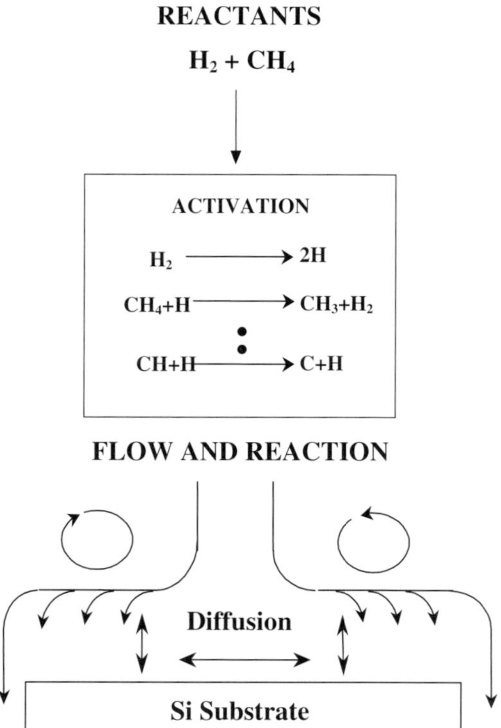

Fig. 1.5.   Schematic reaction mechanisms of conventional diamond deposition [1].

or important precursor species always dominating the deposition, several reactions, whose importance depends on experimental conditions such as temperature, pressure, gas composition, residence time, activation mode, and reactor geometry, happen inside the chamber at the same time [1].

In order to understand diamond growth behavior, general kinetics have to be considered. Regardless of specific growth mechanisms, there appears to be a growing consensus on two general features of diamond growth reaction kinetics. First, the principal element of diamond growth is the formation of an active site — a surface carbon sp^3 radical — followed by the addition of a carbonaceous gaseous species to the surface radical formed. Second, the density of carbon surface radicals is determined primarily by the balance between the abstractions of hydrogen from surface C–H bonds by gaseous hydrogen atoms,

$$C_dH + H \rightarrow C_d\bullet + H_2, \qquad (1.2)$$

and the combination of the carbon surface radicals with free gaseous hydrogen atoms,

$$C_d\bullet + H \rightarrow C_dH, \qquad (1.3)$$

first revealed by Frenklach and Wang [1]. The notations C_dH and $C_d\bullet$ denote the surface sites of C–H bond and surface radicals, respectively. The reverse of the reaction (1.2),

$$C_d\bullet + H_2 \rightarrow C_dH + H, \qquad (1.4)$$

is too slow to compete with the reaction (1.3) under typical diamond conditions. The density of the surface radicals is determined by the addition of growth species to the surface radicals, and by thermal deposition of sp^3 diamond surface radicals to sp^2 surface carbons, and vice versa. Typical reaction conditions include: highly diluted gas mixtures of hydrocarbons in hydrogen, substrate temperatures of about 800 to 1000°C, and a large super-equilibrium of H atoms. With a deficit of H atoms, the density of active sites is dependent on the concentration of H atoms.

The following are some major observations of conventional diamond growth chemistry:

a. Activation of gas is required for achieving an appreciable diamond growth rate. Activating the gas prior to the deposition increases the diamond growth rate from Å/h to μm/h.
b. The carbon-containing precursor is needed in the deposition.

c. Atomic hydrogen has to be present in the gas phase in a super-equilibrium concentration. The activated hydrogen etches graphite at a much higher rate than it does on diamond.

d. Oxygen added in small amounts to hydrocarbon precursor mixtures enhances the quality of diamond deposits.

e. Graphite and other non-diamond carbons are usually deposited simultaneously with diamonds.

f. Deposition parameters such as temperature, pressure, and gas combination are related to diamond growth rate and quality.

g. A substrate pretreatment of substrate surface with a diamond powder is common in order to enhance nucleation rates and densities. Electrical biasing of substrates can also influence nucleation rates.

1.6. *Growth Mechanisms of Nanocrystalline Diamond (NCD) Films*

With the development of nanomanufacturing and nanomaterials, nanotechnology is now a next-generation technique with huge influence on modern science and technology. For diamond, with the ability to reduce grain size to the nanometer range, the nanocrystalline diamond is now considered for a wide variety of structural, biomedical, and microelectronical applications due to its unique electrical, chemical, physical, and mechanical properties. Thus, the fabrication of NCD films by CVD is of great interest because of its potential experimental and theoretical applications. The drastic changes in the mechanical and electrical properties of phase-pure diamond films, such as changing from an insulating to an electrically conducting material as a result of reductions in crystallite size (grain boundaries appear to conduct, and because their numbers dramatically increase with decreasing crystallite size, the entire film becomes electrically conducting and functions as an excellent cold cathode [14]), are largely due to the fact that the grain boundary carbon is π-bonded. Also, because of the small crystallite size of NCD films, film surfaces become smooth and they can function extremely well in tribological applications.

In conventional CVD diamond film growth, hydrocarbon–hydrogen gas mixtures consist typically of 1% CH_4 in 99% H_2. Atomic hydrogen is generally believed to play a key role in various diamond CVD processes. As mentioned in the previous section, atomic hydrogen is generated from the collision process inside the microwave plasma. With a hydrocarbon precursor such as CH_4 used in MCD deposition, gas-phase hydrogen abstraction reactions lead to the generation of methyl radical $-CH_3$ and additional hydrogen abstraction reactions form carbon–carbon bonds resulting in diamond lattice. Methyl radical CH_3 and Acetylene molecule C_2H_2 are the dominant growth species in CH_4/H_2 systems [15]. However, hydrogen is not the only thing that can be used to produce diamond. Deposition of carbon films from 5% CH_4, 95% Ar microwave plasma has been shown by several groups of researchers to result in graphite or amorphous carbon deposition with only small amounts of diamond inclusions [16]. The presence of carbon dimer, C_2, in the plasma is believed to be responsible for the formation of graphite or amorphous carbon. Phase-pure diamond films have been grown from C_{60}/Ar microwave plasmas and it was found that the sizes of diamond grains are in the nanometer range. The deposition was carried out under the following conditions: total argon pressure of 98 Torr, C_{60} partial pressure of 10^{-2} Torr, flow rate 100 sccm, and microwave power of 800 W. A new growth mechanism has been proposed that C_2 dimer is a growth species to form diamond films with nanocrystalline microstructure [14].

Carbon dimer is known to be a highly reactive chemical species. It is produced in microwave plasma under highly non-equilibrium conditions. High C_2 concentrations in the plasma, achieved by using C_{60} as the precursor, appear to increase heterogeneous nucleation rates and result in small diamond grain size. Emission spectroscopy study shows that the transition in crystallite size and morphology is correlated with C_2 concentrations in hydrogen-rich or hydrogen-poor plasmas [17]. A schematic diagram of growth of diamond lattice by insertion of C_2 into C–H bands on the (110) surface is shown in Fig. 1.6. It can be seen that the insertion of C_2 into C–H bonds on the hydrogenated (110) surface has small activation barriers of less than 5 kcal/mol and is highly exothermic. Subsequent steps involving linkage of C_2 units also have low barriers and lead to growth [18]. Further, nanocrystalline diamond films have also been grown from CH_4/Ar plasma [19]. It was also observed that the diamond grain size decreased as more argon was included to replace hydrogen. Further characterization will be discussed in the next part of this chapter. Optical emission studies of the plasma have shown intensive green Swan-band radiation, indicating the presence of a large number of C_2 dimers in CH_4/Ar plasma [17]. Interestingly, C_2 concentration decreases by about an order of magnitude upon the addition of 20% H_2, which is from the reaction of C_2 with H_2 to form hydrogen carbons. Therefore, this reaction will drastically lower C_2 concentrations in high hydrogen-content plasmas.

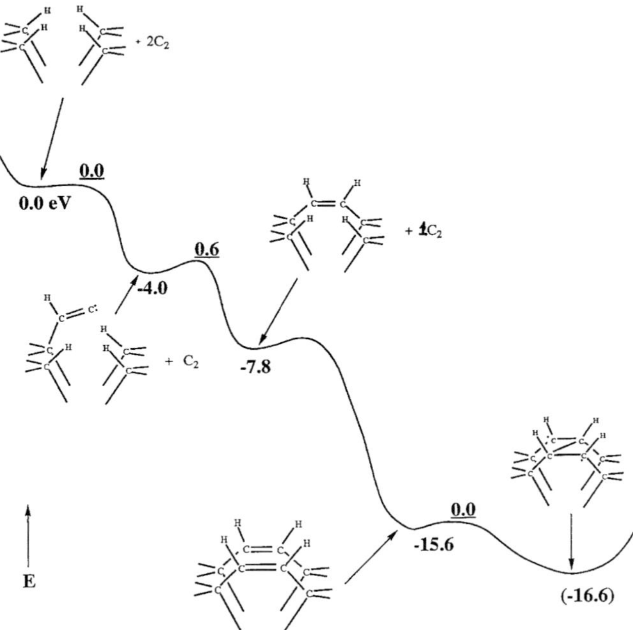

Fig. 1.6. Potential energy surface for C_2 addition to diamond (110) surface (barriers are underlined) [14].

The million-fold difference in crystallite size means that secondary nucleation rates associated with growth in hydrogen-poor plasmas are 10^6 times higher than conventional hydrogen-rich plasmas. The basic idea of secondary nucleation is that the unattached carbon atom can react with other C_2 molecules from the gas phase to nucleate a new diamond crystallite. Extensive theoretical work has been done over the past years to understand the growth mechanisms from $CH_4/H_2/Ar$ plasma. Sternberg *et al.* [20] demonstrated that stable C_2 adsorption configurations which lead to diffusion-assisted chain growth have been found on hydrogen-free (110) surfaces. Diamond (100) surfaces are the most typical growth surfaces and a density-functional-based tight-bonding study has been carried out on a non-hydrogenated diamond (100)-(2×1) surface by Sternberg *et al.* [21]. They have simulated diamond growth steps by studying isolated and agglomerated C_2 molecules adsorbed on a reconstructed non-hydrogenated diamond surface. Their calculations suggest that the most stable configuration is a bridge between two adjacent surface dimers along a dimer row, and there are many other configurations with adsorption energies differing by up to

2.7 eV. The (110) face is the fastest growing one in the C_2 regime while (100) face is slow growing. However, from the simulation, both types of growth result in same growth of small grain sizes.

Overall, diamond can nucleate and grow using carbon dimers as the growth species without the intervention of hydrogen. This new method of diamond synthesis results in very small diamond grains ranging from 5 to 20 nm and very smooth surface. Moreover, the cost of diamond CVD is dramatically reduced since hydrogen is no longer necessary for growth.

2. Characterization of Nanocrystalline Diamond Films

The most apparent difference in films grown by hydrogen-poor plasmas compared to conventional hydrogen-rich plasmas is the smaller diamond grain sizes in the nanometer range. The nanocrystallinity of the films grown by CH_4/Ar plasma has been extensively characterized by various techniques. Surface morphologies have been studied by scanning electron microscopy (SEM), transmission electron microscopy (TEM), and atomic force microscopy (AFM). Raman spectroscopy and near edge X-ray absorption fine structure (NEXAFS) have been used to characterize the bonding structures of films. Nanoindentation has been employed to measure the mechanical and tribological properties. The results of characterization studies on the nanocrystalline diamond films are discussed below.

2.1. *Scanning Electron Microscopy (SEM)*

Scanning electron microscopy is a very useful technique in the characterization of diamond thin films. Conventional light microscopes use a series of glass lenses to bend light waves and create a magnified image. SEM creates magnified images by using electrons instead of light waves. In a SEM system, an electron gun emits a beam of high energy electrons, and this beam travels downward through a series of magnetic lenses designed to focus the electrons to a very fine spot. As the electron beam hits each spot on the sample, secondary electrons are knocked loose from its surface and a detector counts these electrons and sends the signals to an amplifier. The final image is built up from the number of electrons emitted from each spot on the sample. SEM can show very detailed three-dimensional images at much higher magnifications than is possible with a light microscope.

In the previous section, we discussed that the microstructure of diamond films changes dramatically with the continued addition of Ar to reacting gas mixtures during CVD. The transition from micro- to nanocrystallinity

Table 2.1. Summary of reactant gases used for diamond deposition in microwave plasma-enhanced CVD system.

Sample No.	Ar (vol. %)	H_2 (vol. %)	CH_4 (vol. %)
1	2	97	1
2	20	79	1
3	40	59	1
4	60	39	1
5	80	19	1
6	90	9	1
7	97	2	1
8	99	0	1

by systematically adding argon to hydrogen-rich plasma has been characterized by SEM micrographs as a function of argon content [19]. Different combinations of gas mixtures are listed in Table 2.1. Note that when Ar gas was added, the CH_4 was kept at 1% all the time and the volume percentages of H_2 were changed accordingly.

Plan-view and cross-section SEM micrographs are shown in Fig. 2.1 and Fig. 2.2. It can be seen that the surface morphology of diamond films changes dramatically with the continued addition of Ar. Figure 2.1(a) shows the surface morphology of a diamond film grown with 2% Ar. Well-faceted microcrystalline diamond grains with grain sizes ranging from 0.5 to 2.0 μm have been deposited. (111) and (110) planes can be found and the surface is very rough. The plan-view SEM micrograph of a film grown with 20% Ar, 79% H_2, and 1% CH_4 mixture shows that small diamond particles start to form during the deposition process. For films grown with 40% and 60% Ar, the number density of the small crystals increases significantly and most of these small crystals nucleate and grow between boundaries of large diamond crystals. At 80% Ar, large diamond crystals begin to disappear and well-faceted microcrystalline diamond grains no longer exist. Diamond crystals change from microcrystalline to nanocrystalline when 90% Ar is added to the $Ar/H_2/CH_4$ plasma, shown in Fig. 2.1(f). Faceted microcrystalline diamond crystals have disappeared and diamond crystals at nanometer scale have been formed. For films deposited with 97% and 99% Ar, the crystals are further reduced to less than 20 nm. Therefore, it is demonstrated that the microstructure of the diamond film can be controlled by varying the gas mixtures. The more the content of Ar in the reactant gases, the smaller the diamond grain sizes and the smoother the surfaces.

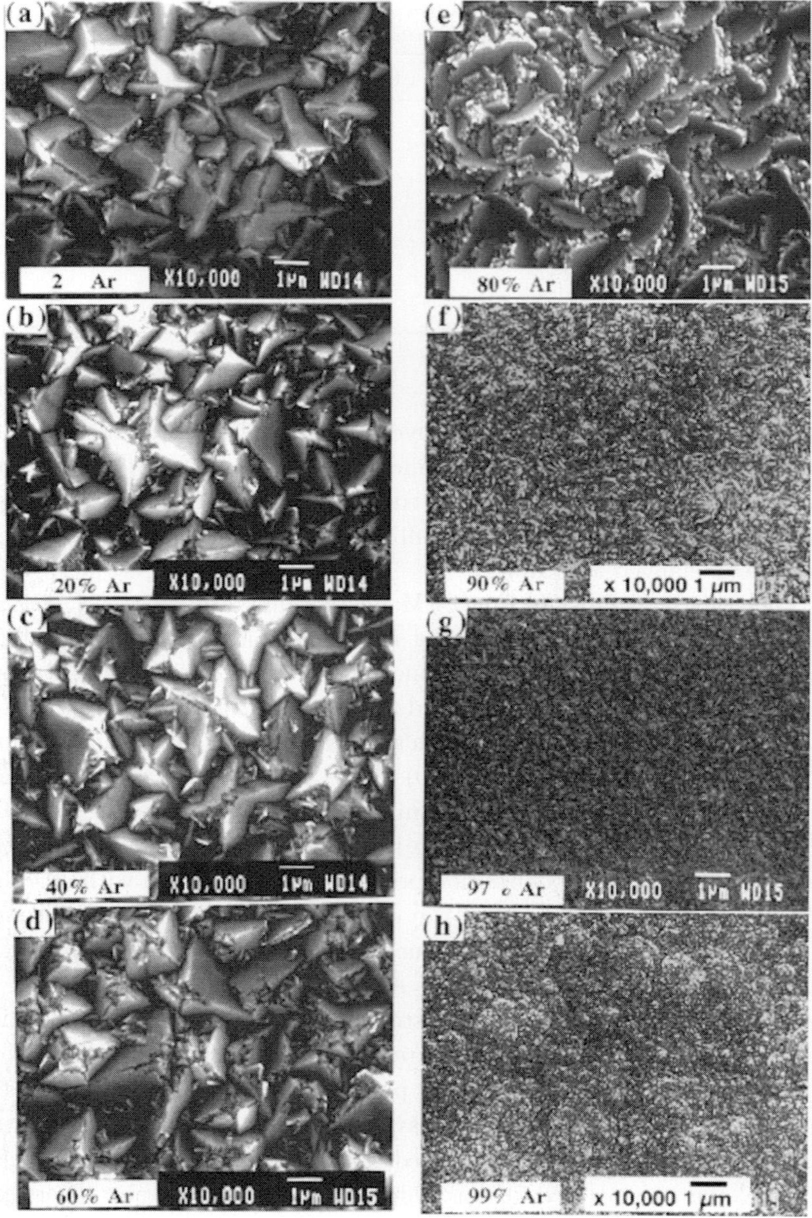

Fig. 2.1. Plan-view SEM images of the as-grown films prepared from microwave plasmas with different mixtures of Ar, H_2, and CH_4 as the reactant gases (listed in Table 2.1) showing the transition of microcrystalline to nanocrystalline diamond films: (a) film 1; (b) film 2; (c) film 3; (d) film 4; (e) film 5; (f) film 6; (g) film 7; (h) film 8 [19].

In order to obtain information on the development of the growth morphology, cross-section SEM micrographs of diamond films with different Ar volume percentages are shown in Fig. 2.2. Columnar growth structures can be observed in film grown with 2% Ar, shown in Fig. 2.2(a). This type of surface morphology can be explained by the van der Drift

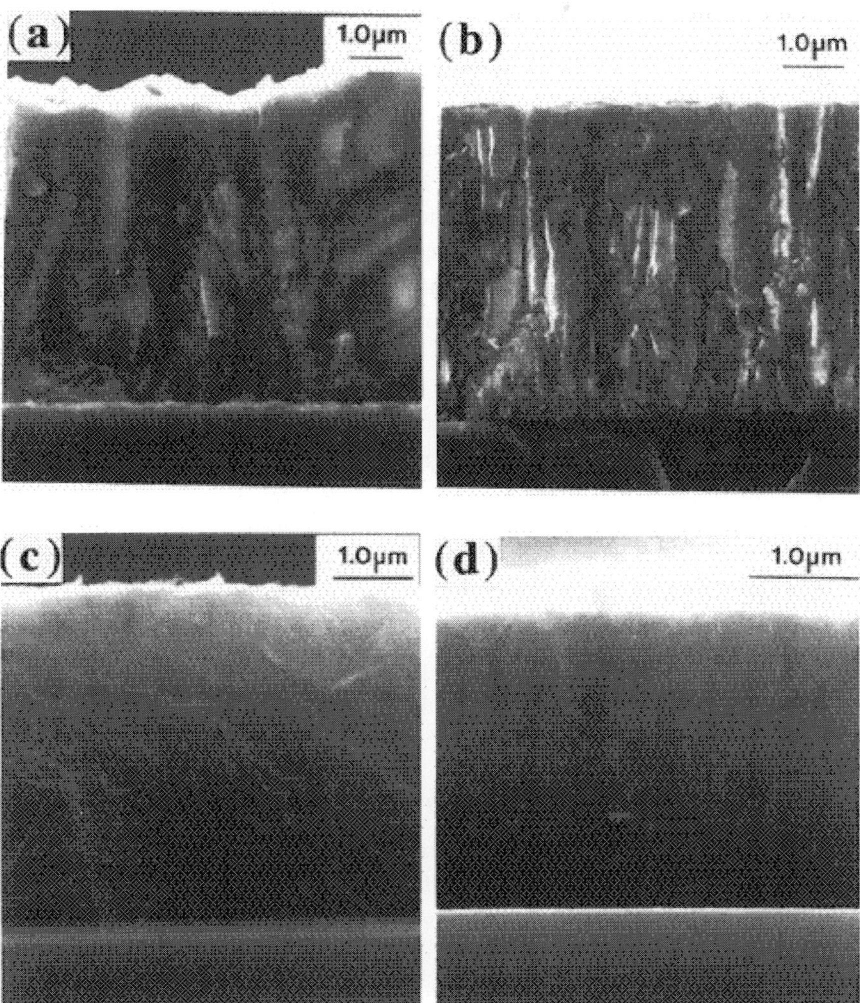

Fig. 2.2. Cross-section SEM images of the as-grown films prepared from microwave plasmas with different mixtures of Ar, H_2, and CH_4 as the reactant gases showing different growth phenomena of microcrystalline and nanocrystalline diamond films: (a) film 1; (b) film 2; (c) film 3; (d) film 4 [19].

growth mechanism [22]. During conventional diamond growth, hydrogen prevents diamond crystals from secondary nucleation by regasfying small or non-diamond phase. Therefore, after a period of time, only large diamond crystals can survive to form columnar crystallites, which are the so-called "evolutionary selection of crystallites". The columnar structures become much smaller and narrower for films grown with 80% Ar. For films deposited with more than 90% Ar, the columnar structure totally disappears, as shown in Fig. 2.2(c). Very fine and smooth surfaces are observed from the SEM micrographs, suggesting that the nanocrystalline diamond does not grow from the initial nuclei at the substrates during nucleation but from the second nucleation with C_2 as the nucleation species. So far it has been demonstrated that the changes in diamond microstructure are reflections of the changes in the plasma chemistry. During NCD growth, the C_2 dimmer and atomic hydrogen concentrations in the microwave discharges are monitored. Increasing the Ar content in the reactant gas mixtures significantly promotes the concentration of C_2 dimer, which is the nucleation species for NCD growth. Besides the growth mechanism, surface morphology and microstructure, the growth rate of diamond film has also been found to depend on the ratio of Ar to H_2 in the plasma. Zhou *et al.* [19] have demonstrated that the growth rate doubles in value for films grown with 60% Ar and then decreases rapidly in the 80–97% Ar range. So for NCD growth, although the concentration of C_2 dimer increased with the inclusion of more Ar, the growth rate decreased. They also established the relationship between the growth rate and the reactant gas pressure. When pressure was less than 40 Torr, no C_2 emission was detected and there was no NCD growth. As the reactant gas pressure increased, both the emission intensity of C_2 and the growth rate of NCD films increased significantly.

2.2. *Transmission Electron Microscopy (TEM)*

TEM is a powerful technique due to its high spatial resolution ($\sim$0.2 nm) in imaging. A TEM works much like a slide projector. A projector transmits a beam of light through the slide, as the light passes through it is affected by structures and objects on the slide. This transmitted beam is then projected onto the viewing screen, forming an enlarged image of the slide. TEM works the same way except that it transmits a beam of electrons through the specimen. The transmitted part is projected onto a phosphorous screen for the user to see. Materials for TEM must be specially prepared to thicknesses which allow electrons to be transmitted through. As the wavelength of electrons is much smaller than that of light, the optimal resolution of TEM images is many orders of magnitude better than that of a light microscope. Thus, TEM can reveal the finest details of internal structure as small as individual atoms.

The first detailed electron microscopy examination of nanocrystalline diamond films showed that various planar defects, common in diamond synthesis by conventional method, such as twinning boundaries and stacking faults, were hardly found in the films [23]. Zhou *et al.* [24] performed TEM analysis for diamond films grown with different Ar/H_2 ratio plasma with 1% CH_4. Films prepared with 90% Ar consisted of grains ranging from 30 to 50 nm. However, films prepared with 97% Ar had grain sizes in the range of 10 to 30 nm. Moreover, film prepared without hydrogen had smallest grain sizes ranging from 3 to 20 nm. Thus, the grain size of the diamond films decreased strongly with the continuous addition of Ar to the microwave plasma. Figure 2.3 shows a bright field TEM image revealing

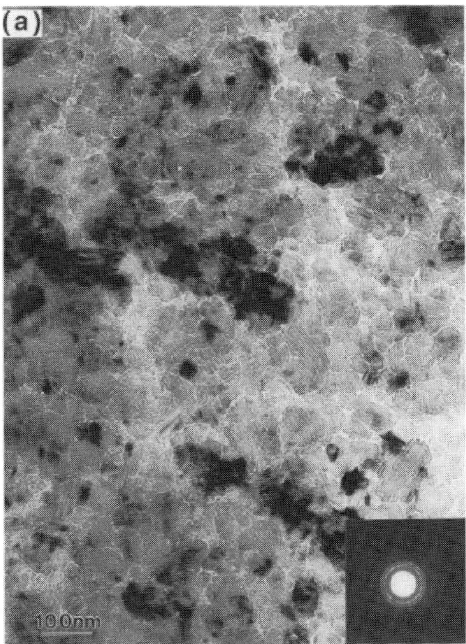

Fig. 2.3.   A plan-view TEM image of the diamond film prepared from an Ar/CH_4 plasma at 100 Torr showing that the diamond film consists of nanocrystalline grains ranging from 3 to 20 nm. The inset shows a sharp ring pattern of a selected area electron diffraction, indicating that the diamond grains have a random orientation. No graphite or amorphous carbon reflections can be observed from indicating the film is phase-pure diamond [24].

the plan-view morphology of the NCD thin film. The grain boundary width estimated from the TEM micrograph is 0.2 to 0.5 nm. The insert image shows a selected area (over 10 μm in diameter) of the electron diffraction pattern. Sharp Bragg reflections are located in concentric circles, indicating the random orientation of nanocrystalline diamond grains.

Further TEM investigations of NCD films have been performed by Jiao *et al.* [25] to compare the differences between NCD films grown with Ar/1% CH_4 and Ar/1% CH_4/5% H_2. Figure 2.4 shows two dark field images of NCD films. Dark field imaging provides better contrast than bright field imaging and can be used for determining the grain sizes with least effect by grain overlapping. The median and average size for NCD films produced with an Ar/1% CH_4 plasma are 4 and 7 nm respectively, and 5 and 8 nm, respectively, for films produced with Ar/1% CH_4/5% H_2 plasma.

However due to the limitation of the small objective aperture, results obtained by tradition TEM might be incorrect as grain size decreases to the nanometer scale diamond grains will overlap each other [26]. High resolution electron microscopy (HREM) on the other hand offers high resolution and more reliable results. Jiao *et al.* developed an interesting method to identify the grain boundaries by taking HREM images using different focusing conditions. By adjusting the focusing conditions, the contrast could change along grain boundaries from over-focus to under-focus. Figure 2.5 shows a series of HREM images of an Ar/1% CH_4 NCD film taken under different focusing conditions. These images show Fresnel fringes on grain boundaries.

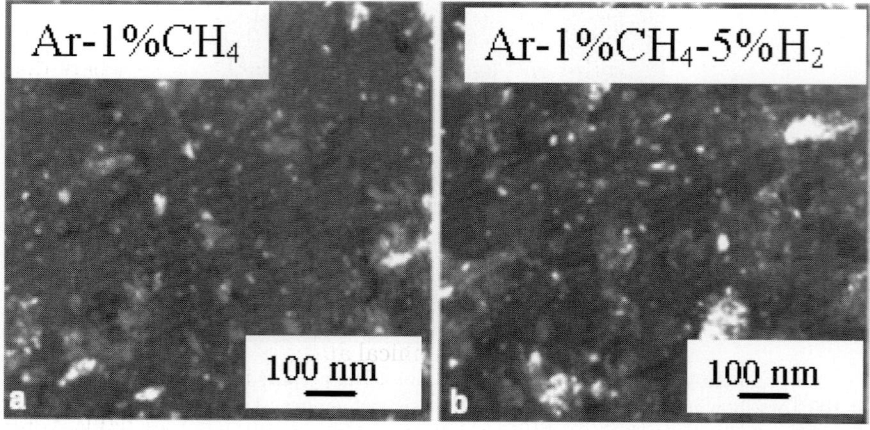

Fig. 2.4.   Plan-view TEM dark field images of NCD films grown using (a) Ar/1% CH_4 and (b) Ar/1% CH_4/5%H_2 microwave plasma CVD [25].

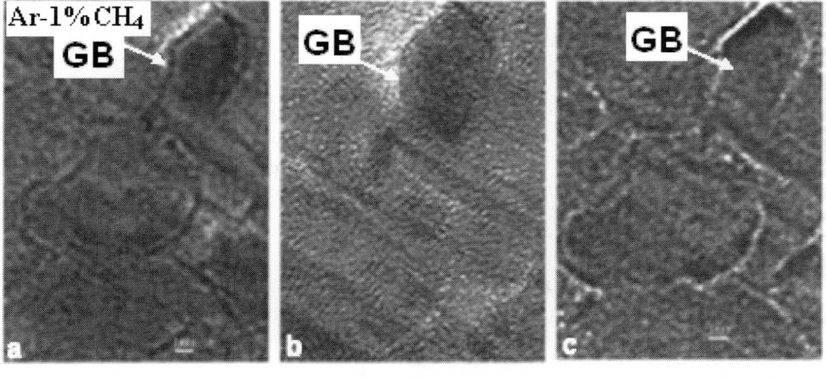

Δf=170 nm, over-focus　　　　Δf=0, in-focus　　Δf=100nm, under-focus

Fig. 2.5.　HREM images of an Ar/1% CH₄–NCD film taken using different focusing conditions: (a) over-focus, Df5170 nm; (b) in focus, Df50; and (c) under-focus, Df52100 nm. Note the contrast change along grain boundaries from over-focus to under-focus, and the effect of grain overlapping on fringes at GB [25].

In the over-focus image, grain boundaries appear dark while they appear white in the under-focus one. Very little contrast is shown in the in-focus image and lattice fringes are clearly visible. Therefore, it is estimated that the grain boundary thickness ranges from 0.2 to 0.4 nm [25].

Cross-section TEM analysis have been performed by Jiao *et al.* [25] and two different nucleation mechanisms have been proposed for NCD film grown with and without a small amount of hydrogen in Ar/CH_4 plasma. Figure 2.6 shows the NCD/Si interfaces TEM bright field images of two NCD films grown with Ar/1% CH_4 and Ar/1% CH_4/5% H_2. For film grown without any hydrogen, an amorphous carbon layer was formed where diamond grains start to nucleate. No such layer can be observed for samples grown with 5% H_2, possibly due to the etching effect of atomic hydrogen. Therefore, diamond grains nucleate directly on the Si substrate with the aid of diamond seeds introduced during the seeding process.

2.3. *Raman Spectroscopy*

Raman spectroscopy is a method of chemical analysis that enables real-time reaction monitoring and characterization of compounds in a non-contact manner. The sample is illuminated with a laser and scattered light is collected. The wavelengths and intensities of the scattered light can be used to identify functional groups in a molecule. Raman Spectroscopy has found

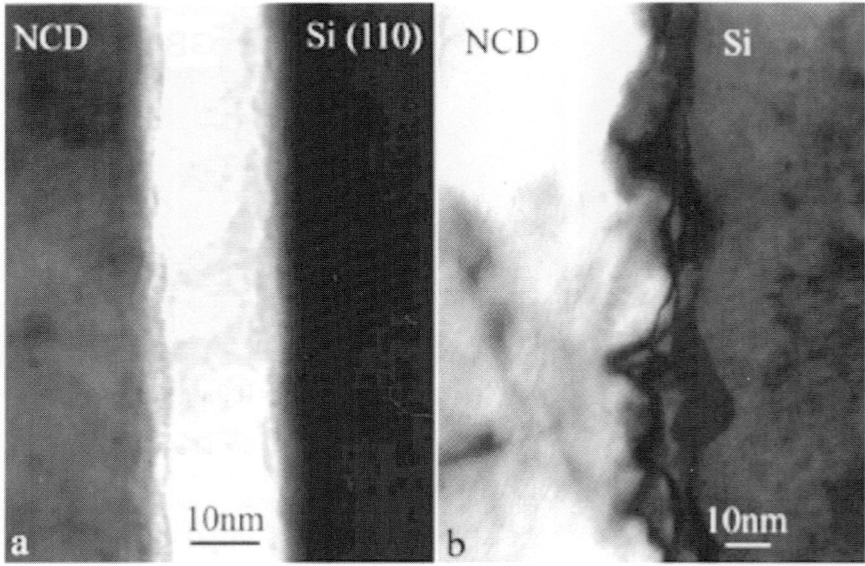

Fig. 2.6.   Cross-section TEM bright field images of (a) Ar/1% CH$_4$–NCD and (b) Ar/1% CH$_4$/5% H$_2$–NCD films showing the NCD/Si interface [25].

wide application in chemical, polymer, semiconductor, and pharmaceutical industries because of its high information content and ability to avoid sample contamination.

When light is scattered from a molecule most photons are elastically scattered. The scattered photons have the same energy (frequency) and, therefore, wavelength, as the incident photons. However, a small fraction of light (approximately 1 in 10^7 photons) is scattered at optical frequencies different from, and usually lower than, the frequency of the incident photons. The process leading to this inelastic scatter is termed the Raman Effect. Raman scattering can occur with a change in vibrational, rotational or electronic energy of a molecule.

Raman scattering is about 50 times more sensitive to π-bonded amorphous carbon and graphite than to the phonon band of diamond. Hence, Raman spectroscopy could be used to establish the crystalline quality of diamond thin films by estimating the amount of sp^2-bonded carbon in the films. Lin *et al.* [27] performed Raman analysis of diamond films grown with Ar/CH$_4$/H$_2$ plasmas with different gas mixtures. Figure 2.7 shows the plots of micro-Raman spectra obtained of diamond films grown by different ratio of Ar/H$_2$ with fixed 1% CH$_4$ in Ar/H$_2$/CH$_4$ plasma. For the

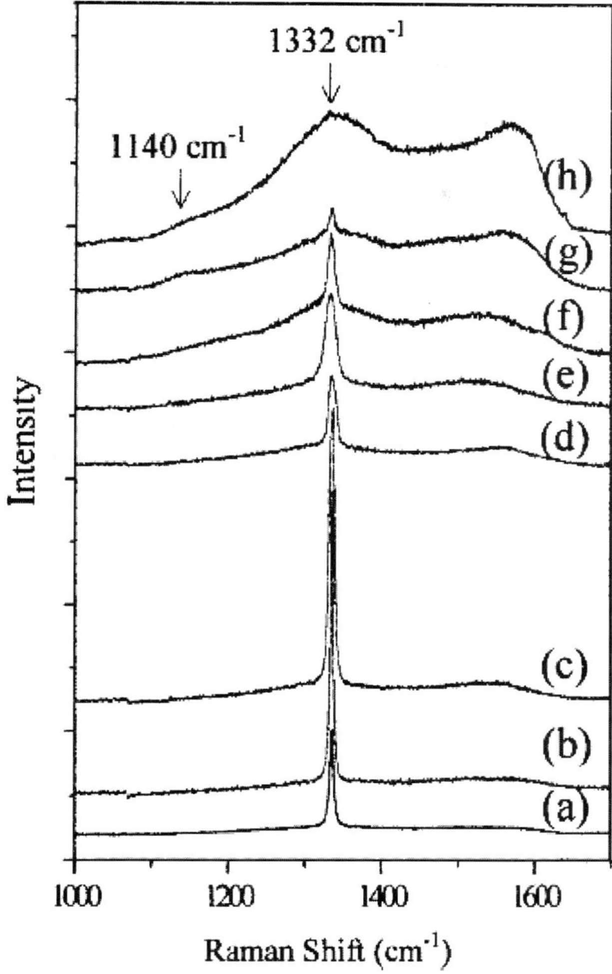

Fig. 2.7.   Micro-Raman spectra of HFCVD diamond grown at increasing Ar volume fraction: (a) 0%; (b) 20%; (c) 50%; (d) 80%; (e) 90%; (f) 92%; (g) 94%; and (h) 95.5% for growth mixture using Ar–CH$_4$–H$_2$ [27].

film grown without Ar, a sharp diamond characteristic peak is observed at 1332 cm^{-1}. No Raman scattering can be found in the range from 1400 to 1600 cm^{-1} suggesting that the diamond film contains very little sp^2-bonded carbon. With addition of argon to the reactant gas up to 92%, a sharp diamond peak still exists indicating the presence of microcrystalline diamond grains. However, the diamond peak shrinks and the full width at half maximum (FWHM) of the peak increases. For films grown with more than 92% Ar, the spectra change dramatically compared to previous samples. The sharp diamond peak at $1332\,\text{cm}^{-1}$ no longer exists, and it is significantly broadened as a result of decreasing the grain size to the nanometer scale. A broad peak in the 1400–$1600\,\text{cm}^{-1}$ is observed due to the presence of sp^2-bonded carbon at the grain boundaries in the nanocrystalline diamond films.

Further studies have been carried out to study the transition of diamond films from micro- to nanocrystalline by UV Raman spectroscopy. For visible Raman, when a laser with a wavelength in the visible region is used, the energy of the incident photons is much lower than the energy of the bandgap for sp^3-bonded carbon. The spectra are thus completed dominated by Raman scattering from sp^2-bonded carbon [28]. However, UV Raman can provide higher photon energy that is close to local gap of sp^3-bonded carbon at $\sim 5.5\,\text{eV}$ [29]. Thus, UV Raman allows us to further interpretation of some nanocrystalline diamond Raman signatures. Birrell *et al.* [30] interpreted the Raman spectra of NCD films, especially the peak at $1120\,\text{cm}^{-1}$ using UV Raman spectroscopy. Figure 2.8(a) shows the visible Raman spectra of NCD films grown with different amounts of H_2 in $Ar/H_2/CH_4$ plasma. There are three peaks: a sharp diamond peak at $1332\,\text{cm}^{-1}$, a shoulder at $1190\,\text{cm}^{-1}$, and a broad peak at $1532\,\text{cm}^{-1}$. As less H_2 is included in reactant gas, a shoulder at $1125\,\text{cm}^{-1}$ and peaks at 1400–$1600\,\text{cm}^{-1}$ start to appear while the diamond peak shrinks. For the sample grown with 0 and 1% H_2, a shoulder at $1125\,\text{cm}^{-1}$, a D-bank peak at $1330\,\text{cm}^{-1}$, a poorly defined peak at $1450\,\text{cm}^{-1}$ and a peak at $1560\,\text{cm}^{-1}$ are observed. Figure 2.8(b) shows Raman spectra taken with UV excitation with a laser wavelength of 266 nm. All spectra show the same three peaks: a sharp peak at $1332\,\text{cm}^{-1}$ due to diamond, a broad peak at $1580\,\text{cm}^{-1}$ due to the presence of sp^2-bonded carbon at the grain boundaries, and a $1546\,\text{cm}^{-1}$ feature caused by the excitation of oxygen molecules. It is widely believed that the peak around $1120\,\text{cm}^{-1}$ originated from the presence of confined phonon modes in diamond. Nemanich *et al.* [31] suggested that this feature was due to the regions of microcrystalline or amorphous diamond. Fayette *et al.* [32] concluded that peak at $1140\,\text{cm}^{-1}$ was one of the characteristics of CVD diamond films. However, Birell *et al.* suggested that the peaks at 1120 and $1450\,\text{cm}^{-1}$ were due to carbon–hydrogen bonds in the grain boundaries, and the peaks at 1330 and $1560\,\text{cm}^{-1}$ were due to the D-band and G-band sp^2-bonded carbon. Although NCD contains nearly 95% sp^3-bonded carbon, as diamond thin films change from microcrystalline to nanocrystalline,

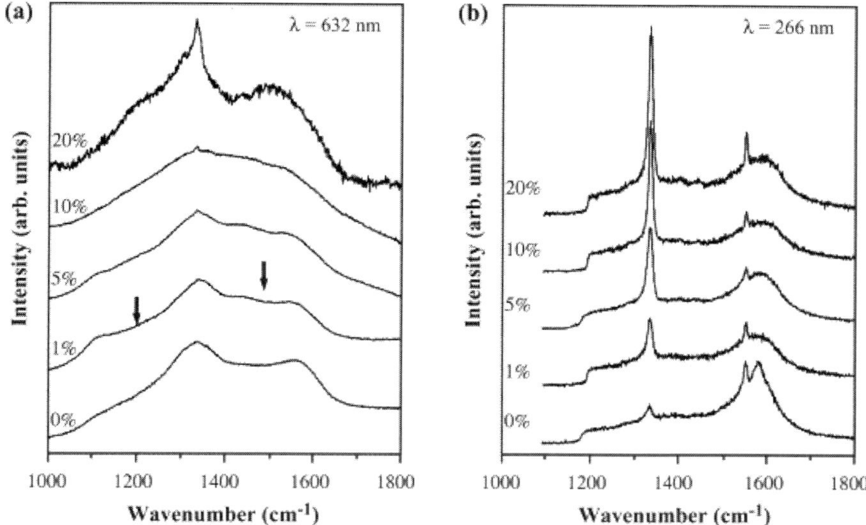

Fig. 2.8. (a) Visible and (b) UV Raman spectra of UNCD thin films grown with successive amounts of hydrogen added to the plasma [30].

the visible Raman signal was due entirely to the sp^2-bonded grain boundary materials in the sample. For nanocrystalline diamond films, the sp^3-bonded carbon can only be characterized UV Raman spectroscopy instead of regular visible Raman spectroscopy [30].

2.4. *Near Edge X-Ray Absorption Fine Structure (NEXAFS)*

A technique that can distinguish between sp^2-bonded carbon (graphite) and sp^3-bonded carbon (diamond) would be very useful for characterizing the nanocrystalline diamond films. X-ray core level reflectance and absorption spectra are site and symmetry selective and the magnitudes and energy positions of spectral features contain bonding information of the films. Particularly, the near-edge region of the core-level photoabsorption has been used to relative quantity of sp^2- or sp^3-bonding in BN powders and thin films [33]. Its sensitivity to the local bond order in a material arises from the dipole-like electronic transitions from core states, which have well-defined orbital angular momenta, into empty electronic (e.g. antibonding) states [34]. Therefore, the symmetry of the final state can be determined and the difference between sp^2 and sp^3 can be distinguished. Besides, NEXAFS with a photon beam of the size of millimeter has the ability to investigate the film over a large area.

Capehart *et al.* [35] first demonstrated that NEXAFS studies on CVD diamond films could be used to calculate the fraction of any sp^2-hybridized graphitic content inside the film. The characterization of NCD films by core-level photoabsorption was first studied by Gruen *et al.* [34]. NEXAFS measurements have been carried out on NCD films as well as two reference samples, a highly oriented pyrolytic graphite (HOPG) that is 100% sp^2-bonded, and a single crystal diamond that is 100% sp^3-bonded. Figure 2.9 shows the C (1s) photoabsorption data from four different samples: natural diamond, graphite, microcrystalline diamond deposited with Ar/CH$_4$, and nanocrystalline diamond deposited with Ar/C$_{60}$ [34]. From previous studies, we have found that amorphous carbon shows the characteristic transition to sp^2-bonded π^* orbitals at $\sim$285 eV and bulk diamond shows a

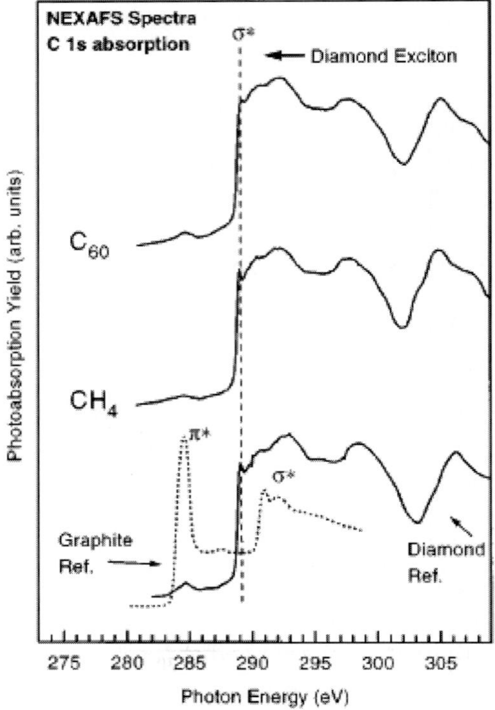

Fig. 2.9. The C (1) photoabsorption data from four different samples: C$_{60}$, nanocrystalline diamond film grown using the C$_{60}$-based deposition method; CH$_4$, diamond film with micron-sized grains grown using a conventional CH$_4$-based deposition method; diamond, gem quality natural diamond; graphite, highly oriented pyrolytic graphite [34].

sharp peak at $\sim$290 eV due to the bulk C-1s core exciton and characteristic transitions to sp^3-bonded σ^* orbitals [36]. These two features are directly related to the bonding structures of diamond and graphite and could be used to determine the sp^2 to sp^3 ratio in a film. From Fig. 2.9 we can clearly see that all of the qualitative features of sp^3-bonding are presented in NCD film. Based on the qualitative line shape of the spectra, no more than 1 to 2% of sp^2-bonded carbon exists in these NCD films.

Chang *et al.* [37] studied the quantum-size effect on the exciton and energy gap of NCD films. Exciton energy and conduction/valence band energies were observed generally to shift to higher/lower energies when the decrease of the crystal size. Their results showed that when the diamond grain size decreased from 5 μm to 3.6 nm, the C k-edge NEXAFS of NCD shifted toward higher energies, especially for crystals smaller than 18nm, as shown in Fig. 2.10. The inset of this figure reveals a similar trend to excitonic binding energy, ΔE_{ex}, where $\Delta E_{\text{ex}} = |E_{\text{ex}} - E_{\text{CB}}|$, the energy separation between the E_{ex} (exciton state) and the E_{CB} (conduction band

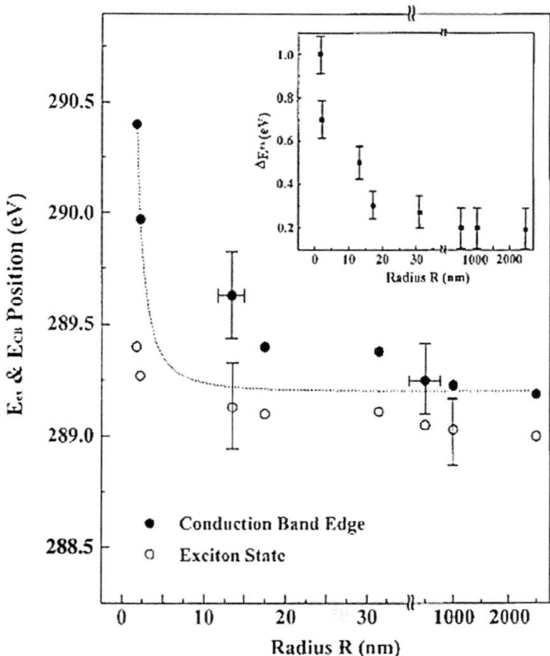

Fig. 2.10. The exciton state and conduction band edge of the NCD are plotted as a function of the ~crystallite radius. The dashed line is the least square fitting of all conduction band edge data. The inset is the exciton binding energy shift of the NCD films plotted as a function of the crystallite radius [37].

edge). NCD films prepared with $H_2/1\%$ CH_4 by HFCVD have also been characterized by NEXAFS [38]. The spectrum of NCD film exhibits a sharp spike at 289 eV (an excitonic transition), and a secondary band at 302 eV, both characteristics of pure diamond spectrum. A shift of exciton energy around 0.25 eV has been observed for 10 nm NCD.

2.5. *X-Ray Diffraction (XRD)*

XRD is sensitive to the presence of crystalline carbons such as diamond or graphite instead of amorphous carbon. Hence, it is frequently used to characterize CVD diamond films. Figure 2.11 shows that NCD films exhibit a similar pattern compared to conventional synthesized MCD films [24]. Three peaks, related to (111), (220) and (311) crystalline diamond peaks are observed. Compared to MCD films, the diffraction peaks of NCD are significantly broadened due to the very small diamond grain sizes.

2.6. *Characterization of Mechanical Properties of NCD*

Diamond has exceptional mechanical, electrical, thermal, and optical properties. In particular, the high hardness and stiffness of diamond

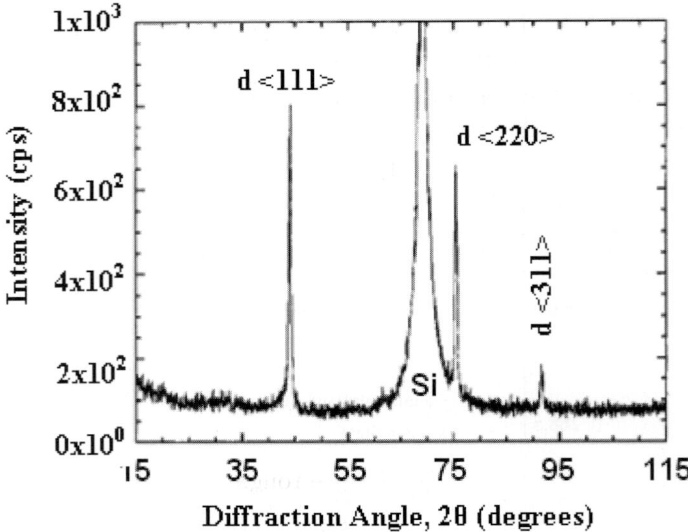

Fig. 2.11.   X-ray diffraction of the as-grown nanocrystalline diamond film prepared from an Ar–CH_4 plasma at 100 Torr. The labels show the diffraction peaks from different planes of the cubic diamond [24].

makes it desirable for many mechanical applications. Moreover, interest in diamond-based applications is increasing as nanocrystalline diamond has been successful synthesized due to its reduced intrinsic roughness and tribological properties. However, nanocrystalline diamond films have grain size dependent mechanical properties that may significantly differ from those of microsized diamond films. The changes in the mechanical properties of NCD films are due to the presence of larger percentages of grain boundaries that contain sp^2-bonded carbon, broken bonds and amorphous regions [39].

The mechanical properties of nanocrystalline diamond films can be measured by nanoindentation. Indentation tests are perhaps the most commonly applied means of testing the mechanical properties of materials. In such a test, a hard tip, typically a diamond, is pressed into the sample with a known load. After some time, the load is removed. The area of residual indentation in the sample is measured and the hardness, H, can be calculated from the maximum load, P, divided by the residual indentation area, A.

Figure 2.12 shows nanoindentation test results of NCD films. The average hardness of the NCD film is $\sim 88\,$GPa, close to the value of bulk diamond. Note that the measured hardness is independent of indentation depth, indicating that the results are reliable [40]. Guillen *et al.* [41] demonstrated that the fracture strength, between 3.2 and 3.9 GPa, and the Young's modulus, between 800 and 980 GPa, of the NCD films were independent of the film thickness. They suggested that the possibility to control and adjust the absolute value of the stress inside the diamond film with respect to Si substrate allowed one to use the built-in stress as a design parameter of MEMS devices for bistable membrane configurations. Philip *et al.* [42] suggested that the Young's modulus of the NCD films was affected by the nucleation rate. For films grown with lower nucleation density (with lower nucleation density, $10^{10}\,cm^{-2}$ or lower), the Young's modulus was roughly half of those films grown with much higher nucleation density ($>10^{12}\,cm^{-2}$). For the latter case, the Young's modulus (1120 GPa) was close to the values measured from natural diamond.

Microcrystalline diamond films have relatively high surface roughness that precludes the immediate application of diamond films in machining and tool wear applications. High surface roughness results in high friction and severe wear losses on mating surfaces. Polish of these MCD are often applied to smooth the film surface although it is costly and impractical in some circumstances such as complex geometries. However, with the development of growing smooth NCD films, low coefficient of friction (COF)

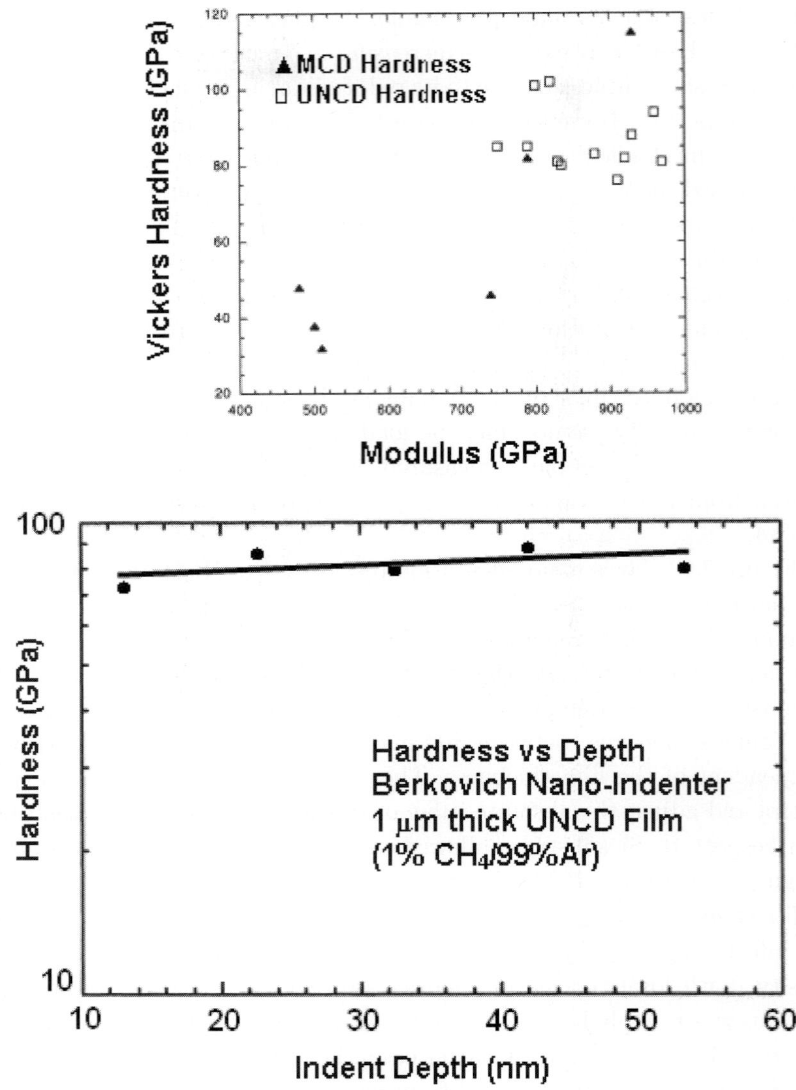

Fig. 2.12. (a) Nanoindenter data (Berkovich indenter) for MCD and UNCD films; (b) measured hardness versus indentation depth [40].

~0.05–0.1 and fine finish of the sample surface (20–40 nm, rms) have been achieved. Espinosa *et al.* [43] studied the surface roughness of the NCD films with different seeding techniques. Higher surface roughness (250–300 nm) was obtained for film seeded by mechanically polish, while lower surface

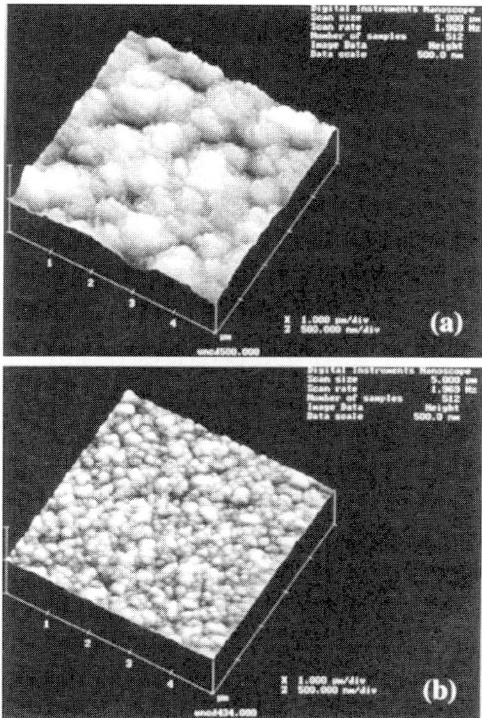

Fig. 2.13.   Roughness analysis of NCD surfaces using AFM. (a) The mechanically-seeded sample has a root mean square root (rms) roughness of 107 nm and a distance from peak to valley of 250–300 nm. (b) The ultrasonically-seeded one has a rms of 20 nm with a distance from peak to valley of $\sim$70 nm [43].

roughness (20 nm) was obtained for film seeded ultrasonically with fine diamond powders, as shown in Fig. 2.13. Clearly, ultrasonic seeding results in much better nucleation and growth, no obvious porosity, no scratches, and enhanced surface smoothness.

Several microscale testing techniques have been employed to investigate fracture strength of thin films. Sharpe *et al.* [44] have performed microtension tests to study the fracture toughness of SiC and polysilicon. Espinosa *et al.* [45] have performed the membrane deflection experiment (MDE) to investigate the strength of submicron freestanding NCD thin films. Table 2.2 summarizes the fracture strength of some hard materials.

NCD films have been grown on Si films and pin-on-disk tests have been carried out to evaluate the MCD and NCD films' friction behavior against Si_3N_4 balls in open air and dry N_2 [46]. The friction results are shown in

Table 2.2.   Fracture strength of some hard materials [43].

Material	Fracture strength (GPa)
Silicon	0.30
Microcrystalline diamond	0.88 ± 0.12
SiC	$1.2 \ \pm 0.5$
Polysilicon	$1.5 \ \pm 0.25$
NCD	4.13 ± 0.90
Single Crystal Diamond	2.8

Fig. 2.14. After the initial stage during which both films exhibit relatively high values, the COF of NCD film decreases rapidly to 0.1 whereas the COF of the MCD film remains high and unsteady in both air and dry N_2 environments. NCD films have also deposited on tungsten carbide substrates to slide against Al-alloy, steel and Al_2O_3 [47]. Similar results have been obtained to show that the COF of NCD film decreases to 0.1 after the break-in period while the COF of uncoated WC sample are much higher (0.6–0.9)

2.7. *Electron Energy Loss Spectroscopy (EELS)*

The electron energy loss spectroscopy (EELS) in TEM is a powerful technique to study the electronic structure of materials [48, 49]. The energy loss near-edge structure (ELNES) is sensitive to the crystal structure. For sp^2-bonded carbon the π^* states can be observed between 282 eV and 288 eV, and σ^* states, between 290 and 320 eV. For sp^3-bonded carbon, σ^* states can be observed between 289 and 320 eV [50]. Therefore, diamond and non-diamond carbon can be easily separated by EELS and ELNES. Figure 2.15 shows electron energy loss spectra taken from a single crystal diamond and nanocrystalline diamond film, respectively. Identical spectra have been obtained, suggesting that NCD film is phase-pure sp^3-bonded [26]. Only an EELS edge at 289 eV corresponding to diamond characteristic has been shown and no other amorphous or graphite phases presented. The results have also been confirmed by Zhou *et al.* [19] where no energy loss feature at 284 eV has been observed, indicating the absence of sp^2-bonded carbon. Okada *et al.* [51] studied the sp^2-bonding distributions in nanocrystalline diamond particles by EELS. The EEL spectrum showed a peak at 290 eV and a slight peak at 285 eV, which is similar to that of nanocrystalline diamond thin films.

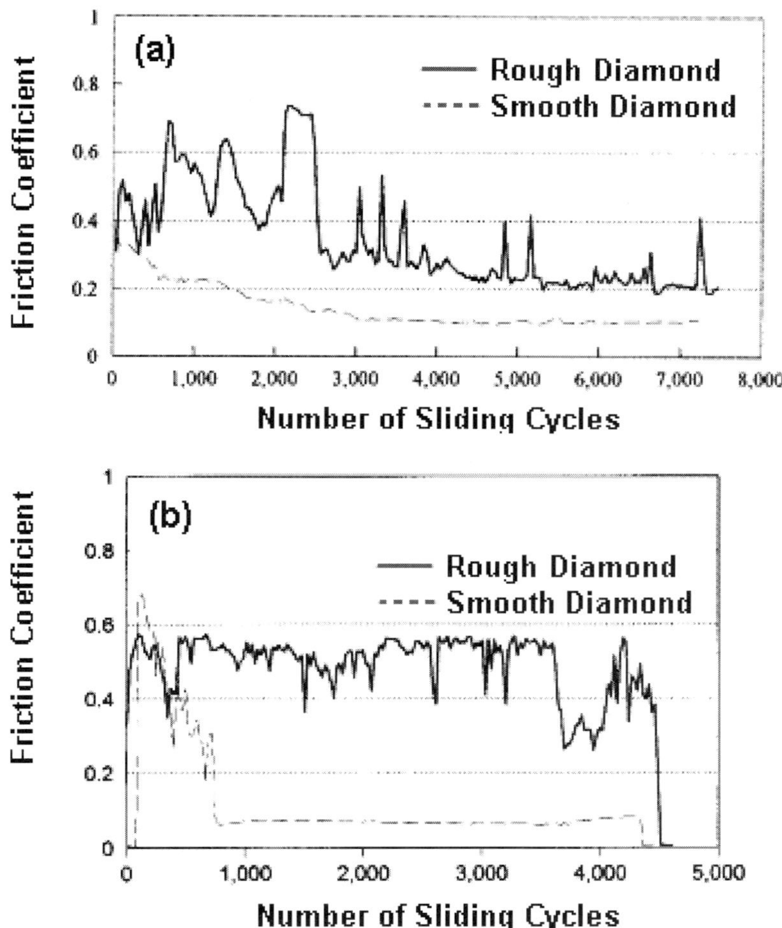

Fig. 2.14. Friction coefficients of rough MCD and smooth NCD films against Si_3N_4 balls in (a) open air and (b) in dry N_2. (Test conditions: load, 2 N; velocity, $0.05\,\mathrm{m\,s^{-1}}$; relative humidity, 37%; sliding distance, 40 m; ball diameter, 9.55 mm [46].)

2.8. *Characterization of Electrical Properties of Doped NCD Films*

The use of diamond for electrical applications has attracted much attention with the development of CVD techniques. Most doped diamond depositions system use microwave-assisted CVD, which minimizes film contamination with other materials. A special aspect of diamond is its capability of revealing a negative electron affinity (NEA) [52–55]. NEA means that electrons

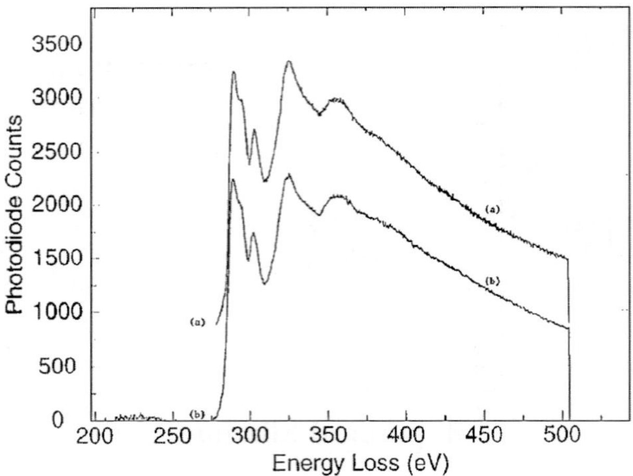

Fig. 2.15. PEELS spectra of (a) CVD-grown nanodiamond thin film and (b) single-crystal diamond. A weak peak at 284 eV is identified in the nanodiamond film spectrum, indicating that sp^3-bonded carbons are present at the grain boundaries, which accounts for about 5–10% of the total atoms in the nanocrystalline material [26].

can exit from conduction band edge into vacuum without further energy barrier. The only way to take advantage of this NEA is to form ohmic or tunnel contacts by n-type doping diamond. So far, nitrogen [56], sulfur [57] and phosphorus [58] have been used to produce n-type diamond.

Chen *et al.* [59] reported that nitrogen-doped NCD thin films exhibited semimetal-like electronic properties and could be used as electrochemical electrodes over 4 eV potential range. Bhattacharyya *et al.* [56] studied the synthesis and characterization of highly conductive nitrogen-doped NCD films. Up to 0.2% total nitrogen content was observed with a $CH_4(1\%)/Ar$ gas mixture and 1% to 20% nitrogen gas included. The electrical conductivity of the film increased by five orders of magnitude (up to 143 $\Omega^{-1}cm^{-1}$) compared to that of undoped film. As N_2 gas was added, the density of C_2 and CN radicals as well as their relative density increased substantially. Conductivity and Hall measurements made as a function of temperature showed that these films had the highest n-type conductivity and carrier concentration for phase-pure diamond thin films. Figure 2.16(a) shows secondary ion mass spectroscopy (SIMS) data of the total nitrogen content in the films as a function of N_2 gas in the plasma along with the conductivities measurements. Temperature-dependent conductivity data in the range of 300–4.2 K are shown in the Arrhenius plot in Fig. 2.16(b). Finite conduction has been observed for temperatures as low as 4.2 K. Nonlinear curves

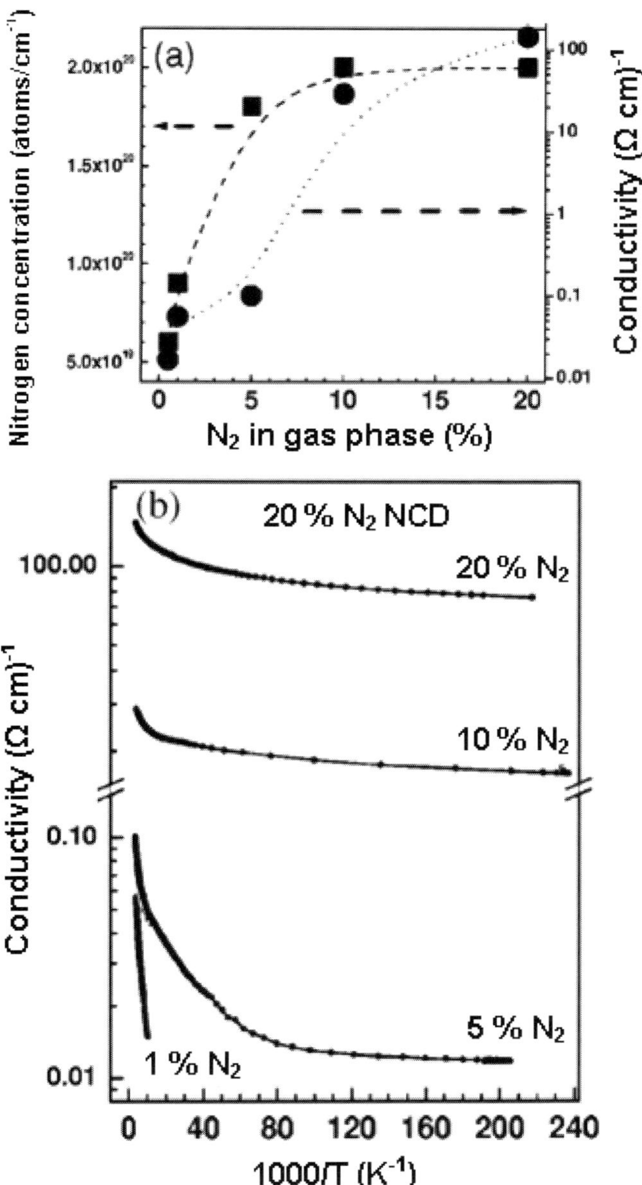

Fig. 2.16. (a) Total nitrogen content (left axis) and room-temperature conductivity (right axis) as a function of nitrogen in the plasma. (b) Arrhenius plot of conductivity data obtained in the temperature range 300–4.2 K for a series of films synthesized using different nitrogen concentrations in the plasma as shown [56].

have been observed, indicating thermally-activated conduction mechanisms with different activation energies [60].

The bonding structures of nitrogen-doped NCD films haven been investigated by NEXAFS measurements [61]. It is revealed that the globe amount of sp^2-bonded carbon in these films increased slightly with nitrogen doping. Nitrogen exists primarily in the tetrahedrally coordinated sites while the grains remain phase-pure diamond. The increased conductivity is attributed to the nitrogen impurities incorporated into grain boundaries, providing more midgap states in the bandgap which allows hopping conduction or other thermally activated conduction mechanisms to occur.

The results of the total electron yield (TEY) NEXAFS experiments on nitrogen-doped NCD are shown in Fig. 2.17(a), and a detailed set of spectra of the π^*-bonding peak and the second bandgap feature are shown in Figs. 2.17(b) and 2.17(c), respectively. The nitrogen-doped NCD films show nearly identical spectra as single crystal diamond, indicating that they are phase-pure diamond containing only a very small amount of sp^2-bonded carbon. As nitrogen is added to the plasma, the sp^2-bonding increases, as indicated by the increase of the π^*-bonding peak and the reduced diamond exciton peak [61].

Fujimori *et al.* [62] indicated that doping with phosphorous produces n-type diamond films, but the resistivity of the phosphorous doped diamond films was too high for practical use. Koizumi *et al.* [63] reported that P-doped CVD (111) films produced using PH_3 showed n-type conductivity by Hall-effect measurement. However, the mobility of the film was only $23\,cm^2\,V^{-1}\,s^{-1}$. Lightly P-doped diamond films have been synthesized and the highest mobility values currently, $660\,cm^2\,V^{-1}\,s^{-1}$, have been achieved at room temperature by Katagiri *et al.* [58]. The phosphorous concentration in the film was controlled at a low doping level of the order of $10^{16}\,cm^{-3}$. Diamond films doped with sulfur have also been deposited by introducing hydrogen sulfide into the CVD process [57]. The mobility of electrons at room temperature was $597\,cm^2\,V^{-1}\,s^{-1}$ and the conductivity was $1.3\times10^3\,\Omega^{-1}\,cm^{-1}$. The donor level introduced by sulfur was at $0.38\,eV$ below the conduction band. Koeck *et al.* [64] reported sulfur-doped NCD films for field emission measurements. The sulfur-doped NCD films were deposited using a 50 ppm hydrogen sulfide in hydrogen mixture. The films showed diminished threshold fields at elevated temperatures as well as a low effective work function. This emission behavior makes this material a prime candidate for vacuum thermionic energy conversion where efficient electron sources are requisite for the transformation of thermal into electrical energy.

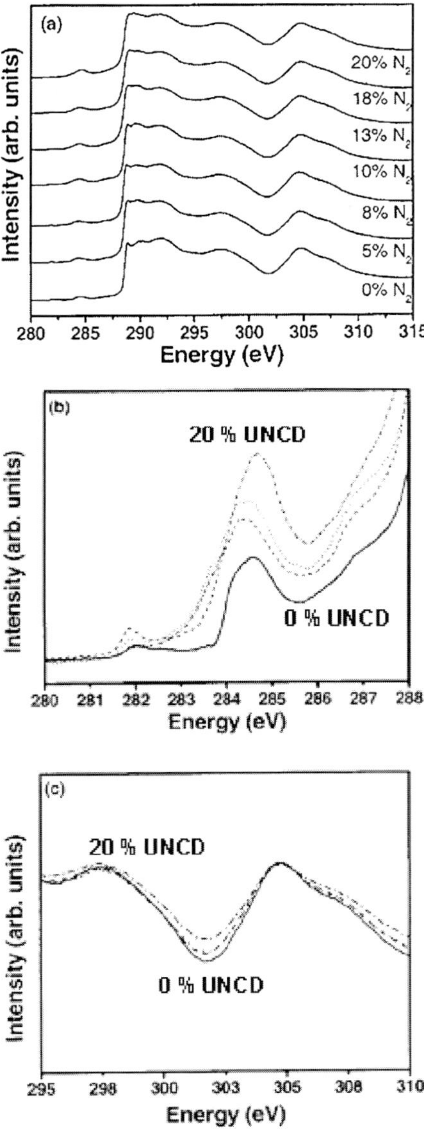

Fig. 2.17. TEY NEXAFS scans of nitrogen doped UNCD thin films: (a) full scan, showing the relevant features for diamond, (b) detailed scan of the π^*-bonding peak, and (c) detailed scan of the second bandgap feature. As the amount of nitrogen in the growth plasma increases, the height of the π^*-bonding peak at 285 eV as shown in (b) increases, the diamond exciton at 289.5 eV illustrated in (a) becomes less distinct, and the depth of the second bandgap decreases, as shown in (c) [61].

3. Applications of NCD

3.1. *MEMS/NEMS Applications of NCD Films*

Micro/Nano-electro-mechanical systems (MEMS/NEMS) devices are currently fabricated mainly using silicon. Complex shapes and devices, such as gears, pinwheels, micromotors, etc. have been produced using stand lithographic patterning and etching techniques [65, 66]. However, a major problem with the Si-based MEMS devices is that Si has poor mechanical and tribological properties due to its hydrophilicity, high friction, poor brittle fracture strength and poor wear properties. Rotating devices such as microturbines [67] has been fabricated, but are frequently subjected to stiction problems or wear-related failure after a few minutes of operation [66]. Until now, there are no commercially available MEMS devices that involve sliding interfaces. With the rapid development of micro- and nanosystems (MEMS/NEMS) such as optical switches, radio frequency resonators, pressure sensors, a new type of material with significantly improved mechanical and tribological properties is of great need for MEMS applications involving rolling or sliding contact in harsh and hostile environment.

Diamond is a super hard material with extreme mechanical strength, outstanding thermal stability and conductivity, and chemical inertness. Relevant mechanical properties of Si, SiC and diamond are shown in Table 3.1. Moreover, diamond has a much lower coefficient of friction (COF) than Si and SiC. The COF of Si has a relatively high and constant value of $\sim$0.6 throughout the temperature ranges; whereas the COF of diamond is an exceptionally low $\sim$0.05–0.1 at room temperature. The COF of Si is high at room temperature $\sim$0.5, which drops to a value of $\sim$0.2 as the temperature exceeds 200°C. The COF of diamond drops to 0.01–0.02 due to the formation of a stable adsorbed oxygen layer on the diamond surface that

Table 3.1.   Selected mechanical and tribological properties of Si, SiC, diamond (at room temperature).

Property	Si	SiC	Diamond
Lattice Constant (Å)	5.43	$\sim$4.35	3.57
Cohesive Energy (eV)	4.64	6.34	7.36
Young's Modulus (Gpa)	130	450	1,200
Shear Modulus (Gpa)	80	149	577
Hardness (kg/mm^2)	1,000	3,500	10,000
Fracture Toughness	1	5.2	5.3
Flexural Strength (Mpa)	127.6	670	2,944
Coefficient of Friction (COF)	0.6	0.5	0.01–0.02

saturates the diamond dangling bonds and minimizes bonding between contacting diamond surface.

3.1.1. *Fabrication of NCD MEMS/NEMS Devices*

The new nanocrystalline and ultrananocrystalline diamond developed from Ar/CH_4 gas mixtures is emerging as one of the most promising materials for MEMS/NEMS. In previous discussion of mechanical and tribological properties of NCD film, we have mentioned that the average nanocrystalline diamond film grain size is less than 50 nm and the surface roughness, 20 nm.

Numerous NCD-based MEMS components and devices have been developed in recent years. There are several ways to fabricate these devices by taking advantage of the extraordinary properties of nanocrystalline diamond. Among them, one of the most apparent ways is to form a conformal NCD coating on existing Si products. Modern Si fabrication technology can produce all kinds of devices in the nanometer scale. The basic idea is to form a thin, continuous, and conformal coating with low roughness on the Si component. However, this could not be realized until the recent development of nanocrystalline deposition technique. Conventional diamond CVD deposition results in discontinuous films with large grains and rough surface. Figure 3.1(a) shows Si microwhiskers coated with (a) conventional CVD diamond and (b) nanocrystalline diamond. Separated diamond grains can

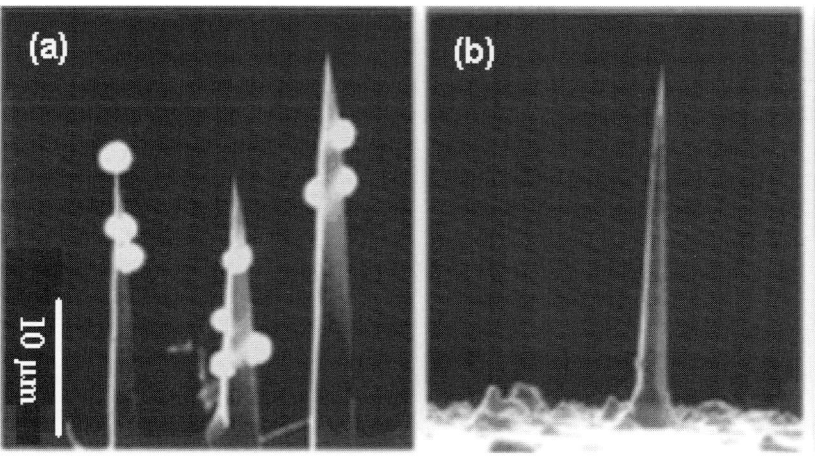

Fig. 3.1. (a) Si microwhisker coated with conventional CVD diamond; (b) single Si microwhisker coated with NCD [40].

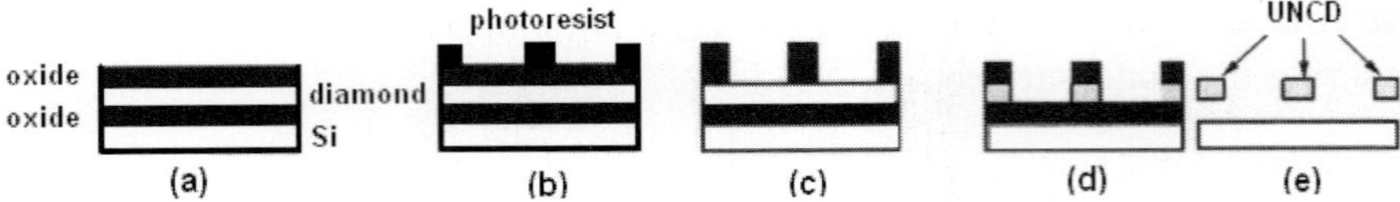

Fig. 3.2.   Etching scheme for producing a freestanding NCD film [40].

be observed for MCD coating, whereas a very smooth conformal coating was obtained for the NCD-coated microwhisker.

Selective deposition has been used for fabrication of NCD microdevices, shown in Fig. 3.2 [40]. Nanocrystalline diamond deposition process needs a nucleation layer, usually achieved by mechanical seeding with fine diamond particles or by bias-enhanced nucleation. So NCD films can be selectively deposited by selective seeding the substrates. Selective seeding can be achieved by: (1) removing part of the seeded layer; (2) masking the film during seeding to produce a patterned seeded layer. Note that it is hard to achieve high feature resolution by selective seeding since precise control of seeding is hard to define. Instead, a more advanced method has been developed by the lithographic patterning technique. Continuous NCD films are deposited on the SiO_2 layer, followed by a sacrifice SiO_2 layer deposited by CVD. Photoresist is deposited onto the top SiO_2 layer and it is patterned by RIE. The NCD film is then etched by oxygen plasma.

3.1.2. *NCD Cantilever*

Free standing NCD cantilever structures, shown in Fig. 3.3, have been fabricated for microscale mechanical testing. The detailed procedures are summarized as follows [45]:

1. Seeding the Si wafer and NCD growth (0.5–0.6 μm).
2. Deposition of 300 nm Al by sputtering. Al is used as mask material due to its resistance to oxygen RIE.
3. Photoresist spin-coating with S 1805; exposure with Karl Suss MA6; developing; postbaking.
4. Wet chemical etching of Al.
5. O_2 reactive ion etching (RIE), 50 mTorr, 200 W, until the exposed NCD is etched away. During the etching, the photoresist is also removed. Removal of Al mask is performed using wet etching.
6. Si wafer KOH etching from the front side (90 min, KOH30% at 80°C) using the NCD pattern as a masking layer. Cantilevers are released.

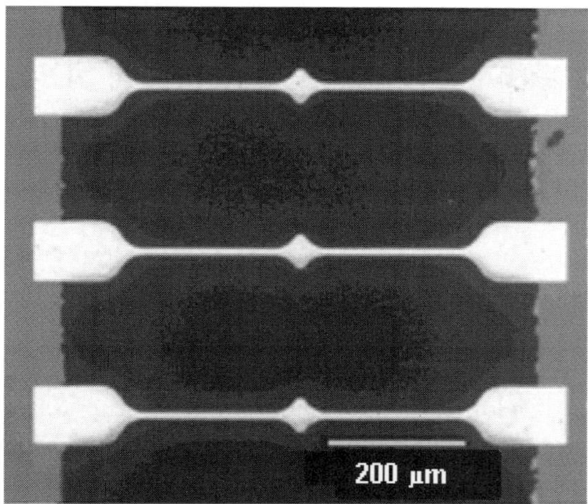

Fig. 3.3. Optical image of three as-microfabricated NCD membranes [45].

Underetching of NCD cantilevers is possible due to a slight misalignment of the Si wafer $\langle 110 \rangle$ direction with respect to the cantilever structure.

3.1.3. *SAW Devices*

Surface acoustic wave (SAW) devices are critical components in wireless communication systems. Diamond exhibits high acoustic phase velocity which makes it an attractive material for high frequency (GHz range) SAW devices. However, high surface roughness and large grain sizes prevent the application of diamond for SAW applications. Time-consuming post deposition is required to smooth the MCD surface in order to obtain free-standing smooth surface. Moreover, MCD-based SAW devices suffer from relatively low performances due to propagation losses in polycrystalline structures. It is reported that the propagation losses decreases while the diamond grain size diminishes [68, 69]. Therefore, NCD based SAW devices are expected to have better performance than MCD films.

A model has been established by Elmazria *et al.* [70] based on AlN/NCD/Si layered structure to determine the acoustic phase velocity V and the electromechanical coupling coefficient k^2 as a function of NCD thickness. Figure 3.4 shows a schematic drawing of an NCD-based SAW device. NCD films are grown on (100)-oriented silicon substrate and smooth piezoelectric AlN films with columnar structures and (002) orientations are

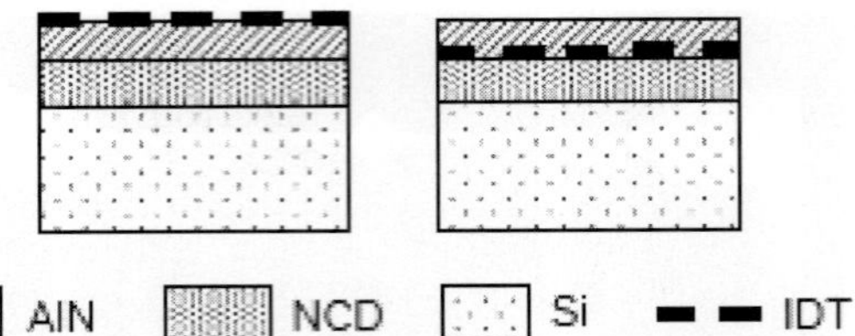

Fig. 3.4. Schematic cross-sections of the two possible IDT configurations in the AlN/NCD/Si layered structure [70].

deposited on top of NCD. Aluminium IDE of 16 and 32 um wavelengths on AlN/NCD/Si is developed by photolighography. Figure 3.5 shows the phase velocity of measured NCD SAW device as a function of NCD and AlN film thickness, where kh_{NCD} is the normalized NCD thickness and kh_{AlN} is the normalized AlN thickness. It is shown that devices working at frequency $f_0 = 2.5\,\text{GHz}$ can be achieved with acoustic velocity as high as $9500\,\text{m/s}$ and $K^2 = 1.4\%$. Note that the experimental values and the theoretical values agree very well and the phase velocity strongly increases with the NCD normalized thickness.

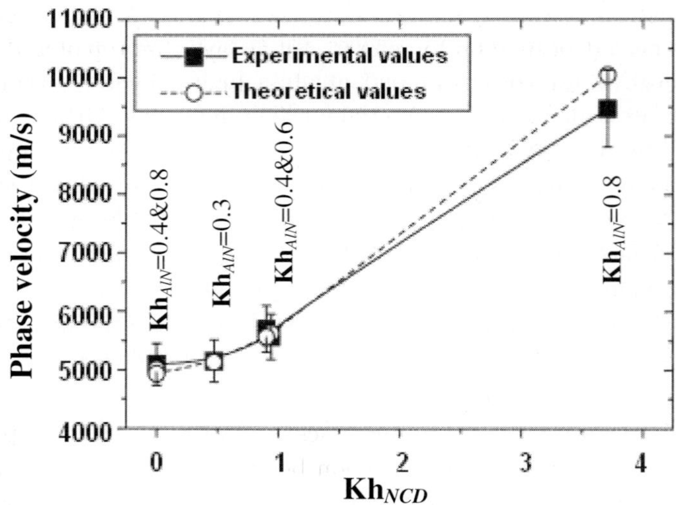

Fig. 3.5. Phase velocity measured from SAW device high frequency characterization for several kh_{NCD} and kh_{AlN} values along with theoretical data calculated for the AlN/NCD/Si layered structure [70].

3.1.4. *NCD RF MEMS Devices*

The wide bandgap, small (or negative) electron affinity, high breakdown voltage, high saturation drift velocity, high carrier mobility, radiation hardness and high thermal conductivity make diamond a promising material for the fabrication of high power, high frequency and high temperature solid state microelectronic devices, sensors/MEMS and vacuum microelectronic devices [71]. Polycrystalline diamond-based RF electronics have been fabricated by various groups. Gildenblat *et al.* [72] have reported high temperature Schottky diode with thin film diamond. Diamond field effect transistors (FETs) have been investigated by *Aleksov et al.* [73]. Prins reported the first diamond bipolar junction transistor (BJT) in 1982 on natural, p-type single crystal diamond [74]. Aleksov *et al.* [75] presented the p–n diodes and p^+np BJT structures using nitrogen and boron doping during CVD. It was thought that only polycrystalline diamond films could be used for RF MEMS devices due to the difficulty of producing large areas of NCD films. Surface area is needed for RF structures including active devices, passive components, waveguide circuits and heat dissipation. Until recently, large area deposition of nanocrystalline diamond becomes possible by Ar/CH_4 reactants with microwave chemical vapor deposition technique. Due to the unique properties of nitrogen-doped nanocrystalline diamond, a nanocrystalline/dingle crystal diamond heterostructure diode has been fabricated and characterized [76]. As shown in Fig. 3.6, N-type doped NCD films has been deposited on top of a p^- active layer (on top of a p^+ contact layer) where all the structures are built based on a single crystal diamond substrate. The diode characteristics show rectification and high temperature stability. Figure 3.7 shows that the device has been successfully tested at $1050°C$ which represents the highest temperature measurement of any semiconductor to date. Low forward losses combined with high breakdown strength have been achieved, which is not possible for any other polycrystalline diamond diodes.

Kusterer *et al.* [77] demonstrated a bistable microactuator based on stress-engineered nanocrystalline diamond. The NCD film was compressively pre-stressed controlled by different growth conditions. After release from Si substrate, NCD film could buckle into two quasistable positions, downwards or upwards. The transition between these two positions could be realized by combining the diamond film with Ni/Cu strips acting as thermal bimetal elements. Thus, a $10\,\mu m$ deflection could be achieved with film thickness $\sim 2\,\mu m$ resulting in a $12\,V$ threshold voltage to switch

 Z. Xu and A. Kumar

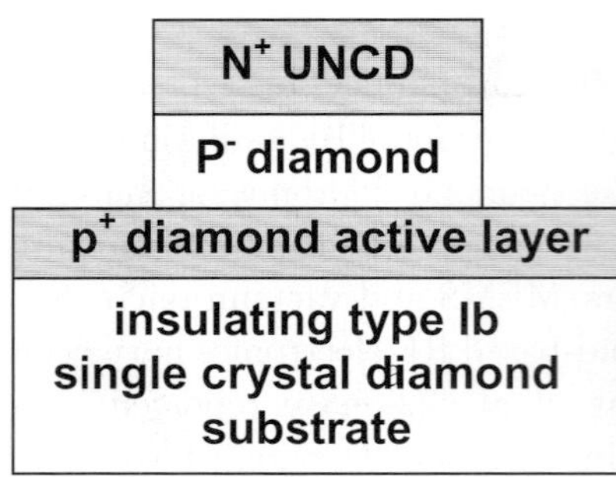

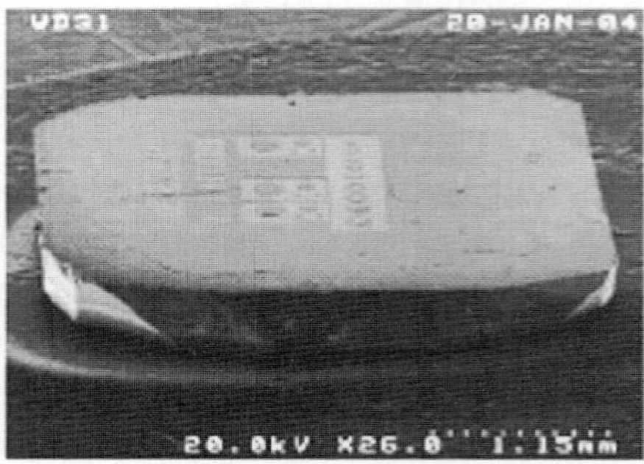

Fig. 3.6.   (a) Schematic drawing and (b) SEM micrograph of a NCD/single crystal diamond heterostructure diode [76].

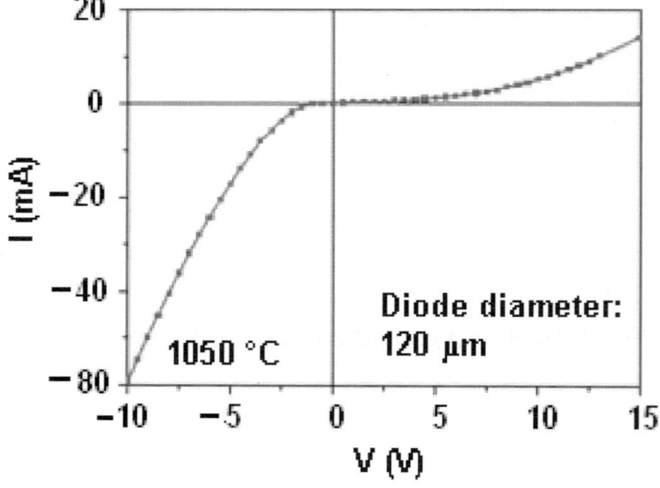

Fig. 3.7.   IV characteristics of a diode with 120 µm diameter at 1050°C [76].

between two stable positions. The first hybrid structure based on NCD as a RF switch was fabricated recently [77]. Figure 3.8 shows the principle of device structure and hybrid integration of NCD actuator and base plate. NCD actuator is fixed on Al_2O_3 substrate including aninterrupted co-planar waveguide. The actuator is fixed in the center of the waveguide, and moves upwards and downwards during operation between two stable positions. In the down-state position, the microbridge acts as signal contact connecting the segments of the line. Numerical simulations show that the diamond-based actuator and base plant can deliver large forces in the

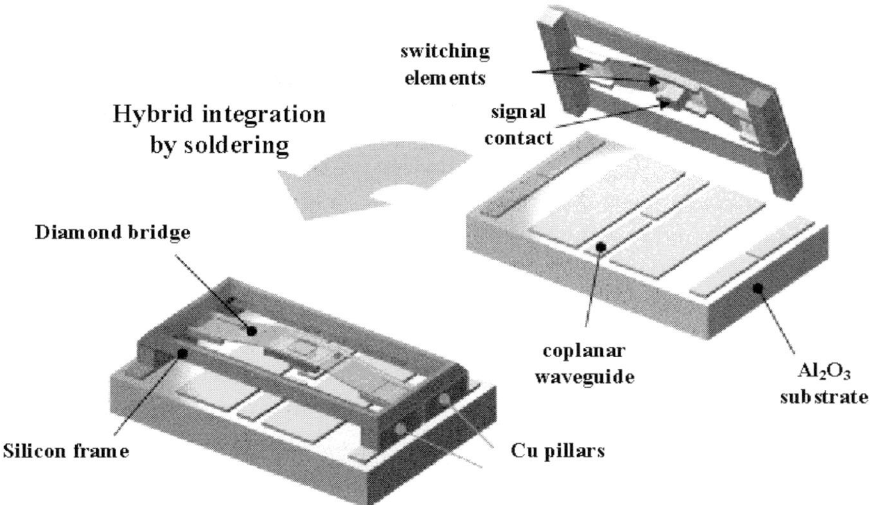

Fig. 3.8.   Principle of device structure and hybrid integration of diamond actuator and base plate. Here the microtool acts as signal contact to close the break of a coplanar waveguide on a low loss Al_2O_3-substrate [77].

range of hundred μN. Therefore, this NCD actuator could also be used for numerous purposes such as microsqueezer, micropunch or microanvil.

3.2. Electrochemistry Applications of NCD Films

3.2.1. History of the Electrochemistry of Diamond Films

Diamond was first reported for electrochemical applications by Iwaki *et al.* [78]. Conducting diamond were obtained by ion implantation of natural diamond with argon and zinc. With the development of CVD technology, CVD diamond electrodes have been widely used in electrochemistry. Pleskov *et al.* [79] and coworkers first extensively studied the electrochemical properties of diamond electrodes. The diamond electrodes grown by CVD were normally undoped, but had sufficient conductivity by using specific growth conditions during deposition to introduce defects in diamond films for electrochemical measurements. Moreover, boron-doped diamond electrodes have been studied by Swain [80, 81] and coworkers. Diamond electrodes exhibit low capacitance and featureless background current which are highly desirable for electroanalytical and sensor applications.

3.2.2. *Basic Diamond Properties in Electrochemistry*

The electrochemical response of diamond electrodes is determined by the nature of the diamond film as well as the surface termination of diamond. The chemical inertness of diamond is remarkable compared to that of other electrode materials. However, the diamond electrodes are not completely electrochemically inert. The electrode surface can change from hydrophobic to hydrophilic and chemically-bound oxygen is found on a diamond surface after anodic polarization [82, 83]. Furthermore, cyclic voltammetry shows that polycrystalline diamond has a redox couple at 1.7 V versus standard hydrogen electrode (SHE) whereas single crystal diamond does not [82].

As-deposited diamond electrodes that have been cooled down in a hydrogen-rich environment are terminated with hydrogen. These hydrogen terminated surfaces result in a wide potential range, which is illustrated in Fig. 3.9. Note that various types of electrodes have been evaluated. A wide potential range has been obtained for high quality diamond. Low quality diamond electrodes have a potential window similar to glass carbon and graphite. With this wide "window", other electrochemical reactions

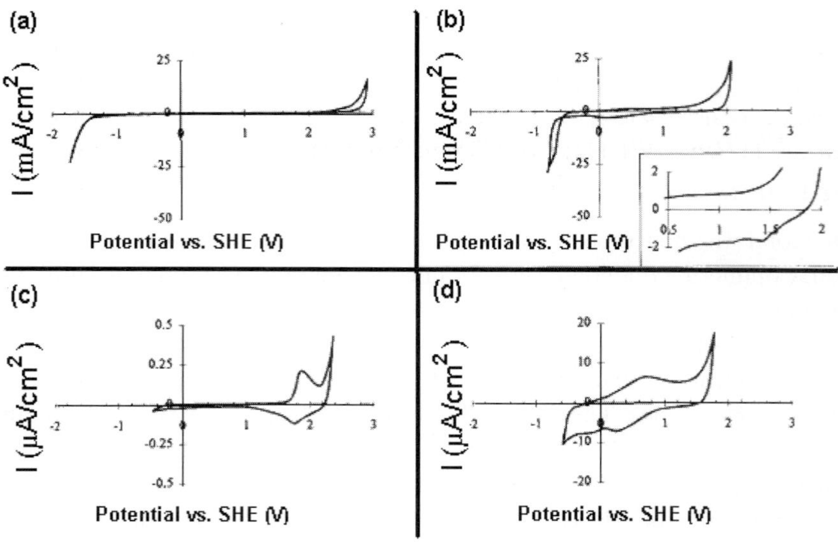

Fig. 3.9. Cyclic voltammograms for diamond electrodes in 0.5 M H_2SO_4. Scan rate was 200 mV/s; reference electrode was Pt/H_2. (a) High quality diamond at mA/cm^2 current densities; (b) low quality diamond at mA/cm^2 current densities, expanded view for low quality diamond for potentials +0.5 to +2.0 V (inset); (c) high quality diamond at μA/cm^2 current densities; and (d) low quality diamond mA/cm2 current densities [83].

can be observed which makes diamond electrodes attractive for sensor applications [83].

Most of the previous electrochemical applications of diamond use boron-doped diamond electrodes. The electrical conductivity induced by boron has been widely studied and its reviews can be found elsewhere [84]. During deposition diamond changes its nature from a dielectric, to a wide gap semiconductor and finally a quasimetal when boron is doped to diamond electrodes. Thus, it is possible to tune the electrical properties of diamond by changing the doping concentrations. It has been observed that as the doping levels are increased, the potential window of water stability decreases and the crystalline quality decreases [85]. Nitrogen and phosphorus are two deep donors to make the diamond n-type. Recently, sulfur has been reported to produce n-type diamond. Eaton *et al.* [86] have doped diamond with sulfur and small quantities of boron to produce diamond electrodes with n-type conductivity in the near surface region.

The electrochemical properties of nanocrystalline diamond were first investigated by Fausett *et al.* [87]. The electrochemical activity of the NCD electrodes was probed using $Fe(CN)_6^{-3/-4}$, $Ru(NH_3)^{+2/+3}$, methyl viologen and 4-tert-butylcatechol. It was suggested by Fausett that the advantages of using NCD electrodes in electrochemistry compared to microcrystalline diamond included: (1) the ability to deposit continuous films at nanometer thicknesses rather than micrometer thicknesses leading to time and cost savings, (2) easier coating of irregular geometry substrates like fibers and high surface area meshes, and (3) different film morphologies and characteristic electronic properties might lead to unique electrochemical behaviors [87].

3.2.3. *Basic Principles of Electrochemical Measurements*

The electrochemical measurement involves the measurement of an electrical signal (e.g. potential, current or charge) associated with the oxidation or reduction of a redox analyte dissolved in solution, and relating this signal to the analyte concentration [88]. Diamond electrodes possess a number of unique electrochemical properties compared to other sp^2-bonded carbon electrodes, including linear dynamic range, limit of quantitation, fast response time and response stability.

Figure 3.10 shows the design of a typical cell used for electrochemical measurements [89]. The three-neck cell is constructed of glass. Three electrodes, diamond working electrode, counter electrode (Pt) and reference electrode (Ag/AgCl), are clamped to the bottom of the glass cell. Diamond

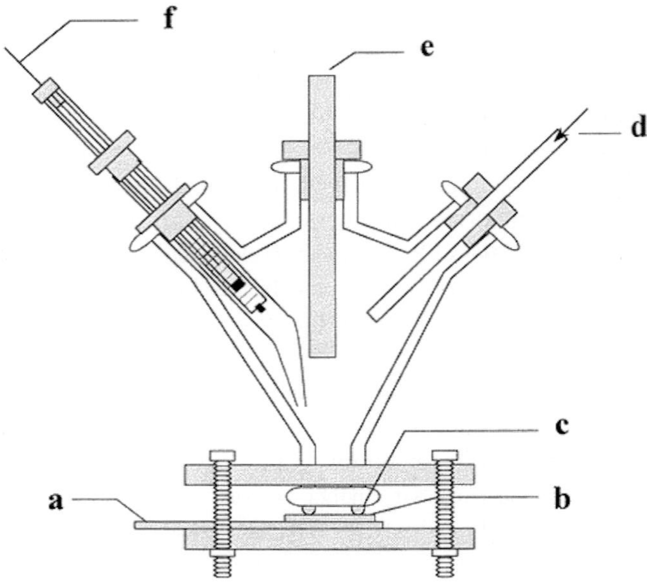

Fig. 3.10. Diagram of the single-compartment, glass electrochemical cell used. (a) Cu or Al metal current collecting plate, (b) diamond film electrode, (c) Viton O-ring seal, (d) input for nitrogen purge gas, (e) carbon rod or Pt counter electrode, and (f) reference electrode inside a glass capillary tube with a cracked tip [89].

film is placed on the bottom of the cell and ohmic contact is made on the back side of the substrate for measurement.

3.2.4. *Electrochemical Properties of NCD*

Boron-doped diamond exhibits very low background current density and a wide working window in voltammetric and amperometric measurements. The comparison of nanocrystalline diamond has been studied, and it appears that NCD films deposited from C_{60}/Ar gas mixture have basic electrochemical properties similar to boron-doped microcrystalline diamond films. Low voltammetric background current (one order of magnitude lower than glassy carbon), a wide working potential window (3 V) and a high degree of electrochemical activity for several redox systems have been obtained. Figure 3.11 shows cyclic voltammetric i–E curves for a NCD film and a boron doped MCD film. Figure 3.11(a) shows voltammograms for NCD and MCD films between -0.5 and 1.0 V (versus SCE). The curve of NCD film is basically featureless within this potential window.

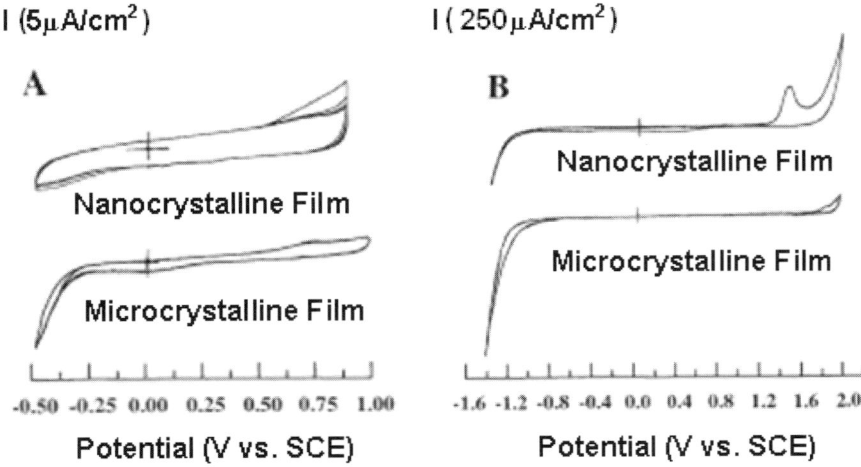

Fig. 3.11. Cyclic voltammetric i–E curves for a nanocrystalline diamond film deposited from a 1% C_{60}/Ar gas mixture and a boron-doped microcrystalline diamond film deposited from a 0.3% CH_4/H_2 gas mixture over (A) a potential range from -0.5 to 1.0 V (versus SCE) and (B) a potential range from -1.6 to 2.0 V (versus SCE). The electrolyte was 1 M H_2SO_4 and the potential sweep rate for (A) was 50 mV/s and 25 mV/s for (B) [87].

Figure 3.11(b) shows voltammograms for NCD and MCD films between -1.6 and 2.0 V (versus SCE). The response shows a large anodic peak at 1.5 V and a smaller cathodic peak at 0.4 V. The anodic peak at 1.5 V is likely due to redox-active carbon in the grain boundaries. A similar peak has been observed in MCD films previously and been attributed to the presence of reactive sp^2-bonded grain boundary carbon [87].

Further experiments have been carried out for NCD films in different redox analytes. Figure 3.12 shows cyclic voltammetric i–E curves for (A) 0.1 mM $Fe(CN)_6^{-3/-4}$, (B) 0.1 mM $Ru(NH_3)_6^{+2/+3}$, (C) 0.1 mM methyl viologen, and (D) 0.1 mM 4-tert-butylcatechol at the nanocrystalline diamond film. The active responses have been observed for all these analytes without any doping or surface treatment for NCD films.

Nitrogen-doped electrically conductive NCD films has been obtained by adding N_2 gas to Ar/CH_4 during deposition. The film deposited form Ar/N_2/CH_4 mixtures are a little rougher morphologically with an average cluster size of 150 nm. These kinds of NCD films also possess good electrochemical activity without any conventional pretreatment. They exhibit semimetal-like electronic properties over a wide potential range from at least 0.5 to -1.5 V versus SCE, shown in Fig. 3.13. It is clear that the responses

     Z. Xu and A. Kumar

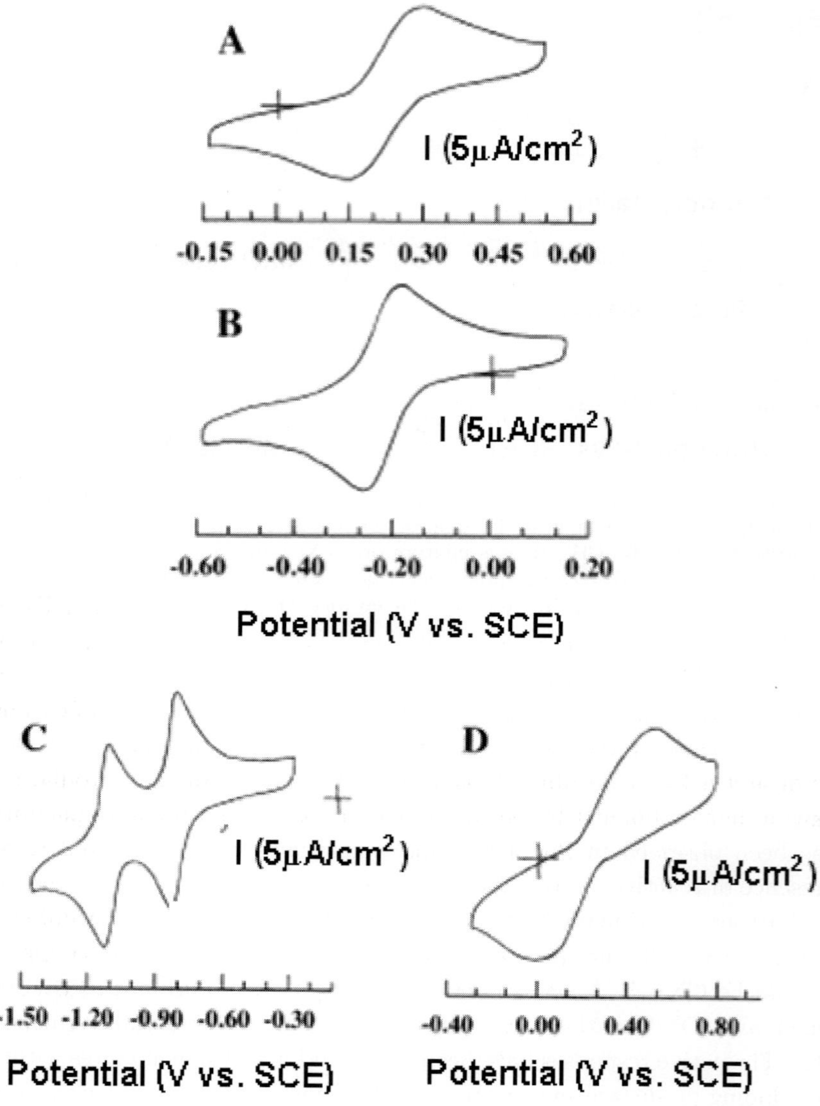

Fig. 3.12. Cyclic voltammetric i–E curves for (A) 0.1 mM $Fe(CN)_6^{-3/-4}$, (B) 0.1 mM $Ru(NH_3)_6^{+2/+3}$, (C) 0.1 mM methyl viologen, and (D) 0.1 mM 4-tert-butylcatechol at the nanocrystalline diamond deposited from a 1% C_{60}/Ar gas mixture. The analyte concentrations were all 0.1 mM and the electrolyte was 1 M KCl. Potential sweep rate: 50 mV/s [87].

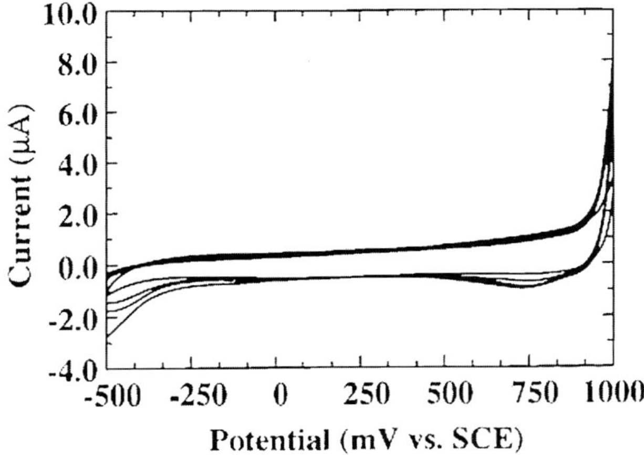

Fig. 3.13. Cyclic voltammetric *i–E* curves for nanocrystalline diamond thin films deposited from 1% CH_4/1% N_2/98% Ar, 1% CH_4/2% N_2/97% Ar, 1% CH_4/4% N_2/95% Ar and 1% CH_4/5% N_2/94% Ar. Electrolyte: 1 M KCl. Scan rate: 0.1 V/s [59].

of the electrodes are independent of the level of nitrogen doping. For all the films the background currents are mostly featureless within this potential range. Generally, disordered carbon electrodes exhibit oxidation and reduction peaks between −200 and 500 mV due to the formation of carbon–oxygen functionalities. But only very small anodic and cathodic peaks can be observed for NCD films. It is suggested that the low background current of diamond is attributed to two factors: a reduced pseudocapacitance from the absence of redox-active and/or ionizable surface carbon–oxygen functional groups, and a slightly lower internal-charge carrier concentration, due to the semimetal-like electronic properties [84]. Wide potential working windows have also been obtained for nitrogen-doped NCD films: 0, 2, 4, 5% N_2 doped films have working potential windows of 4.27, 4.05, 3.85, and 3.87 V [59], respectively. These windows are 1 V greater than what is usually observed for glassy carbon.

Boron-doped nanocrystalline diamond have also been successfully obtained by Swain [84]. The boron-doped NCD films consist of clusters of diamond grains around 100 nm in diameter and surface roughness of 34 nm [90]. The electrical conductivities of nitrogen doped NCD films are dominated by the high fraction of sp^2 bonded grain boundaries. However, the electrical properties of boron-doped NCD films are dominated by the charge carriers. Good electrochemical properties have been observed for these boron-doped samples. A large potential window of around 3.1 V has

                                    Z. Xu and A. Kumar

been observed for both types of nitrogen- and boron-doped films for cyclic
voltammetric i–E curves in 1 M KCl. Well-defined symmetric curves are
observed with peak separations of 65 and 73 mV. This low separation value
indicates relatively rapid response of diamond electrodes [89].

3.2.5. *Electroanalytical Applications*

Nanocrystalline diamond provides a new type of electrochemical electrodes
with fast responsiveness, low background current, wide working potential
window and stability for a wide range of applications. The following are
some examples of functionalization of NCD films based on electrochemical
techniques.

Wang and Carlise [91] have developed a method to covalent immobilize
the glucose oxidase on conducting ultrananocrystalline diamond thin films
for glucose detection. Electrochemical glucose detection is based on the
enzyme glucose oxidase. A side product of the reactions between glucose
oxidase and glucose is hydrogen peroxide (H_2O_2), whose electroactivity
can be used to obtain a measurable current signal. Glucose oxidase was
attached to the NCD surface via tethered aminophenyl functional moieties
that were previously grafted to NCD surface by electrochemical reduction
of aryl diazonium salt. Figure 3.14 shows a schematic drawing of covalent
immobilization of glucose oxidase on diamond surface.

The functionalization of NCD film was monitored by XPS measurements of C (1s), O (1s) and N (1s) core level spectra shown in Fig. 3.15.
The as-grown NCD film showed noticeable O (1s) signal due to chemisorp-
tion of oxygen or water left on the NCD surface. No N (1s) signals could
be observed. After surface functionalization with aminophenyl groups, an
intense O (1s) signal (O/C ration about 18.4%) was observed which arise
from the tethered carboxyl group. Enhanced N (1s) signal was observed

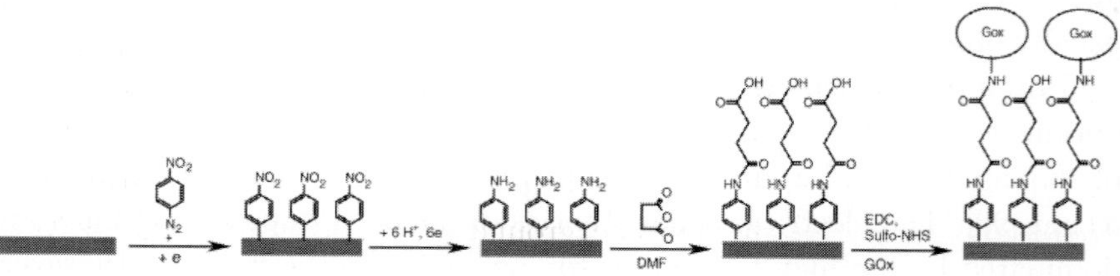

Fig. 3.14.   Schematic drawing of covalent immobilization of glucose oxidase on diamond
surface [91].

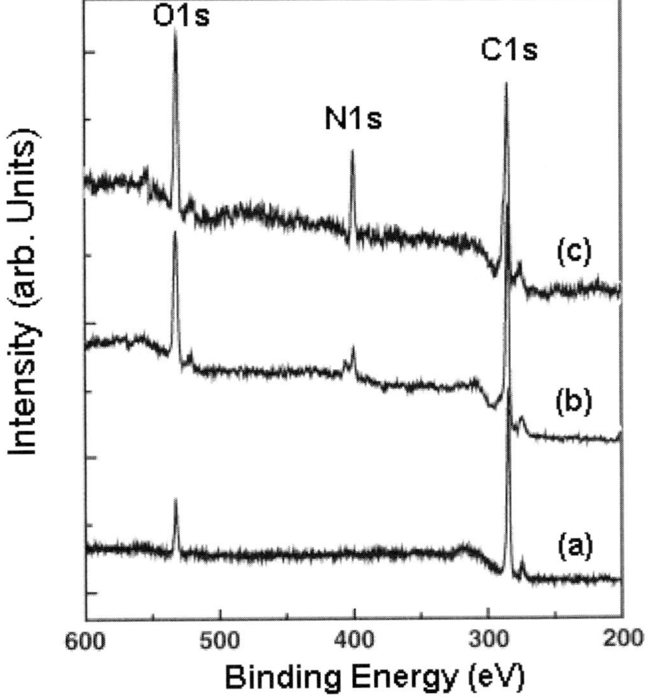

Fig. 3.15. XPS spectra for (a) bare, (b) succinic acid modified, and (c) glucose oxidase modified UNCD thin films [91].

after the succinic acid modification with glucose oxidase due to the successful attachment of protein molecules onto NCD surface.

Figure 3.16 shows the linear sweep voltammograms recorded in 0.1 M phosphate buffer solution (PBS) with and without the presence of glucose (10 mM). A significant increase of anodic peak in the presence of glucose can be observed, corresponding to the anodic oxidation of hydrogen peroxide released from enzyme–glucose oxidation reaction. Oxidation current density at 1.1 V is plotted as a function of glucose concentration. Low detection limit about 0.1 mM and linear response of glucose concentration have been achieved due to the low background current characteristic of NCD electrodes.

Compared with surface oxidation approach in which oxygen functionalities are introduced to diamond surface, the electrochemical approach causes

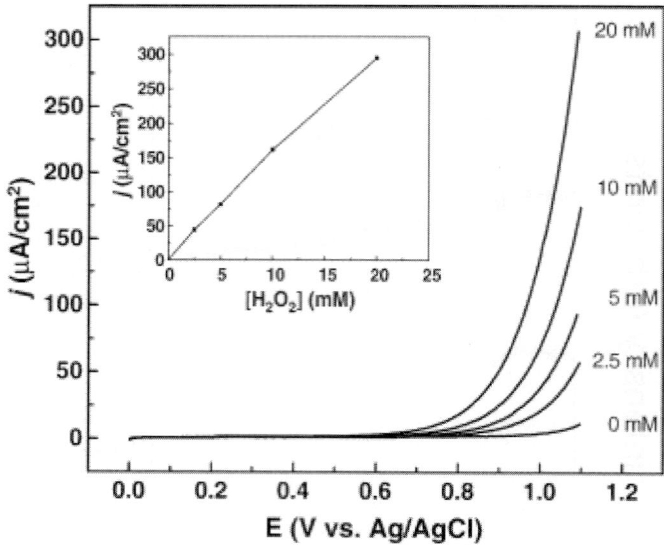

Fig. 3.16.   Linear sweep voltammograms for a glucose oxidase modified UNCD electrode in oxygen-saturated 0.1 M PBS solution (pH = 7.2) with (solid line) and without (dashed line) the presence of 10 mM glucose. Scan rate = 50 mV/s. The inset shows the plot of the anodic oxidation current density recorded at 1.1 V versus the concentration of glucose [91].

less damage to diamond surface and results in more stable C–C linkage at the interface, providing better detection limit and reproducibility.

3.3. *Biomedical Applications of NCD Films*

Diamond is known as a biocompatible material which is a good candidate as a biointerface with soft materials for biomedical applications. However, NCD biomedical applications require control of its surface properties such as surface chemistry, wettability, optical properties, etc. The surface modification of synthetic diamond has been investigated by several methods: (1) direct chemical functionalization of hydrogen-terminated surface [93]; (2) generation of oxygen-containing surface functionalities [94]; (3) electrochemical modification of NCD surface by functional groups [92].

The first DNA modification of NCD thin film was reported by Yang *et al.* [93]. A photochemical modification scheme is shown in Fig. 3.17. In order to chemically modify clean, H-terminated NCD surfaces grown on silicon substrates, the H-terminated substrates were photochemically reacted with a long-chain ω unsaturated amine, 10-aminodec-1-ene, that

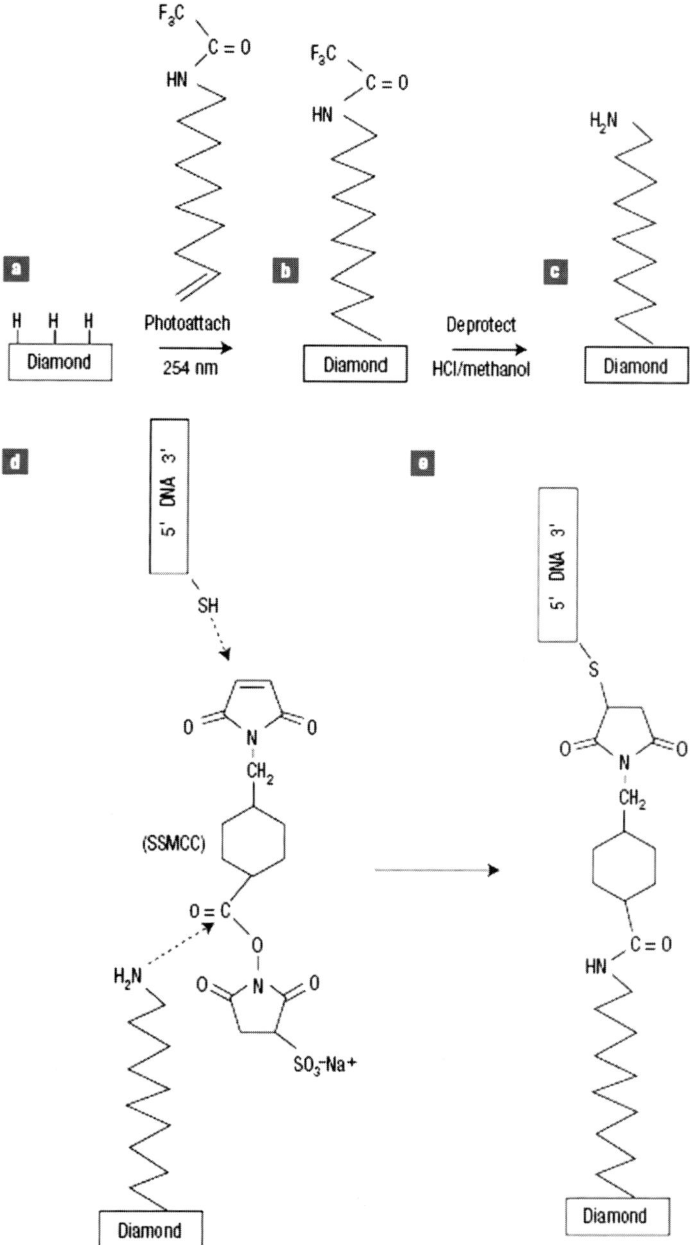

Fig. 3.17.   Sequential steps in DNA attachment to diamond thin films [93].

has been protected with the trifluoroacetamide functional group (TFAAD), producing a homogeneous layer of amine groups for DNA attachment. The protected amine was then de-protected, leaving behind a primary amine [Fig. 3.17(c)]. The primary amine was reacted with a heterobifunctional crosslinker sulphosuccinimidyl-4-(N-maleimidomethyl) cyclohexane-1-carboxylate (SSMCC) and finally reacted with thiol-modified DNA [Fig. 3.17(d)] to produce the DNA-modified diamond surface [Fig. 3.17(e)]. Hybridization reactions with fluorescently tagged compelementary and non-complementary oligonucleotides showed good selectivity of NDA-modified NCD films with matched and mismatched sequences. Comparison of NCD with other substrates including gold, silicon, glass, and glassy carbon showed that diamond surfaces exhibited extremely good stability, and without loss of selectivity (Fig. 3.18). No decrease in intensity from a complementary sequence, and no increased background after exposure to a non-complementary sequence has been found by the end of 30-cycle stability test [93].

Not only DNA but other proteins have been covalently attached to NCD films by chemical functionalization [95]. H-terminated NCD films were modified by using a photochemical process and green fluorescent protein was covalently attached. A direct electron transfer between the enzyme's redox center and the diamond electrode was detected. An enzyme-based amperometric biosensor was fabricated by functionalizing the surface of a NCD film with the enzyme catalase (from bovine liver). The device was fabricated on top of the hydrogenated NCD film. Two small Ti/Au coated contacts were oxidized on top of the diamond surface. The contacts and the hydrogenated area in between were isolated from the rest of the sample surface by second oxidation step. Then another layer of gold was evaporated to establish good electrical contact between the hydrogenated area and the contacts. A schematic drawing of the device is shown in Fig. 3.19. The sample was then functionalized by a serious biofunctionalization process. Detailed functionalization steps can be found in the paper published by Hartl *et al.* [95]. They have demonstrated that biomolecules can be covalently immobilized on NCD surfaces and these immobilized biomolecules retain their functionality, shown in Fig. 3.20. The strong C–C covalent bond at the diamond/biolayer interface provides an important advantage over normally used metal electrodes such as gold.

Yang and Hamers [96] have reported the fabrication and characterization of a biologically sensitive field-effect transitor (Bio-FET) using a NCD thin film. Biomolecular recognition capability was provided by linking

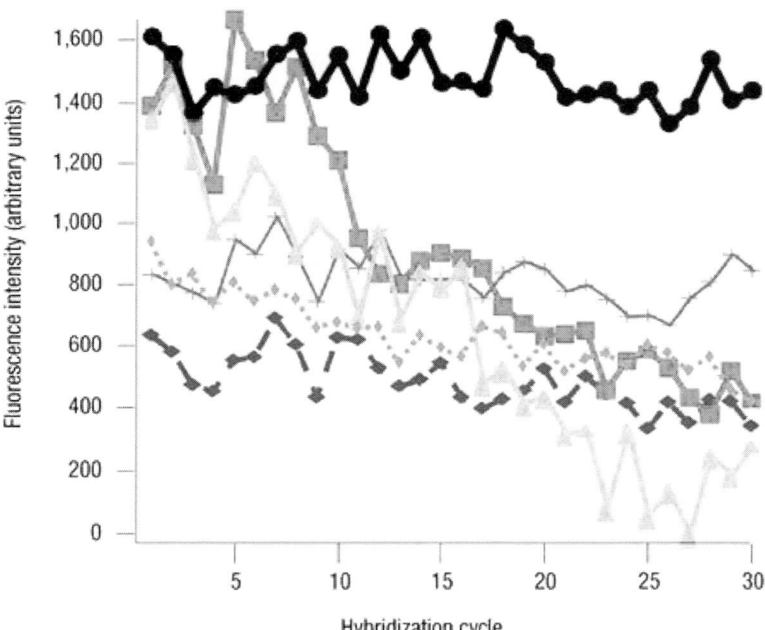

Fig. 3.18. Stability of DNA-modified NCD and other materials during 30 successive cycles of hybridization and denaturation [93].

human immunoglobulin G (IgG) to the diamond surface. The biomolecular recognition and specificity characteristics were tested using two different antibodies anti IgM and anti IgG. The results showed that the bio-FET device responded only to the anti-IgG antibody. The chemical functionalization process is shown in Fig. 3.21(a) and a schematic illustration of diamond bio-FET device is shown in Fig. 3.21(b). The diamond surfaces were first terminated with hydrogen in a radio frequency plasma system. An organic protection monolayer film was covalently linked (step 1) to the surface via photoexcitation at 254 nm. De-protection in 0.36 M HCl/methanol (step 2) flowed by reaction with 3% solution of glutaraldehyde in a sodium cyanoborohydride coupling buffer for 2 h (step 3) then produced a diamond

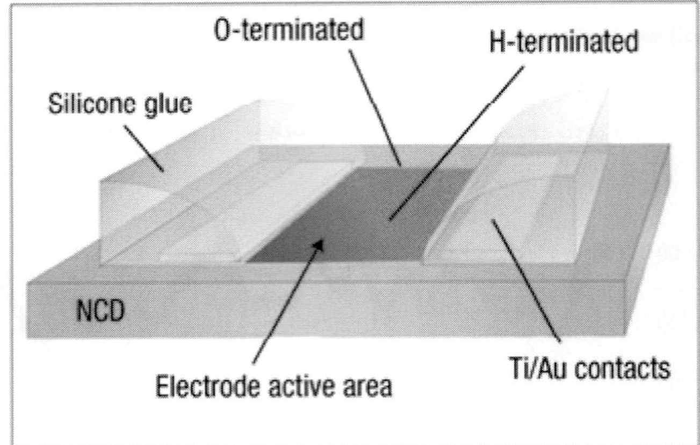

Fig. 3.19.   Schematic drawing of a biofunctionalized NCD-based bio sensor [95].

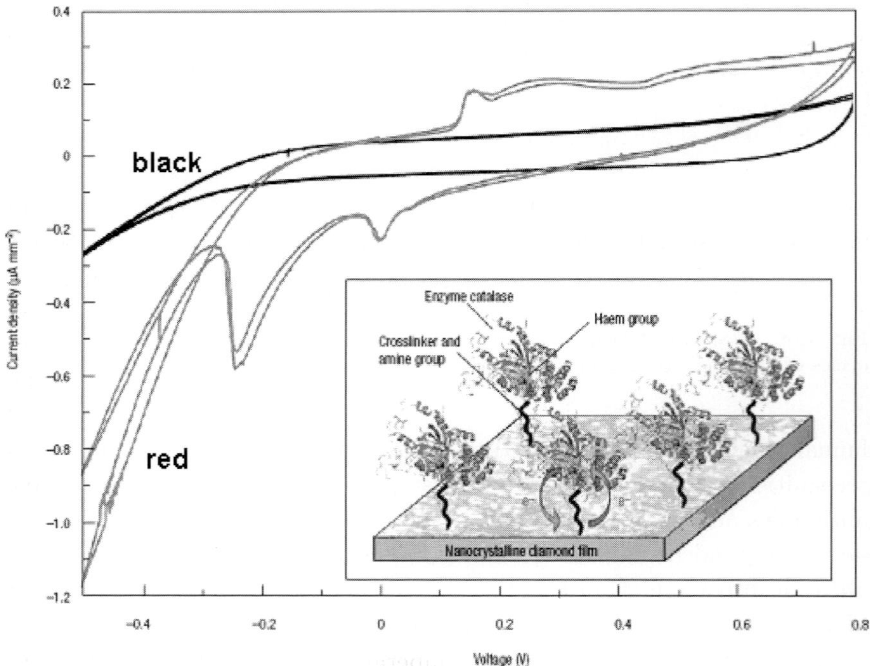

Fig. 3.20.   Cyclic voltammograms of an untreated NCD electrode (black) and a catalase-modified NCD electrode (red), measured in pure phosphate buffer. The inset shows a schematic view of the functionalized electrode, illustrating direct electron transfer between the diamond electrode and the haem group of the catalase [95].

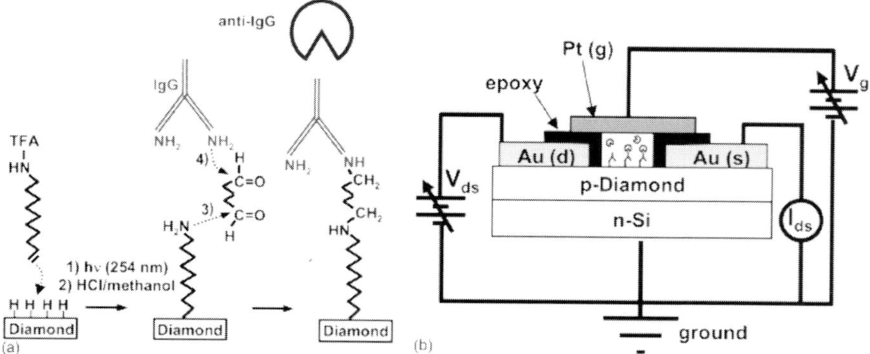

Fig. 3.21. (a) Overview of the chemistry for linking human IgG to diamond surfaces. (b) Schematic illustration of diamond Bio-FET [96].

surface with some exposed aldehyde groups. Gold was then sputtered on the diamond surfaces to form ohmic source and drain contacts. The metal channel was coved with epoxy to insulate them from the solution. The modified diamond surfaces were immersed in human immunoglobulin G (IgG) solution for 8 h (step 4) which binds the IgG to the surface. The top electrodes were finally deposited by sputtering a thin Pt layer. The results show that the modification of diamond surface with IgG yields the ability to selectively recognize and detect the corresponding anti IgG. The sensitivity of the sensor is estimated to be $\sim 7\,\mu g/ml$ of anti IgG [96].

Lu *et al.* [97] reported a newly developed DNA-modified diamond surface that showed excellent chemical stability compared to modified gold and silicon surfaces using similar chemical attachment under high temperature (60°C). Detection sensitivity was improved by a factor of ~ 100 by replacing the DNA-modified gold surface with a more stable DNA-modified diamond surface. Experiments by Remes *et al.* [98] demonstrated the aminofunctionalization of NCD diamond surface in RF plasma of vaporized silane coupling agent *n*-(6-aminohexyl) aminopropyl trimethoxysilane. NCD films were grown on low alkaline glass substrates at lower temperatures below 700°C. Surface functionalization was performed for times ranging from 3 to 180 min in two RF reactors with a standard excitation frequency of 13.56 MHz. It was found that low temperature and low RF plasma power were two key factors in successfully functionalizing the NCD film. Uniform coverage with a high density of the primary amines have been achieved. Adsorption and immobilization of cytochrome C on nanodiamonds have been achieved by Huang and Chang [99]. The immobilization started with carboxylation/oxidization of diamonds with strong acids, followed by coating the surfaces with poly-L-lysine (PL) for covalent attachment of proteins using heterobifunctional cross-linkers. The functionalization was proved with fluorescent labeling of the PL-coated diamonds by Alexa Fluor 488 and subsequent detection of the emission using a confocal fluorescence microscope.

3.4. *Field Emission Devices*

Metal field emission diode and triode are the first developed field emission devices [100–102]. However, these field emission devices have limited applications because of their high work functions and high turn-on electrics. Furthermore, impurity adsorption on metal surfaces leads to instable current and high turn-on voltage. Several approaches have been proposed, such as Si and W emitters, although no significant improvement has been achieved [103–105]. With the discovery of diamond negative electron affinity (NEA) surfaces, diamond and related materials have attracted extensive investigation of their electron field emission properties [54]. NEA means that electrons can exit from conduction band edge into vacuum without further energy barrier. A NEA occurs when the vacuum level lies below the conduction band minimum at the semiconductor/vacuum interface. The presence of a NEA for a semiconductor allows electrons in the conduction band to be freely emitted into vacuum without a barrier. Experimental results show that electron field emission from diamond or diamond-coated surface yield large currents at low electric fields. The chemical inertness of diamond allows diamond-based field emitters to work in a relatively low vacuum compared to other materials. In addition, strong crystal structure ensures the devices to be operated with long life. Furthermore, due to its high electrical breakdown field and high thermal conductivity, diamond can operate at high temperature or at high power. Thus, diamond emitters possess promising performances for potential applications, such as flat panel displays or other field emission devices.

Wang *et al.* [106] were the first to report field emission from diamond surfaces. Zhu *et al.* [107] demonstrated a correlation between the defect density present in undoped and *p*-type doped diamond films and the required field for emission. It was proposed that defects in the film created sub-bands just below the conduction band and facilitated the promotion of electrons to the conduction band for subsequent emission. Although diamond has negative electron affinity surface, it plays a minor role in enhancing the field emission in previous mentioned models for undoped or *p*-type doped polycrystalline diamond films. All the electron emissions do not involve the conduction band for these films [107]. However, *n*-type doping could solve this problem. The electrons could be supplied to the diamond conduction band via a low resistance ohmic contact and they would be emitted at the NEA surfaces. From previous discussions, we know that *n*-type doping could be incorporated into nanocrystalline diamond films and made *n*-type, which would result in a large number of researches on NCD-based field devices.

Nitrogen has a high solubility in diamond and is found in both natural and synthetic diamonds. Geis *et al.* [108] reported enhanced field emission properties of nitrogen-doped single crystal doped diamond. Okano *et al.* have reported threshold fields less than $1\,\mathrm{V}/\mu\mathrm{m}$ for nitrogen-doped diamond films [109]. Zhou *et al.* [110] studied the field emission properties of nanocrystalline diamond prepared from C_{60} precursors. Their results showed that the field emission characteristics strongly depended on the film's microstructure which could be controlled by growth conditions. The field emission properties were investigated by using a flat probe configuration apparatus shown in Fig. 3.22. A 1.8 mm diameter stainless steel electrode was used as an anode. The gap between the anode (probe) and the cathode (diamond) was controlled by a computer system through a stepping motor. The emission data were collected and the emission current

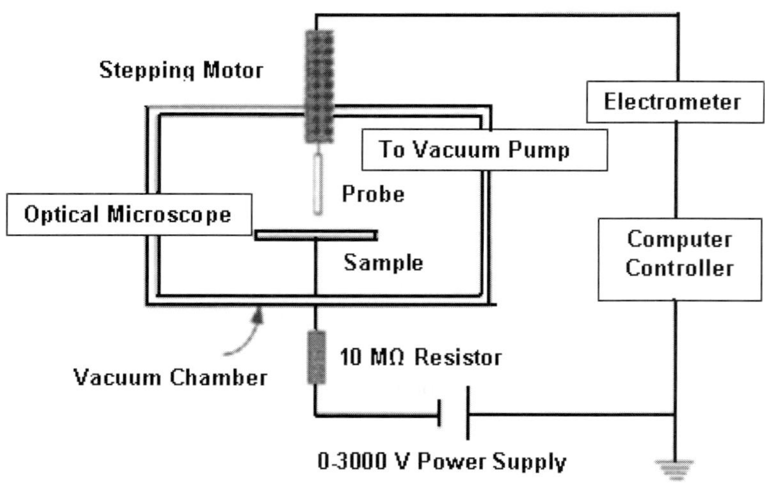

Fig. 3.22. Schematic diagram of the electron field emission test apparatus [110].

was then plotted as a function of applied field at $0\sim3000\,\mathrm{V}$ at various gaps. The measurements were performed at ultrahigh vacuum environment at $\sim10^{-8}$ Torr. NCD films were grown with Ar/C_{60} gas mixtures with additions of 2, 5, 10% H_2. It was found that emission properties improve with successive hydrogen addition. They proposed that unlike single crystal diamond, although electrons cannot be freely transported, nitrogen-doped nanocrystalline diamond films contain sp^2-bonded grain boundaries and defects that may be sufficient to form a conduction channel to transport electrons through the bulk to the surface. The microstructure analysis suggested that film grown with 20% H_2contained highest density of lattice defects, providing an alternative mechanism of transporting electrons to the surface and enhancing the electron emission at the surface under a low field [110].

Krauss *et al.* [111] have investigated the field emission properties of UNCD film coated Si tips. The NCD were grown by CH_4-Ar plasma on sharp single Si microtip emitters. Field emission results show that CND-coated tips exhibit high emission current (60–100 μm/tip). The emission turn-on voltage for diamond coated tips decreased dramatically compared to uncoated Si tips. For uncoated Si tips, the threshold voltage was proportional to the tip radius. For NCD-coated tips, the threshold voltage becomes nearly independent of the diamond coating thickness, the tip radius and the surface topography. A model which field emission occurs at the grain boundaries of NCD has been proposed.

Subramanian *et al.* [112] have reported enhanced electron field emission properties of NCD tip array prepared from $CH_4/H_2/N_2$ gas mixtures. NCD films grown by this protocol have grain sizes as small as $\sim5\,\mathrm{nm}$. The improved field emission behavior was attributed to their better geometrical shapes, increased sp^2-bonded carbon content and high electrical conductivity by incorporation of nitrogen impurity in the diamond. The NCD field emission cathode array was fabricated by FIB nanomold transfer technique [113]. The fabrication procedures are illustrated in Fig. 3.23. N-type silicon was nanomolded by high precision milling using a focused ion beam, where a focused beam of gallium ions is used to form arrays of high-respect ratio conical cavities of controlled diameter and depth. The mold can also be created by anisotropic etching of Si with KOH solution although the aspect ratio of this method is relatively low. NCD films were then deposited with $CH_4/H_2/N_2$ plasma on top of the mold followed by thick MCD layer ($\sim 30\,\mu\mathrm{m}$) growth with conventional CH_4/H_2 plasma. A metal layer such as Ti/Ni (1 μm) was deposited on diamond film followed by silicon etching using KOH solution to yield diamond nanotips with high aspect ratio.

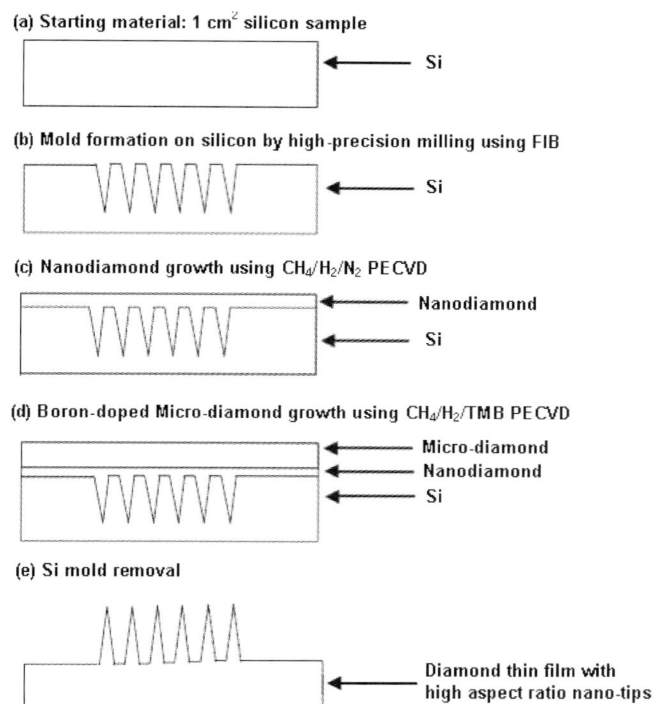

Fig. 3.23.   Fabrication process of the nanodiamond vacuum field emission cathode by the focused ion beam nanomold transfer technique [113].

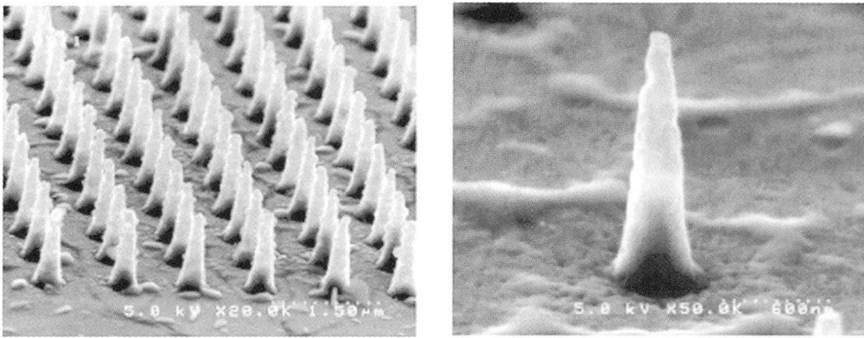

Fig. 3.24.   SEM micrographs of arrays of high aspect ratio diamond nanotips fabricated using the FIB molds: (a) diamond nanotips with $1\,\mu$m space; (b) a single diamond nanotip [113].

Figure 3.24 shows the SEM micrographs of the nanotip array molds created on a silicon substrate. The radius of curvature of the diamond nanotips is about 25 nm and the aspect ratio of the nanotips is found to be as high as $\sim$120.

The effect of gases on the field emission properties of NCD films has been studied by Hajra *et al.* [114]. A significant reduction of turn-on voltage and an increase in the emission current has been observed with the exposure of hydrogen. Tips exposed to Ar and N_2 showed less emission current at pressure $\sim 10^{-5}$ Torr while the current reached its original value after the chamber was pumped down to 8×10^{-10} Torr.

The field emission properties of NCD nanotip array cathode has been studied by Subramanian *et al.* [112]. The conductivity measurements at NCD emitter tip surface showed low electrical resistance $\sim 2\,\Omega$. The $CH_4/H_2/N_2$ nanotips exhibit significantly enhanced field emission characteristics compared to undoped NCD films grown by CH_4/H_2 method. A low threshold electric field of $1.6\,V/\mu m$ and $10\,\mu A$ emission current at $\sim 3.3\,V/\mu m$ has been obtained for $CH_4/H_2/N_2$ film, whereas CH_4/H_2 tip has a high turn-on field of $14\,V/\mu m$ and $10\,\mu A$ emission current at $\sim 23\,V/\mu m$. Figure 3.25 shows the I–J–E plot of the doped nanotips. A high emission current of 19 mA at a low electric field of $\sim 6\,V/\mu m$ can be observed from the plot. Note that the observed current is solely from field emission phenomenon due to the linearity of the corresponding F–N plot

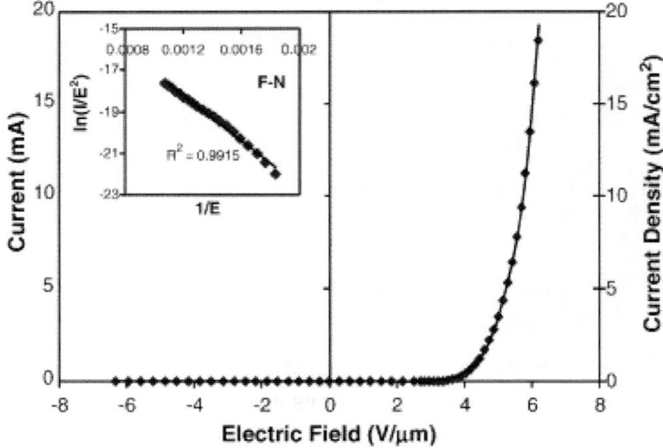

Fig. 3.25. I–J–E behavior of the $CH_4/H_2/N_2$-nanodiamond microtip array cathode, demonstrating $\sim 19\,mA/cm^2$ at $\sim 6\,V/\mu m$, current density being $\sim 19\,mA/cm^2$; inset: F–N plot with shallow slope [112].

and the absence of reverse leakage. Current stability tests indicated that over a time period of 30 min, nanotip showed stable emission current with a fluctuation of $\sim$4%.

3.5. *Other Applications of NCD Films*

NCD films have recently attracted considerable interest with their outstanding properties. In addition, NCD in an amorphous carbon matrix can be realized where diamond and matrix could be combined to a nanocomposite coating with improved properties [115, 116]. The NCD/a-C films were deposited by using a CH_4/N_2 gas mixture with a high methane concentration of 17%. The composite contains a mixture of sp^2-bonded carbon (20–30%) and sp^3-bonded carbon. 10% hydrogen was found inside the film which is bonded to sp^3-bonded carbon atom. Four point Hall measurements showed the NCD/a-C film is p-type conducting ($\rho = 0.14\,\Omega\,cm$) with a carrier concentration of $p = 1.9 \times 10^{17}$ cm^{-3} and a Hall mobility of $250\,cm^2/Vs$. This composite showed different electrical properties as compared to the n-type NCD films grown from 1% $CH_4/N_2/Ar$ plasma. No growth mechanism has been proposed although high nitrogen content within the film may play a role. The NCD/a-C film composite film showed that good mechanical properties with hardness of 39.7 GPa and Young's modulus of 387 GPa. The tribological tests showed that the coefficient of friction of the films is less than 0.1 after break-in period. The simulated body fluid (SBF) test has shown that no hydroxyl apatite could be observed after the film was exposed to the SBF, suggesting the NCD/a-C films are bioinert.

Yang *et al.* [117] successfully deposited NCD films on quartz glass substrate at a low temperature of 500°C. An optimal transmittance of 65% in the visible light range was achieved with coating thickness of $1.0\,\mu$m. The Viker's hardness tests showed hardness of the film was between 61 and 95 GPa. Thus, this NCD film could be used as a transparent protective coating for optical components.

The outstanding tribological and mechanical properties make NCD film a promising coating for cutting tools and inserts. Composites materials and high silicon content aluminum alloys are extremely abrasive and corrosive which are difficult to be machined with conventional high speed steel or tungsten carbide cutting tools. As a result, in order to machine these kinds of materials, efficient and energy saving cutting tools are required to reduce tool downtime, increase cutting productivity and improve the quality of the machined surface. Nanocrystalline diamond coating has been developed to improve the performance of cemented carbide tools due to its outstanding mechanical properties. Along with great wear resistance, the advantages of NCD coating include high surface hardness, high thermal conductivity, reduced friction, better corrosion protection, low coefficient of friction and improved optical properties. Jian *et al.* [118] have developed ultrafine diamond composite coatings on tungsten carbide. The mechanical and tribological properties of the films have been studied. Low COF of the film has been measured using pin on disk tests as 0.122, 0.143, 0165 for SiC PRAMC, Cu-1, and Al-0 respectively.

Thermionic energy converters have been fabricated based on nitrogen-doped NCD films by Koeck and coworkers [119, 120]. Thermionic energy converters transfer thermal energy into more useful electrical energy. At an elevated temperature, an electron emitter and a collector are separated by a vacuum gap and a voltage is generated due to the temperature difference between the emitter and collector. The emission current density J can be described in terms of the work function, the Richardson constant A, the temperature and the Boltzmann constant. In order to obtain high current, low work function materials such as tungsten and tungsten impregnated with barium or thorium have been utilized earlier. However, all the existing converters such as metallic electrodes require very high temperatures, $\sim$1900 K, to provide sufficient emission current. With the unique negative electron affinity property of diamond surfaces, a low temperature thermionic electron emitter could be achieved. The vacuum level is located below the conduction band so that the electron could be discharged from the solid due to the absence of a surface barrier for emission. The NEA surface could be induced by exposing diamond film to hydrogen plasma. Koeck and Nemanich [119] demonstrated that thermionic electron emission has been observed for nitrogen-doped NCD films at temperatures less than 900 K with relatively low working function ranging from 1.5–1.9 eV. Figure 3.26 shows that thermionic electron emission commences at temperatures as low as $\sim$520 K and increases with temperature. The thermionic emission from boron-doped NCD films has been characterized by Robinson *et al.* [121]. Hydrogen- and nitrophenyl-terminated NCD films have been produced and characterized. Surface termination plays an important role for thermionic emission and it was found that the hydrogenated sample has a more homogeneous work function. Lowest work functions measured for the hydrogen- and nitrophenyl-terminated films were 3.95 and 3.88 eV, respectively. Sulfur-doped NCD films have been grown

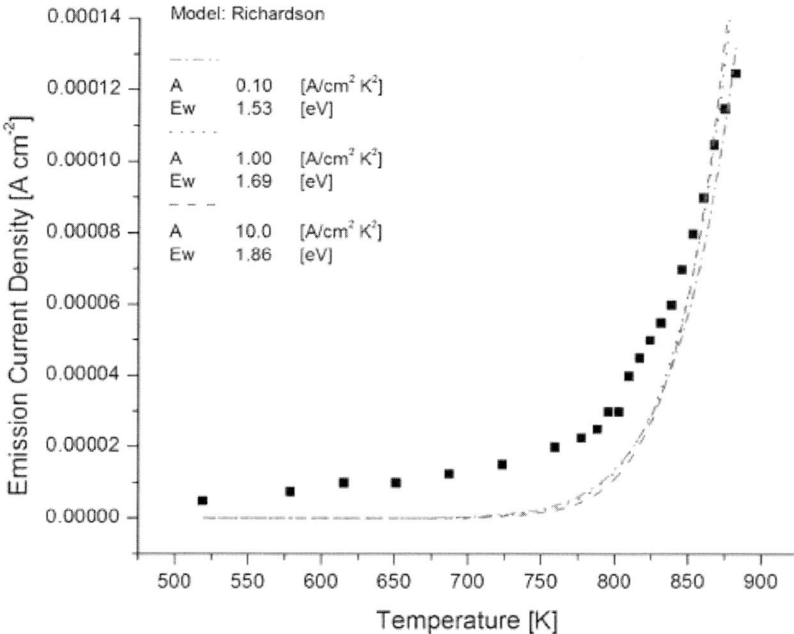

Fig. 3.26. Thermionic electron emission from a nitrogen-doped diamond film and fits to the Richardson equation using three values of the Richardson constant, A [119].

and characterized at various emitter temperatures [122]. Electron emission from these films is non-uniform at room temperature. Enhanced field emission with low work function $\sim 1.7\,$eV has been achieved at an elevated temperature.

4. Conclusions

This chapter has reviewed the recent discovery of a new type of material known as nanocrystalline diamond. Compared to conventional CVD diamond growth from CH_4/H_2 plasma, the nanocrystalline diamond, rather than of microcrystalline, is the result of new growth and nucleation mechanisms. The C_2 dimmer, produced from carbon-containing noble plasmas, is believed to be responsible for high secondary nucleation rates resulting in small grains in the nanometer scale. By adjusting the hydrogen/noble gas ratio, the continuous transition of diamond structure from micro- to nanocrystalline has been achieved.

Extensive characterization studies have been carried out to analyze nanocrystalline diamond films. It has been demonstrated that nanocrystalline diamond films are nearly phase-pure diamond with π-bonded carbon present in grain boundaries. The effects of the grain boundaries on the mechanical, electrical and electrochemical properties of diamond are enormous, especially for nitrogen-doped films. The properties of nitrogen-doped nanocrystalline diamond films as well as their prospective applications have been the focus of a variety of research. In this chapter, we have extensively characterized the nitrogen-doped films to investigate the effect of doping on the bonding structures and the grain boundaries. Although tremendous tasks remain to be undertaken in order to develop n-type diamond films that have properties comparable to p-type boron-doped ones, increased research activities in different fields based on nitrogen-doped diamond films have already begun, including electrochemical electrodes, field emission devices, and biomedical sensors. Some of the other applications of nanocrystalline diamond films, such as MEMS devices and conformal coatings, have also been reviewed. In summary, the nanocrystalline diamond, with its unique properties, is a rapidly developing area for both fundamental research as well as potential nanotechnological applications.

Acknowledgments

The authors would like to thank Dr. Jayanta Banerjee for providing useful comments. They would also like to acknowledge the support from NSF NIRT grant (ECS-0404137, Program Director: Dr. Rajinder Khosla). Part of this diamond research is also supported by General Motors Corp. (Grant # 2105-099L0). Z. Xu would like to thank Drs. Leonid Lev, Y. T. Cheng and other members from the GM R&D center for providing internship opportunity related to diamond research. The authors would also like to thank Dr. Arun Sikder (GE, Bangalore, India) for his contribution to diamond related materials research at the University of South Florida.

References

1. R.F. Davis, *Diamond Films and Coatings* (1993).
2. J.C. Angus, H.A. Will and W.S. Stanko, *J. Appl. Phys.* **39** (1968) 2916.
3. B.V. Derryagin, V.A. Ryabov, D.V. Fedoseev, B.V. Spitsyn, V.M. Lukyanovich and K.S. Uspensakya, Second All Union Symposium Process for Nucleation and Growth of Crystals and Films of Semiconducting Compounds (1969).
4. S. Matosumoto, Y. Sato, M. Kamo and N. Setaka, *J. Mater. Sci.* **17** (1982) 3106–3109.
5. W.G. Eversole, U.S. Patents No. 3,030,187 and 3,030,188 (1962).

6. S. Jin and T.D. Moustakas, *Appl. Phys. Lett.* **63** (1993) 2354.

7. R. Samlenski, C. Huang, R. Breen, C. Wild, R. Locher and P. Koidl, *Appl. Phys. Lett.* **67** (1995) 2798.

8. R.B. Heimann, S.E. Evsvukov and Y. Koga, *Carbon* **35** (1997) 1654–1658.

9. M. Inagaki, New carbons (2000).

10. V.L. Kuznetsov, A.L. Chuvilin and Y.V. Butenko, *Chem. Phys. Lett.* **209** (1994) 72.

11. Banhart and P.M. Ajayan, *Nature* **382** (1996) 433.

12. J.A. Viecelli, S. Bastea, J.N. Glosli and F.H. Ree, *J. Chem. Phys.* **115** (2001) 2730–2736.

13. Zhenqing Xu., Master Thesis: Fabrication and optimization of MPECVD system for growing diamond films (*University of South Florida*, 2003).

14. D.M. Gruen, *Annu. Rev. Mater. Sci.* **29** (1999) 211–259.

15. S.J. Harris and D.G. Goodwin, *J. Phys. Chem.* **97** (1993) 23.

16. O. Matsumoto and T. Katagirl, *Thin Solid Films* **146** (1987) 283.

17. D.M. Gruen, S. Liu, A.R. Krauss, J. Luo and X. Pan, *Appl. Phys. Lett.* **64** (1994) 1502.

18. P.C. Redfern, D.A. Horner, L.A. Curtiss and D.M. Gruen, *J. Phys. Chem.* **100** (1996) 11654.

19. D. Zhou, D.M. Gruen, L.C. Qin, T.G. McCauley and A.R. Karuss, *J. Appl. Phys.* **84** (1998) 1981–1989.

20. M. Sternberg, M. Kaukonen, R.M. Nieminen and T. Frauenheim, *Phys. Rev. B* **63** (2001) 165414.

21. M. Sternberg, P. Zapol and L.A. Curtiss, *Phys. Rev. B* **68** (2003) 205330.

22. A. Drift, *Philips Res. Rep.* **22** (1967) 267.

23. J.S. Luo, D.M. Gruen, A.R. Karuss, X.Z. Pan and S.Z. Liu, *Recent Advances in the Chemistry and Physics of Fullerenes and Rel. Materials*, eds. R.S. Ruoff and K.M. Kadish (1995), pp. 43–50.

24. D. Zhou, T.G. McCauley, L.C. Qin, A.R. Krauss and D.M. Gruen, *J. Appl. Phys.* **83** (1998) 540–543.

25. S. Jiao, A. Sumant, M.A. Kirk, D.M. Gruen, A.R. Krauss and O. Auciello, *J. Appl. Phys.* **90** (2001) 118–122.

26. L.C. Qin, D. Zhou, A.R. Krauss and D.M. Gruen, *Nanostructured Mater.* **10** (1998) 649–660.

27. T. Lin, G.Y. Yu, T.S. Wee, Z.X. Shen and K.P. Loh, *Appl. Phys. Lett.* **77** (2000) 2692–2694.

28. A. Gicquel, F. Silva and K. Hassouni, *J. Electrochem. Soc.* **147** (2000) 2218.

29. V.I. Merkulov, J.S. Lannin, C.H. Munro, S.A. Asher, V.S. Veersamy and W.I. Milne, *Phys. Rev. Lett.* **78** (1997) 4869.

30. J. Birrell, J.E. Gerbi, O. Auciello, J.M. Gibson, J. Johnson and J.A. Carlisle, *Diam. Relat. Mater.* **14** (2005) 86–92.

31. R.J. Nemanich, J.T. Glass, G. Lucovsky and R.E. Shroder, *J. Vac. Sci. Technol.* **A6** (1998) 1783–1787.

32. L. Fayette, B. Marcus, M. Mermoux, G. Tourillon, K. Laffon, P. Parent and F.L. Normand, *Phys. Rev. B* **57** (1998) 14123–14132.

33. A. Chaiken, L.J. Terminello, J. Wong, G.L. Doll and C.A. Taylor, *Appl. Phys. Lett.* **63** (1993) 2112.

34. D.M. Gruen, A.R. Krauss, C.D. Zuiker, R. Csencsits, L.J. Terminello, J.A. Carlise, I. Jimenez, D.G.J. Sutherland, D.K. Shuh, W. Tong and F.J. Himpsel, *Appl. Phys. Lett.* **68** (1996) 1640.

35. T.W. Capehart, T.A. Perry, C.B. Beetz, D.N. Belton and G.B. Fisher, *Appl. Phys. Lett.* **55** (1989) 957.

36. Y. Wang, H. Chen and R.W. Hoffman, *J. Vac. Sci. Technol.* **A9** (1990) 1154.

37. Y.K. Chang, H.H. Hsieh, W.F. Pong, M.H. Tsai, F.Z. Chien, P.K. Tseng, L.C. Chen, T.Y. Wang, K.H. Chen, D.M. Bhusari, J.R. Yang and S.T. Lin, *Phys. Rev. Lett.* **82** (1999) 5377–5380.

38. Y.H. Tang, X.T. Zhou, Y.F. Hu, C.S. Lee, S.T. Lee and T.K. Sham, *Chem. Phys. Lett.* **372** (2003) 320–324.

39. R.W. Siegal and G.E. Fougere, *Nanophase Materials*, eds. G.C. Hadjipanayis and R.W. Siegel (1994), pp. 233–261.

40. A.R. Krauss, O. Auciello, D.M. Gruen, A. Jayatissa, A. Sumant, J. Tucek, D.C. Mancini, N. Moldovan, A. Erdemir, D. Ersoy, M.N. Gardos, H.G. Busmann, E.M. Meyer and M.Q. Ding, *Diam. Relat. Mater.* **10** (2001) 1952–1961.

41. F.J.H. Guillen, K. Janischowsky, J. Kusterer, W. Ebert and E. Kohn, *Diam. Relat. Mater.* **14** (2005) 411–415.

42. J. Philip, P. Hess, T. Feygelson, J.E. Butler, S. Chattopadhyay, K.H. Chen and L.C. Chen, *J. Appl. Phys.* **93** (2003) 2164–2171.

43. H.D. Espinosa, B. Peng, B.C. Prorok, N. Moldovan, O. Auciello, J.A. Carlise, D.M. Gruen and D.C. Mancini, *J. Appl. Phys.* **94** (2003) 6076–6084.

44. W.N. Sharpe, K.M. Jackson, K.J. Hemker and Z. Xie, *J. Microelectromech. Syst.* **10** (2001) 3.

45. H.D. Espinosa, B.C. Prorok, B. Peng, K.H. Kim, N. Moldovan, O. Auciello, J.A. Carlise, D.M. Gruen and D.C. Mancini, *Experimental Mechanics* **43** (2003) 256–268.

46. A. Erdemir, G.R. Fenske, A.R. Krauss, D.M. Gruen, T. McCauley and R.T. Csencsits, *Surf. Coat. Technol.* **120** (1999) 565–572.

47. F. Deuerler, N. Woehrl and V. Buck, *Int. J. Refract. Met. Hard Mater.* **5** (2006) 392–398.

48. K. Kimoto, T. Sekiguchi and T. Aoyama, *J. Electron Microsc.* **46** (1997) 369.

49. R. Brydson, *Electron Energy Loss Spectroscopy* (Springer, New York, 2001).

50. R.F. Egerton and M.J. Whelan, *J. Electrochem. Soc.* **3** (1974) 232.

51. K. Okada, K. Kimoto, S. Komatsu and S. Matsumoto, *J. Appl. Phys.* **93** (2003) 3120–3122.

52. J. Weide, Z. Zhang, P.K. Baumann, M.G. Wemnsell, J. Bernhold and R.J. Nemanich, *Phys. Rev.* **B50** (1994) 5803.

53. F.J. Himpsel, J.A. Knapp, V. Vechten and D.E. Eastman, *Phys. Rev.* **B20** (1979) 624.

54. I.L. Krainsky, V.M. Asnin, G.T. Mearini and J.A. Dayton, *Phys. Rev.* **B53** (1996) 7650.
55. C. Bandis and B.B. Pate, *Appl. Phys. Lett.* **69** (1996) 366.
56. S. Bhattacharyya, O. Auciello, J. Birrell, J.A. Carlise, L.A. Curtiss, A.N. Goyette, D.M. Guren, A.R. Krauss, J. Schlueter, A. Sumant and P. Zapol, *Appl. Phys. Lett.* **79** (2001) 1441–1443.
57. I. Sakaguchi, M.N. Gamo, Y. Kikuchi, E. Yasu and H. Haneda, *Phys. Rev.* **B60** (1999) R2139–R2142.
58. M. Katagiri, J. Isoya, S. Koizumi and H. Kanda, *Appl. Phys. Lett.* **85** (2004) 6365–6367.
59. Q. Chen, D.M. Gruen, A.R. Krauss, T.D. Corrigan, M. Witek and G.M. Swain, *J. Electrochem. Soc.* **148** (2001) E44.
60. K. Okano, S. Koizumi, S.R. Silva and G.A.J. Amaratnga, *Nature (London)* **398** (1996) 140.
61. J. Birrell, O. Auciello, J.M. Gibson, D.M. Guren and J.A. Carlisle, *J. Appl. Phys.* **93** (2003) 5606–5612.
62. N. Fujimori, T. Imai, H. Nakahata, H. Shiomi and Y. Nishibayashi, *Mater. Res. Soc. Sym. Proc.* **162** (1990) 23.
63. S. Koizumi, M. Kamo, Y. Sato, H. Ozaki and T. Inuzuka, *Appl. Phys. Lett.* **71** (1997) 1065–1067.
64. F.A.M. Koeck, M. Zumer, V. Nemanic and R.J. Nemanich, *Diam. Relat. Mater.* (2006), in press.
65. A.P. Lee, A.P. Pisano and M.G. Lim, *Mater. Res. Soc. Sym. Proc.* **276** (1992) 67.
66. Sniegowski, *Tribology Issues and Opportunities in MEMS*, ed. B. Bushan, (1998), p. 325.
67. L. Tong, M. Mehregany and L.G. Matus, *Appl. Phys. Lett.* **60** (1992) 2992.
68. T. Uemura, S. Fujii, H. Kitabayashi, K. Itakura, A. Hachigo and H. Nakahaata, *J. Appl. Phys.* **41** (2002) 3476.
69. O. Elmazria, M. Hakiki, V. Mortet, M.B. Assouar, M. Nesladek and M. Vanecek, *Ferroeletr. Freq. Control* **51** (2004) 1704.
70. O. Elmazria, F. Benedic, M. Hakiki, H. Moubchir, M.B. Assouar, M. Nesladek and F. Silva, *Diam. Relat. Mater.* **15** (2005) 193–198.
71. Y. Gurbuz, O. Esame, I. Tekin, W.P. Kang and J.L. Davison, *Solid State Electron.* **49** (2005) 1055–1070.
72. G.Sh. Gildenblat, S.A. Grot, C.W. Hatfield, A.R. Badzian and T. Badzian, *IEEE Electron. Device Lett.* **11** (1990) 371–372.
73. A. Aleksov, M. Kubovic, M. Kasu, P. Schmid, D. Grobe, S. Ertl, M. Schreck, B. Strizker and E. Kohn, *Diam. Relat. Mater.* **13** (2004) 233–240.
74. J.F. Prins, *Appl. Phys. Lett.* **41** (1982) 950–952.
75. A. Aleksov, A. Denisenko and E. Kohn, *Solid State Electro.* **44** (2000) 369–375.
76. T. Zimmermann, M. Kubovic, A. Denisenko and K. Janischowsky, *Diamond Relat. Mater.* **14** (2005) 416–420.
77. J. Kusterer, F.J. Hernandez, S. Haroon, P. Schmid, A. Munding, R. Muller and E. Kohn, *Diam. Relat. Mater.* (2006), in press.

78. M. Iwaki, S. Sato, K. Takahashi, K. Takahashi and H. Sakairi, *Nucl. Instrum. Methods Phys. Res.* **209** (1983) 1129–1133.

79. Y.V. Pleskov, A.Y. Sakharova, M.D. Krotova, L.L. Bouilov and B.V. Spitsyn, *J. Electroanal. Chem.* **228** (1987) 19–27.

80. G.M. Swain and R. Ramesham, *Anal. Chem.* **65** (1993) 345–351.

81. G.M. Swain, A.B. Anderson and J.C. Augus, *M. R. S. Bull* **23** (1998) 56–60.

82. H.B. Martin, J.C. Angus and U. Landau, *J. Electrochem. Soc.* **146** (1999) 2959–2964.

83. H.B. Martin, A. Argoitia, U. Landau, A.B. Anderson and J.C. Angus, *J. Electrochem. Soc.* **143** (1996) L133–L136.

84. G.M. Swain, Electrochemistry of Diamond, in *Thin-Film Diamond II, Semiconductors and Semimetals* **77** (2004) 97–120.

85. C. Levy-Clement, F. Zenia, N.A. Ndao and A. Deneuville, *New Diamond Frontier Carbon Tech.* **9** (1999) 189–206.

86. S.C. Eaton, A.B. Anderson, J.C. Angus, Y.E. Evstefeeva and Y.V. Pleskov, *Electrochem Solid-State Lett.* **5** (2002) G65–G68.

87. B. Fausett, M.C. Granger, M.L. Hupert, J. Wang, Swain M. Greg and D.M. Gruen, *Electronanlysis* **12** (1999) 7–15.

88. G. M. Swain, Electroanalytical Applications of Diamond Electrodes. *Thin-Film Diamond II*, in *Semiconductors and Semimetals* **77** (2004) 121–148.

89. M.C. Granger, M. Witek, J. Xu, J. Wang, M. Hupter, A. Hank, M.D. Koppang, J.E. Butler, G. Lucazeau, M. Mermoux and J.W. Strojek, *Anal. Chem.* **72** (2000) 3793.

90. Y. Show, M.A. Witek, P. Sontahlia and G.M. Swain, *Chem Mater.* **15** (2003) 879.

91. J. Wang and J.A. Carlise, *Diam. Relat. Mater.* **15** (2005) 279–284.

92. T. Matrab, M.M. Chehimi, J.P. Boudou, F. Benedic, J. Wang, N.N. Naguib and J.A. Carlise, *Diam. Relat. Mater.* **15** (2006) 639–644.

93. W. Yang, O. Auciello, J. Butler, W. Cai, J.A. Carlise, J.E. Gerbi, D.M. Gruen, T. Knickerbocker, T.L. Lasseter, J.N. Russell, L.M. Smith and J. Hamers, *Nature Mater.* **1** (2002) 253–257.

94. P.H. Chung, E. Perevedentseve, J.S. Tu, C.C. Chang and C.L. Cheng, *Diam. Relat. Mater.* **15** (2006) 622–625.

95. A. Hartl, E. Schmich, J.A. Garrido, J. Hernando, S.C.R. Catharino, S. Walter, P. Feulner, A. Kromka, D. Steinmuller and M. Stutzman, *Nature Mater.* **3** (2004) 736–742.

96. W. Yang and R.J. Hamers, *Appl. Phys. Lett.* **85** (2004) 3626–3628.

97. M. Lu, T. Knickerbocker, W. Cai, W. Yang, R.J. Hamers and L.M. Smith, *Biopolymers* **73** (2003) 606–613.

98. Z. Remes, A. Choukourov, J. Stuchlik, J. Potmesil and M. Vanecek, *Diam. Relat. Mater.* **15** (2006) 745–748.

99. L.C.L. Huang and H.C. Chang, *Langumir* **20** (2004) 5879–5884.

100. C.A. Spindt, A Thin Film Field Emission Cathode, *J. Appl. Phys.* **39** (1968) 3504.

101. C.A. Spindt, C.E. Holland, A. Rosengreen and I. Brodie, *IEEE Trans. Electron. Devices* **38** (1991) 2355.

102. T. Sakai, T. Ono, M. Nakamoto and N. Sakuma, *J. Vac. Sci. Technol.* **B16** (1998) 770.

103. J. Liu, V.V. Zhirnov and A.F. Myers, *J. Vac. Sci. Technol.* **B13** (1994) 422.

104. J. Liu, V.V. Zhirnov and G.J. Wojak, *Appl. Phys. Lett.* **22** (1994) 2842–2844.

105. F.Y. Chuang, C.Y. Sun, H.F. Cheng, C.M. Huang and I.N. Lin, *Appl. Phys. Lett.* **68** (1996) 1666.

106. C. Wang, A. Garcia, D.C. Ingram, M. Lake and M.E. Kordesch, *Electron. Lett.* **27** (1991) 1459.

107. W. Zhu, G.P. Kochanski, S. Jin and L. Seibles, *J. Appl. Phys.* **78** (1995) 2707.

108. M.W. Geis, N.N. Efremow, K.E. Krohn, J.C. Twichell, T.M. Lyszczara, R. Kalish, J.A. Greer and M.D. Tabat, *Nature (London)* **393** (1998) 431.

109. K. Okano, A. Hiraki, T. Yamada, S. Koizumi and J. Itoh, *Ultramicroscopy* **73** (1998) 43.

110. D. Zhou, A.R. Krauss, T.D. Corrigan, T.G. McCauley, R.P.H. Chang and D.M. Guren, *J. Electrochem. Soc.* **144** (1997) L224–228.

111. A.R. Krauss, O. Auciello, D.M. Gruen, Y. Huang, V.V. Zhirnov, E.I. Givargizov, A. Breskin, R. Chenchen, E. Shefer, V. Konov, S. Pimenov, A. Karabutov, A. Rakhimov and N. Suetin, *J. Appl. Phys.* **89** (2001) 2958–2967.

112. K. Subramanian, W.P. Kang, J.L. Davidson, R.S. Takalkar, B.K. Choi, M. Howell and D.V. Kerns, *Diam. Relat. Mater.* **15** (2006) 1126–1134.

113. K. Subramanian, W.P. Kang, J.L. Davidson and W.H. Hofmeister, *Diam. Relat. Mater.* **14** (2005) 404–410.

114. M. Hajra, C.E. Hunt, M. Ding, O. Auciello, J. Carlise and D.M. Gruen, *J. Appl. Phys.* **94** (2003) 4079–4083.

115. S. Barnett and Madam, *Phys. World* **11** (1998) 45.

116. C. Popov, W. Kulisch, M. Jelinek, A. Bock and J. Strnad, *Thin Solid Film* **494** (2006) 92–97.

117. W.B. Yang, F.X. Lu and Z.X. Cao, *J. Appl. Phys.* **91** (2002) 10068–10073.

118. X.G. Jian, L.D. Shi, M. Chen and F.H. Sun, *Diam. Relat. Mater.* **15** (2006) 313–316.

119. F.A.M. Koeck and R.J. Nemanich, *Diam. Relat. Mater.* **15** (2006) 217–220.

120. F.A.M. Koeck, J.M. Garguilo and R.J. Nemanich, *Diam. Relat. Mater.* **13** (2004) 2052–2055.

121. V.S. Robinson, Y. Show, G.M. Swain, R.G. Reifenberger and T.S. Fisher, *Diam. Relat. Mater.* **15** (2006) 1601–1608.

122. F.A.M. Koeck and R.J. Nemanich, *Diam. Relat. Mater.* **14** (2005) 2051–2054.

CHAPTER 5

PROPERTIES OF HARD NANOCOMPOSITE THIN FILMS

J. Musil

Department of Physics, Faculty of Applied Sciences
University of West Bohemia
Univerzitní 22, 306 14 Plzeň, Czech Republic

Institute of Physics, Academy of Sciences of the Czech Republic
Na Slovance 2, 182 21 Praha 8, Czech Republic
musil@kfy.zcu.cz

1. Introduction

Nanocomposite coatings represent a new generation of materials. Nanocomposite films are composed of at least two separate phases with nanocrystalline and/or amorphous structure. The nanocomposite materials, due to (1) very small ($\leq$10 nm) size of grains from which they are composed and (2) a significant role of boundary regions surrounding individual grains, behave in a different manner compared to that of the conventional materials with grains greater than 100 nm, and so they exhibit completely new properties. New unique physical and functional properties of the nanocomposite films are a main driving force stimulating a huge development of these materials [1–49]. This chapter reviews the state-of-the-art in the field of hard and superhard nanocomposite films. At present, it is accepted that hard and superhard films are films with hardness $H \geq 20$ GPa and $H \geq 40$ GPa, respectively.

2. Present State of Knowledge

At present, it is known that

1. There are two groups of hard and superhard nanocomposites: (i) nc-MeN/hard phase and (ii) nc-MeN/soft phase.
2. Nanocrystalline and/or X-ray amorphous films are created in transition regions between (i) the crystalline phase and the amorphous phase,

281

(ii) two crystalline phases and/or (iii) two different crystallographic orientations of grains of the same material.

3. There are huge differences in the microstructure of single- and two-phase films; here nc- denotes the nanocrystalline phase and Me = Ti, Zr, Ta, Mo, W, Cr, Al, etc. are elements forming nitrides.

Using these findings a complete concept of nanocomposites with enhanced hardness was developed [49]. This concept is based on the geometry of nanostructured features, i.e. on the size of grains and the shape of crystallites.

3. Enhanced Hardness

3.1. *Origin of Enhanced Hardness*

Main mechanisms responsible for the enhanced hardness H of hard films are: (1) dislocation-dominated plastic deformation, (2) cohesion forces between atoms, (3) nanostructure and (4) compressive macrostress σ generated in the film during its formation. The magnitude of material hardness H depends on the deformation processes dominating in a given range of the size d of grains, see Fig. 3.1. There is a critical value of the grain size $d_c \approx 10\,\mathrm{nm}$ at which a maximum hardness $H_{\max}$ can be achieved. A region around $H_{\max}$ corresponds to a continuous transition from the activity of

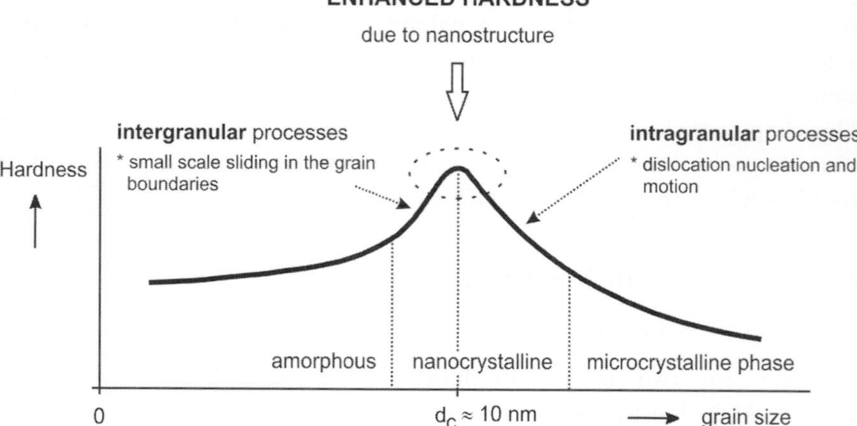

Fig. 3.1. Schematic illustration of material hardness H versus size d of grains. Adapted from [45, 49].

intragranular processes at $d > d_c$, dominated by the dislocation activity and described by the Hall–Petch law ($H \sim d^{-1/2}$), to that of *intergranular* processes at $d < d_c$, dominated by a small-scale sliding in grain boundaries. The high macrostress σ in the film is detrimental and so it is usually suppressed by proper control of the deposition parameters.

3.2. *Formation of Nanocomposite Films*

Nanocrystalline films are characterized by broad, low-intensity X-ray reflections. Such films are formed in the transition regions where the film structure (crystallinity) strongly changes. There are three groups of transitions: (1) transition from the crystalline phase to the amorphous, (2) transition between two phases of different materials and (3) transition between two preferred orientations of grains of the same material, see Fig. 3.2.

3.3. *Microstructure of Films Produced in Transition Regions*

3.3.1. *Transition Region from Crystalline to Amorphous Phase*

The transition from the crystalline phase to the amorphous occurs in the case when the second element B added to the AN compound forms an amorphous phase $B_x N_y$ in the $A_{1-x}B_x N$ film. The $A_{1-x}B_x N$ films with $B \approx B_1$ produced in the crystalline region (Fig. 3.2(a)) in front of the transition from the crystalline to the amorphous phase mostly exhibit a strong preferred crystallographic orientation of grains. Such films have a columnar microstructure and the columns perpendicular to the film/substrate

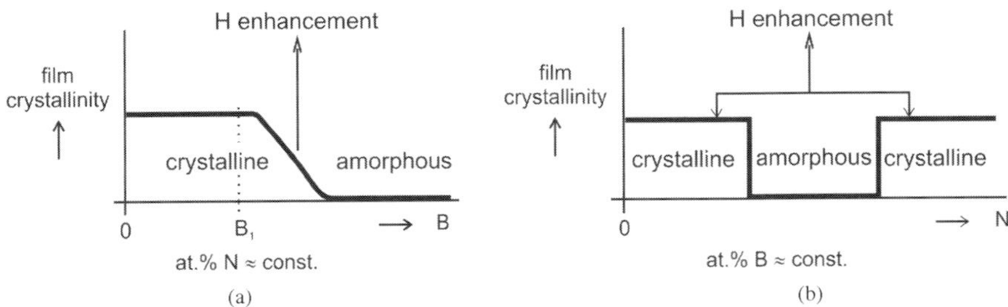

Fig. 3.2. Schematic illustration of three transition regions of $A_{1-x}B_x N$ compounds. (a) Transition region from crystalline to amorphous phase and (b) transition region between two crystalline phases or two preferred crystallographic orientations of grains. Reprinted from [45] with permission of Elsevier.

                    J. Musil

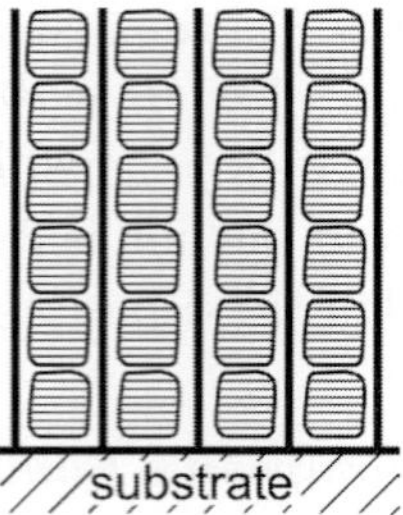

Fig. 3.3. Schematic illustration of the columnar microstructure in the crystalline $A_{1-x}B_xN$ film. The amount of B element in $A_{1-x}B_xN$ film is relatively low (<10 at.%).

interface are composed of small grains oriented in one crystallographic direction, see Fig. 3.3. As an example we can give $Zr(Ni)N_x$ films analyzed in detail in [50].

As the amount of B element in the $A_{1-x}B_xN$ compound increases the content of amorphous B_xN_y phase also increases. At first, the nanocomposite with columnar microstructure (Fig. 3.3) is converted into an $A_{1-x}B_xN$ nanocomposite in which every $A_{1-x}N_{1-y}$ nanograin is surrounded by a thin ($\sim$1–2 ML) tissue B_xN_y phase (Fig. 3.4(a)); here ML is the monolayer. A further increase of the content of a-B_xN_y phase results in the formation of a nanocomposite in which the nanograins are embedded in the amorphous B_xN_y matrix (Fig. 3.4(b)); a- denotes the amorphous phase. The separation distance w between the nanograins (Fig. 3.4(b)) increases and their number decreases with increasing amount of B element in the film up to the formation of a pure amorphous BN phase (Fig. 3.4(c)).

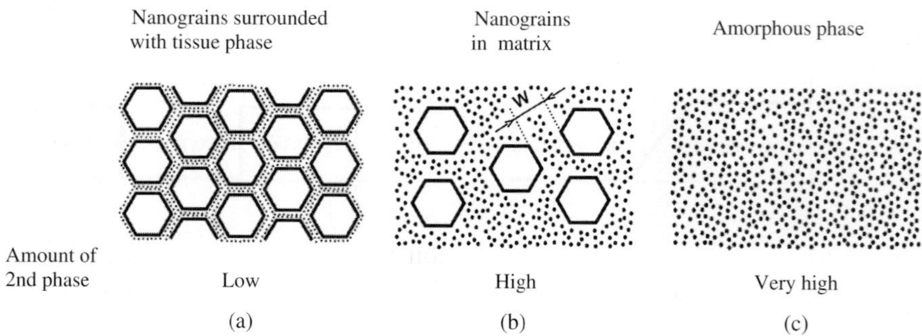

Fig. 3.4. Development of microstructure of films produced in the transition from crystalline to amorphous phase with increasing amount of B element in $A_{1-x}B_xN$ film.

Experiments show that both films with columnar microstructure and those composed of nanograins surrounded by a tissue phase exhibit the enhanced hardness H. To produce films with these microstructures only a relatively low ($\leq$10 at.%) amount of B element is sufficient. This is a reason why the nanocomposites of the type nc-MeN/a-Si$_3$N$_4$ exhibit the enhanced H only in the case when the amount of Si in the nanocomposite is $\leq$10 at.%.

3.3.2. *Transition Between Two Preferred Crystallographic Orientations of Grains or Two Crystalline Phases*

The transition between two preferred crystallographic orientations of grains or two crystalline phases can be easily formed in reactive magnetron sputtering of $A(B)N_x$ and $A_{1-x}B_xN$ nitrides. As an example, Fig. 3.5 displays the structure evolution of $Ti(Fe)N_x$ films reactively sputtered from a TiFe(90/10 at.%) target in $Ar + N_2$ mixture with increasing partial pressure of nitrogen, p_{N2}. This figure clearly shows the transition region between TiN(200) and TiN(220) preferred orientation of grains. Similar transitions between two preferred orientations of grains have also been created in reactive sputtering of other $A(B)N_x$ films when the partial pressure of nitrogen, p_{N2}, increases [50]. If the magnetron discharge is sufficiently strong, two transitions may be created; for details see the reference [51].

The films produced inside the transition region between the preferred crystallographic orientations of grains exhibit an X-ray amorphous structure. This is a reason why the microstructure of $A(B)N_x$ films produced inside the transition region between two preferred crystallographic orientations of grains or two crystalline phases $A(B)N_x$ and $B(A)N_y$, e.g. in the system $Ti_{1-x}Al_xN$, is different from that of the nanocomposite films produced inside the transition from crystalline to amorphous phase. The films formed inside the transition region between two preferred crystallographic orientations of grains exhibit a dense globular microstructure and are composed of a mixture of small grains of different crystallographic orientations. Also, these films exhibit the enhanced H, see Fig. 3.5; more details are given in [50, 51]. This means that a mixture of small grains of the same material but different crystallographic orientation is another microstructure which results in the enhanced hardness H of nanocomposite films. This finding is of great importance because it can explain the enhanced hardness, H, in single phase composite films. Further experiments are, however, necessary to be carried out to demonstrate the validity of the last statement.

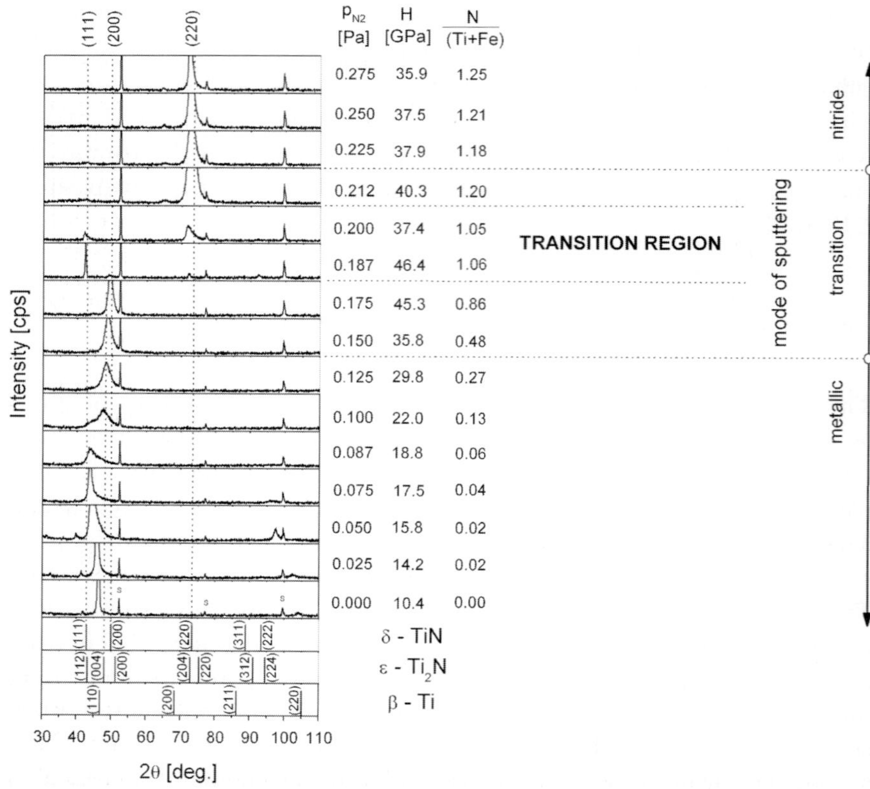

Fig. 3.5.   Evolution of XRD patterns from the reactively sputtered Ti(Fe)N$_x$ films with increasing partial pressure of nitrogen, p_{N2}.
Adapted from [51] with permission of Elsevier.

Particularly, a possibility of the segregation of element B from the nitride of solid solution A(B)N$_x$ must be investigated in detail.

On the contrary, the A(B)N$_x$ films produced near edges (outside) of the transition region between two preferred crystallographic orientations of grains are characterized by strong X-ray reflections. Experiments show that also these films exhibit columnar microstructure. Similarly as in the case of the transition from crystalline to amorphous phase, these films with columnar microstructure also exhibit the enhanced H. The existence of columnar microstructure in the films with strong preferred crystallographic orientation of grains was already demonstrated experimentally [50, 52–55]. A schematic illustration of the microstructure of A(B)N$_x$ films with enhanced hardness, characterized by an X-ray amorphous structure (inside transition)

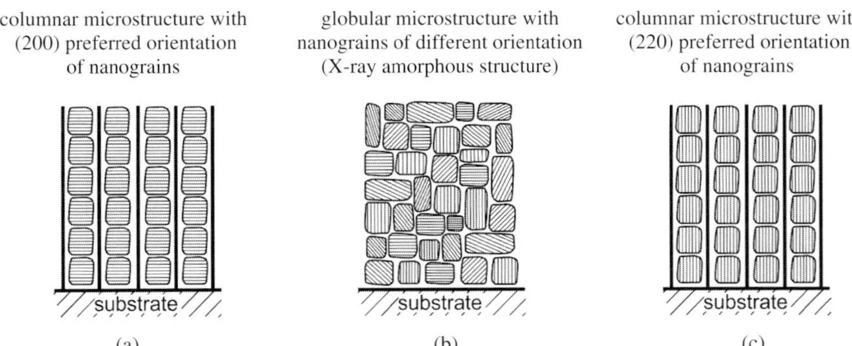

Fig. 3.6. Microstructure of $A(B)N_x$ film produced at edges [(a) and (c)] and inside (b) transition region between two preferred crystallographic orientations of grains or two crystalline phases.

and by a strong preferred crystallographic orientation (at edges and outside transition), is given in Fig. 3.6.

3.3.3. *FE-TEM of Cross-Section of Nanocomposite Films Based on Nitrides*

Many experiments performed to date have demonstrated that the nanostructure of $A(B)N_x$ nitride films strongly depends on the amount of the second B element in the nanocomposite film. If the amount of B element is lower than its solulibility limit a nitride of solid solution $A(B)N_x$ is formed. On the contrary, if the amount of B element exceeds its solulibility limit the excess of B in the compound can segregate to grain boundaries and a two-phase nanocomposite is formed.

Therefore, the microstructure of the nanocomposite film also depends on the amount of B element incorporated in it. If B content is lower than its solulibility limit a dense film is formed. On the contrary, if B content exceeds its solulibility limit and further increases gradually cluster-like, plate-like and columnar microstructures should be formed, see Fig. 3.7.

The development of the microstructure of $A(B)N_x$ films with increasing content of B element schematically shown in Fig. 3.7 has been observed experimentally, see Fig. 3.8. This figure displays the evolution of the microstructure of $Zr(Ni)N_y$ films reactively sputtered from $ZrNi(90/10\,\mathrm{at.\%})$ target in a mixture of $Ar + N_2$ with increasing content of Ni. The obtained experimental results are in a good agreement with the model schematically shown in Fig. 3.7. Selected area electron diffraction (SAED) patterns

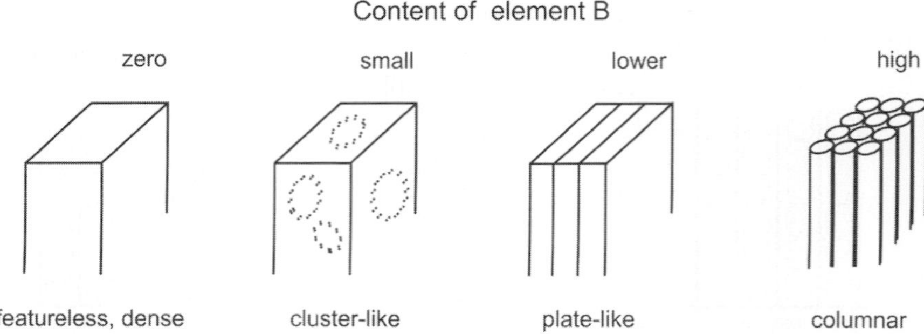

Fig. 3.7.   Schematic illustration of microstructure evolution in $A(B)N_x$ film with increasing content of element B.

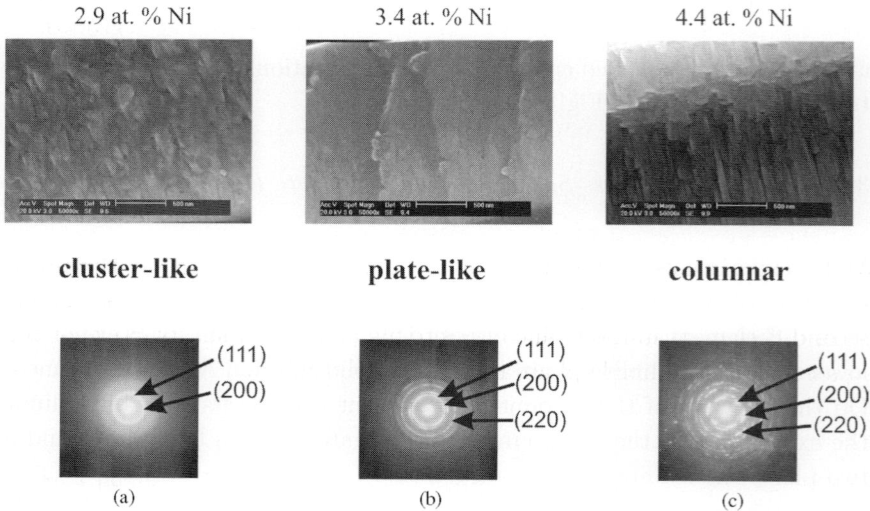

Fig. 3.8.   Effect of Ni content on the microstructure of $Zr(Ni)N_y$ films. (a) globular microstructure at 2.9 at.% Ni, (b) plate-like microstructure at 3.4 at.% Ni and (c) columnar microstructure at 4.4 at.% Ni.
Adapted from [50].

from these $Zr(Ni)N_y$ films clearly show how the film microstructure influences the crystallographic orientation of its grains. A random orientation of nanograins observed in the films with cluster-like microstructure changes relatively rapidly into a preferred crystallographic orientation of grains in the films with columnar microstructure with increasing Ni content. For more details on the microstructure of $Zr(Ni)N_y$ films see [50].

$Zr(N)N_x$ films with globular microstructure were produced inside the transition between two preferred crystallographic orientations of grains and exhibit enhanced hardness $H \approx 29\,\mathrm{GPa}$ [50].

3.4. *Microstructure of Nanocomposites with Enhanced Hardness*

Nanocomposites with enhanced H can exhibit different nanostructures. The analysis given above shows that there are at least three types of microstructure which results in the enhanced H of nanocomposite films: (1) columnar, (2) nanograins surrounded by very thin ($\sim$1–2 ML) tissue phase and (3) mixture of nanograins of different crystallographic orientation, see Fig. 3.9; here ML is the monolayer. According to the film nanostructure, the nanocomposites with enhanced H can be divided into three groups:

1. The nanocomposites with a columnar nanostructure composed of the grains assembled in nanocolumns; there is insufficient amount of the second (tissue) phase to cover all grains, Fig. 3.9(a).
2. The nanocomposites with a dense globular nanostructure composed of nanograins fully surrounded by tissue phase, Fig. 3.9(b).
3. The nanocomposites with a dense globular nanostructure composed of nanograins of different materials (two-phase materials) or nanograins of different crystallographic orientations and/or lattice structure of the same material (single-phase materials), Fig. 3.9(c).

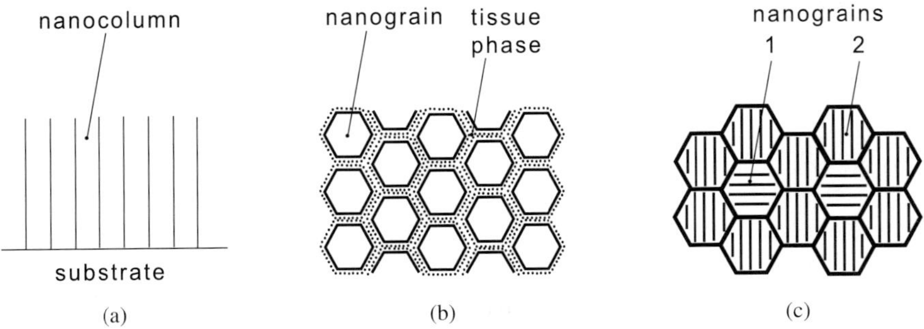

Fig. 3.9. Schematic illustration of different nanostructures of nanocomposites with enhanced H. (a) columnar, (b) nanograins surrounded by tissue phase and (c) mixture of nanograins.
Reprinted from [45], with permission of Springer Science and Business Media.

Columns in the film are perpendicular to the substrate/film interface. The nanocomposites produced at crystalline edges of the transitions (Figs. 3.2(a) and (b)) also exhibit the columnar nanostructure. The nanocomposites composed of nanograins fully surrounded by tissue phase are formed inside the transition from the crystalline to amorphous (Fig. 3.2(a)). The nanocomposites composed of a mixture of small nanograins of different material or nanograins of different crystallographic orientation and/or lattice structure of the same material are formed inside between two crystalline phases or two preferred crystallographic orientations of grains.

The classification given above was already confirmed experimentally, for instance, see Fig. 3.8. It demonstrates that the origin of the enhanced H is closely connected with the size and shape of the building blocks from which the nanocomposite is composed. Based on this finding we can conclude that both the geometry of building blocks and dimensions of grains are the key factors which determine the new unique properties of nanocomposite films.

A very important issue is the finding that the enhanced hardness can include films which are composed of a mixture of nanograins of the same material but of different crystallographic orientations and/or different lattice structures, Fig. 3.9(c). This nanostructure explains the enhanced hardness of single-phase materials. The films with columnar nanostructure or those composed of nanograins surrounded by tissue phase are two-phase materials.

3.5. *New Advanced Materials Composed of Nanocolumns*

Many recent experiments clearly show that the materials composed of nanocolumns perpendicular to the film/substrate interface exhibit not only strongly anisotropic properties but also new unique properties. As an example, we can introduce the enhancement of (1) the strength of materials or (2) the photocalytic acitivity of materials due to the reduction of unwanted electron scattering and recombination in more ordered columnar structures. These effects are main reasons why the materials composed of nanocolumns are expected to be new advanced materials in the very near future and why they have been started to be intensively investigated in many labs.

Materials with nanocolumnar structure can be easily formed by reactive magnetron sputtering. For instance, A(B)N$_x$ nitride solutions with low ($\leq$10 at.%) content of added B element, e.g. Ti(Fe)N$_x$, exhibit nanocolumnar structure. Moreover, low ($\leq$10 at.%) content of the added element B should decrease the melting point T_m of $A(B)N_x$ compounds, i.e. the ratio T_s/T_m will increase and consequently the macrostress σ generated in the film during its growth will decrease due to the thermal recovery [49]; here T_s is the substrate temperature during film deposition. This is another advantage of these new materials which may realize new applications.

Here, it is worthwhile to note that also materials composed of nanotubes belong to this new class of advanced materials. These materials offer further unique properties.

4. Mechanical Properties of Nanocomposite Coatings

Mechanical properties of nanocomposite coatings are well characterized by their hardness, H_f, effective Young's modulus $E_f^* = E_f/(1-\nu_f^2)$ and elastic recovery W_e; here E_f is the Young's modulus and ν_f is the Poisson's ratio. These quantities can be evaluated from loading/unloading curves measured by a microhardness tester. Measured values of H_f and E_f^* permit to calculate the ratio H_f^3/E_f^{*2} which is proportional to the resistance of the material to plastic deformation [56]. The likelihood of plastic deformation is reduced in the materials with high hardness and low Young's modulus E_f^*. In general, a low Young's modulus E_f is desirable as it allows the given load to be distributed over a wider area. All data given in this chapter were measured in our labs using a microhardness tester Fischerscope H 100.

The dependencies $H_f = f(E_f^*)$, $H_f^3/E_f^{*2} = f(H_f)$ and $W_e = f(H_f)$ are basic relations between mechanical properties of thin films [19, 34, 57] because they determine the mechanical behavior of thin films. For selected oxides, carbides and nitrides these dependences are displayed in Fig. 4.1.

As can be seen from Fig. 4.1(a) the dependence $H_f = f(E_f^*)$ can be approximated by a straight line

$$H_f[\text{GPa}] = 0.15E_f^*[\text{GPa}] - 12. \tag{4.1}$$

Similarly, the dependence $H_f^3/E_f^{*2} = f(H_f)$ given in Fig. 4.1(b) can be approximated by a parabola:

$$H_f^3/E_f^{*2} = 4.3 \times 10^{-4}H_f^2. \tag{4.2}$$

Experimental points are quite well distributed along the lines defined by Eqs. (4.1) and (4.2). This finding seems to be of a fundamental importance for the prediction of mechanical behavior of the coating. At first, we see that in all materials displayed in Fig. 4.1 there is a strong relationship between H_f and E_f^*. H_f almost linearly increases with increasing E_f^* (Fig. 4.1(a)). The scatter of experimental points around the straight line

                    J. Musil

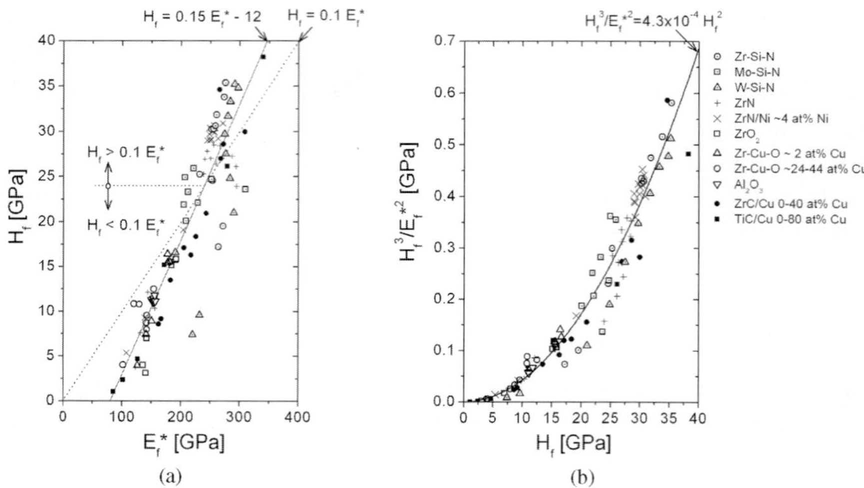

Fig. 4.1. The relationships between (a) H_f and E_f^* and (b) the ratio H_f^3/E_f^{*2} and H_f^* for selected oxides, carbides and nitrides prepared by magnetron sputtering under different deposition conditions.
Reprinted from [58], with permission of Elsevier.

may be of different origin and needs a detailed investigation which is out of the scope of this article. At second, Fig. 4.1 shows that it is possible to control the mechanical behavior of the film, e.g. its resistance to plastic deformation characterized by the ratio H_f^3/E_f^{*2}, because the material hardness H_f is not exactly $(1/10)\,E_f^*$ but $H_f \leq (1/10)E_f^*$ for $E_f^* \leq 240\,\text{GPa}$ and $H_f \geq (1/10)E_f^*$ for $E_f^* > 240\,\text{GPa}$. This fact indicates that the general relationships between H_f, E_f^* and H_f^3/E_f^{*2} defined by Eqs. (4.1) and (4.2) could be used to predict the relation between the cracking of film and its toughness. More details are given below in Sec. 6.

Another important characteristic of the material is its plastic deformation. The plastic deformation W_p of different films as a function of their hardness H_f and the ratio H_f^3/E_f^{*2}, i.e. $W_p = f(H_f)$ and $W_p = f(H_f^3/E_f^{*2})$ is given in Fig. 4.2. As expected W_p decreases with increasing both H_f and H_f^3/E_f^{*2}. The dependence of W_p versus H_f^3/E_f^{*2} exhibits a smaller scatter of experimental points. This is due to the fact that the ratio H_f^3/E_f^{*2} already expresses the combined action of H_f and E_f^* of films on their mechanical behavior. The hard films with $H_f \geq 25\,\text{GPa}$ exhibit relatively low plastic deformation of approximately 30%.

In summary, it is necessary to note that the mechanical properties of the nanocomposite coating strongly depend on (1) elements which form individual phases, (2) kind and relative content of phases, (3) chemical

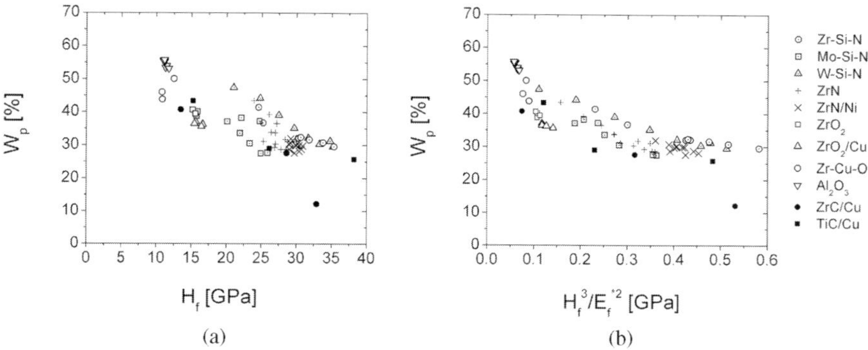

Fig. 4.2. Plastic deformation (a) $W_p = f(H_f)$ and (b) $W_p = f(H_f^3/E_f^{*2})$ of selected oxides, carbides and nitrides prepared by magnetron sputtering under different deposition conditions.

interaction between phases and (4) its microstructure. For details see, for instance [19, 34, 57].

5. High Temperature Behavior of Hard Nanocomposites

5.1. *Thermal Stability of Film Properties*

As was already mentioned above, unique properties of the nanocomposite films are due to their nanostructure. The nanostructure is, however, a metastable phase. This means, if the temperature under which the film is operated achieves or exceeds the crystallization temperature, T_{cr}, the material of the film starts to crystallize. It results in a destruction of its nanostructure and in the formation of new crystalline phases. This is the reason why the nanocomposite films lose their unique properties at temperatures $T > T_{cr}$. Simply said, the crystallization temperature T_{cr}, at which the nanostructure of nanocomposite film is destroyed and new crystalline phases occur, determines the thermal stability of the nanocomposite material. The crystallization temperature, T_{cr}, of the hard nanocomposite films so far produced is lower than 1000°C. However, it is insufficient, for instance, for high speed cutting. Many other applications also require T_{cr} to be higher than 1000°C. Therefore now, it is vitally important to develop new hard nanocomposite materials which will be thermally stable against crystallization and oxidation at operating temperatures, T, considerably exceeding 1000°C.

High thermal stability and high-temperature (high-T) oxidation resistance are very desirable properties of the hard nanocomposite coatings.

These properties strongly depend on the phase composition and thermal stability of individual phases of which the coating is composed. Therefore, it is not surprising that in the development of hard coatings with high-T oxidation resistance their phase composition has been varied from TiC ($\sim$400°C) [59] through TiN ($\sim$650°C) to (Ti, Al)N ($\sim$850°C) [60] and recently also to (Ti, Al, Y)N [61] and Me–Si–N nanocomposites with a low ($\leq$10 at.%) Si content ($\sim$950°C) [62]; here Me = Ti, Zr, Cr, W, Ta, Mo, Nb, etc. A common feature of all these coatings is their polycrystalline structure. These coatings are composed of grains surrounded by boundaries directly connecting their surfaces, exposed to an external oxidizing atmosphere, with the substrates. Mainly this fact is responsible for a relatively low (less than 1000°C) oxidation resistance of the hard protective coatings.

5.2. Si_3N_4/MeN_x Composites with High ($\geq$ 50 vol.%) of a-Si_3N_4 Phase

In principle, there is only one efficient way to increase the oxidation resistance of hard coatings. It is based on the suppression of coating crystallization, i.e. on the elimination of grains, and so on the removal of a continuous connection between the coating surface and the substrate along the grains through boundaries surrounding them. This means that the hard coating with high-T oxidation resistance should be amorphous, see Fig. 5.1.

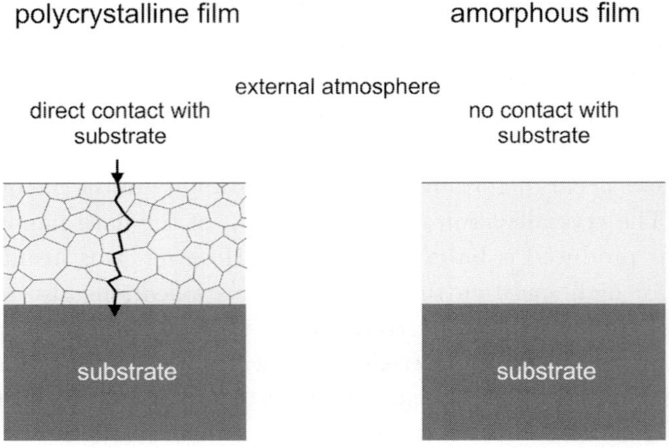

Fig. 5.1.   Schematic illustration of the contact between external atmosphere and substrate through the protective film.

Recently, it was shown that a new family of Si_3N_4/MeN_x composites with a high (≥ 50 vol.%) content of a-Si_3N_4 phase can fulfil this requirement; a- denotes the amorphous phase [63–70]. These composite coatings are amorphous and some of them exhibit two unique properties (1) excellent oxidation resistance considerably exceeding 1000°C and (2) the high hardness ranging from 20 to 40 GPa. Unique properties of these new hard amorphous composites can lead to new applications not mastered so far.

The structure, microstructure, surface morphology, mechanical properties and high-T oxidation resistance of the a-Si_3N_4/MeN_x composites with high (≥ 50 vol.%) content of a-Si_3N_4 phase are analyzed in detail below. Special attention is devoted to the effect of (1) the amount of a-Si_3N_4 phase in the composite, (2) the thermal decomposition of the MeN_x nitride phase, (3) the interdiffusion of elements from the substrate into the coating and (4) the elemental composition of the annealing atmosphere on the oxidation resistance of hard Si_3N_4 based amorphous composites. Examples of hard Si_3N_4 based amorphous composite coatings deposited on Si substrates with high-T oxidation resistance (no increase in mass $\Delta m = 0$ in oxidation test) achieving up to 1300°C are given below.

5.2.1. *Film Preparation and Measurement of Their Oxidation Resistance*

The Si_3N_4/MeN_x composite films were reactively sputtered with an unbalanced DC magnetron equipped with an alloyed $MeSi_2$ target of diameter 100 mm in a mixture of $Ar + N_2$; here Me = Ta, Zr, Mo, W. Films were deposited under the following conditions: discharge current $I_d = 1$ and 2 A, substrate bias, U_s, ranging from U_{fl} to -500 V, substrate ion current density, i_s, ranging from 0.5 to 1 mA/cm^2, substrate temperature, T_s, ranging from 100 to 750°C, substrate-to-target distance $d_{s-t} = 50$ and 60 mm, partial pressure of nitrogen, p_{N2}, ranging from 0 to 0.7 Pa and total pressure $p_T = p_{Ar} + p_{N2} = 0.5$ and 0.7 Pa. Films were sputtered on polished 15330 CSN steel substrates (disk Ø25 mm, 5 mm thick) and Si(100) wafers ($20 \times 7 \times 0.42$ mm^3). Typical thickness, h, of the Si_3N_4/MeN_x composite films ranged from 3000 to 5000 nm.

The high-temperature oxidation resistance of films was measured in pure air with a flow rate of 1 l/h using a symmetrical high resolution Setaram thermogravimetric system TAG 2400. The heating rate was 10°C/min and heating temperature ranged from room temperature (RT) to (1) 1300°C for

Si substrate (temperature limit for Si substrate) and (2) 1700°C for sapphire substrate. Changes in the weight of measured sample caused during the oxidation test by the oxidation of Si or Al_2O_3 substrate were eliminated. More details on deposition conditions, sputtering process and characterization of sputtered Si_3N_4/MeN_x composite films are given in the following papers: Si_3N_4/TaN_x [63], Si_3N_4/ZrN_x [64], Si_3N_4/MoN_x [65] and Si_3N_4/WN_x[66].

5.2.2. *Elemental and Phase Composition*

Based on the fact that the Si_3N_4 phase should be formed preferentially due to (1) a higher affinity of N to Si ($\Delta H_{f\,Si3N4} = -745\,\mathrm{kJ/mol}$, $\Delta H_{f\,ZrN} = -365\,\mathrm{kJ/mol}$ [71]) and (2) lower affinity of Si to Me (e.g. $\Delta H_{f\,ZrSi2} = -159\,\mathrm{kJ/mol}$ [72]), we assume that only the number of N atoms exceeding that necessary to form the Si_3N_4 phase will be joined with Me atoms and will form the MeN_x phase [73]. On the contrary, in the case when the amount of N is not sufficient to form Si_3N_4 with all Si atoms a $MeSi_x$ phase is formed. This assumption enables us to convert the elemental composition of the film into its phase composition.

As an example, Fig. 5.2 illustrates the conversion of the elemental composition (Fig. 5.2(a)) of the film into its phase composition (Fig. 5.2(b)) for Zr–Si–N films. The Zr–Si–N films were reactively sputtered from a $ZrSi_2$ target using a DC unbalanced magnetron under the following deposition conditions: magnetron discharge current $I_d = 1$ A, substrate bias

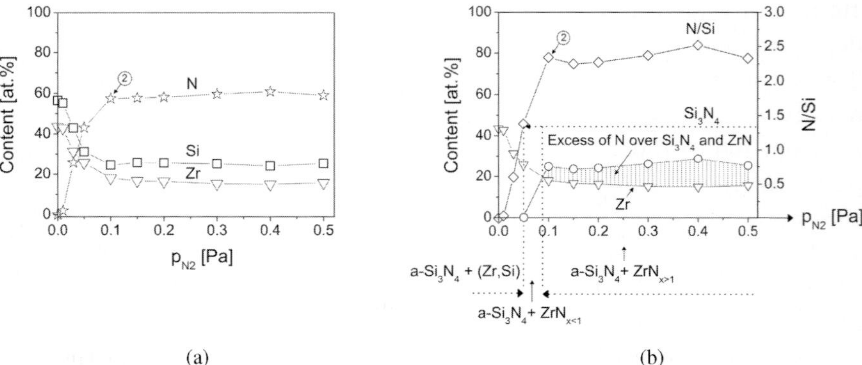

Fig. 5.2. Conversion of the elemental composition of Zr–Si–N films into the phase composition based on known values of heat of formation ΔH_f of individual compounds. Reprinted from [64] with permission of Elsevier.

$U_s = -100\,\mathrm{V}$, substrate ion current density $i_s = 1\,\mathrm{mA/cm^2}$, substrate temperature $T_s = 500°\mathrm{C}$, partial pressure of nitrogen, p_{N2}, ranging from 0 to $0.5\,\mathrm{Pa}$, substrate-to-target distance $d_{s-t} = 50\,\mathrm{mm}$ and total sputtering gas pressure $p_{\mathrm{T}} = p_{\mathrm{Ar}} + p_{\mathrm{N2}} = 0.7\,\mathrm{Pa}$. As can be seen from Fig. 5.2(b), the phase composition of Zr–Si–N film gradually varies from $\mathrm{ZrSi_2}$ through a-$\mathrm{Si_3N_4} + (\mathrm{Zr}, \mathrm{Si})$ to a-$\mathrm{Si_3N_4} + \mathrm{ZrN}_{x<1}$ and next to a-$\mathrm{Si_3N_4} + \mathrm{ZrN}_{x>1}$. Here, it is necessary to note that the phase composition of the Zr–Si–N film strongly depends on the combination of process parameters (I_d, U_s, i_s, T_s, p_{N2}, p_{T}, d_{s-t}) under which the film is produced. More details are given in [64].

The correctness of conversion of the elemental composition into the phase composition was verified by measurement of the structure evolution of the Zr–Si–N films using X-ray diffraction. The development of XRD patterns from the Zr–Si–N films with increasing p_{N2} is displayed in Fig. 5.3. In this figure, values of the film microhardness H and the ratio N/Si are also given. Under the assumption that at first the $\mathrm{Si_3N_4}$ phase is formed, all sputtered films with N/Si ≥ 1.33 contain a stoichiometric $\mathrm{Si_3N_4}$ nitride with N/Si $= 1.33$. From this experiment the following important issues can be drawn

1. The $\mathrm{ZrSi_2}$ films sputtered at $p_{\mathrm{N2}} = 0\,\mathrm{Pa}$ and Zr–Si–N films sputtered at $p_{\mathrm{N2}} \leq 0.03\,\mathrm{Pa}$ are crystalline and characterized with a strong (200) reflection ($p_{\mathrm{N2}} = 0\,\mathrm{Pa}$) and many weak, low-intensity (111), (131), (081) and (280) reflections ($p_{\mathrm{N2}} \leq 0.03\,\mathrm{Pa}$) from $\mathrm{ZrSi_2}$ grains; the intensity of (200) reflection decreases with increasing ratio N/Si.
2. The films sputtered at $p_{\mathrm{N2}} > 0.03\,\mathrm{Pa}$ are X-ray amorphous and are characterized by one very broad, low-intensity reflection line with a maximum located at $2\theta \approx 37.5°$.
3. The amorphous structure of Zr–Si–N films is created in consequence of a very efficient formation of a-$\mathrm{Si_3N_4}$ phase.
4. The transition from the crystalline to amorphous phase is caused by a strong increase of the amount of N incorporated into the Zr–Si–N film and by the formation of large ($\geq 50\,\mathrm{vol.\%}$) amount of $\mathrm{Si_3N_4}$ phase.
5. The microhardness H of amorphous Zr–Si–N films is high and only slightly increases with increasing p_{N2} from approximately $30\,\mathrm{GPa}$ at $p_{\mathrm{N2}} = 0.1\,\mathrm{Pa}$ to $35\,\mathrm{GPa}$ at $p_{\mathrm{N2}} = 0.5\,\mathrm{Pa}$.

Based on measurements of the evolution of XRD patterns from the Zr–Si–N films with increasing p_{N2} (Fig. 5.3) we can conclude that the phase composition of the Zr–Si–N film changes from crystalline $\mathrm{ZrSi_2}$ through a-$\mathrm{Si_3N_4} + \mathrm{ZrSi_2}$ to a-$\mathrm{Si_3N_4} + \mathrm{ZrN}_x$ phase with increasing p_{N2}. Similar

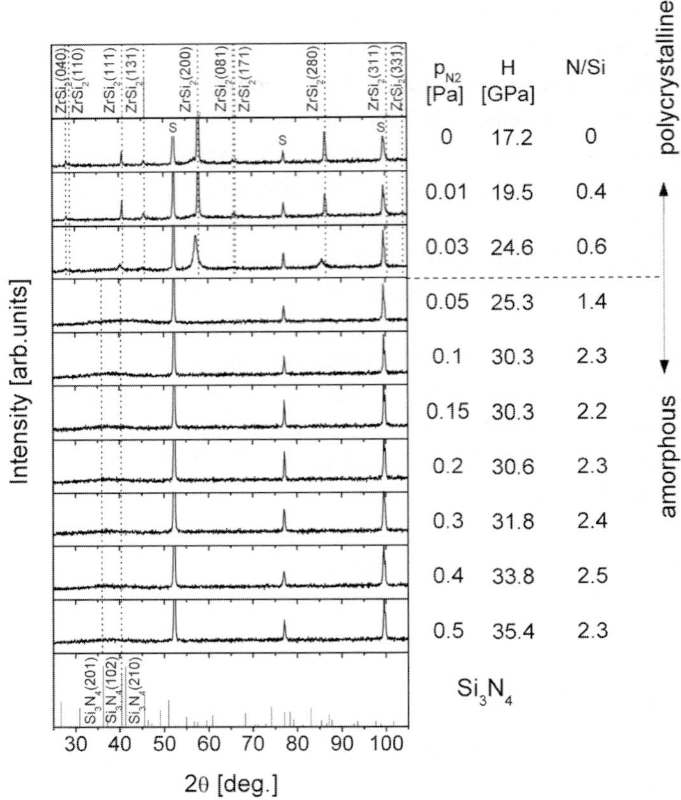

Fig. 5.3. Development of structure of Zr–Si–N films sputtered at $I_d = 1$ A, $U_s = -100$ V, $i_s = 1$ mA/cm^2, $T_s = 500°$C, $d_{s-t} = 50$ mm, $p_T = 0.7$ Pa on steel substrates with increasing p_{N2}.

evolution of the XRD patterns from the Me–Si–N films with increasing p_{N2} was found also for the films with Me = Ta, Mo, W and Ti.

The physical properties and thermal stability of Zr–Si–N film depend on its phase composition. For instance, the Zr–Si–N films deposited at $p_{N2} \leq 0.1$ Pa and $p_{N2} > 0.1$ Pa strongly differ in the electrical conductivity and the optical transparency. While the films sputtered at $p_{N2} \leq 0.1$ Pa are electrically conductive and optically opaque, those sputtered at $p_{N2} > 0.1$ Pa are electrically insulating and optically transparent. This difference in the film properties correlates well with strong changes in its phase composition with increasing p_{N2}. The Zr–Si–N films sputtered at $p_{N2} \leq 0.1$ Pa are composed of three a-Si$_3$N$_4$ + ZrN$_{x<1}$ $\equiv$ a-Si$_3$N$_4$ + Zr + ZrN phases. Free Zr atoms are responsible for the electrical conductivity and nontransparency (opaqueness) of these films. As soon as a sufficient amount of N is available, i.e. at $p_{N2} > 0.1\,Pa$, all Si atoms are converted into the Si_3N_4 phase and the nanocomposite is composed of two $a\text{-}Si_3N_4 + ZrN_{x>1}$ phases. These films are (1) amorphous as shown in the XRD patterns given in Fig. 5.3, (2) optically transparent and (3) electrically insulating.

5.3. *Thermal Stability of Amorphous Me–Si–N Nanocomposites*

The thermal stability of amorphous films is determined by its crystallization temperature, T_{cr}. T_{cr} depends on three factors: (1) the thermal stability of individual components (phases) from which Me–Si–N film is composed, (2) the elemental composition of annealing atmosphere and (3) the interdiffusion of the substrate elements into the film during annealing. The effect of these three factors on the film crystallization is further illustrated on an example of the crystallization of Zr–Si–N films during their thermal annealing performed under different conditions.

5.4. *Crystallization of Amorphous Zr–Si–N Films During Post-Deposition Thermal Annealing*

5.4.1. *Crystallization of Amorphous Zr–Si–N Film on Si(100) Substrate*

A. Thermal Annealing in Vacuum

The thermal annealing of as-deposited amorphous Zr–Si–N films was investigated in vacuum at selected annealing temperatures T_a for 30 minutes. The annealing temperature T_a was gradually increased in 200°C steps from 500 to 700°C and in steps 50°C from 800 to 1150°C. After each annealing cycle the film structure was investigated by means of XRD measurements. Results of this investigation are summarized in Fig. 5.4 and Table 5.1.

Figure 5.4 clearly shows that changes in the structure of a-Zr–Si–N film strongly depend on their phase composition in as-deposited state. The phase composition of as-deposited films, determined on the basis of known values of the heat of formation of compounds from which the film is composed, i.e. on $\Delta H_{f\,Si3N4} = -743\,kJ/mol$, $\Delta H_{f\,ZrN} = -365\,kJ/mol$, and $\Delta H_{f\,ZrSi2} = -159\,kJ/mol$ [71, 72], is given in Table 5.1.

Differences in properties of both films can be briefly described as follows.

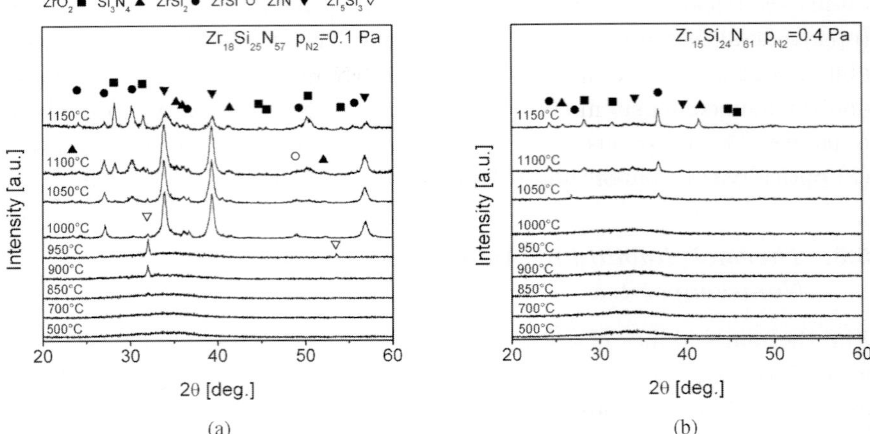

Fig. 5.4. Evolution of the structure of two as-deposited amorphous Zr–Si–N films, magnetron sputtered at $I_d = 1$ A, $U_s = -100$ V, $i_s = 1$ mA/cm^2, $T_s = 500°$C, $p_T = 0.7$ Pa and (a) $p_{N2} = 0.1$ and (b) 0.4 Pa, during thermal annealing in vacuum with increasing annealing temperature T_a at heating rate 10 K/min and cooling rate 30 K/min [69].

Table 5.1. Crystallization temperature T_{cr} and phase composition of amorphous Zr–Si–N films sputtered on Si(100) substrate.

Film	$Zr_{18}Si_{25}N_{57}$	$Zr_{15}Si_{24}N_{61}$
Phase composition		
of as-deposited film	a-Si$_3$N$_4$ + ZrN$_{x=1.3}$	a-Si$_3$N$_4$ + ZrN$_{x=1.9}$
T_{cr1} [°C]	850	~1050
First c-phase	Zr$_5$Si$_3$	ZrSi$_2$
T_{cr2} [°C]	1000	1100
Next c-phases	ZrN + ZrSi$_2$	ZrSi$_2$ + ZrO$_2$
T_{cr3} [°C]	1150	1150
Next c-phases	ZrN + Si$_3$N$_4$ + ZrO$_2$ + ZrSi$_2$	ZrSi$_2$ + ZrO$_2$ + Si$_3$N$_4$ + ZrN

c- is the crystalline phase, T_{cr1}, T_{cr2} and T_{cr3} are the crystallization temperature of the first phase and next phases, respectively.

$Zr_{18}Si_{25}N_{57}/Si(100)$ System

The $Zr_{18}Si_{25}N_{57}$ film is *electrically conductive and optically opaque*. In spite of these facts, it exhibits an X-ray amorphous structure up to approximately 800°C. During thermal annealing in vacuum the following crystalline phases occur: (1) Zr$_5$Si$_3$ at $T_a \approx 850°$C, (2) ZrN + ZrSi$_2$ at $T_a \approx 1000°$C, (3) ZrN + Si$_3$N$_4$ + ZrO$_2$ + ZrSi$_2$ at $T_a \approx 1100°$C, see Fig. 5.4(a). It is worthwhile to note that the reflections from some phases decrease with increasing T_a. The reflection from Zr$_5$Si$_3$ grains almost disappear at $T_a \approx 950°$C. The

Zr_5Si_3 phase is probably converted into $ZrSi_2$ phase at $T_a \approx 1000°C$ in consequence of the enrichment of the film with Si from the Si(100) substrate. On the contrary, the reflections from the ZrN phase are strongly reduced at $T_a \approx 1150°C$. This may be due to the decomposition of or desorption of N from the ZrN phase which can result in the formation of a new ZrO_2 phase, see increasing of ZrO_2 reflections with increasing T_a. The increase of $ZrSi_2$ reflection is probably connected with an enhanced interdiffusion of Si from the substrate into the film as T_a increases.

$Zr_{15}Si_{24}N_{61}/Si(100)$ System

The $Zr_{15}Si_{24}N_{61}$ film with a strongly overstoichiometric $ZrN_{x=1.9}$ phase is electrically insulating and optically transparent as well. It exhibits even higher thermal stability, almost up to 1050°C, compared with $Zr_{18}Si_{25}N_{57}$ film. During thermal annealing in vacuum the following crystalline phases occur: (1) $ZrSi_2$ at $T_a \approx 1050°C$, (2) $ZrSi_2 + ZrO_2$ at $T_a \approx 1100°C$ and (3) $ZrSi_2 + ZrO_2 + Si_3N_4 + ZrN$ at $T_a \approx 1150°C$, see Fig. 5.4(b). This investigation shows that the crystallization of the $ZrN_{x>1}$ phase at $T_a \approx 1150°C$ is very weak and is accompanied also with very weak crystallization of the Si_3N_4 phase. Free Zr created after ZrN decomposition is converted into the ZrO_2 phase.

This experiment clearly shows that the thermal stability of Zr–Si–N film increases with increasing content of N incorporated in the film and is maximum when the overstoichiometric metal nitride $MeN_{x>1}$ is formed. It can be expected that the thermal stability achieves a maximum value when the columnar microstructure of the film is fully eliminated, i.e. nanocrystalline structure of the film is converted into amorphous one. This hypothesis, however, needs to be demonstrated in next experiments.

B. Thermal Annealing of Zr–Si–N Film with Overstoichiometric $ZrN_{x>1}$ Phase in Flowing Argon

Thermal annealing of the Zr–Si–N films in flowing argon and vacuum strongly differ. A comparison of experiments performed in the vacuum and flowing argon clearly shows a strong effect of the elemental composition of annealing atmosphere on the film crystallization. It is worthwhile to note that already a small amount of O_2 in the annealing atmosphere strongly affects the crystallization process. This fact demonstrates the comparison of the structure of $Zr_{15}Si_{25}N_{60}$ film annealed in vacuum (10^{-2} Pa) and argon (atmospheric pressure) at $T_a = 1100°C$ for 30 min, see Fig. 5.5.

          J. Musil

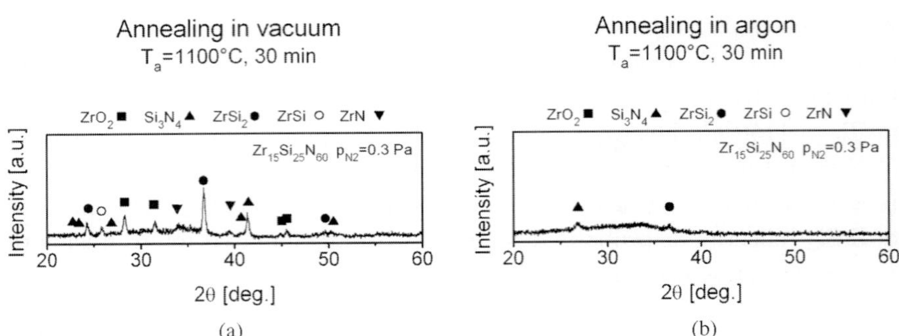

Fig. 5.5.   Structure of optically transparent and electrically insulating as-deposited amorphous $Zr_{15}Si_{23}N_{60}$ film sputtered on Si(100) substrate after its annealing in (a) vacuum and (b) argon at $T_a = 1100°C$ for 30 min.

From Fig. 5.5 it is clearly seen that while the film annealed in the vacuum is well crystallized at $T_a = 1100°C$ (Fig. 5.5(a)), the film annealed at the same temperature T_a in argon exhibits only a very weak crystallization of $ZrSi_2$ and Si_3N_4 grains from amorphous phase (Fig. 5.5(b)). Particularly, $ZrSi_2$, Si_3N_4 and ZrO_2 reflections are quite strong when the film is annealed in vacuum. This means that even a very small amount of O_2 contained in the residual atmosphere (vacuum of 10^{-2} Pa) is sufficient to influence strongly the crystallization of amorphous material.

C.   Thermal Annealing of Zr–Si–N Film with Almost Stoichiometric $ZrN_{x≈1}$ Phase in Flowing Air

The thermal annealing of the electrically conductive and optically opaque as-deposited amorphous Zr–Si–N film with almost stoichiometric $ZrN_{x≈1}$ phase at $T_a = 1300°C$ in both the flowing air and argon results in its strong crystallization, see Fig. 5.6. In spite of strong crystallization in both atmospheres there are, however, huge differences in crystalline phases created during annealing. The film annealed in Ar exhibits strong reflections from ZrN and $ZrSi_2$ grains and small reflections from Si_3N_4 grains. It shows that the film annealed in Ar is composed of many $ZrSi_2$ grains of different crystallographic orientations but contains no ZrO_2 grains (no ZrO_2 reflections). On the contrary, in the film annealed in air the crystallization of ZrN and Si_3N_4 phases is strongly suppressed and $ZrSi_2$, $ZrSiO_4$, ZrO_2 reflections dominates. This indicates that the reaction of O_2 with the film is very strong. The oxidation process is now under detailed study in our labs. The

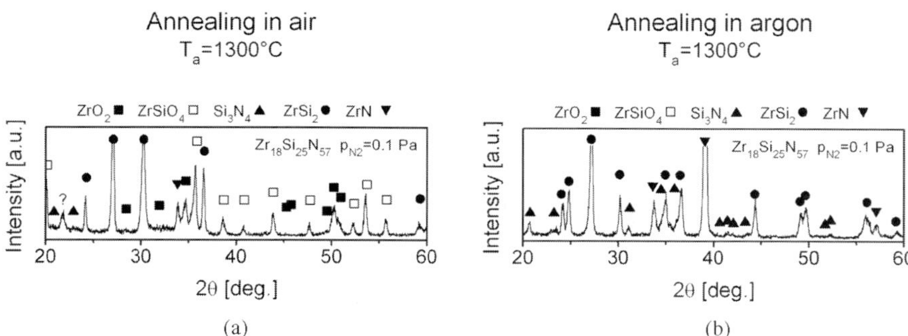

Fig. 5.6. Structure of electrically conductive and optically opaque as-deposited amorphous $Zr_{18}Si_{25}N_{57}$ films sputtered on Si(100) substrate after its annealing in flowing (a) air and (b) argon at $T_a = 1100°C$ for 30 min.

slowdown of the crystallization of Si_3N_4 phase during thermal annealing in air seems to be an interesting finding.

In summary we can conclude that the stoichiometry, x, of MeN_x nitride phase in Si_3N_4-based composites is a very important parameter that determines the crystallization of Me–Si–N films. The amorphous Zr–Si–N films with a strongly overstoichiometric $ZrN_{x\gg1}$ phase are more resistant to crystallization during thermal annealing compared to Zr–Si–N films with nearly stoichiometric ($ZrN_{x\approx1}$) or even substoichiometric ($ZrN_{x\leq1}$) phase. This fact is demonstrated by the following experiments in which the film was separated from the substrate and the interdiffusion of the substrate elements into the film during thermal annealing is excluded.

5.4.2. *Crystallization of Amorphous Zr–Si–N Films Separated from Substrate in Flowing Argon*

Changes in the film structure during thermal annealing can also be strongly influenced by the interdiffusion of substrate elements into the film. To avoid this undesirable effect, the Zr–Si–N film was separated from the substrate prior to thermal annealing and then only the film material was investigated using differential scanning calorimetry (DSC). The crystallization temperature T_{cr} is determined from the exothermic peaks (increase in heat flow) created on the DSC curve. This procedure completely excludes the influence of the substrate elements on the evolution of the Zr–Si–N film structure with increasing annealing temperature T_a.

As shown above the crystallization of as-deposited amorphous Zr–Si–N film which is composed of two phases $Si_3N_4 + ZrN_x$ strongly depends on the stoichiometry $x = N/Zr$ of ZrN_x phase. The Zr–Si–N film containing understoichiometric $ZrN_{x\leq1}$ phase should crystallize at a lower value of T_a compared to that containing overstoichiometric $ZrN_{x>1}$ phase. To confirm the correctness of this statement the crystallization of two as-deposited amorphous Zr–Si–N films were investigated in detail. The elemental and phase composition of these two films are summarized in Table 5.2.

Results of annealing experiments are given in Figs. 5.7 and 5.8. Figure 5.7 displays the DSC measurements and Fig. 5.8 displays the evolution of XRD patterns from Zr–Si–N films with increasing T_a. From Fig. 5.7 it is clearly seen that while the $Si_3N_4 + ZrN_{x\approx0.8}$ composite film crystallizes at a relatively low temperature $T_{cr} \approx 1130°C$ (Fig. 5.7(a)), the $Si_3N_4 + ZrN_{x\approx1.2}$ composite film crystallizes at much higher temperature $T_{cr} \approx 1530°C$ (Fig. 5.7(b)). Lower value of T_{cr} for the $Si_3N_4 + ZrN_{x\approx0.8}$ is

Table 5.2.   Elemental and phase composition of two Zr–Si–N films used in DSC measurements.

Film	Zr	Si	N at.%	N for Si_3N_4 N_{Si3N4}	N for Zr ΔN	$\Delta N/Zr$	Phase composition	T_{cr} [°C]
$Zr_{18}Si_{29}N_{53}$	18	29	53	38.7	14	0.8	$Si_3N_4 + ZrN_{0.8}$	1130
$Zr_{16}Si_{28}N_{56}$	16	28	56	37	19	1.2	$Si_3N_4 + ZrN_{1.2}$	1530

$\Delta N/Zr = x$ is the stoichiometry of ZrN_x phase.

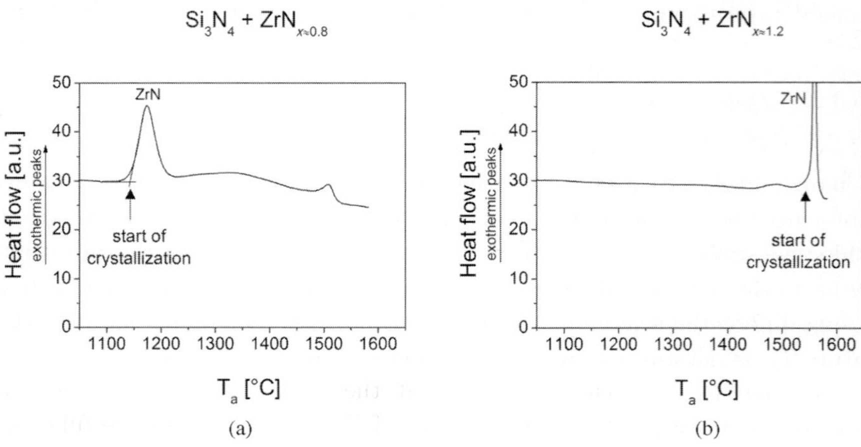

Fig. 5.7.   DSC curves of as-deposited amorphous Zr–Si–N films containing (a) understoichiometric $ZrN_{x\leq1}$ and (b) overstoichiometric $ZrN_{x>1}$ phase. Thermal annealing was carried out in flowing argon at atmospheric pressure with heating rate 10 K/min.

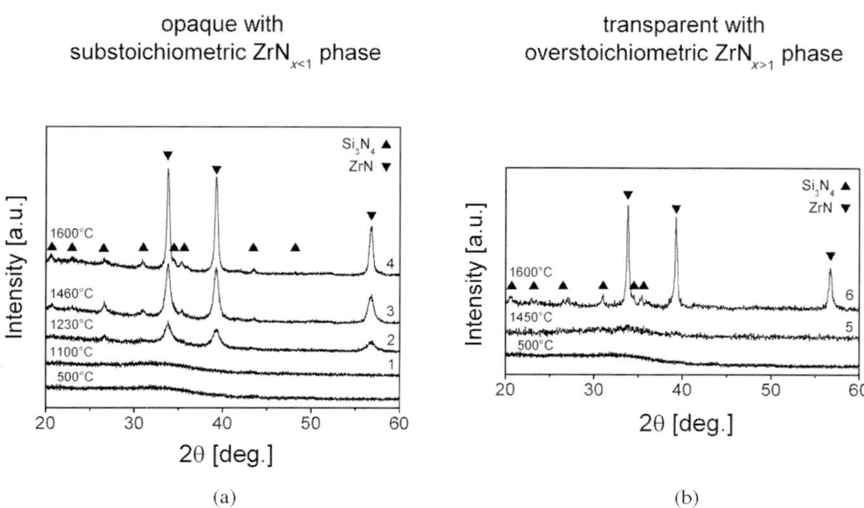

(a)   (b)

Fig. 5.8.   Development of XRD patterns from as-deposited amorphous Zr–Si–N films containing (a) understoichiometric ZrN$_{x\leq1}$ and (b) overstoichiometric ZrN$_{x>1}$ phase.

due to the fact that ZrN$_{x<1}$ is Zr-rich compound, composed of a mixture of ZrN + Zr, and free Zr atoms easily form new compounds.

XRD patterns from thermally annealed Zr–Si–N films fully confirm the DSC measurements. Besides, the XRD measurements show clear differences in the crystallization of the Zr–Si–N films with understoichiometric ($x = $ N/Zr < 1) and overstoichiometric ($x > 1$) ZrN$_x$ phase. The formation of ZrN grains is shifted to higher values of T_a of at least 1450°C for the Zr–Si–N film with overstoichiometric ZrN$_{x>1}$ phase compared to the Zr–Si–N film with substoichiometric ZrN$_{x<1}$ phase. Moreover, it is worthwhile to note that the formation of ZrN grains is accompanied by the crystallization of the Si$_3$N$_4$ phase. This means that the presence of free Zr or the deficiency of N in the film accelerates the crystallization of a-Si$_3$N$_4$ phase. This result seems to be of great practical importance.

In summary, we can conclude that the highest thermal stability of the Zr–Si–N film can be achieved if the following conditions are fulfilled: (1) the Zr–Si–N film should contain overstoichiometric ZrN$_{x>1}$ phase, (2) the annealing atmosphere should be inert and (3) the film has to be separated from the substrate by a barrier interlayer to avoid the interdiffusion of substrate elements into the film, i.e. to avoid undesirable and uncontrolled changes of its phase composition.

5.5. *Oxidation of Amorphous Me–Si–N Films in Flowing Air*

Recent experiments show that the oxidation resistance of the Me–Si–N composite films depends on (1) thermal stability of their components, i.e. MeN_x and Si_3N_4 phases, (2) interdiffusion of the substrate elements into the thermally annealed film and (3) type of oxide, MeO_x, which can be either a solid or volatile phase. The thermal stability of the Si_3N_4 phase is higher ($\approx 1530°C$ in argon when $MeN_{x>1}$, see Fig. 5.7(b)) than that of the MeN_x phase; also the thermal stability of $MeN_{x>1}$ is higher than that of $MeN_{x\leq1}$. The overstoichiometric $MeN_{x>1}$ nitrides improve the oxidation resistance. The interdiffusion of elements into the film from the substrate should be avoided because this process changes the elemental composition of the film and, therefore, its phase composition. After the decomposition of MeN_x nitride ($MeN_{x1} \to MeN_{x2} + N_g + Me$), free Me atoms form a metal oxide MeO_x according to the following reaction: $MeN_{x1} + O_2 \to MeN_{x2} + N_g + MeO_s$ or $MeN_{x2} + N_g + MeO_g$; here the indexes $x1$ and $x2$ are the stoichiometry of MeN_x prior to and after the nitride decomposition, i.e. $x1 > x2$, s and g denotes the solid and gas phase, respectively. Different elements (Me) form nitrides with different thermal stability (resistance to crystallization) and different types (solid or volatile) of oxides also with diferent thermal stability. Therefore, the selection of the element Me in the Me–Si–N composite is of the key importance when the Me–Si–N composite films with the highest oxidation resistance are required to be developed.

5.5.1. *SEM Cross-Section Images of Amorphous Me–Si–N Films After Thermal Annealing in Flowing Air*

The scanning electron microscope (SEM) cross-section images of 2500 to 4000 nm thick Ta–Si–N [63], Mo–Si–N [65] and W–Si–N [66] films with high (≥ 50 vol.%) content of Si_3N_4 phase deposited on Si(100) substrates after the thermal annealing in flowing air at $T_a = 1300°C$ are given in Fig. 5.9. The film was heated with the rate $10\,K/min$ and immediately upon reaching the temperature $T_a = 1300°C$ was cooled down with rate $30\,K/min$.

From Fig. 5.9 the following issues can be drawn:

1. Ta–Si–N film is the best. The surface of film is covered by a thin ($\sim$100 to 400 nm) compact surface oxide layer. On the contrary, the bulk of the

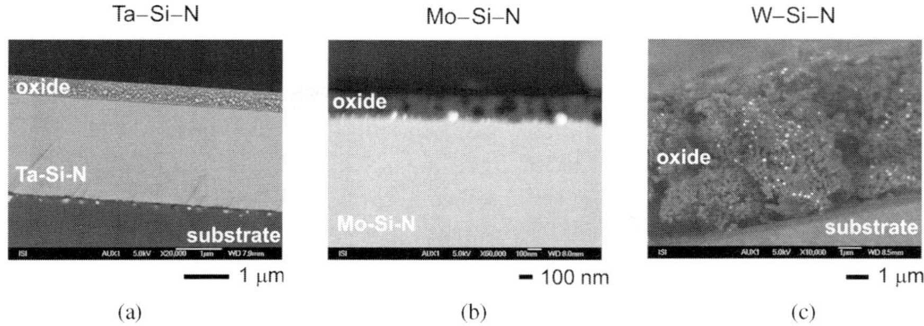

Fig. 5.9.   Comparison of SEM cross-section images of amorphous (a) Ta–Si–N, (b) Mo–Si–N and (c) W–Si–N films deposited on Si(100) substrates after thermal annealing in flowing air at $T_a = 1300°$C.

film under the surface oxide layer is amorphous. This means that there is no direct contact between the external atmosphere and the substrate so the oxidation resistance is excellent. Similar behavior exhibits also Zr–Si–N films [64].

2. Mo–Si–N film is covered by ~300 nm thick porous surface layer. The porous surface layer creates due to the formation of volatile MoO_x oxides at $T_a \approx 800 - 1000°$C. The release of volatile oxides from surface layer results in the mass decrease ($\Delta m < 0$) in thermogravimetric measurements. The bulk of the film remains amorphous. Therefore, the oxidation resistance of film is good. However, there is an open question what happens when the time of annealing at $T_a = 1300°$C will increase.

3. W–Si–N film is oxidized through the whole thickness of the film due to very strong oxidation of free W which creates in consequence of a strong decomposition of $WN_x \rightarrow W + N$ at $T_a \geq 1100°$C. The WO_x is formed in consequence of the following reaction $W + O_2 \rightarrow WO_x$. Because WO_x is volatile it escapes from the film. Therefore, its microstructure is porous and its mass reduces ($\Delta m < 0$). The oxidation resistance of W–Si–N film at $T_a \geq 1100°$C is very bad. The stronger oxidation of W compared to Mo is also supported by values of the formation enthalpy ($\Delta H_{WO3} = -838$ kJ/mol compared with $\Delta H_{MoO3} = -746$ kJ/mol [71]).

Here, it is necessary to note that results given above can be influenced by the interdiffusion of Si from the substrate into annealed film.

All experiments described in this paragraph clearly show that the high (≥ 60 vol.%) content of a-Si_3N_4 phase in Me–Si–N films is not a sufficient condition to produce the amorphous Si_3N_4/MeN_x composite films with the oxidation resistance considerably exceeding 1000°C. To reach high values of the oxidation resistance exceeding 1000°C the MeN_x phase of Si_3N_4/MeN_x composites must exhibit (1) the highest temperature of its decomposition and (2) the lowest ability of Me element to form oxide, i.e. the lowest negative value of the heat of oxide formation. Experiments performed till now demonstrated that the highest oxidation resistance is exhibited by $Si_3N_4/ZrN_{x>1}$, $Si_3N_4/TaN_{x>1}$ and $Si_3N_4/TiN_{x>1}$ composite films with high (≥ 50 vol.%) content of a-Si_3N_4 phase.

5.6. *Summary of Main Issues*

The experiments described in this chapter show that the a-Si_3N_4/MeN_x composite films with high (≥ 50 vol.%) content of a-Si_3N_4 phase exhibit high crystallization temperature $T_{cr} > 1000$°C, high oxidation resistance considerably exceeding 1000°C and good protection of the substrate if (1) the metal nitride $MeN_{x>1}$ phase is (i) overstoichiometric and (ii) resistant to its decomposition during the thermal annealing, and (2) free Me created during the decomposition of the MeN_x phase forms the dense solid state oxide; the formation of a volatile oxides needs to be avoided. The crystallization of the Me–Si–N film strongly depends also on (1) the elemental composition of the annealing atmosphere and (2) the interdiffusion of elements from the substrate into the film. An interlayer barrier is necessary to be included between the film and the substrate to suppress the latter effect. At present, the a-Si_3N_4/MeN_x composite films containing Zr, Ta and Ti exhibit best thermal behavior. The Zr–Si–N nanocomposite deposited on Si(100) substrate shows no increase in mass ($\Delta m = 0$) in thermogravimetric measurements up to 1300°C (thermal limit of Si substrate). A similar behavior is reported also for the amorphous Si–B–C–N films on Si(100) [74]. The a-$Si_3N_4/ZrN_{x>1}$ film separated from Si(100) substrate is stable during thermal annealing in flowing argon up to 1530°C (Fig. 5.7). Very recent experiments show that the a-$Si_3N_4/TiN_{x>1}$ film deposited on c-Al_2O_3 (sapphire) substrate exhibits no oxidation ($\Delta m = 0$) during thermal annealing in flowing air up to 1400°C and the strong oxidation ($\Delta m \geq 0.025$ mg/cm^2) starts at $T_a = 1500$°C, see Fig. 5.10 [75]. Figure 5.10 clearly shows a dramatic effect of the structure (crystalline or amorphous) of the as-deposited film on its oxidation resistance.

The present status in the oxidation resistance of hard coatings is summarized in Fig. 5.11. Here, a weight gain Δm as a function of the annealing temperature T_a is displayed. The temperature T_a corresponding to the

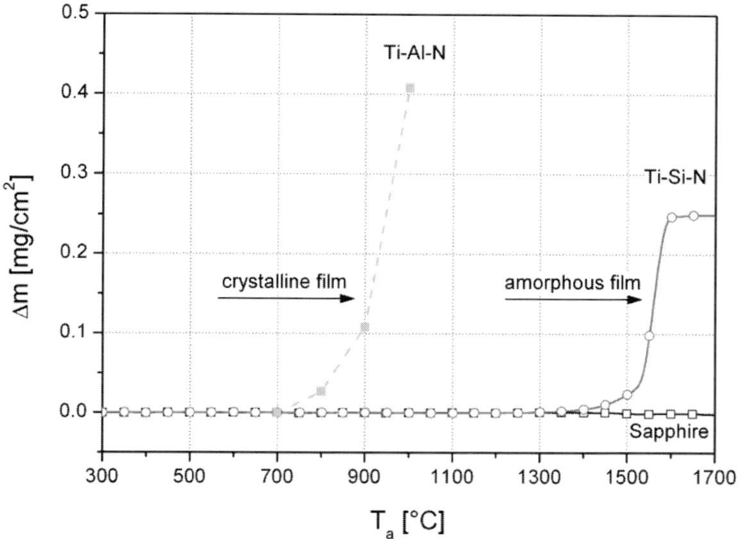

Fig. 5.10. Comparison of oxidation resistance of as-deposited polycrystalline Ti–Al–N film [60] and as-deposited X-ray amorphous Ti–Si–N film [75].

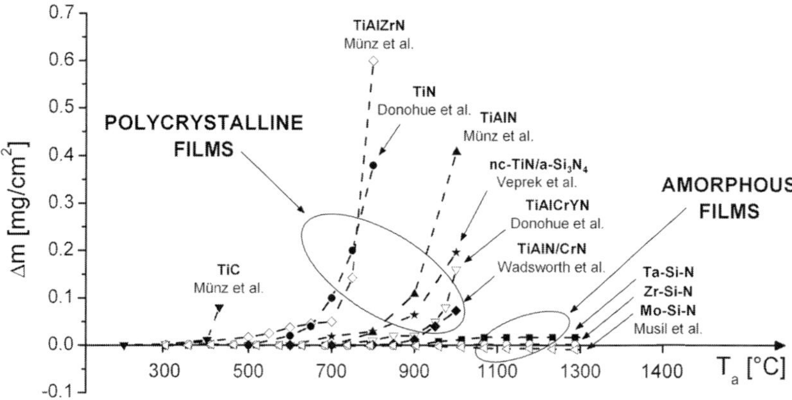

Fig. 5.11. Oxidation resistance of selected hard binary, ternary, quaternary nitrides and hard amorphous Si_3N_4/MeN_x composite films represented as $\Delta m = f(T_a)$ [70].

sharp increase of the film mass Δm, is defined as a maximum temperature $T_{\max}$ which still avoids the oxidation of the film. The oxidation resistance is higher as $T_{\max}$ becomes higher. All films with a sharp increase in Δm given in Fig. 5.11 are crystalline or nanocrystalline. For all these films the oxidation resistance is lower than 1000°C. This fact is not surprising because the films composed of grains always allow the direct contact of the external atmosphere at the film surface with the substrate via grain boundaries. This phenomenon dramatically decreases the oxidation resistance of the bulk of film and so its barrier action. Some improvement could be, however, expected to be achieved if an intergranular glassy phase is used. The best high-T oxidation resistance is exhibited by the films which are in the as-deposited state X-ray amorphous.

6. Toughness of Thin Nanocomposite Coatings

Up to recently, main attention was concentrated on the hardness H of materials (coatings), ways of H enhancement and on the achievement of H approaching or even exceeding that of diamond. New advanced nanocomposite films based on nitrides, particularly the composites of the type nc-MeN/a-Si$_3$N$_4$ with low ($\leq$10 at.%) Si content, were successfully developed. These nanocomposites exhibit enhanced H up to 50–70 GPa but none of them exhibited H approaching that of diamond. This means that the diamond still remains the hardest material.

Simultaneously it was recognized that the hard materials are often very brittle. The high brittleness of hard coatings strongly limits their practical utilization. It concerns mainly hard ceramic materials based on oxides with a very wide application range from protective to functional coatings. This is the main reason why now many labs all over the world try to develop new advanced ceramics with a low brittleness and simultaneously with sufficiently high ($\geq$20 GPa) hardness [44, 47, 76]. In spite of the fact that hardness is one of the most important mechanical properties of the material it is not sufficient to use hardness alone to select the material for a given application. The hardness H must be combined with a sufficient toughness because the film toughness can be for many applications more important than its hardness. Therefore, it is vitally important to master the formation of hard films with high toughness. The hard films with high toughness represent a new class of the advanced nanocomposite coatings, see Fig. 6.1.

6.1. *Toughening Mechanisms*

According to the definition, the toughness of a material is its ability to absorb energy during deformation up to its fracture. This means that the toughness can be enhanced if crack initiation and propagation are hindered or at least reduced.

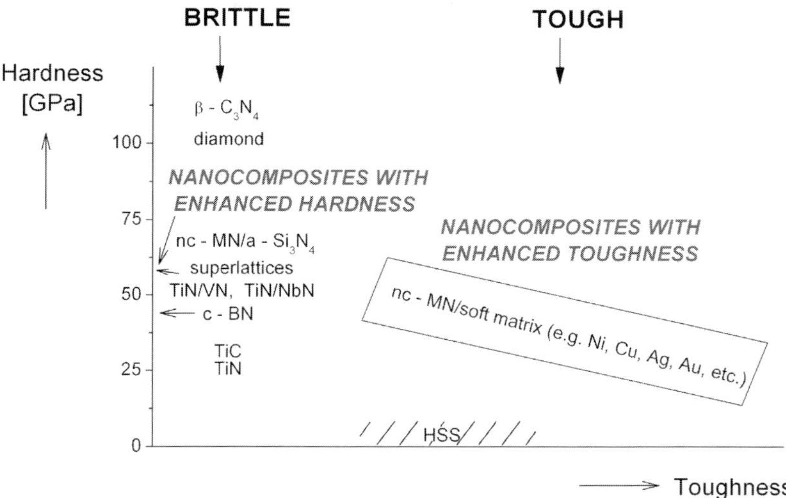

Fig. 6.1. Classification of nanocomposites according to their hardness and toughness. Adapted from [77].

There are several ways to reach this goal: (1) ductile phase toughening, i.e. the addition of certain ductile phases (metals) to ceramic matrices, (2) nanograin toughening based on crack deflection or branching along grain boundaries or grain boundary sliding, (3) multilayer structure toughening based on alternation of many brittle and ductile thin layers, (4) fiber or nanotube toughening based on bridging or deflection of cracks, (5) phase transformation toughening based on the extraction of the fracture energy and consuming it for the phase transformation and (6) compressive stress toughening which prevents the initiation of cracks by their closing [47]. For illustration, the principles of three toughening mechanisms are schematically displayed in Fig. 6.2.

6.2. *Fracture Toughness of Bulk Materials and Thin Films*

The length of cracks is commonly used for the determination of the fracture toughness of bulk materials. Under the plane strain conditions, the fracture toughness is related to the rate of strain energy release by the following formula [76]:

$$K_c = \sigma_f[\pi a/(1 - \nu^2)]^{1/2}, \tag{6.1}$$

                    J. Musil

Toughening
using

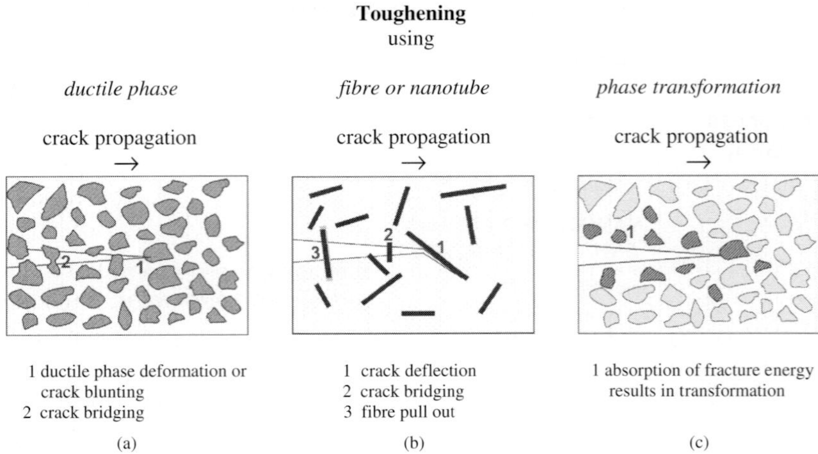

Fig. 6.2. Schematic illustration of three toughening mechanisms: (a) ductile phase toughening, (b) fibre or nanotube toughening and (c) phase transformation toughening. Adapted from [47], Copyright (2005) with permission from Elsevier.

where σ_f is the fracture strength, a is the length of the crack and ν is the Poisson's ratio. K_c is called the critical stress intensity factor and fracture toughness increases with K_c. Unfortunately, this formula can be used only for thick films with a minimum thickness $h_{\min} = 2.5(K_c/\sigma_y)^2$ [78]; here σ_y is the yield stress. For brittle materials with $(K_c/\sigma_y)^2 \approx 0.1\,\mathrm{mm}$ [79] and the minimum film thickness $h_{\min} \approx 0.25\,\mathrm{mm}$. This means that the toughness of thin ($\leq 10\,\mu\mathrm{m}$) films cannot be calculated from the formula derived for bulk materials.

At present, the determination of the toughness of material is still a difficult task. There are some attempts to assess the toughness of thin films using bending, indentation and scratch test measurements [47]. However, no systematic study devoted to the determination of (1) the toughness of thin films and (2) main factors influencing the toughness of thin films has been carried out so far. Moreover, it is not clear if a relation between the cracking of a thin film and its toughness really exists. Therefore, the determination of key factors influencing the film cracking is the main aim of next few paragraphs. The assessment of the toughness of thin films is based on a detailed analysis of correlations between the formation of cracks, mechanical properties of both the film and the substrate, structure of film and macrostress, σ, generated in the film during its growth. In this investigation, selected ceramic films prepared by the reactive magnetron sputtering were used.

6.3. *Films and Methods Used for Characterization of Thin Film Toughness*

The Zr–Cu–O, Zr–Cu–C, Ti–Cu–C and Si–Me–N (Me = Ta, Zr, Mo, W) nanostructured films were reactively sputtered using a DC unbalanced magnetron equipped with a round target of diameter 100 mm under different deposition conditions on steel and Si(100) substrates. Typical thickness, h, of films ranged from 2 to 5 μm. The structure of films was characterized by X-ray diffraction (XRD) and their mechanical properties, i.e. microhardness H_f, effective Young's modulus $E_f^* = E_f/(1 - \nu_f^2)$ and elastic recovery W_e, were evaluated from the load versus displacement curves measured using a computer controlled microhardness tester Fischerscope H 100; here E_f is Young's modulus of the film. The mechanical properties of the films were measured at low values of the diamond indenter load $L \leq 50$ mN which ensured that the ratio $d/h < 0.1$ and so the measured values of H_f and E_f^* of the film are correct; here d is the depth of diamond impression. The brittleness of thin films was characterized by (1) *the formation of cracks* during the impression of diamond indenter into the film under high loads $L = 0.5$ and 1 N and (2) the ratio H_f^3/E_f^{*2}, which is proportional to the resistance of the film to plastic deformation [56].

6.4. *Formation of Cracks*

At present, it is well known that the fracture toughness K_c of bulk materials and thick films can be calculated from the length of radial cracks created during diamond impression from Eq. (6.1). Considerably less information is, however, available on factors influencing the formation of cracks and particularly on cracks formed in thin films deposited on different base material (substrate). Main factors that determine the formation of cracks are (1) the structure and mechanical properties of the film, (2) the residual stress σ generated in the film during its growth and (3) the mechanical properties of the substrate. The formation of cracks is the result of a combined action of all these factors. This is the reason why (1) the determination of the toughness of thin films is a very difficult task and (2) the investigation of the effect of individual factors influencing film cracking is needed. The effect of individual factors on the formation of cracks in thin ceramic films is further analyzed in detail.

6.4.1. *Effect of Substrate*

To demonstrate the effect of the substrate on the formation of cracks in the film, the same amorphous Zr–Cu–O film was sputtered onto three substrates: (1) steel, (2) glass and (3) Si(100). These substrates strongly differ in the values of their hardness H_s and effective Young's modulus E_s^* and this fact results in different deformation of the same film under the same load ($L = 1\,\mathrm{N}$) of the diamond indenter, see Fig. 6.3.

The cracks are circular for soft substrate (steel). On the contrary, radial cracks are formed on Si(100) substrate with H_s sligthly higher than H_f of the film. This means that a transition between circular and radial cracks should exist. Such a transition with no cracks was really found on the glass substrate with $H_s = 7.1\,\mathrm{GPa}$ (glass). This experiment clearly shows that the formation of cracks strongly depends on the mechanical behavior of substrate, see Table 6.1. Moreover, it is worthwhile to note that there is no formula which enables the calculation of the film toughness from circular cracks. More details are given in [80].

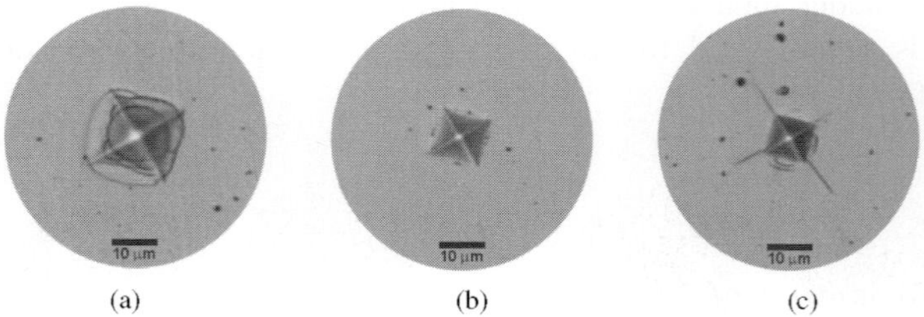

(a)   (b)   (c)

Fig. 6.3.   Micrographs of diamond indenter impressions at load $L = 1\,\mathrm{N}$ into 2 μm thick amorphous Zr–Cu–O film with 38 at.% Cu and tension macrostress $\sigma_{\mathrm{Si(100)}} = 0.3\,\mathrm{GPa}$ deposited on (a) 15330 steel, (b) glass and (c) Si(100) substrate.
Reprinted from [80], with permission of Elsevier.

Table 6.1.   Mechanical properties of amorphous Zr–Cu–O film with 38 at.% Cu and steel, glass and Si(100) used as substrates.

Material	H[GPa]	E^*[GPa]	H^3/E^{*2}[GPa]	W_e[%]	Structure of substrate	d/h [%]	Crack in film
Zr–Cu–O	10	120	0.069	0.50			
Steel	2.9	212	0.00054	0.10	polycrystalline	202	circular
Glass	7.1	69	0.075	0.57	amorphous	177	none
Si(100)	12.6	132	0.11	0.57	single crystal	158	radial

W_e is the elastic recovery, h is the film thickness and d is the depth of diamond indenter impression.

6.4.2. *Effect of Film Structure*

The diamond impression into the amorphous and crystalline film, deposited on the same substrate, strongly differs even in the case if they are produced at the same indenter load L, see Fig. 6.4

Two important facts were found: (1) lower load L is sufficient to produce cracks in polycrystalline film and (2) cracks are (i) circular in the amorphous film and (ii) radial in the polycrystalline film. This indicates that (a) amorphous films exhibit better fracture toughness compared to that of polycrystalline ones and (b) grain boundaries in the crystalline films facilitate the propagation of cracks. Here, it is necessary to note that there is a strong coupling between the film structure and macrostress, σ, generated in the film during its formation. This fact considerably complicates an exact determination of the resistance of film to cracking.

6.4.3. *Effect of Residual Stress in Film*

The macrostress, σ, generated in the film during its formation also influences the formation of cracks in indentation measurements. This fact was

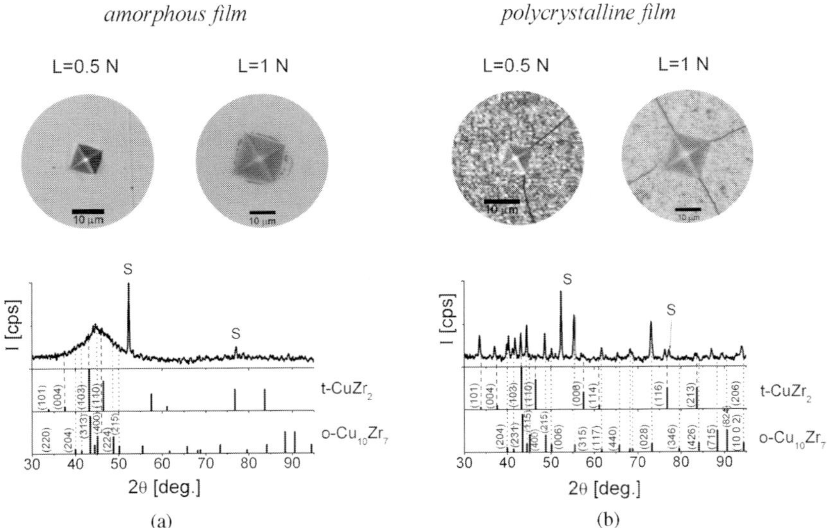

Fig. 6.4. Micrographs of diamond indenter impressions at load $L = 0.5$ and 1 N into (a) amorphous and (b) crystalline ZrCu films with 44 at.% Cu sputtered on Si(100) substrate in pure argon at $I_d = 2$ A, $U_d = 250$ V, $U_s = U_{fl}$, $p_{Ar} = 1$ Pa and $T_s = 300$ and 550°C, respectively [80]. Film structure is documented by XRD patterns. Reprinted from [80] with permission of Elsevier.

Film in: tension (σ>0)   compression (σ<0)
                    σ = 0.3 GPa   σ = -0.3 GPa

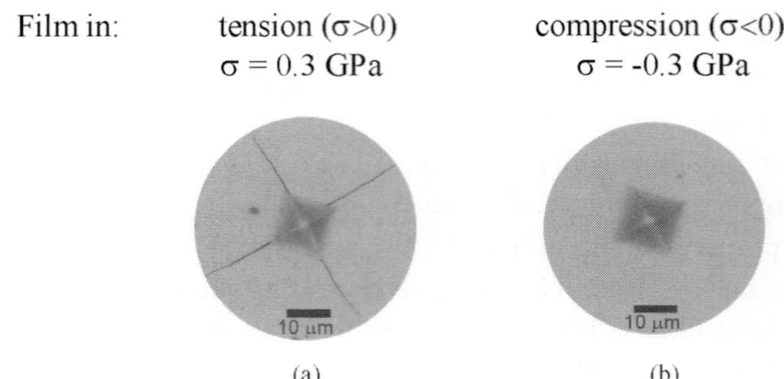

Fig. 6.5.   Effect of macrostress σ on film cracking. Micrographs of diamond indenter impressions into two Zr–Cu–O films with 38 at.% Cu with the same values of H_f, E_f, W_e and with (a) tensile and (b) compressive macrostress σ sputtered on Si(100) substrate at load $L = 1$ N are shown.

demonstrated by the experiments performed with Zr–Cu–O films sputtered on Si(100) substrate. It was found that a compressive ($\sigma < 0$) macrostress prevents the formation of cracks, see Fig. 6.5. Here, micrographs of the diamond indenter impressions into two Zr–Cu–O films with the same values of $H_f \approx 11\,\mathrm{GPa}$, $E_f = 110\,\mathrm{GPa}$, $W_e \approx 0.41$ and $H_f^3/E_f^{*2} \approx 0.1\,\mathrm{GPa}$ but with a different macrostress, σ, (tensile and compressive) at load $L = 1$ N are compared. From Fig. 6.5 it is clearly seen that while the Zr–Cu–O film in tension ($\sigma > 0$) cracks under load $L = 1\,\mathrm{N}$, the Zr–Cu–O film in compression ($\sigma < 0$) exhibits no cracks.

It is worthwhile to note that already a small value of compressive macrostress ($\sigma \leq -0.3\,\mathrm{GPa}$) is sufficient to prevent the formation of cracks in the film during its loading by the diamond indenter. This experiment clearly shows that the compressive σ helps to close the cracks. For more details, see [58, 80].

6.4.4. *Effect of Film Thickness*

The thickness, h, of film also influences its cracking under a given external load L. There is a direct proportionality between h, L and cracking. As expected, no cracks are formed when the ratio $d/h < 1$. On the contrary, the film cracks if the depth, d, of the diamond indenter impression under the same load L approaches the film thickness, h, or is even greater than h, i.e. in the case when $d/h \geq 1$, see Fig. 6.6. Properties of two films of different thickness, h, are compared, see Table 6.2. The cracks are formed

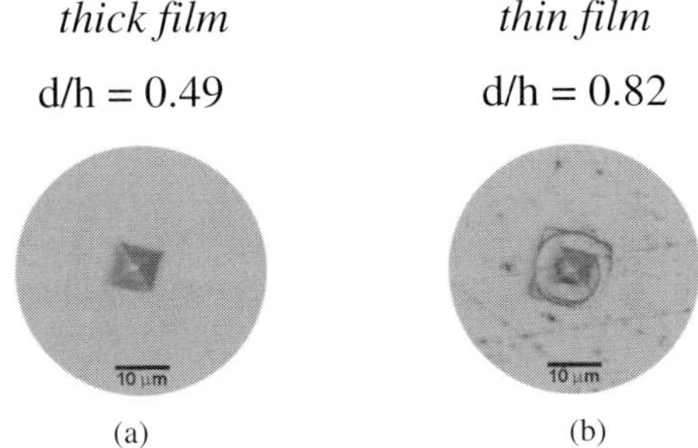

Fig. 6.6.   Effect of film thickness h on film cracking. Comparison of diamond indenter impression into amorphous W–Si–N film of the same composition and two thicknesses (a) $h = 5700$ nm and (b) $h = 4100$ nm on 15330 steel substrate produced at the same load $L = 1$ N.

Table 6.2.   Mechanical properties of two hard amorphous W–Si–N films of different thickness h deposited on 15330 steel substrates. d is the depth of diamond indenter impression into film after its loading at $L = 1$ N.

Film	H_f [GPa]	E_f^* [GPa]	H_f^3/E_f^{*2} [GPa]	h [μm]	d [μm]	d/h	E_f^*/E_s^*	Cracks
a	31.7	280	0.41	5.7	2.8	0.49	1.32	no
b	34.8	297	0.48	4.1	3.3	0.82	1.40	x

x denotes that the crack is formed.

in film b in spite of the fact that the ratios H_f^3/E_f^{*2} and E_f^*/E_s^* are greater than those of film a.

This experiment clearly shows that the ratio d/h is also important for film cracking. To avoid the cracking, the ratio d/h should be 0.5 or lower. Therefore, the thickness, h, of the protective coating must increase with increasing load L to ensure that $d/h \leq 0.5$.

6.5. *Assessment of Toughness of Thin Films*

Recently, Zhang *et al.* reported that the toughness of thin film can be assessed from the depth, d, of the impression of the diamond indenter created after the characterization of its mechanical properties using the microindentation [81], i.e. from the depth of impression created in the film after its loading by the diamond indenter at a small load L ensuring that the ratio $d/h \leq 0.1$. The plasticity of the film, measured as the ratio of the plastic displacement, d, over the total displacement in the nanoindentation test, is considered as the first approximation of the film toughness. According to this definition the toughness is the higher the greater is d. This seems to be valid for the material of thin film or the self-sustained film but it is not a sufficient condition to prevent the cracking of the thin film deposited on the substrate when the film is exposed to an external load. This fact was confirmed experimentally.

6.5.1. *Cracking of Hard Films with* $E_f^* \leq E_s^*$

Experiments indicate that a resistance of the thin film/substrate system against cracking increases with the ratio H_f^3/E_f^{*2}. At present, there is a question as to what is the maximum value of the ratio H_f^3/E_f^{*2} at which no cracks in the film are formed. It is necessary to note that a maximum resistance of the film against cracking is important for good protection of the substrate but it cannot be achieved by a high toughness of the film alone. Therefore, instead of the film toughness the ratio H_f^3/E_f^{*2} should be used to assess the protective efficiency of the thin film exposed to the external load.

As an example, we present the mechanical behavior of Zr–Cu–O films with high (≥ 30 at.%) amount of Cu, see Table 6.3. The Zr–Cu–O films were prepared by reactive magnetron sputtering from ZrCu(90/10) target in Ar + O$_2$ mixture at $T_s = 400°$C, $p_T = p_{Ar} + p_{O2} = 1\,$Pa and different values of p_{O2}.

From Table 6.3 it is seen that (1) W_p increases with increasing ratio d/h and decreases with increasing ratio H_f^3/E_f^{*2}, (2) the films with $W_p \geq 50\%$ very easily crack during indentation test and (3) no cracks are formed in

Table 6.3. Mechanical properties of (i) nanocrystalline Zr–Cu–O films with high (≥ 30 at.%) amount of Cu and (ii) substrates and formation of cracks during their loading at high diamond indenter load $L = 1$ N.

p_{O2}	H_f	E_f^*	H_f^3/E_f^{*2}	W_p	σ		Cracks	
[Pa]	[GPa]	[GPa]	[GPa]	[%]	[GPa]	d/h	Steel	Si(100)
0.15	10.7	109	0.10	42	−0.1	0.45	no	x
0.2	10.5	117	0.09	46	0.3	0.48	no	small
0.3	8.9	127	0.04	56	0.4	0.57	x	x
0.4	8.0	125	0.03	60	0.3	0.61	x	x
Si(100)	12.6	132	0.115					
Steel	2.9	212	0.0005					

$W_p = 1 - W_e$ is the plastic deformation and x denotes that the cracks are formed.

Increase in resistance to plastic deformation

$$\rightarrow$$

$$H_f^3/E_f^2 = \qquad 0.03 \qquad\qquad 0.04 \qquad\qquad 0.09 \qquad\qquad 0.10$$

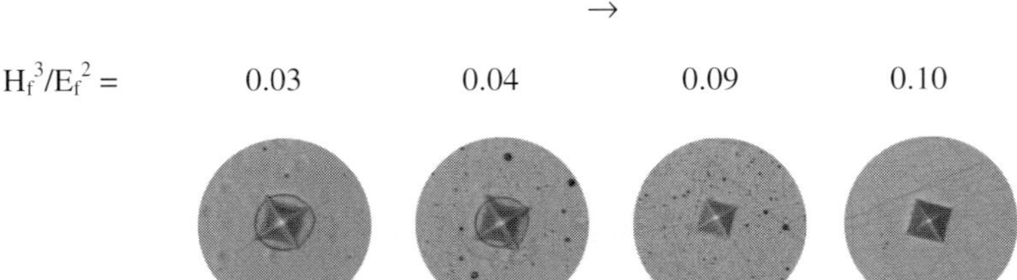

Fig. 6.7. Development of micrograph of diamond impressions into $\sim$5000 nm thick Zr–Cu–O films with large ($\geq$24 at.%) content of Cu and $E_f/E_s < 1$, sputtered on 15330 steel at $T_s = 400^\circ$C, with increasing ratio H_f^3/E_f^2. More details are given in [80].

films with $H_f^3/E_f^{*2} \geq 0.1$ and $E_f^* \leq E_s^*$. These results indicate that the films with $H_f^3/E_f^{*2} \geq 0.1$ and $E_f^* \leq E_s^*$ should exhibit a maximum toughness of the thin film/substrate system.

The resistance of the film to the formation of cracks increases with increasing ratio H_f^3/E_f^{*2}, see Fig. 6.7. Therefore, the resistance of the film to plastic deformation, i.e. the ratio H_f^3/E_f^{*2}, should be maximized to improve the film elastic recovery W_e and its toughness.

6.5.2. *Cracking of Hard Films with $E_f^* > E_s^*$*

It is well known that the film hardness H_f is determined by its elemental and phase composition, chemical bonding, structure (crystalline, amorphous) and microstructure (geometry of grains and building blocks). The effect of the elemental composition of film on its hardness, H_f, is illustrated in Fig. 4.1. This figure shows that while the films based on oxides are softer with hardness, H_f, values up to 15 GPa only, the films based on nitrides and carbides exhibit much higher H_f up to 35 GPa and also high values of the ratio H_f^3/E_f^{*2} up to 0.6 GPa. This means that the films based on nitrides exhibit considerably higher resistance to plastic deformation compared to the films based on oxides. Here, it is also necessary to note that the effective Young's modulus E_f^* increases almost linearly with increasing H_f, and, for $H_f > 20$ GPa the value of E_f^* is, for the majority of materials, greater than E_s^*. Therefore, there is an open question of whether hard films with $H_f \geq 20$ GPa can be resistant to cracking.

Experiments show that even very hard thin films with $H_f \approx 30$ GPa can exhibit no cracks in the case when (1) the film is (i) amorphous and (ii) in compression ($\sigma < 0$) and (2) the substrate is hard ($H_s \approx 0.5H_f$) and

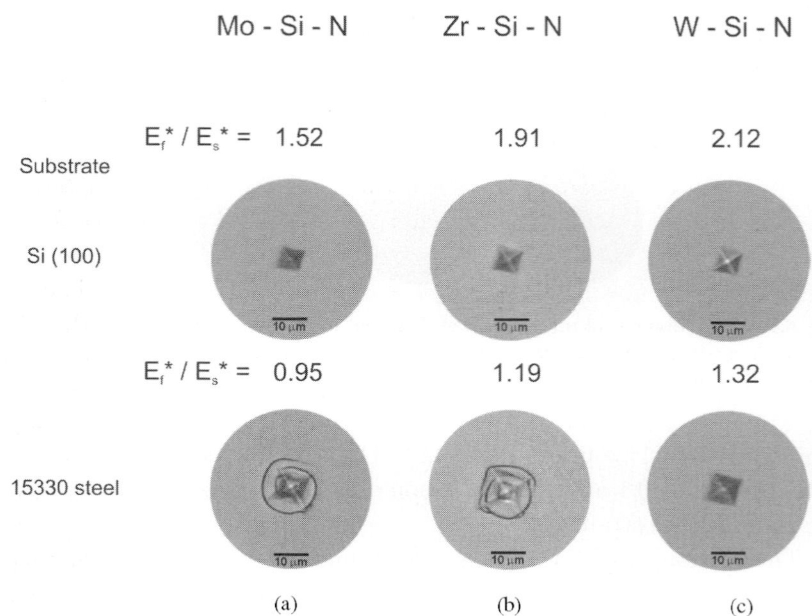

Fig. 6.8. Comparison of diamond indenter impression into amorphous (a) Mo–Si–N, (b) Zr–Si–N and (c) W–Si–N films with high hardness $H_f \geq 25$ GPa deposited on 15330 steel and Si(100) substrates.

its Young's modulus $E_f^* > E_s^*$. This fact is illustrated in Fig. 6.8 for thick, amorphous Zr–Si–N, Mo–Si–N and W–Si–N films with $H_f \geq 25$ GPa which exhibit a compressive macrostress, σ, see Table 6.4.

From Fig. 6.8 it is seen that no cracks are formed in the films deposited on the hard Si(100) substrate because $E_f^*/E_s^* \geq 1.5$. On the contrary, cracks are formed in the films deposited on the soft steel substrate if $E_f^*/E_s^* \leq 1.3$.

Table 6.4. Physical and mechanical properties of selected amorphous Me–Si–N films with high (>50 vol.%) of Si_3N_4 phase [64–66]. These films were used in experiment whose results are given in Fig. 6.8.

Film	h [μm]	d [μm]	H_f [GPa]	E_f^* [GPa]	H_f^3/E_f^{*2} [GPa]	σ [GPa]
Mo–Si–N	3.1	2.5	25.4	201	0.41	−2.2
Zr–Si–N	5.2	3.3	30.3	252	0.44	−1.1
W–Si–N	5.3	2.4	31.7	280	0.41	−1.6
Si(100)			12.6	132	0.115	
15330 steel			2.9	212	0.0005	

This means that the ratio $E_f^*/E_s^* \geq 1.3$ is the necessary condition to avoid film cracking.

6.6. *Summary of Main Issues*

The toughness of a material is defined as a resistance of the material to cracking under loading by an external load L. There is a simple formula which allows the calculation of the toughness of bulk materials from the length of radial cracks produced at a given load L. This method cannot, however, be used for the determination of the toughness of thin films because (1) the formula was derived under the assumption that the material thickness $h \geq 0.25\,\mathrm{mm}$ and cracks are radial, (2) cracking of thin film material is strongly influenced by the substrate and (3) geometry of cracks can be different (radial or circular) and strongly depends on the mechanical properties (H_s and E_s^*) of the substrate. Therefore, the thin film/substrate system must be considered as one unit if one wants to find conditions under which cracking of the protective film can be avoided. The determination of the toughness of the thin film alone is not sufficient to achieve this goal.

Experiments described in this chapter show that the toughness of thin films should be assessed from the resistance of the thin film to cracking. The toughness increases with increase in the resistance of the film to cracking. This resistance depends on (1) the film structure (crystalline, amorphous), (2) mechanical properties of both the film (H_f, E_f^*) and the substrate (H_s, E_s^*) and (3) the macrostress, σ, (tensile, compressive) in the film. It was found that:

1. Crystalline films are more brittle than amorphous films.
2. No cracks are formed in films which exhibit a compressive macrostress, σ; even a small ($\approx -0.1\,\mathrm{GPa}$) compressive σ is sufficient to prevent the formation of cracks.
3. Cracks are formed in (a) crystalline films when the diamond indenter load L is sufficiently high and (b) thin films with the ratio $d/h \geq 0.5$.
4. The geometrical form of cracks depends on the substrate hardness H_s. The cracks are radial for hard substrates ($H_s \geq 0.5 H_f$). On the contrary, the cracks are circular for soft substrates ($H_s < 0.5 H_f$), e.g. 15330 steel with $H_s = 2.9\,\mathrm{GPa}$.
5. The resistance of films to cracking increases with increasing ratio H_f^3/E_f^{*2} similarly as H_f increases with increasing ratio H_f^3/E_f^{*2}. This

also means that hard amorphous films in compression can be resistant to cracking when they are exposed to high loads ($L \geq 1$ N). The maximum hardness $H_{f\,max}$, however, depends on the elemental and phase composition of the film. This is the reason, why $H_{f\,max}$ of individual composite films strongly differ, see Fig. 4.1. It was found that the a-Si_3N_4/MeN_x composites with (1) high content of Si_3N_4 phase, (2) relatively high (15 to 35 GPa) hardness H_f and (3) no cracking can be prepared. This fact indicates that the plasticity (W_p) of the film must be accompanied by a certain elasticity (W_e) which prevents the formation of cracks. Since $W_p + W_e = 1$ and W_p decreases with increasing H_f, the amorphous hard films resistant to cracking must also exhibit W_p decreasing with increasing H_f.

6. There are general interrelationships between basic mechanical properties of material, i.e. H_f, E_f^*, H_f^3/E_f^{*2} and W_e, which can be described by simple empirical formulas enabling the prediction of the mechanical behavior of thin films under loading.

In summary, we can conclude that the resistance of thin protective films against cracking is not determined by the toughness of thin film alone but by the combined action of both the thin film and the substrate. The resistance of the film to cracking increases with increasing ratio H_f^3/E_f^{*2}. This means that the resistance of film to cracking can be easily assessed using microindentation techniques.

7. Future Trends

Further research activity in the field of nanocomposite films will be concentrated mainly on the following problems: (1) the development of films with controlled size of grains in the range from 1 to 10 nm with the aim (a) to investigate size-dependent phenomena in nanocomposite films and (b) to develop new advanced coatings with unique physical and functional properties, (2) nanocrystallization of amorphous materials, (3) electronic charge transfer between nanograins with different chemical composition and different Fermi energies again with the aim to produce films with new functional properties, (4) development of protective coatings with oxidation resistance exceeding 2000°C, (5) formation of crystalline films on unheated heat sensitive materials such as polymer foils and polycarbonate and (6) development of new PVD systems for the production of nanocomposite coatings under new physical conditions. Also, it can be expected that in the very near future the thin nanocomposite films will be used as experimental models for the design of nanocomposite bulk materials with prescribed properties.

Acknowledgments

The author would like to thank Prof. RNDr. Jaroslav Vlček, CSc., head of the Department of Physics at the University of West Bohemia in Plzeň, Czech Republic, for many stimulating discussions and to all his Ph.D. students for their hard and enthusiastic work on this project. He would also like to thank Dipl. -Ing. Jan Šůna, Ph.D., for careful preparation of all figures used in this chapter. This work was supported by the Grant Agency of the Czech Republic under Project No. 106/96/K245 (1996–2000) and by the Ministry of Education of the Czech Republic under Projects No. MSM 235200002 (1999–2004) and MSM 4977751302 (2005–2010).

References

1. H. Gleiter, Nanocrystalline materials, *Prog. Mater. Sci.* **33** (1989) 223–315.
2. R. Birringer, Nanocrystalline materials, *Mater. Sci. Eng.* **A117** (1989) 33–43.
3. R.W. Siegel, Cluster-assembled nanophase materials, *Annu. Rev. Mater. Sci.* **21** (1991) 559–579.
4. S.A. Barnett, Deposition and mechanical properties of superlattice thin films, in *Physics of Thin Films*, eds. M.H. Fracombe and J.A. Vossen (Academic Press, N.Y., 1993), pp. 1–73.
5. R.W. Siegel, What do we really know about the atomic-scale structures of nanophase materials?, *J. Phys. Chem. Solids* **55**(10) (1994) 1097–1106.
6. R.W. Siegel and G.E. Fougere, Grain size dependent mechanical properties in nanophase materials, in *Mat. Res. Soc. Symp.* Proc. **362**, eds. H.J. Grant, R.W. Armstrong, M.A. Otooni and K. Ishizaki (Materials Research Society, Warrendale, PA, 1995), pp. 219–229.
7. S. Vepřek and S. Reiprich, A concept for the design of novel superhard coatings, *Thin Solid Films* **265** (1995) 64–71.
8. H. Gleiter, Nanostructured materials: State of the art and perspectives, *Nanostruct. Mater.* **6** (1996) 3–14.
9. K. Lu, Nanocrystalline metals crystallized from amorphous solids: Nanocrystallization, structure, and properties, *Mater. Sci. Eng.* **R16** (1996) 161–221.
10. B.X. Liu and O. Jin, Formation and theoretical modelling of non-equilibrium alloy phases by ion mixing, *Phys. Stat. Sol.(a)* **161** (1997) 3–33.
11. F. Vaz, L. Rebouta, M.F. da Silva and J.C. Soares, Thermal oxidation of ternary and quaternary nitrides of titanium, aluminium and silicon, in

Protective Coatings and Thin Films, eds. Y. Pauleau and P.B. Barna (Kluwer Academic Publisher, 1997), pp. 501–510.

12. S. Yip, The strongest size, *Nature* **391** (1998) 532.

13. S. Vepřek, P. Nesládek, A. Niederhofer, F. Glatz, M. Jílek and M. Šíma, Recent progress in the superhard nanocrystalline composites: Towards their industrialization and understanding of the origin of the superhardness, *Surf. Coat. Technol.* **108–109** (1998) 138–147.

14. A.A. Voevodin and J.S. Zabinski, Superhard, functionally gradient, nanolayered and nanocomposite diamond-like carbon coatings for wear protection, *Diam. Relat. Mater.* **7** (1998) 463–467.

15. J. Musil and J. Vlček, Magnetron sputtering of films with controlled texture and grain size, *Mater. Chem. Phys.* **54** (1998) 116–122.

16. H.S. Kim, A composite model for mechanical properties of nanocrystalline materials, *Scripta Mater.* **39**(8) (1998) 1057–1061.

17. S. Veřek, The search for novel, superhard materials, *J. Vac. Sci. Technol.* **A17** (1999) 2401–2420.

18. A. Niederhofer, K. Moto, P. Nesládek and S. Veřek, Diamond is not the hardest material anymore: Ultrahard nanocomposite nc-TiN/a- & nc-TiSi$_2$ prepared by plasma CVD, in Proc. 14th Int. Symp. Plasma Chem., Vol. 3, Prague, Czech Republic, August 2–6, 1999, eds. M. Hrabovský, M. Konrád, and V. Kopecký (Institute of Plasma Physics AS CR, Prague, Czech Republic, 1999), pp. 1521–1525.

19. J. Musil, Hard and superhard nanocomposite coatings, *Surf. Coat. Technol.* **125** (2000) 322–330.

20. S. Veprek, A. Niederhofer, K. Moto, T. Bolom, H.-D. Mannling, P. Nesladek, G. Dollinger and A. Bergmaier, Composition, nanostructure and origin of the ultrahardness in nc-TiN/a-Si$_3$N$_4$/a- and nc-TiSi$_2$ nanocomposites with $H_V = 80$ to ≥ 105 GPa, *Surf. Coat. Technol.* **133–134** (2000) 152–159.

21. A.A. Voevodin and J.S. Zabinski, Supertough wear-resistant coatings with "chameleon" surface adaptation, *Thin Solid Films* **370** (2000) 223–231.

22. H. Conrad and J. Narayan, On the grain size softening in nanocrystalline materials, *Scripta Mater.* **42** (2000) 1023–1030.

23. H. Gleiter, Nanostructured materials: Basic concepts and microstructure, *Acta Mater.* **48** (2000) 1–29.

24. R. Hauert and J. Patscheider, From alloying to nanocomposites — Improved performance of hard coatings, *Adv. Mater.* **2**(5) (2000) 247–259.

25. L. Hultman, Thermal stability of nitride thin films, *Vacuum* **57**(1) (2000) 1–30.

26. J. Musil and J. Vlček, Magnetron sputtering of hard nanocomposite coatings and their properties, *Surf. Coat. Technol.* **142–144** (2001) 557–566.

27. J. Patscheider, T. Zehnder and M. Diserens, Structure-performance relations in nanocomposite coatings, *Surf. Coat. Technol.* **146–147** (2001) 201–208.

28. H. Gleiter, Tuning the electronic structure of solids by means of nanometer-sized microstructures, *Scripta Mater.* **44** (2001) 1161–1168.

29. S. Veřek and A.S. Argon, Mechanical properties of superhard nanocomposites, *Surf. Coat. Technol.* **146–147** (2001) 175–182.

30. A. Badzian, Diamond challenged by hard materials: A reflection on developments in the last decade, *Mater. Chem. Phys.* **72** (2001) 110–113.

31. J. Sung, The design of exotic superhard materials, *Mater. Chem. Phys.* **72** (2001) 141–146.

32. V.V. Brazhkin, A.G. Lyapin and R.J. Hemley, Harder than diamond: Dreams and reality, *Philosophical Magazine* **A82**(2) (2002) 231–253.

33. S. Vepřek and A.S. Argon, Towards the understanding of mechanical properties of super- and ultrahard nanocomposites, *J. Vac. Sci. Technol.* **B20**(2) (2002) 650–664.

34. J. Musil, J. Vlček, F. Regent, F. Kunc and H. Zeman, Hard nanocomposite coatings prepared by magnetron sputtering, *Key Engi. Mater.* **230–232** (2002) 613–622.

35. H. Gleiter and M. Fichtner, Is enhanced solubility in nanocomposites an electronic effect?, *Scripta Mater.* **46** (2002) 497–500.

36. S. Zhang, D. Sun and Y. Fu, Superhard nanocomposite coatings, *J. Mater. Sci. Technol.* **18**(6) (2002) 485–491.

37. R.A. Andrievskii, Thermal stability of nanomaterials, *Russ. Chem. Rev.* **71**(10) (2002) 853–866.

38. J. Patscheider, Nanocomposite hard coatings for wear protection, *M.R.S. Bulletin* **28**(3) (2003) 180–183.

39. S. Zhang, D. Sun, Y. Fu and H. Du, Recent advances of superhard nanocomposite coatings: A review, *Surf. Coat. Technol.* **167** (2003) 113–119.

40. S. Veřek, S. Mukherjee, P. Karvanková, H.-D. Mannling, J.L. He, K. Moto, J. Procházka and A.S. Argon, Limits to the strength of super- and ultrahard nanocomposite coatings, *J. Vac. Sci. Technol.* **A21**(3) (2003) 532–544.

41. G.M. Demyashev, A.L. Taube and E. Siores, Superhard nanocomposite coatings, in *Handbook of Organic-Inorganic Hybrid Materials and Nanocomposite*, Vol. 1, ed., H. S. Nalwa (American Scientifuc Publishers, 2003), pp. 1–61.

42. G.M. Demyashev, A.L. Taube and E.E. Siores, Superhard nanocomposites, in *Encyclopedia of Nanoscience and Nanotechnology*, Vol. 10, ed. H. S. Nalwa, (American Scientifuc Publishers, 2003), pp. 1–46.

43. J. Musil, Hard nanocomposite films prepared by magnetron sputtering, Invited lecture at the NATO-Russia Advanced Research Workshop on "Nanostructured Thin Films and Nanodispersion Strengthened Coatings", December 8–10, 2003, Moscow, Russia, in NATO Science Series Volume. *Nanostructured Thin Films and Nanodispersion Strengthened Coatings*, eds. A.A. Voevodin, E. Levashov, D. Shtansky and J. Moore (Kluwer Academic B.V. Publishers Dordrecht, The Netherlands, 2004), pp. 43–56.

44. J.D. Kuntz, G.D. Zhang and A.K. Murherjee, Nanocrystalline-matrix ceramic composites for improved fracture toughness, *M.R.S. Bulletin* January (2004), pp. 22–27.

45. J. Musil and S. Miyake, Nanocomposite coatings with enhanced hardness, in *Novel Materials Processing (MAPEES'04)*, ed. S. Miyake (Elsevier Ltd., Amsterdam, 2005), pp. 345–356.

46. R.A. Andrievski, Nanomaterials based on high-melting carbides, nitrides and borides, *Russ. Chem. Rev.* **74**(12) (2005) 1061–1072.

47. S. Zhang, D. Sun, Y. Fu and H. Du, Toughening of hard nanostructured thin films: A critical review, *Surf. Coat. Technol.* **198** (2005) 2–8.

48. J. Musil, Nanocomposite coatings with enhanced hardness, *Acta Metallurgica Sinica (English Letters)* **18**(3) (2005) 433–442.

49. J. Musil, Physical and mechanical properties of hard nanocomposite films prepared by reactive magnetron sputtering, in *Nanostructured Hard Coatings,* eds. J.T.M. DeHosson and A. Cavaleiro (Kluwer Academic/Plenum Publishers, N.Y., 2006), pp. 407–463.

50. J. Šůna, J. Musil, V. Ondok and J.G. Han, Enhanced hardness in sputtered Zr-Ni-N films, *Surf. Coat. Technol.* **200** (2006) 6293–6297.

51. J. Musil, H. Poláková, J. Šůna and J. Vlček, Effect of ion bombardment on properties of hard reactively sputtered Ti(Fe)N$_x$ films, *Surf. Coat. Technol.* **177–178** (2004) 289–298.

52. M. Nose, W.A. Chiou, M. Zhou, T. Mae and M. Meshii, Microstructure and mechanical properties of Zr-Si-N films prepared by RF reactive sputtering, *J. Vac. Sci. Technol.* **A20**(3) (2002) 823–828.

53. M. Nose, Y. Deguchi, T. Mae, E. Honbo, T. Nagae and K. Nogi, Influence of sputtering conditions on the structure and properties of Ti-Si-N thin films prepared by RF reactive sputtering, *Surf. Coat. Technol.* **174–175** (2003) 261–265.

54. X.D. Zhang, W.J. Meng, W. Wang, L.E. Rehn, P.M. Baldo and R.D. Evans, Temperature dependence of structure and mechanical properties of Ti-Si-N coatings, *Surf. Sci. Technol.* **177–178** (2004) 325–333.

55. Z.G. Li, M. Mori, S. Miyake, M. Kumagai, H. Saito and Y. Muramatsu, Structure and properties of Ti-Si-N films prepared by ICP assisted magnetron sputtering, *Surf. Coat. Technol.* **193** (2005) 345–349.

56. T.Y. Tsui, G.M. Pharr, W.C. Oliver, C.S. Bhatia, R.L. White, S. Anders, A. Anders and I.G. Brown, Nanoindentation and nanoscratching of hard carbon coatings for magnetic disks, *Mater. Res. Soc. Symp. Proc.* **383** (1995) 447–451.

57. J. Musil, F. Kunc, H. Zeman and H. Poláková, Relationships between hardness, Young's modulus and elastic recovery in hard nanocomposite coatings, *Surf. Coat. Technol.* **154** (2002) 304–313.

58. J. Musil and M. Jirout, Toughness of hard nanostructured ceramic thin films, Invited paper at 5th Asian-European Int. Conf. Plasma Surf. Eng., Sept. 12–16, 2005, Qingdao, China, *Surf. Coat. Technol.* **201** (2007) 5148–5152.

59. W.D. Munz, Titanium aluminium nitride films: A new alternative to TiN coatings, *J. Vac. Sci. Technol.* **A4** (1986) 2717–2725.

60. L.A. Donohue, I.J. Smith, W.-D. Munz, I. Petrov and J.E. Greene, Microstructure and oxidation resistance of Ti$_{1-x-y-z}$Al$_x$Cr$_y$Y$_z$N layers grown by combined steered-arc/unbalanced magnetron sputter deposition, *Surf. Coat. Technol.* **94–95** (1997) 226–231.

61. I. Wadsworth, I.J. Smith, L.A. Donohue and W.-D. Munz, Thermal stability and oxidation resistance of TiAlN/CrN multilayer coatings, *Surf. Coat. Technol.* **94–95** (1997) 315–321.

62. S. Vepřek, M. Hausmann and S. Reiprich, Structure and properties of novel superhard nanocrystalline/amorphous composite materials, *Mater. Res. Soc. Symp. Proc.* **400** (1996) 261.

63. H. Zeman, J. Musil and P. Zeman, Physical and mechanical properties of sputtered Ta-Si-N films with a high ($\geq$40 at.%) content of Si, *J. Vac. Sci. Technol.* **A22**(3) (2004) 646–649.

64. J. Musil, R. Daniel, P. Zeman and O. Takai, Structure and properties of magnetron sputtered Zr-Si-N films with a high ($\geq$25 at.%) Si content, *Thin Solid Films* **478** (2005) 238–247.

65. J. Musil, P. Dohnal and P. Zeman, Physical properties and high-temperature oxidation resistance of sputtered Si_3N_4/MoN_x nanocomposite coatings, *J. Vac. Sci. Technol.* **B23**(4) (2005) 1568–1575.

66. J. Musil, R. Daniel, J. Soldán and P. Zeman, Properties of reactively sputtered W-Si-N films, *Surf. Coat. Technol.* **200** (2006) 3886–3895.

67. P. Zeman and J. Musil, Difference in high-temperature oxidation resistance of amorphous Zr-Si-N and W-Si-N films with a high Si content, *Appl. Sur. Sci.* **252** (2006) 8319–8325.

68. P. Zeman, J. Musil and R. Daniel, High-temperature oxidation resistance of Ta-Si-N films with a high Si content, *Surf. Coat. Technol.* **200** (2006) 4091–4096.

69. R. Daniel, J. Musil, P. Zeman and C. Mitterer, Thermal stability of magnetron sputtered Zr-Si-N films: *Surf. Coat. Technol.* **201** (2006) 3368–3376.

70. J. Musil and P. Zeman, Hard amorphous a-Si_3N_4/MeN_x nanocomposite coatings with high thermal stability and high oxidation resistance, Invited paper at the Int. Workshop on Designing of Interfacial Structures in Advanced Materials and their Joints (DIS'06) May 18–20, 2006, Osaka, Japan and *Solid State Phenomena* **127** (2007) 31–36.

71. *Smithells Metals Reference Book*, ed. E.A. Brandes, in association with Fulmer Research Institute, 6th edition (Butterworh, Heinemann, 1992), pp. 8–24.

72. *Phase Diagrams of Ternary Boron Nitride and Silicon Nitride Systems*, eds. P. Rogl and J.C. Schuster (ASM International, Ohio, 1992).

73. C. Louro, A. Cavaleiro and F. Montemor, What is the chemical bonding of W-Si-N sputtered coatings?, *Surf. Coat. Technol.* **142–144** (2001) 964–970.

74. Vlček J., S. Potocký, J. Čížek, J. Houška, M. Kormunda, P. Zeman, V. Peřina, J. Zemek, Y. Setsuhara and S. Konuma, Reactive magnetron sputtering of hard Si-B-C-N films with a high-temperature oxidation resistance, *J. Vac. Sci. Technol.* **A23**(6) (2005) 1513–1522.

75. J. Musil, P. Zeman and P. Dohnal, Ti-Si-N films with a high content of Si, *Plasma Processes and Polymers* **4** (S1) (2007) S574–S578.

76. W.D. Callister Jr., *Materials Science and Engineering, an Introduction*, 6th edn. (Wiley, New York, 2003).

77. J. Musil and J. Vlček, Magnetron sputtering of alloy-based films and its specifity, *Czech. J. Phys.* **48** (1998) 1209–1224.

78. W.F. Brown Jr. and J.E. Srawley, *ASTM STP* **410** (1996) 12.

79. A.D.S. Jayatilaka, *Fracture of Engineering Brittle Materials* (Applied Science Publishers, London, 1979).
80. M. Jirout and J. Musil, Effect of addition of Cu into ZrO_x film on its properties, *Surf. Coat. Technol.* **200** (2006) 6792–6800.
81. S. Zhang, X.L. Bui, Y. Fu, D.L. Butler and H. Du, Bias-graded deposition of diamond-like carbon for tribological applications, *Diam. Relat. Mater.* **13** (2004) 867–871.

CHAPTER 6

NANOSTRUCTURED, MULTIFUNCTIONAL TRIBOLOGICAL COATINGS

John J. Moore[*,†], In-Wook Park[†], Jianliang Lin[†],
Brajendra Mishra[†] and Kwang Ho Kim[‡]

[†]*Advanced Coatings and Surface Engineering Laboratory (ACSEL)*
Colorado School of Mines, Golden, CO 80401, USA
[‡]*School of Materials Science and Engineering*
Pusan National University, Keumjung-Ku
Busan 609-735, South Korea
[*]*jjmoore@Mines.EDU*

1. Introduction

Nanostructured coatings have recently attracted increasing interest because of the possibilities of synthesizing materials with unique physical–chemical properties [1,2]. A number of sophisticated surface-related properties, such as optical, magnetic, electronic, catalytic, mechanical, chemical and tribological property can be obtained by advanced nanostructured coatings [3, 4]. There are many types of design models for nanostructured coatings, such as three-dimensional nanocomposite coatings [2,5], nanoscale multilayer coatings [6, 7], functionally graded coatings [1, 4], etc. The optimized design of nanostructured coatings needs to consider many factors, e.g. ion energy and ion flux of depositing species, interface volume, crystallite size, single layer thickness, surface and interfacial energy, texture, epitaxial stress and strain, etc., all of which depend significantly on materials selection, deposition methods and process parameters [2, 8].

In particular, pulsed reactive magnetron deposition techniques have been investigated, more recently, since it is possible to conduct reactive sputtering without arcing during deposition. Pulsed reactive sputtering can also change and control the plasma constituents, increase the ion energy and ion flux, and microstructural growth of the thin film through ion bombardment [8]. The applications of pulsing in reactive magnetron sputtering

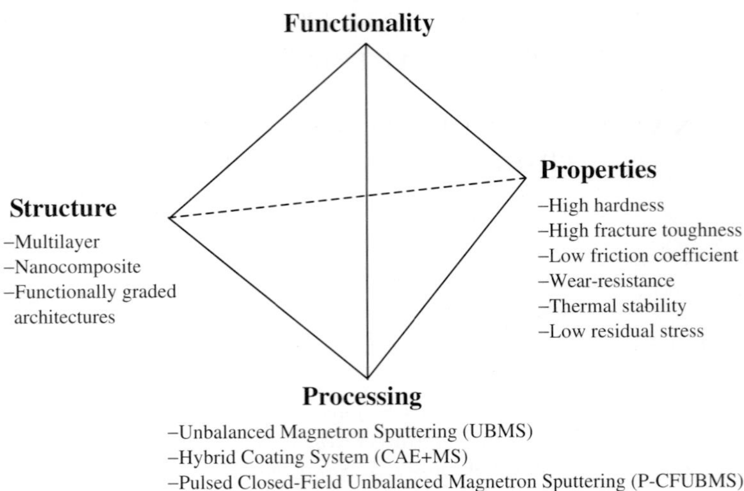

Fig. 1.1.   Tetrahedron representing the relationship among processing, structure, properties, and functionality for nanostructured, multifunctional tribological coatings.

opens up considerable opportunities for the control of ion energy and ion flux to optimize the deposition process and tailor the as-deposited coating structure and properties.

The focus of this chapter is to introduce the relationships between processing, structure, properties, and functionality of nanostructured coatings using various magnetron sputtering deposition processes, such as unbalanced magnetron sputtering (UBMS), hybrid coating system of cathodic arc evaporation (CAE) and magnetron sputtering (MS), pulsed closed-field unbalanced magnetron sputtering (P-CFUBMS), and high-power pulsed magnetron sputtering (HPPMS), as shown in Fig. 1.1.

2. Classification of Nanostructured, Multifunctional Tribological Coatings

2.1. *Nanoscale Multilayer Coatings*

Research on using nanoscale multilayers (i.e. "Superlattices") to increase the hardness and toughness of coatings has provided significant advancements in understanding the advantages of employing this type of coating architecture. Early research by Palatnik with multilayers of metals showed that significant improvements in strength were achieved when layer thickness was decreased below 500 nm [6,9]. In early modeling, Koehler [7] predicted that high shear strength coatings could be produced by alternating layers of high and low elastic modulus. Key elements of the concept are that very thin layers ($\leq$10 nm) inhibit dislocation formation, while differences in elastic modulus between layers inhibit dislocation mobility. Lehoczky [47] demonstrated these concepts on metallic Al/Cu and Al/Ag multilayers and showed that a Hall–Petch type equation could be used to relate hardness to $1/(\text{periodicity})^{1/2}$ in where periodicity is a minimum periodic length between layers in the multilayer coating. Springer and Catlett [10], and Movchan *et al.* [11] reported on mechanical enhancements in metal/ceramic (e.g. Ti/TiN, Hf/HfN, W/WN, etc.) [12] and ceramic/ceramic (e.g. TiN/VN [13], TiN/NbN [14,15], TiN/V$_x$Nb$_{1-x}$N [16, 17], etc.) laminate structures that followed a Hall–Petch relationship. These pioneering works were followed by intensive research in multilayers [18, 19], which has produced coatings significantly harder than the individual components making up the layers. To achieve increased hardness, the layers must have sharp interfaces and periodicity in the 5–10 nm range. The multilayer architectures, as shown in Fig. 2.1, exhibiting high hardness are frequently called superlattices [20]. The different design architectures have been classified and some reports have formalized the multilayer design [4, 21]. Multilayer architectures clearly increase coating hardness and have commercial applications, especially in the tool

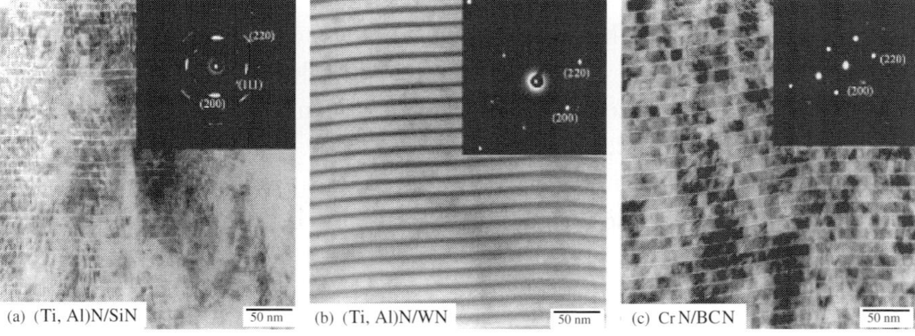

Fig. 2.1. Cross-sectional TEM images and selected area diffraction patterns (SADP) of nanoscale multilayer coatings: (a) (Ti, Al)N/SiN, (b) (Ti, Al)N/WN, and (c) CrN/BCN [20].

industry. However, they can be difficult to apply with uniform thickness on three-dimensional components and rough surfaces. If the layers are not of the correct periodicity, the superlattice effect is lost. Another relatively new technology, nanocomposites, offers the same advantages as multilayers (plus other benefits) and their properties are not critically dependent on thickness or substrate geometry.

2.2. *Nanocomposite Coatings*

Nanostructured composite (i.e. "Nanocomposite") coatings are usually formed from ternary or higher order systems and comprise at least two immiscible phases: two nanocrystalline phases or, more commonly, an amorphous phase surrounding nanocrystallites of a secondary phase. The most interesting and extensively investigated nanocomposite coatings are ternary, quaternary or even more complex systems, with nanocrystalline (nc-) grains of hard transition metal-nitrides (e.g. TiN, CrN, AlN, BN, ZrN, etc.), carbides (e.g. TiC, VC, WC, ZrC, etc.), borides (e.g. TiB_2, CrB_2, VB_2, WB, ZrB_2, etc.), oxides (e.g. Al_2O_3, TiO_2, SiO_2, MgO, TiO_2, Y_2O_3, ZrO_2, etc.), or silicides (e.g. $TiSi_2$, $CrSi_2$, $ZrSi_2$, etc.) surrounded by amorphous (*a*-) matrices (e.g. Si_3N_4, BN, C, etc.). The physical, mechanical, and thermal properties of these hard materials are summarized in Table 2.1 [22]. The synthesis of such nanocomposite (nc-/*a*-) coatings critically depends on the ability to co-deposit both the nanocrystalline and amorphous phases, such as Ti–Si–N (nc-TiN/nc-and *a*-$TiSi_2$/*a*-Si_3N_4) [2], Ti–Al–Si–N (nc-TiAlN/*a*-Si_3N_4) [5], W–Si–N (nc-W_2N/*a*-Si_3N_4) [23], Cr–Si–N (nc-CrN/*a*-Si_3N_4) [24], Ti–B–C–N (nc-TiB_2 and TiC/*a*-BN) [25], TiC/DLC (nc-TiC/*a*-C) [26], WC/DLC (nc-WC/*a*-C) [27], etc. as schematically presented in Fig. 2.2(a). A variety of hard compounds can be used as the nanocrystalline phases, including nitrides, carbides, borides, oxides, and silicides. Veprek *et al.* [28] suggested that the nanocrystalline grains must be 3~10 nm in size and separated by 1~2 nm within an amorphous phase as shown in Fig. 2.2(a). For example, Ti–B–N nanocomposite, which consists of nanocrystalline TiN (~5 nm in size) in an amorphous BN matrix, has been synthesized and observed by Lu *et al.* [29], as shown in Fig. 2.2(b).

2.3. *Functionally Graded Coatings*

In order to counteract brittle failure and improve fracture toughness, two concepts have been explored. The first involves the use of graded interfaces

Table 2.1.   The physical, mechanical, and thermal properties of hard materials.

Phase	Crystal structure	Lattice parameters (nm)	Density (g.cm^{-3})	Melting point ($^{\circ}$C)	Linear thermal expansion, α (10^{-6} K^{-1})	Thermal conductivity, λ (W m^{-1}K^{-1})	Electrical resistivity ($10^{-6}\Omega$cm)	Enthalpy at 298K (kJ mol^{-1})	Young's modulus (10^5 N mm^{-2})	Micro hardness (10 N mm^{-2})	Oxidation resistance ($\times 100^{\circ}$C)
Nitrides											
AlN	hex	0.311/0.498	3.05	2200	6	10	10^{11}	288.9	3.15	1200	13
BN	hex	0.251/0.669	2.25	3000	3.8	284.7	3×10^{14}	252.5	0.9	4400 HV	10
CrN	fcc	0.415	6.1	1050	2.3	11.72	640	118–124	4	1800–2100	7–7.5
	cub-B1	0.4149	5.39–7.75	1450	2.3	11.72	640	123.1	3.236	1100	
Cr$_2$N	Hex	0.4760/0.4438	5.9	1500	9.4			30.8	3.138	2250 HV	86.3–110.3
HfN	Fcc	0.452	13.8	3310	6.9	11.3	26	369.4	3.33–4.8	1700–2000	
Si$_3$N$_4$	hex	0.78/0.56	3.44	1900	2.4	20–24	10^{18}	750.5	2.1	1410 HV	12–14
TaN	hcp	0.52/0.29	13.6–13.8	3000	3.6	8.58	128	225.7	5.756	3240	5–8
Ta$_2$N	hex	0.30/0.493	15.8	3000		10.05	263	270.9		3000	
TiN	cub-B1	0.423	5.21	3220	9.35	30	21.7	336.2	2.512	2400 HV	5
VN	fcc	0.41	6.13	2050	8.1	11.3	85–100	147.8	4.6	1520	5–8
ZrN	fcc	0.46	6.93	3000	6	16.75	13.6	365.5	5.1	2000	12
Carbides											
B$_4$C	rhom	0.5631/1.2144	2.52	2450	6	27.63	10^6	72	4.5	3700	11–14
Cr$_3$C$_2$	ortho	1.146/0.552/0.2821	6.68	1900	10.3	18.8	75	88.8	4	1500–2000	12
NbC	fcc	0.45	7.78	3490	6.65	14.24	35–74	139.8	3.4	2400	11
SiC	α :hex	β : 0.4360	3.2	2200	5.68	15.49	10^5	71.6	4.8	3500	14–17
	β :fcc	α : 0.3–7.3/1–1.5	3.17	2700	5.3	63–155	10^5	73.3	3.9–4.1	1400 HV	13–14
TaC	cub-B1	0.4454	14.65	3877	6.04	22.19	25	159.5	2.91	1490	11–14
TiC	cub-B1	0.429–0.433	4.93	3150	7.4	17–23.5	68	179.6	3.22	3200 HV	11–14
VC	cub-B1	0.4173	5.36	2770	6.55	4.2	156	105.1	4.34	2760 HV	8–11
WC	hex	0.29/0.28	15.7	2600	5.2–7.3	121.42	17	35.2	7.2	2080	8
ZrC	cub-B1	0.4989–0.476	6.51	3400	6.93	20.5	42	181.7	4	2600 HV	12
Borides											
AlB$_2$	hex	0.3006/0.3252	3.17	1975				67			
CrB	ortho	0.2969/0.7858/0.2932	6.05	1550				75.4			14–18
CrB$_2$	hex	0.279/0.307	5.6	2200	11.1		56	94.6	2.15	2250	
HfB$_2$	hex	0.3141/0.3470	11.01	3200	5.3	430	10	336.6		2800	11–17
MoB$_2$	hex	0.3/0.31	7.8	2100			45	96.3		1380 HV	11–14
NbB$_2$	hex	0.31/0.33	6.8	3000	7.1–9.6	16.75	32	150.7	2.6	2600	11–14
SiB$_6$	ortho	1.4470/1.8350/0.9946	2.43	1950	8.3		10^7		3.3	1910	

Table 2.1. (*Continued*)

Phase	Crystal structure	Lattice parameters (nm)	Density (g.cm^{-3})	Melting point (°C)	Linear thermal expansion, α (10^{-6} K^{-1})	Thermal conductivity, λ (W m^{-1}K^{-1})	Electrical resistivity ($10^{-6}\Omega$cm)	Enthalpy at 298K (kJ mol^{-1})	Young's modulus (10^5 N mm^{-2})	Micro hardness (10 N mm^{-2})	Oxidation resistance (×100°C)
TaB$_2$	hex	0.31/0.33		3150	5.1	21.35	68	209.3	2.62	2200	11–14
TiB$_2$	hex	0.3/0.32	4.5	2900	6.39	25.96	9	150.7	3.7	3840	11–17
Ti$_2$B	tet	0.61/0.46		2200						2500	
VB$_2$	hex	0.3/0.31	4.8	2400	5.3		16	203.9	5.1	2080	13
WB	tet	0.31/1.7	15.5	2860						3750	
W$_2$B	tet	0.56/0.47	16.5	2770	4.7		21.43			2350	8–14
ZrB	fcc	0.47	6.5	3000				163.3		3600	
ZrB$_2$	hex	0.32/0.35	6.1	3000	6.83	23.03	9.2	326.6	3.5	2200	11
Oxides											
Al$_2$O$_3$-α	hex	0.5127	3.99	2043	8	30.1	10^{20}	1580.1	4	2100 HV	17
	rhom	0.513	3.9	2030	7.2–8.6	4.2–16.7	10^{20}	1678.5	3.6	2100 HV	20
BeO	hex	0.2699/0.4401	3	2450	9	264	10^{23}	569	3	1230–1490 HV	17
CrO$_2$	tet	0.441/0.291	4.8					582.8			
CrO$_3$	ortho	0.573/0.852/0.474	2.81	170–198				579.9			
Cr$_2$O$_3$	rhom	0.536	5.21	2440	6.7		10^{13}	1130.4		1000 HV	
	hex	0.495876/1.35942	5.21	2400	5.6			1130.4		2300 HV	
HfO$_2$	mono	0.512	9.7	2900	10	3	5×10^{15}	1053.4		900	
MgO	cub	0.4208	3.6	2850	11.2	36	10^{12}	568.6	3.2	745 HV	17
SiO$_2$	quartz	0.4093/0.5393	2.33	1703–1729	0.4	1.38	10^{22}	911	0.5–1.0	1130–1260	
	trigonal	0.421/0.539	2.2	1713	0.5–0.75	1.2–1.4	10^{21}	911.5	1.114	1200	
ThO$_2$	fcc	0.5859	10.05	3250	10	10	10^{16}	1173.1	1.38	950 HV	17
TiO	cub-B1	0.417	4.88	1750	7.6	11		520		1300	
TiO$_2$	tet	0.4593/0.2959	4.19	1900	4.21–4.25	8	1.2×10^{10}	945.4	2.05–2.80	767–1000 HV	
	cub-B1	0.45933/0.29592	4.25	1867	9				0.8–2.0	1000 HV	
Ti$_2$O$_3$	rhom	0.5454	4.6	2130			10^{19}–10^{24}	1433.1		980 HV	
Ti$_3$O$_5$	mono	0.9828/0.3776/0.9898		1780				2461			
ZrO$_2$	cub	0.511	5.6	2750	7.5–10.5	0.7–2.4	10^{16}	1035	1.63	1200	17

Table 2.1. (*Continued*)

Phase	Crystal structure	Lattice parameters (nm)	Density (g.cm^{-3})	Melting point (°C)	Linear thermal expansion, α (10^{-6} K^{-1})	Thermal conductivity, λ (W m^{-1}K^{-1})	Electrical resistivity ($10^{-6}\Omega$cm)	Enthalpy at 298K (kJ mol^{-1})	Young's modulus (10^5 N mm^{-2})	Micro hardness (10 N mm^{-2})	Oxidation resistance ($\times 100$°C)
Silicides											
CrSi	cub	0.4629	5.38	1550				53.2		1000	
CrSi$_2$	hex	0.442/0.655	4.91	1630				100.5		1100	14–18
Cr$_3$Si	cub	0.455	6.52	1710				105.5		900–980	
MoSi$_2$	tet	0.32/0.786	6.3	2050	8.4	221.9	21	108.9	3.84	1290	17
NbSi$_2$	hex	0.48/0.66	5.5	1950	8.4		6.3	50.2		700	8–11
TaSi$_2$	hex	0.4773/0.6552	9.2	2200	8.9/8.8		38	150.7		1410	11
TiSi$_2$	ortho	0.8236/0.4773/0.8523	4.39	1520	11.5		18	134.4	2.556	892	11
VSi$_2$	hex	0.46/0.64	4.5	1650	11		9.5	95		960	
WSi$_2$	tet	0.321/0.788	9.5	2165	6.5	45	12.5	92.1	5.3	1090	16
ZrSi$_2$	ortho	0.372/1.416/0.367	4.87	1700	9.7		161	159.4	2.348	1030 HV	8–11

Legend:

HV: Vickers hardness, HK: Knoop hardness

hcp: hexagonal closed-packed, fcc: face-centered cubic

ortho: orthorhombic, hex: hexagonal, cub: cubic, cub-B1: cubic NaCl-type, rhom: rhombohedral/trigonal, tet: tetragonal, mono: monoclinic, tri: triclinic

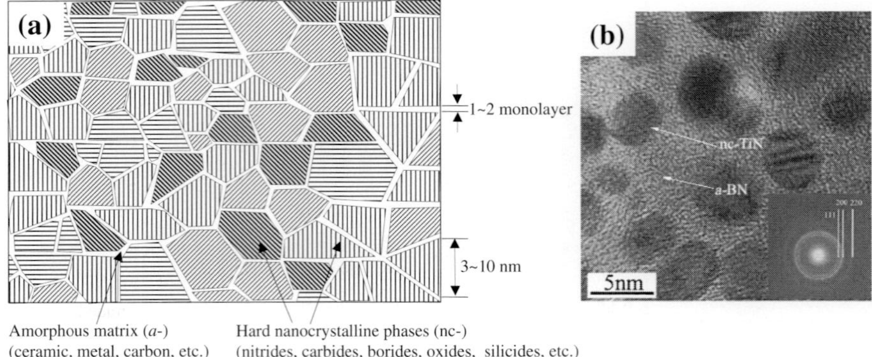

Fig. 2.2. (a) Schematic diagram of a nanostructured nanocomposite coating proposed by Veprek *et al.* [28], and (b) HRTEM image and selected area diffraction pattern (SADP) of nanocomposite Ti–B–N (nc-TiN/*a*-BN) [29].

between the coating and substrate and between layers in the coating. For example, a WC–TiC–TiN (outside layer) graded coating for cutting tools was reported by Fella *et al.* [30], which showed considerably less wear than single layer hard coatings used in the cutting of steels. This type of coating is functionally and chemically graded to achieve better adhesion, oxidation resistance, and mechanical properties. One example of how functionally graded architectures improve coating performance is the adhesion of DLC to steels. DLC, and especially hydrogen-free DLC, has a very high hardness and generally has a large residual compressive stress. The coatings are relatively inert, and adhesion failures of coated steel surfaces were a roadblock to success. This problem was solved through designing and implementing a graded interface between the coating and the substrate. Examples of effective gradient compositions are Ti–TiN–TiCN–TiC–DLC for hydrogenated DLC [31] and Ti–TiC–DLC for hydrogen-free DLC [32]. In the development of the later composition, the importance of a graded elastic modulus through the substrate coating/interface was highlighted as shown in Fig. 2.3. The gradual build-up in material stiffness from the substrate with $E = 220\,\text{GPa}$ to the DLC layer with $E = 650\,\text{GPa}$ avoids sharp interfaces that can provide places for crack initiation, good chemical continuity, and load support for the hard DLC top-coat. This functionally graded approach can be combined with multilayer and nanocomposite architectures to further enhance tribological properties.

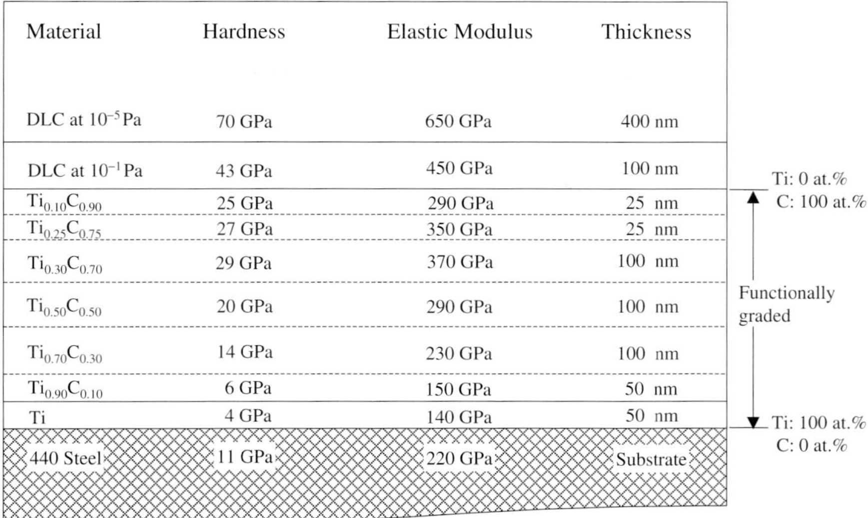

Material	Hardness	Elastic Modulus	Thickness	
DLC at 10^{-5} Pa	70 GPa	650 GPa	400 nm	
DLC at 10^{-1} Pa	43 GPa	450 GPa	100 nm	Ti: 0 at.%
$\text{Ti}_{0.10}\text{C}_{0.90}$	25 GPa	290 GPa	25 nm	C: 100 at.%
$\text{Ti}_{0.25}\text{C}_{0.75}$	27 GPa	350 GPa	25 nm	
$\text{Ti}_{0.30}\text{C}_{0.70}$	29 GPa	370 GPa	100 nm	
$\text{Ti}_{0.50}\text{C}_{0.50}$	20 GPa	290 GPa	100 nm	Functionally graded
$\text{Ti}_{0.70}\text{C}_{0.30}$	14 GPa	230 GPa	100 nm	
$\text{Ti}_{0.90}\text{C}_{0.10}$	6 GPa	150 GPa	50 nm	
Ti	4 GPa	140 GPa	50 nm	Ti: 100 at.%
440 Steel	11 GPa	220 GPa	Substrate	C: 0 at.%

Fig. 2.3. Schematic diagram of a functionally gradient Ti–TiC$_x$–DLC coating, where chemistry and elastic moduli are transitioned from metallic substrate to hard DLC top layer [32].

3. Background of Nanostructured Superhard Coatings

Hardness is defined as the resistance to plastic deformation. Plastic deformation of crystalline materials occurs predominantly by dislocation movement under applied load. Therefore, a higher resistance to dislocation movement of a material will generally enhance its hardness. One approach to obtain high resistance to dislocation movement and plastic deformation is to preclude the formation of stable dislocations. "Superhard" coatings, with a hardness value in excess of 40 GPa, have attracted significantly increasing interest during the past 10–15 years [33]. A concept for superhard nanocomposite coatings was suggested by Veprek and Reiprich [34]. The strength and hardness of engineering materials are orders of magnitude smaller than the theoretically predicted values. They are determined mainly by the microstructure which has to be designed in such a way as to efficiently hinder the multiplication and movement of dislocations and the growth of microcracks. This can be achieved in various ways known from metallurgy, such as solution, work, and grain boundary hardening [35, 36]. In this way, the strength and hardness of a material can be increased by a factor of 3–7 times, i.e. superhard material should form when such enhancement can be achieved starting from a hard material ($HV > 20$ GPa). Solution and work hardening do not operate in small nanocrystals of about $<10\,\text{nm}$ because solute atoms segregate to the grain boundary and there are no dislocations. Therefore, we consider the possibilities of extending the grain boundary hardening in poly- and microcrystalline materials, described by the Hall–Petch relationship [37,38], Eq. (3.1), down to the range of a few nanometers:

$$\sigma_C = \sigma_0 + \frac{k_{\text{gb}}}{\sqrt{d}}. \tag{3.1}$$

Here σ_C is the critical fracture stress, d is the crystallite size, and σ_0 and k_{gb} are constants. Many different mechanisms and theories describe Hall–Petch strengthening [37,38]. Dislocation pileup models and work hardening yield the $d^{-1/2}$ dependence but different formula for σ_0 and k_{gb}, whereas the grain–grain boundary composite models also give a more complicated dependence of σ_C on d. The strength of brittle materials, such as glasses and ceramics, is determined by their ability to withstand the growth of microcracks. Brittle materials do not undergo any plastic deformation up to their fracture. Their strength or hardness is proportional to the elastic modulus. The critical stress which causes the growth of a microcrack of size a_0 is given by the general Griffith criterion:

$$\sigma_C = k_{\text{crack}}\sqrt{\frac{2E\gamma_s}{\pi\,a_0}} \propto \frac{1}{\sqrt{d}}. \tag{3.2}$$

Here E is the Young's modulus, γ_s is the surface cohesive energy, and k_{crack} is a constant which depends on the nature and shape of the microcrack and on the kind of stress applied [35]. Thus, the $d^{-1/2}$ dependence of the strength and hardness in a material can also originate from the fact that the size a_0 of the possible flaws, such as voids and microcracks that are formed during the processing of the material, also decreases with decreasing grain size. For these reasons, the Hall–Petch relationship, Eq. (3.1), should be considered as a semi-empirical formula which is valid down to a crystallite size of 20–50 nm (some models predict an even higher limit [39, 40]). With the crystallite size decreasing below this limit, the fraction of the material in the grain boundaries strongly increases which leads to a decrease of its strength and hardness due to an increase of "grain boundary sliding" [40, 41]. A simple phenomenological model (i.e. rule of mixtures) describes the softening in terms of an increasing volume fraction of the grain boundary material, f_{gb}, with the crystallite size decreasing below 10–6 nm [42]:

$$H(f_{\text{gb}}) = (1 - f_{\text{gb}})H_c + f_{\text{gb}}H_{\text{gb}}. \tag{3.3}$$

Here $f_{\mathrm{gb}} \propto (1/d)$. Due to the flaws present, the hardness of the grain boundary material, H_{gb}, is smaller than that of the crystallites H_c. Thus the average hardness of the material decreases with d decreasing below 10 nm. The first report of a "Reverse" (or "negative") Hall–Petch relationship was by Chokshi *et al.* [43] Later on it was the subject of many studies, with controversial conclusions regarding the critical grain size where "Normal" Hall–Petch relationship changes to the reverse one [39, 44]. Various mechanisms of grain boundary creep and sliding were discussed and are described by deformation mechanism maps in terms of the temperature and stress [45, 46]. Theories of grain boundary sliding were critically reviewed in [39]. Recent computer simulation studies [41] confirm that the negative Hall–Petch dependence in nanocrystalline metals is due to the grain boundary sliding that occurs due to a large number of small sliding events of atomic planes at the grain boundaries without thermal activation. Although many details are still not understood, there is little doubt that grain boundary sliding is the reason for softening in this crystallite size range. Therefore, a further increase of the strength and hardness with decreasing crystallite size can be achieved only if grain boundary sliding can be blocked by appropriate design of the material. This is the basis of the concept for the design of superhard nanocomposites, suggested by Veprek and Reiprich [34].

As mentioned in Sec. 2.1, another possible way to strengthen a material is based on the formation of nanoscale multilayers consisting of two different materials with large differences in elastic moduli, sharp interface, and small bilayer thickness (lattice period) of about 10 nm [7]. Because this design of nanostructured superhard materials was suggested and experimentally confirmed before superhard nanocomposites were developed, we will deal with superlattice coatings in Sec. 3.1.

3.1. *Nanoscale Multilayer Coatings*

In a theoretical paper published in 1970, Koehler [7] suggested a concept for the design of strong solids which are nowadays called superlattices. Originally he suggested depositing epitaxial multilayers of two different metals, A and B, having as different elastic constants as possible $E_A < E_B$ but a similar thermal expansion and strong bonds. The thickness of the layers should be so small that no dislocation source could operate within the layers. If under applied stress a dislocation, which would form in the softer layer A, would move towards the A/B interface, elastic strain induced in the second layer B with the higher elastic modulus would cause a repulsing force that would hinder the dislocation from crossing that interface. Thus, the strength of such multilayers would be much larger than that expected from the rule of mixture. Koehler's prediction was further developed and experimentally confirmed by Lehoczky who deposited Al/Cu and Al/Ag superlattices and measured their mechanical properties [47]. According to the rule of mixtures, the applied stress which causes elastic strain is distributed between the layers proportional to their volume fractions and elastic moduli. Lehoczky has shown that the tensile stress–strain characteristics measured on multilayers consisting of two different metals displayed a much higher Young's modulus and tensile strength than that predicted by the rule of mixtures, and both of which increased with decreasing thickness of the double layer (lattice period). For layer thicknesses of $<70\,$nm, the yield stress of Al/Cu multilayers was 4.2 times larger and the tensile fracture stress was 2.4–3.4 times higher than the values given by the rule of mixture [47]. This work was followed by work of a number of researchers who confirmed the experimental results on various metal/metal, metal/ceramic, and ceramic/ceramic multilayer systems (see Sec. 2.1). In all these cases, an increase in hardness by a factor of 2–4 was achieved when the lattice period decreased to about 5–10 nm. For a large lattice period, where the dislocation multiplication source can still operate, the increase of the hardness and the tensile strength (most researchers measured the hardness because it is simpler than the measurement of tensile strength conducted by Lehoczky) [47] with decreasing layer thickness is due to the increase in the critical stress, which is dependent on multiplicate dislocations such as a Frank–Read dislocation source,

$$\sigma_C = \frac{Gb}{l_{\mathrm{pp}}} \propto \frac{1}{l_{\mathrm{pp}}}, \tag{3.4}$$

where G is the shear modulus, b the Burgers vector, and l_{pp} the distance between the dislocation pinning sites [35]. Usually one finds strengthening dependence similar to the Hall–Petch relationship but with a somewhat different dependence on the layer period (λ^{-n}), instead of the crystallite size d in Eq. (3.1), with $n = 1/2$ for layers with different slip systems and $n = 1$ for layers with a similar slip systems [48]. A more recent theoretical discussion of the Hall–Petch relationship for superlattices was published by Anderson and Li [49]. According to their calculations, strong deviations from continuum Hall–Petch behavior occur when the thickness of the layers is so small that the pileup contains only one to two dislocations.

In a remark added in the proof, Koehler mentioned that the ideas described in his paper would also be valid if one of the layers is amorphous. Recently, several papers appeared in which one of the layers consists of an amorphous CN_x and a transition metal nitride such as TiN [50] or ZrN [51]. However, with decreasing layer thickness the layered structure vanished and a nanocrystalline composite (i.e. "Nanocomposite") structure appeared [50,51]. Such films also exhibit a high hardness of 40 GPa or more.

3.2. *Single Layer Nanocomposite Coatings*

Using similar ideas for restricting dislocation formation and mobility as used in multilayer approaches to "hardening", nanocomposite coatings can also be superhard [25,28]. These composites have 3–10 nm crystalline grains embedded in an amorphous matrix and the grains are separated by 1–3 nm. This designs a "Architecture" which leads to ultrahard (hardness above 100 GPa) coatings as reported by Veprek *et al.* [2]. The nanocrystalline phase may be selected from the nitrides, carbides, borides, and oxides, as shown in Table 2.1, while the amorphous phase may also include metals and diamond-like carbon (DLC) as shown in Fig. 2.1(a). The initial model proposed by Veprek to explain hardness in nanocomposites is that dislocation operation is suppressed in small grains (3–5 nm) and that the narrow space between them (1 nm separation) induces incoherence strains [34,52]. The incoherence strain is likely increased, when grain orientations are close enough to provide interaction between matched but slightly misoriented atomic planes. In the absence of dislocation activity, Griffith's equation, as shown in Eq. (3.2), for crack opening was proposed as a simple description of the nanocomposite strength. This equation suggests that strength can be increased by increasing elastic modulus and surface energy of the combined phases, and by decreasing the crystalline grain sizes. It is noted that elastic modulus is inversely dependent on grain sizes that are in the nm size range due to lattice incoherence strains and the high volume of grain boundaries [2]. In practice, grain boundary defects always exist, and a 3 nm grain size was found to be close to the minimum limit. Below this limit, a reverse Hall–Petch effect has been observed and the strengthening effect disappears because grain boundaries and grains become indistinguishable and the stability of the nanocrystalline phase is greatly reduced [34,43,53].

Nanocomposites with metal matrixes are in a special category for this discussion. They have been demonstrated to increase hardness, but also have good potential for increasing toughness. Mechanisms for toughening within these systems are discussed in the next section, while mechanisms for hardening are discussed here. The composite strength of ceramic/metal nanocomposites may be described by the following form of the Griffith–Orowan model [54] when the dimensions of the metal matrix permit operation of dislocations:

$$\sigma_C = \sqrt{\frac{2E(\gamma_s + \gamma_p)}{\pi a} \frac{r_{\text{tip}}}{3d_a}},$$ (3.5)

where γ_p is the work done during plastic deformation, r_{tip} is the curvature of the crack tip, and d_a is the interatomic distance. It is noted that crack tip blunting and the work of plastic deformation considerably improve material strength, while the lower elastic moduli of metals cause a reduction in strength as compared to ceramics. However, in nanocomposites, dislocation operation may be prohibited because the separation of grains is very small. For example, the critical distance, l_{pp}, for a Frank–Read dislocation source is very small. Matrix dimensions in hard nanocomposite coatings are from 1–3 nm, which is well below the critical size for dislocation source operation, even in very soft metal matrices. Therefore, the mechanical behavior of such nanocomposites can be expected to be similar to that of ceramic matrix composites.

Composite designs that increase elastic modulus and hardness, do not necessarily impart high toughness. First, dislocation mechanisms of deformation are prohibited and crack opening is the predominant mechanism for strain relaxation when stresses exceed the strength limit. Second, Griffith's equation does not take into account the energy balance of a moving crack, which consists of the energy required to break bonds and overcome friction losses, potential energy released by crack opening, and kinetic energy gained through crack motion [55]. From crack energy considerations, a high amount of stored stress dictates a high rate of potential energy release in the moving crack. In such conditions, a crack can achieve the self-propagating (energetically self-supporting) stage sooner, transferring into a macrocrack and causing brittle fracture. However, nanocomposites contain a high volume of grain boundaries between crystalline and amorphous phases. This type of structure limits initial crack sizes and helps to deflect, split, and blunt growing cracks.

4. New Directions for Nanostructured Supertough Coatings

While superhard coatings are very important, quite notably for protection of cutting tools, most tribological applications for coatings either require, or would receive significant benefit from increased toughness and lower friction. In particular, "High Fracture Toughness" is necessary for applications where high contact loads and, hence, significant substrate deformations, are encountered [27]. A material is generally considered tough if it possesses both high strength and high ductility. High hardness (H) is directly related to high elastic modulus (E) and high yield strength, but it is very challenging to add a measure of ductility to hard coatings. For example, the superhard coating designs, as stated earlier, prevent dislocation activity, essentially eliminating one common mechanism for ductility. Therefore, designs that increase ductility must also be considered to provide tough tribological coatings. Pharr [56] has suggested that an indication of fracture toughness (i.e. H/E ratio) can be obtained by examining the surface radial cracks created during indentation, described by the equation.

$$K_c = \alpha_1 \left(\frac{E}{H} \right)^{1/2} \left(\frac{P}{c^{3/2}} \right), \tag{4.1}$$

where P is the peak indentation load, c the radial crack length and α_1 an empirical constant related to the indenter geometry. K_c describes the "critical stress intensity" for crack propagation, but it is not an intrinsic parameter that can be used to measure fracture toughness directly. However, it is proportional to fracture toughness. Thus, fracture toughness of coatings would appear to be improved by both a high hardness and a "Low Elastic Modulus". In this work, the H/E values were calculated and discussed relatively for each coating [57].

4.1. *Functionally Graded Multilayer Coatings*

An effective route for improving toughness in multilayers is the introduction of ductile, low elastic modulus alternate layers into the coating structure to relieve stress and allow crack energy dissipation by plastic deformation in the crack tip. This approach will result in a decreased coating hardness, but the gain in the fracture toughness improvement may be more important in many tribological applications, excluding coatings for the cutting tool industry. For example, $[Ti/TiN]_n$ multilayer coatings on cast iron piston rings relaxed interface stress and improved combustion engine performance [58]. Figure 4.1(a) shows a schematic of a multilayer $[Ti/DLC]_n$ coating on a graded load support foundation, where the ductile Ti layers in the multilayer stack were graded at every DLC interface to avoid brittle fracture [4]. A cross-sectional photograph of this coating with 20 [Ti/DLC] pairs is shown in Fig. 4.1(b). The ductile Ti layers reduced the composite

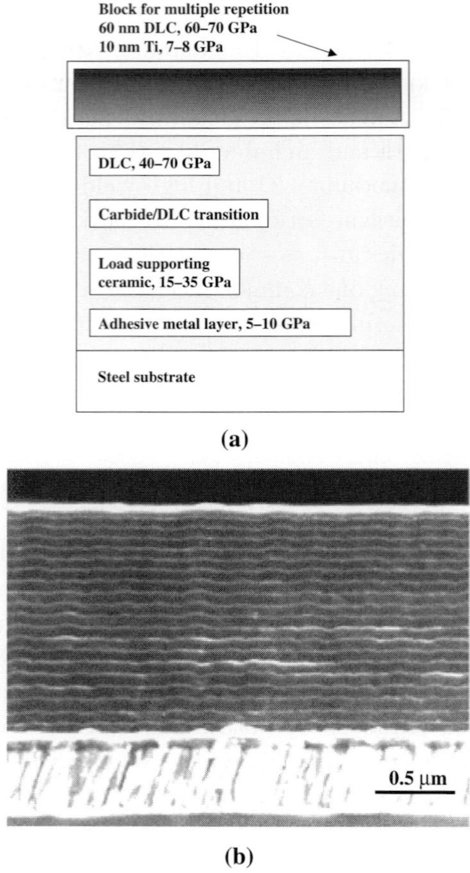

(a)

(b)

Fig. 4.1. A multilayer coating with multiple Ti/DLC pairs on top of a functionally gradient layer for an optimum combination of cohesive and adhesive toughness: (a) design schematic, and (b) cross-sectional photograph of the coating produced with 20 Ti/DLC pairs [4].

coating hardness to 20 GPa as compared to a single layer DLC coating, which has a hardness of about 60 GPa. However, due to dramatic improvement in toughness the multilayer coating design permitted operation during sliding friction at contact pressures as high as 2 GPa without fracture failure compared to 0.6–0.8 GPa for single layer DLC.

In general, the combination of multilayer and functionally gradient approaches in the design of wear protective coatings produces exceptionally tough wear protective coatings for engineering applications. One potential drawback slowing the wide spread use of new coatings was the need for reliable process controls to ensure that the correct compositions, structures, and properties are implemented during growth. However, modern process instrumentation and control technologies are able to meet the challenge and permit successful commercialization [59]. Thus, functionally gradient and multilayer designs are commonly utilized in the production of modern tribological coatings.

4.2. *Functionally Graded Nanocomposite Coatings*

An alternative to employing multilayers to toughen coatings is embedding grains of a hard, high yield strength phase into a softer matrix allowing for high ductility. This approach has been widely explored in macrocomposites made of ceramics and metals which are known as cermets [60]. It was recently scaled down to the nanometer level in thin films made of hard nitrides and softer metal matrices [28]. Combination of the nanocrystalline/amorphous designs with a functionally graded interface, as shown in Fig. 4.2(a), provides high cohesive toughness and high interface (adhesive) toughness in a single coating. Several examples of tough wear-resistant composite coatings have been reported. Two of them combined nanocrystalline carbides with an amorphous DLC matrix designated as TiC/DLC and WC/DLC composites. In another example, nanocrystalline MoS_2 was encapsulated in an Al_2O_3 amorphous matrix as shown in Fig. 4.2(b). In

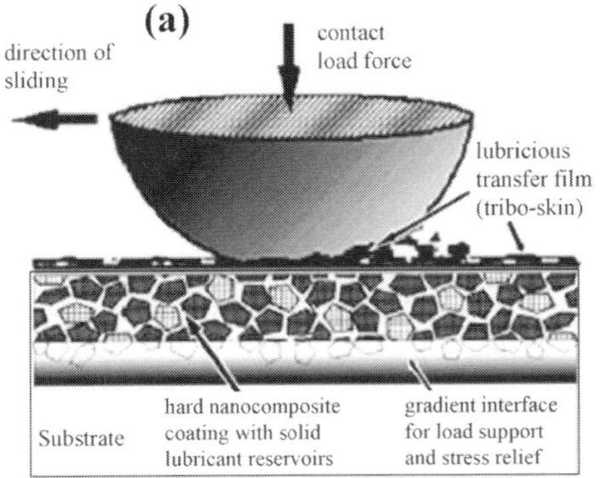

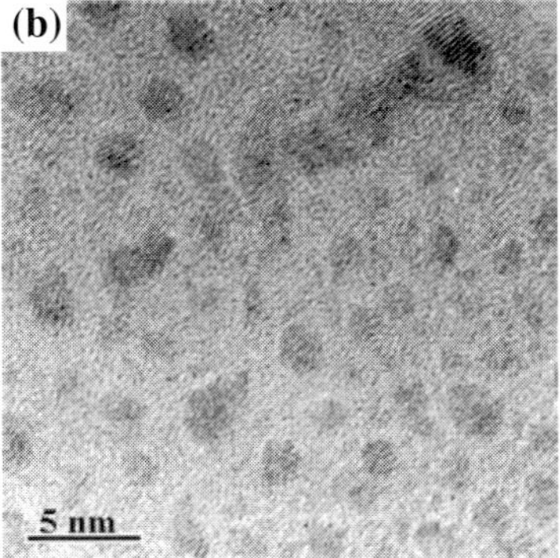

Fig. 4.2.   (a) Schematic representation of a tough nanocomposite coating design, combining a nanocrystalline/amorphous structure with a functionally gradient interface and (b) TEM image of an Al_2O_3/MoS_2 nanocomposite coating consisted of an amorphous Al_2O_3 ceramic matrix encapsulating 5–10 nm inclusions of nanocrystalline MoS_2 grains [61].

all cases, the large fraction of grain boundary phase provided ductility by activating grain boundary slip and crack termination by nanocrack splitting. This provided a unique combination of high hardness and toughness in these coatings. Novel nanocomposite designs suggested by Voevodin and Zabinski [61] for tough tribological coatings are very promising and provide a very attractive alternative to multilayer architectures. One of these novel designs incorporates the "Chameleon" approach in which components are added to the coating that provide a low friction coefficient under extreme environmental conditions such as low and high humidity and low and high temperatures [61]. Nanocomposite coatings are more easily implemented, since they do not require precise control in the layer thickness and frequent cycling of the deposition parameters, as is required for fabrication of multilayer coatings. They are however relatively recent developments, and suitable scale-up of deposition techniques is currently under intense study.

5. Other Possible Properties of Nanostructured Coatings

To prevent tribological failures, there are additional requirements related to the normal (load) and tangential (friction) forces. In general terms, a tough "Wear-resistant" coating must support high loads in sliding or rolling contact without failure by wear, cohesive fracture, and loss of adhesion (delamination). A "Low Friction Coefficient" reduces friction losses and may increase load capability. Tribological coatings where a low friction coefficient is also required may be obtained by producing nanocomposite coatings with a mix of hard and lubricating phases, in which a hard primary phase (e.g. nitrides, carbides, or borides, etc.) provides wear-resistance and load-bearing capability and a lubricating secondary phase (e.g. a-C, a-Si$_3$N$_4$, a-BN, etc.) reduces the friction between two contacting components. Finally, "Thermal Stability" is required to optimize coating performance and lifetime. The amorphous phases in grain boundaries can act as diffusion barriers (e.g. a-Si$_3$N$_4$, a-SiO$_2$, etc.) for improved thermal stability. For instance, nc-TiN/a-Si$_3$N$_4$ coatings with amorphous Si$_3$N$_4$ matrix did not show grain growth at temperatures up to 1050°C as well as superhardness of about 45 GPa [62]. Moreover, silicon nitride acts as an efficient barrier against oxygen diffusion at the grain boundaries and also by forming an oxidation resistant SiO$_2$ surface layer, and thus resulting in excellent thermal stability.

In summary, in addition to high hardness, other aspects such as, high toughness, low friction coefficient, and high thermal stability are decisive characteristics of nanostructured coatings for their potential as protective tribological coatings.

6. New Processes for Industrial Applications of Multifunctional Tribological Coatings

6.1. *Hybrid Coating System of Cathodic Arc Ion Evaporation (CAE) and Magnetron Sputtering (MS)*

For more than a decade, cathodic arc evaporation (CAE) has been widely used to deposit various kinds of refractory and wear-resistant coatings such as Ti–Al–N [63] and its variants [64] for cutting tools, and CrN [65] coatings for automotive applications. The cathodic arc evaporation process is characterized by a combination of a high deposition rate and a high degree of ionization ($\geq$90 %) of evaporated species and high ion energies (20–120 eV), which makes this process a versatile deposition technology for many industries. Due to the nature of the high-current vacuum arc discharge, however, only target materials with good electro-conductivity can be used as evaporation sources. Also materials with a too high or low melting point or poor mechanical strength cannot be used. On the other hand, the sputtering process has fewer restrictions concerning the target material. Various kinds of materials, which are hard to use in the arc process, are usable in the sputtering process. One of the drawbacks of the sputtering process for the deposition of hard coatings is its low degree of ionization, which is in the range of a few percent. However, recently, this has been amended by the introduction of unbalanced magnetron sputtering (UBM) [66] and, more recently, by high-power pulsed magnetron sputtering (HPPMS) [67], which enhances plasma irradiation during the deposition or ionization of sputtered species. Recently, more complex coatings in composition and structure such as nanocomposite coatings and multilayer coatings have been under development to increase the service life of the coated tools and machine parts. Considering this situation, combining the arc and the sputtering process can be a way to enhance the possibility of depositing a coating system with a more complex structure and composition for improved performance. In this chapter, the microstructure and mechanical properties of the nanocomposite coating system will be discussed: Ti–Al–N and Ti–Al–Si–N coatings deposited by a hybrid coating system of CAE and sputtering will be introduced and discussed in Sec. 7.1; and Ti-B-C-N and Ti–Si–B–C–N coatings deposited by UBMS will be discussed in Sec. 7.2.

6.2. *Pulsed Closed-Field Magnetron Sputtering (P-CFUBMS)*

DC magnetron sputtering is a state-of-art deposition technique to produce coatings for diverse industrial applications, such as hard coatings, semiconductor, optical and decorative films etc. In DC magnetron sputtering, a glow discharge is created in low-pressure inert gas by applying a negative voltage on metal targets under a vacuum condition. The use of a magnetic field in the magnetrons behind the metal target allows sputtering to be performed at low pressures and voltages and at high deposition rates [68,69].

In recent years, there have been many improvements on DC magnetron sputtering. The development of unbalanced magnetron sputtering (the inner magnet and the outer magnet magnetic strength is unbalanced) has significantly increased the plasma generation area away from the target surface, see Fig. 6.1 [70]. With the development of closed-field unbalanced magnetron sputtering, a closed magnetic field is created by applying opposite magnetic poles in adjacent magnetrons, thereby resulting in much higher ion current densities and dense, hard well-adhered coatings by enhanced chemical reaction at the substrate [70]. A comparison of the magnetic configuration and plasma confinement in conventional, unbalanced and dual-magnetron closed-field systems is presented in Fig. 6.1. Closed-field systems can be configured using any even number of magnetrons. The use of multiple magnetrons in conjunction with a two-axis or three-axis substrate rotation system allows for the uniform deposition of large and complex shaped components in the closed-field plasma. Applying a negative bias voltage to the substrate can increase the energy of the ions bombarding the substrate. The energetic ions increase the bombardment of the substrate surface therefore producing coatings that exhibit high hardness and good wear-resistance and good adhesion to the substrate. One of the most important improvements in magnetron sputtering has been to reactively deposit ceramic and compound coatings of various components. Under normal magnetron sputtering, the sputtered targets generate a range of species that include energetic ions (1–10%), and neutral atoms (Ti, Cr, Ar, etc.). These species can chemically react with gas atoms that impinge on the substrate surface to form a compound film. Reactive sputtering can be used to deposit most nitride, carbide and oxide thin films with desired composition as a controlled monolithic or compositionally graded structure.

The deposition of hard ceramic coatings such as TiN [71], CrN [72], TiC [73], Ti–Al–N [74], Cr–Al–N [75], Al_2O_3 [76] etc., is achieved using

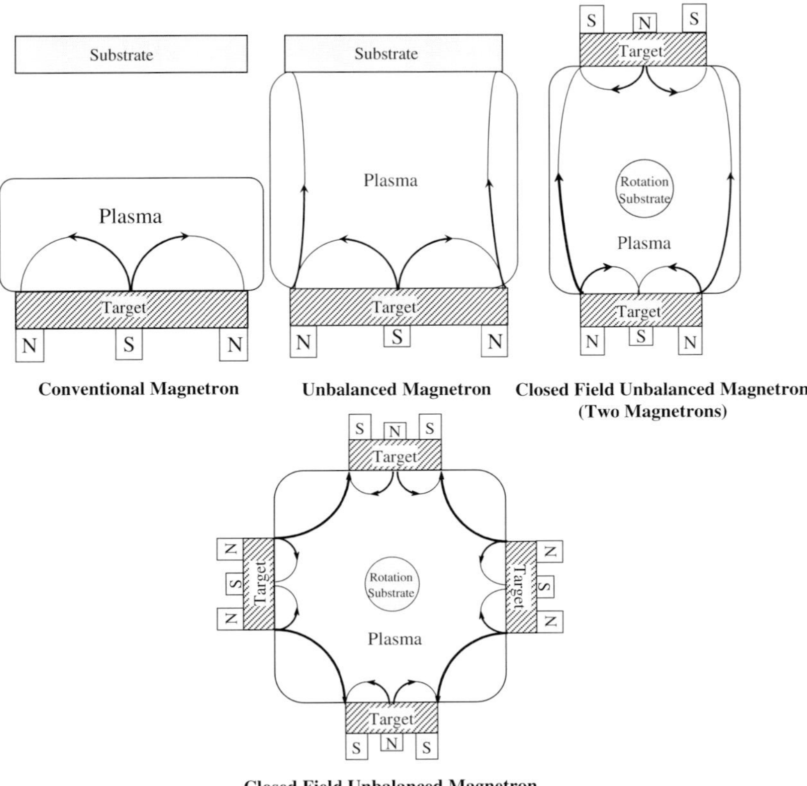

Fig. 6.1. A comparison of the magnetic configuration and plasma confinement in conventional, unbalanced and dual-magnetron closed-field systems [70].

reactive sputtering from the metal targets (Ti, Cr, Al, etc.) in a reactive gas atmosphere, e.g. N_2/Ar, CH_4/Ar or O_2/Ar gas mixture. However, for deposition of insulating films, the insulating film builds up on the surface of the chamber and/or anode. When the insulating layer on the anode becomes thick, the sputtering discharge becomes unstable. This phenomenon is called "disappearing anode". The target can also charge up quickly due to a non-conductive layer formed on the target at which point the target is said to "poisoned" which makes sputtering more difficult, decreases the sputtering rate and increases the power in the target. The charges will generate arcing problems in which the microparticles will be ejected from the target and incorporated into the deposited coatings.

This condition leads to non-uniform deposition, and the inhomogeneity and defects in the coatings, as well as reduced deposition rate [77].

In order to overcome this problem, the use of radio frequency (RF) to sputter material was developed in the 1960s. However, RF sputtering is not used extensively for commercial magnetron sputtering due to its low deposition rate, high cost, and the generation of a high temperature from the self-bias voltage associated with the RF power. In recent years, an alternative technique using a pulsed DC plasma has been developed and is called pulsed DC magnetron sputtering. Pulsed DC magnetron sputtering utilizes a pulsed potential to neutralize the positive charge on the target surface and eliminate the arcing by appropriately controlling the pulsing parameters of the target potential. Figure 6.2 exhibits some typical target voltage potential waveforms used in pulsed DC magnetron sputtering. The continuous target voltage in DC mode is either turned off periodically in the unipolar mode [Fig. 6.2(b)], or more commonly, switched to a positive voltage [Figs. 6.2(b) and (c)] in pulsed DC magnetron sputtering. During the normal pulse-on (target-on) period ($\tau_{\rm on}$), the negative sputtering voltage is applied to the target as in conventional DC sputtering. However, the target negative potential is periodically interrupted by a positive pulse voltage with a period of $\tau_{\rm rev}$. The reversed positive voltage is variable depending on the power supply allowance, which is reversed to a smaller positive voltage than the nominal negative pulse voltage in the asymmetric bipolar mode [Fig. 6.2(c)] or reversed to the same magnitude of positive voltage as the nominal negative pulse voltage in the symmetric bipolar mode [Fig. 6.2(d)]. The duty cycle is defined as the negative pulse time divided by the period of the pulsing cycle ($\tau_{\rm cycle}$). For a given positive pulse width, the full range of frequencies may not be available due to the power supply limitation, e.g. for an Advanced Energy Pinnacle Plus power supply, the smallest duty cycle is 50% [78]. In practice, the waveforms are virtually never as intended due to nonlinearities of either the plasma or the power supply circuitry. Therefore, the shapes of the resulting power waveforms are complex [79].

In the closed-field configuration, two or more magnetrons are commonly used for reactive sputtering. All magnetrons can be run in pulsing condition, thereby resulting in different combinations of pulsing modes. Normally, the magnetron potential can be pulsed in either asynchronous bipolar mode [Fig. 6.3(a)], in which the two target voltage waveforms are out of phase, or in synchronous bipolar mode [Fig. 6.3(b)], in which the two target voltage waveforms are in phase. The degree of out of phase in asynchronous mode is totally dependent on the frequencies and reverse time on each target.

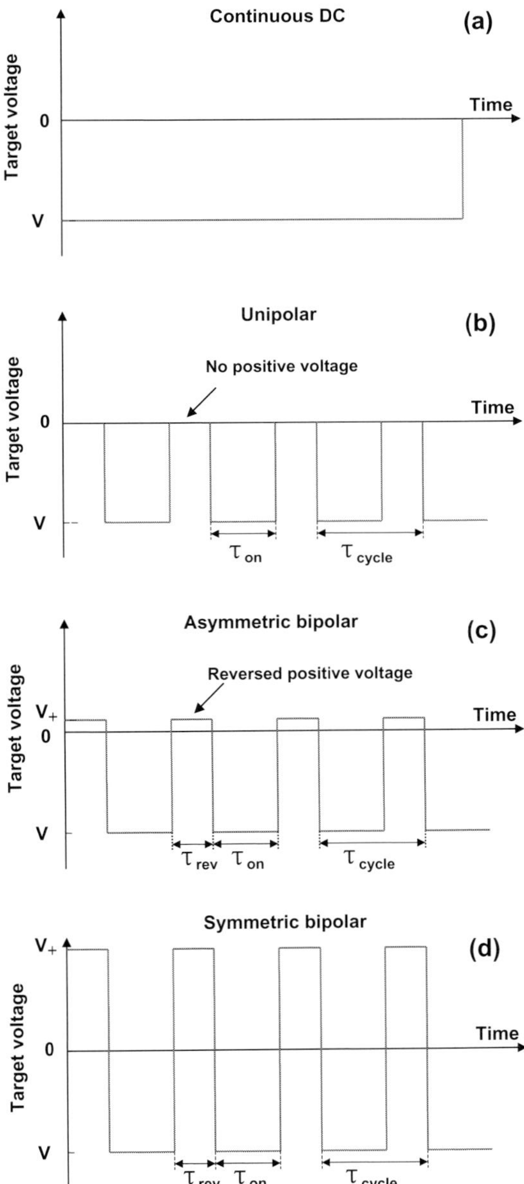

Fig. 6.2. The target voltage waveform when operated in (a) continuous DC, (b) unipolar pulsed mode (c) asymmetric bipolar pulsed mode, and (d) symmetric bipolar pulsed mode, (τ_{rev}: the reversed positive pulse period, τ_{on}: the normal negative target potential period, τ_{cycle}: the whole pulse period).

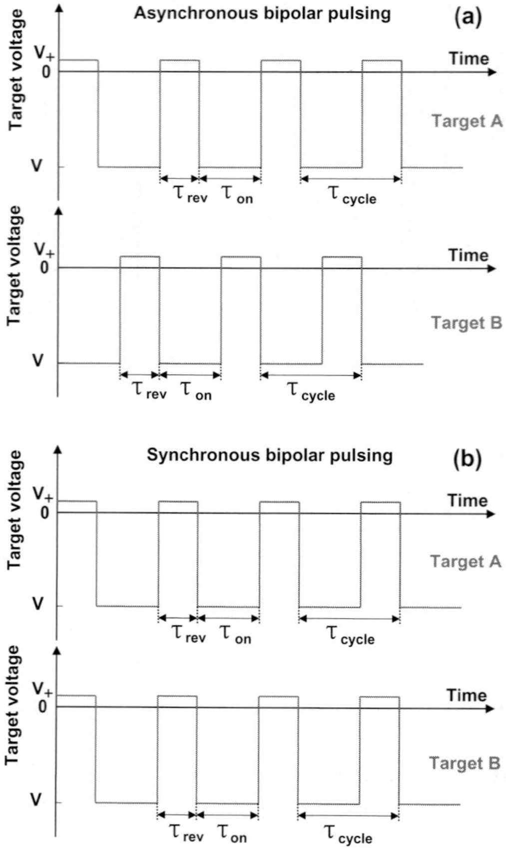

Fig. 6.3. Different target voltage waveforms in a dual-magnetron pulsed mode: (a) asynchronous bipolar pulsing, and (b) synchronous bipolar pulsing.

During the reversed positive pulse, the charge built up on the insulating material during the negative pulse period is discharged to eliminate breakdown and arcing. Therefore, a stable deposition process and smooth coating structure can be obtained. Moreover, with precise control of the reactive gas, high deposition rate can be achieved in depositing insulating films. Kelly *et al.* [80] compared the microstructure of Al_2O_3 films deposited by P-CFUBMS and normal DC magnetron sputtering. A fully dense and smooth surface Al_2O_3 film was produced by P-CFUBMS. On the other hand, the film deposited by DC magnetron sputtering exhibited a porous structure with microparticles covering the film surface (Fig. 6.4).

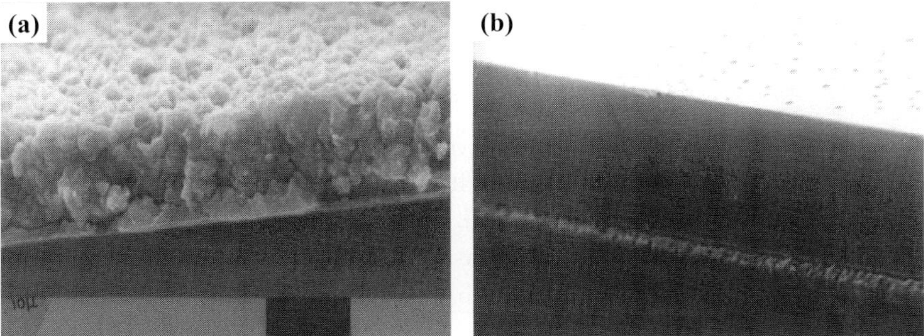

Fig. 6.4. SEM photomicrographs of fracture sections of aluminum oxide coatings deposited by (a) DC reactive sputtering, and (b) pulsed reactive sputtering [80].

6.3. *High-Power Pulsed DC Magnetron Sputtering (HPPMS)*

Recently, high-power pulsed magnetron sputtering (HPPMS), operating in an unipolar mode, has attracted wide attention. In this new method, a high density plasma is created in front of the sputtering source by using pulsed high target power density, ionizing a large fraction of the sputtered atoms [81,82]. In normal DC magnetron sputtering, the degree of ionization is low (typically less than 10%) which is due to the low power density (e.g. 3 Wcm^{-2}) limited by the target overheating from the ion bombardment [83]. However, in high-power pulsed magnetron sputtering, the average thermal load on the target is low by operating the target at high-power density (e.g. 3000 Wcm^{-2}) in a pulsed condition; therefore a considerably large fraction of ionized species of the target material can be created by the high probability for ionizing collisions between the sputtered atoms and energetic electrons. It was reported that the fraction of ionized species in high-power pulsed DC magnetron sputtering can be considerably increased up to and beyond 70% [84]. The high degree of ionized ions with high energy makes it possible to control the film growth behavior and thereby produce high quality films. The potential of this emerging new high-power pulsed magnetron sputtering technique includes depositing fully dense films with equiaxed structure, controlling the orientation of the film, and depositing thick (maybe up to 20 μm or more) films with low residual stress. However, this new technique is in its developing stage, and the deposition rate and cost of the power supply are the most important challenges that need to be considered.

7. Preparation–Microstructure–Properties of Nanostructured Coatings

7.1. *Hybrid Coating System of Ti–Al–Si–N Coatings*

Quaternary Ti–Al–Si–N coatings have been prepared by a hybrid coating system, where CAE was combined with a magnetron sputtering technique. Various analyses (e.g. HRTEM, XPS, XRD) revealed that the synthesized Ti–Al–Si–N coatings exhibited nanostructured composite microstructures consisting of solid–solution (Ti, Al, Si)N crystallites and amorphous Si_3N_4. The Si addition caused the grain refinement of (Ti, Al, Si)N crystallites and its uniform distribution with percolation phenomenon of amorphous silicon nitride similar to that of the Si effect in TiN films [2]. Figure 7.1 shows dark-field TEM images of Ti–Al–Si–N coatings containing different amounts of Si. The (Ti, Al)N crystallites [in Fig. 7.1(a)] appear to be large grains with a columnar structure. The (Ti, Al)N crystallites became finer with a uniform distribution as the Si content was increased. In Fig. 7.1(b), the crystallites were embedded in an amorphous matrix. These crystallites were revealed to be solid–solution (Ti, Al, Si)N phases of typical fcc crystal structure from the electron diffraction patterns. The (Ti, Al, Si)N crystallites showed partly aligned microstructure penetrated ("percolated") by an amorphous phase, but were not distributed homogeneously in the amorphous matrix. The microstructure, however, changed to that of a nanocomposite, having much finer (Ti, Al, Si)N crystallites (approximately 10 nm) and uniformly embedded in an amorphous matrix as the Si content in films increased to 9 at.% [Fig. 7.1(c)]. Such microstructure as shown in Fig. 7.1(c) was in agreement with the concept of a nanocomposite architecture suggested by Veprek and Reiprich [34]. Therefore, the Ti–Al–Si–N coatings with Si content of 9 at.% exhibited a maximum hardness among the experimental conditions. On the other hand, at a higher Si content of 19 at.% [Fig. 7.1(d)], (Ti, Al, Si)N crystallites decreased ($\sim$3 nm) and the film consisted mainly of the amorphous phase.

Figure 7.2(a) shows the nanohardness and Young's modulus of Ti–Al–Si–N coatings with various Si contents and average grain sizes. As the Si content increased, the nanohardness and Young's modulus of the Ti–Al–Si–N coatings steeply increased, and reached maximum values of approximately 55 GPa and 650 GPa at Si content of 9 at.%, respectively, and then dropped again with further increase of Si content. The reason for the large increases in hardness and Young's modulus of Ti–Al–N with Si addition is due to the grain boundary hardening both by strong cohesive

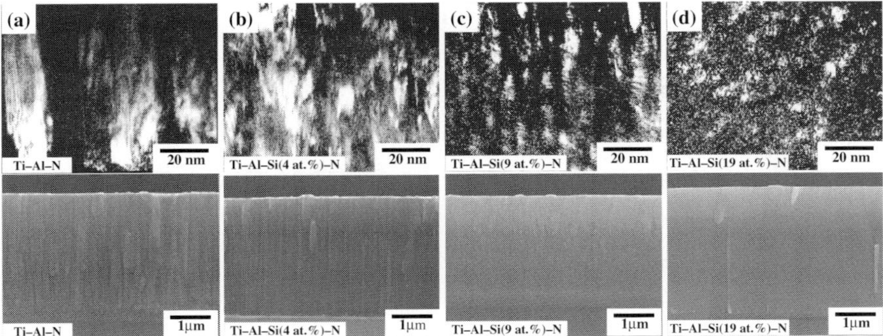

Fig. 7.1. Dark-field TEM and SEM images of Ti–Al–Si–N coatings containing (a) 0, (b) 4, (c) 9, and (d) 19 at.% Si, respectively [5].

energy of interphase boundaries and by Hall–Petch strengthening derived from crystal size refinement, as mentioned in Sec. 4, which were simultaneously caused by the percolation of amorphous Si_3N_4 ($H = \sim 22$ GPa, $E = \sim 250$ GPa) [85] into the Ti–Al–N film. Another possible reason would be due to solid–solution hardening of crystallites by Si dissolution into Ti–Al–N. The maximum hardness value at the silicon content of about 9 at.% results from the nanosized crystallites and their uniform distribution embedded in the amorphous Si_3N_4 matrix as shown in Fig. 7.1(c). On the other hand, the hardness reduction with further increase of Si content after maximum hardness as shown in Fig. 7.2(a) has been explained with the thickening of amorphous Si_3N_4 phase with increase of Si content [86]. When the amorphous Si_3N_4 is increased, the ideal interaction between nanocrystallites and the amorphous phase is lost, and the hardness of the nanocomposite becomes dependent on the property of the amorphous phase. On the other hand, Young's modulus, which must be related with density and atomic structure of the film, also largely increased from 470 GPa to 670 GPa with Si addition. This latter result was attributed to the densification of Ti–Al–Si–N films by filling the open structure of the Ti–Al–N grain boundaries with amorphous silicon nitride. Young's modulus reduction with further increase of Si content above 9 at.% was explained by the increase of volume fraction of amorphous Si_3N_4 phase, which has a lower atomic density than crystalline (Ti, Al, Si)N phase. Figure 7.2(b) shows the friction coefficients of the Ti–Al–N, Ti–Al–Si(9 at.%)–N, and Ti–Al–Si(31 at.%)–N films against a steel ball counterpart. The average friction coefficient of the film decreased from 0.9 to 0.6 with increasing Si content. This result is caused by tribochemical reactions, which often take

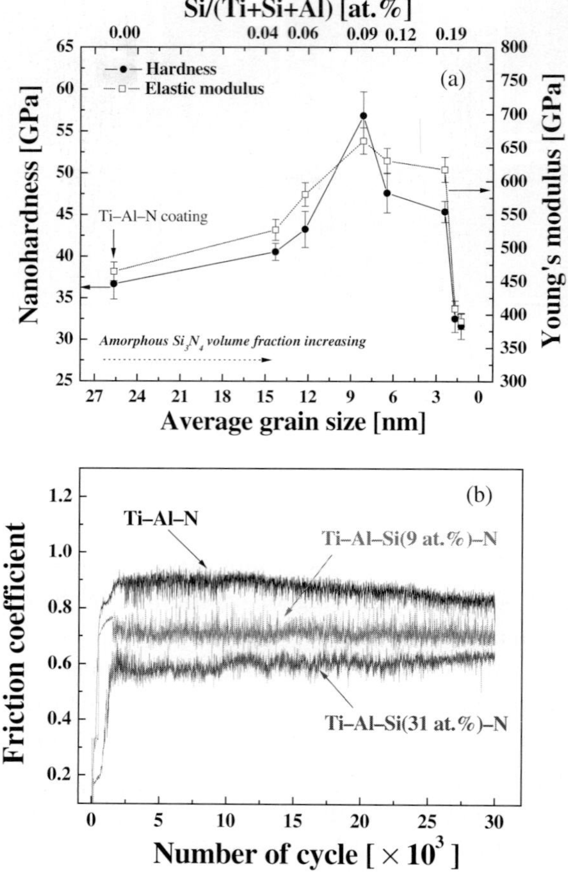

Fig. 7.2. (a) Nanohardness and Young's modulus, and (b) friction coefficients of nanocomposite Ti–Al–Si–N [nc-(Ti, Al)N/a-Si$_3$N$_4$] coatings as a function of Si content.

place in many ceramics [87], e.g. Si$_3$N$_4$ reacts with H$_2$O to produce SiO$_2$ or Si(OH)$_2$ tribolayer [88]. The products of SiO$_2$ and Si(OH)$_2$ are known to play a role of a self-lubricating layer.

Figure 7.3 shows the surface morphologies of the wear track and composition analyses for the wear debris after a sliding test. The surface morphology of the wear track for Ti–Al–N film was rough, and the width of the wear track relatively narrow as shown in Fig. 7.3(a), whereas, the surface morphology for the Ti–Al–Si(9 at.%)–N film was relatively smooth and the width of wear track was wide [Fig. 7.3(b)]. This result is due to

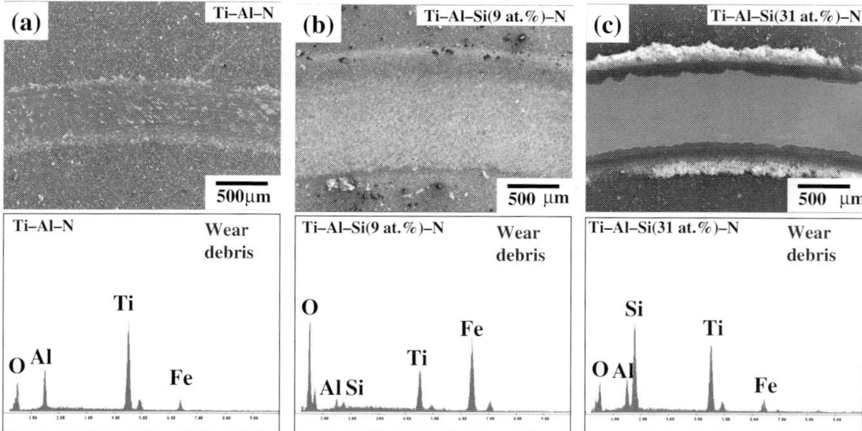

Fig. 7.3.   SEM morphologies of wear track and composition analyses for the wear debris after sliding test: (a) Ti–Al–N, (b) Ti–Al–Si(9 at.%)–N, and (c) Ti–Al–Si(31 at.%)–N film.

the adhesive wear behavior between the hard film ($\sim$55 GPa) and relatively soft steel ($\sim$700 $H_{v0.2}$). Thus, the steel ball is worn and smeared onto the Ti–Al–Si(9 at.%)–N film having higher hardness ($\sim$55 GPa). On the other hand, the surface morphology of wear track for Ti–Al–Si(31 at.%)–N film was very smooth, and the width of wear track narrowed again as shown in [Fig. 7.3(c)]. This reflects that the formation of a self-lubricating tribolayer such as SiO_2 or $Si(OH)_2$ was activated on increasing the Si content. From EDS analyses of the wear debris [Figs. 7.3(a) and 7.3(b)], the peak intensity of Fe for Ti–Al–Si(9 at.%)–N film was higher than those for Ti–Al–N film, and the peak intensities of Ti and Al were reversed for the two films. Similar EDS results were found between Ti–Al–Si(9 at.%)–N and Ti–Al–Si(31 at.%)–N films. This indicates that the harder film was more wear-resistant against the steel counterpart.

7.2. *Unbalanced Magnetron Sputtering of Ti–Si–B–C–N Coatings*

Figure 7.4 shows the X-ray diffraction patterns of Ti–B–C–N and Ti–Si–B–C–N films on AISI 304 stainless steel substrates with various Si target powers at fixed TiB_2–TiC composite target power of 700 W. At the Si target power of 0 W, the diffraction pattern of Ti–B–C–N film exhibited crystalline hexagonal TiB_2 phase with preferred orientations of (001) or (002)

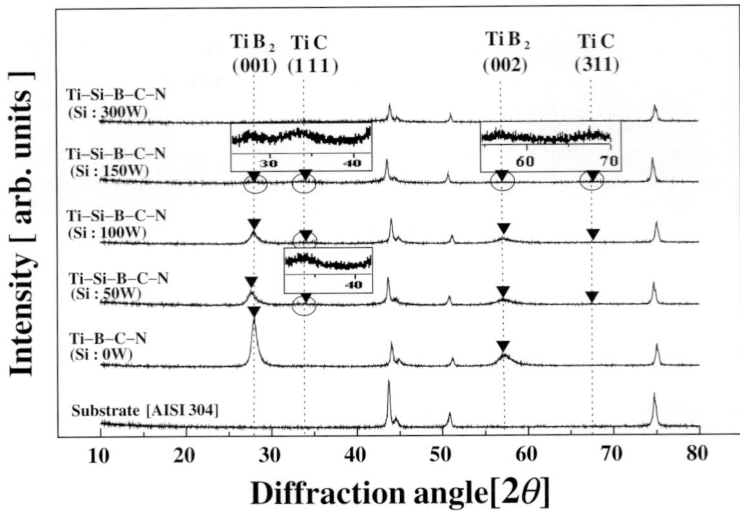

Fig. 7.4.   XRD patterns of Ti–B–C–N and Ti–Si–B–C–N films on AISI 304 stainless steel substrates with various Si target powers at fixed TiB$_2$–TiC composite target power of 700 W.

crystallographic planes. Any XRD peaks corresponding to crystalline TiC or TiN phase were not observed from the Ti–B–C–N diffraction pattern. As the Si target power was increased, the diffraction peak intensities of TiB$_2$ (001) and (002) gradually reduced and completely disappeared at the Si target power of 300 W. At the Si target power of 50 W and 100 W, the TiB$_2$ peaks corresponding to the same (001) and (002) planes as well as a small TiC peaks for (111) and (311) crystallogaphic planes, were present. And, at the Si target power of 150 W, very small diffraction TiB$_2$ peaks for (001) and (002) as well as TiC peaks for (111) and (311) were observed. Furthermore, at the highest Si target power of 300 W, the XRD pattern presented no diffraction peaks for the film, indicating that the film is comprised mainly of an amorphous phase. The gradual changes in the XRD patterns of Ti–Si–B–C–N films with Si additions into Ti–B–C–N are similar to the case of the N addition into Ti–B–C, as previously reported by the authors for the Ti–B–C–N nanocomposite system [25].

In the report, it was revealed that the crystallites in Ti–B–C–N films were composed with solid–solution (Ti, C, N)B$_2$ and Ti(C, N) crystallite ($\sim$10 nm in size). Addition of nitrogen into the Ti–B–C film led to grain refinement of (Ti, C, N)B$_2$ and Ti(C, N) crystallites and their distribution is coupled with a percolation phenomenon of amorphous BN and carbon phase. In order to determine the chemical composition and to investigate the bonding status of Ti–Si–B–C–N coating, X-ray photoelectron spectroscopy (XPS) was performed on Ti–B–C–N and Ti–Si–B–C–N coatings deposited by unbalanced magnetron sputtering from a TiB_2–TiC composite target and a Si target at different Si target powers. Figure 7.5(a) provides the content of each element in the Ti–Si–B–C–N coating with two different Si target powers and a fixed TiB_2–TiC target power of 700 W. As the Si target power was increased, the Si content was increased in the Ti–Si–B–C–N film from 0 to 14.2 at.%. Boron content steeply decreased. Figs. 7.5(b) to 7.6(d) present the XPS spectra binding energies of Si [Fig. 7.5(b)], C [Fig. 7.5(c)], and O [Fig. 7.5(d)] for the Ti–B–C–N and Ti–Si(14.2 at.%)–B–C–N coatings. For Si 2p region [Fig. 7.5(b)], $TiSi_2$ is clearly present with smaller amount of SiC and SiB_4. C 1s region [Fig. 7.5(c)], confirms the

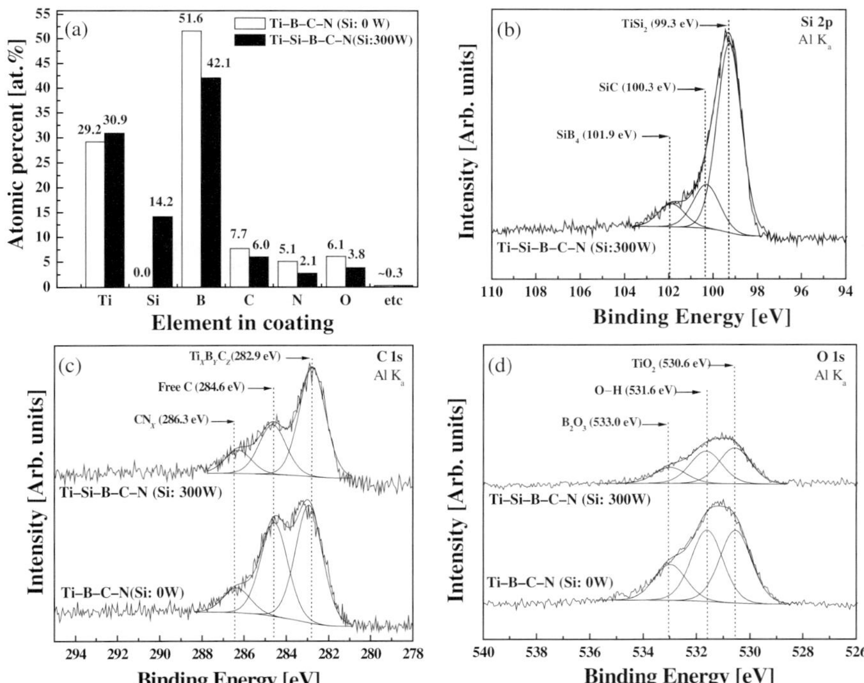

Fig. 7.5.   XPS data for: (a) Content of Ti, Si, B, C, N, and O, and XPS spectra of (b) Si 2p, (c) ls, and (d) O ls for Ti–B–C–N with Si target power of 0 W and Ti–Si–B–C–N coating with Si target power of 300 W (14.2 at.% Si in film).

presence of $Ti_xB_yC_z$ components in amorphous free carbon and CN_x. With addition of Si into the Ti–B–C–N coating, the free carbon peak intensity significantly decreased. The large decrease in the free carbon peak intensity of Ti–B–C–N with the 14.2 at.% Si addition is most likely the result of formation of SiC as shown in Fig. 7.5(b).

Based on the results from the XRD and XPS analyses, it is concluded that the Ti–Si–B–C–N coatings are nanocomposites consisting of nanosized $(Ti, C, N)B_2$ and $Ti(C, N)$ crystallites embedded in an amorphous $TiSi_2$ and SiC matrix including some carbon, SiB_4, BN, CN_x, TiO_2, and B_2O_3 components.

Figure 7.6 presents the nanohardness and H/E values of the Ti–Si–B–C–N coatings as a function of Si target power. The hardness of the Ti–Si–B–C–N coating decreased from $\sim$42 GPa at 0 W Si target power to $\sim$36 GPa at 50 W Si target power. The hardness was constant at about 35 GPa from 50 W to 150 W Si target power and decreased again with the further increase in Si target power to about 25 GPa at Si target power of 300 W (14.2 at.% Si in coating). The decrease in hardness of Ti–B–C–N with 50 W Si target power is most likely due to a reduction in hard TiB_2 based crystallites. The supporting evidence is shown in Fig. 7.5(a), where boron content decreases with increasing silicon content. However, the Ti–Si–B–C–N

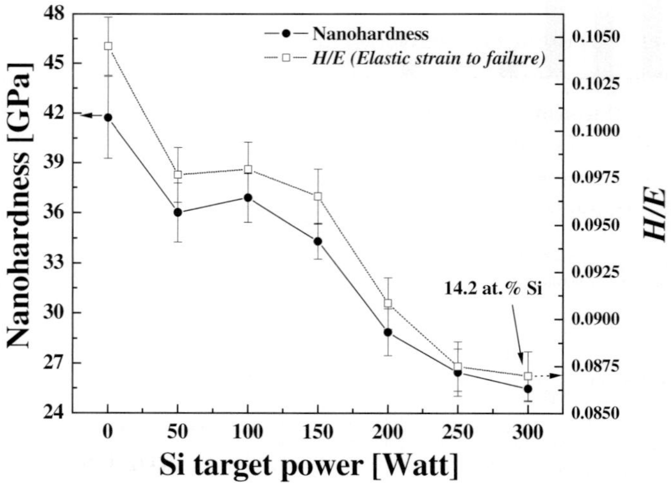

Fig. 7.6.   Nanohardness and H/E values of Ti–Si–B–C–N coating as a function of Si target power.

coating with Si target power of from 50 W to 150 W exhibited a high hardness of about 35 GPa. The reason for maintaining the high hardness ($\sim$35 GPa) in Ti–Si–B–C–N coatings with a small amount of Si is most likely due to percolation of amorphous $TiSi_2$ and SiC in the grain boundaries. Veprek *et al.* [28] have found that ultrahardness (80 GPa $\leq H_v \leq$ 105 GPa) is achieved when the nanosized and/or amorphous $TiSi_2$ precipitate in the grain boundaries in their Ti–Si–N (nc-TiN/a-Si_3N_4/a- and nc-$TiSi_2$) nanocomposites. On the other hand, the hardness reduction with further increase in Si target power above 200 W can be explained by either an increase in the soft amorphous $TiSi_2$, SiC, and SiB_4 phases or reduction of hard TiB_2-based crystallites. When the amount of amorphous phase is increased the ideal interaction between nanocrystallites and the amorphous phase can be lost, and the hardness of the nanocomposite becomes dependent on the property of the amorphous phase [89]. In addition, H/E values, the so-called "elastic strain to failure criterion", were calculated from the obtained hardness (H) and Young's modulus (E). Recently, Leyland and Matthews [57] have suggested that a high H/E value is often a reliable indicator of good wear-resistance. In Fig. 7.6, the H/E values exhibited a similar tendency with hardness. As the Si target power increased, the H/E value of Ti–Si–B–C–N coatings decreased from $\sim$0.105 to 0.087. From the results of hardness and H/E, it can be suggested that the Ti–Si–B–C–N coating with Si target power up to 150 W could provide a better wear-resistance with higher fracture toughness than that of Ti–Si–B–C–N coating with Si target power above 200 W.

Figure 7.7(a) shows the friction coefficients of the Ti–Si–B–C–N coating against a WC–Co ball as a function of Si target power. The average friction coefficient of Ti–Si–B–C–N coating rapidly decreased by increasing the Si target power and showed a minimum value of approximately 0.15 at Si target power of 50 W, and then rebounded with the further increase in the Si target power above 100 W. This large decrease in friction coefficient of Ti–Si–B–C–N coating with the 50 W Si target power is most likely due to the formation of smooth solid-lubricant tribolayers formed by tribochemical reactions during the sliding test. For example, silicon compounds such as $TiSi_2$, SiC, or SiB_4in the coating react with ambient H_2O and oxygen to produce SiO_2 or $Si(OH)_2$ tribolayers. These by-products of SiO_2 and $Si(OH)_2$have been known [87] to play a role as a self-lubricating layer. This Si effect on tribological behavior in nanocomposites, has also been found by other authors [90]. In addition, the carbon

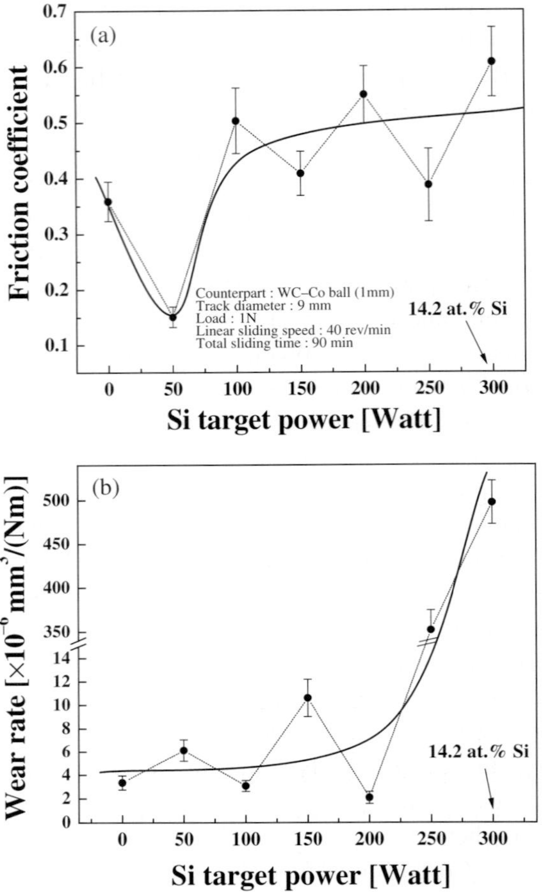

Fig. 7.7.   (a) Friction coefficients, and (b) wear rates of Ti–Si–B–C–N coating against WC–Co ball as a function of Si target power.

and hydroxide phases in the coating, as shown in Figs. 7.5(c) and (d), with a small Si content may also contribute to the minimum friction coefficient value.

However, the friction coefficient slightly increased with further increase of Si content in the coating. The increase in friction coefficient with increase target power is most likely due to a reduction in free carbon and hydroxide phase, which are self-lubricating phases, in the Ti–Si–B–C–N coating as shown in Figs. 7.5(c) and (d). Figure 7.7(b) represents the wear rates of Ti–Si–B–C–N coating on AISI 304 substrates as a function of Si target power. The wear rate of the Ti–Si–B–C–N coating slightly increased from $\sim 3 \times 10^{-6}\,\mathrm{mm^3/(N\cdot m)}$ at 0 W to $\sim 10 \times 10^{-6}\,\mathrm{mm^3/(N\cdot m)}$ at Si target power around 200 W. These very low wear rates would be due to the adhesive wear behavior between the hard coating (~ 35 GPa) and relatively soft WC–Co (~ 22 GPa) ball. On the other hand, at the Si target power above 250 W, the wear rate of the Ti–Si–B–C–N coating steeply increased to about $500 \times 10^{-6}\,\mathrm{mm^3/(N\cdot m)}$ at Si target power of 300 W. This large increase in wear rate would be due to the abrasive wear behavior between the relatively soft coating (~ 25 GPa) and the WC–Co ball. Combining the results of the H/E [Fig. 7.6] values and wear rates [Fig. 7.7(b)], the Ti–Si–B–C–N coating with a higher H/E of above ~ 0.090 had a better wear-resistance against WC–Co ball in agreement with Leyland and Matthews [57].

7.3. Pulsed Closed-Field Magnetron Sputtering of Cr–Al–N Coatings

7.3.1. Characterization of Ion Energy and Ion Flux Change in the P-CFUBMS

As described in the previous section, pulsed DC magnetron sputtering has been used to eliminate arcing problems during the reactive sputter deposition of insulating films, stabilize the discharge and reduce the formation of defects in the film. Besides its primary goal of eliminating arcing, it has been found that applying pulsing in DC magnetron sputtering has significant influence on the ion energy and ion flux change within the plasma. In addition, coating structure and properties will be benefit from additional ion bombardment by the increased ion energy and ion flux. In recent years, time and space-resolved Langmuir probe and electrostatic quadrupole plasma mass spectrometer measurements have been intensively used to investigate the plasma condition in pulsed DC magnetron sputter deposition [91, 92]. Bartzsch *et al.* [93] found that the ion current density was increased at the substrate during pulsed bipolar magnetron sputtering. The ion current increased nearly linearly with sputtering power for both bipolar and unipolar pulsing modes. In addition, the ion current density increased with a decrease in the duty cycle at a constant frequency and sputtering power. Muratore *et al.* [91] compared the ion energy distributions (IED) of pulsed DC magnetron sputtering and conventional DC magnetron sputtering during deposition of TiO films using a Hiden Analytical Ltd., electrostatic quadrupole plasma analyzer (EQP). An increased ion energy up to 140 eV was observed in pulsed DC magnetron sputtering compared with a 17 eV in the DC discharge. Increase of crystallographic texture and an 11% increase in hardness were observed in the TiO films processed with the pulsed magnetron sputtering. Backer *et al.* [94] calculated the ion energy impinging on the pulsed DC magnetron sputtering of Ti and TiO_2 films from the measurements of plasma parameters using a planar Langmuir probe and found that the ion energy values during the positive pulse period increased more than tenfold the time-average value.

The use of high ion energy and ion flux from the pulsed plasma for the reactive sputtering deposition is still a new field. Only a very limited number of coating materials, such as Cr–Al–N [95], TiC [96], TiO_2 [91,97], BN [98], Al_2O_3 [99] have been studied in some detail. More investigation in this field is needed in understanding the nature of the pulsed plasma and its influence on the film structure and properties. One approach to understanding the nature of these states employs both laboratory measurements and theoretical studies aimed at using an EQP to characterize the positive ion energies and the flux change within the plasma. Then it is possible to control ion bombardment energy and ion flux to deposit nanostructure films during deposition of Cr–Al–N thin films using P-CFUBMS. This research program will be described here in an effect to explain using P-CFUBMS to produce nanostructured Cr–Al–N films.

Lin *et al.* [100] first investigated the ion energy and ion flux change in the pulsed closed-field dual-magnetron sputtering of Cr–Al–N films. To demonstrate how the dual-magnetron pulse influences the energy spectra of the ions, the evolution of the IED in response to pulsing only one target and pulsing both targets (Cr and Al) asynchronously and synchronously is considered, see Fig. 7.8(a). A significant increase in ion energy up to 150 eV and 170 eV were observed when both targets were pulsed asynchronously and synchronously at 350 kHz and 1.4 μs reverse time respectively. On the other hand, when only one target is pulsed, the ion energy remained at 20 eV which is similar to that of the threshold of conventional DC sputtering. The low ion energy region also increases up to 30–40 eV when both targets are pulsed compared with the 10 eV value when only one target is operated in pulse mode [Fig. 7.8(a)].

It was also found that a target positive voltage overshoot is developed at the beginning of the positive pulse period and the positive voltage overshoot can result in an increase in the ion energy, which can greatly improve the film structure and properties due to the increased ion bombardment of the substrate [91, 100]. The voltage waveforms on the Cr and Al targets measured when only pulsing one target and pulsing both the Cr and Al

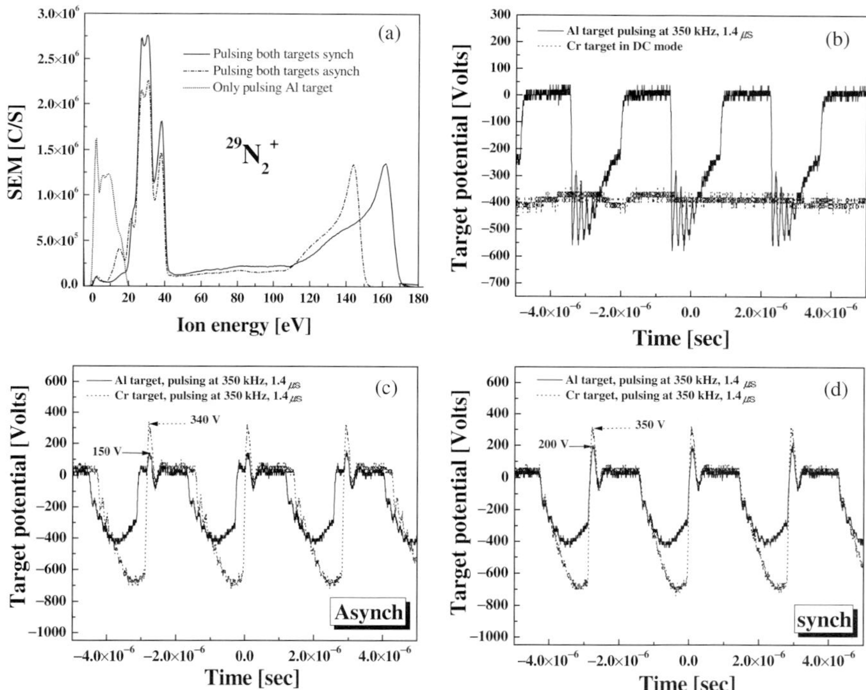

Fig. 7.8. (a) The ion energy distributions of $^{29}N_2^+$ during pulsed discharge for pulsing only Al target and pulsing both targets synchronously and asynchronously at 350 kHz, 1.4 µs, (b) The voltage waveforms of targets for pulsing the Al target at 350 kHz, 1.4 µs and keeping the Cr target in a DC discharge, (c) The voltage waveforms of targets for pulsing both targets asynchronously at 350 kHz, 1.4 µs, and (d) The voltage waveforms of targets for pulsing both targets synchronized at 350 kHz, 1.4 µs [100].

targets asynchronously and synchronously at 350 kHz and 1.4 µs are presented in Figs. 7.8(b), (c) and (d) respectively. In Fig. 7.8(b), when only the Al target is pulsed, the Cr target potential has a constant DC negative value reaching −400 V, and the Al target negative potential pulse is periodically interrupted by the positive pulse voltage, which is about 10% of the nominal negative pulse voltage, 40 V in this case. On the other hand, when both the Cr and Al targets are running in the pulse mode [Figs. 7.8(c) and (d)], steep and large amplitude positive voltage overshoots appear at the start of the positive pulse period in both targets and collapse to the nominal positive pulse voltage value after 200–300 ns. Misina *et al.* [101] demonstrated that the total ion energy at the substrate is mostly determined by the plasma potential. The fast voltage overshoot was observed at the beginning of the positive pulse period and contributed to an increase in the plasma potential and the ion energy. Some physical models have been proposed to explain the effect of the positive voltage overshoot on the increased plasma energy [92]. Moiseev and Cameron [102] proposed that at the beginning of the positive pulse time, a high positive potential is developed on the targets, and therefore a large electron current is extracted from the plasma bulk. Due to the slow ion movement, compared with that of electrons, the equilibrium cannot be reached in a short time period, therefore a negative charge density gradient and a positive charge density gradient will be created in the plasma bulk resulting in an electron field potential being established under this charge gradient. Ions will be accelerated by the electron field potential and will gain energy from it.

In pulsed closed-field dual-magnetron sputtering, the pulsed discharge can not only change the ion energy but also vary the ion flux in the plasma. The frequency and duty cycle of the applied pulses play an important role in determining the ion energy and ion flux because the frequency sets the timescale of the pulsing cycle, while the duty cycle establishes the time within the cycle the pulse is set to positive and negative voltage [103]. As a result, different combinations of pulsing frequencies and duty cycles can change the ion flux and the ion energy values in the plasma, which has a great influence on the deposition process and the film structure and properties. Figures 7.9(a) and (b) exhibit the maximum ion energy evolution ($^{29}N_2^+$) as a function of duty cycle at different pulsing frequencies in the synchronized and asynchronized pulsing modes, respectively. The corresponding integrated ion flux values in the two pulsing modes are plotted in Figs. 7.9(c) and (d) [100].

The ion energy will have higher values when both targets are pulsing at higher frequencies in most cases. In the asynchronized mode, the maximum ion energy decreases with increasing the duty cycle at most frequencies and drops suddenly to the low energy region at certain duty cycles due to the total separation of the two target positive pulse periods [100], see Fig. 7.9(b). The ion flux decreases with an increase in the duty cycle at all frequencies in both the synchronized and asynchronized pulsing modes [Figs. 7.9(c) and (d)]. As the duty cycle decreases (the positive pulse time increases), the cathode is switched proportionally to a positive voltage for a longer period of time, which in turn provides more time for positive ions to stream away from the target. This increased escape time results in a higher number of ions gaining the additional kinetic energy available from the

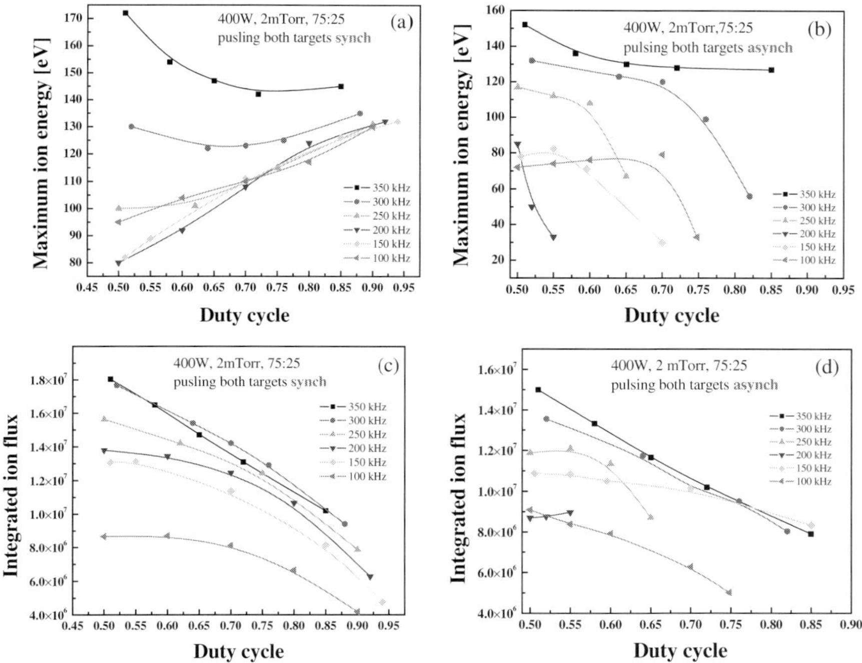

Fig. 7.9.   The maximum ion energy $(^{29}N_2^+)$ as a function of the duty cycle at different pulsing frequencies in (a) synchronized mode, and (b) asynchronized mode; The integrated ion flux as a function of the duty cycle at different pulsing frequencies in (c) synchronized mode, and (d) asynchronized mode [100].

positive potential switch, thereby increasing the flux of higher energy ions at the substrate. The negative voltage overshoot developed on the targets at the beginning of the negative pulse period also contributes to the ion flux change within the plasma. Bradley *et al.* [92] investigated the time evolution of the electron density, the electron temperature, and the plasma potential using a Langmuir probe in pulsed DC magnetron sputtering. It was found that both the effective electron temperatures and densities were elevated during pulse periods compared to the DC equivalent discharge. During the positive pulse time of the cycle, electrons are drawn to the positive potential on the target. When the target voltage switches back to negative, these electrons are now reflected and accelerated back into the plasma, causing a spike in the electron temperature. Under these latter conditions electrons will gain higher kinetic energy from the higher negative voltage overshoot and may also contribute to ionization.

The change of the plasma composition in pulsed reactive magnetron sputtering plays an important role in determining the as-deposited coating composition. During the deposition of Cr–Al–N thin films using P-CFUBMS, the change in the Cr–Al–N coating composition was found to be influenced by the Cr^+, Al^+, and N^+ ion fluxes in the plasma, resulting in improvements in film density, hardness, and stoichiometry [95]. In addition, some other studies [104, 105] have also demonstrated the importance of applying a pulsed DC substrate bias to control the film properties, such as adhesion, texture, morphology, and density of the film.

7.3.2. *Microstructure and Properties of Cr–Al–N Coatings*

The increased ion energy and ion flux in P-CFUBMS as discussed in the previous section will strongly affect the structure and properties of the films. The high ion energy and ion flux will apply additional ion bombardment, increase the mobility of the atoms on the substrate, and reduce the shadowing effect of the columnar structure, thereby changing the film growth microstructure. Figure 7.10 shows the cross-section SEM images of Cr–Al–N films deposited at two ion energy levels by changing the pulsing parameters in P-CFUBMS. For ion energy at 20 eV, which is obtained by pulsing the Cr and Al targets asynchronously at 100 kHz and 1.0 μs, a typical columnar structure with grain size of 60 nm is observed [Fig. 7.10].

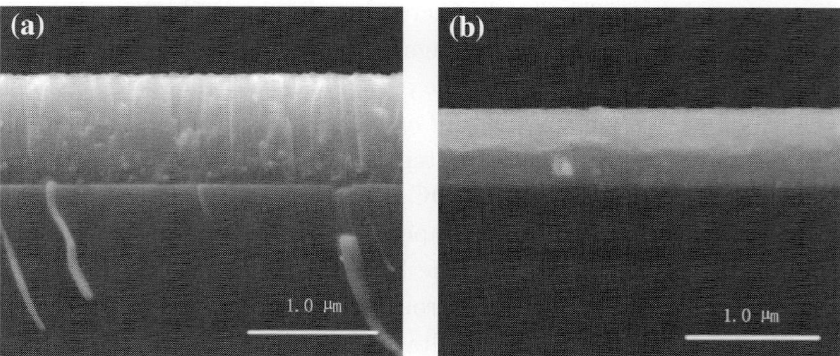

Fig. 7.10. Scanning electron micrographs of cross section of Cr–Al–N films deposited asynchronously at: (a) 100 kHz and 1.0 μs (20 eV ion energy), and (b) 350 kHz and 1.4 μs (150 eV ion energy) [100].

On the other hand, the film deposited at ion energy of 150 eV, which is obtained by pulsing both targets asynchronously at 350 kHz and 1.4 μs, exhibits a denser structure with nanocrystallized and equiaxed grains of less than 60 nm [Fig. 7.10(b)] [100].

In the case when high ion energy and ion flux bombardment are used in the film growth, the grain size and the surface roughness of the films will be reduced. For example, a comparison of the grain size and surface roughness of Cr–Al–N films deposited by P-CFUBMS through an increase in ion flux by sevenfold near the substrate region is presented in Fig. 7.11. Figures 7.11(a) and (b) are the 2D and 3D AFM images of Cr–Al–N film surfaces deposited at relatively low ion flux. The morphology of film indicates separated mounds of grains of relatively large lateral and vertical sizes which indicate a large average grain size and porous structure with a surface roughness of about 23.45 nm. On the other hand, the mounds become smaller (<100 nm), more continuous when the ion flux is increased by sevenfold in the plasma, thereby showing more nuclei and a denser microstructure [Figs. 7.11(c) and (d)] [105].

The high ion energy bombardment created by pulsing in reactive sputtering deposition can result in different film surface growth processes. The most dominant energy impact to the film is the direct implantation of the energetic ions into the near-surface region. The ions carrying high energy can be directly implanted into the voids of the growing film. This process can increase the film density as well as the compressive stress of the film. The energy imparted to the film also results in increase in the film growth temperature, and can thus aid in surface diffusion and relaxation, which can result in a higher defect annealing rate than the defect generation rate in the film, thereby lowering the film internal stress. The temperature at which this occurs depends on several factors such as film material, impinging ion energy. However, as the temperature is increased the thermal stress will also increase [106]. Another high ion energy effect on the film growth is the knock-on implantation, which takes place when the incoming particles collide with the surface atoms of the film. The ions with high energy can knock the surface atoms out of their original lattice positions and move them deep into the voids (interstitial) of the lattice structure of the film, thereby increasing film density. The already bonded atoms in the near surface region of the film may be removed again due to the high energy bombardment, resulting in re-sputtering, which may change the film growth orientation and decrease the deposition rate [Fig. 7.10(b)]. A schematic representation of the surface processes during the ion bombardment in Cr–Al–N films deposition is presented in Fig. 7.12 [107].

7.3.3. *Properties of Cr–Al–N Coatings*

The improved microstructure of films at controlled ion energy and ion flux levels by pulsing the targets results in significant improvement of the mechanical and tribological properties. Figure 7.13 shows the increase in hardness obtained by increasing ion energy from 20 eV to 150 eV in

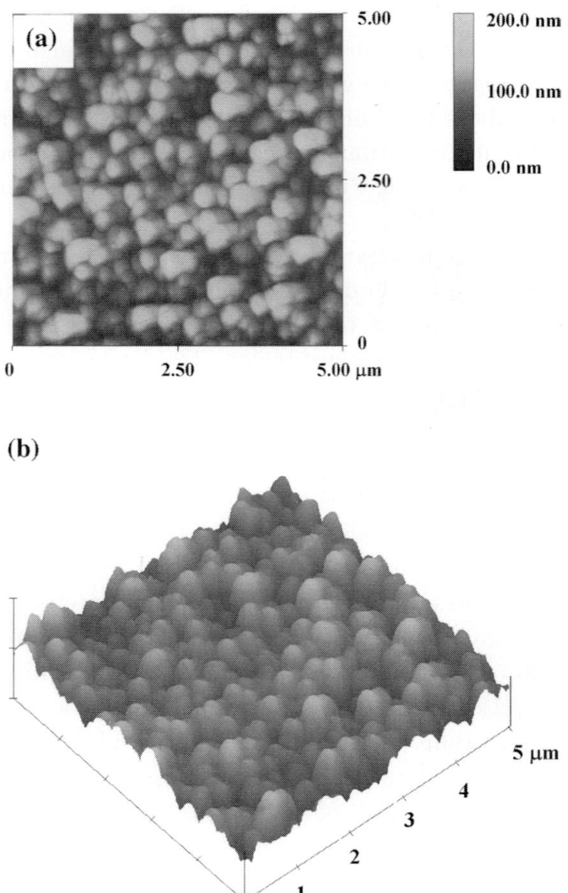

Fig. 7.11. Two- and three-dimensional AFM images of Cr–Al–N films deposited at: (a) and (b) relatively low ion flux, (c) and (d) high ion flux [sevenfold higher than in (a)] [105].

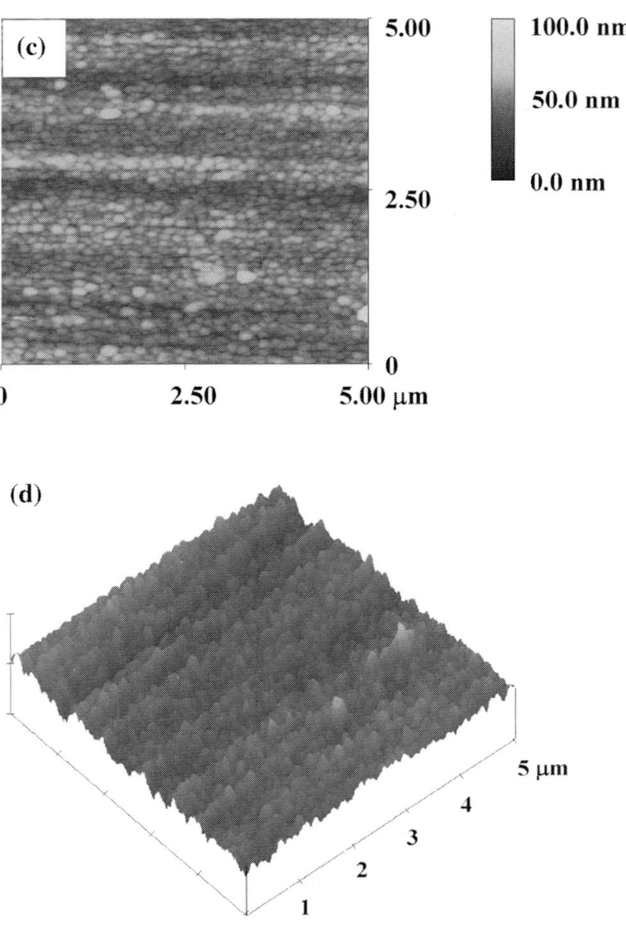

Fig. 7.11. (*Continued*)

Cr–Al–N films deposited using P-CFUBMS system. The different maximum ion energy was achieved by varying the pulsing parameters asynchronously on the Cr and Al targets [100]. The film hardness increased from 34 GPa to 48 GPa when the maximum ion energy in the discharged plasma increased from 20 eV to 150 eV accordingly. The Young's modulus of film exhibits a similar trend.

The improved hardness of the films as using the high ion energy and ion flux bombardment is directly related to the dense microstructure and decreased (nanocrystalline) grain size. According to the Hall–Petch relationship [37, 38], the hardness of the material increases with decreasing

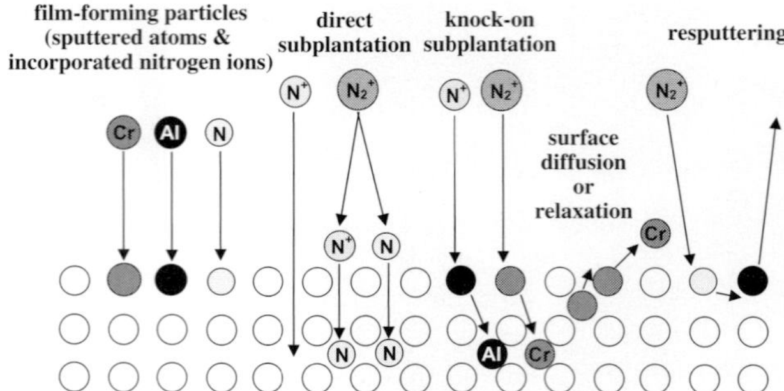

Fig. 7.12.   Schematic drawing of surface processes during film growth [107].

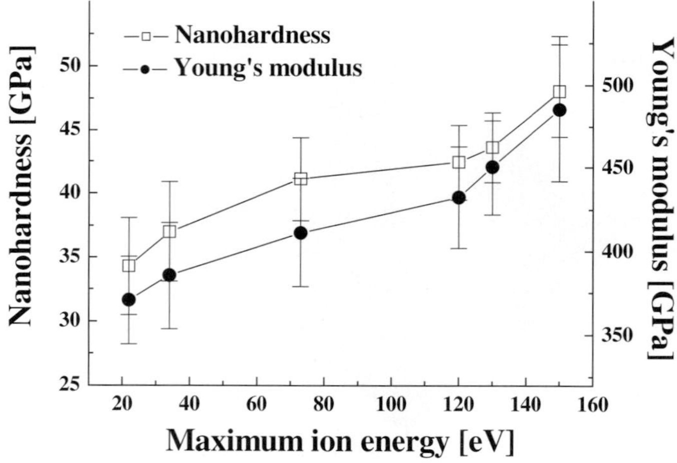

Fig. 7.13.   The hardness and Young's Modulus values of Cr–Al–N films deposited in asynchronized mode as a function of maximum ion energy in the plasma.

crystallite size. The contribution of development of internal residual stress also needs to be considered. The bombarding particles will be implanted in the growing film resulting in increased defect incorporation, provided their energy is high enough. The density of defects such as dislocations, point defects, trapped gases, etc., will strongly affect the stress state of the film. The variation in the in-plane residual stress as a function of the maximum ion energy in the discharged plasma is shown in Fig. 7.14 for P-CFUBMS Cr–Al–N films. The internal stress increases with increased maximum ion energy in the plasma and develops rapidly when the ion energy is between 72 eV to 130 eV. The film deposited with 150 eV ion energy exhibits a high

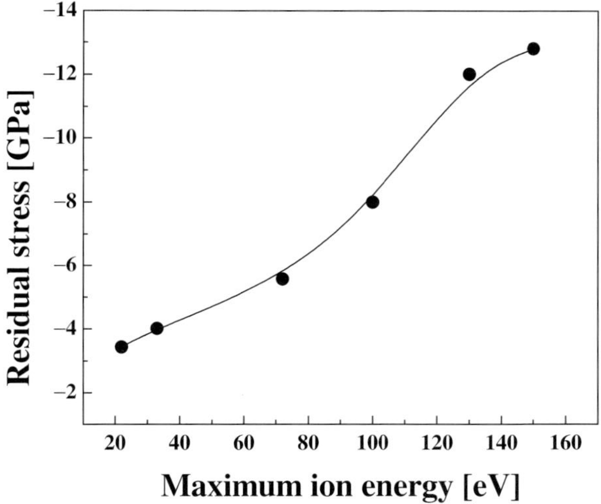

Fig. 7.14. The internal residual stress in Cr–Al–N films as a function of maximum ion energy incorporated in deposition by P-CFUBMS.

residual stress of $\sim$13 GPa. Since stresses result from a change in the average distance between atoms in the lattice away from their equilibrium positions, the elastic response of the material will change. It is understandable that highly compressive stressed materials will show large hardness values. Also, a high defect concentration in a compressively stressed material will restrict the plastic flow, and thus, be a contributing factor in enhancing the hardness [106].

The wear-resistance of hard films is determined by many parameters such as contact geometry, surface roughness, applied load, adhesion, etc. Usually for hard coatings, with a combination of good surface finish and adhesion strength, the wear mechanism is mainly governed by the surface interaction between the tip and the coating itself. On the other hand, if there is a low adhesion strength between the film and the substrate, the adhesion of the film will totally control the wear properties. The steady-state coefficient of friction (COF) value and calculated wear rate of Cr–Al–N films deposited in asynchronized mode with different maximum ion energy are plotted in Fig. 7.15. The COF and wear rate of Cr–Al–N films increased with an increase in the maximum ion energy.

Figure 7.16 are optical photomicrographs of the wear track of Cr–Al–N films deposited at different ion energy values after sliding against 1 mm WC ball at a load of 3 N for 100 meters travel length. The films deposited

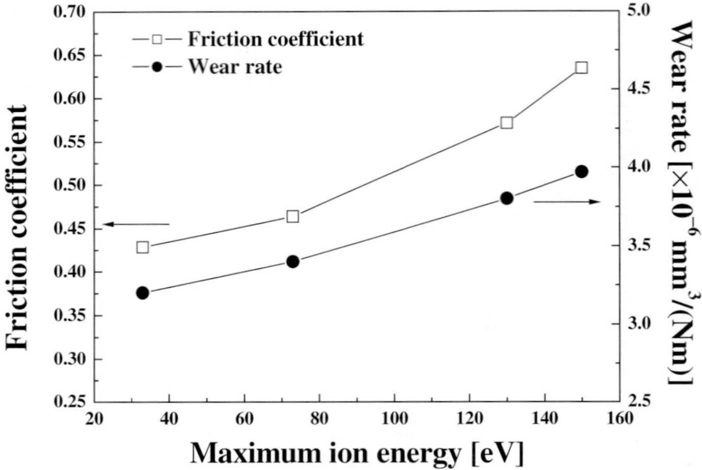

Fig. 7.15.   Friction coefficient and wear rate of Cr–Al–N films deposited in asynchronized mode with different maximum ion energy.

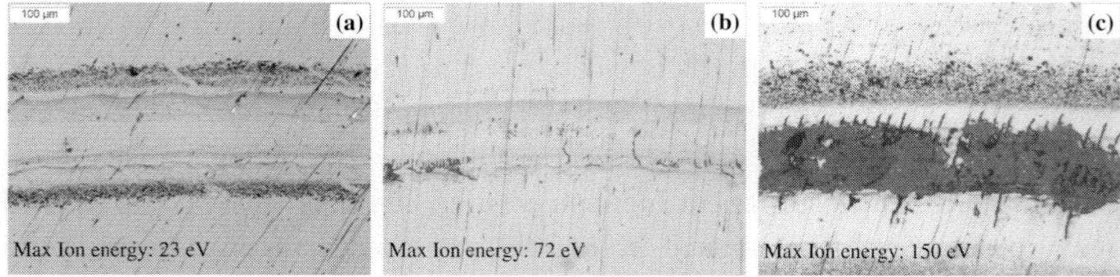

Fig. 7.16.   Optical photomicrographs of wear tracks of Cr–Al–N films deposited at different ion energies after sliding against 1 mm WC ball at a load of 3 N for 100 meters travel length.

with maximum ion energy of 23 eV and 72 eV exhibit good wear-resistance, and there are no obvious defects inside the wear tracks [Figs. 7.16 (a) and (b)]. On the other hand, the films deposited with 150 eV ion energy bombardment exhibit extensively abrasive wear (black area) and cracking in the wear track [Fig. 7.16(c)].

In the case of high ion energy and ion flux assisted deposition, the increased ion energy and ion flux can facilitate the ion bombardment on the substrate, increase the mobility of the atoms on the substrate surface, resulting in a denser film with decreased grain size, and films with super-hardness (Fig. 7.13). The superhard Cr–Al–N films deposited at high ion energy and ion flux (hardness above 40 GPa) can change the wear behavior between the film and the WC ball. Since the hardness of the WC ball (16–20 GPa) is much lower than the superhard Cr–Al–N films (48 GPa in this case), the main abrasive wear body can switch from the coating to the WC ball, and this was confirmed from SEM observations of the WC ball after the wear test. Therefore, the increased contact surface between the WC ball and the films due to the ball wear can result in an increase in the COF of the films, nevertheless the film can still have a low wear rate due to the superhardness of the film. However, if the film receives excessive ion energy and flux bombardment, the internal residual stress and concentrations of point and line defects in the films will increase. Excessive high residual stress and point defect concentration can reduce the film toughness and increase the brittleness of the film, thereby reducing the film wear-resistance. Therefore, care should be taken to control both the ion energy and ion flux during film deposition in P-CFUBMS to achieve both high hardness and wear-resistance in the film. The high in-plane stress built up with excessive ion bombardment can be converted to a stress gradient by depositing compositionally graded structure between the substrate and film, and varying the pulsing conditions throughout the deposition processes.

8. Concluding Remarks

A number of multicomponent, nanostructured coatings have been prepared and summarized for a range of tribological applications. The relationship among processing, structure, properties, and performance of nanostructured coating systems: Ti–Al–N, Ti–Al–Si–N, Ti–B–C–N, Ti–Si–B–C–N, Cr–Al–N produced by various deposition processes, such as unbalanced magnetron sputtering (UBMS), hybrid coating system of cathodic arc evaporation (CAE) and magnetron sputtering (MS), pulsed closed-field unbalanced magnetron sputtering (P-CFUBMS), and high-power pulsed DC magnetron sputtering (HPPMS) was discussed with concepts of design for their industrial application in detail. In each case, the coating system needs to exhibit a range of tribological properties to meet the required application, such as high wear-resistance, low friction coefficient, self-lubrication, high oxidation and/or corrosion resistance. In particular, the effect of the film chemistry, pulsing the magnetron and utilizing a closed-field configuration was discussed as suitable control parameters in tailoring the structure and properties of the coatings to meet specific tribological applications.

References

1. A.A. Voevodin, J.S. Zabinski and C. Muratore, *Tsinghua Sci. Technol.* **10**(6) (2005) 665.
2. S. Veprek, A. Niederhofer, K. Moto, T. Bolom, H.D. Mannling, P. Nesladek, G. Dollinger and A. Bergmaier, *Surf. Coat. Technol.* **133–134** (2000) 52.
3. D. Pilloud, J.F. Pierson and L. Pichon, *Mater. Sci. Eng.* **B131** (2006) 36.
4. A.A. Voevodin, S.D. Walck and J.S. Zabinski, *Wear* **203–204** (1997) 516.
5. I.W. Park, K.H. Kim, J.H. Suh, C.G. Park and M.H. Lee, *J. Kor. Phys. Soc.* **42**(6) (2003) 783.
6. L.S. Palatnik, A.I. Il'inskii and N.P. Sapelkin, *Sov. Phys. Solid State* **8** (1967) 2016.
7. J.S. Koehler, *Phys. Rev.* **B2** (1970) 547.
8. P.J. Kelly, R. Hall, J.O. Brien, J.W. Bradley, P. Henderson, G. Roche and R.D. Arnell, *J. Vac. Sci. Technol.* **A19** (2001) 2856.
9. L.S. Palatnik and A.I. Il'inskii, *Doklady Sov. Phys. Techn. Phys.* **9** (1964) 93.
10. R.W. Springer and D.S. Catlett, *Thin Solid Films* **54** (1978) 197.
11. B.A. Movchan, A.V. Demchishin, G.F. Badilenko, R.F. Bunshah, C. Sans, C. Deshpandey and H.J. Doerr, *Thin Solid Films* **97** (1982) 215.
12. K.K. Shih and D. B. Dove, *Appl. Phys. Lett.* **61** (1992) 654.
13. U. Helmersson, S. Todorova, S.A. Barnett, J.E. Sundgren, L.C. Markert and J.E. Greene, *J. Appl. Phys.* **62** (1987) 481.
14. M. Larsson, P. Hollman, P. Hedenqvist, S. Hogmark, U. Wahlström and L. Hultman, *Surf. Coat. Technol.* **86–87** (1996) 351.
15. K.M. Hubbard, T.R. Jervis, P.B. Mirkarimi and S.C. Barnett, *J. Appl. Phys.* **72** (1992) 4466.
16. P.B. Mirkarimi, L. Hultman and S.A. Barnett, *Appl. Phys. Lett.* **57** (1990) 2654.
17. P.B. Mirkarimi, S.A. Barnett, K.M. Hubbard, T.R. Jervis and L. Hultman, *J. Mater. Res.* **9** (1994) 1456.
18. H. Holleck and H. Schulz, *Thin Solid Films* **153** (1987) 11.
19. H. Holleck and V. Schier, *Surf. Coat. Technol.* **76–77** (1995) 328.
20. K. Yamamotoa, S. Kujimeb and K. Takahara, *Surf. Coat. Technol.* **200** (2005) 435.
21. P. Robinson, A. Matthews, K.G. Swift and S. Franklin, *Surf. Coat. Technol.* **62** (1993) 662.
22. R. Riedel, *Handbook of Ceramic Hard Materials* (Wiley, New York, 2000).
23. A. Cavaleiro and C. Louro, *Vacuum* **64**(3) (2002) 211.
24. E. Martinez, R. Sanjinés, A. Karimi, J. Esteve and F. Lévy, *Surf. Coat. Technol.* **180–181** (2004) 570.
25. I.W. Park, K.H. Kim, A.O. Kunrath, D. Zhong, J.J. Moore, A.A. Voevodin and E.A. Levashov, *J. Vac. Sci. Technol.* **B23**(2) (2005) 588.
26. M. Stuber, H. Leiste, S. Ulrich, H. Holleck and D. Schild, *Surf. Coat. Technol.* **150** (2002) 218.

27. A.A. Voevodin, J.P. O'Neill, S.V. Prasad and J.S. Zabinski, *J. Vac. Sci. Technol.* **A17**(3) (1999) 986.

28. S. Veprek, P. Nesladek, A. Niederhofer, F. Glatz, M. Jilek and M. Sima, *Surf. Coat. Technol.* **108–109** (1998) 138.

29. Y.H. Lu, P. Sit, T.F. Hung, Haydn Chen, Z.F. Zhou, K.Y. Li and Y.G. Shen, *J. Vac. Sci. Technol.* **B23**(2) (2005) 449.

30. R. Fella, H. Holleck and H. Schulz, *Surf. Coat. Technol.* **36** (1988) 257.

31. A.A. Voevodin, J.M. Schneider, C. Rebholz and A. Matthews, *Tribol. Int.* **29** (1996) 559.

32. A.A. Voevodin, M.A. Capano, S.J.P. Laube, M.S. Donley and J.S. Zabinski, *Thin Solid Films* **298** (1997) 107.

33. L. Shizhi, S. Yulong and P. Hongrui, *Plasma Chem. Plasma Proc.* **12**(3) (1992) 287.

34. S. Veprek and S. Reiprich, *Thin Solid Films* **268** (1995) 64.

35. R.W. Hertzberg, *Deformation and Fracture Mechanics of Engineering Materials*, 3rd edn. (Wiley, New York, 1989).

36. A. Kelly and N.H. MacMillan, *Strong Solids*, 3rd edn. (Clarendon, Oxford, 1986).

37. E.O. Hall, *Proc. Phys. Soc. London, Sect.* **B64** (1951) 747.

38. N.J. Petch, *J. Iron Steel Inst., London* **174** (1953) 25.

39. A. Lasalmonie and J. L. Strudel, *J. Mater. Sci.* **21** (1986) 1837.

40. E. Arzt, *Acta Mater.* **46** (1998) 5611.

41. J. Schiotz, E.D. Di Tolla and K.W. Jacobsen, *Nature (London)* **391** (1998) 561.

42. J.E. Carsley, J. Ning, W.W. Milligan, S.A. Hackney and E.C. Aifantis, *Nanostruct. Mater.* **5** (1995) 441.

43. A.H. Chokshi, A. Rosen, J. Karch and H. Gleiter, *Scr. Metall.* **23** (1989) 1679.

44. R.W. Siegel and G.E. Fougere, *Nanostruct. Mater.* **6** (1995) 205.

45. M.F. Ashby, *Acta Metall.* **20** (1972) 887.

46. S.C. Lim, *Tribol. Int.* **31** (1998) 87.

47. S.L. Lehoczky, *J. Appl. Phys.* **49** (1978) 5479.

48. S.A. Barnett, Mechanics and Dielectric Properties, *Phy. Thin Films* **17** (1993) 2.

49. P.M. Anderson and C. Li, *Nanostruct. Mater.* **5** (1995) 349.

50. W.D. Sproul, *Surf. Coat. Technol.* **86–87** (1996) 170.

51. M.L. Wu, X.W. Lin, V.P. Dravid, Y.W. Chung, M.S. Wong and W.D. Sproul, *J. Vac. Sci. Technol.* **A15** (1997) 946.

52. S. Veprek, *Thin Solid Films* **317** (1998) 449.

53. W.D. Sproul, *J. Vac. Sci. Technol.* **A12** (1994) 1595.

54. G.E. Dieter, *Mechanical Metallurgy* (McGraw Hill, New York, 1976).

55. M. Marder and J. Finberg, *Phys. Today* **49** (1996) 24.

56. G.M. Pharr, *Mater. Sci. Eng.* **A253** (1998) 151.

57. A. Leyland and A. Matthews, *Wear* **246** (2000) 1.

58. V.V. Lyubimov, A.A. Voevodin, A.L. Yerokhin, Y.S. Timofeev, I.K. Arkhipov, *Surf. Coat. Technol.* **52** (1992) 145.

59. A.A. Voevodin, P. Stevenson, J.M. Schneider and A. Matthews, *Vacuum* **46** (1995) 723.

60. J. Musil, P. Zeman, H. Hruby and P.H. Mayrhofer, *Surf. Coat. Technol.* **120–121** (1999) 179.

61. A.A. Voevodin and J.S. Zabinski, *Compo. Sci. Technol.* **65** (2005) 741.

62. S. Veprek, *Surf. Coat. Technol.* **97** (1997) 15.

63. O. Knoteck and H.G. Prengel, *Surf. Modif. Technol.* **4** (1991) 507.

64. Y. Tanaka, N. Ichimiya, Y. Onishi and Y. Yamada, *Surf. Coat. Technol.* **146–147** (2001) 215.

65. A.M. Merlo, *Surf. Coat. Technol.* **174** (2003) 21.

66. S.C. Seo, D.C. Ingram and H.H. Richardson, *J. Vac. Sci. Technol.* **13** (1995) 2856.

67. A.P. Ehiasarian, P.Eh. Hovsepian, L. Hultman and U. Helmersson, *Thin Solid Films* **457** (2004) 270.

68. W.D. Sproul, P.J. Rudnik, M.E. Graham and S.L. Rohde, *Surf. Coat. Technol.* **43–44** (1990) 270.

69. P.J. Kelly and R.D. Arnell, *Surf. Coat. Technol.* **108–109** (1998) 317.

70. R.D. Arnell and P.J. Kelly, *Surf. Coat. Technol.* **112** (1999) 170.

71. L. Combadiere and J. Machet, *Surf. Coat. Technol.* **88** (1997) 17.

72. E. Forniés, R. Escobar Galindo, O. Sánchez and J.M. Albella, *Surf. Coat. Technol.* **200** (2006) 6047.

73. S. Inoue, Y. Wada and K. Koterazawa, *Vacuum* **59** (2000) 735.

74. R. Cremer, K. Reichert and D. Neuschütz, *Surf. Coat. Technol.* **142–144** (2001) 642.

75. A. Sugishima, H. Kajioka and Y. Makino, *Surf. Coat. Technol.* **97** (1997) 590.

76. A. Belkind, A. Freilich and R. Scholl, *Surf. Coat. Technol.* **108–109** (1998) 558.

77. A. Anders, *Proc. 5th ICCG*, Saarbruecken, 2004.

78. Advanced Energies Inc., Pinnacle^{TM} Plus User Manual, 2002.

79. Advanced Energies Inc., Power supplies for pulsed plasma technologies: State-of-the-art and outlook, Whitepaper.

80. P.J. Kelly, O.A. Abu-Zeid, R.D. Arnell and J.Tong, *Surf. Coat. Technol.* **86–87** (1996) 28.

81. V. Kouznetsov, K. Macak, J.M. Schneider, U. Helmersson and I. Petrov, *Surf. Coat. Technol.* **122** (1999) 290.

82. K. Macak, V. Kouznetsov, J.M. Schneider, U. Helmersson and I. Petrov, *J. Vac. Sci. Technol.* **A18** (2000) 1533.

83. C. Christou and Z.H. Barber, *J. Vac. Sci. Technol.* **A18** (2000) 2897.

84. A.P. Ehiasarian, R. New, W.D. Munz, L. Hultman, U. Helmersson and V. Kouznetsov, *Vacuum* **65** (2002) 147.

85. M. Diserens, J. Patscheider and F. Levy, *Surf. Coat. Technol.* **120–121** (1999) 158.

86. S.H. Kim, J.K. Kim and K.H. Kim, *Thin Solid Films* **420–421** (2002) 360.

87. S. Wilson and A.T. Alpas, *Wear* **245** (2000) 223.

88. J. Takadoum, H. Houmid-Bennani and D. Mairey, *J. Eur. Ceram. Soc.* **18** (1998) 553.

89. J. Patscheider, T. Zehnder and M. Diserens, *Surf. Coat. Technol.* **146** (2001) 201.

90. J. Xu and K. Kato, *Wear* **245** (2000) 61.

91. C. Muratore, J.J. Moore and J.A. Rees, *Surf. Coat. Technol.* **163–164** (2003) 12.

92. J.W. Bradley, H. Backer, P.J. Kelly and R.D. Arnell, *Surf. Coat. Technol.* **142–144** (2001) 337.

93. H. Bartzsch, P. Frach and K. Goedicke, *Surf. Coat. Technol.* **132** (2000) 244.

94. H. Backer, P.S. Henderson, J.W. Bradley and P.J. Kelly, *Surf. Coat. Technol.* **174–175** (2003) 909.

95. K. Bobzin, E. Lugscheider and M. Maes, *Surf. Coat. Technol.* **200** (2005) 1560.

96. J.M. Anton, Investigation of structure and property modifications of titanium carbide-carbon thin films produced by pulsed DC closed field unbalanced magnetron sputtering, Doctor Thesis, Colorado School of Mines, 2005.

97. P.S. Henderson, P.J. Kelly, R.D. Arnell, H. Backer and J.W. Bradley, *Surf. Coat. Technol.* **174–175** (2003) 779.

98. X.Z. Ding, X.T. Zeng and H. Xie, *Thin Solid Films* **429** (2003) 22.

99. A. Belkind, A. Freilich and R. Scholl, *J. Vac. Sci. Technol.* **A17**(4) (1999) 1934.

100. J. Lin, J.J. Moore, B. Mishra, W.D. Sproul and J.A. Rees, *Surf. Coat. Technol.* **201** (2007) 4640.

101. M. Misina, J.W. Bradley, H. Backer, Y.A. Gonzalvo, S.K. Karkari and D. Forder, *Vacuum* **68** (2003) 171.

102. T. Moiseev and D.C. Cameron, *Surf. Coat. Technol.* **200** (2005) 5306.

103. E.V. Barnat and T.M. Lu, *Pulsed and Pulsed Bias Sputtering* (Kluwer Academic Publishers, 2003).

104. A. Mitsuo, S. Uchida and T. Aizawa, *Surf. Coat. Technol.* **186** (2004) 196.

105. J. Lin, B. Mishra, J.J. Moore, W.D. Sproul and J.A. Rees, *Surf. Coat. Technol.* **201** (2007) 6960.

106. R.F. Bunshah, *Handbook of hard Coatings: Deposition technologies, properties and applications*, (William Andrew Publishing, LLC, 2001).

107. S. Ulrich, H. Holleck, J. Ye, H. Leiste, R. Loos, M. Stüber, P. Pesch and S. Sattel, *Thin Solid Films* **437** (2003) 164.

CHAPTER 7

NANOCOMPOSITE THIN FILMS FOR SOLAR ENERGY CONVERSION

Yongbai Yin

Solar Energy Group, Physics School
University of Sydney, NSW 2006, Australia
yyin@Physics.usyd.edu.au

1. Introduction

Nanocomposite thin films are widely used in solar energy conversion applications. There are several important technologies that have been developed for converting solar radiation energy into other conventional energy forms, such as heat, electricity, and hydrogen. Until now, the most promising technologies for solar energy conversion include solar thermal conversion and solar cells electricity generation. For example, in 2004 installed solar thermal systems were equivalent to 9 GWTH. Solar hot water systems counted for most of the solar thermal market. Other applications in solar thermal conversion include solar air conditioning, solar thermal electricity generation, etc. Both solar thermal and solar cells electrical conversion products have had an annual growth rate of approximately 30% in the last 5 years. The growth trend is expected to continue in the next two decades, owing to the momentum to reduce the greenhouse effect and in response to the pressure of increasing energy demand worldwide. In this chapter we will take solar thermal and solar electrical energy conversion as application examples of nanocomposite thin film materials.

2. Solar Thermal Energy Conversion Nanocomposite Thin Films

2.1. *Solar Thermal Energy Conversion Thin Films*

The principle of solar thermal conversion of thin films or coatings relies on the properties of high absorption of solar radiation and low thermal emission at their operating temperatures. Solar radiation is in the wavelength region between 0.3 and 2.5 μm. The thermal radiation region can be generally expressed as an analogue to the well-known black body radiation spectrum. For example, for surfaces less than 300°C the radiation spectrum is mainly in the wavelength range between 5 and 100 μm. The well-separated wavelength regions enable the possibility to absorb solar radiation selectively and to emit much less thermal radiation than a blackbody at the operating temperature, resulting in increase of net thermal energy output.

All surfaces emit thermal radiation unless they are at absolute zero temperature. Figure 2.1 illustrates the solar radiation spectrum and emission spectra of a blackbody surface at different temperatures. In Fig. 2.1 the solar radiation distribution has air mass of 1 (AM1.5), and the blackbody radiation spectra are for temperatures of 100, 300 and 400°C respectively. Ideal solar thermal surfaces absorb the most of incident solar radiation while simultaneously suppressing emittance itself, corresponding to the step function of the reflectance spectrum (dashed line to indicate a step function for absorption spectrum) as shown in Fig. 2.1. The feature of selectively absorbing solar radiation and suppressing thermal radiation is referred commonly as "solar thermal selective surfaces" for the solar thermal energy conversion surfaces.

Mathematically, we can use the performance parameters of solar absorptance and thermal emittance to quantify the selectivity. The solar absorptance and thermal emittance may depend on incident or emitting angle,

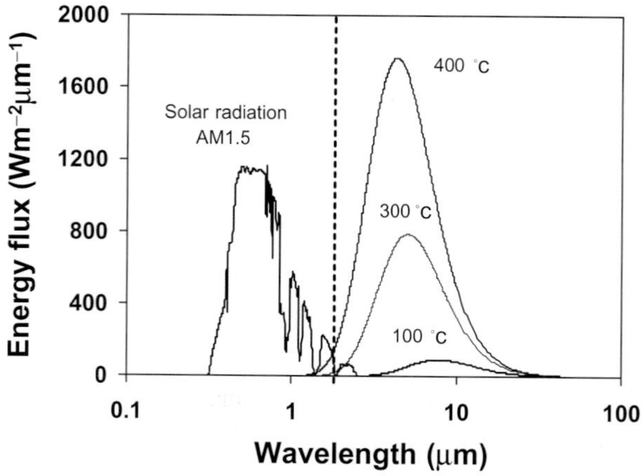

Fig. 2.1. The solar radiation distribution air mass 1.5 (AM1.5) and the blackbody spectra for the temperatures of 100, 300 and 400°C.

wavelength, and operating temperature. The angular dependent solar absorptance α and thermal emittance ε are defined, respectively, by integration of the whole wavelength region:

$$\alpha(\theta) = \int_0^\infty A(\lambda)[1 - R(\theta, \lambda)]d\lambda \Big/ \int_0^\infty A(\lambda)d\lambda, \qquad (2.1)$$

and

$$\varepsilon(\theta, T) = \int_0^\infty E(T, \lambda)[1 - R(\theta, \lambda)]d\lambda \Big/ \int_0^\infty E(T, \lambda)d\lambda. \qquad (2.2)$$

In Eqs. 2.1 and 2.2, $A(\lambda)$ is the solar spectral irradiance, and $E(T, \lambda)$ is the spectral blackbody emissive power. $R(\theta, \lambda)$ is the angular dependent spectral reflectance. Integrating the angular emittance, the hemispherical emittance ε_h is given by

$$\varepsilon_h(T) = \int_0^{\pi/2} \sin(2\theta)d\theta \int_0^\infty E(T, \lambda)[1 - R(\theta, \lambda)]d\lambda \Big/ \int_0^\infty E(T, \lambda)d\lambda. \qquad (2.3)$$

The solar absorptance in general depends on solar radiation angle distribution. Solar radiation at the surface of the Earth consists of mainly two components: direct radiation and diffused radiation. The direct radiation has very little angular dependence, commonly less than $1°$. The diffused radiation is approximately independent of incident angle. The intensity ratio of the direct radiation to the diffused radiation depends mainly on the condition of clouds, particles in the atmosphere, humidity, and the objects on the ground surface. For simplicity, the absorptance of solar selective surface can be calculated by integrating the wavelength dependent absorptance in the wavelength region from 0.3 to 3.0 μm without considering the angular dependence, and the emittance is calculated in the range of 1–50 μm for a given temperature. The photothermal conversion efficiency η can be calculated using the following formula,

$$\eta = \alpha - \varepsilon_\eta \sigma T^4/(CI), \qquad (2.4)$$

where σ, T, C, ε_η, and I are the Stefan–Boltzmann constant, operating temperature, concentration factor, hemispherical emittance, and solar radiation intensity, respectively. The concentrate factor is different from unity when a reflector or mirror is used to concentrate or de-concentrate solar radiation for the solar selective surface. A higher value of C gives higher solar thermal conversion efficiency.

The use of the concept of solar selective surfaces can be traced back as early as the 1950s [1]. Selective solar absorbing surfaces were commonly made of composite thin film materials, which were realized later on to most likely be nanocomposite thin film materials. The first researchers who discovered or reported the nanocomposite feature of solar selective surfaces may have been Niklasson and Granqvist [2]. On the other hand, the optical theory of nanocomposite materials was established early last century. In Sec. 2, we will discuss the theories, development, and current topics of interest in the community of nanocomposite thin films in solar thermal energy conversion.

2.2. *Theories of Nanocomposite and Nanoparticles in Solar Thermal Energy Conversion*

The optical properties of an inhomogeneous (or composite) medium can be described by a complex dielectric function and a complex magnetic permeability, which are both a function of position. In a two-component metal and insulator mixture, the dielectric function has two possible values, ε_m at the metal as the Drude dielectric function with ω_p as the plasma frequency and τ for the relaxation time for electrons, and ε_i at the insulator, which is a frequency independent real number.

The characterization of the inhomogeneous nanocomposite medium (Fig. 2.2) by these two dielectric functions is not straightforward, and requires knowledge of the exact geometrical arrangement of the constituents of the material. However, when the wavelength of the electromagnetic radiation is much larger than the metallic particle size, classical theories of inhomogeneous media presume that the material can be treated as a homogeneous substance with an effective dielectric function and effective

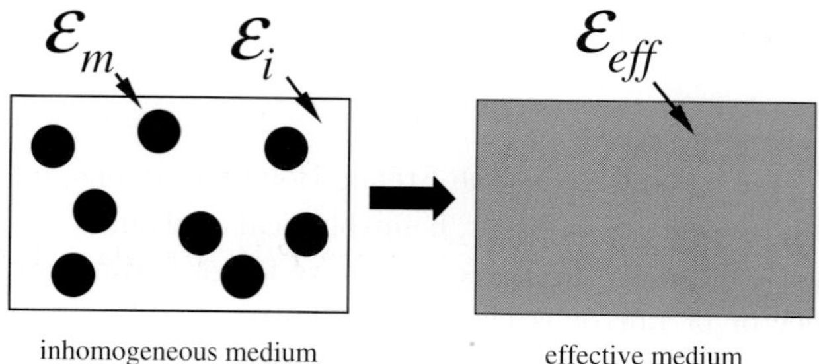

Fig. 2.2.   An inhomogeneous system containing nanoparticles is replaced by an (virtual) effective medium.

magnetic permeability. These quantities depend upon the properties of the constituents, as well as their volume fractions and sizes [3].

The most commonly used effective medium theory (EMT) is the Maxwell–Garnett Theory (MGT), which is also known as the Clausius–Mosotti relation [4]. The MGT theory was developed about 100 years ago. In the derivation of the Maxwell–Garnett theory, the particle size of metal was assumed to be approximately less than one tenth of the wavelength. Solar radiation is peaked at a wavelength of about 550 nm, so that it requires metal particles sizes which are less than 55 nm. For solar thermal energy conversion applications, wavelengths can be as short as about 300 nm. Taking one tenth of the wavelength as the size of the particles to satisfy the requirement of "much larger", the size of metal particles would be less than 30 nm, which means it must be considered as a nanocomposite medium.

The terminology of the "nanocomposite" thin films for solar selective surfaces was not used until quite recently. In the last 30 years, "metal particles", "small particles", and "ultrafine particles" were frequently used, which basically had the same definition as "nanoparticles" today, and the concept of composite medium thin films has effectively the same definition to the concept of "nanocomposite thin films" we are using today and discuss in this book.

Essentially, the MGT theory is a modification of the Lorentz–Lorenz Formula for small particles as the first approach to consider the local field. It resulted from averaging the electric fields and polarizations induced by the applied electric field in the composite medium. The Maxwell–Garnett geometry is shown in Fig. 2.3 and visualizes: the quasi-static approximation — static with respect to the interaction of light with particles if $2R/\lambda \ll 1$; and dynamic with respect to the dielectric properties of the free electrons in the inclusions. The circle with radius R shows the Lorentz cavity.

Another commonly used theory is a symmetric self-consistent model known as the effective medium approximation (EMA) or Bruggeman Theory [5]. At low volume fractions of metallic particles, MGT and EMA lead to very similar results of the effective dielectric constants, but EMA ensures validity at higher volume fill fractions of metal particles since it treats both constituents symmetrically [6]. Below we introduce the approaches that lead to the MGT theory.

For a linear dielectric, the polarization $\bar{P}$ is proportional to the total macroscopic field $\bar{E}_{\text{total}}$:

$$\bar{P} = \varepsilon_0 \chi_e \bar{E}_{\text{total}}. \tag{2.5}$$

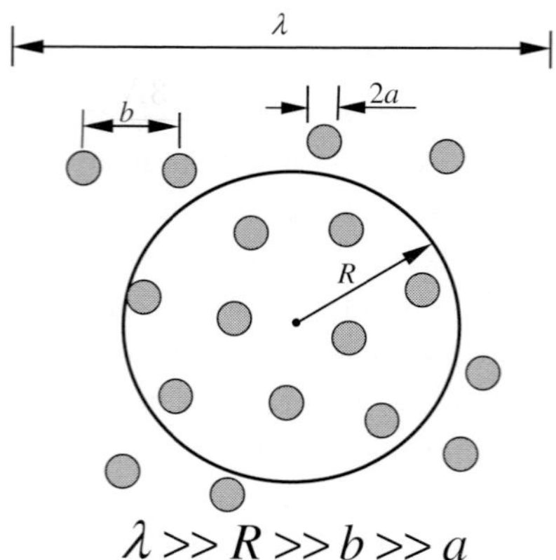

Fig. 2.3.   The geometry of a Maxwell–Garnett composite material.

The polarization of the dielectric is equal to the vector sum of the polarization $\bar{P}$ of the individual atoms or molecules:

$$\bar{P} = N\bar{p}, \tag{2.6}$$

where N is the number of atoms or molecules per unit volume. The polarization of an individual atom or molecule is proportional to the microscopic field at the position of the atom or molecule due to everything except the particular atom or molecule under consideration:

$$\bar{p} = \alpha \bar{E}_{\text{else}}. \tag{2.7}$$

The dipole moment of the atom or molecule will generate an electric field at its center equal to

$$\bar{E}_{\text{self}} = -\frac{1}{4\pi\varepsilon_0}\frac{\bar{p}}{R^3}, \tag{2.8}$$

where R is the radius of the atom or molecule. The total macroscopic field seen by the atom or molecule is therefore equal to

$$\bar{E}_{\text{total}} = \bar{E}_{\text{self}} + \bar{E}_{\text{else}} = -\frac{1}{4\pi\varepsilon_0}\frac{\alpha\bar{E}_{\text{else}}}{R^3} + \bar{E}_{\text{else}} = \left(1 - \frac{1}{4\pi\varepsilon_0}\frac{\alpha}{R^3}\right)\bar{E}_{\text{else}}$$

$$= \left(1 - \frac{N\alpha}{3\varepsilon_0}\right)\bar{E}_{\text{else}}, \tag{2.9}$$

where N is the number of atoms per unit volume. The total polarization of the dielectric is thus equal to

$$\bar{P} = N\bar{p} = N\alpha\bar{E}_{\text{else}} = \frac{N\alpha}{\left(1 - \frac{N\alpha}{3\varepsilon_0}\right)}\bar{E}_{\text{total}} = \varepsilon_0\chi_e\bar{E}_{\text{total}}. \tag{2.10}$$

Therefore,

$$\chi_e = \frac{\frac{N\alpha}{\varepsilon_0}}{\left(1 - \frac{N\alpha}{3\varepsilon_0}\right)} = \frac{3N\alpha}{3\varepsilon_0 - N\alpha}. \tag{2.11}$$

This equation can be rewritten in terms of the dielectric constants K and α as

$$K - 1 = \frac{3N\alpha}{3\varepsilon_0 - N\alpha}, \tag{2.12a}$$

or

$$\alpha = \frac{3\varepsilon_0}{N} \frac{K-1}{K+2}. \tag{2.12b}$$

This equation shows that a measurement of the macroscopic parameter K can be used to obtain information about the microscopic parameter α. This equation is known as the Clausius–Mossotti formula or the Lorentz–Lorenz equation.

The different effective medium theories available all have the same origin, i.e. Maxwell's equations for the static limit. For example, the Bruggeman theorem [5] simply resulted from this fact. The theorem connects the effective dielectric function of a two-phase composite with the geometry information about the composite, described by the spectral function $G(L)$, and the dielectric functions of the two components:

$$\varepsilon_{\text{eff}} = \varepsilon_2 \left(1 - f \int_0^1 \frac{G(L)}{t - L} dL\right), \tag{2.13}$$

where f is the volume filling factor of component 1, and $t = \frac{\varepsilon_2}{\varepsilon_2 - \varepsilon_1}$.

In words, the theorem states that all possible geometric resonances of a two-phase composite occur for real values of the variable t in the interval $[0, 1]$; with the integration over L one scans all possible resonance positions, and whether a resonance occurs or not is determined by the spectral function $G(L)$ which carries all geometry information.

The spectral representation clearly distinguishes between the influence of the geometrical quantities and that of the dielectric properties of the components on the effective behavior of the system. The spectral representation generally holds as long as the quasistatic approximation is valid. No further restriction has to be made.

Though the spectral function $G(L)$ is basically unknown for an arbitrary two-phase nanocomposite, it is analytically known that it can be numerically derived for the existing mixing rule. As each spectral function has to be non-negative, normalized to unity in the interval $[0, 1]$, it should obey (for isotropic systems) the first moment equation

$$\int_0^1 LG(L)dL = \frac{1-f}{3}. \tag{2.14}$$

The derivation of $G(L)$ and the checking of whether these restrictions are fulfilled or not is useful to validate any mixing rule. Note, there are some mixing rules in the literature, which are not correct with respect to the Bergman spectral representation. For example, the spectral function for the mixing rule by C. Böttcher [7].

For

$$\frac{\varepsilon_{\text{eff}} - \varepsilon_2}{3\varepsilon_1} = f\frac{\varepsilon_1 - \varepsilon_2}{\varepsilon_1 + 2\varepsilon_2}, \quad G(L) = 3\delta(L) - 2\delta\left(L - \frac{1}{3}\right) \tag{2.15}$$

which fulfils only the normalization restriction.

In the following, we give some examples of the mixing rules:

1. Bruggeman model [5]:

$$f\frac{\varepsilon_1 - \varepsilon_{\text{eff}}}{\varepsilon_1 + 2\varepsilon_{\text{eff}}} = (f - 1)\frac{\varepsilon_2 - \varepsilon_{\text{eff}}}{\varepsilon_2 + 2\varepsilon_{\text{eff}}}. \tag{2.16}$$

2. Maxwell Garnett model [4]:

$$f\frac{\varepsilon_1 - \varepsilon_2}{\varepsilon_1 + 2\varepsilon_2} = \frac{\varepsilon_{\text{eff}} - \varepsilon_2}{\varepsilon_{\text{eff}} + 2\varepsilon_2}. \tag{2.17}$$

3. Looyenga model [8]:

$$\varepsilon_{\text{eff}}^{1/3} = f\varepsilon_1^{1/3} + (1 - f)\varepsilon_2^{1/3}. \tag{2.18}$$

Experimentally determining the optical constants of nanocomposite thin films is always desirable. Though theoretically the optical constants of nanocomposite thin films can be determined experimentally, many issues were become significant for accurately determining the optical constants both technically and theoretically. Ellipsometry is primarily used to measure film thickness and optical constants n and k. In recent years, the development and use of spectroscopic ellipsometry has made this technique attractive. Both n and k vary with wavelength and temperature, and are needed to describe real materials as the solar thermal selective surfaces are used at different temperatures. Here n is the refractive index of a sample, which gives us the propagation speed of the wave through the sample and direction of propagation. The k is the extinction coefficient which relates how much of the energy of the wave is absorbed in a material. These two values vary with wavelength and together form the complex refractive index:

$$\tilde{n}(\lambda) = n(\lambda) + i \cdot k(\lambda). \tag{2.19}$$

The values n and k describe the material's responses to light. There is an equivalent definition that describes the interaction of light with the material. Alternatively, it can be described using the complex dielectric function:

$$\varepsilon(\lambda) = \varepsilon_1(\lambda) + i \cdot \varepsilon_2(\lambda). \tag{2.20}$$

ε_1 is the volume polarization term for induced dipoles and ε_2 is the volume absorption related to carrier generation. These equations are just different ways of defining the same information. One can convert between the two forms by using the equations below:

$$n = \sqrt{\frac{\left[\varepsilon_1 + \sqrt{(\varepsilon_1^2 + \varepsilon_2^2)}\right]}{2}}, \quad k = \sqrt{\frac{\left[-\varepsilon_1 + \sqrt{(\varepsilon_1^2 + \varepsilon_2^2)}\right]}{2}}. \tag{2.21}$$

Optical absorption of nanocomposite materials is a function of their optical constants. In general, n is a non-zero constant and k can be close to or equal to zero. Increasing the k value from zero, nanocomposite materials are firstly becoming more and more absorbing in the solar radiation region, then the absorption decreases in this region when further increasing the k value. A similar trend can be found in the infrared region with a different k value for the maximum absorption. Figure 2.4 illustrates the dependence, using a nanocomposite material 30 nm thick and with $n = 2.0$.

Applying Bruggeman theory, Yin and Collins [9] simulated solar thermal selective surfaces using molybdenum and aluminum oxide composite materials. Figure 2.5 shows the simulated spectra for two optimized cases of metal volume fraction profile: graded profile and two-layer profile. The graded profile has a profile where the metal volume fraction increases continuously from the top surface to the substrate. The two-layer profile has two uniform cermet layers with an antireflection dielectric layer at the top; the first cermet layer deposited near the substrate has a higher metal volume fraction than that of the layer deposited sequentially. Excellent solar absorption selectivity is shown in the figure.

Solar thermal selective surfaces using Al_2O_3/Mo nanocomposite material were deposited using the co-evaporation of electron beam technique [10]. Examples of high performance solar selectivity surfaces were obtained as shown in Fig. 2.6.

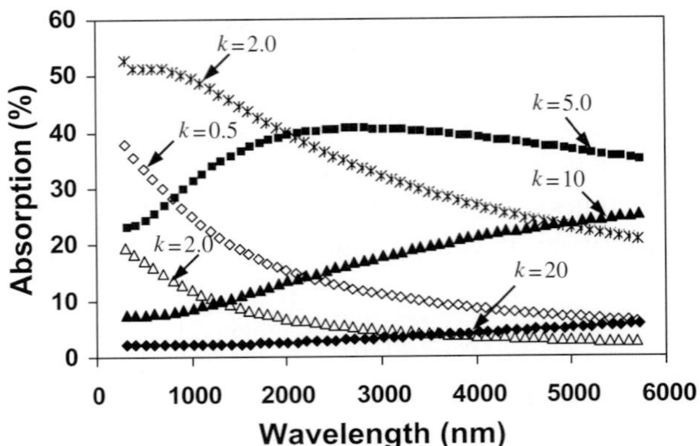

Fig. 2.4.   Absorption spectra of a free-standing 30 nm thick nanocomposite thin film as a function of extinction coefficient, assuming the refractive index n is a constant and equal to 2.0.

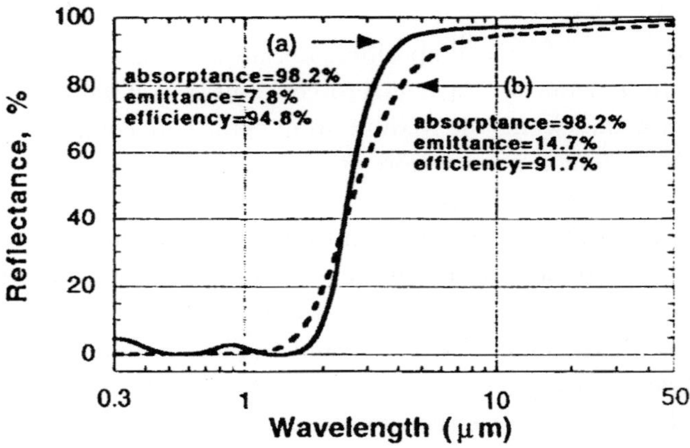

Fig. 2.5.   The simulation results of optimized reflectance spectra for the Al_2O_3/Mo selective surfaces (a) with two-layer of nanocomposite materials; and (b) with a graded nanocomposite profile [9].

2.3. Complications in Nanocomposite Thin Film Materials in Solar Thermal Selective Surfaces: The Effects of Particle Size, Shape, and Orientation

Many researchers have investigated the optical properties of materials by incorporating nanoparticles of metals, such as gold, silver, and nickel [11, 12]. It appears that the first paper to discuss or to address the concept of

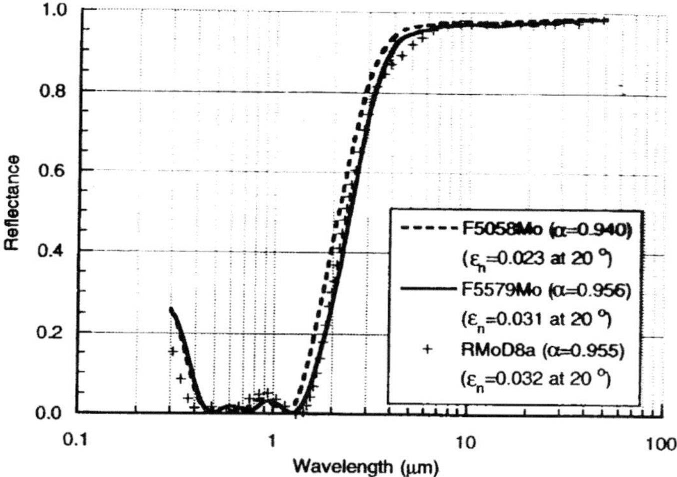

Fig. 2.6.   A near normal reflectance spectrum (+) for a deposited film (FMoD8a) with a two-layer of Mo-Ai203 cermet structure. Two calculated normal reflectance spectra are also included for comparison, the dashed line (F5058Mo) and the solid line (F5579Mo), with the same basic film structure: a low Mo metal volume fraction ($f = 0.34$) layer on a high Mo metal volume fraction ($f = 0.53$) layer with a Cu metal thermal reflector and a 80 nm Al203 anti-reflection coating [10].

nanocomposite materials in solar thermal selective surfaces was by Niklasson and Granqvist [2]. Using transmission electron microscope, they measured chrome particle size, embedded in chrome oxide, ranging between 4 and 15 nm. Similar diameters of chrome nanoparticles in chrome oxide matrix were also reported by Fan [13]. The dimension of nanocrystalline particles, similar to their findings, in solar selective surfaces was commonly known in other solar selective materials, for example, nickel particle in nickel oxide matrix [14].

Both MGT and EMA assume the embedded particles are very small and spherical. This simple approach was widely adopted since the beginning of the development of solar selective surfaces, and the complication of nanocomposite thin films in solar thermal surfaces was not investigated until recently. The apparent driving force for investigating these effects was the discrepancy between prediction and experimental observation of metal volume fraction.

Most notably it was found that the predicted metal volume fraction was often lower than the actual fraction observed in experiments, as reported for example, by Yin and Collins [9] and Zhang *et al.* [10]. In this experiment, the molybdenum metal particles were co-evaporated and embedded in aluminium oxide dielectric. Quartz thickness monitors were used to monitor deposition rates of molybdenum and aluminium oxide respectively by using a partition in the vacuum chamber to prevent seeing the other counterpart evaporation source during the co-evaporation. The measured value of metal volume fraction was found to be as much as 50% higher than the predicted value using the Bruggeman model. Similar phenomena were also observed by the authors using other metals.

It was well understood that solar cermet selective surfaces were based on the optical properties of nanocomposite materials. However, for a long period of time the understanding of the optical proprieties of nanocomposite materials was limited to the effect of metal particle volume fraction, as originally expressed in the Maxwell–Garnett equation, for example, and to the assumption that the particles consisted of very small and spheres. In recent years, a number of attempts were made to determine the effects of particle size and shape on the optical properties. We will now discuss the progress made below in this direction.

When the size of the embedded metal particles becomes larger (more than 50 nm), it is necessary to consider the effect of multiple scattering of light in the materials in the visible and near infrared. Arancibia-Bulness *et al.* [15] studied the effect of particle size theoretically, using the Lorenz–Mie scattering theory for particle sizes from 50 to 130 nm. They found when the particle size increases, the absorption and scattering cut-off shifts to higher wavelengths, which is accompanied with a reduction in the maximum of the scattering and absorption efficiencies.

Scaffardi and Tocho [16] studied the size dependence of the refractive index of gold nanoparticles in the range of a few nanometers, which can be regarded as a nanocomposite material of gold in the medium of air or as a low refractive index medium such as a solvent. They obtained the extinction spectra of spherical gold nanoparticles suspended in a homogeneous media. This work discussed the separated contribution of free and bound electrons on the optical properties of particles and their variation with size for gold nanoparticles. The effects of dielectric function and its changes with size on extinction spectra near plasmon resonance were considered. For the bound electron contribution, two different models were analyzed to fit the extinction spectra. On the one hand, the damping constant for the inter-band transitions and the gap energy were used as fitting parameters, and on the other, the electronic density of states in the conduction band was made size-dependent. For the first case, extinction spectra corresponding to particles with radius $R = 0.7\,\mathrm{nm}$ were fitted using two sets of values of the energy gap and damping constant: $E_g = 2.3\,\mathrm{eV}$ and $\gamma = 158\,\mathrm{meV}/\hbar$ or $E_g = 2.1\,\mathrm{eV}$ and $\gamma = 200\,\mathrm{meV}/\hbar$. For the second case, a simple assumption for the electronic density of states and its contribution to the dielectric function in terms of size allowed adjustment of extinction spectra for all samples studied (from 0.3 to 1.6 nm radius). This last case uses only one parameter, a scale factor $R_0 = 0.35\,\mathrm{nm}$, that controls the contribution of the bound electrons of the nanoparticles in the composite. Contrast between the maximum and the minimum in the extinction spectra near the resonance at 520 nm or alternatively the broadening of the plasmon band can be used to determine the size of gold nanoparticles with radii smaller than 2 nm.

Simulation was conducted using the dynamical Maxwell–Garnett expression, which is based on the simple Maxwell–Garnett equation with an empirical parameter for convenience of consideration of particle size, shape, and orientation [17, 18]. G.B. Smith *et al.* [19] analyzed the effect of columnar structured nanocomposite cermet materials on solar spectral response. They concluded that the simple MG model and its variety of extensions may be erroneous for particle shape predictions. It is possible in nanocomposite cermet to model data with these expressions that give similar appearance and mathematical structure. However, the parameters used in fitting are not those commonly used models. However, since then little progress has been made in understanding in details of the effect of nanoparticle shape.

The effects of particle size and shape may not be able to account for all these differences between the simple MGT model as well as its extensions and experimental findings. There are a few important factors worth great attention. Firstly, the microstructure and crystalline orientation of the nanocrystals or nanoparticles may depend on deposition parameters and the host dielectric medium. The microstructure and crystalline orientation can affect the optical properties of the nanoparticles. Secondly, the interface between the embedded particles and the host dielectric medium may not be ideal as assumed by Maxwell–Garnett theory. Take the example of molybdenum metal particles co-evaporated and embedded in aluminium oxide dielectric, the interface layer between the molybdenum and aluminium oxide may actually be a layer of oxygen deficient molybdenum oxide and oxygen deficient aluminium oxide respectively. Thirdly, when metal volume fraction becomes large, the separation between particles is small, eventually being connected to each other, and the range of particle size will likely become much wider.

The contacts between adjacent nanoparticles form larger "particles". When this occurs, the physical mechanisms which govern the optical constants become more complicated. Recently, a numerous of studies were conducted on this matter. For example, in [20] the dependence of optical constants on crystal size was studied in well-connected metallic thin films. This is an important direction for understanding optical properties of nanocomposite thin film materials.

It is a challenge to study the optical constants of the nanoparticles in composite thin films. For this reason, in the MGT model and its extensions, the optical constants of metallic particles are often assumed to be those of bulk materials, in particular, in the wide wavelength range for solar thermal selective surfaces. Due to the recent availability of infrared spectroscopic ellipsometry, characterization of the effect of nickel film thickness on its optical properties in the infrared region was attempted [20]. In that work, the infrared optical constants of magnetron sputtered nickel thin films were studied as a function of film thickness from a few to a few tens of nanometers, nanocrystalline size, orientation, roughness, and electrical properties. Transmission electron microscopy (TEM) analysis was also conducted, in which it was found that a thin layer of amorphous nickel film was present at the interface with the silicon (001) substrate. The authors [20] concluded that this thin layer of amorphous nickel contributed to the difference in optical constants from that of thick nickel films. The TEM photo is shown in Fig. 2.7.

The band near the interface between the silicon and nickel film is about 6 nm thick. Similar sized nanocrystals on the top of the band are easily identified. The 6 nm thick band of the nickel amorphous layer indicated a strong surface interaction at the initial deposition of the thin layer. In solar thermal cermet selective surfaces, the metallic particles may be as small as a few nanometres. The results from the infrared ellipsometry and TEM analysis thus suggested that it is questionable to use optical constants of bulk nickel for nanosized nickel thin films or particles in solar thermal selective surfaces.

It is advantageous for solar thermal selective surfaces to operate at medium or high temperatures in air to simplify solar energy system design and to reduce the cost. Prolonged heating of solar selective surfaces in air at medium or high temperatures can result in changes of particle size, orientation, metal volume fraction, particle shape, etc. Further understanding of the effects of such changes on the optical and mechanical properties is required.

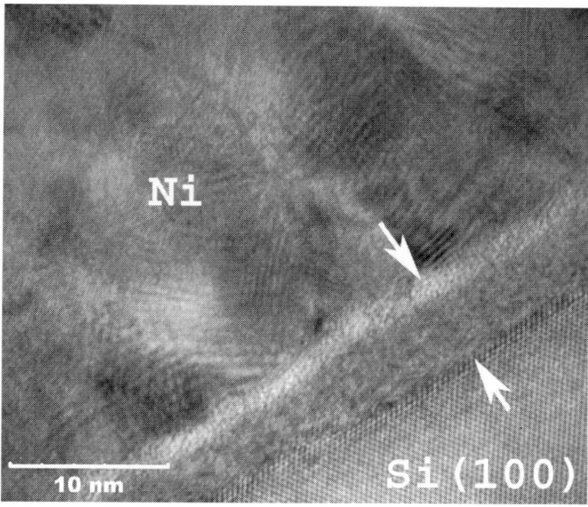

Fig. 2.7.   TEM cross-section image of nickel film on a silicon substrate (the amorphous band near the silicon surface is about 6 nm thick).

3. Nanocomposite Thin Films in Solar Electrical Energy Conversion

3.1. *Photovoltaic Solar Electricity Generation*

Solar electricity generation is based on one of the most important discoveries of the last century on photoelectrical phenomenon, which is called photovoltaic (PV) effect and is the principle mechanism of solar cells. The first conventional photovoltaic silicon cells were produced in the late 1950s, and throughout the 1960s were principally used to provide electrical power for earth-orbiting satellites. In the 1970s, improvements in manufacturing, performance and quality of PV modules helped to reduce costs and opened up a number of opportunities for powering remote terrestrial applications, including battery charging for navigational aids, displays or signs, telecommunications equipment and other critical low power needs. In the 1980s, photovoltaics became a popular power source for consumer electronic devices, including calculators, watches, radios, lanterns and other small battery charging applications. Following the energy crisis of the 1970s, significant efforts began to develop PV power systems for residential and commercial uses both for stand-alone, remote power as well as for utility-connected applications. During the same period, international applications for PV systems to power rural health clinics, refrigeration, water pumping, telecommunications, and off-grid households increased dramatically, and remain a major portion of the present world market for PV products. In the last few years, the PV industry had an average annual growth rate over 30%. Over the last decade, there has been a rapid improvement in solar electrical energy conversion efficiencies as shown in Fig. 3.1.

A typical silicon PV cell is composed of a thin wafer consisting of a layer of phosphorus-doped (N-type) silicon on top of a thicker layer of boron-doped (P-type) silicon, as shown in Fig. 3.2. An electrical field is created near the top surface of the cell where two materials are in contact, which is called the P–N junction. When sunlight radiates the PV cell, this electrical field provides momentum and direction to light-stimulated electrons and holes, resulting in a flow of net current when the solar cell is connected to an electrical load.

A typical crystalline silicon solar cell produces about 0.5–0.6 Volts at open-circuit or no-load conditions. The current and power output depends on the efficiency and size (surface area exposed to solar radiation) of the solar cell, and is proportional to the intensity of sunlight at the surface of

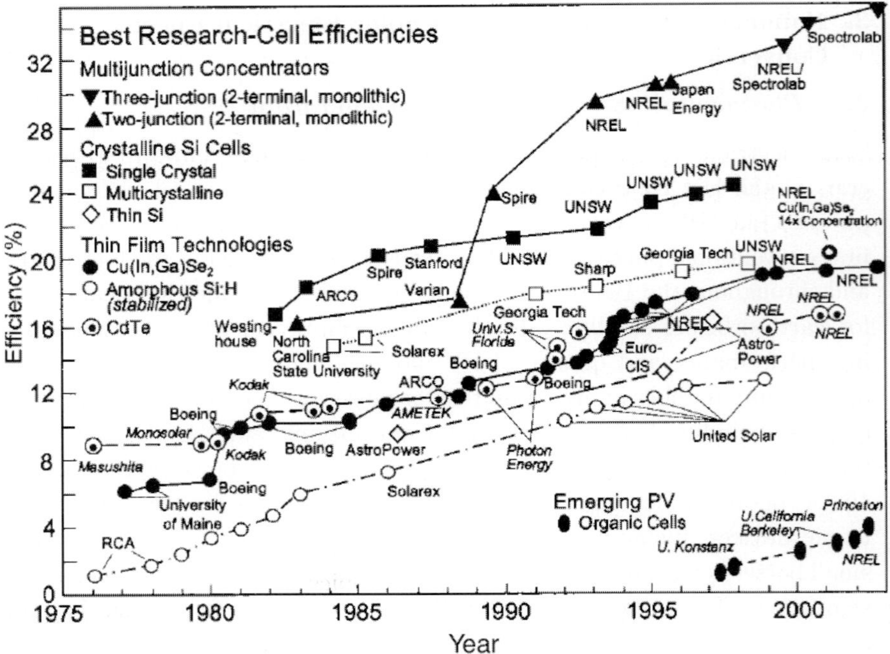

Fig. 3.1. Historical events of reported best solar electrical energy conversion efficiencies of different type of solar cells. The data labels are the organisations, which reported the data [21].

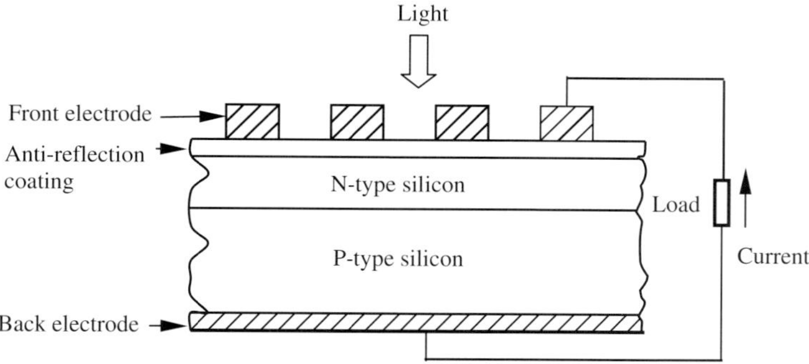

Fig. 3.2.   Schematic diagram of a silicon solar cell.

the cell. For example, under peak sunlight conditions a typical commercial PV cell with a surface area of $100\,cm^2$ and an efficiency higher than 20% will produce more than 2 watts.

Solar cells can be classified as crystalline (or multicrystalline) solar cells, thin film solar cells, and hot-carrier junction solar cells. The crystalline (or multicrystalline) solar cells generally do not consist of nanocomposite materials except for antireflection thin film coatings and metallic contact materials. Thin film solar cells are usually based on nanocomposite thin films, although it is desirable to grow larger sized crystals in order to reduce recombinations caused by boundary defects of smaller crystals. This is a well-studied topic for some solar cell materials. For example, simulation was done by Kazmerski *et al.* [22] on the effects of grain boundaries in $CuInSe_2$ thin film solar cells, showing that minority carrier lifetime reduces at about the same order of magnitude as the diameter of grain size.

However the advantage of a very low recombination coefficient in larger crystal size solar cells is often a trade-off with their commonly lower absorption coefficient so that a thicker film is needed in order to achieve the same solar absorption. For example, in bulk silicon solar cells the required silicon material is about two orders of magnitude more than thin film silicon solar cells, which increases the cost and creates the issue of market availability of silicon material for large quantity production of bulk silicon cells. For this reason, one of the important research topics in thin film solar cells is to nanocrystallize the solar absorbing thin film materials, which we will discuss further in the next section.

Another kind of "thin film" solar cells are dye-sensitized solar cells (DSSC), as noted in Fig. 2.7 as organic solar cells. Dye-sensitized solar cells are relatively new, and are also called bulk heterjunction solar cells. The heterjunction of these kind of cells is not a flat or planar interface as in thin film or bulk solar cells. The materials are prepared from nanoparticles and connected to an electrode while the counter electrode is connected through a medium contacting the surfaces of these nanoparticles, taking advantage of the large surface area. For example, a dye-adsorbed nanoporous TiO_2 (formed using nanoparticles) is prepared on transparent conductive oxide, then immersed in an electrolyte containing a redox couple and placed on a platinum counter electrode. Here, the energy conversion efficiency is determined by (1) light harvesting efficiency, (2) charge injection efficiency, (3) electron transport and collection efficiency in TiO_2 electrodes, and (4) hole transport and collection efficiency in electrolytes. As for light harvesting efficiency and charge injection efficiency, a series of ruthenium dyes with carboxylic groups show wide absorption and high injection efficiency.

Hot-carrier junction solar cells were proposed recently, but do not yet have a prototype cell produced. A hot-carrier junction solar cell splits the cells into many layers to selectively convert photons into electrons or holes with approximately the same energy of the photons. The activated "hot" electrons or holes pass through the interfaces of the layers or between "quantum" dots by quantum hopping without loss of energy in the interfaces or the thin layers.

The advantages of nanocrystals or nanocomposite materials in photovoltaic devices are still far from well understood. The natural advantage in thin film solar cells is that nanocrystalline thin film provides better electronic transportation properties than amorphous materials. For this reason, efforts were made on obtaining larger sized crystals than "nano" size. The compromises for this approach include the thickness limitation and reduction of absorption coefficient of large sized crystals. In the following section of this chapter we will discuss the nanocomposite properties to cover thin film solar cells, dye-sensitized cells, and the "hot carrier" concepts.

3.2. Nanocomposite Materials in Thin Film Solar Cells

3.2.1. Hydrogenated Nanocrystalline Silicon Solar Cells

There are many different thin films solar cells, among them, amorphous hydrogenated silicon solar cells (a-Si:H) are one of the most well-studied solar energy conversion devices. a-Si:H based solar cells are successfully mass produced in production of thin film photovoltaic solar panels. The advantages of a-Si:H based thin film solar cell over conventional crystalline silicon solar cell are: (1) use of large area inexpensive substrates such as glass, polyimide, and stainless steel foils, (2) use of very thin a-Si:H or silicon germanium alloy (a-SiGe:H) layers deposited by glow discharge methods, (3) capability of automatic controlled mass production such as roll-to-roll deposition, and (4) feasibility of building integration. As this technology is not affected by the shortage of silicon raw material faced by the crystalline silicon (c-Si) solar cell manufacturing, the manufacturing capacity of a-Si:H solar panels has increased dramatically in the last a few years. However, the shortcomings of lower efficiency and light-induced degradation caused by the Staebler and Wronski effect [23] in a-Si:H based solar cells may limit the reduction of manufacturing cost and further penetration of a-Si:H solar panels in the market. With current technologies, a-Si:H single-junction solar panels show an initial aperture–area efficiency of $\sim$8–8.5% with about 30–40% light-induced degradation in conversion efficiency. A spectrum split triple-junction cell structure with a-SiGe:H in the middle and bottom cells can increase the initial cell efficiency and improve the stability against light soaking [24, 25]. Currently, the a-Si:H/a-SiGe:H/a-SiGe:H triple-junction solar panels in the market shows an initial aperture-area efficiency of $\sim$10% with $\sim$15% light-induced degradation.

In order to improve the conversion efficiency and the stability of a-Si:H based solar cells, hydrogenated microcrystalline (μc-Si:H) or nanocrystalline (nc-Si:H) silicon solar cells have attracted significant attention in the last ten years since their first invention in the early 90s by the group at the University of Neuchâtel in Switzerland [26–29]. In the literature, the terms μc-Si:H and nc-Si:H are used interchangeably without clearly defining their difference. In the early stage, μc-Si:H was more popular for indicating small or tiny crystallites included in the amorphous matrix or tissues. In recent times, however, the term nc-Si:H has been gradually accepted and commonly (or correctly) used in the community due to crystalline size of the order of a few nanometers to a few tens of nanometers. In addition, other nanometer structures show a lot of unique properties and the study of nanometer related phenomena has become one of the most active fields in science and technology. It appears that use of nc-Si:H is more precise and easy to be aligned with other nanotechnologies. Therefore, we use nc-Si:H in the rest of the chapter. In this section, we will review studies in nc-Si:H fabrication, characterization for material structure and transport properties, and progress in solar cell application.

The first report of nc-Si:H appeared in the late 1960s. Veprek and Marecek used a chemical vapor transport method to deposit nc-Si:H films [30]. Later on, radio frequency (RF) 13.56 MHz glow discharges, called plasma enhanced chemical vapor deposition (PECVD) were used [31, 32]. In the early stage, the nc-Si:H materials showed an n-type character with the Fermi energy level close to the conduction band edge, which is probably due to unintentionally incorporated impurities such as oxygen. Therefore, nc-Si:H was first used as doped layers by adding phosphorus for n layer or boron for p layer [33, 34]. Using nc-Si:H-doped layers, especially nc-Si:H p layer, has significantly increased the open-circuit voltage of a-Si:H solar cells and reduced the loss at the tunnel-junction in multi-junction cell structures [23, 34]. With the reduction of the impurity level by using a clean deposition system and purifying the gases used for the deposition, intrinsic or near intrinsic nc-Si:H can be made for the absorbing layer in solar cells [26–28], which is the most important work done by Neuchâtel's group. Using a very high frequency (VHF) glow discharge with a gas purifier in the gas line, they developed solar cells with intrinsic nc-Si:H as an absorption layer. Comparing with a-Si:H deposition, the key parameter for nc-Si:H is the hydrogen dilution ratio defined by the ratio of hydrogen to active gas (SiH_4 or Si_2H_4) flow rates. It is well known that a high hydrogen dilution reduces the deposition rate significantly. Therefore, developing a method for high rate nc-Si:H deposition is a critical issue for nc-Si:H in solar cell application. Currently, VHF glow discharge under a high pressure regime is the most promising deposition technique for nc-Si:H solar cells due to its high deposition rate and good material quality [35, 36]. Conventional RF glow discharge has also been used in deposition of nc-Si:H solar cells, but the rate is not as high as VHF glow discharge for high quality material even in the high pressure regime. Other techniques such as hot wire chemical vapor deposition and microwave glow discharge have also been investigated [37–40]. However, many technical issues with these two techniques still need to be addressed before consideration in large area manufacturing plants.

The structure of nc-Si:H is a mixture of nanometer size crystallites, grain boundaries, and some residual amorphous components. X-ray diffraction and TEM revealed grain sizes in the range from a few nanometers up to 30 nm [41]. Most nc-Si:H films deposited close to the amorphous/nanocrystalline transition show (220) preferential orientation as shown in Fig. 3.3 and others show a (111) orientation or random distribution depending on the deposition condition, especially the hydrogen dilution and ion bombardment to the growth surface during the film deposition. Raman spectra normally show a sharp peak close to $520 \, \text{cm}^{-1}$ from

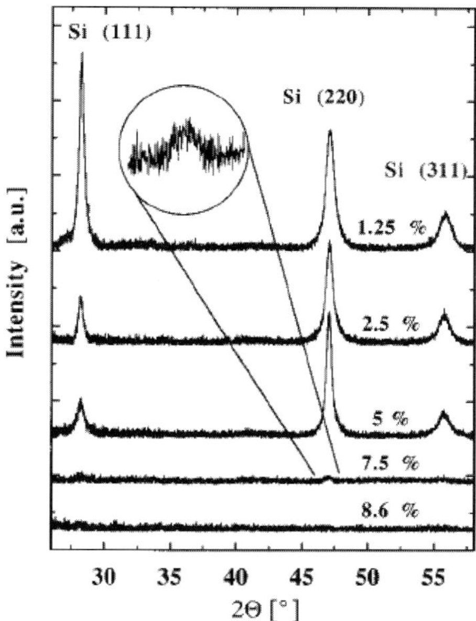

Fig. 3.3.   X-ray diffraction spectra of nc-Si:H as a function of dilution ratio. Reprint from [41].

the nanograins, an inter-medium peak at 490–500 cm^{-1} from the grain boundaries or intermedium orders, and a broad peak at 480 cm^{-1} from the amorphous component [42, 43]. Figure 3.4 shows an example of a Raman spectrum from a device grade nc-Si:H film. From the de-convolution of a Raman spectrum the consideration of the Raman cross sections of each component, in principal, one can obtain the volume fractions for each component. In practice, however, the difference in the Raman cross sections is often ignored due to the complexity and uncertainty in the calculation of the Raman cross sections. The most important phenomena observed by Raman measurement is nanocrystalline evolution [44]. It has been found that the crystalline volume fraction increases with the film thickness. The nanocrystalline evolution has had a remarkable effect on the nc-Si:H solar cell performance and a hydrogen dilution profiling method for controlling this effect has been developed [44].

As in a-Si:H, hydrogen is also incorporated into nc-Si:H films. The average hydrogen content in nc-Si:H films is lower than in a-Si:H films deposited at the same temperature. High quality a-Si:H normally contains 10–15 at.% hydrogen, while high quality nc-Si:H has 4–8 at.%. Since hydrogen atoms

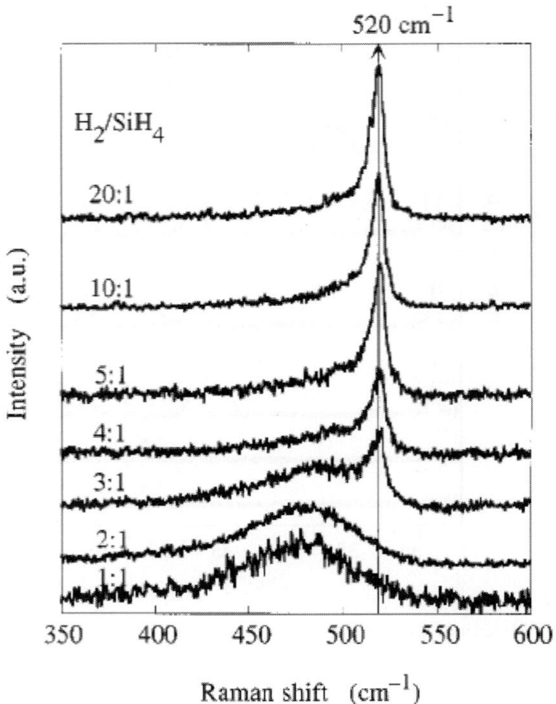

Fig. 3.4. Raman spectra of nc-Si:H as a function of hydrogen dilution ratio.

are mainly in the amorphous phase and at the grain boundaries, the average hydrogen content in these regions could be higher than in conventional a-Si:H films. The hydrogen bonding configuration in nc-Si:H silicon is also different from a-Si:H. In high quality a-Si:H, the main infrared absorption is at 2000 cm^{-1}. A small shoulder at 2100 cm^{-1} corresponding to a dihydride or polyhydride has been proposed as an indication of deterioration in material quality [28]. In nc-Si:H, the main IR absorption peak shifts to 2100 cm^{-1}, which is believed to result from hydrogen atoms bonded in the crystallite surfaces. The hydrogen bonding in the grain boundaries is very important for nc-Si:H quality. In most situations, a poor nc-Si:H quality results from a poor grain boundary passivation [28].

The bandgap of nc-Si:H is slightly larger than in c-Si (1.1 eV) but much smaller than in a-Si:H (1.7–1.8 eV). The optical bandgap obtained from absorption spectra is around 1.2–1.3 eV depending on the crystalline volume fraction. The mobility bandgap estimated from dark current–voltage characteristics of nc-Si:H solar cells is close to the optical bandgap. Since nc-Si:H is a mixture of nanocrystallites and amorphous tissues, the local bandgap may vary from site to site. The most important optical characteristic, the absorption coefficients, are much larger in nc-Si:H than in a-Si:H in the long wavelength region; and are even larger than in c-Si:H [27]. The enhanced long wavelength absorption mainly results from the light trapping effect due to the surface texture and the inhomogeneity inside the material. The high long wavelength absorption coefficients result in higher long wavelength quantum efficiency in nc-Si:H solar cells than in a-Si:H solar cells. However, comparing the absorption in the region above the bandgap of a-Si:H, the absorption coefficients are lower in nc-Si:H than in a-Si:H, which is due to the nature of indirect band absorption in the nanocrystallites, as in c-Si. Therefore, nc-Si:H cells need a much thicker intrinsic layer than a-Si:H cells for a sufficient absorption and a high current density.

The carrier transport properties of nc-Si:H are also important for solar cell applications. The most important parameter is the hole drift mobility (μ_h), which is one of the limiting factors for solar cell performance. In a-Si:H, μ_h is normally 10^{-2}–10^{-3} cm^2/V·s. The increased μ_h (1–10 cm^2/V·s) in nc-Si:H [45] allows a much thicker intrinsic layer in nc-Si:H cells for a high short-circuit current without loss in the fill factor, which is an important parameter for energy conversion efficiency. On the other hand, the μ_h in nc-Si:H is still much lower than in c-Si, which is believed to be due to the trapping in the band-tail states as in a-Si:H. Although most carriers transport in the path of the nanocrystalline phase, the photo-generated carriers still have opportunities to be trapped in or recombined through the tail states and defects in the grain boundaries. The hole mobility measurements show that there is indeed a band tail distribution in nc-Si:H, but the width is smaller than in a-Si:H.

Since its invention, the nc-Si:H cell has attracted remarkable attention for its improved stability against light soaking and its high current capability, which is a good substitute for the a-SiGe:H alloy bottom cell in multijunction solar cells. Basically, two types of cell structure are used, *p-i-n* structure or substrate structure, and *n-i-p* structure or substrate structure. The *p-i-n* or *n-i-p* structure is a sandwich layer formed by placing an intrinsic layer between a *p*-type and a *n*-type layer. The *p-i-n* structures are normally deposited on transparent conductive oxide (TCO), such as doped textured ZnO, which is improving light trapping. The doped *p*-type or *n*-type layer can be a-Si:H or nc-Si:H. A ZnO layer between the *n*-type layer and the back metal (Al or Ag) contact can also reduce the metallic diffusion into the semiconductor layers. The *n-i-p* structures can also be deposited on glass substrates, but are more commonly deposited on flexible substrates such as stainless steel foils and polyimide, on which a textured back reflector with Ag/ZnO or Al/ZnO is used as the back contact to enhance the light trapping.

The dominate factor for nc-Si:H cell performance is the intrinsic nc-Si:H quality. Firstly, very low impurity level is an essential. Because oxygen is a weak *n*-type dopant, most nc-Si:H materials show *n*-type conduction, which causes an electric field collapse in thick nc-Si:H solar cells and results in lower quantum efficiency in the long wavelength region. Secondly, the nc-Si:H should have a compact structure with a very low porosity. It has been found that unoptimized nc-Si:H films with high crystalline volume fraction have a high density of microvoids and microcracks, which causes impurity diffusion into the materials after deposition and exposure to air. Experimentally, an ambient degradation in nc-Si:H cell performance was observed without intentional light soaking, which was contributed to by post impurity diffusion [46]. In addition, nc-Si:H materials with a high crystallinity also show deficiency for hydrogen passivation of grain boundary defects. Therefore, theoretically, nc-Si:H materials with a high crystallinity might produce high current density, but in reality, the best nc-Si:H solar cells are made close to the transition region from amorphous to nanocrystalline [47]. Under this condition the nc-Si:H materials contain a low crystalline volume fraction and small grains with a compact structure.

As mentioned previously, the crystallinity in nc-Si:H increases with the film thickness when a constant deposition condition is used. This means that the initial layer has a very high amorphous volume fraction, and is normally called the incubation layer; while, the final layer has a high crystalline volume faction. The amorphous incubation layer causes problems for carrier transport. To solve this problem, a very high hydrogen dilution is used to produce a so called seeding layer for promoting the crystallite growth. To suppress the nanocrystalline evolution, a hydrogen dilution profiling method has been demonstrated as an effective way for improving the cell performance [44].

In recent years, nc-Si:H cell performance has been improved dramatically. Recently, United Solar Ovonic Corporation in the USA has shown an initial active-area efficiency over 15% using an a-Si:H/a-SiGe:H/nc-Si:H triple-junction structure [48], which is higher than the champion cell with an a-Si:H/a-SiGe:H/a-SiGe:H triple-junction structure [24]. In Japan, Canon and Kaneka Corporation have demonstrated large-area a-Si:H/nc-Si:H/nc-Si:H triple-junction and a-Si:H/nc-Si:H double-junction modules with initial aperture-area efficiency over 13% [49, 50]. It appears that nc-Si:H based solar cells will become an important technology in thin film solar cell manufacturing.

3.2.2. *Other Nanocomposite Thin Film Solar Cells*

Other research-active and promising thin film solar cells may include $CuInS_2$, CuInGaSe, CdTe, TiO_2, and CdS solar cells [51, 52]. All of these use semiconductor nanocomposite materials to form a *p-n* junction.

Marian Nanu *et al.* [53] suggested a so-called three-dimension DSSC. Figure 3.5 illustrates the concept of the nanocomposite $CuInS_2$ solar cell. In this type of nanocomposite, a wide bandgap *n*-type semiconducting oxide and a *p*-type visible light sensitive semiconductor are mixed on a nanometer scale. They employed atomic-layer chemical vapor deposition technique for infiltration of $CuInS_2$ inside the pores of the nanostructured TiO_2. First, a dense film of TiO_2 ($\sim$100 nm) was applied onto the TCO (transparent conducting oxide). On top of this, a 10 μm thick nanoporous TiO_2 film was applied with primary particles between 10 and 50 nm in diameter. Next, these substrates were infiltrated with $CuInS_2$ using atomic-layer chemical vapor deposition. The process conditions were: 2 mbar reactor pressure, temperature between 350 and 500°C, and CuCl, $InCl_3$, and H_2S as precursors. A major concern with these of solar cells is their poor stability if operating in full sunlight. The cells must be sealed rigorously to avoid reaction with oxygen and water.

The interrelation between particle size, crystal structure and optical properties in semiconductor quantum dots has elicited widespread interest. Banerjee *et al.* reported [54] a change in the bandgap due to the effect of the size-induced transformation from a hexagonal to a cubic structure in CdS nanoparticles. CdS nanoparticles with particle size in the 0.7–10 nm

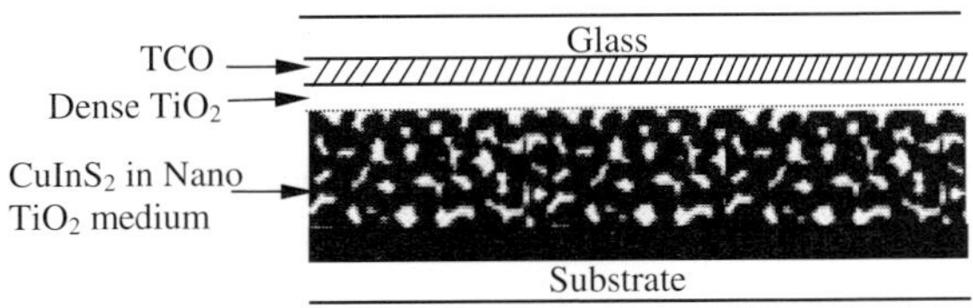

Fig. 3.5. Illustration of a nanocomposite $CuInS_2$ solar cell. The active *p-n* junction is folded and fills the space at the nano TiO_2 and $CuInS_2$ contacts.

range were prepared by chemical precipitation using thiophenol as a capping agent. Whereas the bandgap for bulk hexagonal CdS is about 2.5 eV, for 1 nm cubic CdS nanoparticles it was found to be almost 3.9 eV.

The blue-shift in the bandgap in nanoparticles, due to quantum confinement has the quantitative form [55, 56]:

$$\Delta E_g \equiv E_g^{\mathrm{nano}} - E_g^{\mathrm{bulk}} = h^2/8MR^2, \qquad (3.1)$$

where R is the radius of the particle and M is the effective mass of the system. For hexagonal CdS, $M = 1.919 \times 10^{-31}$ kg. Since CdS-I exhibits a hexagonal structure, Eq. (3.1) could be used to estimate the particle size, using the observed blue-shift in the bandgap, $\Delta E_g = 0.09$ eV. The corresponding particle diameter is 9 nm, which is in very good agreement with the X-ray size (10 nm).

The bulk bandgap and the effective mass are unknown for the cubic Z phase of CdS. Therefore, the optical absorption data for CdS-III and CdS-IV, both of which consist of single phase cubic CdS, were fitted to a simplified form of Eq. (3.2):

$$\Delta E_g = A/R^2, \qquad (3.2)$$

where A is a constant which depends on the effective mass, M. The resulting value of the bandgap for the *bulk* cubic Z phase of CdS is 3.53 eV. This is only a notional value since the cubic phase may never exist in the bulk.

It needs to be pointed out that the optical bandgap of nanosized semiconductor materials does not scale with Eq. (3.2) when the crystal size becomes very small or approaches zero. In amorphous silicon, the optical bandgap is usually about 1.7 eV or slightly higher. This is understood to be due to disordered structures inducing energy states near the edges of the conducting and valence band edges. The measured optical bandgap is determined by the optical excitation and recombination of these states. This topic is beyond the subject of this book. Interested readers may find some useful details in other references, for example, in [57].

Another type of important thin film solar cells is made of CdTe. The bandgap of cadmium telluride has the ideal value of about 1.5 eV and also a high absorptivity, making it as one of the prime materials to use in solar cells. A layer of a few micro-metres can absorb 90% of incident photons. When compared to amorphous silicon thin-film cells, CdTe has the advantage of having high chemical and thermal stability even though a polycrystalline form is used in fabrication. CdTe solar modules are already on the market. CdTe production is fast and robust using an established technology and automated processes. Additionally the production has a high material yield due to strict control of module parameters. There are two main problems associated with CdTe solar cells. The first is that p-type CdTe films tend to have a high electrical resistance. This means that, although an effective absorber of photons, conversion efficiency is compromised. The second problem stems from the fact that cadmium is a heavy metal. There are, therefore, environmental concerns regarding its use both during fabrication and with it end-of-life disposal.

Nanotechnology may provide solutions for CdTe solar cells. Both the high resistance and environmental concerns may be minimized by introducing much less CdTe material. Ernst *et al.* [58] suggested a textured structure by using a layer of CdTe nanocrystal thin film of about 150 nm thickness instead of the conventional thickness of more than $2\,\mu$m. The textured structure is formed first by depositing a conducting nanoporous anatase-phase TiO_2 layer on a flat and commercially available SnO_2-coated glass substrate. The porous TiO_2 was formed using pyrolysis of 3.5×10^{-3} M Ti-tetra isopropolylate in isopropanol at a temperature of $160°$C and a substrate temperature of $180°$C. After the spray-deposition, a final annealing step at $450°$C was applied. The CdTe absorber layer is deposited in a standard three-electrode electrochemical cell from a solution of 0.5 M $CdSO_4$ and 2.5×10^{-4} M TeO_2 in H_2O at a pH 1.6 at $90°$C. The optimum deposition potential for stoichiometric films is $-400\,$mV versus a Pt reference electrode. After deposition, the CdTe film is exposed to hot $CdCl_2$ vapor in methanol for 3 min and then annealed at $400°$C in air for 1 hour. The CdTe grain size is typically $\sim 100\,$nm. Alloying with Hg is achieved by adding $HgCl_2$ to the electrolyte while keeping all other deposition parameters and procedures the same. The back contact of the cell is made by vacuum evaporation of a 150 nm thick Au layer. The efficiency of this type of cell is still relatively low, which is understandable given that the nanostructural properties of this type of solar cell require further investigation.

3.3. *Dye-Sensitized Solar Cells*

The work of Hermann W. Vogel in Berlin in the 19th century can be considered as the first significant study of dye-sensitization of semiconductors, where silver halide emulsions were sensitized by dyes to produce black and white photographic film. Interested readers may go to [59] for useful historical background information. The use of dye-sensitization in photovoltaics remained unsuccessful until a breakthrough in the early 1990s in the Laboratory of Photonics and Interfaces in the EPFL, Switzerland. Combining nanostructured electrodes with efficient charge injection dyes, Grätzel and his coworkers successfully developed a dye-sensitized solar cell with energy conversion efficiency exceeding 7% in 1991 [60] and 10% in 1993 [61]. In contrast to the all-solid conventional semiconductor solar cells, the dye-sensitized solar cell is a photoelectrochemical solar cell, which uses a liquid electrolyte or other ion-conducting phase as a charge transport medium. Due to their significant discovery, the research interest in this technology grew rapidly during the 1990s.

Figure 3.6 is a schematic diagram of a dye-sensitized solar cell, which is comprised of a transparent conducting glass electrode coated with porous nanocrystalline TiO_2 (nc-TiO_2), dye molecules attached to the surface of the nc-TiO2, an electrolyte containing a reduction-oxidation couple such as I^-/I_3^- and a catalyst coated counter-electrode. Under solar radiation, the cell produces voltage over and current through an external load. The absorption of light in the cell occurs by dye molecules and the charge separation by electron injection from the dye to the TiO_2 at the semiconductor electrolyte interface. If non-porous crystalline TiO_2 is used, a single layer of dye molecules at the interface between the TiO_2 and electrolyte is formed, which can absorb only less than one percent of the incoming light. Developed by the Grätzel group, a porous nanocrystalline TiO_2 material was used in order to increase the internal surface area of the electrode to allow a large enough amount of dye to be contacted at the same time by the TiO_2 electrode and the electrolyte [60, 61]. The porous nanocrystalline

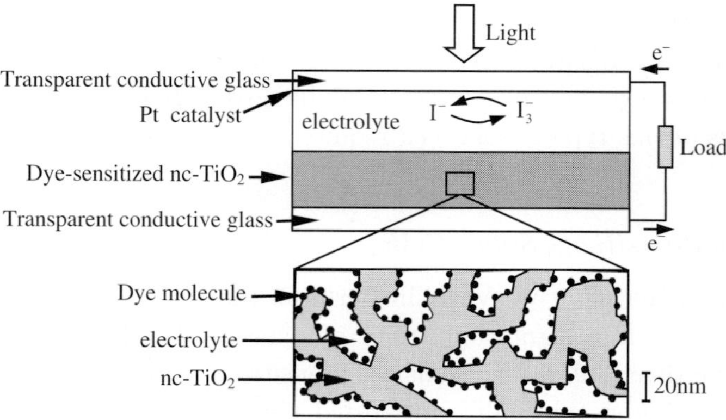

Fig. 3.6. A schematic diagram of the structure and components of the dye-sensitized solar cell.

TiO_2 electrode is typically 10 μm thick with an average particle (as well as pore) size typically in the order of 20 nm and an internal surface area thousands of times greater than the dimension of the cell.

The regenerative working cycle of the dye-sensitized solar cell is demonstrated in Fig. 3.7, showing the relative energy levels of the cell. The incoming photon is absorbed by the dye molecule that is adsorbed on the surface of the nanocrystalline TiO_2 particle. The sequence of the photoelectronic chemistry is: (a) an electron from a molecular ground state S^0 is then excited to a higher energy state S^*; (b) the excited electron is injected to the conduction band of the TiO_2 particle leaving the dye molecule in an oxidized state S^+; (c) the injected electron passes through the porous nanocrystalline material and reaches the transparent conducting oxide layer (anode); (d) at the counter-electrode the electron is transferred to the tri-iodide in the electrolyte to yield iodine; (e) finally the cycle is closed by reduction of the oxidized dye by the iodine in the electrolyte. In this operation cycle, no chemical substances are either consumed or produced. The electron state transfer cycle is summarized below:

$$\text{At anode:} \quad S + h\upsilon \rightarrow S^* \qquad \text{photon absorption} \qquad (3.3)$$

$$S^* \rightarrow S^+ + e^-(TiO_2) \quad \text{electron injection} \qquad (3.4)$$

$$2S^+ + 3I^- \rightarrow 2S + I_3^- \qquad (3.5)$$

$$\text{At cathode:} \quad I^- + 2e^-(Pt) \rightarrow 3I^- \qquad (3.6)$$

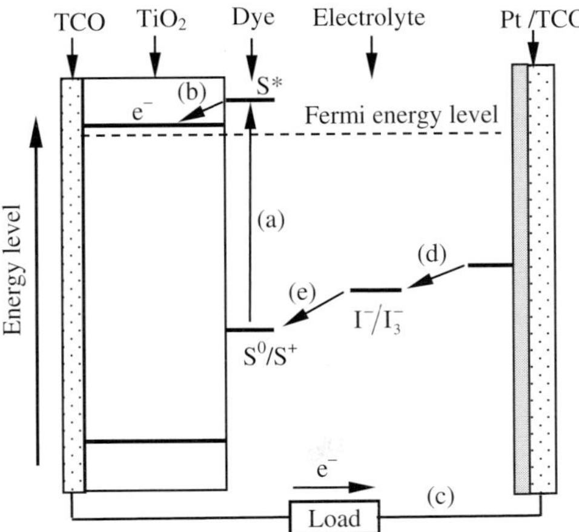

Fig. 3.7. The schematic diagram of the photo-electronic chemistry process of the dye-sensitized solar cell under solar radiation.

The oxide nanocrystalline electrodes in dye-sensitized solar cells are very important because oxide semiconductors are preferential in photoelectrochemistry due to their exceptional stability against photo-corrosion. Furthermore, the large bandgap of oxide semiconductors is advantageous as it provides transparency to a large part of the solar spectrum. In addition to TiO_2, semiconductors used in porous nanocrystalline electrodes in dye-sensitized solar cells include ZnO, CdSe, CdS, WO_3, Fe_2O_3, SnO_2, Nb_2O_5, and Ta_2O_5 [62]. However, using nanocrystalline TiO2 is the main approach for dye-sensitized solar cells.

Two crystalline forms of TiO_2, anatase and rutile, are important for dye-sensitized solar cells. Anatase is a pyramid-like structure and is stable at low temperatures. Rutile structure is needle-like and is preferentially formed in high temperature processes. Anatase has been the main subject of study in dye-sensitized solar cells as it is the primary structure formed in the usual colloidal preparation method of the nanostructured TiO_2 electrodes. Recently it was revealed that dye-sensitized nanostructured TiO_2 electrodes with pure rutile structure showed only fractionally smaller short circuit photocurrents than pure anatase nanocrystalline films with equal open circuit photo voltages [63], indicating further investigation of the effect of TiO_2 nanocrystalline structure on solar energy conversion is needed.

3.4. *Hot-Carrier Junction Nanocomposite Solar Cells*

This is still at this stage fundamental study being done for development of low-cost, thin film solar cells. This research explores the fabrication of semiconductor nanocomposites for photovoltaics using nanostructured materials. The focus is on cells built by high-throughput techniques where multiple junctions, ultrathin layers, and nanoporous structures are used to achieve good energy conversion efficiencies. There is a strong need for the development of photovoltaic cells with low cost, high efficiency, and stability. The broad energy distribution of the solar spectrum creates a fundamental challenge for the development of devices capable of efficient photovoltaic conversion. A multiple junction approach uses stacked cells of different materials so that each different cell converts a portion of the spectrum with an efficiency approaching 70% for single wavelength conversion. In thin film technologies, there exists a common problem with conversion efficiency where the photo-generated electrons and holes can recombine and are hence lost for power conversion. If the solar cell can be made using nanoscale heterojunctions, then the problem of recombination through traps can be greatly reduced. Atomic layer deposition is particularly well suited for this application since it can allow for deposition on complex non-planar structures at the nanoscale level with controllable thickness. With nanoscale diffusion lengths, the material's constraints can be relaxed, and low cost deposition routes become acceptable.

Bulk wafer technologies using the p-n junction design are regarded as first generation solar cells. The manufacturing cost is dominated by the wafer cost and the efficiency is limited by the bandgap that controls the energy available to separate an electron-hole pair. The second generation photovoltaic cell is based on thin film technology in order to reduce the material cost in production and to adjust the bandgap with flexibility of introducing stacked multi-junctions. Both first and second generation solar cells have un-ideal theoretical conversion efficiency. For the next (third) generation photovoltaic cell, higher conversion efficiency and lower production cost with physical principles extending or different from the single p-n junction cell has been targeted. Many new photovoltaic conversion ideas regarded as third generation photovoltaic are discussed elsewhere [64, 65]. Tandem cells, hot carrier cells, multiple electron-hole pairs per photon (such as by up and down conversion), impurity photovoltaic (multiband) cells and thermophotovoltaic cells are some candidates. With the hot carrier cell approach, the theoretical limit can reach the same limit as the infinite tandem approach with an efficiency of 86.8% for direct sunlight and 68.2% for global sunlight [64]. Figure. 3.8 is a schematic diagram of a hot carrier solar cell using quantum dots for the required energy selective contacts as elaborated in [64].

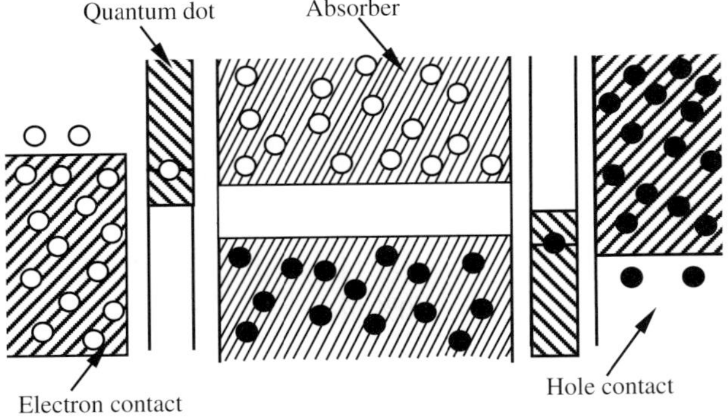

Fig. 3.8.   Schematic diagram of a hot carrier cell.

Hot carrier cells consist of an absorber designed to reduce the carrier cooling and an energy selective contact to collect carriers through a narrow energy range. The major challenge for hot carrier cells is to extract those carriers in this way and to reduce their cooling rate in the absorber, for example, resonant tunneling through defects to produce the required energy selective contacts, the application of silicon quantum dots as engineered defects for such a resonant tunneling device and the extension of silicon quantum dots to a super-lattice structure for all-silicon tandem solar cells are on the list of topics in this area.

Currently, there is also great interest in the optical and transport properties of nano-sized semiconductor particles or quantum dots [66]. They show significant departures from bulk optical and electronic properties when the scale of confinement approaches the Bohr radius, which sets the length scale for optical processes [67]. Quantum dots of II–VI semiconductors have attracted particular attention, because they are relatively easy to synthesize in the size range required for quantum confinement.

The advantage of using nano-sized semiconductor crystals or quantum dots can be schematically illustrated in Fig. 3.9. For 3D bulk semiconductor materials, the dependence of energy states on energy is a smooth and continuous function. In a 2D semiconductor structure, the dependence is split as a few separated smooth and continuous regions. In a 1D structure, the energy states are semi-discrete. When semiconductor materials are limited to zero dimensions (nanocrystals or quantum dots), the energy states become truly

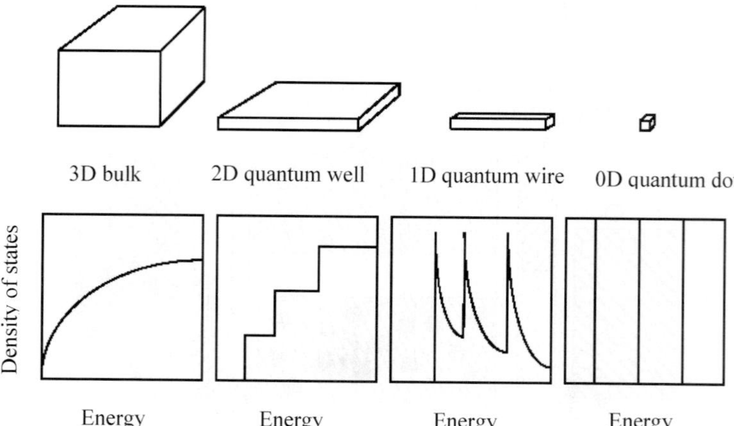

Fig. 3.9.   Schematic diagram of density distribution of energy states for different dimension of a system.

discrete. For this reason, in nanocrystalline silicon particles, the optical and electrical properties depend strongly on the nanoparticle size.

In the case of silicon in an amorphous matrix, quantum dots have the advantage of lower surface area compared to 2D layers and of being single crystalline. An approach based on depositing alternating layers of stoichiometric oxide followed by silicon-rich oxide appears promising in producing silicon quantum dot super-lattices. The approach allows the control of dot diameter and of one spatial coordinate using a potentially low cost process [65]. Figure 3.10 shows a high resolution TEM image of a silicon quantum dot super-lattice [65].

The main factors controlling the per watt price of PV solar power are cost of materials and device efficiency. The high cost that currently prevents widespread application of this technology invariably arises from a shortfall in one of these factors. The vast majority of solar panels today are made of silicon, the workhorse of the semiconductor industry. These devices are called first generation, and make for highly stable and efficient solar cells. Unfortunately, because of the material processing necessary to make these devices, first generation solar cells are inherently expensive.

The Carnot limit on the conversion of sunlight to electricity is 95% as opposed to the theoretical upper limit of 33% for a standard solar cell. This suggests that the performance of solar cells could be improved 2–3 times if different concepts were used to produce a third generation of high

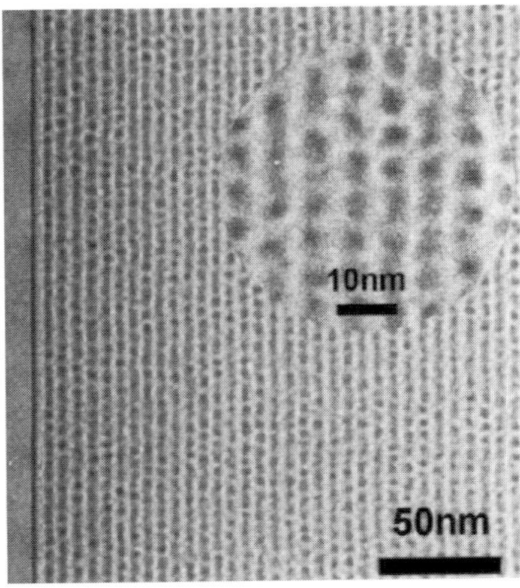

Fig. 3.10.   High resolution image of a silicon quantum dot super-lattice. From [65].

efficiency, low-cost solar cell technologies. A variety of advanced approaches to so-called third generation solar cells are under investigation, but the best scenario would involve a low-cost semiconductor material that could have its bandgap tuned for optimal performance allowing the manufacturer to control the absorptive properties of the solar cell.

4. Summary

Nanocomposite and nanocrystalline thin films have been widely used in solar energy conversion applications. In solar thermal energy conversion applications, the corresponding optical effective medium theory and experimental observation of nanocomposite nature were understood a long time before the recent or current nanoscience revolution. In solar electricity energy conversion applications, we can comfortably conclude that research on nanocomposite and nanocrystalline thin films in this field was one of the main driving forces that resulted in the nanoscience and nanoengineering revolution.

In both solar thermal and solar electricity energy conversions, there are still many unknowns on the nature and the significance of nanocomposite and nanocrystalline thin films. It is apparent that there is a long way to go in order to understand the science of nanocomposite films, to control the formation process of nanocomposite films, and to create new nanocomposite materials for solar energy conversions. In the immediate future, the most important research topics include the effect of nanocrystalline geometric parameters and interface properties on the optical and electronic properties of nanocomposite thin films, methods to control the separation, size and shape of nanoparticles, and to achieve desired properties of nanocomposite thin films by modifying the interface or surface of nanocrystallines. Traditionally, solar thermal thin films and solar electricity conversion solar cells were regarded as different research fields. However, as both fields move into the nanoscale, the gap between the two fields is narrowing.

Acknowledgments

The author greatly appreciates discussion with and assistance from Professor D.R. McKenzie, Professor M.M.M. Bilek, Dr B. Yan, Dr. C. Zhang, Mr S. Allan and Mr Y. Pan.

References

1. H. Tabor, *Sol. Energy Res.* **2** (1958) 3–6.

2. D.A. Niklasson and C.G. Granqvist, Selective absorption of solar energy by ultrafine metal particles, Vol. 2, Proc. Intern. Solar Energy Congr., New Delhi, Jan 16–21 (1978), pp. 870–874.

3. T.C. Choy, *Effective Medium Theory, Principles and Applications* (Oxford University Press, 1999).

4. J.C.M. Garnett, *Phil. Trans. R. Soc. Lond.* **203** (1904) 385–420.

5. D.A.G. Bruggeman, *Ann. Phys. (Leipzig)* **24** (1935) 636–679.

6. B. Wendling, *Preparation and Optical Properties of Mixed Dimensional Gold-Nanostructures*, Masters Thesis (University of Michigan, 2001).

7. C.J.F. Böttcher, *Theory of Electric Polarization* (Elsevier, Amsterdam, 1952), p. 415.

8. H. Looyenga, *Physica* **31** (1965) 401–406.

9. Y. Yin and R.E. Collins, *J. Appl. Phys.* **77** (1995) 6485–6489.

10. Q.C. Zhang, Y. Yin and D.R. Mills, *Sol. Energy Mat. Sol. Cells*, **40** (1996) 43–53.

11. R.W. Cohen, G.D. Cody, M.D. Coutts and B. Abeles, *Phys. Rev.* **B8** (1973) 3689–3701.

12. C.M. Lampert and J. Warshburn, *Sol. Energy Mater.* **1** (1979) 81–92

13. J.C.C. Fan, *Thin Solid Films* **80** (1981) 125–136.

14. S. Zhao, E. Avendano, K. Gelin, J. Lu and E. Wackelgard, *Sol. Energy Mater. Sol. Cells* **90** (2006) 308–328.

15. C.A. Arancibia-Bulness, C.A. Estrada and J.C. Ruiz-Suarez, *J. Phys.* **D33** (2000) 2489–2496.

16. L.B. Scaffardi and J.O. Tocho, *Nanotechnology* **17** (2006) 1309–1315.

17. C.A. Foss, G.L. Hornyak Jr. and C.R. Martin, *J. Phys. Chem.* **98** (1994) 2963–2971.

18. G.L. Hornyak, C.J. Patrissi, C.R. Martin, J. Valmalette, J. Dutta and H. Hofmann, *Nanostruct. Mater.* **9** (1997) 575–578.

19. G.B. Smith, A.V. Radchik, A.J. Reuben, P. Moses, Skryabin I. and S. Dligatch, *Sol. Energy Mater. Solar Cells* **54** (1998) 387–396.

20. Y. Yin, Y.Q. Pan, M.M.M. Bilek, D.R. McKenzie and S. Rubanov, *Opt. Mater.*, in press.

21. T. Surek, Progress in S.S. photovoltaics: Looking back 30 years and looking ahead 20, *Proc. 3rd World Conf. Photovoltaic Energy Conversion* (2003), pp. 2507–2512.

22. L.L. Kazmerski, P. Sheldon and P.J. Ireland, *Thin Solid Films* **58** (1979) 95–99.

23. D. Staebler and C. Wronski, *Appl. Phys. Lett.* **31** (1977) 292–294.

24. J. Yang, A. Banerjee and S. Guha, *Appl. Phys. Lett.* **70** (1997) 2975–2977.

25. A. Banerjee, J. Yang and S. Guha, *Proc. Mat. Res. Soc. Symp.* **557** (1999) 743.

26. A. Shah, J. Dutta, N. Wyrsch, K. Prasad, Curtins, F. Finger, A. Howling and Ch. Hollenstein, *Proc. Mat. Res. Soc. Symp.* **258** (1992) 15–26.

27. J. Meier, R. Fluckiger, H. Keppner and A. Shah, *Appl. Phys. Lett.* **65** (1994) 860–862.

28. J. Meier, P. Torres, R. Platz, S. Dubail, U. Kroll, J.A. Selvan, P.N. Vaucher, Ch. Hof, D. Fischer, H. Keppner, A. Shah, K.D. Ufert, P. Giannoules and J. Koehler, *Proc. Mat. Res. Soc. Symp.* **420** (1996) 3–14.

29. A. Shah, J. Meier, E. Vallat-Sauvain, N. Wyrsch, U. Kroll, C. Droz and U. Graf, *Sol. Energy Mater. Sol. Cells* **78** (2003) 469–491.

30. S. Veprek and V. Marecek, *Solid State Electron.* **11** (1968) 683–684.

31. W.E. Spear, G. Willeke, P.G. LeComber and A.G. Fitzgerald, *J. Phys. (Paris)* **42** (1981) C4–257.

32. G. Lucovsky, C. Wang, R.J. Nemanich and M.J. Williams, *Sol. Cells*, **30** (1991) 419–434.

33. S. Nishida, H. Tasaki, M. Konagai and K. Takahashi, *J. Appl. Phys.* **58** (1985) 1427–1431.

34. S. Guha, J. Yang, P. Nath and M. Hack, *Appl. Phys. Lett.* **49** (1986) 218–219.

35. L. Guo, M. Kondo, M. Fukawa, K. Saitoh and A. Matsuda, *Jpn. App. Phys. Part 2*, **37** (1998) L1116–L1118.

36. Y. Mai, S. Klein, R. Carius, J. Wolff, A. Lambertz, F. Finger and X. Geng, *J. Appl. Phys.* **97** (2005) 114913–114924.

37. S. Klein, F. Finger, R. Carius, B. Rech, L. Houben, M. Luysberg and M. Stutzmann, *Proc. Mat. Res. Soc. Symp.* **715** (2002) 617–622.

38. R.E.I. Schroop, Y. Xu, E. Iwaniczko, G.A. Zaharias and A.H. Mahan, *Proc. Mat. Res. Soc. Symp.* **715** (2002) 623–628.

39. B. Yan, G. Yue, J. Yang, K. Lord, A. Banerjee and S. Guha, *Proc. 3rd World Conf. Photovoltaic Energy Conversion*, May 11–18, 2003, Osaka, Japan, pp. 2773–2776.

40. H. Jia, H. Shirai and M. Kondo, *Proc. Mat. Res. Soc. Symp.* **910** (2006) A13.3.

41. E. Vallat-Sauvain, U. Kroll, J. Meier and A. Shah, *J. Appl. Phys.* **87** (2000) 3137–3142.

42. G. Yue, J.D. Lorentzen, J. Lin and D. Han, *Appl. Phys. Lett.* **75**, (1999) 492–494.

43. D. Han, J.D. Lorentzen, J. Wenberg-Wolf, L.E. McNeil and Q. Wang, *J. Appl. Phys.* **94** (2003) 2930–2936.

44. B. Yan, G. Yue, J. Yang, S. Guha, D.L. Williamson, D. Han and C.S. Jiang, *Appl. Phys. Lett.* **85** (2004) 1955–1957.

45. T. Dylla, F. Finger and E.A. Schiff, *Appl. Phys. Lett.* **87** (2005) 032103–032105.

46. B. Yan, K. Lord, J. Yang, S. Guha, J. Smeets and J.M. Jacquet, *Proc. Mat. Res. Soc. Symp.* **715** (2002) 629–634.

47. T. Roschek, T. Repmann, J. Müller, B. Rech and H. Wagner, *Proc. 28th IEEE Photovoltaic Specialists Conf.*, Sept. 15–22, 2000, Anchorage, AK, USA (IEEE, 2000), p. 150.

48. B. Yan, G. Yue, J.M. Owens, J. Yang and S. Guha, *Proc. 4th World Conf. Photovoltaic Energy Conversion*, May 7–12, 2006, Hawaii, USA.

49. K. Saito, M. Sano, S. Okabe, S. Sugiyama and K. Ogawa, *Sol. Energy Mater. Sol. Cells* **86** (2005) 565–575.

50. K. Yamamoto *et al.*, *Proc. 15th Int. Photovoltaic Sci. Eng. Conf.* Vol. 1, Oct. 10–15, 2005, China, pp. 529–530.

51. C.Y. Kwong, W.C.H. Choy, A.B. Djurisic, P.C. Chui, K.W. Cheng and W.K. Chan, *Nanotechnology* **15** (2004) 1156–1161.

52. V.P. Singh, R.S. Singh, G.W. Thompson, V. Jayaraman, S. Sanagapalli and V.K. Rangari, *Sol. Energy Mater. Sol. Cells* **81** (2004) 293–303.

53. M. Nanu, J. Schoonman and A. Goossens, *Adv. Mater.* **16** (2004) 453–456.

54. R. Banerjee, R. Jayakrishnan and P. Ayyub, *J. Phys. Condens. Mat.* **12** (2000) 10647–10654.

55. Y. Kayanuma, *Phys. Rev.* **B38** (1988) 9797–9805.

56. S.K. Mandal, S. Chaudhuri and A.K. Pal, *Thin Solid Films* **350** (1999) 209–213.

57. H. Fritzsche, *Amorphous Silicon and Related Materials* (World Scientific, Singapore, 1989).

58. K. Ernst, A. Belaidi and R. Konenkamp, *Semicond. Sci. Technol.* **18** (2003) 475–479.

59. A.J. McEvoy and M. Grätzel, *Sol. Energy Mater. Sol. Cells* **32** (1994) 221–227.

60. B. O'Regan and M. Grätzel, *Nature* **353** (1991) 737–740

61. M.K. Nazeeruddin, P. Pechy and M. Gratzel, *Chem. Comm.* **18** (1997) 1705–1706.

62. A. Hagfeldt and M. Grätzel, *Chem. Rev.* **95** (1995) 49–68.

63. N.G. Park, J. van de Lagemaat and A.J. Frank, *J. Phys. Chem.* **B104** (2000) 8989–8994.

64. M.A. Green, *Third Generation Photovoltaics: Advanced Solar Conversion* (Springer-Verlag, 2003).

65. M. Zacharias, J. Heitmann, R. Scholz, U. Kahler, M. Schmidt and J. Bläsing, *Appl. Phys. Lett.* **80** (2002) 661–663.

66. A.P. Alivisatos, *J. Phy. Chem.* **100** (1996) 13226–13239.

67. L.E. Brus, *J. Chem. Phys.* **80** (1984) 4403–4409.

CHAPTER 8

APPLICATION OF SILICON NANOCRYSTAL
IN NON-VOLATILE MEMORY DEVICES

T. P. Chen

School of Electrical and Electronic Engineering
Nanyang Technological University
Singapore 639798
EChenTP@ntu.edu.sg

1. Introduction

Semiconductor memory devices, which are usually fabricated on silicon wafers with the complementary metal-oxide semiconductor (CMOS) technology, can be divided into the two categories:

1. The volatile memories, such as static random access memory (SRAM) and dynamic random access memory (DRAM), which are very fast in writing and reading but lose their data content when the power supply is off.

2. The non-volatile memories (NVMs), such as UV-erasable programmable read-only memory (EPROM), electrically erasable programmable read-only memory (EEPROM), and flash memory, which maintain their data content even without power supply [1, 2].

The NVMs were first introduced in the late 1960s. During the early development of the NVMs, the floating gate metal-oxide semiconductor (MOS) memory device and the metal-nitride-oxide semiconductor (MNOS) memory device were proposed [1]. The floating gate memory device quickly became a dominant design, and today it is still the most prevailing NVM implementation. The most widespread memory array organization is the so-called flash memory. The term "flash" refers to the fact that the contents of the whole memory array, or a memory block (sector), is erased in one step. Although a huge commercial success, conventional floating gate memory devices have their limitations. The most prominent one today is the limited potential for continued scaling of the device structure. At present, the floating gate structure is believed to have run out of steam beyond 65 nm technology [3]. The limitations of the scaling are mainly due to impossibility of scaling the dielectrics (i.e. the tunnel oxide and control oxide) without losing data retention or degrading reliability, to the increase of floating gate interference in ultra-dense NAND devices, and to the impossibility of reducing the drain turn on effect, which limits the channel length scaling of NOR memory cells [4].

NVMs based on silicon nanocrystal (nc-Si), which were first introduced in the early 90s [5, 6], are promising candidate to overcome the above mentioned limitations and to continue the scaling of NVM devices. In a nanocrystal-based NVM device, the continuous floating gate of the conventional NVM device is replaced by an ensemble of Si nanocrystalline islands. In other words, charge is not stored on a continuous floating gate polycrystalline-silicon layer, but instead on a layer of discrete, mutually isolated, crystalline Si nanocrystals or "dots" [3–7]. As compared to conventional floating gate NVM devices, nanocrystal-based NVM devices offer several advantages. First, nanocrystal-based NVM devices provide the possibility to reduce the thickness of the tunnel oxides without sacrificing nonvolatility. Reducing the tunnel oxide thickness is key to lowering operating voltages and/or increasing operating speeds. Secondly, nanocrystal-based NVMs use a more simplified fabrication process as compared to conventional floating gate NVMs by avoiding the fabrication complications and costs of a dual-poly process. Further, due to the absence of drain to floating gate coupling, nanocrystal-based NVMs suffer less from drain-induced barrier lowering (DIBL) and therefore have intrinsically better punchthrough characteristics [3]. On the other hand, the nanocrystals with sub-5 nm dot diameter offer the possibility of single-electron memory devices operating at room temperature [8].

2. Conventional Floating Gate Non-Volatile Memory Devices

Before the discussion on nanocrystal-based memory devices, we should have a brief introduction to the conventional floating gate flash memory [1, 2] because they are very similar to each other. A flash memory device is basically a floating gate metal-oxide semiconductor field effect transistor (MOS-FET) (see Fig. 2.1), i.e. a transistor with a gate completely surrounded by dielectrics such as silicon dioxide (SiO_2), the floating gate (a polysilicon layer), and electrically governed by a capacitively coupled control gate (another polysilicon layer). In other words, a floating gate memory device

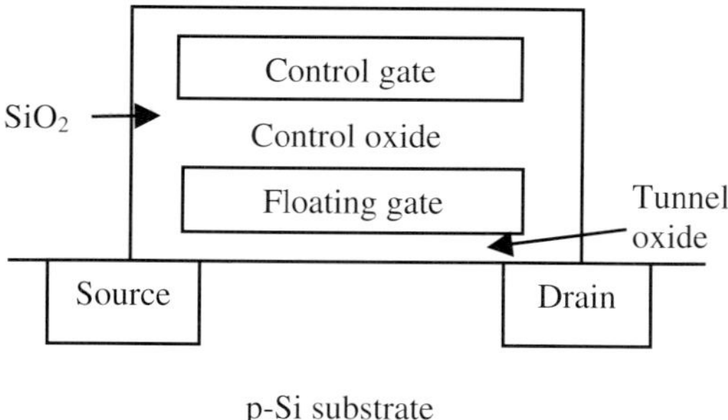

Fig. 2.1. Schematic cross-section of a floating gate flash memory device.

has an additional gate (i.e. the floating gate) as compared to a conventional MOSFET. Being electrically isolated by the surrounding dielectrics (i.e. the tunnel oxide and control oxide), the floating gate acts as the charge-storing node for the memory device; charge that is injected into the floating gate is maintained there, leading to the shift of the threshold voltage of the MOSFET. Obviously the quality of the dielectrics guarantees the non-volatility, while the thickness of the dielectrics allows the possibility to program or erase the memory cell by electrical pulses. Usually the gate dielectric, i.e. the one between the MOSFET channel and the floating gate, is an oxide in the range of several nanometers and is called "tunnel oxide" since Fowler–Nordhein (FN) electron tunneling occurs through it during the programming and erasing operations. The dielectric that separates the floating gate from the control gate is called "control oxide" with a thickness in the range of 15–20 nm [2].

Figure 2.2 shows the energy band diagram for the memory device shown in Fig. 2.1. It can be seen that the floating gate acts as a potential well for charge stored in it as it is surrounded by SiO_2 (the control oxide and the tunnel oxide). Under the influence of appropriate electric fields electrons can tunnel into the floating gate from the channel of the MOSFET during the programming operation or tunnel out from the floating gate to the channel through the tunnel oxide during the erasing operation. The neutral (i.e. uncharged) state or positively charged state (i.e. electrons tunnel out from the floating gate leaving positive charges there) is associated with the logical state "1", and the negatively charged state, corresponding to electrons stored in the floating gate, is associated with the logical "0".

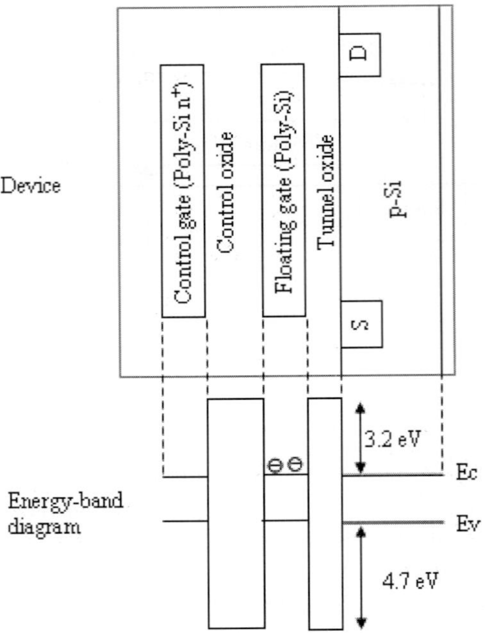

Fig. 2.2.   Schematic energy band diagram of a floating gate memory device.

The threshold voltage (V_{th}) of the floating gate MOSFET for an amount of charge (Q_{FG}) stored in the floating gate can be written as [1]

$$V_{\text{th}} = \phi_{\text{ms}} + 2\phi_{\text{F}} - \frac{Q_{\text{ox}}}{C_{\text{ox}}} - \frac{Q_{\text{D}}}{C_{\text{ox}}} - \frac{Q_{\text{FG}}}{\varepsilon_{\text{ox}}/d}, \tag{2.1}$$

where ϕ_{ms} is the work function difference between the control gate and the Si substrate;

ϕ_{F} is the Fermi level potential;
Q_{ox} is the equivalent fixed oxide charge per unit area;
Q_{D} is the charge in the depletion layer per unit area;
Q_{FG} is the charge stored in the floating gate per unit area;
C_{ox} is the oxide capacitance per unit area;
ε_{ox} is the permittivity of the oxide; and
d is the distance between the floating gate and the control gate.

The threshold voltage shift (ΔV_{th}) due to the charge stored in the floating gate is thus given by [1]

$$\Delta V_{\text{th}} = -\frac{Q_{\text{FG}}}{\varepsilon_{\text{ox}}} d. \tag{2.2}$$

For electron storing in the floating gate, Q_{FG} is negative, and thus ΔV_{th} is positive. This means that the threshold voltage of the MOSFET will increase after electrons are injected into the floating gate.

The data stored in a floating gate memory device can be determined by measuring the threshold voltage of the floating gate MOSFET. The best and fastest way to do that is by measuring the drain current (I_D) of the transistor at a fixed gate bias (V_{G_read}) [2]. As shown in Fig. 2.3, for logic "1" (no charge stored in the floating gate), the transistor has a threshold voltage (V_{th_1}); for logic "0" (electrons are stored in the floating gate), the threshold voltage is increased to $V_{th\_0} = V_{th\_1} + \Delta V_{th}$. Note that $\Delta V_{th} > 0$ for electron storing (i.e. $Q_{FG} < 0$). V_{G_read} is set at a value between V_{th_1} and V_{th_0}, i.e. $V_{th\_1} < V_{G\_read} < V_{th\_0}$. Therefore, for logic "1", the transistor is "ON", and thus there is a large drain current, i.e. $I_D \gg 0$; for logic "0", the transistor is "OFF", and thus there is no drain current flowing, i.e. $I_D = 0$. In this way the two logic states can be easily distinguished. Of course, when the power supply is interrupted, the memory state should remain unchanged in order to provide a nonvolatile device.

Storing charge on a single node (i.e. the floating gate) makes the conventional memory structure particularly prone to failure of the floating gate isolation. One weak spot in the tunnel oxide is sufficient to create a fatal discharge path, compromising long term non-volatility [3]. On the other hand, to allow for long data retention, the tunnel oxide thickness must be maintained a few nanometers to prevent the floating gate charge lose to the channel, source and drain of the MOSFET. A thicker tunnel oxide leads to higher programming, erasing voltage and longer programming, and erasing time. This conflicts with the requirements of low power and high speed

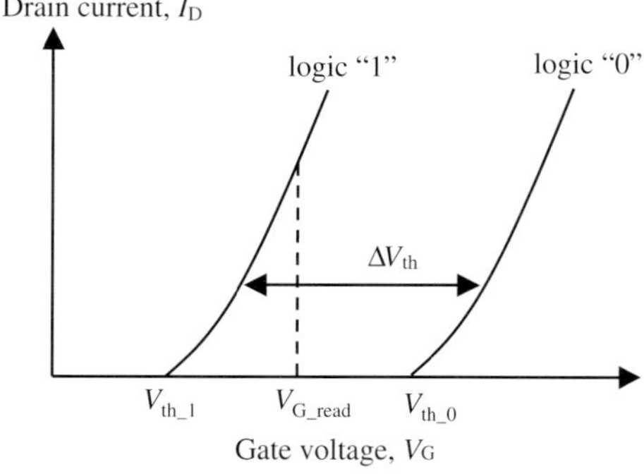

Fig. 2.3.   Distinguishing the two logic states in the read operation.

applications. One way to alleviate the scaling limitations of the conventional floating gate device while still preserving the fundamental operating principle of the memory, is to rely on distributed discrete-trap charge storage instead. Silicon nanocrystal memories, first introduced in the early 90s [5, 6], are one particular implementation of that concept. In silicon nanocrystal memories, the nanocrystals are used as the discrete-trap storage nodes to replace the floating gate of NVMs. With the discrete-trap storage nodes, a single leakage path due to a defect (intrinsic or stress-induced defects) in the oxide can only discharge a single storage node. This allows for a thinner tunnel oxide for continued scaling of device structures.

3. Non-Volatile Memory Devices Based on Si Nanocrystal

3.1. *Device Structure*

In a nanocrystal memory cell, the floating gate of a conventional floating gate memory device is replaced by an array of silicon nanocrystal (nc-Si) of nanometer size (less than 10 nm) separated each other by silicon dioxide. Actually, a nanocrystal memory cell is a MOSFET with silicon nanocrystals embedded in the gate oxide, as shown in Fig. 3.1. The nanocrystal is distributed on a thin tunnel oxide (i.e. a very thin SiO_2 film) with a thickness of several nanometers (typically 2 nm) and covers the entire surface channel region. A control oxide (i.e. a thicker SiO_2 film) with a thickness of a few nanometers (e.g. 7 nm) separates the nanocrystals from the control gate of the MOSFET.

As an example, Fig. 3.2 shows the transmission electron microscopy (TEM) images of a nanocrystal memory cell [4]. The silicon nanocrystals were deposited on top of the tunnel oxide by chemical vapor deposition using SiH_4 as precursor and H_2 as carrier gas. Figure 3.2(c) shows a TEM planar-view with electron energy-loss filtered image of the silicon

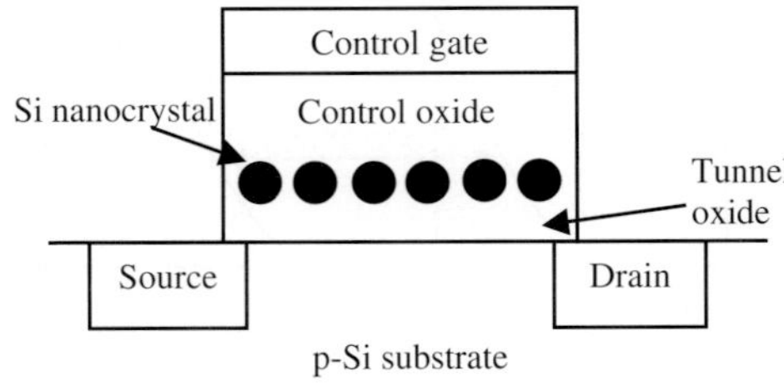

Fig. 3.1.   Schematic of a nanocrystal memory cell.

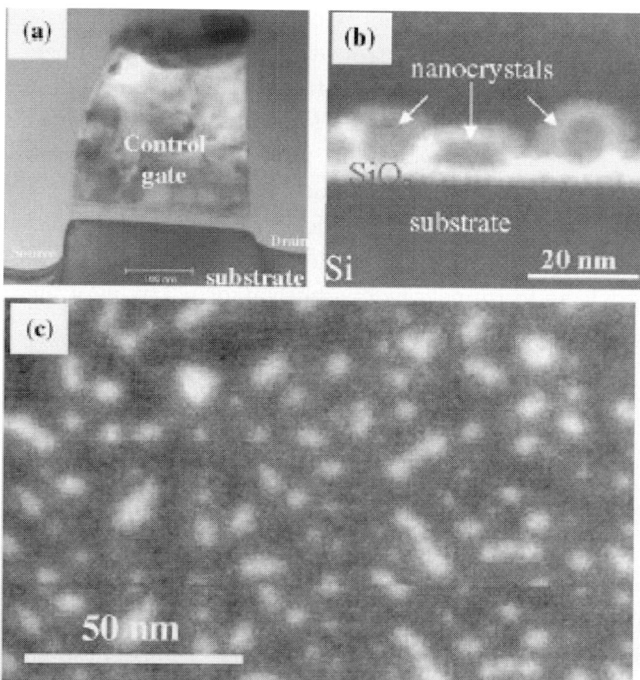

Fig. 3.2. (a) TEM cross-section of the Si nanocrystal memory cell, (b) details of the stack tunnel oxide-Si nanocrystals-control oxide, and (c) energy filtered-TEM plan-view of the Si nanocrystals (white spots) deposited on SiO$_2$ [4].
Reprinted with permission of Elsevier.

nanocrystals nucleated on the SiO$_2$. Tuning the process conditions it is possible to obtain nanocrystal densities in the range 0.1–$1 \times 10^{12}\,\mathrm{cm}^{-2}$ [4].

3.2. *Operation Mechanisms*

The operation principle of nanocrystal memory device is very similar to that of conventional floating gate NVM devices. The electrons stored in the nanocrystals screen the gate charge and reduces the conduction in the inversion layer of the MOSFET, i.e. it effectively shifts the threshold voltage of the device to be more positive, whose magnitude is approximately given by [5, 6, 9]

$$\Delta V_{\mathrm{th}} = \frac{npq}{\varepsilon_{\mathrm{ox}}} \left(t_{\mathrm{ctl}} + \frac{\varepsilon_{\mathrm{ox}} t_{\mathrm{nc}}}{2\varepsilon_{\mathrm{Si}}} \right), \tag{3.1}$$

where n is the nanocrystal number density, p is the average number of electrons stored per nanocrystal, q is the electronic charge, $\varepsilon_{\mathrm{ox}}$ and $\varepsilon_{\mathrm{Si}}$

represent the permittivity of the oxide and Si, respectively, and t_{ctl} represents the thickness of the control oxide, and t_{nc} is the linear dimension of the nanocrystal well. For nanocrystals that are 5 nm in dimension, 5 nm apart, i.e. a nanocrystal density of $1 \times 10^{12}\,cm^{-2}$, with a control oxide thickness of 7 nm, the threshold shift is nearly 0.36 V for one electron per nanocrystal, a large value that in a good transistor changes the subthreshold current by five orders of magnitude. This is easily current-sensed [5, 6]. Clearly, for storing a given number density of electrons, np, susceptibility to isolated defects in tunnel oxide is mitigated by having a higher density, n, of nanocrystals and minimizing the number, p, of electrons stored per nanocrystal. For nanocrystals to be sufficiently electrically isolated with respect to tunneling transport, their typical separation must be greater than about 4 nm from one another [9].

3.2.1. *Programming Operation*

There are two important programming mechanisms for charge injection into the nanocrystals. They are: Fowler–Nordheim (FN) tunneling and channel hot electron (CHE) injection [1]. The former is based on the quantum mechanical tunneling through the tunnel oxide, whereas the latter is based on injection of carriers that are heated in a large electric field in the silicon, followed by injection over the energy barrier of the SiO_2/Si substrate.

A. Fowler–Nordheim Tunneling [1]

When a large positive voltage is applied to the gate, the energy bands of the structure of polysilicon gate, control oxide, nanocrystal, tunnel oxide and Si substrate will be influenced as shown in Fig. 3.3. Due to the high electric field, electrons in the conduction band of the Si substrate see a triangular energy barrier with a width dependent on the applied field. When the applied field is sufficiently high, the width of the barrier becomes small enough that electrons can tunnel through the barrier from the conduction band of the Si substrate into the oxide conduction band. Some of the electrons tunneling from the Si substrate are trapped in the nanocrystals, leading to an increase in the threshold voltage of the n-channel MOSFET.

B. Hot Electron Injection [1]

In n-channel MOSFET, at a large drain voltage, electrons that are injected into the depletion region of the drain junction are accelerated by the high lateral field, and they may gain enough energy to cause impact ionization in the depletion region near the drain. The impact ionization generates many

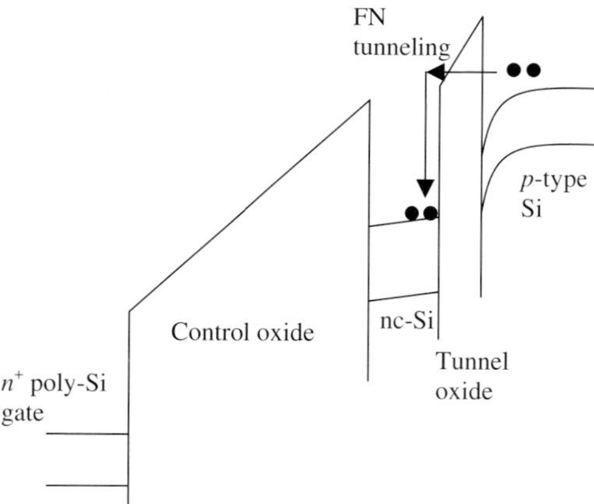

Fig. 3.3.　Energy band diagram for a nanocrystal memory device based on n-channel MOSFET in the regime of FN tunneling under a positive gate voltage.

electrons and holes. Both the generated electrons and holes can also gain energy from the electric field. These carriers have much higher energy than the thermal energy and are called hot carriers. Normally, the maximum hot carrier effect occurs when the gate voltage is approximately half of the drain voltage. The generated holes are normally collected at the substrate and form the substrate current. On the other hand, most of the generated electrons are collected by the drain. But some of them may have sufficient energy to surmount the SiO_2/Si energy barrier to inject into the gate oxide. Some of the injected electrons reach the gate forming a very small gate current, and some of them are trapped in the nanocrystals leading to a shift in the threshold voltage ($\Delta V_{th} > 0$). The energy band diagram under channel hot electron injection is shown in Fig. 3.4.

3.2.2. *Erasing Operation*

In the erasing operation, the electrons trapped in the nanocrystals can be removed by means of the FN electron tunneling through the tunnel oxide from the nanocrystals to the channel of the MOSFET. This process is opposite to the FN programming discussed above. With a sufficiently large negative gate voltage, the electric field will pull electrons out of the nanocrystals to the channel by tunneling. The memory cell is then erased to a low threshold voltage state. Note that both programming and erasing operations can be carried out with the FN tunneling mechanism. However, with the mechanism of hot electron injection, it is only possible to

T. P. Chen

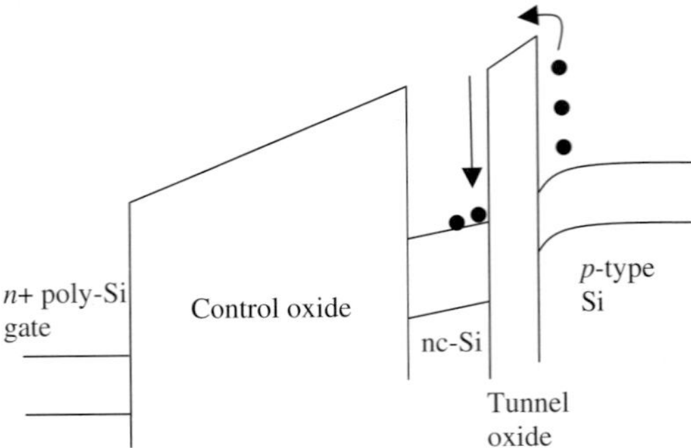

Fig. 3.4. Energy band diagram under hot electron injection. Both the drain voltage and the gate voltage should be positive for n-channel MOSFET. The maximum hot electron injection normally occurs when the gate voltage is approximately half of the drain voltage. The hot electrons could have enough energy to overcome the energy barrier at the interface of tunnel oxide/Si substrate.

bring electrons onto the nanocrystals. But they cannot be removed from the nanocrystals with the same mechanism (i.e. hot electron injection).

3.2.3. *Reading Operation*

The reading operation of nanocrystal NVMs is very similar to that of conventional floating gate NVMs. As shown in Fig. 2.3, logic "1" and "0" exhibit the same transconductance but are shifted by a quantity — the threshold voltage shift (ΔV_{th}) — that is proportional to the stored electron charge in the nanocrystals. Note that logic "1" represents no electron trapping in the nanocrystals while logic "0" represents sufficient electron trapping in the nanocrystals. Therefore, it is possible to fix a reading voltage (i.e. an appropriate gate voltage) in such a way that the drain current of the "1" cell is very high, while the current of the "0" cell is zero (very small actually). In this way, it is possible to define the logical state "1" from a microscopic point of view as no electron charge (or positive charge) stored in the nanocrystals and from a macroscopic point of view as large reading current (i.e. the drain current of the MOSFET). Vice versa, the logical state "0" is defined, respectively, by electron charge stored in the nanocrystals and zero reading current.

The above principle can be clearly illustrated by the example shown in Fig. 3.5. The threshold voltages corresponding to logic "1" (the erased state) and "0" (the programmed state) are $\sim 0.7\,\text{V}$ and $\sim 2.5\,\text{V}$, respectively. If the gate voltage (i.e. the reading voltage) is set at $1.5\,\text{V}$, the drain current

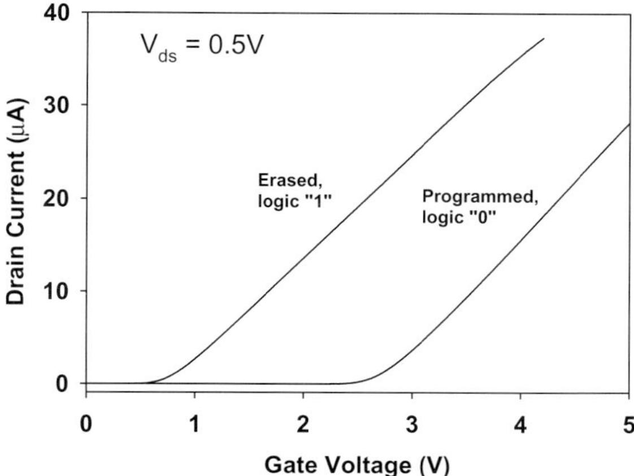

Fig. 3.5. Transfer characteristics of a nanocrystal memory device at the programmed and erased states. The programming and erasing operations are carried out with the FN tunneling at the gate voltage of $+12$ V and -12 V, respectively, for 1 μs. The drain voltage (V_{ds}) is fixed at 0.5 V for the drain current measurement.

of logic "1" cell is 8×10^{-6} A, but the current of logic "0" cell is 1×10^{-13} A. In other words, at the reading voltage, the MOSFET of logic "1" cell is on, while the MOSFET of logic "0" cell is off. The current ratio of logic "1" to logic "0" is 8×10^{7}. Therefore, in this way, the two memory states can be easily distinguished.

4. Synthesis and Characterization of Si Nanocrystal

4.1. *Synthesis of Si Nanocrystal*

The nanocrystal layer embedded in the gate oxide of a MOSFET memory device has to meet certain specifications in order to support properly functioning memory devices [3]. To have a sufficient memory window, the density of nanocrystals should be high enough. A typical density is 1×10^{12} cm^{-2}. This is equivalent to approximately 100 particles controlling the channel of a memory MOSFET with a 100×100 nm active area, and requires nanocrystal diameters of 5–6 nm and below [3]. The nanocrystals should be isolated from each other to allow for long data retention. This is particular important because nanocrystal memory devices rely on the concept of distributed discrete-trap charge storage. With the discrete-trap storage nodes, a single leakage path due to a defect (intrinsic or stress-induced defects) in the oxide can only discharge a single storage node (i.e. a single nanocrystal only). Generally, good process control on the size and size distribution of nanocrystals and uniformity of nanocrystal density is required. Of course, the fabrication process should be simple and compatible with main-stream CMOS process.

Several nanocrystal fabrication processes have been suggested [3], and the most important of them are chemical vapor deposition (CVD) [7, 9–13] and ion implantation [14, 15]. Here we will discuss only the CVD and ion implantation techniques.

4.1.1. *CVD*

CVD technique is a process that offers good process control on the size and density of nanocrystals, adequate planarity of the nanocrystal layer and excellent compatibility with CMOS technology. Si nanocrystal growth during CVD on amorphous dielectrics such as SiO_2 and Si_3N_4 is believed to proceed by atomistic nucleation, with a critical size of between 1 and 4 atoms [9]. Rao *et al.* have studied the growth of silicon nanocrystals with CVD [9]. Some of their results and discussions are reproduced in the following. Figure 4.1(a) shows a schematic nucleation and growth curve and the scanning electron microscopic (SEM) images of the surface during various phases of the curve for Si nanocrystal formation during CVD [9]. During the initial incubation phase, Si adatoms are formed on the SiO_2 surface. When there are enough Si adatoms formed on the surface, nucleation occurs to form nanocrystals, and the number of nanocrystals increases rapidly

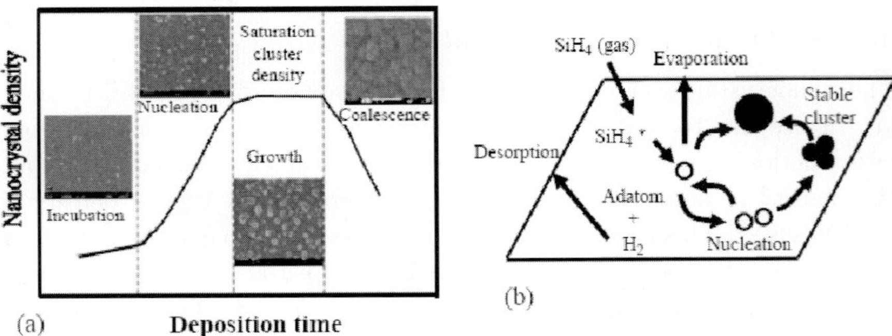

Fig. 4.1.   (a) Typical nucleation and growth curve along with SEM images showing the evolution of nanocrystals on SiO_2 surface during CVD of Si. (b) Atomistic nucleation of Si nanocrystals on a dielectric surface. Major processes include SiH_4 adsorption, adatom re-evaporation and diffusion as well as H_2 desorption [9].
Reprinted with permission of Elsevier.

with deposition time as fresh nuclei are formed continuously during the nucleation phase. When the nucleation phase is completed, the nanocrystals formed by nucleation grow by adatom attachment, and nanocrystal density becomes saturated during this growth phase. Later the growing clusters merge with adjacent ones by coalescence leading to a decrease in the nanocrystal density. Figure 4.1(b) shows the major processes occurring on the surface during formation of Si nanocrystals by SiH_4 CVD [9]. A SiH_4 molecule from the gas phase is first adsorbed on the surface. The adsorbed SiH_4 molecule then dissociates at elevated temperatures, resulting in the formation of a Si adatom on the surface and the desorption of a H_2 molecule. The Si adatoms produced by the dissociation form fresh nanocrystals through atomistic nucleation or are consumed by existing nanocrystals through surface diffusion. Note that some of the Si adatoms may be evaporated at the elevated temperatures.

The size and density of nanocrystals can be controlled in terms of CVD deposition conditions including the partial pressure of precursor and surface temperature. A high nanocrystal density could be obtained with a high adatom flux, low surface diffusion, fast surface H_2 desorption and low adatom evaporation [9]. Figure 4.2(a) shows the effect of precursor partial pressure on Si nanocrystal deposition in CVD at a fixed temperature [9]. As shown in Fig. 4.2(a), when the precursor partial pressure is increased from 0.4 to 0.78 Torr, the time required for the peak nanocrystal density is reduced significantly. This indicates that the nucleation and growth process is accelerated due to the higher adatom flux at the higher

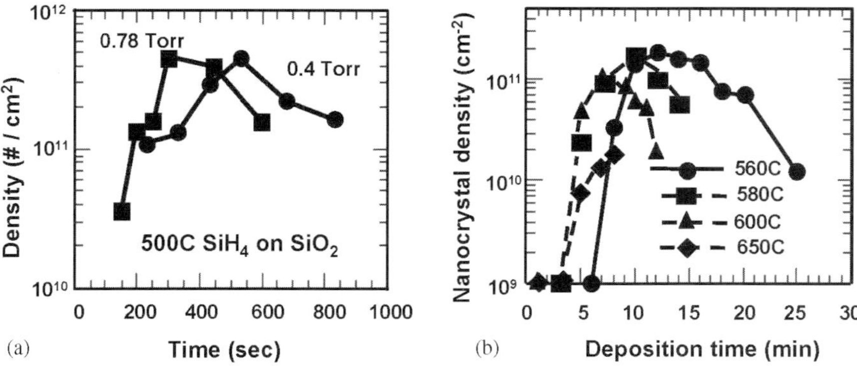

Fig. 4.2. (a) Effect of precursor partial pressure on nucleation curve for Si nanocrystal deposition by SiH_4 CVD on SiO_2 surface. (b) Effect of surface temperature on nucleation curve for Si nanocrystal deposition by SiH_4 CVD on SiO_2 surface [9]. Reprinted with permission of Elsevier.

partial pressure. However, the peak nanocrystal density itself does not show significant dependence on partial pressure. On the other hand, the adatom flux is strongly dependent on the surface temperature, and thus the nucleation process will be affected by the temperature. Figure 4.2(b) shows the effect of the surface temperature on the nucleation curves [9]. As the temperature increases, the dissociation of adsorbed precursor molecule on the surface increases resulting in higher adatom flux and shorter incubation and nucleation timescales. However, surface diffusion of adatoms also increases with temperature. This results in increase in adatom capture by existing Si nuclei, leading to increase in nanocrystal size. This, at the same time, in turn reduces fresh nanocrystal nucleation. Thus, the peak nanocrystal density decreases when the temperature is increased, as shown in Fig. 4.2(b). Therefore, desired size and density of nanocrystals can be achieved by controlling the precursor partial pressure and the surface temperature.

4.1.2. *Ion Implantation*

Ion implantation is a versatile method to produce embedded semiconductor nanoclusters in SiO_2 films. This technique combines low energy (typically a few keV) ion implantation of Si^+ to fluences of 10^{15}–10^{16} cm^{-2} with a subsequent annealing at a high temperature ($\sim 1000°C$) in N_2 gas for duration of a few minutes to ~ 1 hour. The size of Si nanocrystal synthesized with this technique is usually ~ 3–~ 5 nm, and it does not depend much on the annealing time and annealing temperature [16]. In addition, annealing does not very much change the profile of the implanted Si in the oxide. This is due to the fact that Si atoms have an extremely low diffusion in SiO_2. The advantages of this technique include its simple process and fully compatible with mainstream CMOS process.

First investigations of Si^+ implanted thin gate oxides for memory applications were carried out by Kalnitsky *et al.* in 1990 [17, 18]. Since then numerous studies have been performed concerning the application of nanocrystals synthesized with ion implantation in memory devices [19–24]. The potential of ion implantation technique for nanocrystal memory applications has been recently enhanced through the synthesis in the very low to ultra low energy regime ($\leq \sim 2$ keV) [25–34]. In terms of structural possibilities, a combination of very low energy ion implantation and oxide thickness allows for the formation of Si nanocrystals at a location from the SiO_2/Si interface that can be tailored for specified memory applications.

The profile of excess Si in the SiO_2 film due to Si ion implantation can be obtained from the stopping and range of ions in matter (SRIM) simulation [35]. Figure 4.3 shows two cases of profiles of implanted Si in a 30 nm SiO_2 thin film thermally grown on Si substrate: partial and full distributions of the implanted Si atoms in the oxide achieved with an implantation energy of 2 keV and 8 keV, respectively. For the partial distribution, the oxide can be divided into two regions, the implanted-Si-atoms distributed region and the pure SiO_2 region, as shown in Fig. 4.3(a). For the full distribution, the implanted Si atoms are distributed throughout the entire oxide films, as shown in Fig. 4.3(b). The SRIM simulation shows that for the 30 nm gate oxide the full distribution can be achieved with implantation energies higher than 7 keV. For the case shown in Fig. 4.3(b), some of the Si ions are implanted into the Si substrate under the implantation energy of 8 keV. On the other hand, the actual implanted Si profile in SiO_2 thin film can be experimentally determined from secondary ion mass spectroscopy (SIMS) measurement. Figure 4.4 shows such an example [37].

In the following sections, we will limit our discussions to Si ion implantation for simplicity, namely, in Sec. 4.2 we will examine the structural and physical properties of Si nanocrystals synthesized by Si ion implantation, and in Sec. 5 we will also study the electrical characteristics and performance of memory devices based on Si nanocrystals synthesized by Si ion implantation.

4.2. *Properties of Si Nanocrystal*

High resolution transmission electron microscopy (TEM) is frequently used to investigate the structural properties of silicon nanocrystals embedded in SiO_2 thin film. It can provide the information of nanocrystal size and size distribution. Figure 4.5 shows the TEM image of Si nanocrystal embedded in a SiO_2 matrix synthesized with high energy (100 keV) Si ion implantation. The Si nanocrystal synthesized has a size of $\sim$4–5 nm. As mentioned earlier, for memory device application, very low implantation energy is usually required so that the nanocrystal is confined in a narrow region near the SiO_2/Si interface for a proper device operation. Figure 4.6 shows the TEM images of Si nanocrystal embedded in SiO_2 thin films synthesized with implantation energy of 2 keV (Fig. 4.6(a)) and 10 keV (Fig. 4.6(b)). As can be seen in Fig. 4.6, the Si nanocrystal size is also $\sim$4–5 nm for the both implantation energies. Therefore, it seems that nanocrystal size is not sensitive to the implantation energy.

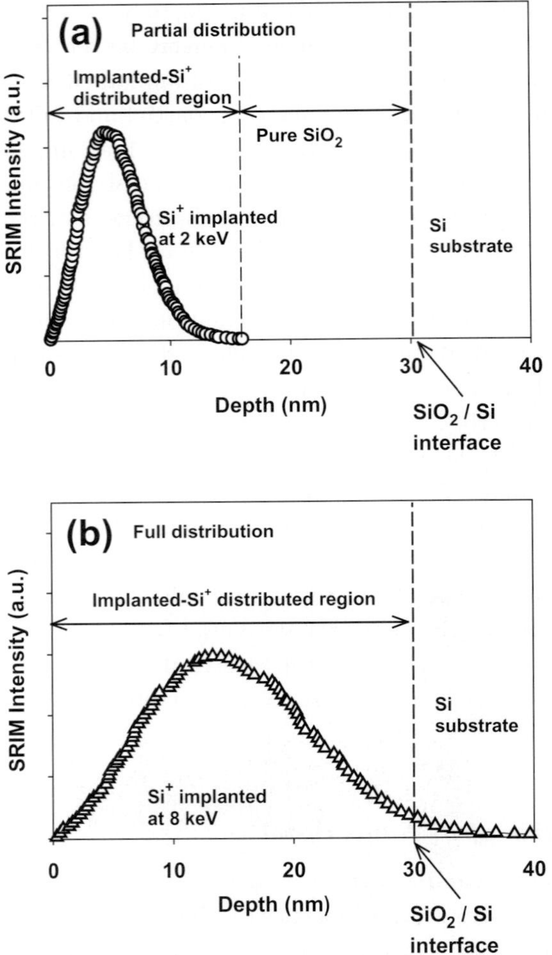

Fig. 4.3. Implanted Si profiles in a 30 nm SiO$_2$ thin film obtained from SRIM simulation. (a) Partial distribution under the ion implantation energy of 2 keV; and (b) full distribution under the implantation energy of 8 keV [36]. (©2006 IEEE).

The information of nanocrystal size can also be obtained from the X-ray diffraction (XRD) measurement. The average size could be estimated from the broadening of the Bragg peak in XRD spectrum based on the Scherrer's equation [39]

$$D = \frac{0.9\lambda}{\Delta\theta \cos(\theta_{\mathrm{B}})}, \tag{4.1}$$

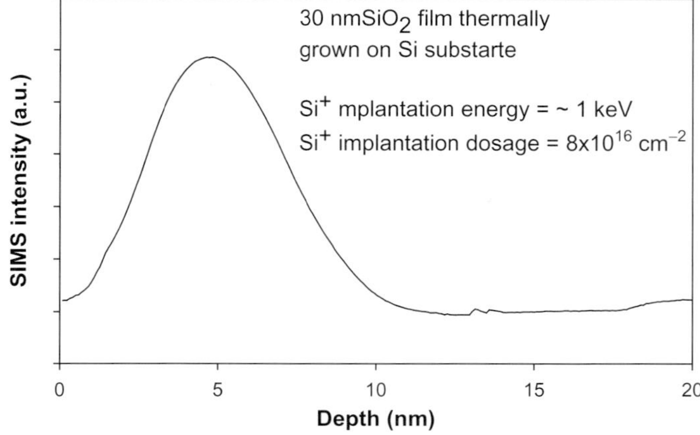

Fig. 4.4.   Implanted Si profile in a 30 nm SiO_2 thin film determined from SIMS measurement. The implantation energy and dosage are 1 keV and 8×10^{16} cm^{-2}, respectively [37]. Reprinted with permission of The Institute of Pure and Applied Physics.

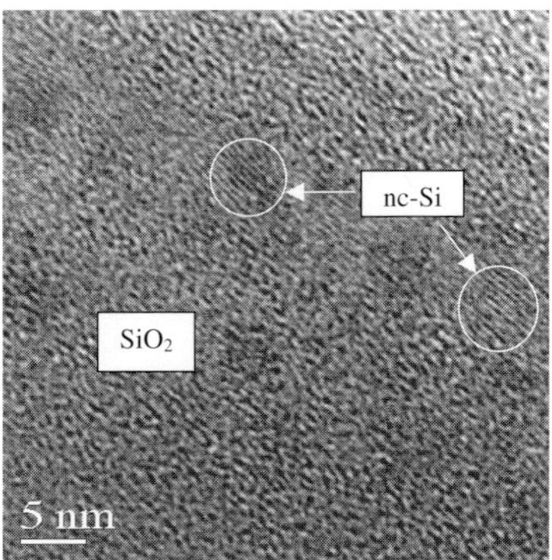

Fig. 4.5.   TEM image of Si nanocrystal embedded in a 550 nm thick SiO_2 film. The SiO_2 film is implanted with a dose of 1×10^{17} atoms/cm^2 of Si^+ at 100 keV, and the sample is annealed at 1000°C in nitrogen gas for duration of 20 min [38].
Reprinted with permission from Ref. 38. Copyright (2003) by the American Physical Society.

where D is the mean size of nanocrystals, λ is the wavelength of the X-ray, θ_B is the Bragg angle and $\Delta\theta$ is the full width of the half maximum (FWHM) of the Bragg peak in radians. Note that a correction for the instrumental broadening should be made when determining the Bragg peak

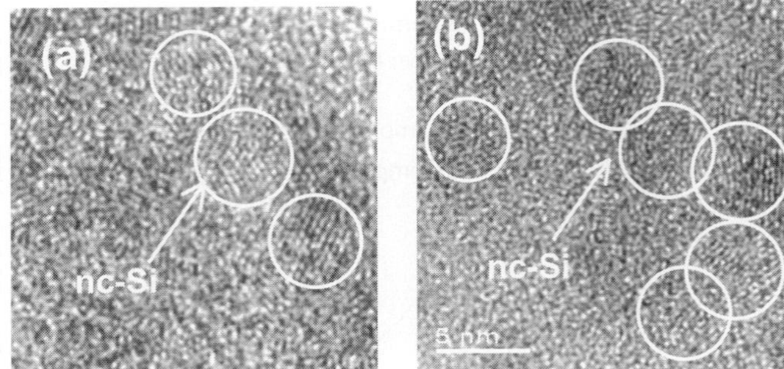

Fig. 4.6. TEM image of Si nanocrystal embedded in a 30 nm gate oxide synthesized with the implantation energy of (a) 2 keV and (b) 10 keV. Thermal annealing is carried out at 1000°C in N_2 ambient for 1 hour to induce nc-Si formation.

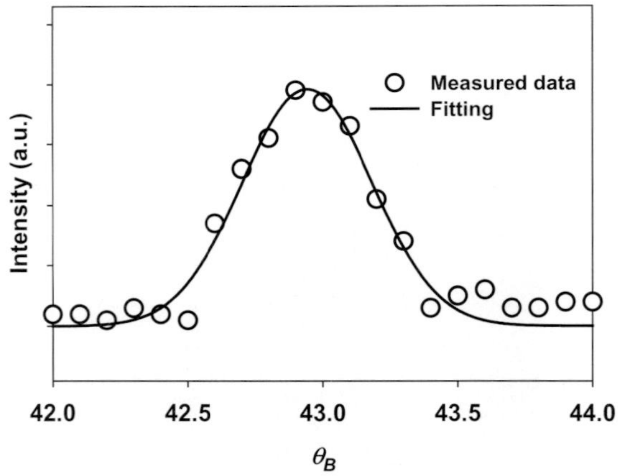

Fig. 4.7. XRD measurement of Si nanocrystal embedded in a 30 nm SiO_2 thin film. The implantation energy and dosage are 2 keV and 1×10^{16} cm^{-2}, respectively. Thermal annealing is carried out at 1000°C in N_2 ambient for 1 hour to induce nanocrystal formation.

broadening. Figure 4.7 shows the XRD measurement on the sample with Si nanocrystal synthesized at very low implantation energy (2 keV), and the nanocrystal size obtained with Eq. (4.1) is ~ 4 nm. For high energy (such as 100 keV) ion implantation, similar XRD result has also been reported [38]. The nanocrystal size is observed to be insensitive to thermal annealing [40, 41]. This is also evident from Fig. 4.8 which shows the nanocrystal size versus annealing time. This insensitivity is due to the very low diffusion coefficient of Si in SiO_2 film [16].

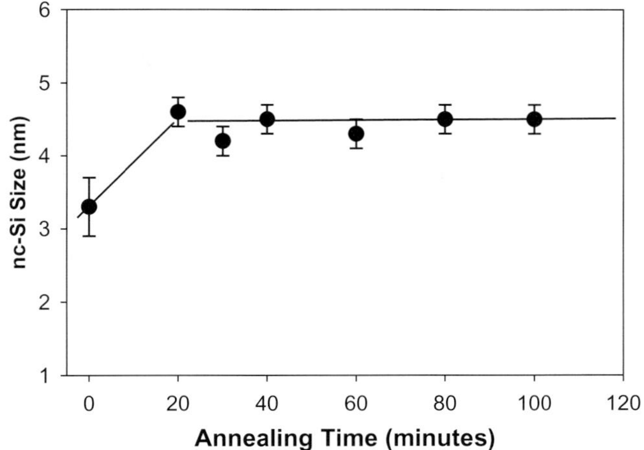

Fig. 4.8.   Nanocrystal size obtained from XRD measurement as a function of annealing time. A dose of 1×10^{17} atoms/cm^2 Si ions were implanted into 500 nm SiO$_2$ film at the energy of 100 keV. Post-implantation thermal annealing was carried out in nitrogen ambient at 1000°C.

The chemical structures of Si$^+$-implanted SiO$_2$ have been recently studied by Liu *et al.* [37, 43] and Chen *et al.* [42] with X-ray photoelectron spectroscopy (XPS). The analysis of the XPS Si 2p peaks shows the existence of the five chemical structures including Si (Si nanocrystals or nanoclusters), Si$_2$O, SiO, Si$_2$O$_3$ and SiO$_2$ corresponding to the Si oxidation states Si^{n+} ($n = 0, 1, 2, 3$, and 4) in the SiO$_2$ films, respectively, as shown in Fig. 4.9 [42]. Figures 4.9(b) and (c) show the XPS Si 2p core level peaks for the as-implanted sample and the sample that was annealed at 1000°C for 60 min, respectively. As the spectra were taken with a sampling depth of ~ 5 nm which is smaller than the SiO$_2$ film thickness (30 nm), the Si substrate did not contribute to the spectra. As shown in Fig. 4.9(b) for the as-implanted sample, the contributions of Si^0, Si^{1+}, Si^{2+} and Si^{3+} are large, showing the coexistence of large amounts of Si nanoclusters/nanocrystals (corresponding to the oxidation state Si^0) and the Si suboxides (corresponding to the three oxidation states Si^{1+}, Si^{2+} and Si^{3+}) in addition to the stoichiometric SiO$_2$ matrix in the SiO$_x$ ($x < 2$) films. However, after the annealing at 1000°C for 60 min, as shown in Fig. 4.9(c), the contributions of the suboxides become small or even insignificant, and Si^{4+} and Si^0 are the dominant components after the annealing, suggesting the formation of a large amount of Si nanocrystals embedded in the stoichiometric SiO$_2$ matrix. The result indicates that the annealing leads to the reductions of the three suboxides

438 *T. P. Chen*

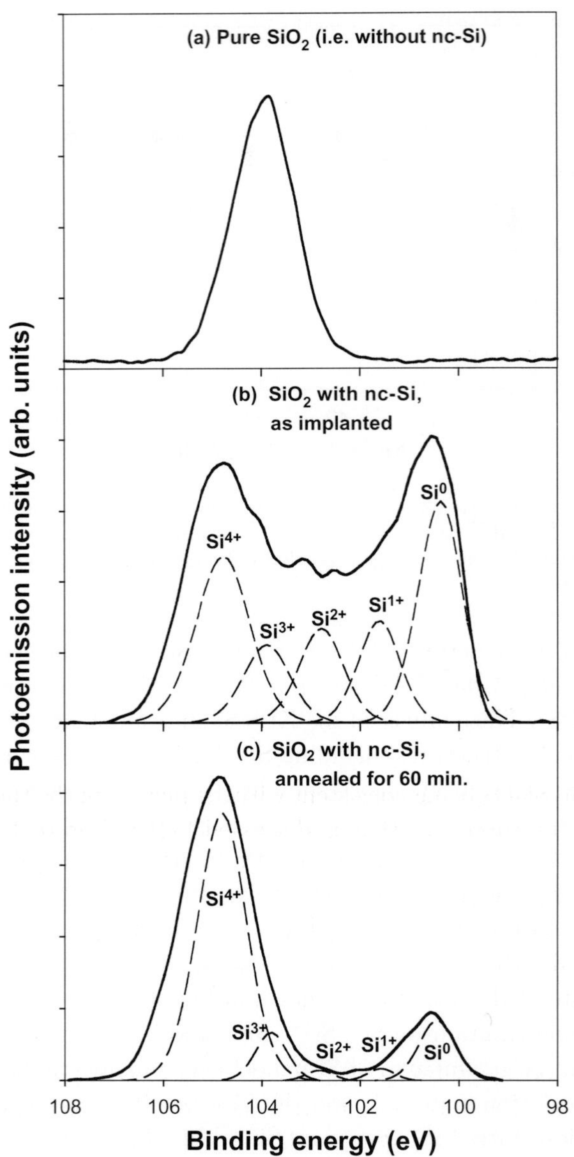

Fig. 4.9. Si 2p spectra for (a) the sample without Si ion implantation (i.e. pure SiO_2 thin film thermally grown on Si substrate); (b) the as-implanted sample; and (c) the sample annealed at $1000°C$ for 60 min. Si ions were implanted into a 30 nm SiO_2 film at the energy of 1 keV with a dose of 8×10^{16} cm^{-2} [42].

Reprinted with permission from Ref. 42. Copyright (2004) by the American Chemical Society.

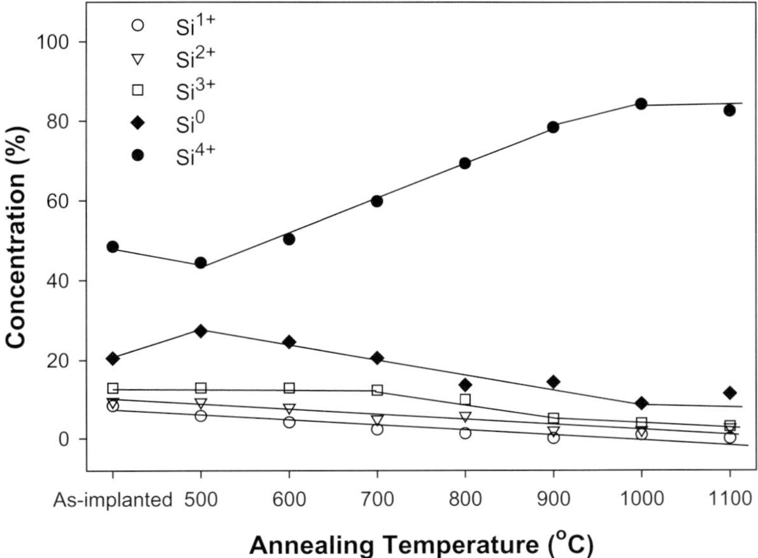

Fig. 4.10. Changes of the concentrations of the five Si oxidation states Si^{n+} ($n = 0, 1, 2, 3$, and 4) with annealing temperature. The annealing time is fixed at 20 min. Si ions were implanted into a 30 nm SiO$_2$ film at the energy of 1 keV with a dose of 8×10^{16} cm^{-2} [43].
Reprinted with permission of IOP Publishing Limited.

Si$_2$O, SiO and Si$_2$O$_3$, being consistent with the picture of the thermal decompositions of the suboxides that is discussed below. Figure 4.10 shows the changes of the concentrations of the five oxidation states with annealing temperature for a fixed annealing time of 20 min [43]. For the low temperature annealing at 500°C, the concentration of Si^0 increases significantly while the concentrations of other oxidation states (Si^{1+}, Si^{2+} and Si^{3+}) decrease. For higher annealing temperatures, with the increase of annealing temperature, the Si^{4+} concentration increases greatly, but the concentrations of all other oxidation states decrease. On the other hand, Fig. 4.11 shows the evolution of the concentrations of the five oxidation states with annealing time for the annealing temperature of 1000°C [43]. As can be seen in this figure, the Si^{4+} concentration has a very large increase after the first 20 min annealing but increases gently after longer annealing, while all other oxidation states Si^{n+} ($n = 0, 1, 2$ and 3) show a decrease in concentration with annealing time.

The above situation is explained in the following. The thermal decomposition (or phase separation) of the Si suboxides Si$_2$O, SiO and Si$_2$O$_3$

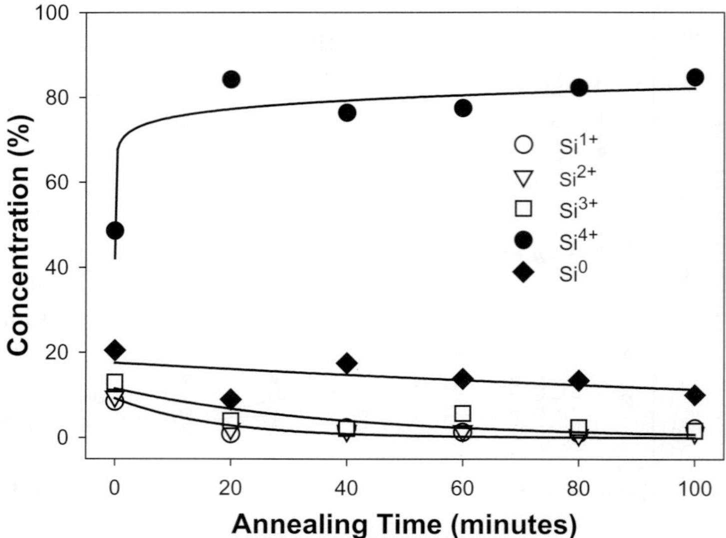

Fig. 4.11. Evolution of the concentrations of the five Si oxidation states Si^{n+} ($n = 0, 1, 2, 3$, and 4) as a function of annealing time. The annealing temperature is fixed at 1000°C [43].
Reprinted with permission of IOP Publishing Limited.

corresponding to the oxidation states Si^{1+}, Si^{2+} and Si^{3+}, respectively, occurs during annealing to form more stable stoichiometric SiO_2 and Si nanocrystals, and the thermal decomposition can be described by [43]

$$SiO_x \rightarrow \frac{x}{2}SiO_2 + \left(1 - \frac{x}{2}\right)Si, \qquad (4.2)$$

where $x = 1/2, 1, 3/2$ for Si_2O, SiO and Si_2O_3, respectively. Obviously, the thermal decomposition of the suboxides reduces the concentrations of the oxidization states Si^{1+}, Si^{2+} and Si^{3+}, and it also leads to the growth of SiO_2 and the formation of Si nanocrystals. On the other hand, thermal oxidation of the implanted Si during the annealing due to the presence of residual oxygen in the nitrogen atmosphere may also lead to the growth of SiO_2 but the reduction of the Si^0 concentration. The thermal decomposition and the oxidation explain the increase of the Si^{4+} concentration with annealing. However, although the thermal decomposition increases the Si^0 concentration, the thermal oxidation reduces the Si^0 concentration. As shown in Fig. 4.10, the low temperature annealing at 500°C leads to an increase in the Si^0 concentration, showing that the contribution from the thermal decomposition is more significant in determining the Si^0 concentration; but the annealing at a higher temperature ($> \sim 800°C$) leads to a decrease in the Si^0 concentration, suggesting that the effect of the thermal oxidation is more significant at higher temperatures. Note that the above discussions are more related to the situation of the surface region in the Si^+-implanted SiO_2 thin films as the XPS probing depth is less than 5 nm. Actually, the concentrations of the five Si oxidation states Si^{n+} ($n = 0, 1, 2, 3$, and 4) vary with the depth, due to the distribution of the implanted Si ions and the difference in the oxidation of the Si atoms between the surface and the regions deep inside the SiO_2 thin films, as shown in Fig. 4.12 [37].

Optical properties of Si nanocrystal embedded in a SiO_2 matrix have been studied with spectroscopic ellipsometry and are found to be well described by the four-term Forouhi–Bloomer (F–B) model and the Lorentz oscillator model [44]. As shown in Fig. 4.13, the Si nanocrystal shows a significant reduction in the dielectric function as compared with bulk crystalline silicon. The bandgap of the Si nanocrystal determined from the

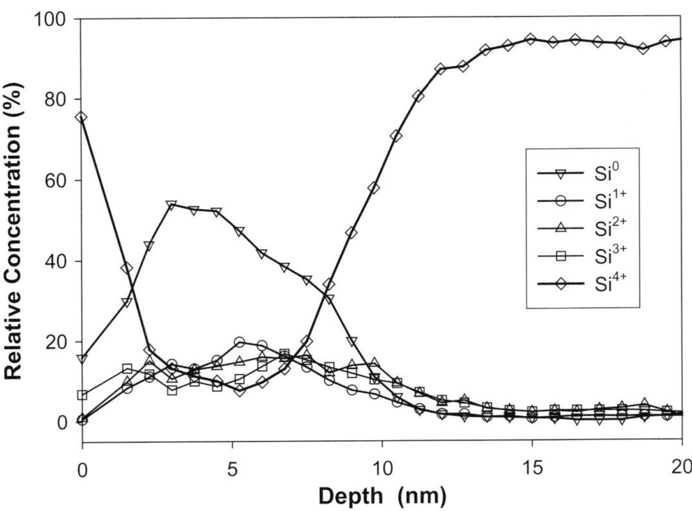

Fig. 4.12.   Depth distributions of the relative concentrations of the five oxidation states Si^{n+} ($n = 0, 1, 2, 3$, and 4) for the samples annealed at 1000°C for 100 min. Si^+ ions were implanted at 1 keV with the dose of 8×10^{16} cm^{-2} into a 30 nm SiO_2 film [37]. Reprinted with permission of The Institute of Pure and Applied Physics.

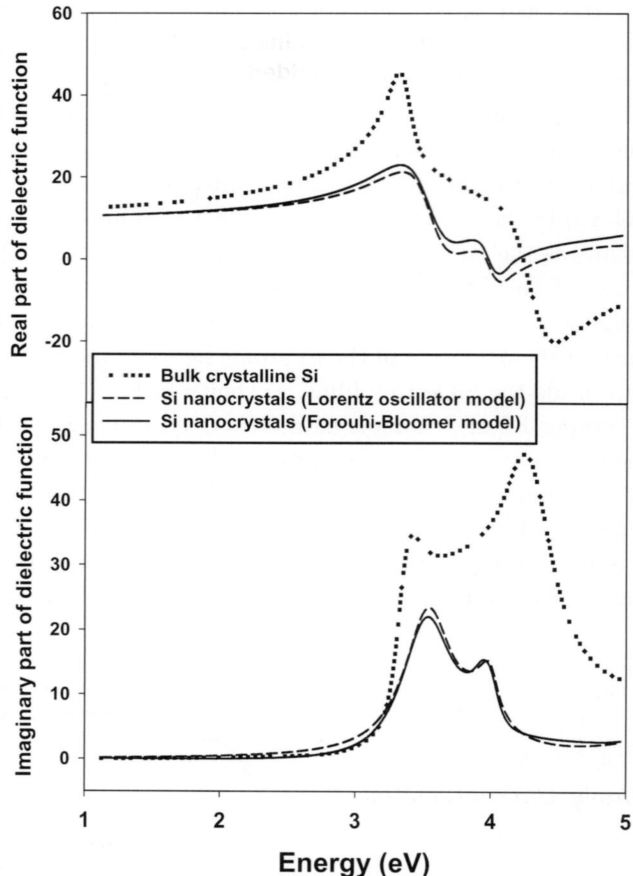

Fig. 4.13. Real (ε_1) and imaginary (ε_2) parts of the complex dielectric function of Si nanocrystal with a size of ~ 4.5 nm obtained from the ellipsometric fittings based on the Lorentz oscillator model and the F–B model. The dielectric function of bulk crystalline silicon is also included for comparison [44].
Reprinted with permission from Ref. 44. Copyright (2005) by the American Physical Society.

study of optical properties is ~ 1.7 eV, showing a large bandgap expansion of ~ 0.6 eV when compared to the bandgap (1.1 eV) of bulk crystalline silicon. The bandgap expansion is consistent with the blueshift of the absorption coefficient, as shown in Fig. 4.14. The bandgap expansion is in very good agreement with the first-principle calculation of the optical gap of silicon nanocrystals based on quantum confinement effect [44].

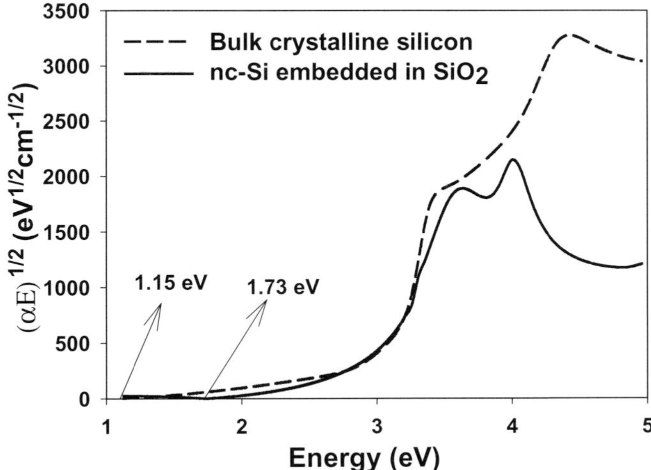

Fig. 4.14. Plot of $(\alpha E)^{\frac{1}{2}}$ versus photon energy (E) for Si nanocrystal and bulk crystalline silicon. α denotes the absorption coefficient and is calculated with $\alpha = \frac{4\pi k}{\lambda}$, where k is the extinction coefficient and λ is the wavelength [44].
Reprinted with permission from Ref. 44. Copyright (2005) by the American Physical Society.

5. Memory Behaviors and Performance of Si Nanocrystal Memory Devices

5.1. *Memory Characteristics*

The charge storage characteristics of n-channel MOS structures with Si^+-implanted gate oxides can be investigated by measuring the shift of the capacitance–voltage (C–V) curves at flatband conditions (the corresponding voltage is the flatband voltage V_{FB} [10, 13]) after applying a voltage to the MOS gate electrode for a given duration. The applied voltage should be positive for programming operation but negative for erasing operation. During the programming operation electrons are injected into the nanocrystals from the Si substrate under the influence of the positive applied voltage; and during the erasing operation the electrons trapped in the nanocrystals are removed by the negative applied voltage. Therefore, the C–V curve after the programming and erasing operations will shift towards the positive and negative sides of the voltage axe, respectively, yielding a flatband voltage difference (ΔV_{FB}) between the two operations, as shown in Fig. 5.1(a). The flatband voltage difference is translated into the threshold voltage difference (ΔV_{th}) of a MOSFET based on the MOS structure between the two

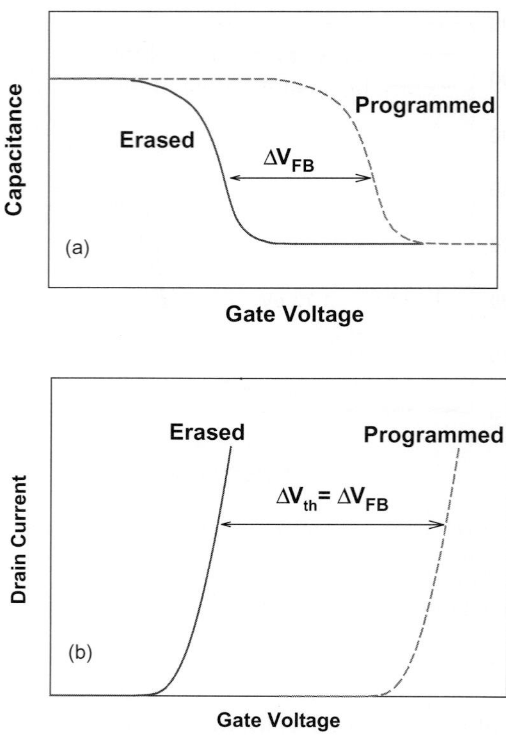

Fig. 5.1. Schematic illustrations of charging-induced shifts in (a) C–V and (b) I–V characteristics of a nanocrystal memory structure.

operations, namely, $\Delta V_{\text{FB}} = \Delta V_{\text{th}}$, as shown in Fig. 5.1(b). The flatband voltage difference or the threshold voltage difference is termed as memory window. The size of the memory window is determined by the number of electrons stored in each nanocrystal (nanocrystal size dependent), the absolute thicknesses of the dielectric layers isolating the nanocrystal from the control gate and the substrate (i.e. thicknesses of the control oxide and tunnel oxide), the capacitive coupling ratios, and the areal density of nanocrystals. Figure 5.2 shows an actual example of C–V shift and the corresponding current–voltage (I–V) (i.e. the transfer characteristic) shift of a nanocrystal memory structure with a tunnel oxide of 7 nm, a control oxide of 20 nm and a Si nanocrystal layer of $\sim$5 nm.

As examples, the performance of some nanocrystal memory devices is discussed in the following. The memory devices are actually *n*-channel MOSFETs with Si nanocrystal confined in a very narrow layer inside the

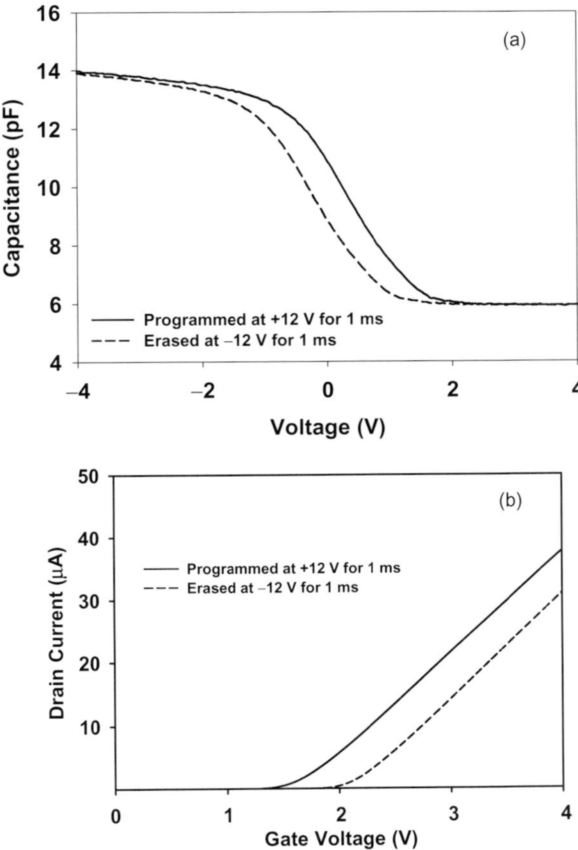

Fig. 5.2. Shifts in (a) C–V and (b) I–V characteristics of a Si nanocrystal memory structure (see the text) as a result of FN programming at $+12\,\mathrm{V}$ for 1 ms and FN erasing at $-12\,\mathrm{V}$ for 1 ms.

gate oxide. They are fabricated with a conventional CMOS process. Initially, a thin SiO_2 layer (e.g. 17 nm) is thermally grown on p-type Si wafers. The Si nanocrystal is synthesized with the ion-implantation technique, i.e. Si^+ ions are implanted into the SiO_2 thin film at a low energy (e.g. 2 keV), and then a high temperature annealing at $1000°C$ in N_2 ambient for one hour is carried out. Another SiO_2 layer (e.g. 20 nm) is deposited on top of the previously grown SiO_2 layer by low pressure CVD (LPCVD) to form the control oxide. The size of the nanocrystal is usually $\sim 4.5\,\mathrm{nm}$ as determined from the high resolution transmission electron microscope (HRTEM) measurement. Figure 5.3 shows the HRTEM cross-section of one memory device.

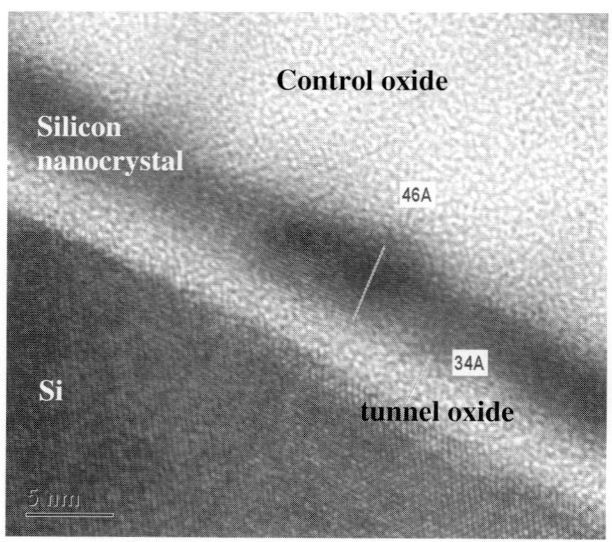

Fig. 5.3.   Cross-sectional TEM image of a memory device [47]. (©2006 IEEE).

Both the programming and erasing operations are carried out with the FN tunneling mechanism.

Figure 5.4 shows the output characteristic (i.e. drain current – drain voltage) of a memory device at the two memory states (i.e. the programmed and erased states) at the gate voltage $V_\mathrm{g} = 1\,\mathrm{V}$. As can be seen in this figure, at $V_\mathrm{d} = V_\mathrm{g} = 1\,\mathrm{V}$, the drain current I_d is $\sim 10^{-6}\,\mathrm{A}$ for the erased state while it is $\sim 10^{-13}\,\mathrm{A}$ for the programmed state, showing a current ratio of $\sim 10^{7}$. Obviously, the two memory states can be easily distinguished through the current sensing.

The memory window strongly depends on the voltage and duration of programming and erasing. As shown in Fig. 5.5, the memory window increases with the programming/erasing voltage. This is because for a fixed programming/erasing time more electrons can tunnel into the nanocrystal at a higher programming voltage while more electrons can be removed from the nanocrystal at a higher erasing voltage. Figure 5.6 shows the dependence of the threshold voltage on the programming time (Fig. 5.6(a)) and the erasing time (Fig. 5.6(b)) at a fixed programming/erasing voltage of $+15/-15\,\mathrm{V}$. The dependence reflects the fact that more electrons can tunnel into the nanocrystal and be removed from the nanocrystals for a longer programming and erasing time, respectively. It can be concluded from Fig. 5.6 that the memory window increases with the programming/erasing time.

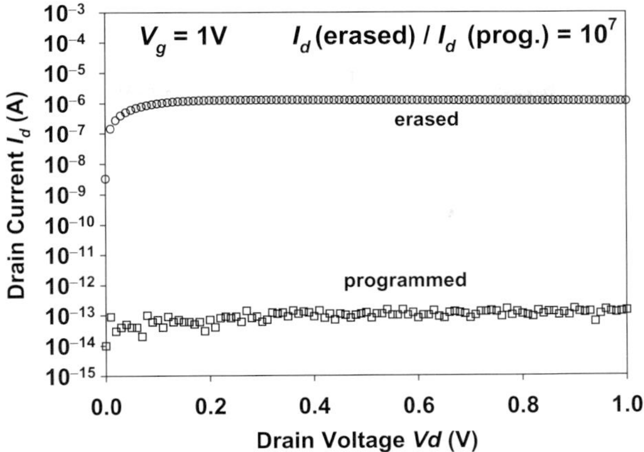

Fig. 5.4. Output characteristics of a memory device in the programmed and erased states.

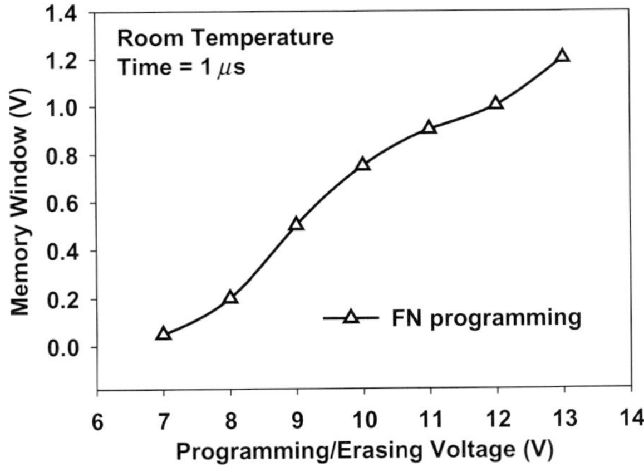

Fig. 5.5. Memory window as a function of programming/erasing voltage. The programming/erasing time is fixed at $1\,\mu$s [47]. (©2006 IEEE).

Figure 5.7 shows the endurance characteristics of a memory device. As can be seen in this figure, the threshold voltage for both programmed and erased states remains unchanged for up to 10^4 programming/erasing (P/E) cycles, and only a small drift up of 0.1 V in the threshold voltage for the both states is observed after 10^5 cycles, showing a good endurance. The drift-up in the threshold voltage is attributed to the electron trapping in the control oxide. Si nanocrystal has a smaller capture cross-section as

 T. P. Chen

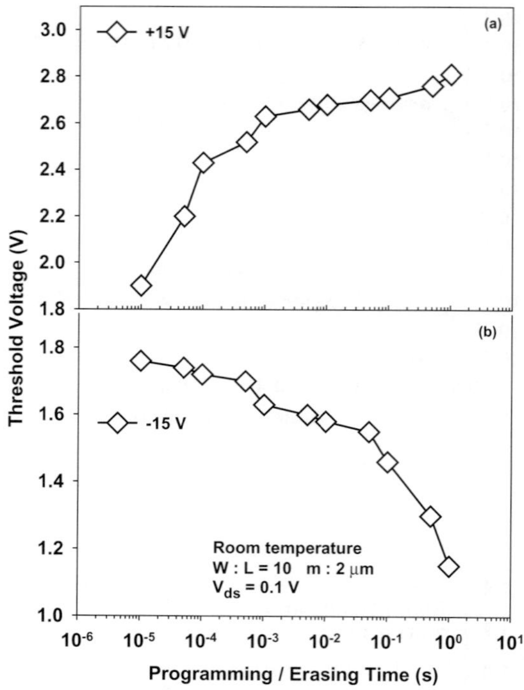

Fig. 5.6. Threshold voltage shift as a function of (a) programming time and (b) erasing time. The FN programming and erasing are carried out at $+15\,$V and $-15\,$V, respectively [48]. (©2006 IEEE).

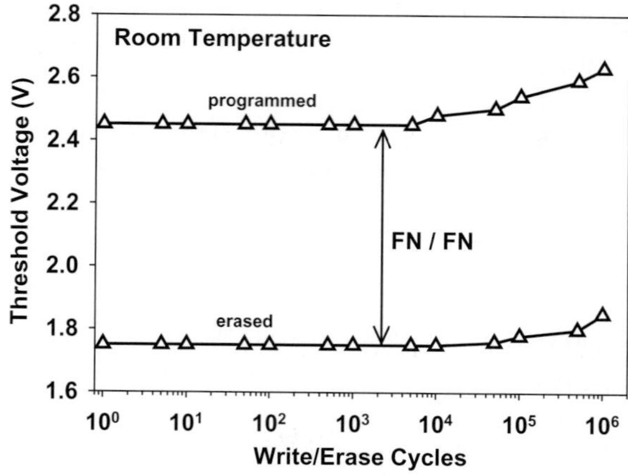

Fig. 5.7. Endurance characteristics of a memory device at room temperature. The programming and erasing operations are carried out at $+12/-12\,$V for $1\,\mu$s [47]. (© 2006 IEEE).

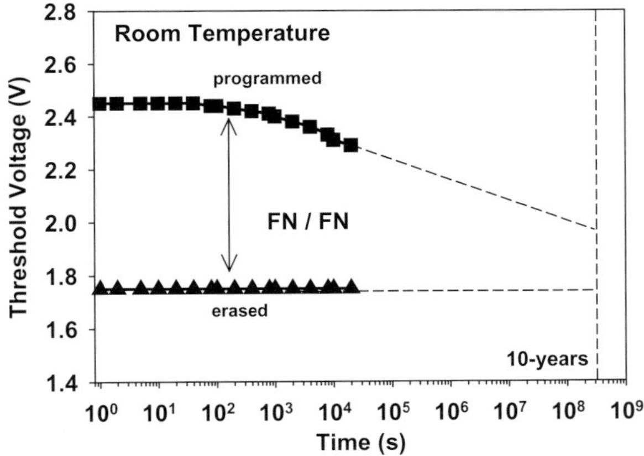

Fig. 5.8. Retention characteristics of a memory device at room temperature. The programming and erasing operation are carried out at $+12/-12\,$V for $1\,\mu$s [47]. (©2006 IEEE).

compared to the polysilicon floating gate and hence only a portion of the injected electrons from the channel are captured by the nanocrystal while the rest are trapped in the control oxide [9]. Incomplete removal of the trapped electrons from the control oxide and the nanocrystal during the erase operation leads to an increase in the flatband voltage.

The retention characteristics of a memory device are shown in Fig. 5.8. For the erased state, there is no obvious change in the threshold voltage with waiting time, indicating that there is no charge lose during the retention measurement. However, for the programmed state, the threshold voltage decreases with time after ~ 1000s, showing that the trapped electrons are moving out from the nanocrystal during the measurement. Based on Fig. 5.8 a memory window of $\sim 0.2\,$V is expected after 10 years. As discussed later, a thicker tunnel oxide and/or channel hot electron injection for programming can improve the charge retention.

5.2. *Effects of Tunnel Oxide Thickness and Programming Mechanism*

5.2.1. *Effect of Tunneling Oxide Thickness*

In the following discussions on the influence of tunnel oxide thickness on the memory performance, we will examine two memory devices: device 1 with a $\sim 7\,$nm tunnel oxide and device 2 with a $\sim 3\,$nm tunnel oxide [45].

The memory devices are fabricated on p-type (100) Si wafers with a conventional 2-μm CMOS process. A SiO_2 film is thermally grown to 20 nm and 17 nm on the p-type Si(100) substrate in dry oxygen at 950°C for device 1 and device 2, respectively. Si^+ ions with a dose of $5 \times 10^{16}\,cm^{-2}$ are then implanted at 2 keV into the oxide. With the implantation energy of 2 keV, the implanted Si ions are distributed from the oxide surface to a depth of $\sim$13 nm. This means that there are no Si ions distributed in the oxide region from the depth of $\sim$13 nm to the interface of SiO_2/Si substrate. In other words, device 1 and device 2 have a tunnel oxide (the oxide between the nanocrystal layer and the interface of SiO_2/Si substrate) of $\sim$7 nm and $\sim$3 nm, respectively. Additional 20 nm SiO_2 is deposited on top of the previously grown oxide by low pressure CVD (LPCVD) method to form the control oxide. Thermal annealing is carried out at 1000°C in N_2 ambient for 1 hour to induce nanocrystal formation. TEM measurement shows the existence of Si nanocrystal with a size of $\sim$4 nm in the gate oxide. The devices have a 20 nm control oxide and a channel width/length (W/L) of 10/2 μm. The programming and erasing operations are carried out with FN tunneling mechanism.

The room temperature low voltage transfer characteristics of the devices are shown in Fig. 5.9 [45]. For device 1, the programmed state is obtained by applying +12 V for 1 ms (Fig. 5.9(a)), while the programmed state for device 2 is obtained by applying +12 V for 1 μs (Fig. 5.9(b)). For device 1, the transistor has an on/off current (I_{On}/I_{Off}) ratio larger than 10^8 and the sub-threshold slope of $\sim$ 110 mV/dec for both the programmed and erased states. For device 2, the transistor has an I_{On}/I_{Off} ratio larger than 10^7 and the sub-threshold slope of $\sim$ 110 mV/dec for the erased state and $\sim$ 130 mV/dec for the programmed state. Obviously, a thicker tunnel oxide leads to a larger I_{On}/I_{Off} ratio. The reduction of sub-threshold slope for the programmed state of device 2 is mainly due to the localized charge effect in the nanocrystal as well as the fringing field effect [46]. However, for device 1, the localized charge effect and fringing field effect are not significant due to the thicker tunnel oxide. As such, the sub-threshold slopes for both the programmed and erased states of device 1 are the same. On the other hand, at the driving current of 1 μA, a memory window of 0.51 V is obtained for device 1 while a memory window of 1 V is obtained for device 2. This indicates that the memory window can be increased by reducing the tunnel oxide.

For the device with the tunnel oxide of $\sim$ 7 nm (device 1), at room temperature, the device exhibits a $\sim$0.5 V memory window under a short

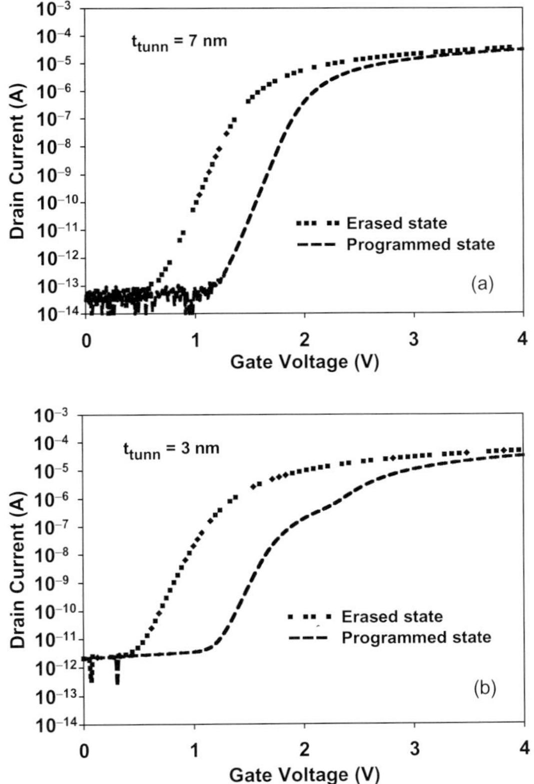

Fig. 5.9. Transfer characteristics of (a) device 1 with tunnel oxide thickness $t_{\text{tunn}} = \sim 7\,\text{nm}$ and (b) device 2 with $t_{\text{tunn}} = \sim 3\,\text{nm}$. The $I_{\text{On}}/I_{\text{Off}}$ ratio is higher for $t_{\text{tunn}} = \sim 7\,\text{nm}$. The programming/erasing is carried out at $+12/-12\,\text{V}$ for 1 ms for device 1 and $1\,\mu\text{s}$ for device 2 [45].
Reprinted with permission of John Wiley & Sons, Inc.

program/erase pulse of $+12/-12\,\text{V}$ and 1 ms. The device shows good endurance of up to 10^5 P/E cycles. However, the threshold voltage starts to drift up after 10^5 P/E cycles as shown in Fig. 5.10 [45]. The drift-up in the threshold voltage is attributed to the electron trapping in the control oxide. During the programming operation, some of the injected electrons from the substrate are trapped inside the control oxide because of the smaller capture cross-section of the nanocrystal layer. During the erasing operation, incomplete removal of the trapped electrons from the control oxide leads to an increase in the threshold voltage. Therefore, as the number of the P/E cycles increases, more and more electrons are trapped in the control oxide,

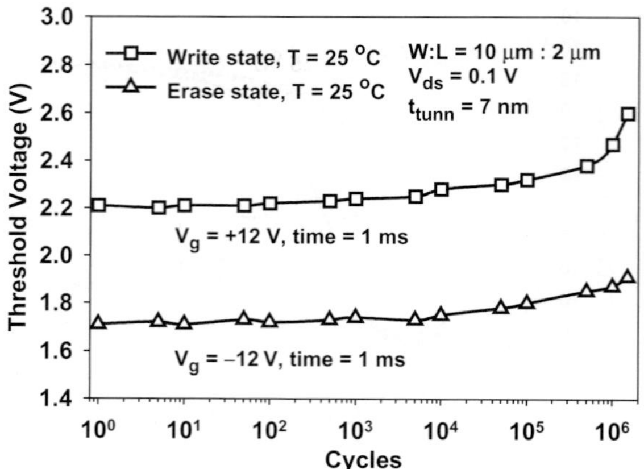

Fig. 5.10. Endurance performance for $t_{\mathrm{tunn}} = {\sim}7\,\mathrm{nm}$ (device 1). Programming/erasing operations are carried out at $+12/-12\,\mathrm{V}$ for 1 ms [45].
Reprinted with permission of John Wiley & Sons, Inc.

and hence the threshold voltage drifts up. With 10^6 P/E cycles, a drift up of ${\sim}0.25\,\mathrm{V}$ in the threshold voltage is observed, indicating a severe electron trapping in the control oxide.

For device 2 with a thinner the tunnel oxide (${\sim}3\,\mathrm{nm}$), the programming voltage and programming time are reduced greatly for a memory window similar to that of device 1. By applying programming/erasing voltage of $+9/-9\,\mathrm{V}$ for $1\,\mu\mathrm{s}$ at the gate terminal, a memory window of ${\sim}0.6\,\mathrm{V}$ can be achieved at room temperature. The device shows a very good endurance with only a slight drift up ($<2\%$) after 10^6 P/E cycles, as shown in Fig. 5.11 [45]. It should be noted that the memory window and the endurance are determined by the programming/erasing voltage. As shown in Fig. 5.12, a memory window of ${\sim}1\,\mathrm{V}$ can be obtained under a programming/erasing pulse of $+12/-12\,\mathrm{V}$ and $1\,\mu\mathrm{s}$ [45]. However, the higher programming/erasing voltage leads to a degradation of the endurance. As shown in Fig. 5.12, the threshold voltage starts shifting up after 10^5 P/E cycles, and a shift of ${\sim}0.15\,\mathrm{V}$ after 10^6 cycles is observed. Increasing the programming voltage leads to more electrons trapped in the control oxide, and thus the threshold voltage drifts up after many P/E cycles.

The retention characteristics of the memory device (device 1) with $t_{\mathrm{tunn}} = {\sim}7\,\mathrm{nm}$ for both the programmed and erased states are plotted in Fig. 5.13 [45]. As shown in this figure, ${\sim}20\,\%$ of the stored charge will be lost after 10 years for the programmed state. However, for the thinner

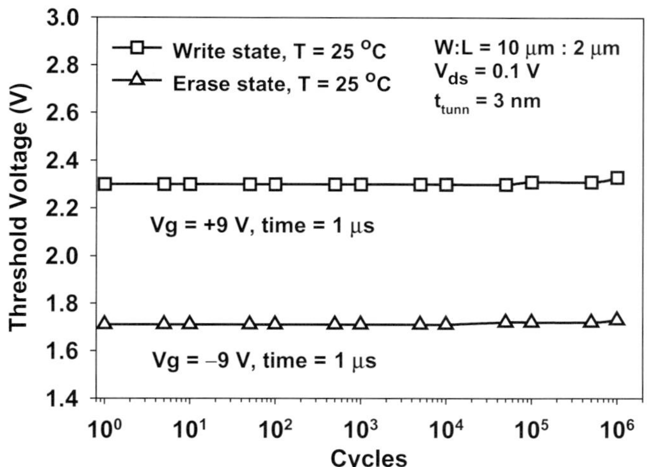

Fig. 5.11. Endurance performance for $t_{tunn} = \sim 3\,nm$ (device 2). The programming/erasing operations are carried out at $+9/-9\,V$ for $1\,\mu s$ [45].
Reprinted with permission of John Wiley & Sons, Inc.

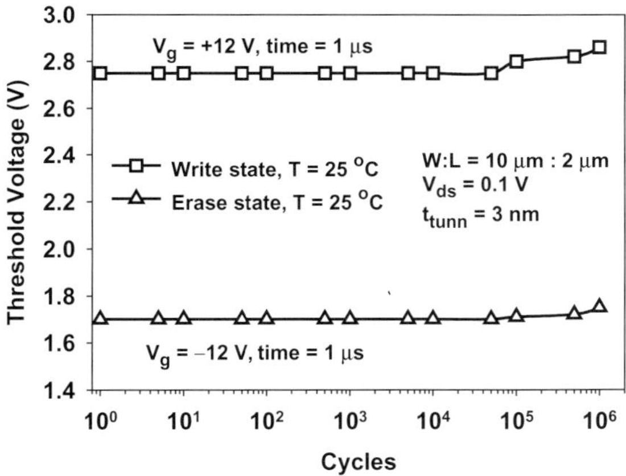

Fig. 5.12. Memory window and endurance performance for $t_{tunn} = \sim 3\,nm$ (device 2) under a higher programming/erasing voltage. The programming/erasing operations are carried out at $+12/-12\,V$ for $1\,\mu s$ [45].
Reprinted with permission of John Wiley & Sons, Inc.

tunneling oxide (device 2), i.e. $t_{tunn} = \sim 3\,nm$, $\sim 70\,\%$ of the charge will be lost after 10 years, as shown in Fig. 5.14. Therefore, the device with $t_{tunn} = \sim 7\,nm$ has a much better data retention ability than the one with $t_{tunn} = \sim 3\,nm$. For the former case, the $\sim 7\,nm$ tunnel oxide is sufficient

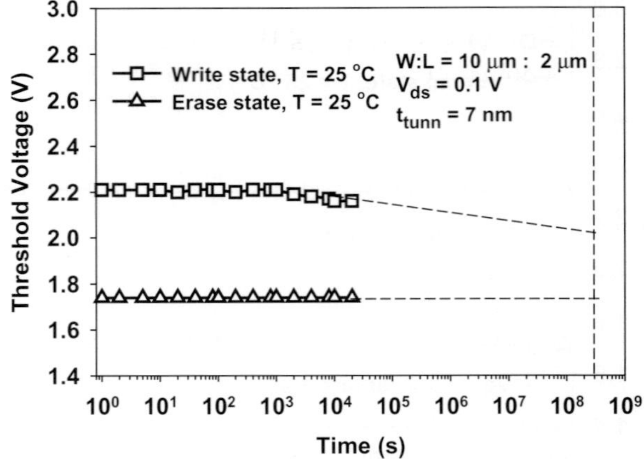

Fig. 5.13. Retention characteristics for $t_{\mathrm{tunn}} = \sim 7\,\mathrm{nm}$ (device 1). $\sim 20\,\%$ of charge loss after 10 years is expected [45].
Reprinted with permission of John Wiley & Sons, Inc.

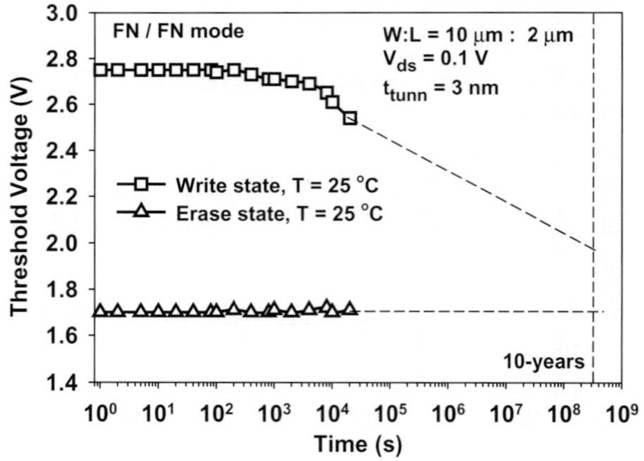

Fig. 5.14. Retention characteristics for $t_{\mathrm{tunn}} = \sim 3\,\mathrm{nm}$ (device 2). $\sim 70\,\%$ of charge loss after 10 years is expected [45].
Reprinted with permission of John Wiley & Sons, Inc.

to block the charge leakage from the nanocrystal to the Si substrate at room temperature. However, for the latter case, the charge retention is not so good due to the tunneling of the trapped electrons through the thin tunnel oxide to the Si substrate. Therefore, the reduction of the tunnel oxide thickness will enhance the charging efficiency but result in poorer retention ability. Nevertheless, a memory window of $\sim 0.3\,$V after 10 years is expected for the both cases, fulfilling the requirement of a minimum of 10-year memory window of $0.3\,$V.

5.2.2. *Effect of Programming Mechanism*

As discussed in Sec. 3.2, nanocrystal memory can be programmed by using channel hot electron (CHE) injection or Fowler–Nordheim (FN) tunneling, but it is usually erased by FN tunneling regardless of the programming mechanisms. In this section, we will examine the influence of the programming mechanisms including the FN and CHE on the memory window, endurance and charge retention of nanocrystal memory. The information presented here is adapted from [47]. The memory devices under study have a tunnel oxide of $\sim 3\,$nm and a control oxide of $\sim 20\,$nm. They have a channel width/length (W/L) of $10/2\,\mu$m. The process of device fabrication is similar to the one described in Sec. 5.2.1.

To program the memory devices, electrons are injected into the nc-Si from the channel by either the CHE or FN mechanisms. For the CHE programming, a positive voltage is applied to both the gate and the drain (i.e. the gate voltage $V_{\mathrm{g}} = $ the drain voltage $V_{\mathrm{d}} > 0$) while both the source and substrate are grounded. For the FN programming, a positive voltage is applied to the gate (i.e. $V_{\mathrm{g}} > 0$) while the other three terminals are grounded. To erase the memory devices, the electrons trapped in the nanocrystal are expelled by the FN mechanism. For the FN erasing, a negative voltage is applied to the gate (i.e. $V_{\mathrm{g}} < 0$) while all the other terminals are grounded. In the following discussions, "FN/FN" and "CHE/FN" denote the operations of FN programming/FN erasing and CHE programming/FN erasing, respectively.

Figures 5.15(a) and (b) show the shift of transfer characteristics of the device at room temperature under the FN/FN ($V_{\mathrm{g}} = +10\,$V for programming and $V_{\mathrm{g}} = -10\,$V for erasing) for $1\,\mu$s and under the CHE/FN ($V_{\mathrm{g}} = V_{\mathrm{d}} = +10\,$V for programming and $V_{\mathrm{g}} = -10\,$V for erasing) for $1\,\mu$s, respectively [47]. As can be seen from Fig. 5.15, the FN/FN leads to a memory window of $\sim 0.75\,$V, while the CHE/FN yields a memory window of $\sim 1\,$V. It is observed that a larger memory window is always achieved by the CHE/FN than by the FN/FN for the same voltage magnitude and time of programming/erasing. This situation is also clearly demonstrated by Fig. 5.16 [47]. Figure 5.16 shows the memory window as a function of the programming/erasing voltage at room temperature for both the CHE/FN

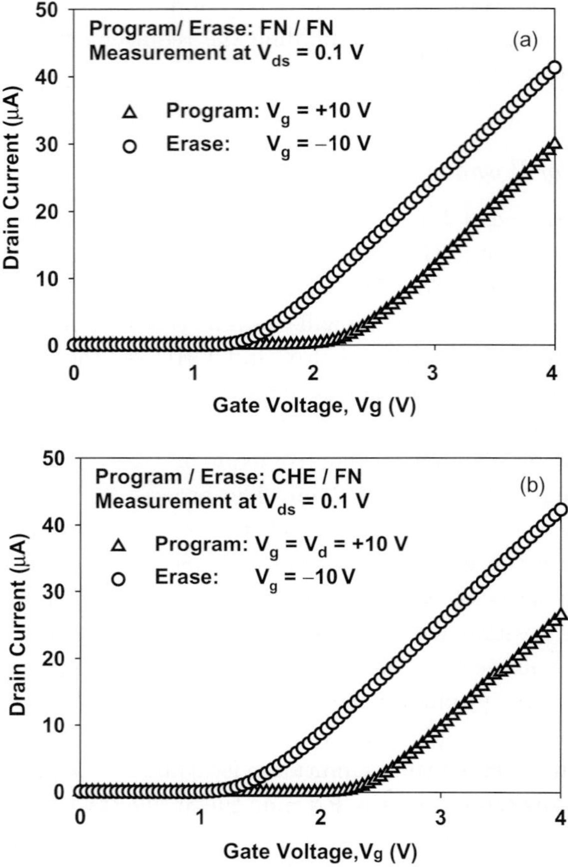

Fig. 5.15.   Transfer characteristics of the programmed and erased states from the programming/erasing operations of (a) FN/FN or (b) CHE/FN. The programming/erasing time is $1\,\mu$s [47]. (©2006 IEEE).

and the FN/FN. The programming/erasing time is fixed at $1\,\mu$s. It is evident from this figure that the CHE/FN leads to a larger memory window as compared to the FN/FN for any programming/erasing voltage. A larger memory window is actually the consequence of a larger threshold voltage increase caused by programming (i.e. a larger $V_{\rm th}$ at the programmed state) and a lower threshold voltage after erasing (i.e. a lower $V_{\rm th}$ at the erased state) in the CHE/FN operation, as discussed in the following.

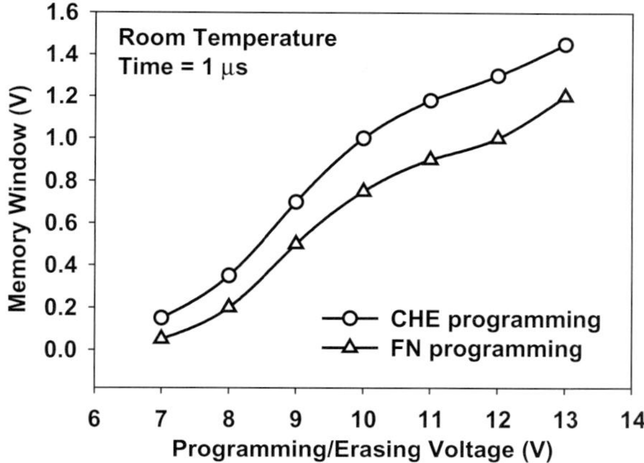

Fig. 5.16. Memory window as a function of magnitude of the programming/erasing voltage for the FN/FN and the CHE/FN operations [47]. (©2006 IEEE).

As can be seen in Fig. 5.17, the CHE programming causes a larger threshold voltage shift as compared to the FN programming [47]. Under the CHE programming, the channel hot electrons are generated by the lateral electric field near the drain junction. The generated hot electrons are injected into the nanocrystals and the gate oxide both near the drain junction, leading to local charge trapping near the drain. This situation is similar to that observed for conventional NVMs. In contrast, under the FN programming, electrons are injected from the Si substrate into the nanocrystals and the gate oxide over the channel, leading to charge trapping with an almost uniform lateral distribution over the whole channel. The difference in the distribution of trapped charges between the two programming mechanisms can explain the larger threshold voltage shift caused by the CHE programming. In the present study, the threshold voltage is measured by reading the drain current with the source grounded (i.e. forward reading). With this measurement configuration, the threshold voltage is sensitive to the charge trapping near the drain, and it is larger for a larger amount of localized charge trapping near the drain. Therefore, the CHE programming can yield a larger threshold voltage shift as it can cause more charge trapping near the drain. On the other hand, as discussed later, the FN programming has more electrons trapped in the control oxide as compared to the CHE programming. The threshold voltage shift due to the electron trapping in the control oxide should be not large because the trapped electrons are far away from the channel. This may be also partially responsible

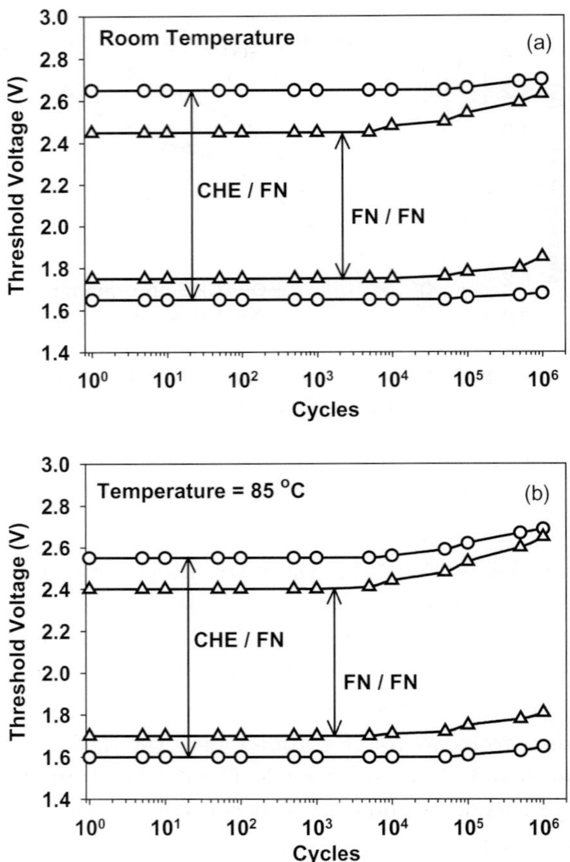

Fig. 5.17. Endurance characteristics for the CHE/FN and FN/FN at (a) room temperature and (b) 85°C [47]. (©2006 IEEE).

for the relatively smaller threshold voltage shift caused by the FN programming.

On the other hand, as shown in Fig. 5.17, the threshold voltage after erasing (i.e. the V_{th} at the erased state) in the CHE/FN operation is lower than that in the FN/FN operation. As discussed above, in the CHE/FN operation, the CHE programming leads to the localized charge trapping near the drain. However, during the FN erasing the entire nanocrystal layer is erased. In other words, the trapped charges in the region near the drain are removed, but at the same time electron tunneling from the nanocrystal of all other regions to the Si substrate can also occur under the influence of the uniform oxide electric field produced by the negative gate bias. This over erase leads to a lower V_{th} at the erased state. Actually, the over erase is evident from the fact that the initial V_{th} before any programming/erasing is $\sim 1.75\,\text{V}$ but the V_{th} decreases to $\sim 1.65\,\text{V}$ at the erased state after a CHE/FN operation. In contrast, in the FN/FN operation, both the charge trapping and the charge erasing are laterally uniform, and thus there is no over erase occurring [47].

The endurance characteristics of the memory device are also shown in Fig. 5.17. For the endurance measurement, the voltage magnitude and duration of programming/erasing are $10\,\text{V}$ and $1\,\mu\text{s}$, respectively, for both FN/FN and CHE/FN. At room temperature, the threshold voltage (V_{th}) starts to drift up after $\sim 10^4$ and $\sim 10^5$ P/E cycles for the FN/FN and the CHE/FN, respectively. After 10^6 P/E cycles, the FN programming shows a V_{th} drift-up of $\sim 0.13\,\text{V}$, while the CHE programming shows a drift-up of only $\sim 0.05\,\text{V}$. At the hot temperature of $85°\text{C}$, the V_{th} drift-up is more significant, and $0.25\,\text{V}$ and $0.13\,\text{V}$ drift-up after 10^6 P/E cycles are observed for the FN/FN and the CHE/FN, respectively. The drift-up in the V_{th} is attributed to the electron trapping in the control oxide [47]. During the erasing process, incomplete removal of the trapped electrons from the control oxide leads to an increase in the threshold voltage, and this effect is enhanced at a higher temperature. The more significant V_{th} drift-up in the FN/FN operation indicates that the FN programming leads to more electrons trapped in the control oxide than the CHE programming.

Figures 5.18(a) and (b) show the retention characteristics of the memory devices at room temperature and at $85°\text{C}$, respectively [47]. The retention measurement was carried out after 1000 P/E cycles. It is observed from the figure that the charge retention for the CHE programming is better than that for the FN programming. For the programmed state at room temperature, $\sim 50\%$ and $\sim 70\%$ charge lost after 10 years are expected for the CHE and FN programming, respectively. The better charge retention is related to the localized charge trapping caused by the CHE programming. As discussed earlier, the trapped charges are distributed uniformly in the entire nanocrystal layer for the FN programming, but the charge trapping is highly localized near the drain for the CHE programming. As such, the cross section (or area) for electron tunneling out from the nanocrystal to the channel during the retention experiment should be much smaller for the latter case. This means that the charge lost is less in the case of CHE programming. For the V_{th} at the erased state at room temperature, for the case of FN/FN it remains unchanged with time, showing no charge

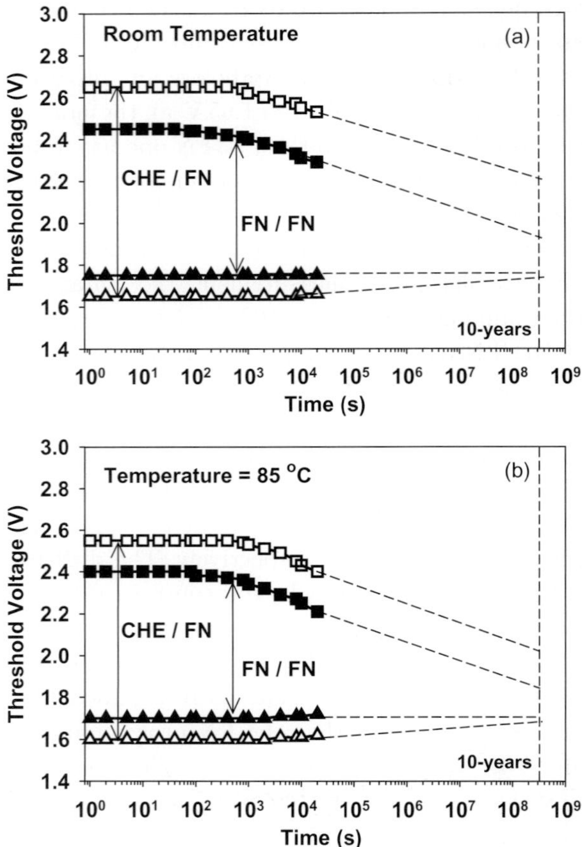

Fig. 5.18. Retention characteristics for the CHE/FN and FN/FN at (a) room temperature and (b) 85°C [47]. (©2006 IEEE).

transfer during the retention experiment. However, for the case of CHE/FN, it starts to increase at the time of 10^4 seconds from $\sim 1.65\,\mathrm{V}$ with a trend to the initial value (i.e. the V_{th} before any programming/erasing) of $\sim 1.75\,\mathrm{V}$, indicating the recovery from the over erase that is discussed earlier. At room temperature, the expected memory window after 10 years is $\sim 0.5\,\mathrm{V}$ for the CHE/FN, which meets the retention requirement of NVMs. However, it is only $\sim 0.2\,\mathrm{V}$ for the FN/FN. At the hot temperature of 85°C, the retention capability degrades significantly for both the CHE/FN and the FN/FN due to the enhancement of thermally-assisted tunneling of electrons from the nanocrystal. However, for the CHE/FN the expected memory window after

10 years is still larger than 0.3 V which could be sufficient for the proper memory functioning.

It is also observed that the CHE programming leads to fewer damages to the gate oxide and the oxide/Si interface. Figure 5.19 shows the gate leakage currents before and after 10^4 P/E cycles for both the FN programming (Fig. 5.19(a)) and the CHE programming (Fig. 5.19(b)) [47]. For the FN programming, a significant stress-induced leakage current (SILC) is observed after 10^4 P/E cycles. The SILC at low gate voltage ($<\sim 3$ V) is attributed to the displacement current associated with the interface traps and oxide traps near the interface that are generated by the FN injection, while the SILC at high gate voltage ($>\sim 3$ V) can be explained in terms of the oxide-trap-assisted tunneling. In contrast to the FN programming,

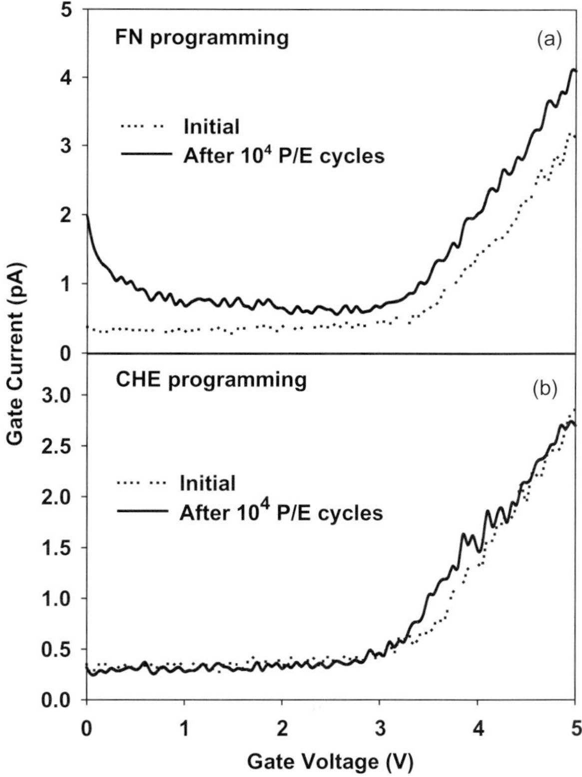

Fig. 5.19. Gate current before and after 10^4 programming/erasing cycles for (a) the FN/FN and (b) the CHE/FN [47]. (©2006 IEEE).

the CHE programming leads to an insignificant SILC after 10^4 P/E cycles. Therefore, it is evident that the CHE programming produces fewer damages to the oxide and the interface and thus has a better reliability than the FN programming.

6. Single-Electron Memory Effect

Single-electron devices, which are based on Coulomb blockade effect [8], have recently attracted much attention. In a single-electron memory device, the addition or subtraction of a small number of electrons to/from a charge storage node can be controlled with one-electron precision. Single-electron memory effect in silicon nanocrystal memory has been demonstrated and studied in the past decade [12, 25, 49–62].

Tiwari *et al.* [49] observed single-electron charging effect in Si nanocrystal memory devices at low temperatures. In their experiment, the memory devices had a nc-Si density of 3–$10 \times 10^{11}\,\mathrm{cm}^{-2}$, a tunnel oxide of 1.5–4.0 nm, a control oxide of 9 nm, a gate length of $0.4\,\mu$m and a gate width of $10.0\,\mu$m. Figure 6.1 shows the threshold voltage as a function of the static voltage between the gate and the source at 300, 77, and 40 K [49]. As the temperature is lowered, the threshold voltage shows plateaus that become pronounced below 120 K and are nearly equidistant. The plateaus

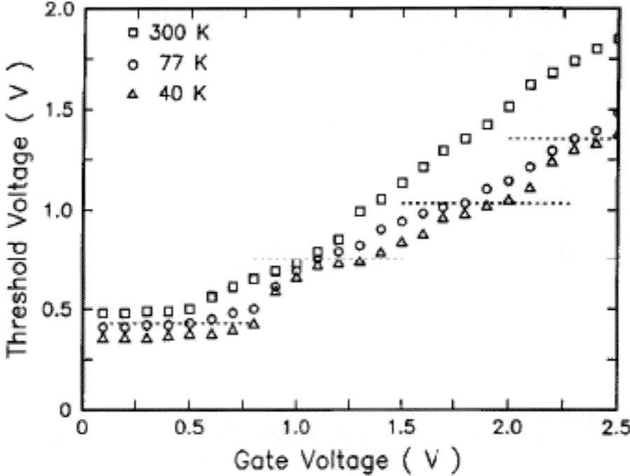

Fig. 6.1.   Pulsed threshold voltage obtained as a function of static gate voltage at 300, 77, and 40 K, for an n-channel nanocrystal memory with a gate length of $0.4\,\mu$m and gate width of $10.0\,\mu$m. The applied drain voltage is 100 mV. This device utilizes an injection oxide of 1.9 nm and a control oxide of ~ 9 nm [49].

Reused with permission from Ref. 49. Copyright 1996, American Institute of Physics.

of the threshold voltage indicate the single-electron charging effect in Si nanocrystal at lower temperatures.

Kim *et al.* [12] demonstrated the first room temperature Coulomb blockade effect in Si nanocrystal memory. The quantized shift in threshold voltage and quantized charging voltage were observed at room temperature, which was attributed to the single-electron charging effect. The Si nanocrystals with a size of about 4.5 nm and a density of $5 \times 10^{11}/\text{cm}^2$ were synthesized by LPCVD technique. With the dot size of 4.5 nm, the energy level splitting due to the quantum effect is calculated to be 128 meV with infinite potential well approximation and the charging energy is calculated to be 82 meV with the assumption that the dot is a spherical-shaped conductor. Therefore, the sum of the quantum energy level spacing and the charging energy is 210 meV. As this value is much larger than the thermal energy of 26 meV at room temperature, single-electron effect can be observed at room temperature. Figure 6.2 shows the room temperature steady-state drain current after static gate bias is applied for a long time sufficient for full programming [12]. The staircase plateauing of steady state drain currents in equidistant gate voltage (V_{GS}) steps with a periodicity of $\Delta V_{\text{GS}} = 1.7\,\text{V}$ shows the single-electron effect at room temperature. The periodicity of the staircase plateaus is consistent with the following

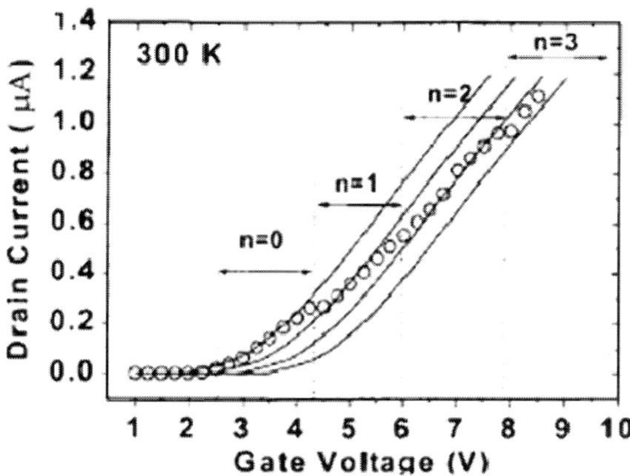

Fig. 6.2.   Saturation drain current versus gate bias applied for a long time — enough for full programming. The drain was biased at 0.1 V for each measurement. The solid lines represent the drain current–gate voltage curves with zero, one, two, and three electrons per quantum dot, respectively. Each region ($n = 0, 1, 2, 3$) represents zero, one, two, and three electron storage per nanocrystal, respectively. This figure shows the single-electron effect at room temperature [12]. (©1999 IEEE).

calculation [12]. The required voltage increase on the control gate for one electron charging onto the dot is calculated by the following equation of $\Delta V_{\mathrm{GS}} = 210\,\mathrm{mV} \times (1 + C_{\mathrm{tt}}/C_{\mathrm{CG}})$. Here, the gate-to-dot capacitance C_{CG} is calculated to be $0.17\,\mu\mathrm{F/cm}^2$, and the dot-to-channel capacitance C_{tt} is calculated to be $1.26\,\mu\mathrm{F/cm}^2$. Then, ΔV_{GS} is calculated to be $1.8\,\mathrm{V}$, being consistent with the measured periodicity of the staircase plateaus (about $1.7\,\mathrm{V}$) mentioned above.

Kapetanakis *et al.* [25] observed room temperature single-electron charging phenomena in large area Si nanocrystal memory obtained by low energy ion implantation synthesis. The evidence revealing a single-electron charging effect is the plateau structure of the drain current (I_{DS}) versus gate voltage (V_{G}) characteristics, shown in Fig. 6.3 for different V_{G} sweep rates [25]. Electron injection into the nanocrystals during each measurement causes a dynamic shift of the threshold voltage V_{Th} of the device, resulting in the observed I_{DS}–V_{G} structures. The time needed for an electron to inject into the nanocrystal, strongly depends on V_{G}. Thus, for the case of a fast sweep, charge injection takes place at relatively high V_{G} resulting in an efficient reduction of the I_{DS} due to the large number of injected electrons. The initial portion of the I_{DS}–V_{G} curve in this case corresponds to the

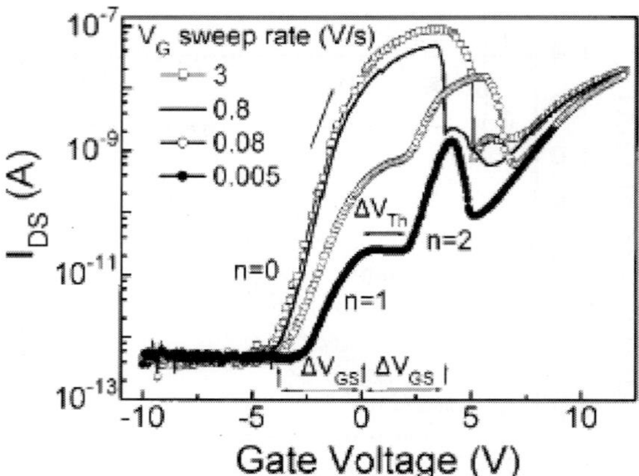

Fig. 6.3. Room temperature transfer characteristics under different V_{G} sweep rates and the drain voltage $V_{\mathrm{DS}} = 0.1\,\mathrm{V}$. With decreasing sweep rate, the I_{DS}–V_{G} curves show a clear staircase structure. The arrows indicate voltage regions ΔV_{GS} with 1 and 2 stored electrons per nanocrystal ($n = 1, 2$), while the initial portion of the fast swept I_{DS}–V_{G} curve corresponds to $n = 0$. ΔV_{Th} is the threshold voltage shift caused by the storage of a single electron per nanocrystal [25].
Reused with permission from Ref. 25. Copyright 2002, American Institute of Physics.

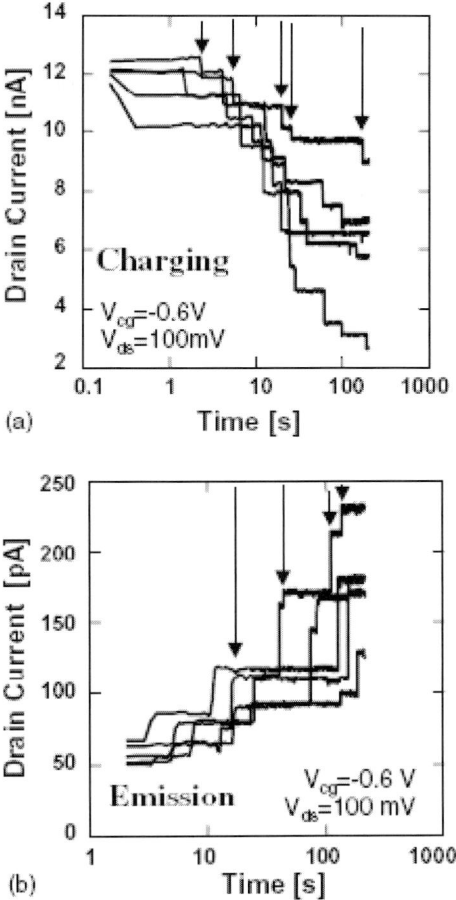

Fig. 6.4. Repeated measurements of the time evolution of the drain current at a constant gate voltage (a) after erasing and (b) after charging. The stepwise behavior clearly appears in both cases. The charging/discharging times are widely spread [58]. Reprinted with permission of Elsevier.

uncharged state of the nanocrystals (i.e. the number of electrons trapped in the nanocrystals $n = 0$). With decreasing sweep rate, V_G for electron injection decreases together with the number of involved electrons, resulting in the appearance of a plateau in the I_{DS}–V_G curves. The plateau structure for the sweep rate of 5 mV/s clearly shows the Coulomb blockade effect at room temperature.

Molas *et al.* [58] also studied the single-electron charging/discharging phenomena in a silicon nanocrystal memory at room temperature. The silicon nanocrystals were formed by depositing and successively annealing (at 1050°C) a 3 nm-thick suboxide layer ($SiO_{0.5}$). To investigate the discrete electron charging phenomena, they measured the time evolution of the drain current at a constant gate voltage, after an erasing stress and after a writing stress, and the result is shown in Fig. 6.4. Clear staircase phenomena appear corresponding to a one-by-one electron trapping/detrapping during the time-domain measurement of the drain current at room temperature. These discrete behaviors show the single-electron charging/discharging effect at room temperature.

7. Summary

Nanocrystal-based NVM devices, which were first introduced in the early nineties, provide a promising alternative to conventional floating gate NVM devices, by allowing a further decrease in the tunnel oxide and reduction in operating voltages. In a nanocrystal memory device, the continuous floating gate of the conventional NVM device is replaced by an ensemble of Si nanocrystal. In other words, charge is not stored on a continuous floating gate poly-Si layer, but instead on a layer of discrete, mutually isolated, crystalline Si nanocrystals embedded in the gate oxide. For the memory application, Si nanocrystal can be synthesized with various techniques such as CVD and Si ion implantion followed by a high temperature annealing. As compared to conventional floating gate NVM devices, nanocrystal memory devices offer several advantages. The main advantage is the potential to use thinner tunnel oxides without sacrificing non-volatility. In addition, nanocrystal memory devices use a more simplified fabrication process as compared to conventional floating gate NVM's.

The operation principle of nanocrystal memory device is very similar to that of conventional floating gate NVM devices. Nanocrystal memory device can be programmed with either the FN tunneling or channel hot electron injection but is usually erased with FN tunneling. However, channel hot electron programming is found to yield a better memory performance and reliability. Promising device results have been presented, demonstrating low voltage operation for comparable memory windows, and good thin tunnel oxide retention performance that suggests to meet long-term non-volatility requirements. An important issue for nanocrystal memory is the charge trapping in the control oxide during the programming operation. The charge trapping leads to an increase in the threshold voltage for both the programmed and erased states during repeated programming/erasing operations. This problem could be overcome by development of high quality oxides and/or device design solutions.

The nanocrystals also offer the possibility of single-electron memory devices operating at room temperature. Single-electron memory effect on silicon nanocrystal memory has been demonstrated. However, there are still many issues and difficulties to be overcome, and the challenges are becoming more and more concrete.

Acknowledgments

This work has been financially supported by the Ministry of Education of Singapore under project No. ARC 1/04. The author wishes to thank his Ph.D. students, Y. Liu, C. Y. Ng, L. Ding, M. Yang and J. I. Wong for their contributions to the results from the group of Nanyang Technological University presented in the review. He also thanks many of the authors cited in this review for discussions.

References

1. G. Groeseneken, H.E. Maes, J. Van Houdt and J.S. Witters, Basics of Non-volatile Semiconductor Memory Devices, in *Nonvolatile Semiconductor Memory Technology*, eds. W.D. Brown and J.E. Brewer (IEEE Press, 1998), pp. 1–88.
2. R. Bez, E. Camerlenghi, A. Modelli and A. Visconti, Introduction to Flash Memory, *Proc. IEEE* **91**(4) (2003) 489–502.
3. J.D. Blauwe, Nanocrystal Nonvolatile Memory Devices, *IEEE Trans. Nanotechnol.* **1**(1) (2002) 72–77.
4. G. Ammendola, V. Ancarani, V. Triolo, M. Bileci, D. Corso, I. Crupi, L. Perniola, C. Gerardi, S. Lombardo and B. DeSalvo, Nanocrystal memories for FLASH device applications, *Solid State Electron.* **48** (2004) 1483–1488.
5. S. Tiwari, F. Rana, K. Chan, H. Hanafi, W. Chan and D. Buchanan, Volatile and Non-Volatile Memories in Silicon with Nano-Cyrstal Storage, in *IEDM Tech. Dig.* (1995), pp. 521–524.
6. S. Tiwari, F. Rana, H. Hanafi, A. Hartstein, E.F. Crabbé and K. Chan, A silicon nanocrystals based memory, *Appl. Phys. Lett.* **68** (1996) 1377–1379.
7. S. Lombardo, B.D. Salvo, C. Gerardi and T. Baron, Silicon nanocrystal memories, *Microelectronic Eng.* **72** (2004) 388–394.
8. K. Yano, T. Ishii, T. Sano, T. Mine, F. Murai, T. Hashimoto, T. Kobayashi, T. Kure and K. Seki, Single-Electron Memory for Giga-to-Tera Bit Storage, *Proc. IEEE* **87**(4) (1999) 633–651.
9. R.A. Rao, R.F. Steimle, M. Sadd, C.T. Swift, B. Hradsky, S. Straub, T. Merchant, M. Stoker, S.G.H. Anderson, M. Rossow, J. Yater, B. Acred, K. Harber, E.J. Prinz, B.E. White Jr. and R. Muralidhar, Silicon nanocrystal

based memory devices for NVM and DRAM applications, *Solid State Electron.* **48** (2004) 1463–1473.

10. G. Ammendola, M. Vulpio, M. Bileci, N. Nastasi, C. Gerardi, G. Renna, I. Crupi, G. Nicotra and S. Lombardo, Nanocrystal metal-oxide semiconductor memories obtained by chemical vapor deposition of Si nanocrystals, *J. Vac. Sci. Technol. B* **20** (2002) 2075–2079.

11. M. Saitoh, E. Nagata and T. Hiramoto, Large memory window and long charge-retention time in ultranarrow-channel silicon floating-dot memory, *Appl. Phys. Lett.* **82** (2003) 1787–1789.

12. I. Kim, S. Han, K. Han, J. Lee and H. Shin, Room Temperature Single Electron Effects in a Si Nano-Crystal Memory, *IEEE Electron. Device Lett.* **20** (1999) 630–631.

13. Y. Shi, K. Saito, H. Ishikuro and T. Hiramoto, Effects of traps on charge storage characteristics in metal-oxide semiconductor memory structures based on silicon nanocrystals, *J. Appl. Phys.* **84** (1998) 2358–2360.

14. T. Gebel, J. von Borany, H.-J. Thees, M. Wittmaack, K.-H. Stegemann and W. Skorupa, Non-volatile memories based on Si^+-implanted gate oxides, *Microelectron. Eng.* **59** (2001) 247–252.

15. J. von Borany, T. Gebel, K.-H. Stegemann, H.-J. Thees and M. Wittmaack, Memory properties of Si^+ implanted gate oxides: From MOS capacitors to nvSRAM, *Solid State Electron.* **46** (2002) 1729–1737.

16. L. Ding, T.P. Chen, Y. Liu, C.Y. Ng, Y.C. Liu and S. Fung, Thermal annealing effect on the band gap and dielectric functions of silicon nanocrystals embedded in SiO_2 matrix, *Appl. Phys. Lett.* **87** (2005) 121903–121905.

17. A. Kalnitsky, A.R. Boothroyd, J.P. Ellul, E.H. Poindexter and P.J. Caplan, Electronic states at Si–SiO_2 interface introduced by implantation of Si in thermal SiO_2, *Solid State Electron.* **33** (1990) 523–530.

18. A. Kalnitsky, A.R. Boothroyd and J.P. Ellul, A model of charge transport in thermal SiO_2 implanted with Si, *Solid State Electron.* **33** (1990) 893–905.

19. T. Gebel, J. von Borany, H.-J. Thees, M. Wittmaack, K.-H. Stegemann and W. Skorupa, Non-volatile memories based on Si^+-implanted gate oxides, *Microelectron. Eng.* **59** (2001) 247–252.

20. J. von Borany, T. Gebel, K.-H. Stegemann, H.-J. Thees and M. Wittmaack, Memory properties of Si^+ implanted gate oxides: From MOS capacitors to nvSRAM, *Solid State Electron.* **46** (2002) 1729–1737.

21. C. Bonafos, M. Carrada, N. Cherkashin, H. Coffin, D. Chassaing, G.B. Assayag, A. Claverie, T. Müller, K.H. Heinig, M. Perego, M. Fanciulli, P. Dimitrakis and P. Normand, Manipulation of two-dimensional arrays of Si nanocrystals embedded in thin SiO_2 layers by low energy ion implantation, *J. Appl. Phys.* **95** (2004) 5696–5702.

22. Y. Liu, T.P. Chen, C.Y. Ng, M.S. Tse, S. Fung, Y.C. Liu, S. Li and P. Zhao, Charging Effect on Electrical Characteristics of MOS Structures with Si Nanocrystal Distribution in Gate Oxide, *Electrochem. Solid State Lett.* **7** (2004) G134–G137.

23. C.Y. Ng, T.P. Chen, Y. Liu, M.S. Tse and D. Gui, Modulation of Capacitance Magnitude by Charging/Discharging in Silicon Nanocrystals Distributed

Throughout the Gate Oxide in MOS Structures, *Electrochem. Solid State Lett.* **8** (2005) G8–G10.

24. C.Y. Ng, T.P. Chen, H.W. Lau, Y. Liu, M.S. Tse, O.K. Tan and V.S.W. Lim, Visualizing charge transport in silicon nanocrystals embedded in SiO_2 films with electrostatic force microscopy, *Appl. Phys. Lett.* **85** (2004) 2941–2943.

25. E. Kapetanakis, P. Normand, D. Tsoukalas and K. Beltsios, Room-temperature single-electron charging phenomena in large-area nanocrystal memory obtained by low energy ion beam synthesis, *Appl. Phys. Lett.* **80** (2002) 2794–2796.

26. E. Kapetanakis, P. Normand, D. Tsoukalas, K. Beltsios, J. Stoemenos, S. Zhang and J. van den Berg, Charge storage and interface states effects in Si-nanocrystal memory obtained using low energy Si^+ implantation and annealing, *Appl. Phys. Lett.* **77** (2000) 3450–3452.

27. P. Normand, D. Tsoukalas, E. Kapetanakis, J.A. Van Den Berg, D.G. Armour, J. Stoemenos and C. Vieu Formation of 2-D Arrays of Silicon Nanocrystals in Thin SiO_2 Films by Very-Low Energy Si^+ Ion Implantation, *Electrochem. Solid State Lett.* **1** (1998) 88–90.

28. P. Normand, P. Dimitrakis, E. Kapetanakis, D. Skarlatos, K. Beltsios, D. Tsoukalas, C. Bonafos, H. Coffin, G. Benassayag, A. Claverie, V. Soncini, A. Agarwal, Ch. Sohl and M. Ameen, Processing issues in silicon nanocrystal manufacturing by ultra low energy ion-beam-synthesis for non-volatile memory applications, *Microelectron. Eng.* **73–74** (2004) 730–735.

29. P. Normand, E. Kapetanakis, P. Dimitrakis, D. Skarlatos, D. Tsoukalas, K. Beltsios, A. Claverie, G. Benassayag, C. Bonafos, M. Carrada, N. Cherkashin, V. Soncini, A. Agarwal, Ch. Sohl and M. Ameen, Effects of annealing conditions on charge storage of Si nanocrystal memory devices obtained by low-energy ion beam synthesis, *Microelectron. Eng.* **67–68** (2003) 629–634.

30. P. Normand, K. Beltsios, E. Kapetanakis, D. Tsoukalas, T. Travlos, K. Stoemenos, J.A. Van Den Berg, S. Zhang, C. Vieu, H. Launois, J. Gautier, F. Jourdan and L. Palun, Formation of 2-D arrays of semiconductor nanocrystas or semiconductor-rich nanolayers by very low energy Si or Ge ion implantation in silicon oxide films, *Nucl. Instr. Meth. Phys. Res.* **B178** (2001) 74–77.

31. P. Normand, E. Kapetanakis, P. Dimitrakis, D. Skarlatos, K. Beltsios, D. Tsoukalas, C. Bonafos, G. Ben Assayag, N. Cherkashin, A. Claverie, J.A. Van Den Berg, V. Soncini, A. Agarwal, M. Ameen, M. Perego and M. Fanciulli, Nanocrystals manufacturing by ultra-low-energy ion-beam-synthesis for non- volatile memory applications, *Nucl. Instr. Meth. Phys. Res.* **B216** (2004) 228–238.

32. P. Normand, E. Kapetanakis, P. Dimitrakis, D. Tsoukalas, K. Beltsios, N. Cherkashin, C. Bonafos, G. Benassayag, H. Coffin, A. Claverie, V. Soncini, A. Agarwal and M. Ameen, Effect of annealing environment on the memory properties of thin oxides with embedded Si nanocrystals obtained by low-energy ion-beam synthesis, *Appl. Phys. Lett.* **83** (2003) 168–170.

33. P. Dimitrakis, E. Kapetanakis, P. Normand, D. Skarlatos, D. Tsoukalas, K. Beltsios, A. Claverie, G. Benassayag C. Bonafos, D. Chassaing, M. Carrada and V. Soncini, MOS memory structures by very-low-energy-implanted Si in thin SiO_2, *Mat. Sci. Eng.* **B101** (2003) 14–18.

34. P. Dimitrakis, E. Kapetanakis, D. Tsoukalas, D. Skarlatos, C. Bonafos, G.B. Asssayag, A. Claverie, M. Perego, M. Fanciulli, V. Soncini, R. Sotgiu, A. Agarwal, M. Ameen, C. Sohl and P. Normand, Silicon nanocrystal memory devices obtained by ultra low energy ion-beam synthesis, *Solid State Electron.* **48** (2004) 1151–1517.

35. J.F. Ziegler, J.P. Biersack and U. Littmark, *The Stopping and Range of Ions in Solids*, Vol. 1 (Pergamon, New York, 1985).

36. C.Y. Ng, T.P. Chen, L. Ding, M. Yang, J.I. Wong, P. Zhao, X.H. Yang, K.Y. Liu, M.S. Tse, A.D. Trigg and S. Fung, Influence of Si nanocrystal distributed in the gate oxide on the MOS capacitance, *IEEE Trans. Electron Dev.* **53**(4) (2006) 730–736.

37. Y. Liu, Y.Q. Fu, T.P. Chen, M.S. Tse, J.H. Hsieh, S. Fung and X.H. Yang, Depth profiling of Si oxidation states in Si-implanted SiO_2 films by X-ray photoelectron spectroscopy, *Jpn J. Appl. Phys.* **42** (2003) L1394–L1396.

38. T.P. Chen, Y. Liu, M.S. Tse, O.K. Tan, K.Y. Liu, P.F. Ho, D. Gui and A.L.K. Tan, Dielectric Functions of Si Nanocrystals Embedded in SiO_2 Matrix, *Phys. Rev.* **B68** (2003) 153301–153304.

39. R. Govindaraj, R. Kesavamoorthy, R. Mythili and B. Viswanathan, The formation and characterization of silver clusters in zirconia, *J. Appl. Phys.* **90** (2001) 958–963.

40. G. Garrido, M. López, O. González, A. Pérez-Rodríguez, J.R. Morante and C. Bonafos, Correlation between structural and optical properties of Si nanocrystals embedded in SiO_2: The mechanism of visible light emission, *Appl. Phys. Lett.* **77** (2000) 3143–3145.

41. M. López, B. Garrido, C. Bonafos, A. Pérez-Rodríguez and J.R. Morante, Optical and structural characterization of Si nanocrystals ion beam synthesized in SiO_2: Correlation between the surface passivation and the photoluminescence emission, *Solid State Electron.* **45** (2001) 1495–1504.

42. T.P. Chen, Y. Liu, C.Q. Sun, M.S. Tse, J.H. Hsieh,Y.Q. Fu, Y.C. Liu and S. Fung, Core-Level Shift of Si Nanocrystals Embedded in SiO_2 Matrix, *J. Phys. Chem.* **B108** (2004) 16609–16612.

43. Y. Liu, T.P. Chen, Y.Q. Fu, M.S. Tse, P.F. Ho, J.H. Hsieh and Y.C. Liu, A Study on Si Nanocrystal Formation in Si-implanted SiO_2 films by X-Ray Photoelectron Spectroscopy, *J. Phys.* **D36** (2003) L97–L100.

44. L. Ding, T.P. Chen, Y. Liu, C.Y. Ng and S. Fung, Optical properties of silicon nanocrystals embedded in a SiO_2 matrix, *Phys. Rev.* **B72** (2005) 125419–125425.

45. C.Y. Ng, T.P. Chen and A. Du, A study of the influence of tunnel oxide thickness on the performance of flash memory based on ion-beam synthesized silicon nanocrystals, *Physica Status Solidi (a)* **203**(6) (2006) 1291–1295.

46. E. Lusky, Y. Shacham-Diamand, G. Mitenberg, A. Shappir, I. Bloom and B. Eitan, Investigation of Channel Hot Electron Injection by Localized

Charge-Trapping Nonvolatile Memory Devices, *IEEE Trans. Electron Dev.* **51** (2004) 444–451.

47. C.Y. Ng, T.P. Chen, M. Yang, J.B. Yang, L. Ding, C.M. Li, A. Du and A. Trigg, Impact of programming mechanisms on performance and reliability of non-volatile memory devices based on Si nanocrystal, *IEEE Trans. Electron Dev.* **53**(4) (2006) 663–667.

48. C.Y. Ng, T.P. Chen, L. Ding and S. Fung, Memory characteristics of MOS-FETs with densely-stacked silicon-nanocrystal layers in the gate oxide synthesized by low-energy ion beam, *IEEE Electron Dev. Lett.* **27**(4) (2006) 231–233.

49. S. Tiwari, F. Rana, K. Chan, L. Shi and H. Hanafi, Single charge and confinement effects in nano-crystal memories, *Appl. Phys. Lett.* **69** (1996) 1232–1234.

50. I. Kim, S. Han, H. Kim, J. Lee, B. Choi, S. Hwang, D. Ahn and H. Shin, Room Temperature Single Electron Effects in Si Quantum Dot Memory with Oxide- Nitride Tunneling Dielectrics, *IEDM Tech. Dig.* (1998), pp. 111–114.

51. I. Kim, S. Han, K. Han, J. Lee and H. Shin, Si Nanocrystal Memory Cell with Room-Temperature Single Electron Effects, *Jpn. J. Appl. Phys. Part I* **40** (2001) 447–451.

52. F. Yun, B.J. Hinds, S. Hatatani and S. Oda, Room Temperature Single-Electron Narrow-Channel Memory with Silicon Nanodots Embedded in SiO_2 Matrix, *Jpn. J. Appl. Phys. Part II* **39** (2000) L792–L795.

53. G. Molas, B. De Salvo, G. Ghibaudo, D. Mariolle, A. Toffoli, N. Buffet, R. Puglisi, S. Lombardo and S. Deleonibus, Single Electron Effects and Structural Effects in Ultrascaled Silicon Nanocrystal Floating Gate Memories, *IEEE Trans. Nanotechnol.* **3** (2004) 42–48.

54. Z. Yu, M. Aceves, J. Carrillo and F. Flores, Single electron charging in Si nanocrystals embedded in silicon-rich oxide, *Nanotechnology* **14** (2003) 959–964.

55. J.S. Sim, J.D. Lee and B.G. Park, The simulation of single-charging effects in the programming characteristics of nanocrystal memories, *Nanotechnology* **15** (2004) S603–S611.

56. A. Thean and J.P. Leburton, 3-D Computer Simulation of Single-Electron Charging in Silicon Nanocrystal Floating Gate Flash Memory Devices, *IEEE Electron Dev. Lett.* **22** (2001) 148–150.

57. A. Thean and J.P. Leburton, Stack effect and single-electron charging in silicon nanocrystal quantum dots, *J. Appl. Phys.* **89** (2001) 2808–2815.

58. G. Molas, B. De Salvo, D. Mariolle, G. Ghibaudo, A. Toffoli, N. Buffet and S. Deleonibus, Single electron charging and discharging phenomena at room temperature in a silicon nanocrystal memory, *Solid State Electron.* **47** (2003) 1645–1649.

59. C. Pace, F. Crupi, S. Lombardo, C. Gerardi and G. Cocorullo, Room-temperature single-electron effects in silicon nanocrystal memories, *Appl. Phys. Lett.* **87** (2005) 182106–182108.

60. S. Decossas, J. Vitiello, T. Baron, F. Mazen and S. Gidon, Few electrons injection in silicon nanocrystals probed by ultrahigh vacuum atomic force microscopy, *Appl. Phys. Lett.* **86** (2005) 033109–033111.
61. J.S. Sim, J. Kong, J.D. Lee and B.G. Park, Monte Carlo Simulation of Single-Electron Nanocrystal Memories, *Jpn. J. Appl. Phys.* **43** (2004) 2041–2045.
62. A. Thean and J.P. Leburton, Geometry and strain effects on single-electron charging in silicon nano-crystals, *J. Appl. Phys.* **90** (2001) 6384–6390.

CHAPTER 9

NANOCRYSTALLINE SILICON FILMS FOR THIN FILM TRANSISTOR AND OPTOELECTRONIC APPLICATIONS

YOUNGJIN CHOI*, YONG QING FU and ANDREW J. FLEWITT

Electrical Engineering Division, Department of Engineering
University of Cambridge
Cambridge CB3 0FA, U.K.
**yjc23@cam.ac.uk*

1. Introduction

Silicon-based materials and devices have made possible the realization of complex electronics for data processing, control, sensing, and communication. The single crystalline silicon (c-Si) electronics achieving submicron dimension and gigascale integration have been developed, and amorphous silicon (a-Si) and nanocrystalline silicon (nc-Si) electronics on inexpensive substrates, such as glass or polymers have been developed in flat panel displays, solar cells and image sensors. In principle, a-Si is distinguished from c-Si by its disordered atomic configurations. Even if it does not have long-range order it still has short range-order, i.e. bond angle, bond length, and coordination number similar to those of c-Si. In contrast to a-Si, nc-Si, including microcrystalline silicon (μc-Si), has a three-dimensional periodic arrangement of atoms (long range order). It has nanoscale grains of crystalline silicon prolonged in the growth direction, which are embedded in a residual amorphous Si matrix. This is different from polycrystalline silicon (poly-Si), which consists of large crystalline silicon grains separated by grain boundaries. Since nc-Si consists of a number of small crystals and has a lower hydrogen concentration, it shows many superior properties over a-Si for thin film transistor applications, such as higher mobility and better stability for active devices. It also displays enchanced light-absorption in the red-infrared range and shows no Staebler–Wronski effect, which leads it to be used in solar cell applications.

473

The advantage of nc-Si is the fact that its technology, the deposition temperature below 250°C, is fully compatible with the conventional a-Si technology. This low deposition temperature is a great potential of nc-Si with respect to poly-Si, which requires much higher deposition temperature above 650°C. Although as-grown nc-Si is highly conductive, it generally cannot achieve the mobility as high as poly-Si has. However nc-Si can be prepared much easier than poly-Si as its primary deposition method is conventional plasma enhanced chemical vapor deposition (PECVD), which are widely used to fabricate a-Si at low temperature. The current driving force in the research and development of nc-Si is focused on enhancing the device quality and increasing the deposition rate of nc-Si films (typically 1.8–18 nm/min), which is inferior to that of a-Si (60–600 nm/min). To realize this, new deposition techniques, for example, very high frequency PECVD, microwave PECVD, electron cyclotron resonance PECVD, and hot-wire CVD, have been investigated and shown much improved properties, such as plasma stability, deposition rate, and device stability .

In this chapter, the mechanisms of nc-Si formation, advances in the deposition techniques and properties of nc-Si films will be introduced, followed by the applications to thin film transistors and solar cells.

2. Deposition Techniques and Growth Models

The characteristics of nc-Si are strongly determined by the deposition conditions and techniques. In this section, reviews on the deposition techniques to grow nc-Si films will be given. Following this, growth mechanisms for the formation of nc-Si and the role of hydrogen will be discussed.

2.1. *Deposition Techniques*

Different growth methods result in a wide range of possible silicon structures available: pure amorphous silicon (a-Si), hydrogenated amorphous silicon (a-Si:H), and crystalline silicon (c-Si). Polycrystalline silicon (poly-Si) in its physical and morphological properties is more related to single crystalline silicon. It contains relatively large grains (> 0.1–$1 \, \mu$m) with a crystalline fraction of close to 100% and only a small percentage of amorphous matrix in the grain boundaries ($<1\%$). Nanocrystalline silicon (nc-Si), which has small crystallites (~ 3–50 nm) embedded into a amorphous silicon matrix, is at the other end of the scale and more closely related to a-Si:H.

There has been considerable effort to produce a low thermal and large area process nc-Si using different techniques. It is of great interest for the semiconductor industry to find a suitable way to grow poly-Si and/or nc-Si for different applications such as solar cells or TFTs, especially when it can be realized for a large substrate, at low deposition temperature and at a relatively low cost. In this section several methods to grow nc-Si films are introduced.

2.1.1. *Radio Frequency-Plasma Enhanced Chemical Vapor Deposition (RF-PECVD)*

Plasma enhanced chemical vapor deposition (PECVD) is also called glow discharge deposition because of the visible luminosity of the plasma glow region, which is mainly the result of the de-excitation of emitting molecular and atomic species contained in the plasma. Glow discharge is sustained by inelastic electron-impact processes that are initiated by electrons which have acquired sufficient energy from the electric field as a result of successive elastic collisions with gas molecules. The field can be direct current (DC), radio frequency (RF), very high frequency (VHF), and microwave frequency.

The PECVD process includes all the technical requirements necessary to obtain nc-Si films at low temperatures and at a relatively low cost. The gas phase and the surface processes are the crucial steps for the film growth in PECVD.

Major steps of a PECVD process include source gas diffusion, electron impact dissociation, gas-phase chemical reaction, radical diffusion, and deposition. Electron impact dissociation of precursor gases in the glow discharge is the primary step for chemical reactions in a PECVD system.

The common PECVD systems used at lower pressures such as DC or RF discharges operate at pressures of 10–100 mTorr. The applied DC or RF potential increases at lower pressure and this gives rise to high ion energy, which causes destructive high-energy ion bombardment during the film growth. Matsuda [1] and Perrin [2] have pointed out the importance of the ion to neutral flux-ratio Φ_{ion}/Φ_o, where very low ion energies ($<50\,eV$) can be used to obtain good quality films for the silicon film deposition. Ions can provide momentum transfer to the surface and thereby they can increase surface mobility [2].

In RF-PECVD system, electron and ion densities are at most $10^{11}\,cm^{-3}$ at a pressure of 0.1 Torr in which the number density of molecules is the order of $10^{15}\,cm^{-3}$. The threshold energy to create neutral fragments by electron impact is known to be close to the threshold of photolytic decomposition, E_p. The electron-impact ionization needs significantly larger threshold energy than E_p. Consequently the predominant flux impinging onto the substrate surface is inferred to be radicals rather than ions, and the deposition process might be controlled by the generation rate of radicals or by the surface reactions among radicals. Plasma-assisted gas decomposition reduces the substrate temperature required for the deposition process. The lower substrate temperature makes it possible for sufficient hydrogen to be incorporated during the deposition.

It is believed that SiH_3 molecule is the main precursor in the growth of nc-Si, as it has a longer lifetime in silane plasma than all the other radicals. This is because the radicals SiH_2 and SiH tend to react especially at higher pressure [3]. Typical reactions are [4]:

$$SiH + SiH_4 \Rightarrow Si_2H_5, \tag{2.1}$$

and

$$SiH_2 + SiH_4 \Rightarrow Si_2H_6. \tag{2.2}$$

The consideration of dissociation and subsequent reactions in the plasma leads to a fragmentation pattern into radicals and ions as a function of the electron temperature. The main contribution of particles to the a-Si:H or nc-Si growth seems to originate from the radicals due to their increased number at low electron energies [5, 6]. However, PECVD deposition of nc-Si proceeds from the dissociation of silicon bearing source gas in the process chamber. This source gas is normally SiH_4 although Si_2H_6 [7] or SiF_4 [8] have also been employed. To deposit nc-Si, it is usually necessary to highly dilute up to silane to hydrogen gas ratio of 1:50 by volume. Another difference in the deposition conditions for nc-Si is that higher deposition power is required since at low power ($28\,mW/cm^2$) a-Si:H films were formed but at high power ($280\,mW/cm^2$) the deposited films were nanocrystalline [9]. It has also been reported that higher deposition pressure leads to nc-Si formation rather than a-Si:H [10]. However conventional PECVD nc-Si films have shown low deposition rate, typically less than 1 Å/s, even at high powers, due to the high hydrogen dilution.

2.1.2. *Very High Frequency-Plasma Enhanced Chemical Vapor Deposition (VHF-PECVD)*

Curtins *et al.* [11, 12] studied the effects of power-source frequency in a silane glow discharge from 25 MHz to 150 MHz. They observed a strong dependence of the deposition rate on the frequency, with deposition rate peaking above 2 nm/s at 70 MHz. Above 70 MHz the deposition rate fell off again. They claimed that there are lower internal stresses in films produced at 70 MHz than those produced at 13.56 MHz. At 70 MHz the plasma glow is self-confined, eliminating the need for other confinement. Chatham and Bhat [13] studied glow discharge deposition of a-Si:H from 13.56 MHz to 110 MHz. They observed that the deposition rate increases monotonically with frequency until 100 MHz. They also found that the optimized operating pressure decreases with increasing frequency for good quality a-Si:H deposition. They believed there are two main benefits with a high frequency deposition: (1) The ion bombardment energy decreases with increasing frequency as the self-bias potential decreases from 30 to 50 V at 13.56 MHz to 10 V or lower at 110 MHz. (2) The flux of low energy ions increases with increased frequency, due to lower operating pressure. This improves the surface mobility at the deposition surface. Wertheimer and Moisan [14] proposed that some observed frequency effects should be interpreted as a function of electron energy distribution functions of the plasma. Oda and Matsumura [15] concluded that frequencies greater than the collision frequency and close to the plasma frequency are useful in plasma processing because of the high efficiency of radical generation and the high feasibility of control of the energy and spatial distribution of electrons.

The explanation of influence of very high frequency (VHF) on the deposition rate is still a topic of discussion. It has been proposed theoretically that the high-energy tail in the electron energy distribution function is increased with an increasing ratio of the excitation frequency and the energy collision frequency [14, 20, 23]. This leads to an increased ionization rate [16–18]. However, J. Bezemer *et al.* [21, 22] reported that the increase of the deposition rate cannot solely be attributed to the enhancement of radical production. It has also been found or deduced that the flux of ions towards the surface is increased with increasing frequency [16, 19, 24, 25] while at the same time the ion energy is decreased. This low energy, high flux ion bombardment enhances surface mobilities of adsorbed species. Due to the fact that the wavelength of the radio frequency signal is of the order of the substrate dimensions (3 m at 100 MHz), it can be expected that uniform deposition is more difficult at these high frequencies [26]. In fact, a practical optimum frequency is used around 60–70 MHz [27, 28], which provides a good compromise between high deposition rate and attainability of uniform deposition.

The use of VHF (125 MHz) led to an increase in the deposition rate, grain size, and Hall mobility of nc-Si films [29]. High precursor flux to the growth surface is the prerequisite of high rate deposition, but methods leading to effective gas dissociation, such as high discharge power, are generally accompanied by high ion energy, which is considered to be detrimental for nc-Si growth [30]. Therefore, optimization of precursor flux versus ion energy might be of importance for high growth rate processes. Reduction of ion energy can be achieved with high excitation frequencies, where sheath voltages are reduced and with high pressure where a reduced mean free path leads to ion energy loss by collision.

2.1.3. *Hot-Wire CVD (HW-CVD)*

Hot-wire (HW) CVD uses a tungsten filament at 1100–1300°C to decompose SiH_4. The filament is placed above a substrate, which is kept at much lower temperature. Substrate temperatures as low as room temperature have been reported for the deposition of poly-Si [31]. The high temperature of the filament leads to a total decomposition of the SiH_4 and H_2 into atomic Si and H. These species then recombine away from the filament to form radicals such as SiH_3, without any ions present in the gas phase. Consequently, the SiH_3 radical is the dominant species responsible for the film growth. As with other deposition methods of nc-Si, HW-CVD is also dependent on the hydrogen dilution to promote crystallinity although the primary deposition parameters are the pressure and filament temperature. The HW-CVD leads to relatively good nc-Si films properties [31–33]. This could solely be attributed to the absence of ions. The evaporation of the tungsten filament is a problem, due to the high temperatures used, resulting in possible contamination. Technological difficulties also arise from the short lifetime of a very fragile tungsten filament and inhomogeneity when large area depositions are needed.

2.1.4. *Electron Cyclotron Resonance CVD (ECR-CVD)*

Electron cyclotron resonance (ECR) CVD systems have attracted interest as a technique of low temperature deposition of a-Si:H, nc-Si [34], and dielectric materials since the energy of the electrons in the plasma, which are responsible for the precursor formation, can be controlled independently of the generation of damaging ion species. The microwave frequency is matched to the cyclotron frequency set by magnetic field in the resonant chamber. When this matching occurs, electrons absorb energy resonantly from the exciting electric field, thereby the dissociation of the source gas is efficiently promoted. ECR plasma is potentially operated under low pressures ~0.01–$1\,Pa$. Therefore, the neutral radicals such as SiH_3 radicals or H atoms are able to reach on the surface of substrate without colliding so many times with the other species. As a result, the formation of higher radicals or particles, which deteriorate the film quality, is considerably suppressed. It has been reported that with a mixture gas of H_2 and SiH_4, highly crystallized films could be grown at 120°C by ECR-CVD [35]. However, in the high-density plasma operating at a low pressure, the ions will attack the growth surface and the damage induced by the ion bombardment prevents the nucleation of nc-Si, because the crystalline structure of nc-Si is very sensitive to the ion bombardment.

2.1.5. *Electron Cyclotron Wave Resonance CVD (ECWR-CVD)*

ECWR works on the principle of wave penetration into the plasma at low magnetic fields [36], providing a better power coupling to the plasma. An important consequence of this is that ECWR plasmas can be used for large area depositions. It can also be used for a PECVD system, where poly-Si can be grown without any dilution of silane being necessary [37]. Compared to the conventional PECVD systems, ECWR affords to control the ion energies between 10–50 eV, independent of the power input. This is essential in a CVD process, where ion bombardment is not desired. The power coupling allows the independent control of the amount of ions and their energy, independent of the electron temperature T_e. The working pressure for the ECWR system depends on the gases used but it usually ranges from $\sim$3.8$\cdot$10^{-2}–38 mTorr ($\sim$5$\cdot$10^{-5}–5$\cdot$10^{-2} mbar).

2.2. **Growth Models**

The periodic atomic structure of crystalline materials leads to Bloch's theory to account for the band structure and electrical characteristics. Amorphous silicon (a-Si:H) is distinguished from crystalline silicon (c-Si) by its disordered atomic configurations. Since nanocrystalline silicon (nc-Si) is a mixed phase material containing small crystalline regions of silicon, up to 500 nm in diameter, embedded within a-Si:H matrix, these nanocrystallites introduce two new structural features: ordered regions of silicon and grain boundaries. There have been many efforts to explain how nc-Si grows and which deposition parameter affects the growth of nc-Si. Most of the growth models for nc-Si are based on how hydrogen reacts and involves in the deposition since hydrogen plays a crucial role in the formation of nc-Si. There are three primary models that can account for the growth of nc-Si.

2.2.1. *Selective Etching (Hydrogen Radical Etching)*

The layer-by-layer (LBL) technique and PECVD process for nc-Si films formation are the main methods used, leading to the model of hydrogen radical etching. It has been suggested that the mechanism of PECVD nc-Si films is a balance between film creation and etching [38–40]. This is described by the reaction equation:

$$\mathrm{SiH}_{n(\mathrm{plasma})} \Leftrightarrow \mathrm{Si}_{(\mathrm{solid})} + n \cdot \mathrm{H}_{(\mathrm{plasma})}. \tag{2.3}$$

The reaction equation is not in an equilibrium state under normal deposition conditions and the reaction on the right side is where the deposition takes place. The reaction can be pushed towards the equilibrium by adding hydrogen, which leads to reduction in the deposition rate. No deposition occurs if the reactions are completely balanced. It is called hydrogen radical etching, if the reaction takes place from the right to the left of Eq. (2.3).

The LBL technique was employed to separate the deposition and etching processes. LBL consists of a sequence of alternating a-Si:H depositions followed by H_2 plasma exposure. Van Oort *et al.* [41] showed with this technique that the hydrogen radical etching model is insufficient to describe the growth of nc-Si films. Fang *et al.* [42] and Solomon *et al.* [43] proposed a more refined etching model to explain the growth mechanism for the LBL nc-Si, the so-called selective etching model. They observed a preferential etching of amorphous matrix and greatly reduced etching of the nc-Si during the growth of nc-Si. The nc-Si also turned out to contain a higher fraction in voids, after the H_2 exposure. The growth rate was expected to change drastically at the transition from amorphous to crystalline structure in the etching model, but this could not be observed. It was also found that etching did not reduce essentially the film thickness [41]. In the initial stage silicon atoms will be adsorbed onto the substrate in all orientations. However, where atoms are arranged with a {112} plane surface, they will be more resistant to hydrogen radicals etching, and then form a crystallite seed. More silicon atoms will bond to these seeds in the same orientation and crystallites will grow in size with film thickness in a conical structure. Therefore the crystalline region will expand laterally as nc-Si thickness increases [44].

2.2.2. *Chemical Annealing (Subsurface Transformation)*

The selective etching model arose from the fact that nc-Si starts to grow with an initial amorphous layer, which is slowly transferred into crystalline structure during growth. However, Abelson *et al.* [5] found the evidence that the subsurface model [45] for a-Si:H growth can describe the nc-Si growth in DC sputtering experiments. The modification of amorphous into crystalline silicon within the bulk of the amorphous material was also observed with ellipsometry [46]. The chemical annealing model [47] explained that the surface and the subsurface volume are in a liquid-like state allowing silicon atoms to reconfigure their bonds and also even change their sites. The liquid-like state is attributed to an increase in the effective surface temperature due to the following exothermic reaction:

$$2n\mathrm{H} + \mathrm{SiH_3} \rightarrow \mathrm{Si_{(solid)}} + \left(n + \frac{3}{2}\right)\mathrm{H_2} + (2 + 4.5n)\,\mathrm{eV}. \qquad (2.4)$$

The rise in temperature gives enough energy to allow the silicon atoms to rearrange to the energetically more favorable crystalline structure. However this model was controversially discussed because it could not account for the fast energy dissipation during deposition. The model also assumes that an amorphous layer should result, after atomic hydrogen treatment, in a crystalline network. The LBL methods showed contrary results: no crystalline growth appears with a cleaned [48] or shuttered cathode during the H-plasma treatment cycle [1], which suggests that LBL in practice often involves chemical transport from cathode to substrate.

2.2.3. *Surface Diffusion (Hydrogen Radical Coverage)*

Surface diffusion model is a different interpretation for the preferential growth of microcrystalline silicon in CVD at lower temperatures ($< 400°$C) [30], considering the strong dependence of the crystalline fraction, grain size and hydrogen content on substrate temperature and ion contribution. The surface diffusion is smaller at lower temperatures and therefore the formation of nc-Si nuclei is reduced or disappeared. This is related with less or no diffusion of SiH_x adsorbates into stable sites. The diffusion coefficient D_s can be written as [30]:

$$D_s = \frac{1}{4} \cdot a_0^2 \cdot \nu \cdot \exp\left(-\frac{E_s}{k_B T}\right), \qquad (2.5)$$

where E_s is the thermal activation energy for surface diffusion jump, a_0 the jumping distance from one site to another and ν the jumping frequency. The SiH_x adsorbates are attached primarily to the hydrogen on the surface. In addition they are mobile corresponding to the particle gradient and their energy until a stronger bond is established. A change in the hydrogen content with T_s can be observed and explained by the influence of D_s and E_s, which both depend on the hydrogen coverage of the surface. When the surface is less covered with hydrogen at high T_s, less diffusion will occur, since E_s is increased and more reactions take place on the surface. The transition from crystalline to amorphous growth can also be observed, when the number of hydrogen ions impinging the substrate surface exceeds a critical value. The grain size is also decreased when the ion bombardment of the film surface is enhanced.

A limitation to this model is that it is hard to explain why crystalline growth occurs at higher substrate temperatures when hydrogen is completely effused from the surface above 500°C. Diffusion experiments performed by Okada *et al.* [49] suggested that hydrogen diffusion on the surface is of minor importance during crystalline growth.

The three major growth models for nc-Si films were developed from results obtained by different deposition techniques. However a model, which combines a couple of growth models, was also suggested as a possible explanation for the growth of nc-Si [50]. A recent investigation showed that different deposition techniques possibly result in different materials, for example when PECVD and HW-CVD are compared [52]. In PECVD the ions contribute to the growth of a-Si:H [51] and nc-Si [52], whereas the ions are not present in HW-CVD.

3. Characterization and Properties of nc-Si Films

Before we discuss specific properties of nc-Si films, it should be noted that the properties of nc-Si generally depend strongly on film thickness. For example, the grain size of nc-Si is proportional to the thickness as the grains grow upward [53]. Most of electrical and optical properties of nc-Si films are strongly correlated to the grain size. Therefore it is important that the properties must be compared with the others only if they have been prepared with similar thickness.

3.1. *Electrical Properties*

3.1.1. *Dark and Photoconductivity*

The conductivity is the most important and commonly measured property of nc-Si. To determine if nc-Si is suitable for device application, it is necessary to measure the dark and the photoconductivity under illumination (generally AM1, 100 mW/cm^2 solar-simulator illumination). The energy of photon of light is given by hv, and the bandgap of Si is denoted by E_g. When the photon energy is higher than the bandgap of Si ($hv > E_g$), the photon energy is absorbed by Si and then an electron is excited to a level higher by hv. In this excited level, the carriers are generated and then the conductivity under illumination (photoconductivity) is increased. The a-Si:H usually shows significant increase in the conductivity when it is exposed to light (several order higher current), but in the transition from a-Si:H to nc-Si the photoconductivity is more and more reduced until the photosensitivity disappears (Fig. 3.1). However, doped (n^+ and/or p^+) nc-Si layers can be deposited by adding doping gases, for example phosphine (PH_3) for n^+ or diborane (B_2H_6) for p^+, during the nc-Si layers deposition.

The conductivity of nc-Si is strongly restricted in the direction of the charge carrier flow because nc-Si is an inhomogeneous material. Since the growth direction of crystallites is more likely to be perpendicular to the substrate surface with conical shape, the number of grain boundaries along the current flow is increased and more amorphous material can be distributed on the surface of the substrate. This may lead to a rise in the conductivity with thickness, causing higher leakage current in devices where the active layer thickness is increased.

The model to describe the electronic transport in a-Si:H relies on the behavior of carriers near the mobility gap which separates localized and

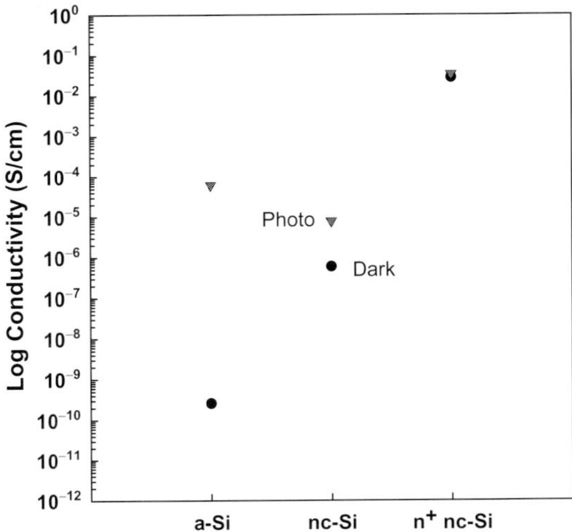

Fig. 3.1. Photo and dark conductivity of a-Si, nc-Si, and n^+ nc-Si. As the transition from a-Si to nc-Si takes place, the photosensitivity drastically decreases.

non-localized states. The carrier transport in localized states is limited where a hopping conduction prevails. Since nc-Si has small grains embedded in a-Si:H, the conduction mechanism is dominated by the amorphous matrix rather than the grain characteristics. However if these gains are sufficiently large, then extended state conduction can be enhanced.

In general, as grown nc-Si is a highly conductive, strongly n-type semiconductor when deposited without special precautions, because oxygen contamination involves and the oxygen atoms act as donors. This can be explained by thermal donors (TDs) in Czochralski Si, which are formed during thermal annealing at 400~500°C [54]. In TD formation, thermal energy causes the oxygen diffusion in solid Si and subsequent oxygen precipitation; however, surface diffusion of the oxygen-related species during the growth must instead play an important role in nc-Si. Since such surface diffusion is a thermally activated process, fewer oxygen aggregates are formed as nc-Si growth temperature decreases. Therefore, a lower growth temperature suppresses oxygen donor formation, which could be an alternative explanation for the intrinsic character of lower temperature (140°C) grown nc-Si [55].

3.1.2. *Thermally Activated Conductivity*

If we measure the temperature characteristics of dark conductivity σ_d and plot it in logarithm scale against the temperature inverse ($\log \sigma_d^{-1/T}$), it forms the straight line shown in Fig. 3.2, i.e. $\sigma_d \propto e^{-\Delta E/kT}$, in which ΔE is the activation energy. In an intrinsic material, if it is possible to assume that the Fermi level of the material is in the middle of bandgap and then ΔE of the material represents the Fermi level E_F, and $E_g = 2\Delta E = 2E_F$. The activation energy is therefore often used as an indication of the Fermi level, even in doped semiconductors. For example, if $\Delta E = 0.4\,\mathrm{eV}$ in nc-Si layer, the Fermi level is $0.5\,\mathrm{eV}$ nearer to the conduction band from the center of the forbidden bandgap (assuming $E_{\mathrm{opt}} \sim E_g = 1.8\,\mathrm{eV}$). The activation energy of device quality a-Si:H has typical values between 0.7–$0.9\,\mathrm{eV}$, whereas that of nc-Si often has values below $0.5\,\mathrm{eV}$ [56].

3.2. *Physical Properties*

3.2.1. *Bandgap*

The optical bandgap of a-Si:H is determined by the wavelength dependence of absorption coefficient $\alpha(\omega)$. In nc-Si the absorption can be enhanced when

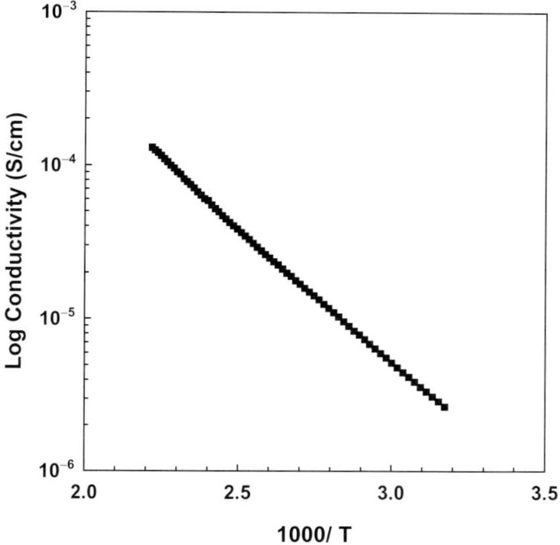

Fig. 3.2.   Dark conductivity of nc-Si plotted against temperature.

the grain size is below 10 nm [57], which is possibly related to quantum confinement effects in small grains, internal reflection in grains or surface roughness. Therefore, the absorption coefficient depends on the grain size [57] and the observed bandgap will increase with smaller grain size. In general, the light frequency v and the optical bandgap E_g are related by

$$\sqrt{\alpha h \nu} = B(h\nu - E_g), \tag{3.1}$$

where h is Planck's constant and B is a proportional constant [58]. Below the Tauc edge there is an exponential region of the absorption edge, then the absorption coefficient can be described by the formula

$$\alpha = \alpha_0 \exp[(E - E_g)/E_0], \tag{3.2}$$

where E_g is the energy to the order of E_g^{opt} and E_0 is the characteristic energy representing the logarithmic slope of the absorption coefficient in this region [58]. The exponential tail of Eq. (3.2) is called the Urbach edge. When $\sqrt{\alpha E}$ is plotted as a function of the energy, it yields a straight line (within a certain energy interval), where the intercept is defined as the E_g (Tauc plot, Fig. 3.3).

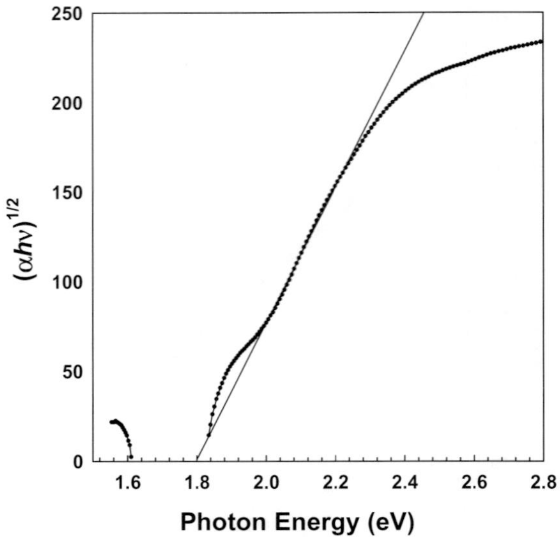

Fig. 3.3.   Tauc plot of nc-Si film. The optical (extrapolated) bandgap is 1.8 eV.

3.2.2. *Hydrogen Concentrations*

The infrared spectra of a-Si:H and nc-Si can be used to estimate the hydrogen content. The infrared absorption spectrum of a-Si:H exhibits the typical fingerprints with peaks at 640, 840, 880, 2000 and 2090 cm^{-1} [59]. On the other hand nc-Si has much narrower peaks at 626, 900, 2000 and 2101 cm^{-1} [60]. These vibration modes were assigned to mono and dihydride bonds on (100) and (111) surfaces of crystalline silicon [60]. They therefore reflect the hydrogen bonded at the surface of crystalline grains. The method of estimating of total hydrogen content by numerical integration of Si-H rocking-wagging peak at 640 cm^{-1} [61]. Generally, infrared spectroscopy can estimate the hydrogen content if the proportionality constant is known. Figure 3.4 shows a simplified example of the characteristic vibration modes of nc-Si. The hydrogen content in a-Si:H can be derived from the absorption strength of the mode centred at 630 cm^{-1} and from the two modes at 2000 cm^{-1} and 2100 cm^{-1} [62, 63]. It has been shown [64] that the hydrogen content in nc-Si is best derived from the mode centered at 630 cm^{-1}. The hydrogen content in a-Si:H is estimated with the modes near 630 cm^{-1} and 2000 cm^{-1} with the following equation:

$$[H] = A \int_{x}^{\Delta x + x} \frac{\alpha(\omega)}{\omega} d\omega, \tag{3.3}$$

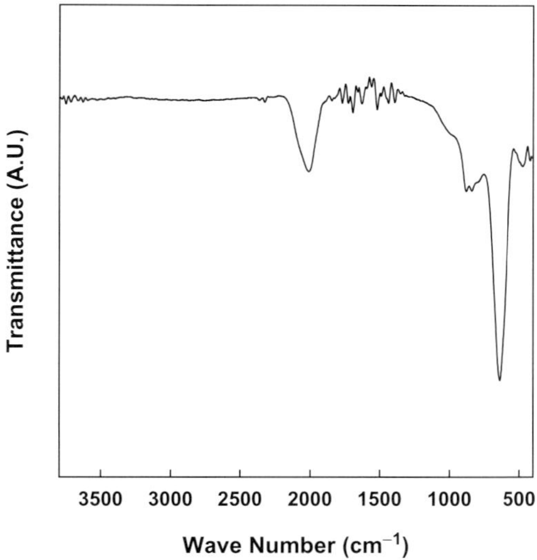

Fig. 3.4.   Fourier transformed infrared spectra for a typical nc-Si film.

where ω is the frequency and α is the absorption coefficient and it is integrated the area of interest (from wavenumber of x to wavenumber of $\Delta x + x$) [59]. "A" is a constant proportional to the inverse oscillator strength and it can be estimated from absolute hydrogen measurements. Langford *et al.* [62] estimated "A" to be at $\sim 630\,\mathrm{cm}^{-1}$ for a-Si:H with $A_{630} = 2.1 \pm 0.2 \cdot 10^{-19}\,\mathrm{cm}^{-2}$. A value of $A_{630} = 3.0 \cdot 10^{-19}\,\mathrm{cm}^{-2}$ was found for nc-Si [63]. The nc-Si infrared spectra again differs from the a-Si:H spectra for the peak around $630\,\mathrm{cm}^{-1}$. In a-Si:H the single peak at $630\,\mathrm{cm}^{-1}$ can be identified for nc-Si as at least two peaks, most likely from the mono and dihydride bonds groups, which have modes at $630\,\mathrm{cm}^{-1}$ and $650\,\mathrm{cm}^{-1}$ [65]. The peaks around $2000\,\mathrm{cm}^{-1}$, which is especially the SiH_2 peak centred at $2084\,\mathrm{cm}^{-1}$, indicates the crystalline phase. This vibration peak has its origin in $Si\text{–}H_2$ bonds, which sit on the surface of the crystalline grains [66]. Moreover, the Si–H bond at nc-Si (111) surfaces have a stretching mode at $2080\,\mathrm{cm}^{-1}$ [67], which may also contribute to the nc-Si spectra.

3.2.3. *Crystalinity from Raman Spectroscopy*

Raman spectroscopy is a further technique to probe the structural composition in nc-Si. Raman scattering is used to probe the vibration spectrum of the nc-Si film. It can be used to obtain the crystalline content X_c in the nc-Si films [68]. The technique is based on the observation that the transverse optical vibration modes of amorphous and crystalline silicon are different. The TO mode of the amorphous phase appears as a broad peak centered at $480\,\mathrm{cm}^{-1}$ while that of crystalline phase appears as a sharp peak at $520\,\mathrm{cm}^{-1}$. Therefore in order to estimate the crystalline fraction from Raman measurements, the intensities of the different peaks are fitted. However the Raman peak, which is assigned to the crystalline phase controlled by the crystalline content, is also influenced by Si–Si bond strain at the grain boundaries and also by mechanical stress in the film [69]. The peaks centered at $520\,\mathrm{cm}^{-1}$ and $480\,\mathrm{cm}^{-1}$ could not be fitted with two single Lorentz functions. An additional peak at approximately $500\,\mathrm{cm}^{-1}$, which was attributed to grain boundaries with possibly different origins must be taken into account [68, 70, 71] as shown in Fig. 3.5. Then X_c in the nc-Si films is given by [72, 73]:

$$X_c = \frac{I_c + I_{\mathrm{gb}}}{I_c + I_{\mathrm{gb}} + y(L)I_a}, \tag{3.4}$$

where I_c, I_a, and I_{gb} are the corresponding integral intensities of the peaks from different phonon bands. $y(L)$ is the ratio of the scattering cross sections for crystalline to amorphous phase, depending on the diameter L as

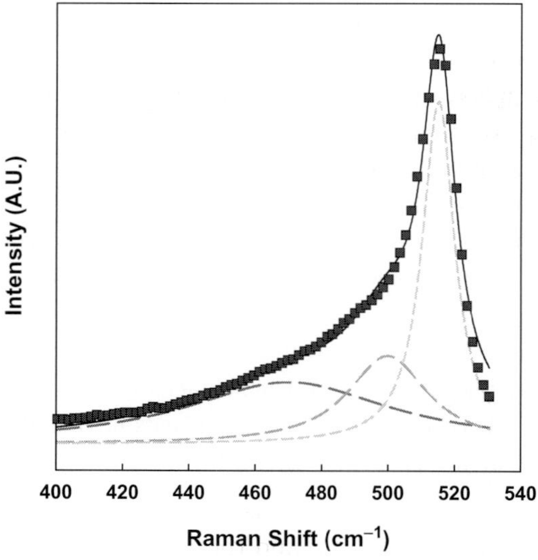

Fig. 3.5. Raman scattering spectra for nc-Si. The crystalline peak at approximately $520\,\mathrm{cm}^{-1}$ with the amorphous phase contribution with a peak at $\sim\!480\,\mathrm{cm}^{-1}$ is shown. A third intermediate peak is at approximately $500\,\mathrm{cm}^{-1}$, which was attributed to grain boundaries with possibly different origin [69].

a measure of the crystallite size. When the size of the crystallites is known, the prefactor can be obtained with [74]:

$$y(L) = 0.1 + \exp(-L/250). \tag{3.5}$$

From Eqs. (3.4) and (3.5), it is also possible to calculate the volume fraction X_{gb} taken by the grain boundary [73]. Additional information about the grain size obtained from X-ray diffraction can be used to correct for the effect of downshift and the broadening of the crystalline peak when the crystalline size decreases [75].

3.2.4. *Crystalline Structure from XRD*

X-rays whose wavelength approaches atomic dimensions ($\sim$0.1 nm) allows the physical structure of the material to be directly probed. Thus, X-ray diffraction (XRD) is an especially useful technique for nc-Si films as it can provide information both on whether crystallites are present, and also on their size to calibrate the grain size obtained from Raman measurements. The average penetration depth of the Cu-K$_\alpha$ radiation (1.54 Å) in silicon is approximately 71 μm. Therefore the absorption of X-ray radiation in thin films with a thickness below $<1\,\mu$m is quite weak, and this results in a long data acquisition time. However it is possible to gain the three prominent diamond structure peaks for the crystalline planes at (111), (220), and (311) in the diffraction pattern (Fig. 3.6). The diffraction patterns can be compared with reference patterns to identify the nature of the preferred orientation present. The full width half maximum (FWHM) is directly related to the grain size and can be calculated using the Scherrer formula [76]:

$$L = \frac{K \cdot \lambda}{\beta \cdot \cos\theta}, \tag{3.6}$$

where L is the grain size, λ is the wavelength of the X-ray, θ is the scattering angle and β is the FWHM of the peak in radian. K is a shape factor which takes into account the shape of the crystallites and is usually set to 1 [77]. The broadening of the peak β is caused by the difference between broadening γ due to the grain size and the broadening of the measurement system b, so that:

$$\beta = \gamma - b. \tag{3.7}$$

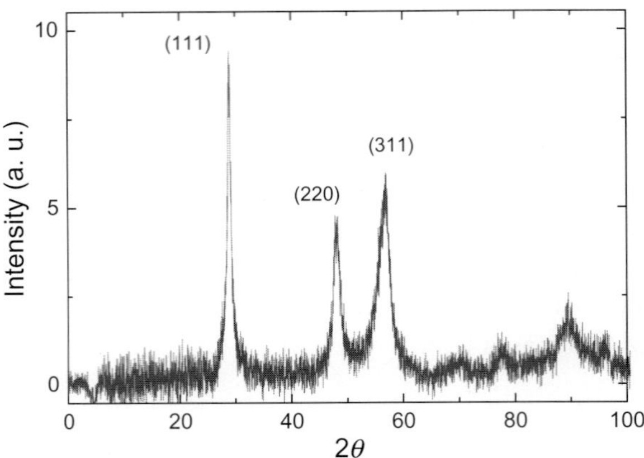

Fig. 3.6. X-ray diffraction pattern for nc-Si with preferential orientation gained. Using the Scherrer formula, it is possible to calculate the grain size of the nc-Si film where $\lambda = 1.54$ Å and $K = 1$. Then the average grain size is 10.4 nm.

3.2.5. *Luminescence*

Generally, a photoluminescence is observed from high quality a-Si:H. When the laser is incident on the a-Si:H film, electron-hole pairs are created. These carriers recombine if they are captured in the localized states. This recombination involves the radiative and non-radiative process. The energy of luminescence is equal to the energy released by the recombination. In direct bandgap materials, the minimum of the conduction band and maximum of the valence band are coincident in the k-space, the probability of radiative recombination is high and the radiative lifetime is typically of the order of a few nanoseconds. Since a-Si:H has an indirect bandgap and a high localized state density due to the defects, the intensity of light emission is very low. Thus, radiative recombination requires an additional phonon, which significantly reduces the probability and rate of that process. The efficiency of light emission can be enhanced by reducing the radiative lifetime or by increasing the non-radiative lifetime. A shorter radiative lifetime is beneficial for the modulation speed of the luminescence. To reduce the radiative lifetime it is necessary to increase the probability of radiative recombination. One possible way to achieve this is by carrier confinement. When a carrier is confined in real space, the k conservation rule may be partially relaxed due to the Heisenberg's uncertainty relation. The majority of the excitons, which is resulted in this localization, would be released and break into free electrons and holes, which can diffuse to fast non-radiative recombination

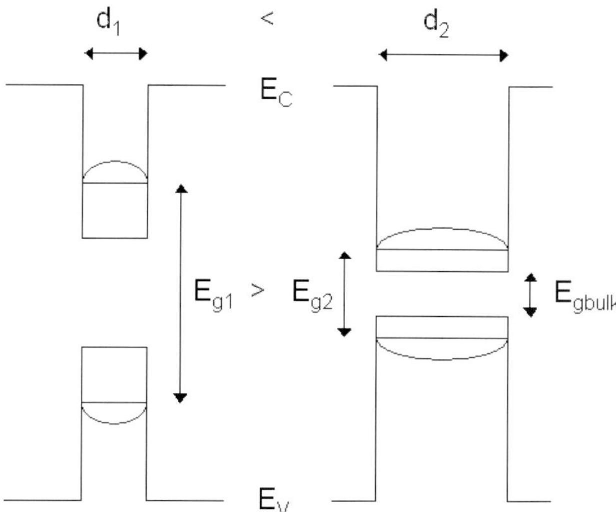

Fig. 3.7.   Quantum confinement influence on the effective bandgap in small structures.

center. It has been reported that silicon nanostructures including nc-Si show a strong luminescence because of quantum confinement [78, 79]. The carriers are confined in silicon nanocrystallites bounded with silicon oxide (SiO_x) regions which has a wider bandgap. A confined carrier has a higher energy than a free carrier (Fig. 3.7) and, therefore, the band structure of Si nanocrystallites is different from that of bulk silicon [80]. It has been also reported that an increase of the bandgap increases as crystallite size decreases due to the carrier confinement when nanocrystallites size becomes smaller than 10 nm [81].

3.3. *Stress Issues in Nanocrystalline Si Films*

It is generally accepted that a large amount of hydrogen dilution into SiH_4 plasma could result in the formation of nanocrystalline Si films, and hydrogen plays a dominant role by enhancing atom mobility, passivating the dangling bond defects, etching disordered phases, reconfigurating the subsurface bonds and modifying Si–Si network [82, 83]. However, all the above effects could cause significant changes in the film stress [84]. Figure 3.8 shows the residual stress of the Si films as a function of SiH_4 gas ratio (i.e. $SiH_4/(H_2 + SiH_4)$). All the films show compressive stress. The film compressive stress increases gradually with a decrease in SiH_4 gas ratio, maximizes at a SiH_4 gas ratio of 2%, and then decreases significantly to a

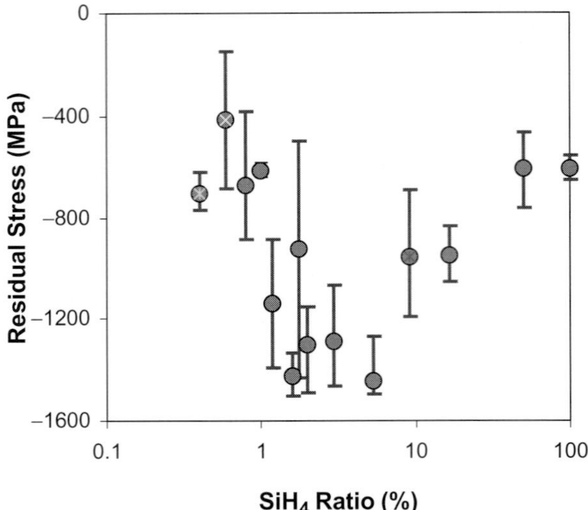

Fig. 3.8.   Residual stress of amorphous and nc-Si films as a function of SiH_4 gas ratio [85].

low value of a few hundred MPa with further decrease in SiH_4 gas ratio. Results clearly showed that there is a maximum value of film stress with the gradual increase in hydrogen content in the gas mixture.

The crystalline volume fraction X_c from Raman analysis (shown in Fig. 3.9) indicates that with SiH_4 gas ratio above 10%, the amorphous phase is dominant, while crystalline structure becomes dominant when SiH_4 gas ratios is lower than 1.75%. This maximizes at about 0.79%, and then decreases slightly with a further decrease in SiH_4 gas ratio. Raman analysis clearly revealed that the H dilution leads to the transition from amorphous to crystalline states. For the films deposited with SiH_4 gas ratio higher than 3%, X-ray diffraction (XRD) results also confirm that with SiH_4 gas ratio at and below 2%, there are three apparent peaks observed at angle of $28.5°$, $47.5°$ and $56.2°$, corresponding to Si(111), (220) and (311) reflections, respectively. The average sizes of the nanocrystals estimated from the (111) diffraction peak (the dominant orientation for all the films) using Scherer's equation are listed in Fig. 3.9. The crystal size of the film increases to a maximum for SiH_4 gas ratio of 1.19%. With further decrease in SiH_4 ratio, grain size decreases, probably associated with decreasing film crystallinity due to hydrogen amorphitization effects. The results indicate that there is a critical hydrogen dilution ratio for crystal size during the film growth [86].

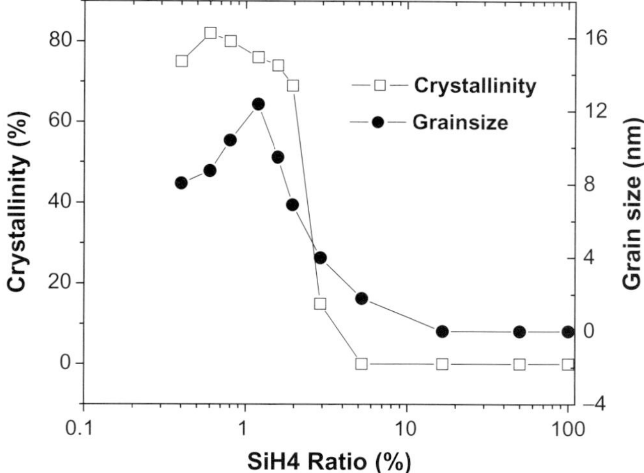

Fig. 3.9.   Crystallinity and grain size of nc-Si films as a function of SiH_4 gas ratio obtained from Raman and XRD tests [85].

As we know, amorphous Si films are formed by short-range ordered Si–Si and/or Si–H_x clusters (with hydrogen atoms inserted into Si–Si bond structures). The hydrogen content in a-Si or nc-Si films has been considered as a dominant effect for residual stress. Amorphous films are formed by H insertion into Si–Si bonds, the increased strained Si–Si bonds in the subsurface region due to high content of H in films can explain the increase in the film intrinsic stress. The hydrogen contents in the films, obtained from FTIR showed that the H content decreases with the decrease in SiH_4 ratio. When the SiH_4 gas ratio is below 2%, the hydrogen content decreases significantly with SiH_4 ratio and then remains at a stable value. Edelberg *et al.* [87] pointed out that film compressive stress could arise from diffusive incorporation of hydrogen or some gaseous impurities in the silicon random network during film growth. If checking the hydrogen content evolution with SiH_4 gas ratio in Fig. 3.10, for the films deposited with high SiH_4 ratio, the hydrogen contents are much higher. Whilst the stress is much lower than those with lower SiH_4 ratio of 2–5%. Therefore, hydrogen in the films cannot be used to explain this stress maximum.

The microstructure of the nc-Si film is that Si nanocrystals are surrounded by grain boundaries and embedded inside an amorphous matrix (Fig. 3.11), which is different from that of amorphous silicon (with seldom order or with short range order) [88]. Comparing results from Figs. 3.8

                    Y. Choi, Y. Q. Fu & A. J. Flewitt

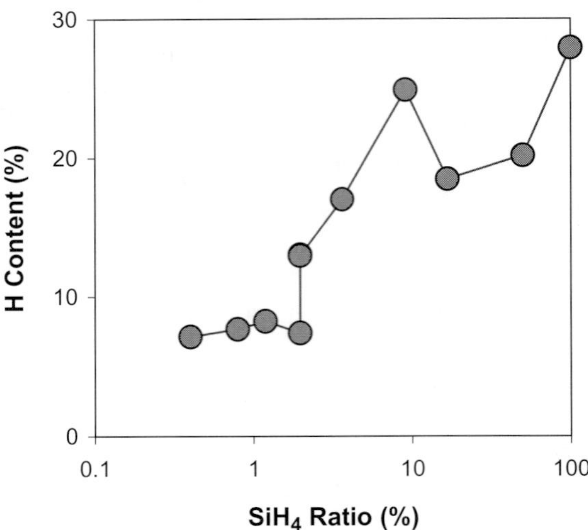

Fig. 3.10.   Hydrogen content evolution for the a-Si and nc-Si films deposited with different SiH$_4$ ratios [85].

Fig. 3.11.   SEM morphology of typical nanocrystalline Si films showing the nanocrystals embedded inside the amorphous matrix.

and 3.9, a maximum compressive stress in the films is observed when nanocrystalline Si forms in the film [89]. When hydrogen gas flow rate is increased to a critical value, hydrogen etching effect becomes significant. Film deposition rate decreases due to rapidly removal of weakly bonded atoms and radicals on the growing surface, thus the hydrogen could have more chances to escape, rather than be trapped inside films. Long range Si–Si order structures (or nanocrystals) could easily form due to hydrogen etching effect. Presence of long-range order in amorphous matrix due to the incorporation of nanometer crystals can greatly increase the film stress due to the significant increase in the interfacial structure. When H_2 gas ratio is further increased, the crystallinity becomes quite high and crystal size also increases, thus the contribution from the interfacial structure or amorphous matrix decreases significantly, resulting in a decrease in stress. In brief, it can be concluded that the film stress maximum is a combination effect of hydrogen insertion into strained Si–Si bonds and the nanocomposite effects.

4. Device Applications

4.1. *Thin Film Transistors (TFTs)*

Since the first proposal by LeComber *et al.* [90] to use a-Si:H TFTs for liquid crystal displays (LCDs), much scientific investigation and technological development has taken place, and a-Si:H thin-film transistor LCDs (TFT-LCDs) are now used in many consumer products. The basic transistor structure, however, has not changed much, and today's mainstream technology uses transistors that, apart from minor variations, are similar to the inverted staggered structures made in the early stages of development [91]. The coplanar top gate structure [Fig. 4.1(b)], for which the gate and the source and drain are located on the same side of the semiconductor layer, is most popular in polycrystalline silicon (poly-Si) TFTs while the inverted staggered structure [Fig. 4.1(c)] is widely used for a-Si:H. However there are several major technical limitations which a-Si:H currently imposes upon the development of TFT-LCDs. These include (1) the poor stability of a-Si:H TFTs to prolonged gate bias stress, which causes the threshold voltage to shift towards the applied gate bias; (2) the low carrier mobility, which limits the switching speed of the devices. A current driving force in the research and development of TFT-LCDs is to integrate the driver circuitry onto the display broad, to improve reliability and reduce costs. These circuits require the use of a material with a higher electron mobility than hydrogenated amorphous silicon (a-Si:H). Presently, polycrystalline

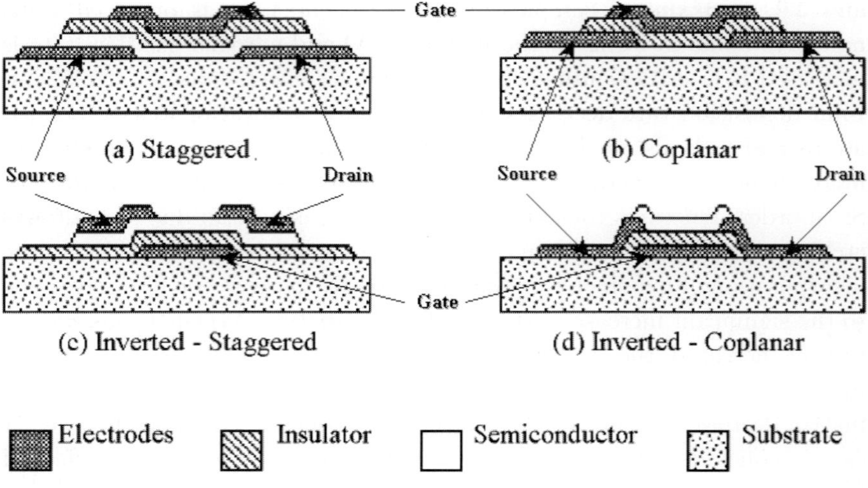

Fig. 4.1. Various a-Si:H TFT structures.

silicon (poly-Si), a potential candidate, is made by laser recrystallization of a-Si:H. Direct PECVD nc-Si would be cheaper, if it could be grown with sufficiently good quality. However the nc-Si TFT exhibits the field-effect mobility of $\sim150\,\mathrm{cm}^2/\mathrm{Vs}$ [145] while the state of art a-Si:H TFT shows $\sim1\,\mathrm{cm}^2/\mathrm{Vs}$ [146].

4.1.1. *Fabrication*

The merits by which the technologies are judged include their static performance (i.e. "On/Off" current ratio, switching speed, field effect mobility, threshold voltage), stability, uniformity, complexity of technology — which will be reflected in the cost — and the potential for further improvement. The a-Si:H TFT structure currently responsible for "state-of-the-art" device performance is the inverted-staggered configuration using PECVD silicon nitride as the gate insulator. The deposition sequence required to manufacture this structure results in a-Si:H/insulator and a-Si:H/source-drain contact interfaces of superior quality. However the grains of nc-Si grow upwards from the bottom layer, so that the grain size is much larger at the top surface. Therefore it is desirable to develop a top gate structure to use with nc-Si, in order to benefit from the larger grain size at the top of this layer.

In order to realize nc-Si top gate TFTs, the most common fabrication process is quite similar to the CMOS process. First, the nc-Si (typically less than 200 nm thick), gate insulator (< 300 nm), and gate metal (~ 100 nm) layers are deposited and then the gate metal and gate insulator layers are patterned. Next, the source and drain regions are doped using ion implantation to form n^+ layer (~ 20 nm in depth). After ion implantation, the annealing process should be followed to activate the dopants. Then source and drain contact metal layer (~ 100 nm) is deposited and patterned. The typical width of TFTs is less than a hundred μm and the length of TFTs is around several μm.

4.1.2. *Operation*

The basic device operation of TFTs is identical to that of metal-oxide semiconductor field-effect transistors (MOSFETs) [92]. A positive voltage applied to the gate will induce $C_i V_g/q$ electrons to accumulate in the intrinsic channel. For small gate voltages, these electrons will be localized in the deep states of the intrinsic nc-Si layer. Above a threshold voltage, a constant proportion of the induced electrons will be mobile, which means that the conductivity increases linearly with gate voltage; the TFT switches on, and a current flows between the source and drain. The n^+ layers in source and drain regions are necessary to reduce the contact resistance between contact metal and intrinsic active layer. If a negative voltage is applied to the gate, hole accumulation will be induced at the interface. This hole channel is, however, not normally observed because the n^+ contacts cannot supply enough holes to sustain a significant current.

4.1.3. *Stability*

The instability in a-Si:H TFTs is the shift in the device threshold voltage that is observed after the prolonged application of gate bias. This is of concern for everyday applications of a-Si:H TFTs such as those found in flat panel displays. This instability manifests itself as a parallel shift of the I–V characteristics to the right when the gate bias voltage is positive and to the left when the gate bias is negative (Fig. 4.2). This causes the threshold voltage (V_{th}) to shift. ΔV_{th} depends strongly on the gate bias, while it varies only slightly with the drain voltage. The annealing process is usually effective in recovering the initial I–V characteristics. After heat treatment at $200°$C for 30–60 minutes, the transfer characteristics return to their initial curves.

The V_{th} shift mechanism has been extensively studied and there are two possible explanations. The first describes the voltage shift by charge

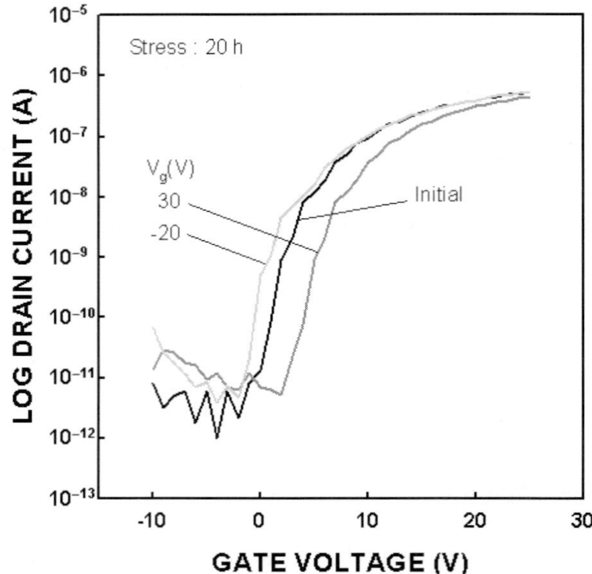

Fig. 4.2. The threshold voltage shifts in the a-Si:H TFT according to the positive or negative gate bias stress.

trapping in slow states located in the SiN_x gate insulator [93–99]. This is supported by the fact that TFTs with silicon-rich SiN_x are less stable than those with stoichiometric SiN_x. The second is that ΔV_{th} arise due to the formation of bias-induced deep trap levels in the amorphous silicon film. Since the channel is formed in the vicinity of the a-Si:H/SiN_x interface, these trap levels may also be associated with the interface [98, 100–102]. In any case, both mechanisms are play roles in the appearance of ΔV_{th}. At higher gate bias, the ΔV_{th} of a-Si:H TFTs is due to charge trapping in SiN_x, mainly through tunneling. The dominant mechanism at low gate bias, on the other hand, is the creation of states or the breaking of bonds in the a-Si:H film at or near the interface.

Figure 4.3(a) shows transfer characteristics of the nc-Si TFT under gate bias stress of 30 V at the room temperature. The transfer characteristic remains essentially unchanged for long bias stress time and no fast carrier induced defect creation is observed. The threshold voltage shifts are plotted against bias stress time in Fig. 4.3(b). The nc-Si TFT drifts at very small rate and exhibits excellent stability characteristics. The main reason why nc-Si TFT shows a high stability is a much lower frequency of attempting to break Si–Si weak bonds. In case of a-Si:H TFTs, electrons are localized

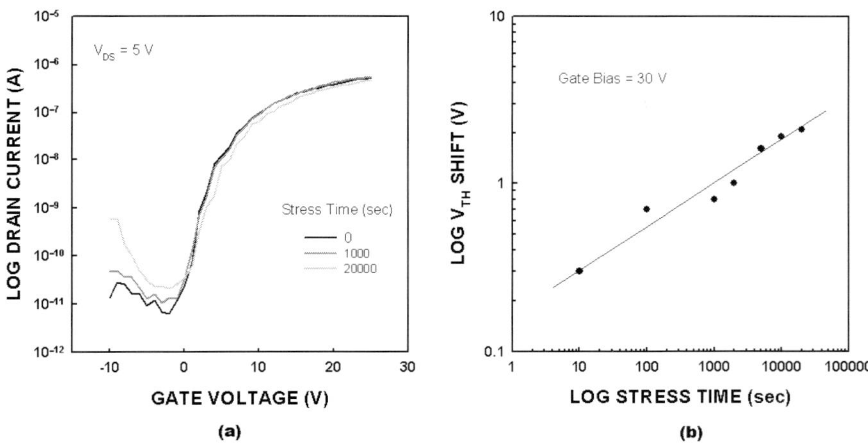

Fig. 4.3. (a) Transfer curves at drain voltage of 5 V under gate bias stress and (b) threshold voltage shifts plotted against bias stress time.

and spending quite long time on single Si atoms, which gives them a high frequency for possible bond breaking. On the other hand, in nc-Si TFTs, electrons are delocalized, which reduces the attempt frequency for bond breaking [103].

4.2. *Solar Cells*

Energy demands are drastically increasing today and at the same time environmental concerns are radically being emphasized. Fossil fuels are currently major energy sources although they are ultimately limited in quantity and causing serious air pollution. Therefore alternative energy sources should be infinite and environmentally friendly. Since solar energy is inexhaustible and clean, solar power seems to be the most probable candidate that is able to fulfill the requirements as alternative energy source. In addition it can be converted into heat, fuels, and electricity. However the greatest disadvantage of solar energy is high generating costs. The high level of generating costs is inevitable because the solar cell materials (mainly c-Si and poly-Si) are expensive. Thin film technologies affords to produce alternative thin film materials, which include a-Si:H, nc-Si, cadmium telluride (CdTe), copper indium selenide (CIS), at lower costs than traditional c-Si based technologies since the low temperature process allows fabrication on inexpensive substrates such as glass, plastic, and stainless steel. Solar cells can be made in a variety of materials such as GaAs, CdTe, $CuInS_2$, and various silicon crystals. A-Si:H is the most common material in commercial solar cells although it has disadvantages of photodegradation. C-Si solar cell shows the efficiencies of $\sim$24%. The efficiency for poly-Si solar cells is $\sim$17%. For nc-Si, it is only $\sim$10%. However, the efficiency for nc-Si solar cells can be enhanced because the grain-boundary defects can be passivated by hydrogen and nc-Si can be used in a tandem structure combined with an a-Si:H solar cell [104]. Along with a higher efficiency, nc-Si solar cells show excellent stability under light absorption [105] and very high deposition rates [106–109]. These are the major attributes of nc-Si as the most promising material for solar cells.

4.2.1. *Fabrication*

To cut manufacturing costs, it is necessary to use thin-film materials rather than c-Si. The deposition of nc-Si at low temperatures is advantageous in that glass substrates can be used. For commercial application, an inexpensive substrate, a high deposition rate and a large area fabrication are essential. Glass is the most obvious substrate for low temperature deposition because it is cheap, transparent, and durable. The drawbacks of low temperature deposition of nc-Si films are that it leads to small grain sizes and low deposition rates. The basic solar cell structures are Schottky barrier, metal insulator semiconductor (MIS), and p–i–n type cells, however the term "thin film solar cell" generally signifies the p–i–n type structure (Fig. 4.4). A thick intrinsic layer of p–i–n cells is needed to separate regions

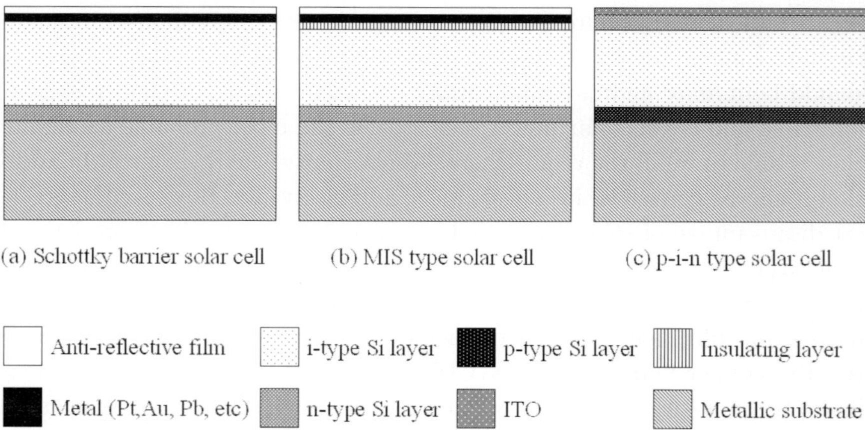

(a) Schottky barrier solar cell   (b) MIS type solar cell   (c) p-i-n type solar cell

Fig. 4.4.   Typical thin film solar cell structures.

of high recombination from the region where charge separation occurs. Since
p- and/or n-type layer is highly doped, dopant diffusion from doped layers
should be minimized to allow majority carriers (holes) to accept minority
carriers (electrons) [110]. In thin film solar cells, a-Si:H is more efficient
to absorb the visible light because of its higher bandgap (1.7–1.8 eV) than
that of c-Si (1.1 eV). The combination of occupied states in the valence
band and empty states in the conduction band determines the dependence
of the absorption coefficient on the photon energy. The photon energies in
the visible range of the spectrum are then absorbed by direct transition,
resulting in a higher absorption of a-Si:H over c-Si. But then a-Si:H fails
to collect the infrared range. As nc-Si has a smaller bandgap than a-Si:H,
when it is stacked with a-Si:H in the tandem structure, the efficiency of
solar cells can be improved by absorbing both visible and infrared range.

4.2.2. *Operation*

Solar cell is a photovoltaic device that absorbs lights and then produces
electrical power. The energy of photon of light is given by hv, and the
bandgap of Si is denoted by E_g. When the photon energy is higher than
the bandgap of Si ($hv > E_g$), the photon energy is absorbed by Si and
then an electron is excited to a level higher by hv. In this excited level,
the electron falls immediately into the conduction band. The covalent bond
where the electron previously was bonded now has one less electron, e.g.
a hole. Therefore the absorption of photons produces electron-hole pairs.
These electron-hole pairs generate an electric field thereby create an open
circuit voltage. When this voltage is applied to a load resistance, it causes
a current flow as the electrons flow from the p-type Si to the n-type Si and
the holes flow from the n-type Si to the p-type Si, the photovoltaic electrical
current flows from the p-type Si to the load. Figure 4.5 shows typical IV
characteristics of solar cell. The optimum operating output power P_{max} can
be expressed as follows.

$$P_{\mathrm{max}} = V_{\mathrm{oc}} \times I_{\mathrm{sc}} \times FF, \tag{4.1}$$

where V_{oc} is the open circuit voltage (current $= 0$) and I_{sc} is the short
circuit current (voltage $= 0$). Here the fill factor (FF) is given by

$$FF = \frac{V_m \times I_m}{V_{\mathrm{oc}} \times I_{\mathrm{sc}}}. \tag{4.2}$$

V_m and I_m are the output voltage and output current respectively at the
optimum operating point.

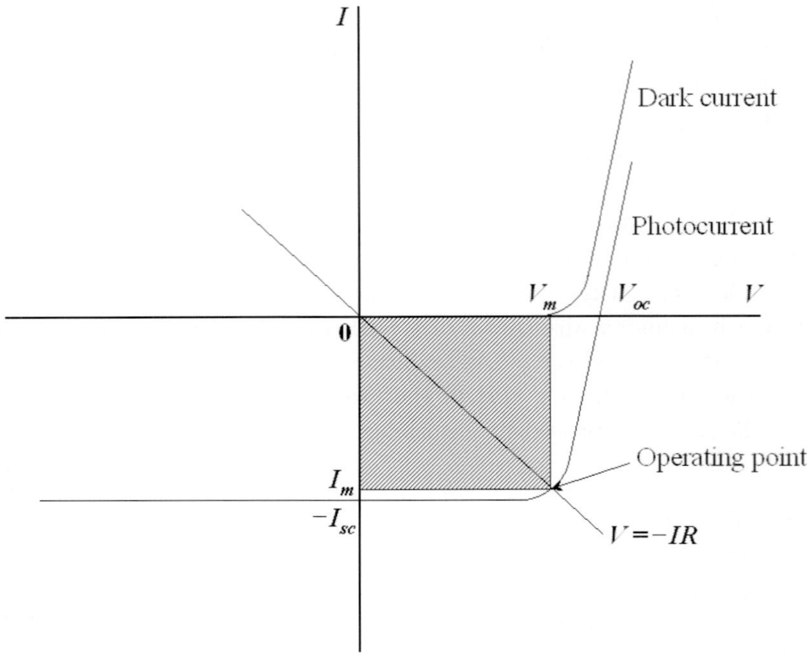

Fig. 4.5.　Current-voltage characteristics of solar cell — dark and under illumination.

B. Rech *et al.* [147] developed series of solar cells to investigate the performances of the a-Si:H single junction, nc-Si (here they called it μc-Si:H) single junction, and a-Si:H/nc-Si tandem junction solar cells. These samples were prepared on commercially available SnO_2:F coated glass. Figure 4.6 shows J–V curves of the solar cells and corresponding photovoltaic parameters. The tandem cell show the highest fill factor (FF), cell efficiency (η) and the lowest short-circuit current density (J_{sc}).

4.2.3. *Photodegradation*

The most troublesome problem inherent to a-Si:H solar cells is photodegradation. Light-induced changes of the properties of a-Si:H films were first reported by Staebler and Wronski [111, 112]. They found that prolonged exposure to light strongly decreases both dark conductivity and photoconductivity of a-Si:H films. The initial state could be restored by annealing above 150°C. The light-induced changes have been attributed to changes in the density and/or occupation of deep-gap states resulting in a shift of the Fermi level toward midgap. These states also act as recombination

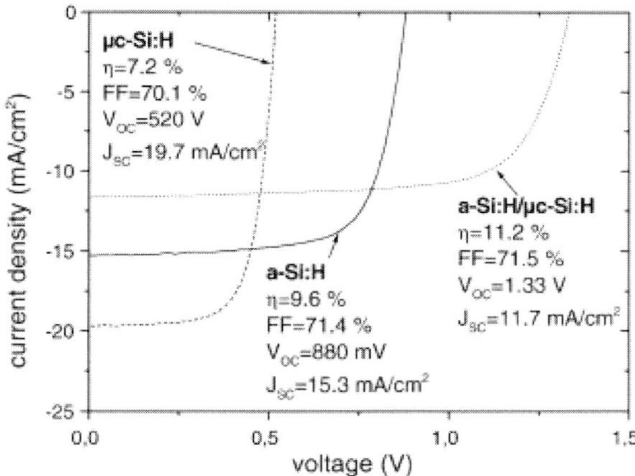

Fig. 4.6. Illuminate current density–voltage (J–V) characteristics and corresponding photovoltaic parameters of an a-Si:H single junction, a nc-Si single junction and an a-Si:H/nc-Si tandem junction solar cells.

centres and decrease the carrier lifetime. Electron spin resonance (ESR) studies identified these states as Si-dangling bonds [113, 114]. The light-induced changes in conductivity are most pronounced when the Fermi level lies about half way between the gap centre and either mobility edge [115]. Dersch *et al.* [113] proposed that non-radiative recombination between the tail states of the conduction and valence bands leads to the breaking of weak Si–Si bonds.

Although the appropriate incorporation of hydrogen is essential in reducing the defect density in hydrogenated amorphous silicon, it is now generally accepted that the ability of hydrogen to move in, out, and within a-Si:H is responsible for access defect creation and thus for degradation of electronic properties of this material [116]. Since the hydrogen is sufficiently mobile above the equilibration temperature and the activation energy for hydrogen motion decreases in the presence of intense illumination [117], both the light-induced degradation and the reversal of the degradation can be explained by hydrogen motion within a metastable defect complex [118]. It has been reported that degradation ratio of the solar cell increases at higher deposition rate while the SiH_2 density also increases correspondingly [119]. The SiH_2 density in the film can be reduced by elevating the deposition temperature because rising deposition temperature can promote surface reactions (desorption, $2Si–H \rightarrow 2Si–Si + H_2$, and structural relaxation). It was found that the higher temperature and hydrogen dilution were effective to suppress the degradation [119]. The observations [120] of stability of PECVD a-Si:H with standard hydrogen content show that the hydrogen content is not directly correlated to higher stability. However, it is still generally accepted that the hydrogen is closely related to the degradation process.

An exact solution to prevent the photodegradation has not been found. However there are few ways to reduce this effect in solar cells. First, using a thinner *i*-layer helps to reduce the degradation since Staebler–Wronski (SW) effect is marked in undoped films but does not occur in heavily doped films. Second, nc-Si has shown extremely low degradation effects because unlike a-Si:H, nc-Si has no SW effect [121]. Therefore, applying nc-Si is presently the most promising solution of the photodegradation.

4.3. *Light Emitting Diodes*

Light emitting diode (LED) is a luminescent device that can emit spontaneous optical radiation as a result of electronic excitation. To emit the light in visible ($hv > 1.8\,\mathrm{eV}$), luminescent materials must have bandgaps larger than this limit. Therefore c-Si whose bandgap is $1.1\,\mathrm{eV}$ is not capable to be as visible LEDs. However the feasibility of all-optical integrated circuits on silicon will depend on a suitable Si light emitter. The discovery of enhanced light emission from porous silicon by Canham [122] has shown that silicon nanostructures may be suitable for Si based LEDs or lasers. Si nanocrystals in the $\sim$10 nm size are efficient light emitters in the visible range due to quantum confinement effects (see Sec. 3.2.5) [78, 79, 122]. Recently silicon seems promising in the LED and light emitter application. First, if Si comes in the form of thin film it has the possibility of low cost fabrication on a large area substrate. Second, it has shown that high efficiencies can be achieved from Si nanoparticles [124]. Lastly, the inherent stability of silicon enables long lifetimes in LEDs.

The fabrication of porous silicon using electrochemical and chemical dissolution method [122] was followed by the discovery of Si nanocrystals in SiO_2 [123]. These Si nanocrystals showed a strong red (1.6 eV) photoluminescence (PL) and a luminescence dependence on size and concentration of the nanoclusters. Recently there have been huge efforts to fabricate Si nanocrystals in a SiO_2 matrix for use in optoelectronic devices by well-established silicon technology. In addition, advanced chemical vapor deposition using SiH_4, which combines deposition with oxidation and annealing, has been used to prepare Si nanocrystals [125]. It has been reported that

Si nanocrystals in Si implanted SiO_2 can be formed by annealing since on annealing the silicon atoms formed clusters, which then ordered into nanocrystals [126]. Photoluminescence in the visible (1.65 eV) has been observed from 2 nm nanocrytal whose bandgap can be increased upto 2.5 eV [127]. Light emission in the visible range from red to blue is observed by reducing the nc-Si diameter from 4 nm to about 1 nm [126]. The red light emitted from nc-Si is based on the radiative recombination through localized states, which forms in the bandgap during surface oxidation in the air [128]. Blue and green luminescence was also obtained from porous Si [129]. However, this luminescence is stable and strong only in a vacuum and/or nitrogen gas but not in the air. Besides blue and green luminescence immediately changes to red after exposing to the air for few seconds and the efficiencies of luminescence are also decreased [129]. The surface oxidation and desorption of hydrogen atoms create the localized states which act as nonradiative recombination centers. These localized states are responsible for the changes in colors and efficiencies. To improve the blue and green luminescence in the air, it is therefore necessary to form a passivation layer on the surface.

Plasma enhanced chemical vapor deposition (PECVD) technique is compatible with the silicon technology which afford to control the crystallite size and the mechanical stress [130–132] as well as the thickness of the grain boundaries [69, 133, 134]. Luminescence has been observed from Si particles synthesized by homogeneous nucleation of reactive SiH_4 fragments in a SiH_4/H_2 plasma or by sputtering [135, 136]. Nc-Si films for photovoltaic applications have also been demonstrated [137–140]. PECVD nc-Si with nanocrystals in 2–15 nm dimensions has shown the room temperature photoluminescence in the visible range [141]. The photoluminescence spectra of the nc-Si films were similar to the spectra from porous silicon by wet method (e.g. anodization and electrochemical dissolution of c-Si). This suggests that the luminescence mechanism of nc-Si films is similar to that of porous silicon.

Erbium (Er) implanted nanocrystal structures have shown the luminescence shift to1.54 μm due to an effective energy transfer from the nanocrystals to the Er^{3+} ions [142, 143]. Since the silica fiber optic exhibits a maximum transmission at 1.54 μm, this particular wavelength is a technological importance for silicon based optoelectronic systems. However, a quantum efficiency of Er-doped Si LED is still insufficient ($<0.01\%$ at room temperature) [142–144] for applications due to high nonradiative recombination.

5. Conclusions

Currently hydrogenated amorphous silicon (a-Si:H) is dominantly used in thin film transistor liquid crystal displays (TFT-LCDs) and solar cells since it can be uniformly deposited over large area at low temperature even though the a-Si:H films show inferior properties of carrier mobility and instability. However, nanocrystalline silicon (nc-Si), which consists of a number of small crystals embedded in an amorphous silicon matrix, can replace a-Si:H in many applications because of its higher mobility and better stability. It has also been shown that device quality nc-Si films can be deposited at low temperature by similar deposition techniques for a-Si:H. In this chapter, the mechanisms of nc-Si formation, advances in the deposition techniques and properties of nc-Si films will be introduced followed by their applications to thin film transistors and solar cells.

References

1. A. Matsuda, *Thin Solid Films*, **337**, 2 (1999).
2. J. Perrin, *Plasma Deposition of Amorphous Silicon-Based Matter* (Academic Press, 1995), p. 177.
3. T. Okada, T. Iwaki, H. Kasahara and K. Yamamoto, *Jpn. J. Appl. Phys.* **24**, 161 (1985).
4. M.J. Kushner, *J. Appl. Phys.* **63** (1988) 2532.
5. J.R. Abelson, N. Maley, J.R. Doyle, G.F. Feng, M. Fitzner, M. Katiyar, L. Mandrell, A.M. Myers, A.Nuruddin, D.N. Ruzic and S.Yang, *Mat. Res. Soc. Symp. Proc.* **219** (1991).
6. J. Perrin, *J. Non-Cryst. Solids* **137–138** (1991) 639.
7. S. Yamamoto, J. Nakamura and M. Migitaka, *Jpn. J. Appl. Phys.* **35** (1996) 3863.
8. J.I. Woo, H.J. Lim and J. Jang, *Appl. Phys. Lett.* **65** (1994) 1644.
9. R.E. Hollingsworth and P.K. Bhat, *Appl. Phys. Lett.* **64** (1994) 616.
10. M.S. Feng and C.W. Liang, *J. Appl. Phys.* **77** (1995) 4771.
11. H. Curtins, M. Favre, N. Wyrsch, M. Brechet, K. Prasad and A.V. Shah, *Proc. 19th IEEE PV Specialists Conf.* (1987), p. 695.
12. H. Curtins, N. Wyrsch and A.V. Shah, *Electron. Lett.* **23** (1987) 228.
13. R.H. Chatham and P.K. Bhat, Annual Subcontractor Report for the period 1 May 1988–30 April 1989, work performed by Glasstech Solar Inc., SERI/STR-211-(1989), p. 3562.
14. M.R. Wertheimer and M. Moisan, *J. Vac. Sci. Tech.* **3** (1985) 2643.
15. S.J. Oda and M. Matsumura, *Jpn. J. Appl. Phys.* **29** (1990) 1889.
16. H. Chatham and P.K. Bhat, *Mater. Res. Soc. Symp. Proc.* **149** (1989) 447.
17. A.A. Howling, J.L. Dorier, C. Hollenstein, U. Kroll and F. Finger, *J. Vac. Sci. Tech.* **10** (1992) 1080.

18. P. Merel, M. Chaker, M. Moisan, A. Ricard and M. Tabbal, *J. Appl. Phys.* **72** (1992) 3220.

19. F. Finger, U. Kroll, V. Viret, A. Shah, W. Beyer, X.-M. Tang, J. Weber, A.A. Howling and C. Hollenstein, *J. Appl. Phys.* **71** (1992) 5665.

20. C.M. Ferreira and J. Loureiro, *J. Phys.* **D17** (1984) 1175.

21. J. Bezemer, W.G.J.H.M. van Sark, M.B. von der Linden and W.F. van der Weg, in *Proc. 12th E.C. Photovoltaic Solar Energy Conf.*, Amsterdam, the Netherlands, 1994, eds. R. Hill, W. Palz and P. Helm (H.S. Stephens & Associates, Bedford, U.K., 1994).

22. J. Bezemer and W.G.J.H.M. van Sark, in Electronic, Optoelectronic and Magnetic Thin films, *Proc. 8th Int. School Condensed Matt. Phy. (ISCMP)*, Varna, Bulgaria, 1994 (Wiley, N.Y., 1995).

23. M.R. Wertheimer, J.E. Klemberg-Sapieha and H.P. Schreiber, *Thin Solid Films* **115** (1984) 109.

24. M. Heintze and R. Zedlitz, *J. Phys.* **D26** (1993) 1781.

25. J. Dutta, U. Kroll, P. Chabloz, A. Shah, A.A. Howling, J.-L. Dorier and C. Hollenstein, *J. Appl. Phys.* **72** (1992) 3220.

26. J. Kuske, U. Stephan, O. Steinke and S. Rohlecke, *Mater. Res. Soc. Symp. Proc.* **377** (1995) 27.

27. U. Kroll, J. Meier, M. Goetz, A. Howling, J.-L. Dorier, J. Dutta, A. Shah and C. Hollenstein, *J. Non-Crystalline Solids* **164** (1993) 59.

28. H. Meiling, J.F.M. Westendorp, J. Hautala, Z. Saleh and C.T. Malone, *Mater. Res. Soc. Symp. Proc.* **345** (1995) 65.

29. F. Finger, P. Hapke, M. Luysberg, R. Carius, H. Wagner and M. Scheib, *Appl. Phys. Lett.* **65** (1994) 2588.

30. A. Matsuda, *J. Non-Cryst. Solids* **59–60** (1983) 767.

31. P. Alpium, V. Chu and J.P. Conde, *J. Appl. Phys.* **86** (1999) 3812.

32. M. Heintze, R. Zedlitz, H.N. Wanka and M.B. Schubert, *J. Appl. Phys.* **79** (1996) 2699.

33. P. Brogueira, J.P. Conde, S. Arekat and V. Chu, *J. Appl. Phys.* **79** (1996) 8748.

34. K.C. Wang, H.L. Hwang, P.T. Leong and T.R. Yew, *J. Appl. Phys.* **77** (1995) 6542.

35. S. Bae, A.K. Kalkan, S. Cheng and S.J. Fonash, *J. Vac. Sci. Technol.* **A16** (1998) 1912.

36. A.J. Perry, D. Vender and R.W. Boswell, *J. Vac. Sci. Technol.* **B9/2** (1991) 310.

37. M. Scheib, B. Schroeder and H. Oechsner, *J. Non-Cryst. Solid* **198** (1996) 895.

38. N. Layadi, P. Roca i Cabarrocas, B. Drévillon and I. Solomon, Phys. Rev. **B52** (1995) 5136.

39. C.C. Tsai, Thompson, C. Doland, F.A. Ponce, G.B. Anderson and B. Wacker, *Mater. Res. Soc. Symp. Proc.* **118** (1988).

40. S. Veprek, *Chimia* **34** (1981) 489.

41. R.C. van Oort, M.J. Geerts, J.C. van den Heuvel and J.W. Metselaar, *Electron. Lett.* **23** (1987) 969.

42. M. Fang, J.B. Chevrier and B. Drevillon, *J. Non-Cryst. Solids* **137–138** (1991) 791.

43. I. Solomon, B. Drevillon, H. Shirai and N. Layadi, *J. Non-Cryst. Solids* **164–166** (1993) 989.

44. M. Heintze, W. Westlake and P.V. Santos, *J. Non-Cryst. Solids* **164–166** (1993) 985.

45. A. Gallagher and J. Scott, *Sol. Cells* **21** (1987) 147.

46. Y.H. Yang, M. Katiyar, G.F. Feng, N. Maley and J.R. Abelson, *Appl. Phys. Lett.* **65** (1994) 1769.

47. T. Akasaka and I. Shimizu, *Appl. Phys. Lett.* **66** (1995) 3441.

48. K. Saitoh, M. Kondo, M. Fukawa, T. Nishimiya, A. Matsuda, W. Fukato and I. Shimizu, *Appl. Phys. Lett.* **71** (1997) 3403.

49. Y. Okada, J. Chen, H. Campbell, P.M. Fauchet and S. Wagner, *J. Appl. Phys.* **67** (1990) 1757.

50. M. Luysenberg, P Hapke, R. Carius and F. Finger, *Phil. Mag.* **A/75** (1997) 31.

51. E.A.G. Hamers, J. Bezemer and W.F. van der Weg, *Appl. Phys. Lett.* **75** (1999) 609.

52. E.A.G. Hamers, A. Fontcuberta i Morral, C. Niikura, R. Brenot and P. Roca i Cabarrocas, *J. Appl. Phys.* **88** (2000) 3674.

53. P. Torres, J. Meier, R. Flückiger, U. Kroll, J.A.A. Selvan, H. Keppner, A. Shah, S.D. Littelwood, I.E. Kelly and P. Giannoules, *Appl. Phys. Lett.* **69** (1996) 1373.

54. See, for example, F. Shimura, *Oxygen in Silicon* (Academic, New York), Ch. 7–8.

55. Y. Nasuno, M. Kondo and A. Matsuda, *Appl. Phys. Lett.* **78** (2001) 2331.

56. N.H. Nickel, W.B. Jackson, N.M. Johnson and J. Walker, *Phys. Stat. Sol.* **A159** (1997) 65.

57. A. Kaan Kalkan and S.J. Fonash, *Mater. Res. Soc. Symp. Proc.* **467** (1997) 319.

58. J. Tauc, R. Grigorovici and A. Vancu, *Physica Status Solidi* **15** (1966) 627.

59. M.H. Brodsky, M. Cardona and J.J. Cuomo, *Phy. Rev.* **B16** (1977) 3556.

60. U. Kroll, J. Meier, A. Shah, S. Mikhailov and J. Weber, *J. Appl. Phys.* **80** (1996) 4971.

61. G. Lucovsky, R.J. Nemanich and J.C. Knights, *Phys. Rev.* **B19** (1979) 2064.

62. A.A. Langford, M.L. Fleet and B.P. Nelson, *Phys. Rev.* **B45** (1992) 13367.

63. H. Shanks, C.J. Fang, L. Ley, M. Cardona, F.J. Demond and S. Kalbitzer, *Phys. Stat. Sol. (b)* **100** (1980) 43.

64. T. Itoh, K. Yamamoto, H. Harada, N. Yamana, N. Yashida, H. Inouchi, S. Nonomura and S. Nitta, *Sol. Energy Mat. Sol. Cells* **66** (2001) 239.

65. W.B. Pollard and G. Lucovsky, *Phys. Rev.* **B26** (1982) 3172.

66. H. Shimizu, S. Mizuno and M. Noda, *Jpn. J. Appl. Phys.* **25** (1986) 775.

67. M. Cardona, *Phys. Status Solidi* **B118** (1983) 463.

68. Z. Iqbal and S. Veprek, *J. Phys. C: Solid State Phys.* **15** (1982) 377.

69. S. Veprek, F.A. Sarott and Z. Iqbal, *Phys. Rev.* **B36** (1987) 3344.

70. L.P. Avakyanta, L.L. Gerasimov, V.S. Gorelik, N.M. Mania, E.D. Obrastsova and Yu.I. Plotnikov, *J. Mol. Struct.* **267** (1992) 177.

71. J. Bandet, J. Frandon, F. Fabre and B. De Mauduit, *Jpn. J. Appl. Phys.* **32** (1993) 1518.

72. V. Paillard, P. Puech, P. Temple-Boyer, B. Caussat, E. Scheid, J.P. Couderc and B. de Mauduit, *Thin Solid Films* **337** (1999) 93.

73. G. Yue, J.D. Lorentzen, J. Lin, D. Han and Q. Wang, *Appl. Phys. Lett.* **75** (1999) 492.

74. E. Bustarret, M.A. Hachicha and M. Brunel, *Appl. Phys. Lett.* **52/20** (1988) 1675.

75. Ch. Ossadnik, S. Veprek and I. Gregora, *Thin Solid Films* **337** (1999) 148.

76. P. Scherrer, *Goett. Nachr.* **2** (1918) 98.

77. H.P. Klug and L.E. Alexander, *X-ray Diffraction Procedures for Polycrystalline and Amorphous Materials* (Wiley & Sons, 1974).

78. T. Takagahara and K. Takeda, *Phys. Rev.* **B46** (1992) 15578.

79. W.L. Wilson, P.F. Szajowski and L.E. Brus, *Science* **262** (1993) 1242.

80. J. Linnros, A. Galeckas, N. Lalic and V. Grivickas, *Thin Solid Films* **297** (1997) 167.

81. C. Delerue, G. Allan and M. Lannoo, *J. Luminescence* **80** (1998) 65.

82. S. Sheng, X. Liao and G. Kong, *Appl. Phys. Lett.* **78** (2001) 2509.

83. S. Miyazaki, N. Fukhara and M. Hirose, *J. Non-Cryst. Solids* **266–269** (2000) 59.

84. U. Kroll, J. Meier, A. Shah, S. Mikhailov and J. Weber, *J. Appl. Phys.* **80** (1996) 4971.

85. Y.Q. Fu, J.K. Luo, S.B. Milne, A.J. Flewitt and W.I. Milne, *Mater. Sci. Eng.* **B124** (2005) 132.

86. G. Viera, S. Huet and L. Boufendi, *J. Appl. Phys.* **90** (2001) 4175.

87. E. Edelberg, S. Bergh, R. Naone, M. Hall and E.S. Aydil, *J. Appl. Phys.* **81** (1997) 2401.

88. P. Roca I Cabarrocas, *J. Non-Cryst. Solids* **266** (2000) 31.

89. S. Ray, S. Mukhopadhyay, T. Jana and R. Carius, *J. Non-Cryst. Solids* **299–302** (2002) 761.

90. P.G. LeComber, W.E. Spear and A. Gaith, *Elec. Lett.* **15**, 179–181 (1979).

91. M. Yamano, H. Ikeda and H. Takesada, *Conf. Rec. Int. Display Res. Conf.* (1983), p. 356.

92. H. Borkan and P.K. Weimer, *R.C.A. Rev.* **24** (1963) 153–165.

93. W.B. Jackson and M.D. Moyer, *Phys. Rev.* **B36** (1987) 6217.

94. M.J. Powell, C. van Berkel, I.D. French and D.H. Nicholls, *Appl. Phys. Lett.* **51** (1987) 1242.

95. N. Nickel, W. Fuhs and H. Mell, *J. Non-Cryst. Solids* **115** (1989) 159.

96. N. Nickel, W. Fuhs and H. Mell, *Philos. Mag.* **B61** (1990) 251.

97. A.V. Galatos and J. Kanicki, *Appl. Phys. Lett.* **57** (1990) 1197.

98. J. Kanicki, C. Godet and A.V. Gelatos, Amorphous Silicon Technology, *M.R.S. Symposia Proc.* **219** (1991) 45.

99. J.P. Kleider, C. Longeaud, D. Mencaraglia, A. Rolland, P. Vitrou and J. Richard, *J. Non-Cryst. Solids* **164** (1993) 739.

100. S.C. Deane, F.J. Clough, W.I. Milne and M.J. Powell, *J. Appl. Phys.* **73** (1993) 2895.

101. P.N. Morgan, W.I. Milne, S.C. Deane and M.J. Powell, *J. Non-Cryst. Solids* **164** (1993) 199.

102. S.C. Deane and M.J. Powell, *J. Appl. Phys.* **74** (1993) 6655.

103. R.B. Wehrspohn, M.J. Powell, S.C. Dean, I. French and P. Roca i Cabarrocas, *Appl. Phys. Lett.* **77** (2000) 750.

104. J. Meier *et al.*, *Sol. Energy Mat. Solar Cells* **49** (1997) 35.

105. W.A. Anderson and C. Ji, *IEEE Photovoltaics* **50** (2003) 1885.

106. U. Graf, J. Meier, U. Kroll, J. Bailat, C. Droz, E. Vallat-Sauvain and A. Shah, *Thin Solid Films* **427** (2003) 37.

107. M. Kondo, M. Fukawa, L. Guo and A. Matsuda, *J. Non-Cryst. Solids* **266–269** (2000) 84.

108. T. Roschek, B. Rech, W. Beyer, P. Werner, F. Edelman. A. Chack, R. Weil and R. Beserman, *M.R.S. Symp. Proc.* **664** (2001).

109. L. Guo, M. Kondo, M. Fukawa, K. Saitoh and A. Matsuda, *Jpn. J. Appl. Phys.* **37** (1998) 1116.

110. Y. Kuwano, S. Tsuda and M. Ohnishi, *Jpn. J. Appl. Phys.* **21** (1982) 235.

111. D.L. Staebler and C.R. Wronski, *Appl. Phys. Lett.* **31** (1977) 292.

112. D.L. Staebler and C.R. Wronski, *J. Appl. Phys.* **51** (1980) 3262.

113. H. Dersch, J. Stuke and J. Beichler, *Appl. Phys Lett.* **38** (1981) 456.

114. M. Stutzmann, W.B. Jackson and C.C. Tsai, *AIP Conf. Proc.* **120** (1984) 213.

115. M.H. Tanielian, N.B. Goodman and H. Fritzsche, *J. Phys. (Paris), C4* **42** (1981) 375.

116. R.A. Street, *Hydrogenated Amorphous Silicon* (Cambridge University Press, Cambridge, 1991).

117. P.V. Santos, N.M. Johnson and R.A. Street, *Phys. Rev. Lett.* **67** (1991) 2686.

118. D.E. Carlson and K. Rajan, *J. Appl. Phys.* **83** (1998) 1726.

119. T. Nishimoto, M. Takai, M. Kondo and A. Matsuda, *J. Non-Cryst. Solids* **292–302** (2002) 1095.

120. S. Guha, *J. Non-Cryst. Solids* **198** (1996) 1076.

121. J. Meier, S. Dubail, R. Fluckiger, D. Fischer, H. Keppner and A. Shah, *Proc. IEEE First World Conf. Photovoltaic Energy Conversion* (New York, 1994), p. 409.

122. L.T. Canham, *Appl. Phys. Lett.* **57** (1990) 1046.

123. L.T. Canham, *Phys. Status Solidi.* **B190** (1995) 9.

124. Y.Q. Wang, Y.G. Wang, L. Cao and Z.X. Cao, *Appl. Phys. Lett.* **83** (2003) 3474.

125. M. Ruckschloss, B. Landkammer and S. Veprek, *Appl. Phys. Lett.* **63** (1993) 1474.

126. S. Guha, M.D. Pace, D.N. Dunn and I.L. Singer, *Appl. Phys. Lett.* **70** (1997) 1207.

127. B. Delley and E.F. Steigmeier, *Appl. Phys. Lett.* **67** (1995) 2370.

128. K. Sato and K. Hirakuri, *J. Appl. Phys.* **97** (2005) 104326.

129. H. Mizuno, H. Koyama and N. Koshida, *Appl. Phys. Lett.* **69** (1996) 3779.

130. M. Rtickschloss, O. Ambacher and S. Veprek, *J. Luminescence* **57** (1993) 1.

131. S. Veprek, *Proc. M.R.S. Europe Symp. Proc.*, eds. P. Pinard and S. Kalbitzer (Les Editions de Physique, Les Ulis, 1984).

132. S. Veprek, F.A. Sarott and M. Ruckschloss, *J. Non-Crystal. Mater.* **137–138** (1991) 733.

133. S. Veprek, H. Tamura, M. Riickschloss and Th. Wirschem, *M.R.S. Symp.* **345** (1994) 311.

134. S. Veprek, Th. Wirschem, M. Ruckschloss, H. Tamura and J. Oswald, *M.R.S. Symp. Proc.* **358** (1995) 99.

135. S. Furukawa and T. Miyasato, *Jpn. J. Appl. Phys.* **27** (1988) L2207.

136. H. Takagi, H. Ogawa, Y. Yamazaki, A. Ishizaki and T. Nakagiri, *Appl. Phys. Lett.* **56** (1990) 2379.

137. I. Solomon, B. Drevillon, H. Shirai and N. Layadi, *J. Non-Cryst. Solids* **164–166** (1993) 989.

138. J.J. Boland and G.N. Parsons, *Science* **256** (1992) 1304.

139. M. Otobe and S. Oda, *J. Non-Cryst. Solids* **164–166** (1993) 993.

140. H.V. Nguyen, I. An, R.W. Collins, Y. Lu, M. Wakagi and C.R. Wronski, *Appl. Phys. Lett.* **65** (1994) 3335.

141. E. Edelberg, S. Bergh, R. Naone, M. Hall and E.S. Aydil, *Appl. Phys. Lett.* **68** (1996) 1415.

142. B. Zheng, J. Michel, F.Y.G. Ren, L.C. Kimerling, D.C. Jacobson and J.M. Poate, *Appl. Phys. Lett.* **64** (1994) 2842.

143. S. Coffa, G. Franzò and F. Priolo, *Appl. Phys. Lett.* **69** (1996) 2077.

144. S. Coffa, G. Franzò, F. Priolo, F.A. Pacelli and A. Lacaita, *Appl. Phys. Lett.* **73** (1998) 93.

145. C.-H. Lee, A. Sazonov and A. Nathan, *J. Vac. Sci. Tech.* **A24** (2006) 618.

146. C.-H. Chen, T.-C. Chang, P.-T. Liu, H.-Y. Lu, K.-C. Wang, C.-S. Huang, C.-C. Ling and T.-Y. Tseng, *IEEE Elec. Dev. Lett.* **26** (2005) 731.

147. B. Rech, T. Repmann, M.N. van den Donker, M. Berginski, T. Kilper, J. Hupkes, S. Calnan, H. Stiebig and S. Wieder, *Thin Sold Films* **511–512** (2006) 548.

CHAPTER 10

AMORPHOUS AND NANOCOMPOSITE DIAMOND-LIKE CARBON COATINGS FOR BIOMEDICAL APPLICATIONS

T. I. T. Okpalugo[*,§], N. Ali[†,¶], A. A. Ogwu[‡], Y. Kousar[†] and W. Ahmed[*]

Northern Ireland Bioengineering Centre (NIBEC)
University of Ulster, Belfast, UK
†*Department of Mechanical Engineering*
University of Aveiro, 3810-193 Aveiro, Portugal
‡*Thin Films Centre, University of Paisley, Paisley, Scotland, UK*
§ *thomas.okpalugo@paisley.ac.uk*
¶ *n.ali@nano-society.org*

1. Introduction

Heart disease is one of the most common causes of death in the world today, particularly in western countries. There are various causes of heart disease, related most commonly to diet and exercise. The failure of heart valves accounts for about 25–30% of heart problems that occur today. Faulty heart valves need to be replaced by artificial ones using sophisticated and sometimes risky surgery. However, once a heart valve has been replaced with an artificial one there should be no need to replace it again and it should last at least as long as the life of the patient. Therefore, any technique that can increase the operating life of heart valves is highly desirable and valuable. Currently, pyrolytic-carbon (PyC) is used for the manufacture of mechanical heart valves. Although, PyC is widely used for heart valve purposes, it is not the ideal material. In its processed form, PyC is a ceramic-like material and like ceramics, it is subject to brittleness. Therefore, if a crack appears, the material, like glass, has very little resistance to the growth/propagation of the crack and may fail under load. In addition, its blood compatibility is not ideal for prolonged clinical use. As a result, thrombosis often occurs in patients who must continue to take anti-coagulation drugs on a regular basis. The anti-coagulation therapy can give rise to some serious side effects, such as birth defects.

513

Occlusive diseases of the arteries that cause life threatening blood flow restrictions can be treated by open heart surgical intervention or by implantation of intracoronary metallic stents. However, in spite of considerable advances in improving the mechanical properties of metallic stents, advances in implantation techniques, and in antithrombotic therapy, the use of stents is still complicated by substantial cases of thrombotic occlusions/stenosis and restenosis [1–5] due to platelet activation resulting from the release of metallic particles/ions, shear forces and blood contacting of the metallic surface [6–9]. Likewise thromboembolism (valve thrombosis and systemic embolism) remains the major drawback in the management of implanted mechanical heart valve prostheses [10, 11]. Patients with these implanted prostheses are faced with life threatening bleeding problems because they are kept under lifelong anticoagulant therapies in order to reduce the risk of thromboembolism. Platelet aggregation in these prostheses is the key factor in thrombus formation and dissemination as emboli which are life threatening. In order to reduce the risk of platelet aggregation/thromboembolism and complications following the lifelong course(s) of anticoagulants, the biomaterials need to be improved in order to achieve better biocompatibility/haemocompatibility [1, 2, 12, 13].

Apart from thrombosis, other problems associated with the failure of medical implants and devices that need to be overcome are problems of mechanical failure, wear, tear and fatigue; the problems of chemical degradation, corrosion and oxidative degeneration; the problem of calcification and the problem of triggering excessive immune response and or infection. Metallic implants may have good fatigue life and may be cheap (e.g. stainless steel) but they can release metallic ions and wear debris into the surrounding tissues, leading to osteolysis, loosening and failure of the implants. All these problems encountered with implanted prostheses and medical devices could be solved with some surface coatings modifications using appropriate durable and biocompatible biomaterial.

It is therefore extremely urgent that new materials, which have better surface characteristics, blood compatibility, improved wear properties, better availability and higher resistance towards breaking are developed. Surface coatings like carbon-based biocompatible coatings are one of the ways of improving both the mechanical, physical and chemical properties of implants in direct contact with the blood and thus improve the biocompatibility and haemocompatibility of implants.

2. Amorphous and Nanocomposite Diamond-Like Carbon Coatings

Diamond is composed of the single element carbon, and it is the arrangement of the C atoms in the lattice that give diamond its amazing properties. Carbon is the sixth element of the periodic table and is grouped into group table IV with silicon. Common natural forms of carbon include charcoal, hydrocarbons ("oils"), graphite, and diamond. Carbon is ubiquitous and unique amongst all the elements in the diversity of its short-range, medium-range and long-range bonding forms with itself and with other elements. It can form double and triple bonds such that it can form networks far more complicated than those of silicon which can only form tetrahedral coordination bonds [14]. It exists in both crystalline and amorphous forms and exhibits both metallic and non-metallic characteristics. Crystalline carbon includes graphite (sp^2, three-fold planar bonding), "cubic" diamond (sp^3, four-fold tetrahedral bonding) and a family of fullerenes [15, 16]. Amorphous forms of carbon include charcoals, hydrocarbons ("oils"), organic materials and synthetic carbon materials. Synthetic amorphous carbon materials include pyrolytic carbon, hydrogenated amorphous carbon (a-C:H) and modified a-C:H-like silicon-doped a-C:H (a-C:H:Si). The bonds that exist in the naturally occurring carbon forms may also dominate in amorphous and synthetic carbon materials and are important dictators of properties like electrical and optical properties, and surface energy.

2.1. *Electronic Structure*

The electronic structure of diamond is based on sp^3 sites. The four valence electrons of the carbon atoms are assigned to a tetrahedrally directed sp^3 hybrid forming σ-bonds (Fig. 2.1). At the sp^2 sites as in graphite, three of the valence electrons are assigned to trigonally directed sp^2 hybrids, which form the σ-bonds such as the strong intralayer σ-bonds of graphite. The remaining one electron, the fourth electron is placed in a π-orbital lying normal to the layer and this forms weaker π-bonds which gives rise to the anisotropic metallic character of graphite. The electronic structure of the amorphous carbons is described in terms of the spectrum of the density of electron states. This is shown in Fig. 2.2 [14, 17, 18].

The σ-bonds form the skeleton of the network. The Fermi level E_F separating the filled and the empty states is at the $E = 0$ level. The valence band of filled bonding σ-states lie at negative energies and is separated

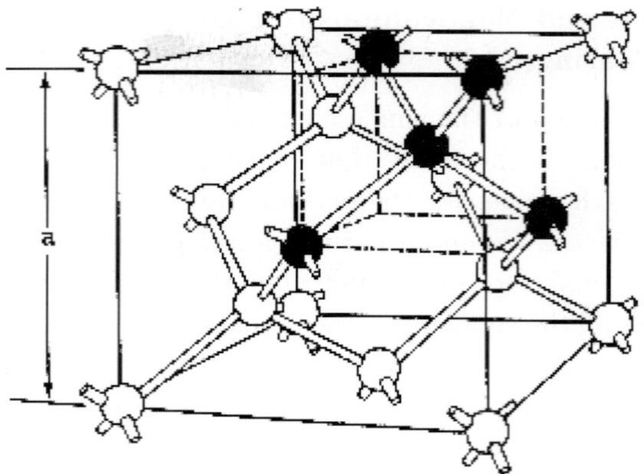

Fig. 2.1.   Schematic representation of the tetrahedral bonding of diamond.

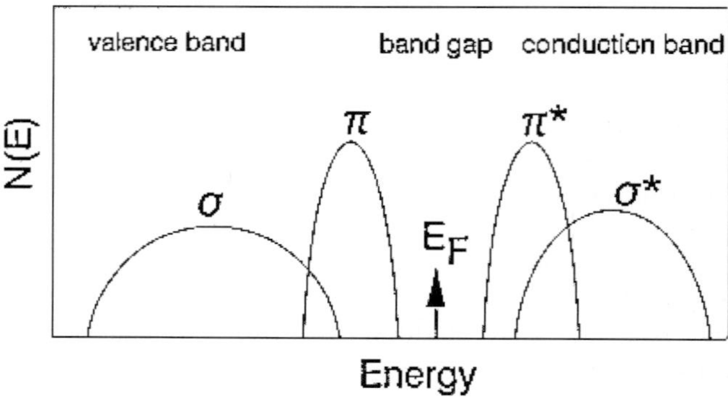

Fig. 2.2.   Schematic electronic band structure of amorphous carbons [14, 17, 18].

from the conduction band of empty anti-bonding σ^*-states by $\sim$5 eV. The π-states consisting of filled π-band and empty π^*-band lie largely within the σ–σ^* gap. The π-states control the electronic structure of the amorphous carbons because they lie closest to the Fermi level and tend to dominate changes in the total energies.

Diamond is an insulator at 0 K due to its crystal structure which also provides the rest of diamond's extreme properties which include the large bandgap (5.48 eV), highest hardness, atom number density, thermal conductivity (26.2 Wcm^{-1}K^{-1}), low conductivity (10^{-16} Scm^{-1}) and lowest chemical reactivity of any substance known [19, 20]. The characteristic properties of graphite are also detected by its electronic and

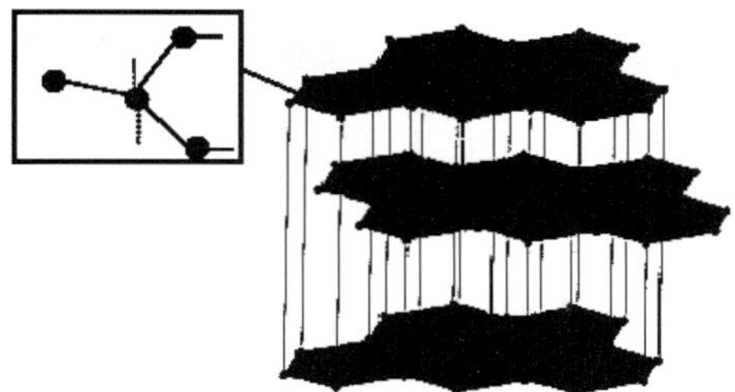

Fig. 2.3. Diagram showing bonding structure of graphite.

crystal structure (Fig. 2.3). These include low hardness, high lubricity, low electrical resistivity $(1\text{–}5 \times 10^{-3}\,\Omega\text{cm}^{-1})$ a thermal conductivity of $\sim 1.5\text{W cm}^{-1}\text{K}^{-1}$ and no optical bandgap. This is due to the type of bonding which leaves one electron of the carbon p-layer, held by weak Van der Waals bonding, available for electrical conduction and/or mixed bonding [19].

2.2. *Plasma-Based Deposition Methods*

DLC is a synthetic thin film and ultrathin film material that is metastable. It is prepared under ion bombardment of the growing film. It is usually synthesized under high vacuum with precursor gas(s), and or target-material using a radio frequency power to generate the capacitatively-coupled plasma or inductively to generate the inductively coupled plasma that forms the material on the substrate.

Physical chemical vapor deposition (P-CVD) processes are the hybrid processes which use glow discharge to activate CVD processes. These are broadly now referred to as plasma-enhanced CVD (PECVD) or plasma-assisted CVD (PACVD). The composition of the films deposited by PECVD is determined to a large extent by the relative fluxes of all the species falling on the substrate.

The process parameters that determine the outcome include [21, 22]:

1. Partial pressure of the feed gas, or the flow rates of the different gases, or the mixing ratio of different gases.
2. Total pressure in the reactor.
3. Substrate temperature and bias.
4. Reactor geometry and material.
5. Electrode material and the distance between the electrodes.
6. Electric power applied to the plasma (self-bias, cathode voltage).

7. Choice of gases.
8. Sample temperature.
9. Sample position with respect to gas flow and the plasma.
10. Frequency of the electric field.
11. Mass flow or gas flow (pumping efficiency).
12. Power distribution (triode).
13. Magnetic field intensity.
14. Pre-treatment of the substrates.

An important relation in terms of these deposition parameters is that due to Bubenzer *et al.* [23] and Catherine [24]:

$$V_{\mathrm{B}} \propto \left(\frac{P_{\mathrm{RF}}}{P} \right)^{1/2} \text{[24]},$$

where V_{B} = substrate bias, P_{RF} = RF power, P = pressure.

$$E_i \propto V_{\mathrm{B}} P^{-1/2} \text{[23]},$$

where E_i = ion energy, V_{B} = substrate bias, P = pressure.

The RF power determines the current and the voltage between the electrodes. The chemical reactions and the properties of the final product may be determined by the composition of gas mixture and the frequency of the electric field. The rates of the reaction may depend on: the composition of the feed gas, the total pressure in the reactor, the electric power, the distance between the electrodes, and the substrate bias or temperature. The structure of the deposited films may be affected by the substrate bias or temperature and the distance between the electrodes. The uniformity of the deposition process may be affected by the flow rate of the gases, the total pressure, and the reactor geometry. The film thickness may be affected by the deposition time, the composition of gas mixture, and the DC voltage and the frequency of the electric field. The electrode material itself is important as well as the building material of the reactor because they can affect the chemistry occurring in the glow discharge. The materials for these should therefore be chemically inert to the precursor gases and the plasma products. The electrodes and the reactor walls are almost always sputtered to a certain degree due to the fact that the surfaces in contact with the plasma are always negative relative to it. These sputtered atoms may hence contaminate the plasma and the processed sample.

Today different forms of PECVD constitute the main method for deposition of DLC. Among the PECVD methods, parallel plate RF reactor systems are the most common type employed. PECVD is commonly used to produce hydrogenated DLC (hydrogenated amorphous carbon, a-C:H) with high film thickness uniformity [25] and coating continuity for over a large substrate area [24]. Filtered cathodic vacuum arc (FCVA) method of DLC deposition has the potential to produce very hard films [26, 27] and to produce hydrogen free DLC [27–29]. However, the macroparticle filters in FCVA used to prevent particle contamination [30, 31] in the film limits the film thickness uniformity on a large area substrates and the repeatability of the deposition. Thus, although, hydrogenated DLC (a-C:H) produced by PECVD are not as mechanically hard as the non-hydrogenated DLC (ta-C) produced by FCVA, the improved tribochemical properties offered by a-C:H leads to improved wear resistance in hard disk interface for example. In principle any hydrocarbon with sufficient vapor pressure can be used for DLC film deposition in PECVD chamber. Commonly acetylene is used. Other precursors include benzene, butane, cyclohexane, ethane, ethylene, hexane, isopropane, methane, pentane, propane, and propylene.

The a-C:H surface is essentially covered with C–H bonds and is therefore chemically passive. Di-radicals and other unsaturated species react strongly and insert directly to the C–C, or C–H bond surfaces, and their sticking coefficient approach ~ 1. Mono-radicals react moderately and cannot insert directly to the surface, but if a dangling bond exists, then they can insert to yield a C–C bond. The dangling bond is created by removal of H from the C–H bond which can occur by an ion displacing H from the bond or by an H atom and or another radical like CH_3 abstracting H from the C–H bond:

$$\equiv C^- + H' \rightarrow \,\equiv C' + H_2$$
$$\equiv C^- + CH_3 \rightarrow \,\equiv C - CH_3.$$

Closed shell neutrals like CH_4 have a very low sticking coefficient of $<10^{-4}$ and a negligible effect. Atomic H' is the most efficient specie for abstraction, about 30 times that of CH_3 [32].

The dissociative attachment, dissociative ionization, and dissociative reactions are a major source for the production of atoms, free radicals and negative ions in the cold plasma [22]. The dissociation of molecules occurs only through their vibrational and electronic excitations. Most dissociation of molecules by slow electrons is induced by electronic excitation and the dissociation occurs only if the molecule is excited above a threshold value. Ionization in discharges taking place in a molecular gas occurs predominantly by electron impact that can produce positive or negative ions, atomic ions and or molecular ions. Positive ions are usually formed in most ionization processes prevalent in cold plasma, however when the atoms or molecules involved in the reaction possess electron affinity, then negative ions can also be formed.

Argon is believed to assist deposition by increasing ionization and ion energies in the plasma [33]. In a radio frequency (RF) discharge, hydrocarbons are partially ionized and the presence of argon in the plasma can be an important additional source of active species. The ionization of acetylene can be enhanced by impact with argon ions as follows:

$$Ar^+ + C_2H_2 \rightarrow Ar + C_2H_2^+.$$

The C_2H_2 ion recombines dissociatively with an electron, with the following possible reaction paths and their associated energy [33]:

$$
\begin{aligned}
C_2H_2^+ + e^- &\rightarrow C_2H + H &&(\Delta H = -10.03\,\text{eV}) \\
&\rightarrow CH + CH &&(\Delta H = -5.77\,\text{eV}) \\
&\rightarrow C_2 + H_2 &&(\Delta H = -9.58\,\text{eV}) \\
&\rightarrow CH_2 + C &&(\Delta H = -6.64\,\text{eV}).
\end{aligned}
$$

Acetylene is the choice gas for low pressure deposition, because its strong $C\equiv C$ bond give a simple dissociation pattern, $C_2H_n^+$ ions [34, 35]. There is a low plasma polymerization when acetylene is used as source gas and it is the preferred source gas for mechanical applications. Acetylene dissociates with less evolution of hydrogen compared to methane and has a moderate growth rate compared to other feed gases. Acetylene, however, is not available in high purity form, and so not good for electronic application due to contamination with substantial amount of nitrogen/air [36] which poses some doping problems especially when used in high density plasma. These dissociations of acetylene and its polymers can also be represented as shown below.

$$
\begin{aligned}
HC &\equiv CH + e^- \rightarrow C_2H^+ + H^+ + 2e^- \\
HC &\equiv CH + e^- \rightarrow C_2H_2^+ + e^- \\
2HC &\equiv CH + e^- \rightarrow (C_2H^+)_2 + H_2 + 2e^-.
\end{aligned}
$$

Methane is a popular choice gas for electronic application because it is available in high purity, though the growth rate is lower than that of acetylene and also it has a disadvantage of high hydrogen evolution, shown below, which is not good for mechanical application.

$$2CH_4 + e^- \leftrightarrow C_2H_2 + 3H_2 + e^-$$
$$2CH_4 + e^- \leftrightarrow C_2H_4 + 2H_2 + e^-$$
$$2CH_4 + e^- \leftrightarrow C_2H_6 + H_2 + e^-.$$

Tetramethysilane (TMS) vapor is one of the vapors used for silicon addition in producing Si-DLCNC films. The possible plasma decomposition paths for the tetramethylsilane (TMS) used in the PECVD deposition proposed by Catherine and Zamouche [37] is by electron impact dissociation/ionization, that is:

$$Si(CH_3)_4 + e \rightarrow Si(CH_3)_3 + CH_3 + e$$
$$Si(CH_3)_4 + e \rightarrow Si(CH_3)_3^+ + CH_3 + 2e$$
$$Si(CH_3)_4 + e \rightarrow Si(CH_3) + 3CH_3 + e$$
$$Si(CH_3)_4 + e \rightarrow Si(CH_3)^+ + 3CH_3 + 2e.$$

Most of the properties of a-C:H depend on the bias voltage and thus, the incident ion energy, E per carbon atom [38]. The depositing specie, usually a molecular ion, $C_mH_n^+$, on hitting the film surface breaks up and shares their energy between individual carbon atoms. Thus the effective energy per carbon atom is given by [35],

$$E = \frac{E_i}{m}.$$

The incident ion energy is usually about 40% of the bias voltage V_B for conventional PECVD systems [39–41]. In contrast to ta-C, the ion flux fraction in a-C:H is much less than 100% and may be typically 10% [24, 39].

Diamond and graphite are two crystalline allotropes of carbon. Both have a similar thermodynamic stability ($\Delta G = 0.04\,eV$ at $300\,K$), though separated by a large kinetic barrier [18]. The large varieties of non-crystalline carbons and hydrocarbons that exist can be classified in terms of their atom density, hydrogen content and their sp^3–sp^2 carbon contents relative to diamond, graphite and hydrocarbon polymers [42, 43].

Diamond has the highest atom density of any known solid at ambient pressure. DLC has atom density much greater than conventional hydrocarbon polymers, and between that of graphite and diamond. The strong directional sp^3-bonding of diamond gives it unique properties like highest elastic modulus, hardness and thermal conductivity at room temperature [43, 44], large optical bandgap and a very low thermal expansion coefficient.

Physical vapor deposition, especially sputtering processes are essential for formation of DLCNC with DLC matrix housing metals islands/clusters such as chromium, titanium, copper, iron, cobalt, nickel, and Mo-DLCNC [45]; Mo, Nb, Ti, W-DLCNC [46]; Cathodic arc evaporation of Ti-, Cr-, Zr-DLCNC [47].

Electrochemical techniques have equally been used for formation of copper nano-particle-DLCNC [48], Ag-DLCNC, Ag-Pt-DLCNC [49].

By fento-second pulsed laser ablation, plasma source ion implantation and others techniques it is possible to produce nanocomposite coatings, such as Ti-, Mo-, W-DLCNC.

2.3. *Characterization*

Atomic force microscopy (AFM) can be used for DLC and DLCNC characterization to investigate the surface morphology of the film which can also give information about the growth process of the film [50–54]. AFM can be used to measure the film thickness, surface force and to study adhesion [55, 56]. Some authors have used AFM for nano-indentation, elastoplastic studies [56], elastic modulus, hardness, microscratch, and microwear studies [57].

Simple eye observation of most synthesized DLC/DLCNC films are smooth, uniform films when the films adheres well to the substrates. However, bucking does tend to appear when the films are not well adherent on the substrates. These features are known to be film thickness dependent, and tend to initiate above several tens of nanometers, and at the edges of the substrates, or other defects, and propagates along the rest of the film surfaces. The generation and propagation of buckling are related to post-deposition environment, especially ambient environment, with buckling increasing mainly in humid environment. This is associated with the diffusion of gaseous species into the film–substrate interface when a high compressive stress exists in the film.

DLCNC do tend to be more adherent to most substrates due to less compressive stress resulting from the incorporation of the metals or dopant particles. As a result, these films are atomically smooth and hardly exhibit buckling. In some films obtained by pulsed laser technique, Narayan [58] observed that microparticulates are exhibited in Ti-DLCNC, and Si-DLCNC films due to "splashing" during pulsed laser deposition.

This splashing was attributed to subsurface boiling or shock wave ejection of the nanoparticulates [58].

In a TEM analysis of some DLCNC films, Narayan [58] reported that Cu-DLCNC and Ag-DLCNC formed Cu-rich inclusions/clusters and Ag-rich inclusions respectively, due to these metals segregating into a second phase within the DLCNC structures; whilst Ti-DLCNC and Si-DLCNC did not form clusters [58]. This is due to the fact that atoms like Ti and Si do form chemical bonds with carbon, and thus are uniformly dispersed within the DLC matrix. According to Narayan, the driving force for the clustering of metals in the DLCNC films is a reduction in surface energy [58]. It is necessary to investigate whether the dispersive or the polar component of the surface energy plays a major role. Investigations by Okpalugo *et al.* [59–65], Si-DLCNC tends to exhibit lower surface energy compared to the DLC counterpart, and in both DLC and Si-DLCNC, more than 90% of the surface energy is dispersive [59–65]. It is interesting to observe that copper in Cu-DLCNC forms Cu-inclusions/clusters and segregate into a second phase within the DLCNC structures, the Cu atoms in DLC-1.20 at.%Cu reported by Narayan have a minimal influence on the radial distribution function (RDF) and sp^3-bonding of DLC as compared to that of diamond [58]. This is expected because the d shell of Cu is fully occupied, thus Cu does not form chemical bonds with carbon and the short range carbon order in DLC matrix is not highly affected.

The surface roughness of DLC has been found to depend on the ion energy [66–70], substrate temperature [52–54, 66], substrate materials [71] and film compositions [72]. Peng *et al.* studied the surface roughness of as-deposited and thermally annealed DLC films deposited under various conditions using a nanoscale tapping mode AFM and identified basic phenomena affecting surface topography (depending on the deposition conditions) to be surface diffusion, ion implantation, sputter removal and etching removal from the deposit [73]. According to Peng *et al.*, surface diffusion, promoted by ion impact energy and by high substrate temperatures, causes surface roughening by promoting the local nucleation and growth of graphitic clusters. They also observed that high ion impingement energies (though causing local and global temperature increases) tend to lead to smoother surfaces. This is because the ions are implanted beneath the surface where the impact energy is less efficient in promoting surface diffusion. The estimated critical ion impingement energy above which smooth surfaces were produced was 50 eV for all the different processes they studied: RF glow discharge from methane, DC magnetron sputtering of graphite target, with substrate RF bias and an ion beam generated from a cathodic arc discharge [73]. Post deposition heat treatment also leads to increased surface roughness though much less pronounced than that resulting from the substrate temperature being high during film growth [73].

The SEM has the ability to image very rough samples, it can measure undercuts. This is due to its large depth of field and large lateral field of view. SEM measurements are destructive due to the energy of the electrons. There is an interaction between the electron beams of the SEM and the weakly bound electrons in the conduction band of the sample. Variation in the surface structure over a large area (e.g. several mms) can be acquired at once in SEM, whereas largest area obtainable in an AFM is about $100\,\mu$m by $100\,\mu$m. The SEM images can be used to infer the consistency of the quantitative AFM results over larger scan area of DLC thin films and both techniques can be complementary [74]. Glancing angle SEM can also be used to obtain topographical information.

The spectrum of visible light has been observed in nature in the form of the rainbow. Newton demonstrated in 1666 that radiation from the sun could be split into the component colors of the rainbow using a glass prism. This has formed the basis for further work and understanding of the electromagnetic spectrum and indeed the subject of spectroscopy.

When a material is exposed to electromagnetic radiation of sufficiently short wavelength (or high photon energy, hv) electronic emission occurs. If a monochromatic (fixed energy) photon of frequency v, is irradiated on the material the energy change in the process is shown by the Einstein relation [75],

$$hv = I_k + E_k,$$

where I_k is the ionization (binding) energy of the kth species of electron

E_k is the kinetic energy of the ejected electron

hv is the energy of the photon.

For carbon XPS, photoelectron excitation is commonly done using Al $K\alpha$ source with photon energy of 1486.6 eV or Mg $K\alpha$ source with photon energy of 1253.6 eV.

The XPS is used to find for example the formation and relative concentration of C–C, C–F, and C–O chemical states of carbon atoms in the near surface region of a sample [76]. The energy of the C_{1s} XPS peak is determined by the bound state of the carbon atom, either incorporated in an adsorbed hydrocarbon species, complexed as carbide or in its graphitic or diamond state [77, 78]. The C 1s peak position for diamond crystal is at

285.50 eV, which is about 1.35 eV higher than the C 1s for graphite (C=C) crystals (284.15 eV) [79]. The XPS peak locations or shifts reflect the chemical states influenced by the environment of the atoms. The chemical shift value E_B (binding energy) is mainly determined by the effective charge of a given atom as a result of the electron charge transfer [80]. In addition, the chemical shift is dependent on the materials lattice field. A higher fraction of sp^3 bonds indicates that the electrons are more tightly bounded to the carbon atom which results in a higher C 1s core binding energy [81]. The Auger line shapes for graphite and diamond have different characteristic features and thus are very sensitive to the concentration of sp^2 or sp^3 states of carbon atoms [76, 82]. Thus the ratio of sp^2 to sp^3 states can be determined by the X-ray excited Auger electron spectroscopy (XAES) experiment [76].

The binding energy for the C 1s core electrons was not changed for Fe-, Co-, and Ni-containing DLCNC films with increasing metal content, metal carbides being formed at higher metal contents of >60 at.% [45].

The C 1s features of a-C:H:Si films can be fit with four peaks representing four carbon bonding configurations (C–C, C=C, C=O, Si–C) [83–86]. In the a-C:H films with low silicon content, the slightly asymmetric C 1s peak is centered at a binding energy of 284.6±0.2 eV [87–89]. An additional peak which increases with increasing silicon content is present at 283.6 eV [87] which is close to the value 283.3 eV expected for Si–C [90, 91]. According to Hauert *et al.* [88] deconvoluting the C 1s signal of the a-C:H film containing 22.5 at.% of silicon into the Gaussian shaped peaks at 284.6 eV (a-C:H) and 283.6 eV (Si–C) splits the total of 77.5 at.% carbon atoms into 27% Si–C–bonds and 50.5% a-C:H bonding structures. The surface of the 22.5 at.% Si containing a-C:H films consists of 44–55 at.% SiC and about the same amount of a-C:H [88].

X-ray diffraction is a good technique for investigating the diffraction patterns of the metal nanocrystals in the amorphous carbon matrix. Under the surface sensitive glazing angle mode, information can be gained on the thin film coatings, from which the thin film density could equally be derived. Baba and Hatada [45] reported a very broad diffraction peak for Fe-, Co-, and Ni-DLCNC at ~ 2 theta value of 42° for 17–40 at.% of metal content in the DLCNC films.

Secondary ion mass spectroscopy (SIMS) is a complementary technique to the other characterization techniques especially XPS. It could be used for identifying the relative intensity of the present ionic species (though is not quantitative) and is good for identifying isotopes.

Raman and infrared (IR) spectra are obtained from transitions between vibrational levels at a range of 10^4–10^2 cm^{-1} of the electromagnetic spectrum. Raman spectroscopy is based on the frequency of the light scattered ($\sim$0.1%) by molecules as they undergo vibrations and rotations. The light scattering mechanisms in Raman and IR spectroscopy are different. In both Raman and IR spectroscopy the sample is irradiated by intense laser beam in the 10^4–10^2 cm^{-1} electromagnetic spectral region. For Raman the scattered ray is usually observed in the direction perpendicular to the incident beam. The vibrational frequency (v_m) is measured as a shift from the incident beam frequency (v_0). Although Raman spectra are observed normally for vibrational and rotational transitions, it is possible to observe Raman spectra of electronic transitions between ground states and low-energy excited states [89]. Rayleigh scattering is strong and has the same frequency as the incident beam, but Raman scattering is very weak $(\sim 10^{-5}\,v_0 \pm v_m)$.

In general, the vibration is Raman-active if the component(s) of the polarizability belongs to the same symmetry species as that of the vibration; the vibration is IR-active if the component(s) of the dipole moment belongs to the same symmetry species as that of the vibration. The symmetry species of normal vibrations can be found by using Herzberg's tables [90].

DLC or a-C:H shows a peak at 1510 ± 20 cm^{-1} [91]. The peak at 1500 cm^{-1} can be attributed to sp^3-bonded carbon plus possible contributions from the D-peak [91]. Nistor *et al.* [93] attributed the 1500 cm^{-1} peak to amorphous graphite. Phonon dispersion relations and the phonon density of state predictions [93–95] are essential in the interpretation of Raman spectrum of carbon because changes in Raman spectra are associated with phonon density of states. Only the phonons with E_{2g} symmetry are predicted to be Raman-active in an infinite graphite crystal. The entire population distribution of phonons as a function of frequency is available from the density of state calculations.

Wang *et al.* [96] investigated Ti-, Cr-, Zr-DLCNC films on biomedical titanium alloy (Ti–6Al–4V) and reported that Cr-DLCNC posses lower Raman I_D/I_G ratio, lower G-peak positions; higher coating hardness, lowest friction coefficient (compared to other films), 0.06 in simulated body fluid (compared to air), and a release of 2.22–10.03 ppm ($<$106.17 ppm from Co–Cr–Mo–alloy) of Cr ions from Cr-DLCNC, indicating that Cr-DLCNC meets the requirement for biomedical application.

Table 2.1. Raman data obtained for the three nanocomposite Cr-DLC samples. The data includes I_D and I_G peak intensities, full-width at half maximum (FWHM) values for D and G bands, $I_\mathrm{D}/I_\mathrm{G}$ ratios and D and G band intensities [97].

Sample	% Cr	I_D (a.u.)	I_G (a.u.)	FWHM (D)	FWHM (G)	$I_\mathrm{D}/I_\mathrm{G}$	D-Peak (cm^{-1})	G-Peak (cm^{-1})
CrDLC1	1	21.833	21.043	363.68	139.37	1.04	1401.6	1538.8
CrDLC2	5	21.628	20.415	361.08	128.51	1.06	1408.4	1540.6
CrDLC3	10	45.588	49.748	389.90	167.24	0.92	1306.8	1511.8

Raman investigation of Cr-DLCNC films are shown in the Table 2.1 [97]. The Cr-DLCNC consisted of nanoscale chromium carbide particles embedded in an amorphous DLC matrix. The G, and D-bands are usually assigned to optically allowed zone center of phonons of E_{2g} symmetry and K-point phonons of A_{1g} symmetry (disorder-allowed zone edge mode of graphite), respectively. The D-band peaks for the three Cr-DLCNC films were positioned at 1401.6, 1408.4 and 1306.8 cm^{-1}, whereas, the G-band peaks were centered at 1538.8, 1540.6 and 1511.8 cm^{-1}. For both D- and G-bands, 5% Cr-DLC sample displayed the smallest values for full width at half maximum (FWHM). The $I_\mathrm{D}/I_\mathrm{G}$ ratio (ratio of D- and G-band intensities) was the least for 10% Cr-DLC and the highest out of the three samples for the 5% Cr-DLC film. This suggests that there are more disordered graphitic phases in sample 5% Cr-DLC, and equally a less compressive stress in this sample (least value for FWHM(G) compared to the other samples).

In Ag-DLCNC and Ag-Pt-DLCNC films [49], the authors report that the D-band increased, whilst the G-band decreased in width (corresponding to a decreased compressive stress within the film) and shifted to a lower wavenumber indicative of a decrease in sp^3 content (increased graphitic cluster size).

Each individual $\mathrm{sp}^m\mathrm{CH}_n$ configuration is characterized by a specific IR absorption peak. It is assumed that one can use these spectral peaks to analyze the relative hybridization $\mathrm{sp}^3/\mathrm{sp}^2$ of the carbon atoms and use the total intensity of the broad peaks centered at about $2900\,\mathrm{cm}^{-1}$ to quantify the hydrogen content of the DLC films [101–106]. The deconvolution of the $2900\,\mathrm{cm}^{-1}$ stretching peak shows that the peak deconvolutes into the individual absorption peaks centered at 2850, 2920, 2970, and $3000\,\mathrm{cm}^{-1}$, with corresponding intensity ratios of 0.6:0.8:0.4:0.1. It appears that the corresponding DLC film contains mostly sp^3 carbon, with only a small fraction of $\sim$5% sp^2 carbon bonds ($\mathrm{sp}^2\mathrm{CH}$ at $3000\,\mathrm{cm}^{-1}$) [104].

2.4. *Doping DLC*

In both crystalline and amorphous materials doping is achieved by introducing atoms which can change the local structure of the bonding of the material interstitially or substitutionally. In amorphous materials large concentration of the dopant is required to compensate the existing traps and dangling bonds. In wide bandgap semiconductors, it is a problem to dope the material to both n- and p-type. This is because the dopant levels are either too deep, or due to low dopant solubility, and/or because of auto-compensation [105]. If the dopant atom is similar in size to the atom of the recipient, then it is likely to be soluble. It is essential for the dopant to have a moderate solubility and a shallow level below the conduction band edge of the recipient for there to be an effective doping and/or for atomic substitution with the dopant to be easily successful. The flexibility of random networks in amorphous semiconductors allows atoms of various sizes to be soluble and incorporated. On the other hand an auto-compensation occurs because network flexibility allows atoms to also exert their chemical preferred valency [105] and to form a non-doping trivalent site with lower energy [106], although, some fractions of the dopant atoms are allowed to exist in the doping sites at equilibrium [106].

DLC has been doped with various elements [107–110]. Silicon doping of DLC has been reported [111–115]. Silicon content was reported to decrease with increasing radio frequency power and to increase with increasing substrate temperature [116]. The surface energy of DLC has been modified by doping with silicon [118, 119]. Silicon incorporation has led to reduction of DLC's typically high internal compressive stress. Silicon incorporation has been used to reduce the etching rate of DLC in oxygen plasma [123]. The valuable effects of silicon doping according to other reports from the literature are the lowering of the friction coefficient to $\sim$0.01 (for Si-DLC) compared to 0.1 of DLC [112, 116, 120–122], improved adhesion and mechanical properties [123–125] with the use of silicon; increased sp^3 to sp^2 ratio and reduction of the size of graphite-like islands [114], because silicon does not form π-bonds. According to Demichelis *et al.*, the effect of a low amount of silicon is to increase the sp^3 to sp^2 ratio, reducing the size of graphitic-like islands. When the amount of silicon is increased over a critical value, the network assumes the characteristics of the semiconductor-type amorphous hydrogenated-SiC, where the properties of silicon are predominant [114]. Speranza *et al.* [126] reported that silicon addition to DLC preserves the diamond-like character of DLC that are lost at higher temperatures (thermal annealing).

Narayan [58] reported an innovative target configuration for incorporating metal atoms into DLC during pulsed laser deposition process in which the composition of these nano-composite films depends on: (a) the scanning radius of the laser beam on the target surface, (b) the laser beam position, (c) the position of the circular target, (d) the size of the metal piece on the target, (e) the laser energy density. The fraction of metal atoms within the film can be estimated by the following relationship [58]:

$$\text{Carbon–metal composite} \ = \frac{\alpha\delta(1 - R_d)}{2\pi\gamma(1 - R_c)},$$

where α = the laser ablation ratio, δ = width of the metal piece on the target,

R_d = the reflectivity of the metal strip, γ = laser beam scanning radius.

Metal nanocomposite integration into carbon matrix is achieved by various techniques as mentioned above (e.g. sputtering, ion-implantation, etc.). Structural information on the architecture formed can be gained for example by Z-contrast or high resolution scanning transmission electron microscopy (TEM), based on observing contrast fields proportional to the squared atomic number of the elements: the darker field corresponding to the higher atomic number region (metal), and the brighter region the lower atomic number (carbon matrix). The metal components usually appear as nanoparticulate nanoclusters or nanoarrays in a carbon matrix. The size and shape of the metal nanoparticles or nanoclusters depends on the metal and the amount incorporated, but usually are few nanometer range clusters/particles (hence nanocluster, and /or nanoparticles). Baba and Hatada [83] reported that the surface features of their magnetron plasma source ion implantation synthesized Fe-, Co-, Ni-, and Mo-DLCNC was featureless and very smooth with the surface roughness of about 0.2–0.3 nm. It is thought that these metal nanoclusters (nanoarrays) may serve as metal reservoirs and or trap for neutral/noble metal and metal-ions in a biologic environment, which could equally prevent risk of metal ion release from the nano-composite films.

The electronic local structures of DLCNC can be studied by the X-ray absorption fine structure techniques (e.g. XANES). Singh *et al.* [127] reported that Cr K-edge XANES spectra of Cr-DLCNC shows that Cr in DLC has a chemical state similar to that of Chromium carbide, at low Cr content (<0.4 at.% Cr), Cr is dissolved in the amorphous DLC matrix forming an atomic scale composite, their simulation studies showing that at this Cr concentration, Cr tends to be present as a very small atomic clusters of 2–3 Cr atoms (dissolved in DLC matrix), implying that the solubility limit of Cr in Cr-DLC films is only between 0.4–1.5%. However, at higher Cr concentration (>1.5 at.% Cr), Cr is present as nanoparticles (<10 nm) of a defect carbide structure forming a nanocomposite [126].

2.5. *Thermal Annealing*

In most crystalline, semi-crystalline materials and/or nanocomposite materials, atomic structural rearrangement could be achieved by thermal activation/annealing. DLC is a metastable material and its structure will change towards graphite-like carbon by either thermal activation or irradiation with energetic photons or particles. DLC films have been altered in UV irradiation [127], in ion beam irradiation [128], and in laser beams [129]. In a-C:H there is evolution of hydrogen as H_2 as well as hydrocarbons unlike the a-Si:H where there is only H_2 evolution. There is loss of hydrogen and CH_x species starting at about 400°C or lower depending on the deposition conditions like the radio frequency power of deposition and the dopant(s) in the film [103, 130]. Nadler *et al.* reported loss of hydrogen starting from 500°C and continued to temperatures over 700°C [132]. Wu *et al.* however, found no detectable loss of hydrogen when DLC films were rapidly annealed until 500°C [133, 134]. As the evolution temperature increases, the fraction of hydrogen evolved as hydrocarbons goes down [134]. This means that the network becomes denser. The loss of hydrogen results in the collapse of the structure into mostly sp^2-bonded network. UV irradiation from a high pressure mercury lamp was found to break the C–C and C–H bonds and cause oxidation of the film with formation of C–O bonds [135]. There was a reduction in the film thickness due to CO_2, CH_4, and H_2 evolving from the film. Hydrogen evolution may vary depending on the type of film (a-C:H or ta-C:H) and the deposition parameters including the source gas(es) used. The ta-C:H deposited from methane is not so different from a-C:H, there is some evolution of hydrocarbon molecule with the main evolution of hydrogen centered on 550°C. The ta-C:H from acetylene is more dense and there is little evolution as hydrocarbon with the main evolution of hydrogen centred on 700°C [136]. Silicon addition to a-C:H to produce Si-DLCNC improves the thermal stability and raises the hydrogen evolution to temperature $\sim$700°C [137].

The development of carbon networks and bonding during thermal annealing has been studied with Raman [136, 138], IR (or FTIR) [135] and optical spectra. With increasing annealing temperature I_D/I_G ratio rises. This starts to rise even before the main H_2 evolution, meaning that once the hydrogen is mobile, though still within the film, the C–C network starts to rearrange. As hydrogen is evolved the bond transforms from sp^3 to more sp^2. The hydrogen motion also allows the density of defects to reduce [139]. There is increase in the number of π-states and a gradual fall of the optical gap [97, 140]. UV irradiation of nitrogen doped DLC films containing about 11 at.% of nitrogen resulted in a decrease in the C–H bonds and an increase in C–C, C=N and C≡N bonds. The sp^2 cluster size in the films became smaller, thus the optical gap increased. Nitrogen-doped DLC (N-DLCNC) with high nitrogen content ($\sim$37 atm.%) did not show any FTIR changes after UV irradiation [135], though their optical gap increased, signifying a reduction of the sp^2 cluster size in the absence of material loss. According to Reyes-Mena *et al.* no Raman signal was observed for annealing carbon films up to 400°C except for the intensity photoluminescence [139].

The onset of thermal relaxation could be as low as 100°C and near full relaxation has been observed at 600°C. About 1–3 eV activation energy has been reported for this thermal relaxation [141]. Residual compressive stress leading to buckling and wrinkles is usually observed in ion bombarded films [142]. The residual stresses usually can be removed by annealing at a suitable temperature for a sufficiently long time. For tetrahedral amorphous carbons vacuum annealing of the films up to 600°C can almost lead to complete elimination of the residual stresses [142, 143]. In hydrogenated amorphous carbon or DLC, attempts to anneal out the residual stress at temperatures above 300°C usually degrade desired properties of DLC [144].

With annealing, the film structure changes from polymer-like to graphite-like and the primary bond changes gradually from sp^3 to sp^2. Graphitization occurs at temperatures greater than 400°C, when the DLC films convert to nanocrystalline graphite with high conductive ability [138, 145]. However, according to Jiu *et al.* the DLC films (in which electrical properties were measured with four point probe) kept the same high resistivity ($>10^7$ Ωcm) at 500°C, the resistivity slightly decreasing to 10^6 Ωcm at 600°C. Jiu *et al.* inferred that DLC film is a mixture of amorphous carbon and very thin graphite particles. The crystallization degree increased at 950°C with clear and intensive diffraction rings seen by use of transmission electron diffraction (TED) technique, which corresponds to those of graphite-like particles [146].

Thermal annealing of the films alters the microstructure of the films and leads to shift in the peak position with annealing temperature.

The formation of DLCNC with metal (or metal alloying to DLC) is a good technique for creating adherent and wear resistant films on various substrates. This is an alternative to thermal annealing, with added advantage that it does not lead to degraded DLC quality that may result from annealing at high temperatures (it is therefore not destructive). The reduction of internal compressive stress by these elements is understood by the effects of metals on the continuous rigid random network (CRN) model of DLC, since these elements/metals are more compliant compared to covalently bonded carbon in DLC, and the metals posses loosely bound electrons that could distort the electron density distribution (short range electronic structure) in the carbon matrix of DLC, thus enabling a relaxation at the film substrate interface. Silicon in Si-DLCNC seems to reduce compressive stress by the vast introduction of disorders/defects at the film–substrate interface since silicon atoms do not form sp^2 bonds (have no pi electrons), but rather form sp^3 bonds with carbon and/or other silicon atoms.

2.6. *Biological Properties and Biocompatibility*

The passive nature of carbon in tissues has been known since ancient times. Charcoal and lampblack were used for ornamental and official tattoos by many. Other forms of carbon have been studied for possible use for biomedical applications stimulated primarily by Gott's original studies [148]: artificial graphite, vitreous or glassy carbons, carbon fibres, pyrolytic carbons, composites, and vacuum vapor deposited carbon coatings [148]. The fundamental nature of these carbon materials and their interactions with the living tissues needs to be explored. Likewise pyrolytic carbon-coated heart valve leaflets have been successfully applied as artificial clinical heart valves [149]. Pyrolytic carbon has the major advantage of being more resistant to thrombus formation, which was the biggest limitation to the earlier generation of stainless steel artificial heart valves. As mentioned earlier, there was always a need to use anticoagulant drugs by patients who had the earlier stainless steel heart valves to prevent clot formation on the stainless steel, but this had the potential to suppress the beneficial effect of the natural blood clotting mechanism in patients. Pyrolytic carbon is an artificial material made of carbon microcrystals with a high density turbostatic structure, originally engineered for use in nuclear reactors and may not be readily available for large scale use. When pyrolytic carbon is alloyed with silicon, it shows excellent thromboresistance [150]. Pyrolytic carbon has also been used as a coating on different types of implant prostheses, such as dental implants, percutaneous devices, and tendon tracheal replacements.

Surface coating modification is essential because it is now known that the outermost layer of a biomaterial (few nanometre scale range) is most crucial in its interfacial interaction *in vivo*. Baier *et al.* have shown that exposure of an organic-free surface to fresh flowing blood for as little as five seconds leads to its complete coating by a very uniform, tenaciously adherent proteinacious thin film [151, 152]. The biomaterials outer surface dictates the configuration of this attached protein film that in turn plays an important role in determining the fate of the biomaterial in the host via a series of cascade interactions. The issue of biocompatibility and haemocompatibility of DLC when used as implants and medical devices will no doubt be expected to stem from possibly favorable tissue–biomaterial surface and interactions. Generally, two main pathways could be feasible, in an effort to create a biocompatible material: creating a material with surfaces that are bioactive (like the host tissue) that can actively support the body's control mechanisms; and/or creating materials that are "inert or passive" to the body's control mechanism in order to avoid triggering an adverse reaction (though there is nothing exactly like absolute inertness in a hostile physiological environment [153]). Moreover, it is impossible to exaggerate though, that among a large number of events occurring during the process of blood coagulation (including, perhaps, thrombus formation and possible thromboembolism) and de-coagulation, the physicochemical adsorption is virtually the main reaction that can be readily regulated unless a bioactive material is used [154]. Thus the present research in the use of DLC is aimed at an approach to modify the physico-chemical properties of DLC and to create a material that may be "passive" and or "bioactive" in the tissue.

The physicochemical surface properties of the outermost interface of bacteria for example and other particles as well as phagocytic cells, can essentially be of only two kinds: (a) interfacial tension and (b) electrical surface potential [155]. When a foreign surface, solid, liquid, or gas, is brought into contact with the body tissue fluid/protein solution, a certain amount of the dissolved protein will be adsorbed to the surface. This process is consistent with Gibbs theory of surface energy and may be described by the adsorption isotherm of Freudlich or Langmuir. The amount of protein adsorbed and the characteristics of the protein monolayer depend mainly on the nature of the foreign surface and the structure and the concentration of the proteins in solution [156]. The protein needs to first approach a distance to the foreign surface that will allow interaction between the molecular forces associated with the foreign surface and the protein — this is governed by diffusion. Then the characteristics of the foreign surface determines the nature of bonds or the type of changes that takes place in the configuration of the protein and the biologic molecule present.

Szent-Gyorgyi [157, 158] suggested that proteins may have electronic structures similar to that of semiconductors. Eley *et al.* [159] reported semiconductivity in certain proteins. This was later corroborated by works from others like that of Postow and Rosenberg [160]. Bruck suggested that intrinsic semiconduction and electronic conduction may be involved in blood compatibility of polymeric systems [161, 162], instead of mere ionic interaction, after the compatibility of the blood with surfaces has been associated chiefly with ionic charges, based on the observation that endothelial wall, platelets, and plasma proteins carry net ionic charges in normal physiologic conditions. Bruck [163–165] observed clotting times six to nine times longer than those observed with non-conducting polymers and also observed little or no platelet aggregation in electroconducting polymers, when compared to non-conducting control samples based on his study with pyrolytic polymers. Bruck concluded that "it is possible that electroconduction and semiconduction is involved in the interaction of surfaces with plasma proteins in the activation of the Hageman factor and the platelets by an unknown mechanism".

The endothelial lining has been reported to be the best non-thrombogenic surface [166]. According to some workers [167–169], various methods including improvement of physico-chemical properties, pretreatment with proteins and incorporation of negative charges have been proposed in order to reduce the surface thrombogenicity of vascular prostheses. Pesakova *et al.* [170] and Hallab *et al.* [171] have also stated that the biocompatibility of materials can be influenced by factors such as surface charge, hydrophobicity, and topography. It has been reported [167, 173], based on surface potential measurements using the vibrating Kelvin probe method, that positively charged surfaces enhance cell adhesion in comparison to neutral or negatively charged surfaces. The hydrophyllic or hydrophobic nature of a surface has also been associated with extent of cell interactions with the surface [167, 173]. Altankov and Grott [174] have reported that wettable (hydrophilic) surfaces tend to be more conducive for cell adhesion. Grinnell [175] also reported observing that cell adhesion seemed to occur preferentially to water wettable surfaces. Van Wachem *et al.* [176, 177] carried out investigation of *in vitro* interaction of human endothelial cells (HEC) with polymers of different wettabilities in culture, and reported observing optimal adhesion of cells with moderately wettable polymers.

The biocompatibility and haemocompatibility of DLC films has been investigated in the literature. Jones *et al.* [178] deposited DLC coatings prepared by PECVD on titanium substrates and tested them for haemocompatibility, thrombogenicity and interactions with rabbit blood platelets. They reported that the DLC coatings produced no haemolytic effect, platelet activation or tendency towards thrombus formation and that platelet spreading correlated with the surface energy of the coatings. Cytotoxicity tests have also been carried out on DLC coatings by several workers [178–180], and they all reported observing no negative effects by DLC coatings on the viability of cells which showed normal metabolic activities like cell adhesion and spreading. Mouse fibroblasts grown on DLC coatings for seven days showed no significant release of lactate dehydrogenase, an enzyme that catalyzes lactate oxidation, often released into the blood when tissue is damaged, compared to control cells, indicating no loss of cell integrity. It has also been reported by Allen *et al.* [184] that mouse macrophages, human fibroblasts and human osteoblast-like cells grown on DLC coatings on various substrates exhibited normal cellular growth and morphology, with no *in vitro* cytotoxicity.

Changes in surface energy, surface charge conditions and electronic conduction have all been suggested to have an effect on the biocompatibility and haemocompatibility of biomaterials. Allen *et al.* [184] implanted DLC-coated cobalt–chromium cylinders in the intramuscular locations in rats and in transcortical sites in sheep and their histological analysis of specimens retrieved 90 days after surgery showed that the DLC-coated specimens were well tolerated in both sites [184]. De Scheerder *et al.* [186] investigated the *in vivo* biointeraction with one particular class of modified DLC coatings: diamond-like nanocomposite coatings (DLN or Dylyn, Bekaert, Kortrijk, Belgium). Either coated or non-coated stents were randomly implanted in two coronary arteries of 20 pigs so that each group contained 13 stented arteries. The pigs underwent a control angiogram at six weeks and were then sacrificed. They performed a quantitative coronary analysis before, immediately after stent implantation, and at six weeks using the semi-automated Polytron 1000 system (Siemens, Erlangen, Germany). They also performed a morphometry using a computerized morphometric program and their angiographic analysis showed similar baseline selected arteries and post-stenting diameters. At six-week followup, they discovered no significant difference in minimal stent diameter and their histo-pathological investigation revealed a similar injury score in the three groups. According to De Scheerder *et al.* [186] inflammatory reactions were significantly increased in the DLN-DLC coating group, thrombus formation was significantly decreased in both coated stent groups and neointimal hyperplasia was decreased in both coated stent groups. However, the difference with the non-coated stents was not statistically significant, and area stenosis was also lower in the DLN-coated stent group than in the control group ($41 \pm 17\%$ versus $54 \pm 15\%$; $p = 0.06$). In their conclusion they indicated that the diamond-like nanocomposite stent coatings are compatible, resulting in decreased thrombogenicity and decreased neointimal hyperplasia; covering this coating with another diamond-like carbon film resulted in an increased inflammatory reaction and no additional advantage compared to the single-layer diamond-like nanocomposite coating [185].

Dowling *et al.* [187] implanted two DLC-coated and uncoated stainless steel cylinders into both cortical bone (femur) and muscle (femoral quadriceps) sites of six adult sheep ($>40\,\text{kg}$), for a period of four weeks (three sheep) and the rest for twelve weeks. According to Dowling *et al.* [187], after explantation of the implants and pathological/histological examination of the implanted cylinders, no macroscopic adverse effect was observed on both the bone and the muscles of the used sheep [186].

Yang *et al.* [188] examined *in vivo* interactions of discs coated with TiN, DLC (deposited on SS316L disc using PVD) and/or Pyrolytic carbon films, implanted into the descending aorta of anaesthesized sheep (six animals) for two hours. They evaluated the three different samples simultaneously in each animal. After explantation they examined the thombus free area on the disc with close-up photography and planimetry, and the test surfaces with SEM. Yang *et al.* found that there were many leucocytes adherent, activated and spread onto PyC and DLC, but on TiN it was the erythrocytes that were mainly adherent [188].

2.7. *Biomedical Applications*

DLC is said to be chemically inert and impermeable to liquids. It could therefore protect biological implants against corrosion as well as serve as a diffusion barriers. DLC films are considered for use as coatings of metallic as well as polymeric biocomponenets to improve their compatibility with body fluids. The potential biomedical applications of DLC and modified DLC (eg. DLC-NC) include surgical prostheses of various kinds: intracoronary stents and prosthetic heart valves. The new prosthetic heart valve designed by FII Company and Pr. Baudet is composed of a Ti6Al4V titanium alloy coated with DLC [179]. When artificial heart organ polymers used for making heart organs are compared to DLC-coated polymers, these polymers seem to show higher complement activation compared to DLC counterpart (polycarbonate substrates coated with DLC, PC-DLC compared with Tecoflex, polyurethane) [189, 190]. DLC and modified DLC can be used in blood contacting devices, e.g. rotary blood pump [189, 190].

DLC-Ag-Pt nanocomposites were reported to exhibit significant antimicrobial efficacy against staphylococcus bacteria, and to exhibit low corrosion rates at the open-circuit potentials in a PBS electrolyte [191].

In orthopedics, DLC can be used as orthopedic pins and coatings in hip implants (e.g. femoral heads). DLC can reduce the wear of the polyethylene cup by a factor of 10–600 when used on metal implants to form a DLC-on-DLC sliding surface. The wear (and the amount of particles causing a foreign body reaction) is 10^5–10^6 lower compared to metal-on-metal pairs. The corrosion of a DLC-coated metal implant can be 100,000 times lower than in an uncoated one. DLC can diminish the bone cement wear by a factor of 500, which can improve the bone cement to implant bonding [187, 192, 193].

In urological dialysis (haemodialysis), DLC-coated microporous polycarbonate and DLC coated dialysis membranes show that DLC imparted an enhanced enzyme electrode performance [194, 195].

DLC has also been reported to do well in both organ [196] and cell culture [197] when compared to the materials conventionally used for this purpose.

DLC can be used as active barrier against attack by micro-organisms and against bio-deterioration of advanced technological devices operating in closed spaces of satellites, aircrafts, and submarines for example [198], and as good protectors against environmental pollutants and atmospheric wastes [199]. In addition nanocrystallite copper modified DLC nanocomposite has been reported to have a fungicidal effect [199].

DLCNC is now being investigated for anti-Prion protective coatings on surgical instruments and as well as anti-MRSA bug in hospital utensils where appropriate due to its known hydrophobic surface properties. DLCNC is equally being investigated as a template/model surface for DNA writing and immobilization for biochemical sensor applications. Specific biomolecules could be easily immobilized on the DLCNC. In Designer

Materials for Nucleic Acid Delivery, titled "Gene delivery by immobilization to Cell-Adhesive Substrates"; surface immobilization of DNA was reported to have several advantages: (1) minimizing the amount of DNA needed to achieve a desired effect and enhancing effective concentration vector; (2) preventing DNA/vector aggregation; (3) reducing toxicity and degradation of delivered particles; and (4) delivering DNA to specific cellular sites [200].

3. Surface Energy of Diamond-Like Carbons

The surface energy can be defined as the algebraic sum of the energy used to break the bonds in a solid or a liquid (for example), to form its surface *in vacuo*, and the energy released, after new bonds are formed on the surface, when the solid or liquid is brought in contact with the second phase . The surface tension of a liquid (or the interfacial tension between a liquid and a fluid) may be measured directly by means of techniques like Wilhelmy plate or capillary rise, but the lack of molecular mobility within a solid prevents the deformation necessary for detecting surface forces in a solid. Thus indirect methods have to be employed while measuring the surface energy of a solid. An example is the "wetting" of a solid by a known reference liquid drop. It is assumed that there will be an interfacial interaction between the solid and the liquid (plus vapor in some cases or vacuum), and that thermodynamic equilibrium is attained. A surface cannot exist on its own: it must be part of an interface between two phases, even if one is a vacuum. The term "surface energy" without qualification strictly applies to the substance concerned in contact with a vacuum [201]. "Surface tension" is a concept closely allied with surface energy. For example the meniscus of mercury can be explained as either due to reduction of the surface area and therefore of surface energy, or of there being a surface tension i.e. a force acting in the surface trying to contact it. The magnitude of the surface energy of a material depends upon the strength of the bonds that have to be broken to form the surface concerned. Thus depending on the types of bonds involved, the surface energy can be given by:

$$\gamma = \gamma^d + \gamma^p + \gamma^h + \cdots,$$

where γ = surface energy, d = dispersion forces, h = hydrogen bonding, γ = liquid surface tension, p = polar forces.

The dispersion forces component is universal and so it is always one of the contributing forces. For non-polar liquids (e.g. alkanes) only dispersion

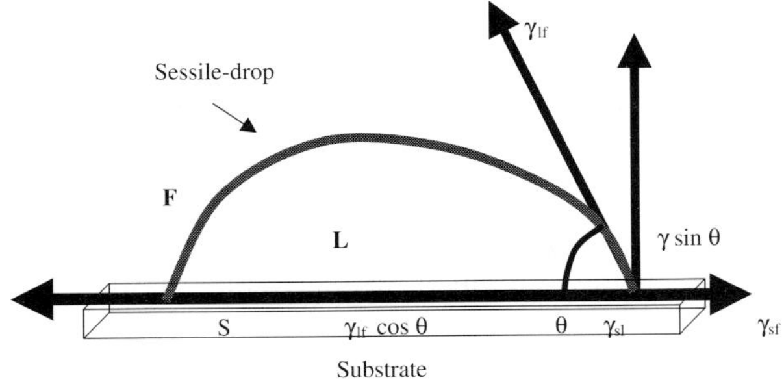

Fig. 3.1.   Liquid (L) exhibiting a contact angle θ on solid (S) with surrounding fluid (F).

forces act between molecules such that, $\gamma = \gamma^d$. A liquid drop on a solid sample creates a contact angle as illustrated below:

Figure 3.1 shows a sessile drop of liquid (L) exhibiting a contact angle θ on solid (S) with surrounding fluid (F) which may be vapor of the liquid (L) or a second liquid. The figure shows a liquid (L) at rest on an ideal solid (S) surface (the triple line of contact where S, L, and fluid (F) meet); to each interface could be attributed a free interfacial energy or interfacial tension γ_{sl}, γ_{sf}, γ_{lf}.

The surface energy (from Young's equation) could be given by:

$$\gamma_{sf} = \gamma_{sl} + \gamma_{lf} \cos \theta.$$

A horizontal force balance is assumed to be achieved. The vertical force component is given by $\gamma \sin \theta$. By thermodynamic work of adhesion (W_a), from equation above:

$$W_a = \gamma_{sf} + \gamma_{lf} - \gamma_{sl}.$$

Eliminating γ_{sl} (the interfacial tension at the solid/liquid phase),

$$W_a = \gamma_{lf}(1 + \cos \theta).$$

The surface energy can be expressed by the addition of the polar (γ_{lf}^p) and the dispersive (γ_{lf}^d) components such that:

$$\gamma_{lf} = \gamma_{lf}^p + \gamma_{lf}^d.$$

The adhesion energy can therefore be taken as the interaction between polar–polar and dispersive–dispersive forces, and using extended Fowkes equation:

$$W_a = \sqrt{\gamma_{lf}^p \gamma_{sf}^p} + \sqrt{\gamma_{lf}^d \gamma_{sf}^d} = \gamma_{lf}(1 + \cos \theta).$$

From the above equation therefore only two measurements with solvents with known polar and dispersive components are essential to solve the equation and to obtain the surface energy for the sample under investigation. Practically surface energy could be determined by a common method known as contact-angle goniometry for both advancing and receding contact angle experiments. The solid–liquid interfacial tension of solids can best be studied by contact angle measurements at gas–liquid–solid interface, using a technique for the investigation of the character of solid surfaces that was first proposed by Thomas Young in 1805 [202]. Today Young's technique has been developed to some degree of sophistication like Zisman's method — the multi-liquid contact angle method for critical surface tension (though the Zisman method is known to have a considerable drawback of exposing cell surfaces to a variety of non-physiological liquids that can interact chemically and physically with them in numerous ways).

The work by van Oss [203], showing the important role played by free energy changes in certain cellular interactions and Tsibouklis *et al.* [204], report of a low surface energy effect in preventing bacterial adhesion onto surfaces point to surface energy consideration in any cellular–biomaterial interactions. Theoretically, for a DLC-coated solid material put in the body fluid, various interfacial interactions are expected to exist depending on the particular application. For fully implanted DLC-coated material, interfacial interactions expected could include solids (DLC, erythrocytes or red-blood cells, platelets, white-blood cells, etc.) interacting with liquid(s) which could be either plasma, interstitial fluid, or even any other "special" body fluid at the same time. When DLC-coated material is partially implanted (for instance when used in a disposable coated syringe) then the third component: "vapor phase" should be considered. The question is, should the cells in the body fluid be considered as "fluid" or should they be considered as "solids" and how is the total number of the cells integrated in the surface energy equations? Moreover, whereas the vertical force component of the assumed equilibrium of forces are infinitesimally small for a "hard solid" like DLC for instance, should it not be considered important for "soft solids" like erythrocytes (red blood cells), platelets, etc.?

McLaughlin *et al.* [205] have indicated that surface energy with surface roughness is crucial in determining haemocompatibility of DLC used for medical guide wires (essentially the hydrophobic nature of DLC indicating a low surface energy). High surface energy has been associated with higher haemolysis, however, titanium substrate used in the study by Jones *et al.* [178] shows that highest surface energy (lowest contact angle) did not produce any haemolytic effect. In their study they concluded that spreading of platelets did not occur on their DLC coating, indicating low surface activation preventing thrombus formation, and that the number of platelets attached to the DLC was similar to those on the other materials that they used. Yu *et al.* [206] examined the surface energy of ta-C prepared by filtered cathodic arc in order to analyze the surface adsorption of human serum albumin (HSA) and human fibrinogen (HFG). The same authors reported that the surface energy and the protein absorption of both ta-C and clinically used LTI-carbon (isotropic pyrolytic carbon) are identical, but differ in haemocompatibility behavior, with the ta-C exhibiting a better anticoagulant property than LTI [206]. They therefore attributed the haemocompatibility behavior to the effective work function instead of the surface energy [206].

The reported surface energy of DLC is 40–44 mNm^{-1} [118, 206], and the reported water contact angle is 55–70° [118, 206–208]. Silicon addition to a-C:H decreases the surface energy from $\sim$42–31 mNm^{-1} [118] and increases the water contact angle. Silicon in a-C:H film reduces mainly the polar component of the surface energy which is due to dipoles at the surface. Silicon does not form π-bonds and therefore increases sp^3 bonding, reducing sp^2 bonding, their polarization potential and the dangling bonds [209–211]. The dispersive component of the surface energy is due to electronic interactions for example "van der Waals" forces. According to Grischke *et al.* [118] an addition of oxygen (critical concentration <35 at.%) to the plasma decomposition of TMS modifying monomer during synthesis of Si-DLCNC do further reduce the dispersive component of the surface energy from 31–23.5 mNm^{-1}, also correlated with a slight increase in the polar component from 2.0–3.7 mNm^{-1}. Decreased density of the network structure has a strong influence on the dispersive component of the surface energy, leading to the lowering of the dispersive component of the surface energy with oxygen addition [118], though a polar binding type like Si–OH is also introduced which results in the slight increase in the polar component of the surface energy. Grischke *et al.* also examined the stability of the surface energy of the Si-DLCNC films to "normal" (the influence of atmospheric pressure, constant $T = 20°C$) and "artificial" (high temperature 80–460°C activation) ageing. Their observations showed an increase in the polar component of the surface energy from 2 mNm^{-1} to a constant value of 3.7 mNm^{-1} during the first 10 days and a slight decrease in the dispersive component of the surface energy from 31–29 mNm^{-1} for the normal ageing process. For the artificial ageing process they observed a significant increase in the polar component of the surface energy for temperatures greater than 260°C and a significant decrease of the dispersive component of the surface energy from 31–23.5 mNm^{-1} [118]. In both types of ageing processes however, Grischke *et al.* detected no incorporation of oxygen to the Si-DLC network.

4. Electrical Conductivity and Conduction Mechanisms

DLC falls into a group of materials called the non-crystalline materials that include liquid metals and semiconductors, glasses, and amorphous films evaporated or deposited by other means. In describing the electrical conductivity of these materials, a theory of non-crystalline array of atoms is of note [212]. DLC conduction is known to be electronic instead of ionic. DLC is a low mobility semiconductor with low electron affinity, a bandgap of 1–4 eV, and room temperature photoconductivity [213]. The sp^2 sites possess π states, and these states form the valence and conduction band edges [213]. DLC has been reported as having high electrical resistivity over a range of values depending on the deposition parameters [213, 214]. The electrical resistivity of DLC can be reduced by doping it with various elements [215–218]. Vogel *et al.* [220] reported ohmic behavior of DLC over range 10E + 02 V/cm to 10E + 06 V/cm electric fields, and Grill and Patel [111] reported resistivity variation with the electric field indicating non-ohmic behavior. Materials could be microscopically homogeneous, for example glasses cooled from the melt (except SiO_2 containing Na_2O) and thin films evaporated under certain conditions. In many materials, however, local variations in density and composition undoubtedly occur. These do produce among other things spatial variations in the bandgap (band edges) and spatial variation in charge density (electrostatic fluctuations). Electrical transport by carriers in the conduction and valence band then will be variable range hopping by electrons with energies at the Fermi energy at low temperatures if the density of state, $N(E_F)$ is finite. At higher temperatures, conduction is thought to be due to thermally excited electrons in non-localized (extended) states [212].

Meyerson and Smith [108, 109] in their experiment (evaporated Mo film on substrate, deposited DLC by PECVD and deposited Mo dot on the DLC) measured the conductivity of DLC using voltage of about 0.1 volts and reported room temperature electrical conductivity of 10^{-16} to 10^{-6} per ohm cm^{-1} as room temperature is increased from 75 to 350°C. They noted strong temperature dependence of conductivity and stated that the expression given by the equation: $\sigma(T) = \sigma_0 \exp(-E_A/k_B T)$, does not apply.

Secondly, at high temperature, e.g. 250°C or 350°C (higher values of $\sigma = 10^{-4}$ to 10^{-2} ohm^{-1}cm^{-1} were observed), $\sigma(T) = \sigma_0 \exp(-E_A/k_B T)$ applies and conductivity is independent of deposition temperature. This conductivity has simple activation form and simple conductivity via extended states in either the valence or conduction band. Thus they concluded that the measured conductivity of DLC is neither simply activated, nor is it consistent with variable range hopping. They tried to explain that the conductivity is instead likely to be an activated hopping conduction in a broad region of localized states (tail states) near one of the band edges, where conduction would occur via thermally activated hopping. Their plots of σ versus $T^{1/4}$ were nonlinear. They suggested that the temperature dependence of σ_0 and E_B (or E_A for the conduction band edge) may be due to the energy dependence of the mobility $\mu(E, T)$. The factor which appears to determine the variation of σ with T_d (deposition temp.) is the decrease of $E_A(250)$ with increasing T_d.

In a four-point probe electrical resistivity measurement of Mo-DLCNC film, a drastic drop in the resistivity was observed at around 10 at.% of Mo and up to about four orders magnitude from 0–60 at. % of Mo-content [83].

The electrical conduction of DLC is not yet fully understood and has been reported as being more complex than mostly reported because of its deviation from both the Mott's relation and the simple conduction by thermal activation [220]. Amorphous semiconductors like a-Ge and a-Si were found to obey Mott's relation when the temperature is low enough, however for evaporated amorphous carbon the hopping conduction seems to occur only in a narrow range of low temperatures: below 8 K and 5 K.

In 1991, it was reported that for an Arrhenius plot of conductivity data of DLC, $\log \sigma$ versus T^{-1} their curves approached a straight line at both low and high temperatures with a transition region in between. They therefore suggested that low temperature conductivity of DLC is due to the excitation of electrons to states lying about 0.3 eV above the Fermi level and then hopping over the sp^3 gap and that at higher temperatures the predominant process is variable range hopping by carriers excited into band tail states (Fig. 4.1).

The main electronic/electrical influence of dopants are their ability to give or take electrons from the host material, to shift the effective Fermi-level towards the conduction band or valence band, and thereby either lowering or increasing the bandgap of the host material. Silicon doping of the a-C:H leads to some microstructural as well as electrical property changes in the film. Silicon does not form π-bonds thus it increases the sp^3 content

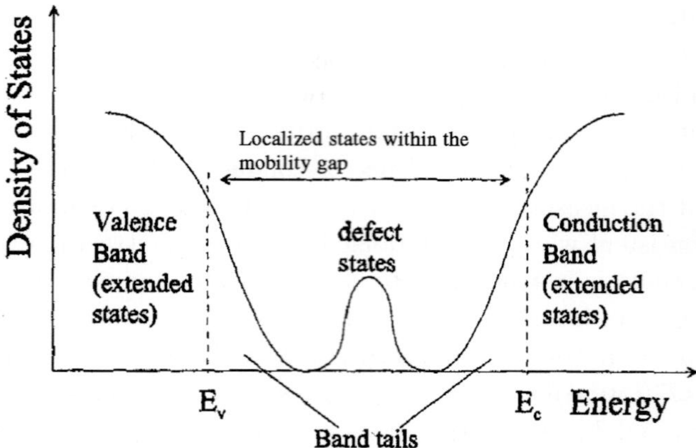

Fig. 4.1.   Diagram of density of states of an amorphous semiconductor [222].

of the film. However, it seems that silicon displaces and causes H_2 exudation from the film. In this chapter, the authors investigate to understand the influence of silicon in DLC thin films, in relation to the way it may affect the biocompatibility and haemocompatibility of the thin films. The knowledge of the conduction mechanisms, the work function, the bandgap/ optical bandgap and the overall density of states and the distribution of the localized states in the mobility gap/edge will be invaluable in understanding the interaction of the material with biological systems.

There is no known report in the literature on the effect of silicon in Si-DLCNC on the electrical properties of amorphous hydrogenated carbon thin films (apart from the papers from this work [62, 63] and the conduction mechanism in Si-DLCNC thin films). However, there are reports on the electrical properties of diamond-like carbon films alloyed with silicon and a third metallic element. Bozhko *et al.* [223] reported on the electro-conductivity of amorphous carbon films containing a mixture of silicon and tungsten with concentrations of tungsten (W) and silicon (Si) $0 < X < 0.15$ and $0 < Y < 0.24$ grown on a dielectric polycrystalline sapphire substrate using plasma decomposition of siloxane vapor. According to Bozhko *et al.* [223] the structure of these films consists of an atomic-scale composite of carbon and silicon random networks in which the carbon network is stabilized by hydrogen and the silicon network stabilized by oxygen forming a self-stabilized C-Si amorphous structures, an ideal matrix for the introduction of metals. They reported that the current–voltage characteristics of their diamond-like films at room temperature are linear on log I–$V^{1/2}$

coordinates, with the applied electric field $E \geq 10^4 \, \text{Vcm}^{-1}$. In the temperature range 150–350 K, the electroconductivity is of thermo-activated Poole–Frenkel nature characterized by two values of activation energy, 0.32 and 0.2 eV, and the observed saturation of the current versus temperature on increasing the electric field is connected with the predominance of tunneling effects, activationless hopping conductivity and direct tunneling at the mobility threshold [222]. Also in another study Bozhko *et al.* [223] reported that in the insulator-rich region (films with little or no amount of tungsten) of their diamond-like films containing tungsten and silicon in a wide range of tungsten concentrations (0–60 at.%) at room temperature, the electric conductivity is that of the Poole–Frenkel model and tends to be activationless following the Shklovskii mechanism at low temperatures [223]. Evolution of the dark room-temperature current–voltage characteristics of hydrogenated amorphous carbon films containing silicon and oxygen with deposition energy growth was also investigated at applied electric fields up to 6×10^5 V/cm by Bozhko *et al.* [224]. They showed that the character of current–voltage dependences is influenced by the deposition energy, which is determined by the value of the self-bias voltage, V_{sb}, varied in the range from -100 to -1400 V, and is described in terms of the space-charge-limited current in the presence of bulk traps, presumably having an exponential energy distribution. In films deposited at moderate values of self bias voltage (-400 to -800 V) they observed a trap-filled limit mode of the Gaussian-distributed deep trap set in the electric fields 5×10^3–10^5 V/cm. At electric fields exceeding 3×10^5 V/cm, phonon-assisted tunneling through the reduced electricfield potential barrier of the trap enhances the space-charge-limited current. The absence of thermal activation of the carriers at the mobility threshold and the Poole–Frenkel mechanism based on the I–V characteristics according to Bozhko *et al.* is possibly caused by the reduction in trap depth in the electric field/deviation of the trap potential from the Coulomb one or by the dipole or multipole character of traps in hydrogenated amorphous carbon films containing silicon and oxygen [224].

5. Work Function/Contact Potential Difference

A potential barrier prevents a valence electron from escaping in order to conduct. To overcome this barrier energy is required. This energy is the work function, the difference between the potential "immediately outside" the metal surface and the electrochemical potential or Fermi energy inside the metal solid [225]. The work function can be defined in various ways:

as the least amount of energy required to remove an electron from the surface of a conducting material, to a point just outside the metal with zero kinetic energy [226]; or for a semiconductor, as the difference in energy between the Fermi level and the vacuum level [227]. It also represents the weighted average of the energies required to remove an electron from the valence and the conduction bands [228]. In the bulk of most non-metallic materials the atoms are bound to their neighbors by covalent bonds. However, at the surfaces of most materials, the atoms are bound to neighbors on only one side, with the other side adjacent to vacuum, thereby creating a dangling bond, a plane of broken bonds consisting of an unpaired electron per suface atom, directed away from the surface. The electron distribution at these surfaces are therefore asymmetrical with respect to the positive ion core. Since the centers of the positive and the negative charges do not coincide there is a double layer of charge (dipole) at the surface. Thus, because the "escaping" electron has to move through the surface region, its energy is influenced by the electrical, optical and mechanical characteristics of the region. The work function is thus, a very sensitive indicator of the surface properties of a material, and is affected by the adsorbed, absorbed and evaporated layers, surface charging, surface imperfections, surface contaminations, oxide layers, and bulk contaminations (for example dopants). From Fig. 5.1, the work function (Φ) is given by:

$$\Phi = \chi + (E_\text{C} - E_\text{F}),$$

where, E_C–E_F is the energy difference between the Fermi level (E_F) and the bottom of the conduction band (E_C) and χ is the electron affinity (the energy difference between an electron at rest outside the surface and an electron at the bottom of the conduction band just inside the surface. Interpretation of work function results can be problematic with amorphous carbon because, unlike other materials, both Φ_Sample and the E_F can vary. The work function of HOP graphite has been measured as 4.65 eV [229], and while exact values for diamond are difficult to obtain, it is known to be less than 1 eV. Ideally E_F should lie mid gap but the presence of traps will cause it to shift closer to the conduction band or valence band. Conversely the CPD cannot readily be used to track E_F changes unless Φ or electron affinity, χ, are known to be constant. For ta-C:H films the Fermi level is reported to lie closer to the valence band, thus making it a p-type semiconductor [140]. Some a-C:H films have been doped n-type and from the observed changes in resistivity, the Fermi-level has moved through the

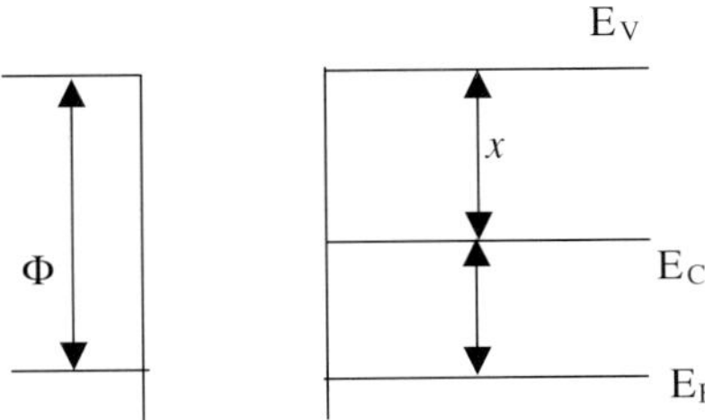

Fig. 5.1.   Band energy diagram for a semiconductor with no surface layer.

mid gap position nearer to the conduction band, hence the initial position of E_F was closer to the valence band, making the as-deposited film p-type. Magill *et al.* [231] reported that the a-C:H films deposited on n-type and p-type silicon also indicates that the films are weakly p-type [231]. Finally there is an additional unknown component, ϕ_{SS}, due to surface states.

In the presence of a surface layer, an additional effect known as "band bending" [232], must be considered because of its contribution to the overall work function. A negatively charged surface layer (e.g. an oxide layer which introduces a corresponding positive space charge) will cause a negative change in the surface potential. This leads to an upward band bending, a lowering of the conduction band and a fall in the Fermi level. Hence the difference between the Fermi level and the vacuum level increases, resulting in an increased work function.

Several techniques have been proposed and used for measuring work function. These are divided mainly into emission and non-emission methods. The most common emission methods are field emission, thermionic emission, and photoelectron emission. Field emission technique requires a high electric field ($\sim$10E $+$ 7 V/cm). The work function is determined from this technique by varying the emission current as a function of the applied voltage and by applying the Fowler–Nordheim equation. The field emission current (J) obtained depends on the, work function (Φ) and the electric field (E) [233] applied in order to enhance the probability of an electron near the Fermi level to tunnel through the surface barrier. The thermionic emission method uses thermal heating to cause electron emission from the surface of the material. It is related exponentially to the work function via the Richardson–Dushman equation [234]. The variation of the thermionic current (I) as a function of the temperature (T) enables the work function to be determined from the slope of $\ln(I/T^2)$ versus I/T which is equal to $e\Phi$. The thermionic technique is only suitable for materials with low vapor pressures at high temperatures (e.g. tungsten, molybdenum, platinium, nickel). In photoelectron emission techniques photons of a specific energy are directed onto the surface of the material. If the electron near the surface absorbs energy from the incident photon with energy close to the work function, it escapes with a kinetic energy which is the difference between the potential energy of the surface (Φ) and the incident photon energy. The threshold energy of emission, hv_0 is related to the work function by the expression:

$$\Phi = \frac{hv_0}{e},$$

where, h = Planck's constant, and v_0 = threshold frequence.

The work function is thus determined by monitoring the photoelectron current as a function of the frequency. The exposure of the material surface to high energy source is the main drawback of the emission techniques, because this will lead to chemical or physical changes in the material surface, thereby altering the actual work function.

The emission methods could be said to be destructive while the alternative, the non-emission method (e.g. contact potential difference, CPD) is not.

W. Thompson, later known as Lord Kelvin, first postulated in 1861 a simple capacitative device consisting of a flat circular electrode (the reference electrode) suspended above a stationary electrode (the specimen). Both electrodes are parallel to each other, thus creating a simple capacitor used to acquire work function difference between two surfaces by measuring the charge flow when the two conducting surfaces are connected. The principle of the Kelvin method was first established by Kelvin in 1898 [235], however his original device produced a once-only measurement due to the surface becoming charged such that the charge had to be dissipated before another measurement could be made. In 1932, Zisman [236] overcame this problem by applying the contact principle in the form of a vibrating capacitor which considerably improved this method in terms of signal detection enhancement. Baikie *et al.* have improved this method with respect to noise rejection [237], signal detection and signal processing [238]. These improvements by Baikie recently led to a Kelvin probe device that is capable of acquiring a work function measurement to a $< 1\,\text{mV}$ accuracy in both air and ultrahigh vacuum environments with increased precision and usage thin film work function studies [239].

6. Protein Adsorption on Biomaterials

It is now known that proteins either present in serum or secreted by the cells play a key role in the adhesion and spreading of the cells on the substrate biomaterial. The adsorption of different types of proteins on bio-surfaces will be discussed in some detail below.

6.1. *Non-Adhesive Proteins*

Albumin and tranferin-like proteins with non-adhesive functions tend to decrease subsequent thromboembolic events [240, 241]. Dion *et al.* [180] examined ^{131}I-labeled albumin plasma protein adhesion on DLC-coated Ti6Al4V and siliconee elastomer, and reported that DLC can adhere more albumin than the medical grade elastomer.

6.2. *Adhesive Proteins*

Adhesive proteins include fibrinogen, fibronectin, VWF and CAM. In general these plasma proteins with adhesive functions tend to increase thrombosis [240–242]. Adhesive proteins and likely increased expression of cell adhesion molecules (CAM) e.g. ICAM-1, VCAM-1, ELAM-1, E-selectin, GMP-140 (P-selectins) and other molecules/ligands from the immunoglobulin and selectin superfamily have been shown to be important in cascade reactions like platelet–leucocyte and leucocyte–endothelial cell adhesion and activation reactions [243]. When expressed on the cell surface the NH_2-terminal lectin-like domains of the selectins bind with their counter-receptors (specific carbohydrate ligands on white blood cells and platelets). Dion *et al.* [180] have also examined ^{125}I-labeled fibrinogen plasma protein adhesion on DLC-coated Ti6Al4V and silicone elastomer and reported that DLC can adhere slightly more fibrinogen than the silicone elastomer.

6.3. *Non-Adhesive/Adhesive Protein Ratios*

It has been shown that platelet adhesion depends on the albumin/fibrinogen ratio: the higher the albumin/fibrinogen ratio, the lower the number of adhering platelets and hence the lower the risks of platelet aggregation and thromboembolism. The albumin/fibrinogen ratio for DLC is 1.24 and 0.76 for silicone elastomer [180]. According to Dion *et al.*, these two ratios allow us to consider that platelet adhesion would be weaker on DLC than on silicone elastomer. However the opposite, in fact, occurred, which they thought could be explained by the large dispersion of results in percentage of platelets retained due to the device concept itself they added [180]. Cui and Li [197] also studied the adhesion of plasma proteins on DLC-coated, CN-coated PMMA, and uncoated PMMA using radioactive targed proteins. They reported the albumin/fibrinogen ratio of 1.008 for DLC, 0.49 for CN and 0.39 for PMMA [197].

7. Endothelial Cell Interactions with Diamond-Like Surfaces

Endothelium is Nature's haemocompatible surface, and the performance of any biomaterial designed to be haemocompatible must be compared with that of endothelium [166]. Endothelial haemocompatibility can be considered under three areas: the interaction between the endothelium and circulating cells (mainly platelets and leucocytes — close interactions between erythrocytes and endothelium are rare); the modulation of coagulation and fibrinolysis by endothelium; and other activities that affect the circulating blood or the vascular wall. Under normal circumstances, platelets do not interact with the endothelial cells — that is, platelet adhesion to the vessel wall and the formation of platelet aggregates do not normally take place except when required for haemostasis. Hence, the surface of endothelial cells does not promote platelet attachment [166]. The formation of platelet aggregates in close proximity to the endothelium is also rendered difficult by prostacyclin (PGI2), a powerful inhibitor of platelet aggregation secreted by the endothelial cells. Prostacyclin can be secreted by endothelial cells in culture as well as by isolated vascular tissue [244]. The vascular endothelium is now known to be a dynamic regulator of haemostasis and thrombosis with the endothelial cells playing multiple and active (rather than passive) roles in haemostasis and thrombosis [245, 246]. Many of the functions of the endothelial cells appear to be anti-thrombotic in nature. Several of the "natural anticoagulant mechanisms", including the heparin–antithrombin mechanism, the protein C-thrombomodulin mechanism, and the tissue plasminogin activator mechanism, are endothelial-associated. Among the proteins on the endothelial surface is antithrombin III [247] which catalyze the inactivation of thrombin by heparin. Endothelial cells also have heparan sulphate and dermatan sulphate (glycosaminoglycans) on their surfaces [248] which are known to have anticoagulant activity. On the other hand, the endothelial cells also appear capable of active pro-thrombotic behavior in some extreme conditions of anticoagulation, because endothelial cells synthesize adhesive co-factors such as the von Willibrand factor [249], fibronectin and thrombospondin [250]. Endothelial cells are now known to play crucial roles in a large number of physiological and pathological processes. Most of these physio-pathologic events take place at the microvasculature (capillary beds) which constitutes the vast majority of the human vascular compartment. Thus it becomes vital to carry out haemocompatibility studies using microvascular endothelial cells. This is also vital because not all endothelial cells are alike. Endothelial cells derived from the microvascular structures of specific tissues differ significantly from large-vessel endothelial cells. The study of human microvascular endothelial cells has been limited due to the fact that these cells are difficult to isolate in pure culture, are fastidious in their *in vitro* growth requirements, and have very short life span undergoing senescence at passages 8–10. Ades *et al.* [251] overcame these problems by the transfection and immortalization of human dermal microvascular endothelial cells (HMEC) [251]. These cells termed CDC/EU.HMEC-1 (HMEC-1) do retain the characteristics of ordinary endothelial cells (HMEC) and could be passaged up to 95 times, grow to densities 3–7 times higher than ordinary HMEC and require much less stringent growth medium [251]. HMEC-1 is just like ordinary endothelial cells and exhibits typical cobblestone (or polyhedral) morphology when grown in a monolayer culture.

Van Wachem *et al.* [177] reported that in their investigation of *in vitro* interaction of human endothelial cells (HEC) and polymers with different wettabilities in culture, optimal adhesion of HEC generally occurred onto moderately wettable polymers. Within a series of cellulose type of polymers, the cell adhesion increased with increasing contact angle of the polymer surfaces [177]. Moderately wettable polymers may exhibit a serum and/or cellular protein adsorption pattern that is favorable for growth of HEC [176]. Van Wachem *et al.* [176] also reported that moderately wettable tissue culture poly(ethylene terephthalate) (TCPETP), contact angle of 44° as measured by captive bubble technique, is a better surface for adhesion and proliferation of HEC than hydrophobic poly(ethylene terephthalate) (PETP), contact angle of 65°, suggesting that vascular prostheses with a TCPETP-like surface will perform better *in vivo* than prostheses made of PETP [176].

In looking at the microstructural changes in Si-DLCNC biomaterial and its effect on its interaction with cells, it is essential to look at the microvascular level of interaction because this is the level at which most of the crucial cellular interaction responsible for haemo-compatibility especially as well as biocompatibility occur. Most of the physio-pathologic events in the body take place at the microvasculature (capillary beds) level, which constitutes the vast majority of the human vascular compartment. It becomes vital therefore to use the human microvascular endothelial cells to investigate haemo-compatibility and biocompatibility of biomaterials.

7.1. *Silicon-Doped Diamond-Like Carbon Nanocomposite Films*

Si-DLCNC films, deposited using plasma-enhanced CVD, were annealed at various temperatures up to 600°C and the attachment of HMEC was counted, Fig. 7.1(a). For amorphous (undoped) DLC films, without annealing, the HMEC attachment was similar to that of the uncoated silicon control wafer. However, with increasing anneal temperature, cell attachment increased up to a maximum at 300°C and fell thereafter. For two values of Si doping, $\sim 5\%$ and $\sim 7.6\%$ (as determined from XPS), a high and approximately constant value of cell attachment occured, irrespective of annealing temperature. In Fig. 7.1(b), the HMEC cell attachment is seen to rise with silicon doping between 0.36% Si and 5% Si, "saturating" thereafter. The contact potential difference (CPD) between the Si-DLCNC and a reference (brass) probe, of known work function, falls in almost inverse relationship to HMEC attachment. The CPD change indicates either a reduction in the Si-DLCNC work function, due to an increased sp^3-fraction or a change in the density of surface charge. *The reduction in work function of the Si-DLC films compared to DLC films could be associated with the creation of new defect states in the bandgap of the Si-DLC films through the formation of clusters like the sp^3-rich Si–H clusters. The presence of silicon and hydrogen in hydrogenated DLC (a-C:H) films, that are more electropositive than carbon, will result in the creation of a surface dipole with an external positively charged side on the silicon-doped DLC film. It is the positive end of the dipole on the surface of the Si-DLC films that is believed to be responsible for anchoring the negatively charged endothelial cells.*

Figure 7.2 shows the effects of % silicon content on HMEC, resistivity and work function. A substantial increase in the number of adherent endothelial cells in the Si-DLCNC compared to the DLC films was observed, as shown in Fig. 7.2, but further silicon addition did not lead to any substantial change in the number of adherent cells on the Si-DLCNC film surface. It has been suggested that the cell adhesion process depends on the sign of the charge carried by the adherent cell. Positively charged surfaces will attract cells with a negative charge or dipole, and vice versa. Also from Fig. 7.2, the work function decreased wih silicon content and then gradually

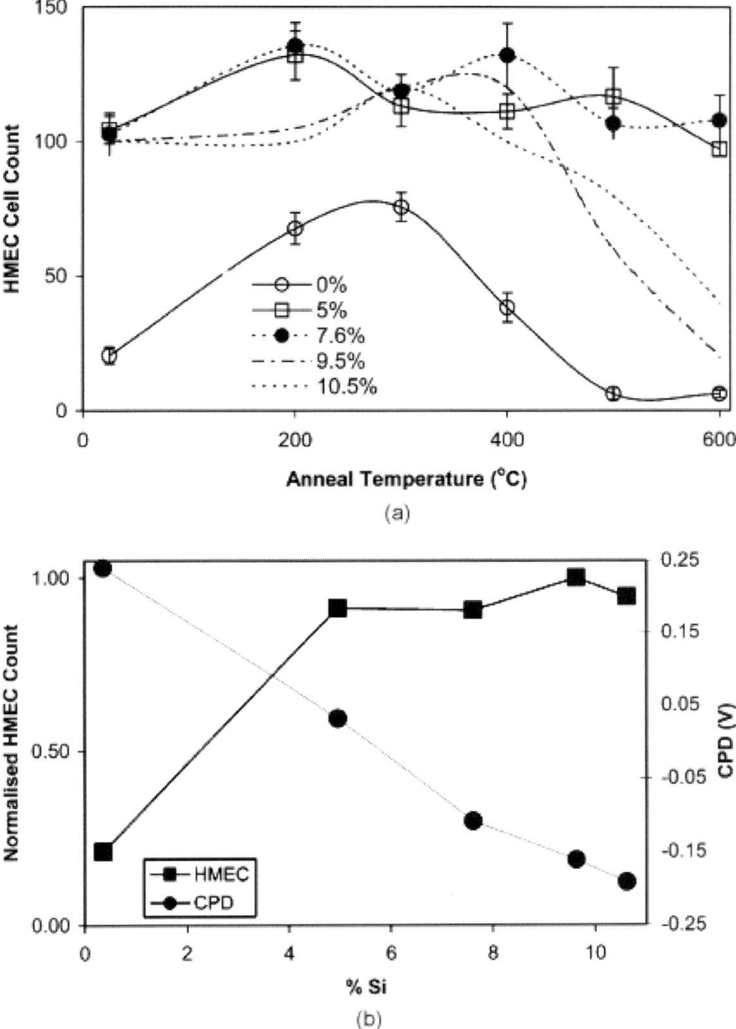

Fig. 7.1. (a) The attachment of HMEC (after 6 h) as a function of anneal temperatures and silicon content. (b) Contact Potential Difference (CPD) variation with silicon content compared with HMEC count (normalized). Cell count area is $600 = 400 \, \mu$m.

leveled off at around 7.6 atomic % of silicon content. The decrease in work function values has also been associated with a reduction in the net surface dipole. It is apparent from Fig. 7.2 that the resistivity gradually decreased with silicon content in the Si-DLCNC films and increased from approximately 9.6 atomic % Si content in the films. The resistivity, obtained from

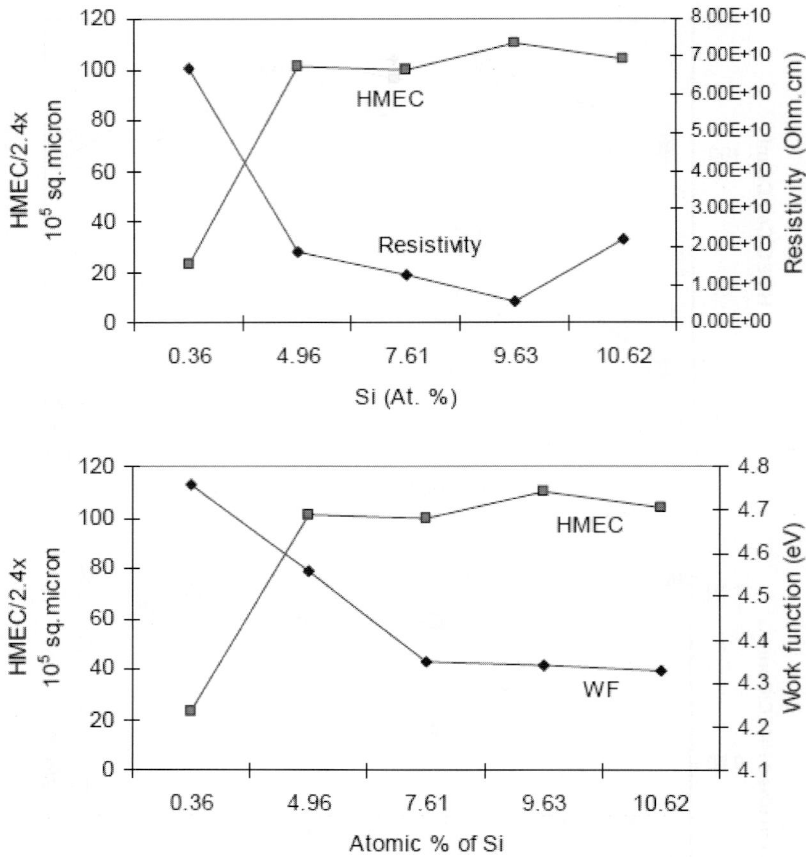

Fig. 7.2. HMEC attachment per $2.4 \times 10^5 \ \mu\text{m}^2$ as a function of the electrical properties of a-C:H (DLC) and a-C:H:Si (Si-DLCNC) thin films.

the slope of the ohmic-like region, indicates up to an order of magnitude decrease in resistivity for silicon-doped films (Si-DLCNC).

Figure 7.3 shows the relationship between atomic % Si content in Si-DLCNC films and HMEC, contact angle and surface energy. The contact angle measurement results obtained using the optical method and the results of the surface energy measured by the Wilhemy plate technique for films deposited on silicon substrates are shown in Fig. 7.3. Generally, silicon doping leads to an increase in the surface contact angles. It is seen that a higher contact angle is good for HMEC seeding, as it follows similar trend when increasing silicon content in the Si-DLCNC films. The introduction of silicon resulted in a lower surface energy with a high dispersive

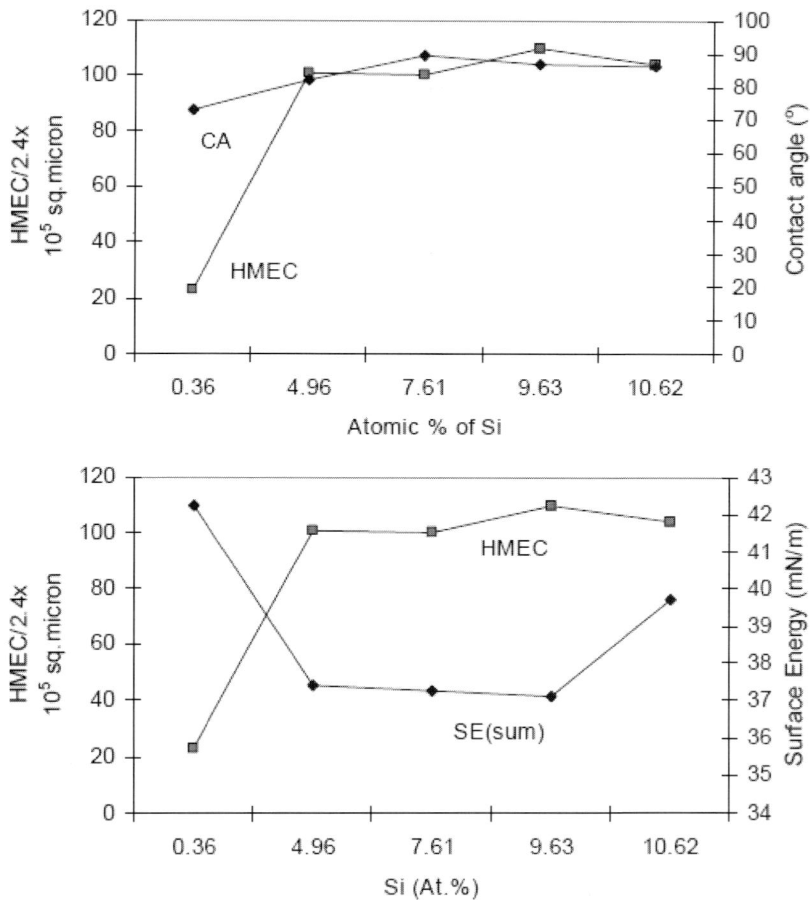

Fig. 7.3. HMEC attachment per 2.4×10^5 μm^2 as a function of the contact angle and the surface energy (sum) of a-C:H (DLC) and a-C:H:Si (Si-DLCNC) thin films.

component, leading to a more hydrophobic film (Fig. 7.3). *A more hydrophobic surface reduces its interaction with water molecules allowing more direct contact with the endothelial cells, rather than an interaction mediated by a water layer. Silicon addition to a-C:H decreases the surface energy from $\sim$42–31 mNm^{-1} [118] and increases the water contact angle. Silicon in a-C:H film reduces mainly the polar component of the surface energy which is due to dipoles at the surface. Silicon does not form π-bonds and therefore increases the sp^3 bonding, reducing the sp^2 bonding, their polarization potential and the dangling bonds [209, 210]. The dispersive component of* *the surface energy is due to electronic interactions, for example "van der Waals" forces. The presence of silicon and hydrogen in the hydrogenated DLC (a-C:H) films result in the creation of a surface dipole on the silicon-doped DLC film.*

Whereas increasing contact angle (or lowering of surface energy) of DLC and increasing the intrinsic electroconduction may lead to increased human microvascular endothelial cell adhesion, graphitization may lead to a decreased adhesion of human microvascular endothelial cells. This implies that an increase in the contact angle as well as a moderate increase in the intrinsic electroconductivity may lead to improved haemo-compatibility of DLC and modified DLC when they are seeded with human microvascular endothelial cells *in vitro*. Thus the electrical properties, the surface energy and the microstructure of DLC play a key role in the interfacial interactions of this biomaterial.

Generally the bulk of the surface energy in both a-C:H (DLC) and a-C:H:Si (Si-DLCNC) consist of the dispersive component. In most cases, over 99% of the surface energy is dispersive. However it seems that where the surface energy values are comparatively lower, the percentage of the polar component seems a bit higher even though the total value of the surface energy is comparatively lower.

7.2. *Chromium-Doped Diamond-Like Carbon Nanocomposite Films*

Chromium-modified DLC (Cr-DLCNC) films with varying %Cr content were deposited on 50 mm circular silicon wafers using magnetron sputtering utilizing the intensified plasma-assisted processing process. Table 7.1 shows the conditions employed to deposit the Cr-DLCNC samples with varying %Cr contents. The as-deposited Cr-DLCNC films were nanocomposite and consisted of nanosized chromium carbide particles embedded in an amorphous DLC matrix.

Figure 7.4 shows the XRD spectrum for a typical Cr-DLC film, with 10%Cr, in the 2-theta range 15–115 degrees. It was found from HRTEM studies that the nanostructured Cr-DLCNC film consisted of nanosized (5 nm) chromium carbide particles. The presence of chromium carbide particles was evident from the XRD peak centered at around 33° 2-theta value. The other major intense peak centered at around 68° 2-theta value corresponds to silicon, which was from the silicon wafer substrate material. We found that the Cr-carbide nanoparticles were embedded deep into the

Table 7.1.   Deposition parameters employed in depositing Cr-DLCNC films using the IPAP process.

Sample	Deposition rate (nm/min)	Magnetron current (mA)
Cr(1%)DLC	6.6	155
Cr(5%)DLC	7.6	220
Cr(10%)DLC	15.2	310

Bias Voltage: -1000 V; flow rate (sccm): CH_4:Ar 7.4:40;
Chamber pressure: 2.66 Pa; processing time: 2 hrs;
sputter cleaning: Ar^+: 3.3 Pa, -1500 V, $\sim$20 minutes.

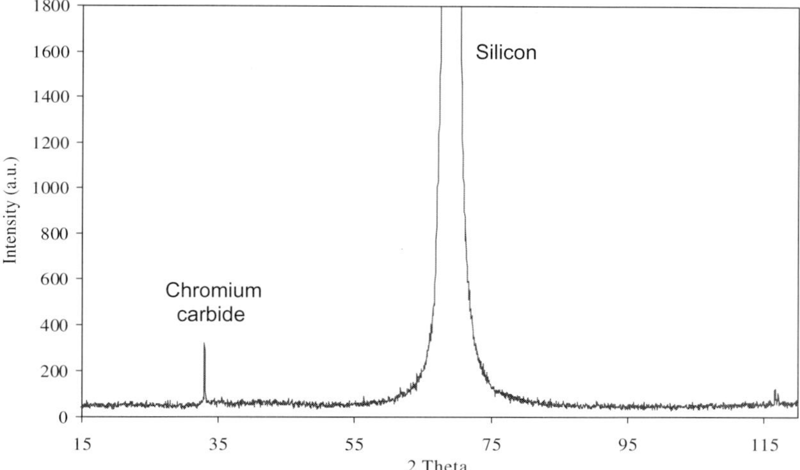

Fig. 7.4.   XRD spectrum for sample Cr(10%)-DLCNC showing the silicon substrate and chromium carbide peaks [97].

amorphous DLC matrix and were not present on the DLC surface. This enabled the nanoparticles to be protected by the amorphous DLC film.

Figure 7.5 displays the high resolution (HRTEM) micrograph representing 5%Cr-DLC and showing the microstructure exhibited by the same as-deposited film. The micrograph shows the presence of nanoclusters (NCs) around 5 nm in diameter surrounded by a $\sim$2 nm thick amorphous matrix. Electron diffraction showed that the dark contrast NCs correspond to Cr carbide, encapsulated by an amorphous matrix.

The contact angle measurements obtained using the optical method are as shown in Fig. 7.6. Chromium doping leads to a gradual increase in the contact angles, as shown in Fig. 7.6. The increase begins to level out with Cr content above 5% where it begins to reach saturation point. The

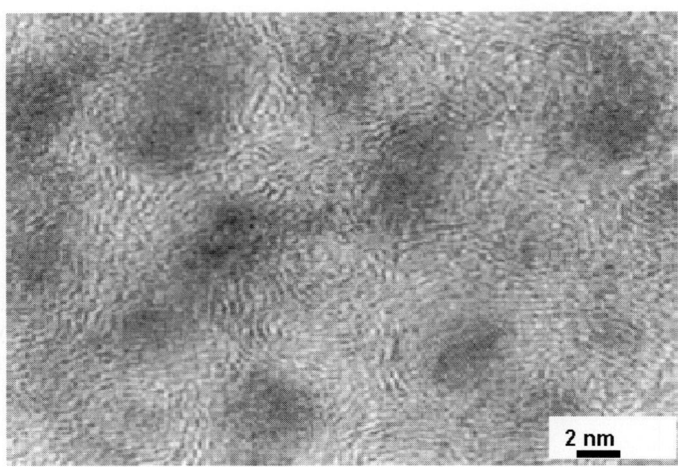

Fig. 7.5. High resolution TEM micrograph of the 5%Cr-DLCNC sample showing the microstructure exhibited by the thin film coating [97].

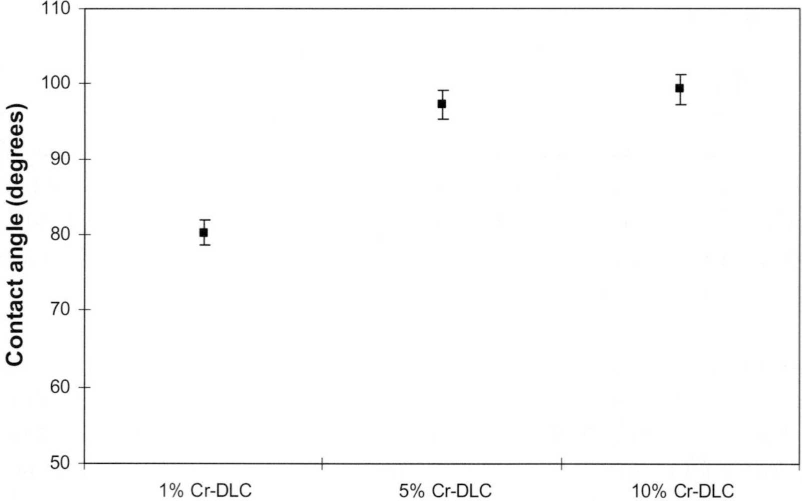

Fig. 7.6. Graph showing the contact angle measurements as obtained using the optical method for the three Cr-DLCNC samples [97].

average contact angles displayed by samples 1%Cr-, 5%Cr- and 10%Cr-DLC samples were calculated to be 80.285, 97.23 and 99.274 degrees, respectively.

Raman spectroscopy was used to characterize the quality of the as-deposited Cr-DLCNC films with different Cr contents, in terms of diamond carbon-phase purity. The data from the Raman spectroscopy studies, including intensities of D (I_D) and G (I_G) bands, full width at half maximum (FWHM) of I_D and I_G bands, I_D/I_G ratio and the positioning of the D and G-band peaks can all be found in Table 2.1. From the Raman investigations, it was found that the Cr-DLCNC films displayed the two D and G bands of graphite. The G and D bands are usually assigned to zone center of phonons of E_{2g} symmetry and K-point phonons of A_{1g} symmetry, respectively. The D-band peaks for the three Cr-DLCNC films were positioned at 1401.6, 1408.4 and 1306.8 cm^{-1}, whereas, the G-band peaks were centered at 1538.8, 1540.6 and 1511.8 cm^{-1}. From both D and G bands, the 5%Cr-DLCNC sample displayed the smallest values for FWHM. The I_D/I_G ratio was the least for 10%Cr-DLCNC, and the highest out of the three samples for the 5%Cr-DLCNC film. This suggests that there are more disordered graphitic phases in sample 5%Cr-DLCNC, and the least similar disorder in 10%Cr-DLCNC film.

The results of the cell count analysis shown in Fig. 7.7, give an indication of the influence of Cr content in Cr-DLCNC films on the adherent cell population of the three samples. 5%Cr-DLCNC provided the best conditions for HMV-EC seeding, while, 10%Cr-DLCNC film resulted in the least population of adherent human endothelial cells onto its surface. It should be noted that sample 1%Cr-DLCNC was a better base material for seeding endothelial cells than 10%Cr-DLCNC. Figure 7.8 displays the SEM micrographs showing the population of endothelial cell attachment onto 1%Cr-, 5%Cr- and 10%Cr-DLCNC film surfaces. All three films displayed smooth surface profiles, which is a key requirement in artificial heart valve applications.

It was noted that there was a direct correlation between the I_D/I_G ratio and the population of endothelial cells attaching to the three Cr-DLCNC films with different Cr contents. We noted that from the three Cr-DLCNC films, the highest value displayed for the I_D/I_G ratio was by 5%Cr-DLCNC film, which also gave the highest adherent cell population onto its surface. The lowest I_D/I_G ratio value was for 10%Cr-DLCNC, which showed the least, from the three samples investigated in this study, population of cell attachment to its surface after conducting the cell seeding procedures.

It is apparent that increased Cr content into the growing DLC films alters the microstructure of the deposited films. Furthermore, the density of nanosized Cr-carbide particles produced during film growth is expected to be the highest in 10%Cr-DLCNC and the least in 1%Cr-DLCNC sample.

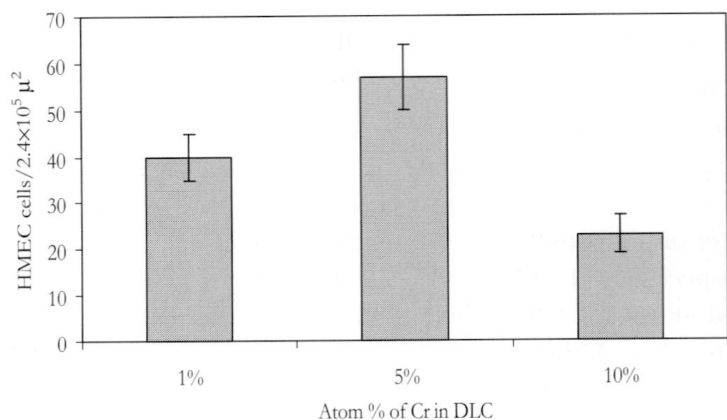

Fig. 7.7. Graph showing the cell count results obtained after seeding the human microvascular endothelial cells onto the Cr-DLCNC surfaces [97].

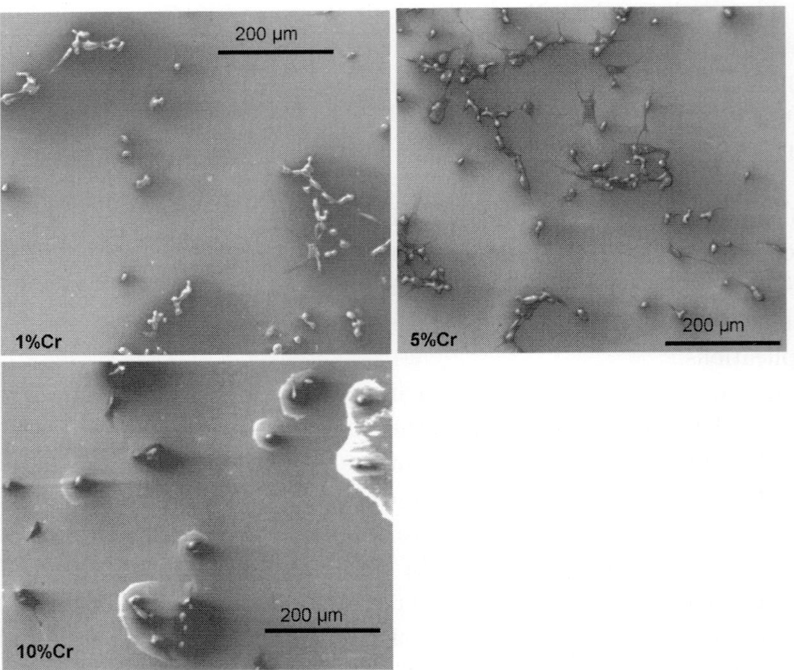

Fig. 7.8. SEM micrographs showing endothelial cell seeding on the three types of Cr-DLC film surfaces [97].

This difference in the Cr-carbide content in the three films is sure to influence the surface chemistry of the DLC films. It was difficult to correlate the water contact angle results with the cell seeding efficiency of the films.

8. Summary

Amorphous and diamond-like carbon nanocomposite films have been discussed with respect to their potentials for applications in biomedical areas, such as mechanical heart valves. Traditional amorphous DLC films have been modified by doping with silicon and chromium to produce DLCNC films. Human microvascular endothelial cells were seeded on amorphous DLC, thermally annealed and unannealed Si-DLCNC and Cr-DLCNC film surfaces. It was found that the atomic %Si and %Cr content in the deposited DLCNC films had a significant effect on the seeding of endothelial cells on the surface of such films. Thermal annealing also influenced the cell seeding behavior on DLC and Si-DLCNC films. It was found that the following factors had a noticeable effect on the seeding behavior of endothelial cell on diamond-like surfaces: (a) contact angle, (b) surface energy, (c) resistivity, (d) work function, (e) contact potential difference and (f) atomic %dopant concentration (silicon and chromium). The results were correlated with one another in order to enhance overall understanding in producing an endothelial lining (endothelium) on diamond-like coatings for applications in heart valves. This work has promising future applications involving possible *in vitro* growth of endothelial cells on DLCNC-coated devices before the implantation of such devices in the human body.

References

1. J.A. Bittl, Advances in coronary angioplasty, *N. Engl. J. Med.* **335** (1996) 1290–1302.
2. J.A. Bittl, Subacute stent occlusion: Thrombus horribilis, *JACC* **28** (1996) 368–370.
3. M. Gawaz, F.J. Neumann, I. Ott, A. May and A. Schomig, *Circulation* **94** (1996) 279–285.
4. T. Inoue, Y. Sakai, T. Fujito, K. Hoshi, T. Hayashi, K. Takayanagi, S. Morooka and R. Sohma, *Circulation* **94** (1996) 1518–1523.
5. J. Lahann, D. Klee, H. Thelen, H. Bienert, D. Vorwerk and H. Hocker, *J. Mater. Sci. Mater.* **10** (1999) 443–448.
6. C.L. Haycox and B.D. Ratner, *J. Biomed. Mater. Res.* **27** (1993) 1181–1193.
7. J.M. Courtney, N.M.K. Lamba, S. Sundaram and C.D. Forbes, *Biomaterials* **15** (1994) 737–744.

8. C.L. Klein, P. Nieder, M. Wagner, H. Kohler, F. Bittinger, C.J. Kirkpatrick and J.C. Lewis, *J. Pathophysiol.* **5** (1994) 798–807.

9. K. Gutensohn, C. Beythien, R. Koester, J. Bau, T. Fenner, P. Grewe, K. Padmanaban, P. Schaefer and P. Kuehnl, *Infusionstherapie und Transfusionmedizin* **27**(4) (2000) 200–206.

10. Y. Yang, S.F. Franzen and C.L. Olin, *Cells and Materials* **6**(4) (1996) 339–354;.

11. Y. Yang, S.F. Franzen and C.L. Olin, *J. Heart Valves Dis.* **5** (1996) 532–537.

12. K. Mark, G. Belli, S. Ellis and D. Moliterno *J. Am. Coll. Cardiol.* **27** (1996) 494–503.

13. A. Colombo, P. Hall, S. Nakamura, Y. Almagor, L. Maiello, G. Martini, A. Gaglione, S.L. Goldberg and J.M. Tobis, *Circulation* **91** (1995) 1676–1688.

14. J. Robertson, *Adv. Phys.* **35**(4) (1986) 317–374.

15. B. Bhushan, *Diam. Relat. Mater.* **8** (1999) 1985.

16. Y. Lifshitz, *Diam. Relat. Mater.* **8** (1999) 1659.

17. J.L. Bredas and G.B. Street, *J. Phys.* **C18** (1985) L651.

18. J. Robertson, *Surf. Coat. Technol.* **50** (1992) 185–203.

19. C.L. Mantell, "Carbon and Graphite Handbook" (Interscience, 1968).

20. R.E. Clausing, L.L. Horton, J.C. Angus and P. Koidl, *Diamond and Diamond-Like Films and Coatings* (Plenum Press, 1990).

21. K. Enke, *Mater. Sci. Forum* **52–53** (1989) 559.

22. A. Grill, *Cold Plasma in Material Fabrication: From Fundamentals to Applications* (IEEE Press, 1993).

23. A. Bubenzer, B. Dischler, G. Brandt and J. Koidl, *J. Appl. Phys.* **54** (1983) 4590.

24. Y. Catherine, in Diamond and Diamond like films and coatings, *NATO-ASI series B: Physics*, Vol. 266, eds. R.E. Clausing, L.L. Horton, J.C. Angus and P. Koidl (Plenum Pub., 1991), pp. 193–227.

25. S.M. Ohja, *Physics of Thin Films* (1982), p. 237.

26. B.K. Gupta and B. Bhushan, *Wear* **190** (1995) 110–122.

27. X. Shi, D. Flynn, B.K. Tay, S. Prawer, K.W. Nugent, S.R.P. Silva, Y. Lifshitz and W.I. Milne, *Phil. Mag.* **B76**(3) (1997) 351–361.

28. P.J. Fallon, V.S. Veerasamy, C.A. Davis, J. Robertson, J.A. Amaratunga, W.I. Wilne and J. Koskinen, *Phys. Rev.* **B48**(7) (1993) 4777.

29. M. Chhowalla, M. Weiler, C.A. Davis, B. Kleinsorge and G.A.J. Amaratunga, *Appl. Phys. Lett.* **67**(7) (1995) 894–896.

30. S. Anders, A. Anders and I. Brown, *J. Appl. Phys.* **75**(10) (1994) 4894–4899.

31. D.R. McKenzie, Y. Yin, E.G. Gerstener and M.M.M. Bilek, *IEEE Trans. Plasma Sci.* **25**(4) (1997) 653–659.

32. A. Von Keudell, T. Schwarz-Sellingber and W. Jacob, *J. Appl. Phys.* **89** (2001) 2979.

33. J.W.A.M. Gielen, Ph.D. Thesis (Eindhoven University, 1996).

34. R. Kleber, M. Weiler, A. Kruger, S. Sattel, G. Kunz, K. Jung and H. Ehrhardt, *Diam. Relat. Mater.* **2** (1993) 246.

35. M. Weiler, S. Sattel, K. Jung, H. Ehrhardt, V.S. Veerasamy and J. Robertson, *Appl. Phys. Lett.* **64** (1994) 2797.

36. N.M.J. Conway, A.C. Ferrari, A.J. Flewitt, J. Robertson, W.I. Milne, A. Tagliaferro and W. Beyer, *Diam. Relat. Mater.* **9** (2000) 765.

37. Y. Catherine and A. Zamouche, *Plasma Chem. Plasma Process* **5** (1985) 353.

38. J. Robertson, *Diam. Relat. Mater.* **3** (1994) 361.

39. P. Koidl, C. Wagner, B. Dischler and M. Ramsteiner, *Mater. Sci. Forum* **52** (1990) 41.

40. C. Wild and P. Koidl, *Appl. Phys. Lett.* **64** (1989) 505.

41. C. Wild and P. Koidl, *J. Appl. Phys.* **69** (1991) 2909.

42. J.C. Angus, *Proc. Eur. MRS* **17** (1987) 179.

43. J.C. Angus and Hayman, *Science* **241** (1988) 913.

44. M.W. Geis, *J. Vac. Sci. Technol.* **A6** (1988) 1953.

45. Baba and Hatada, *Surf. Coat. Technol.* **196**(1–3) (2005) 207–210.

46. C. Corbella, G. Oncins, M.A. Gomez and M.C. Polo *et al.*, *Diam. Relat. Mater.* **14**(3–7) (2005) 1103–1107.

47. D.-Y. Wang, Y.-Y. Chang, C.-L. Chang and Y.-W. Huang, *Surf. Coat. Technol.* **200** (2005) 2175–2180.

48. L. Huang, H. Jiang, J. Zhang, Z. Zhang and P. Zhang, Synthesis of copper nanoparticles containing diamond-like carbon films by electrochemical method. *Electrochem. Commun.* **8**(2) (2006) 262–266.

49. M.L. Morrison, R.A. Buchanan, P.K. Liaw, C.J. Berry, R.L. Brigmon, L. Riester, H. Abernathy, C. Jin and R.J. NarayanMaizza *et al.* (1999); Electrochemical and antimicrobial properties of diamondlike carbon-metal composite films, *Diam. Relat. Mater.* **15**(1) (2006) 138–146.

50. Y. Lifshitz, G.D. Lempert and E. Grossman, *Phys. Rev. Lett.* **72**(17) (1994) 2753.

51. Y. Lifshitz, G.D. Lempert, E. Grossman, I. Avigal, C. Uzansaguy, R. Kalish, J. Kulik, D. Marton and J.W. Rabalais, *Diam. Relat. Mater.* **4** (1995) 318.

52. S. Sattel, H. Ehrhardt, J. Robertson, Z. Tass, D. Wiescher and M. Scheib, *Diam. Relat. Mater.* **6**(2–4) (1997) 255–260.

53. S. Sattel, J. Robertson and H. Ehrhardt, *J. Appl. Phys.* **82** (1997) 4566–5476.

54. S. Sattel, T. Giessen, H. Roth *et al.*, *Diam. Relat. Mater.* **5** (1996) 425–428.

55. J.B. Pethica and D. Tabor, *Surf. Sci.* **89** (1979) 182.

56. N.A. Burnham and R.J. Colton, *J. Vac. Sci. Technol.* **A7**(4) (1989) 2905.

57. B. Bhushan, *Diam. Relat. Mater.* **8** (1999) 1985.

58. R.J. Narayan, *Appl. Surf. Sci.* **245** (2005) 420–430.

59. T.I.T. Okpalugo, A.A. Ogwu, P.D. Maguire, J.A. McLaughlin and D.G. Hirst, In-vitro blood compatibility of a-C:H:Si and a-C:H thin films, *Diam. Relat. Mater.* **13**(4–8) (2004) 1088–1092.

60. T.I.T. Okpalugo, A.A. Ogwu, P.D. Maguire and J.A. McLaughlin, Platelet adhesion on silicon modified hydrogenated amorphous carbon films, *Biomaterials* **25**(3) (2004) 239–245.

61. T.I.T. Okpalugo, A.A. Ogwu, P.D. Maguire, J.A. McLaughlin and R.W. McCullough, Human microvascular endothelial cellular interaction with atomic *n*-doped compared to Si-doped DLC, *J. Biomed. Mater. Res. Part B; Appl. Biomater.* **78B**(2) (2006) 222–229.

62. T.I.T. Okpalugo, A.A. Ogwu, P. Maguire and J.A.D. McLaughlin, Technology and Health Care, *Int. J. Health Care Eng.* **9**(1–2) (2001) 80–82.

63. T.I.T. Okpalugo, *Ph.D. Thesis*, University of Ulster, Belfast, UK.

64. T.I.T. Okpalugo, E. McKenna, A.C. Magee, J.A. McLaughlin and N.M.D. Brown, The MTT assays of bovine retinal pericytes and human microvascular endothelial cells on DLC and Si-DLC-coated TCPS, *J. Biomed. Mater. Res. Part A* **71A**(2) (2004) 201–208.

65. T.I.T. Okpalugo, P.D. Maguire, A.A. Ogwu and J.A. McLaughlin, The effect of silicon doping and thermal annealing on the electrical and structural properties of hydrogenated amorphous carbon thin films, *Diam. Relat. Mater.* **13**(4–8) (2004) 1549–1552.

66. Y. Lifshitz, *Diam. Relat. Mater.* **5** (1996) 388–400.

67. C. Park, H. Park, Y.K. Hong, J. Kim and J.K. Kim, *Appl. Surf. Sci.* **111** (1997) 140–144.

68. K.K. Hirakuri, T. Minorikawa, G. Friedbacher and M. Grasserbauer, *Thin Solid Films* **302** (1997) 5–11.

69. A. Ali, K.K. Hirakuri and G. Friedbacher, *Vacuum* **51** (1998) 363–368.

70. M.K. Fung, W.C. Chan, K.H. Lai *et al.*, *J. Non-Cryst. Solids* **254** (1999) 167–173.

71. G.J. Vandentop, P.A.P. Nascente, M. Kawasaki, D.F. Ogletree, G.A. Somorjai and M. Salmerson, *J. Vac. Sci. Technol.* **A9** (1991) 2273–2278.

72. Z.Y. Rong, M. Abraizov, B. Dorfman *et al.*, *Appl. Phys. Lett.* **65** (1994) 1379–1381.

73. X.L. Peng, Z.H. Barber and T.W. Clyne, *Surf. Coat. Technol.* **138** (2001) 23–32.

74. P. Lemoine, R.W. Lamberton, A.A. Ogwu, J.F. Zhao, P. Maguire and J. McLaughlin, Complementary analysis techniques for the morphological study of ultrathin amorphous carbon films, *J. Appl. Phys.* **86**(11) (1999) 6564–6570.

75. A. Einstein, *Ann. Phys.* **17** (1905) 132.

76. A.P. Dementjev and M.N. Petukhov, *Diam. Relat. Mater.* **6** (1997) 486.

77. K. Miyoshi and D.H. Buckley, *Appl. Surf. Sci.* **10** (1982) 357.

78. T. Mori and Y. Namba, *J. Appl. Phys.* **55** (1984) 3276.

79. Y. Taki and O. Takai, *Thin Solid Films* **316** (1998) 45.

80. M.V. Kuznetsov, M.V. Zhuravlev, E.V. Shalayeva and V.A. Gubanov, *Thin Solid Films* **215** (1992) 1.

81. L. Li, H. Zhang, Y. Zhang, K.P. Chu, X. Tian, L. Xia and X. Ma, *Mater. Sci. Eng.* **B94** (2002) 95–101.

82. A.P. Dementjev and M.N. Petukhov, *Surf. Interface Anal.* **24** (1996).

83. Baba and Hatada, Preparation and properties of metal-containing diamond-like carbon films by magnetron plasma source ion implantation, *Surf. Coat. Technol.* **196** (2005) 207–210.

84. R.D. Evans, G.L. Doll, P.W. Morrison Jr, J. Bentley, K.L. More and J.T. Glass, *Surf. Coat. Technol.* **157** (2002) 197–206.

85. J.F. Zhao, P. Lemoine, Z.H. Liu, J.P. Quinn, P. Maguire and J.A. McLaughlin, *Diam. Relat. Mater.* **10** (2001) 1070–1075.

86. M. Ramm, M. Ata, K.W. Brzezinka, T. Gross and W. Unger, *Thin Solid Films* **354** (1999) 106–110.

87. W.K. Choi, T.Y. Ong, L.S. Tan, F.C. Loh and K.L. Tan, *J. Appl. Phys.* **83**(9) (1998) 4968–4973.

88. R. Hauert, U. Muller, G. Francz, F. Birchler, A. Schroeder, J. Mayer and E. Wintermantel, *Thin Solid Films* **308–309** (1997) 191–194.

89. J.R. Ferraro and K. Nakamoto, *Introductory Raman Spectroscopy* (Academic Press Inc., 1994).

90. G. Herzberg, IR and Raman Spectra of Polyatomic Molecules, *Molecular Spectra and Molecular Structure*, Vol. II (Van Nostrand, Princeton, NJ, 1945).

91. Y. Wang, D.C. Alsmeyer and R.L. McCreery, *Chem. Mater.* **2** (1990) 557–563.

92. J. Wagner, M. Ramsteiner, Ch. Wild and P. Koidl, Resonant Raman scattering of amorphous carbon and polycrystalline diamond films, *Phys. Rev.* **B40**(3) (1989) 1817–1824.

93. L.C. Nistor, J. Van Laudugt, V.G. Ralchenko, T.V. Kononen-ko, E.D. Obraztsova and V.E. Strelnitsky, *Appl. Phys.* **58** (1994) 137.

94. R. Nicklow, N. Watabayashi and H.G. Smith, *Phys.Rev.* **B5** (1972) 4951.

95. D.B. Chase, *J. Am. Chem. Soc.* **108** (1986) 7485.

96. D.-Y. Wang, Y.-Y. Chang, C.-L. Chang and Y.-W. Huang, *Surf. Coat. Technol.* **200** (2005) 2175–2180.

97. N. Ali, Y. Kousar, T.I. Okpalugo, V. Singh, M. Pease, A.A. Ogwu, J. Gracio, E. Titus, E.I. Meletis and M.J. Jackson. Human micro-vascular endothelial cell seeding on Cr-DLC thin films for mechanical heart valve applications, *Thin Solid Films* **515** (2006) 59–65.

98. B. Dischler, A. Bubenzer and P. Koidl, *Appl. Phys. Lett.* **42**(8) (1983) 636–638.

99. B. Dischler, in Amorphous Hydrogenated Carbon Films, *Proc. European Mat. Res. Soc. Symp.* **17**, eds. P. Koidl and P. Oelhafen, Les edition de physique, Paris (1987), p. 189.

100. M. Shimozuma, G. Tochitani, H. Ohno and H. Tagashira, in *Proc. 9th Int. Symp. Plasma Chem*, ed. R. d'Agostino, IUPAC, Bari, Italy, 1462.

101. W. Dworschak, R. Kleber, A. Fuchs, B. Scheppat, G. Keller, K. Jung and H. Erhardt, *Thin Solid Films* **189** (1990) 257.

102. P. Couderc and Y. Catherine, *Thin Solid Films* **146** (1987) 93.

103. M.A. Baker and P. Hammer, *Surf. Interface Anal.* **25** (1997) 629–642.

104. A. Grill and B. Meyerson, in *Synthetic Diamond: Emerging CVD Science and Technology*, eds. K.E. Spear and J.P. Dismukes (Wiley, NY 1994), p. 91.

105. Robertson, *Mater. Sci. Eng.* **R37**(4–6) (2002) 129–281.

106. J. Robertson and C.A. Davies, *Diam. Relat. Mater.* **4** (1995) 441.

107. M. Golden, M. Knupfer, J. Fink *et al, J. Phys. Condens. Mater.* **7** (1995) 8219.

108. B. Meyerson and F.W. Smith, *J. Non-Cryst. Solids* **35–36** (1980) 435–440.

109. B. Meyerson and F. Smith, *Solid State Commun.* **34** (1980) 531.

110. O. Amir and R. Kalish, *J. Appl. Phys.* **70** (1991) 4958.

111. A. Grill and V. Patel, *Diam. Relat. Mater.* **4** (1994) 62.

112. T. Hioki *et al.*, in Structure-Property Relationships in Surface Modified Ceramics, ed. C.J. McHargue (Kluwer Academic Pub., Dordrecht, 1989), p. 303.

113. K. Oguri and T. Arai, *J. Mater. Res.* **5**(11) (1990) 2567.

114. S. Miyake, R. Kaneko, Y. Kikuya and I. Sugimoto, *Trans. ASME J. Tribol.* **113** (1991) 384–389.

115. F. Demichelis, C.F. Pirri and A. Tagliaferro, *Mater. Sci. Eng.* **B11** (1992) 313–316.

116. W.-J. Wu and M.-H. Hon, *Thin Solid Films* **307** (1997) 1–5.

117. W.-J. Wu, T.-M. Pai and M.-H. Hon, *Diam. Relat. Mater.* **7** (1998) 1478–1484.

118. M. Grischke, K. Bewilogua, K. Trojan and H. Dimigen, *Surf. Coat. Technol.* **74–75** (1995) 739–745.

119. M. Grischke, A. Hieke, F. Morgenweck and H. Dimigen, *Diam. Relat. Mater.* **7** (1998) 454–458.

120. A. Grill, V. Patel, K.L. Saenger, C. Jahnes, S.A. Cohen, A.G. Schrott, D.C. Edelstein and J.R. Paraszczak, *Mater. Res. Soc. Symp. Proc.* **443** (1997) 155.

121. K. Oguri and T. Arai, *Surf. Coat. Technol.* **47** (1991) 710.

122. C.D. Martino, F. Demichelis and A. Tagliaferro, *Diam. Relat. Mater.* **3** (1994) 547–550.

123. A.K. Gangopadhyay, P.A. Willermet, M.A. Tamor and W.C. Vassell, *Tribology Int.* **30** (1997) 9–18, 19–31.

124. B. Oral, K.H. Ernse and C.J. Schmutz, *Diam. Relat. Mater.* **5** (1996) 932–937.

125. A. Matthews and S. S. Eskildsen, *Diam. Relat. Mater.* **3** (1994) 902.

126. G. Speranza, N. Laidani, L. Calliari and M. Anderle, *Diam. Relat. Mater.* **8** (1999) 517–521.

127. V. Singh, V. Palshin, R.C. Tittsworth and E.I. Meletis, *Carbon* **44** (2006) 1280–1286.

128. M. Zhang and Y. Nakayama, *J. Appl. Phys.* **82** (1997) 4912.

129. D.G. McCulloch, D.R. McKenzie, S. Prawer, E. Merchant, G. Gerstner and R. Kalish, *Diam. Relat. Mater.* **6** (1997) 1622.

130. R.W. Lamberton, S.M. Morley, P.D. Maguire and J.A. McLaughlin, *Thin Solid Films*, **333**(1–2) (1998) 114–125.

131. A. Grill, V. Patel and C. Jahnes, *J. Electrochem. Soc.* **145** (1998) 1649.

132. M.P. Nadler, T.M. Donovan and A.K. Green, Structure of carbon films formed by the plasma decomposition of hydrocarbons, *Appl. Surf. Sci.* **18**(1–2) (1984) 10–17.

133. R.L.C. Wu *et al.*, *Surf. Coat. Technol.* **54/55** (1992) 576.

134. R.L.C. Wu, Synthesis and characterisation of diamond-like carbon films for optical and mechanical applications, *Surf. Coat. Technol.* **51**(1–3) (1992) 258–266.

135. J. Ristein, R.T. Stief, L. Ley and W. Beyer, *J. Appl. Phys.* **84** (1998) 3836.

136. M. Zhang and Y. Nakayama, *J. Appl. Phys.* **82** (1997) 4912.

137. N.M.J. Conway, A.C. Ferrari, A.J. Flewitt, J. Robertson, W.I. Milne, A. Tagliaferro and W. Beyer, *Diam. Relat. Mater.* **9** (2000) 765.

138. S.S. Camargo Jr., R.A. Santos, B.A.L. Neto, R. Carius and F. Finger, *Thin Solid Films* **332** (1998) 130–135.

139. A. Reyes-Mena, J. González-Hernández, R. Asomoza and B.S. Chao, *J. Vac. Sci. Technol.* **A8**(3) (1990) 1509.

140. N.M.J. Conway, A. Ilie, J. Robertson and W.I. Milne, *Appl. Phys. Lett.* **73** (1998) 2456.

141. J. Fink, T. Muller-Heinzerling, J. Pfluger, B. Scheerer, B. Dischler, P. Koidl, A. Bubenzer and R.E. Sah, *Phys. Rev.* **B30** (1984) 4713.

142. S. Reinke and W. Kulisch, *Surf. Coat. Technol.* **97** (1997) 23.

143. J.P. Sullivan, *Electr. Mater.* **26** (1997) 1022.

144. T.A. Friedman *et al.*, *Proc. MRS* Fall Meeting (Warrendale PA) Abstracts, Contribution Cb 15.5 (1996).

145. A.A. Ogwu, R.W. Lamberton, S. Morley, P. Maguire and J. McLaughlin, *Physica* **B269** (1999) 335–344.

146. D.R. Tallant, J.E. Parmeter, M.P. Siegal and R.L. Simpson, The thermal stability of Diamond like carbon, *Diam. Relat. Mater.* **4**(3) (1995) 191–199.

147. J.-T. Jiu, H. Wang, C.-B. Cao and H.-S. Zhu, *J. Mater. Sci.* **34** (1999) 5205–5209.

148. V.L. Gott, D.E. Koepke, R.L. Daggett, W. Zarnstorff and W.P. Young, The coating of intravascular plastic prostheses with colloidal graphite, *Surgery* **50** (1961) 382–389.

149. A. Haubold, *Annals N.Y. Aca. Sc.* **283** (1977) 383.

150. S.L. Goodman, K.S. Tweden and R.M. Albrecht, *J. Biomed. Mater. Res.* **32** (1996) 249–258.

151. R.E. Baier, *Bull N. Y. Aca. Med.* **48** (1972) 273.

152. R.E. Baier, *Bulletin N. Y. Acad. Med.* **48**(2) (1972) 257–272.

153. D.F. Williams, *J. Biomed. Eng.* **11** (1989) 185.

154. E. Salzman (ed.), Interaction of blood with natural and artificial surfaces (Marcel Dekker, 1981).

155. C. Van Oss, *J. Ann. Rev. Microbiol.* **32** (1978) 19–39.

156. S. Kochwa, R.S. Litwak, R.E. Rosenfield and E.F. Leonard, *Ann. N.Y. Acad. Sci.* **283** (1977) 37.

157. A. Szent-Gyorgyi, *Bioenergetics* (Academic Press, NY, 1957).

158. A. Szent-Gyorgyi, *Nature* **157** (1946) 875.

159. D.D. Eley, G.D. Parfitt, M.J. Perry and D.H. Taysum, *Trans. Faraday Soc.* **49** (1953) 79.

160. E. Postow and B. Rosenberg, *Bioenergetics* **1** (1970) 467.

161. S.D. Bruck, *Nature* **243** (1973) 416.

162. S.D. Bruck, *Polymer* **16** (1975) 25.

163. S.D. Bruck, *Nature* **243** (1973) 416.

164. S.D. Bruck, *Polymer* **16** (1975) 25.

165. S.D. Bruck, *Polym. Sci.* **C17** (1967) 169.

166. J.L. Gordon, 1986 in *Blood-Surface Interactions: Biological Principles Underlying Haemocompatibility with Artificial Materials*, eds. J.P. Cazenave, J.A. Davies, M.D. Kazatchkine and W.G. van Aken (Elsevier Science Publishers, 1986), p. 5.

167. E. Cenni, C.R. Arciola, G. Ciapetti, D. Granchi, L. Savarino, S. Stea, D. Cavedagna, T. Curti, G. Falsone and A. Pizzoferrato, *Biomaterials* **16** (1995) 973–976.

168. M.B. Herring, A. Gardner and J.A. Gloves, *Surgery* **84** (1978) 498.

169. M. Remy, L. Bordenave, R. Bareille, F. Rouais, Ch. Baquey, A. Gorodkov, E.S. Sidorenko and S.P. Novikova, *J. Mater. Sci. Mater. Med.* **5** (1994) 808–812.

170. V. Pesakova, Z. Klezl, K. Balik and M. Adam, *J. Mater. Sci. Mater. Med.* **11** (2000) 797.

171. N.J. Hallab, K.J. Bundy, K. O'Connor, R. Clark and R.L. Moses, *J. Long-Term Effects of Medical Implants* **5** (1995) 209.

172. A. Ahluwalia, G. Basta, F. Chiellini, D. Ricci and G. Vozzi, *J. Mater. Sci. Mater. Med.* **12** (2001) 613–619.

173. G.L. Bowlin and S.E. Rittger, *Cell Transplantation* **6** (1997) 623.

174. G. Altankov and T. Grott, *J. Biomater. Sci. Polymer Edn.* **8** (1997) 299.

175. F. Grinnell, *Int. Rev. Cytol.* **53** (1978) 65.

176. P.B. Van Wachem, J.M. Schakenraad, J. Feijen, T. Beugeling, W.G. van Aken, E.H. Blaauw, P. Nieuwenhuis and I. Molenaar, *Biomaterials* **10** (1989) 532–539.

177. P.B. Van Wachem, T. Beugeling, J. Feijen, A. Bantjes, J.P. Detmers and W.G. van Aken, *Biomaterials* **6** (1985) 403–408.

178. M.I. Jones, I.R. McColl, D.M. Grant, K.G. Parker and T.L. Parker, *Diam. Relat. Mater.* **8** (1999) 457–462.

179. T.L. Parker, K.L. Parker, I.R. McColl, D.M. Grant and J.V. Wood, *Diam. Relat. Mater.* **3** (1994) 1120–1123.

180. I. Dion, X. Roques, C. Baquey, E. Baudet, B. Basse Cathalinat and N. More, *Biomed. Mater. Eng.* **3** (1993) 51–55.

181. A. O'Leary, D.P. Bowling, K. Donnelly, T.P. O'Brien, T.C. Kelly, N. Weill and R. Eloy, *Key Eng. Mater.* **99–100** (1995) 301–308.

182. M.J. Allen, B.J. Myer, F.C. Law and N. Rushton, *Trans. Orthop. Res. Soc.* **20** (1995) 489.

183. M. Allen, F.C. Law and N. Rushton, *Clin. Mater.* **17** (1994) 1–10.

184. M. Allen, R. Butter, L. Chandra, A. Lettington and N. Rushton, *Biomed. Mater. Eng.* **5**(3) (1995) 151–159.

185. M. Allen, B. Myer and N. Rushton, *J. Biomed. Mater. Res.* **58**(3) (2001) 319–328.

186. I. De Scheerder, M. Szilard, H. Yanming, X.B. Ping, E. Verbeken, D. Neerinck, E. Demeyere, W. Coppens and F. Van de Werf, *J. Invasive Cardiol.* **12**(8) (2000) 389–394.

187. D.P. Dowling, P.V. Kola, K. Donnelly, T.C Kelly, K. Brumitt, L. Lloyd, R. Eloy, M. Therin and N. Weill, *Diam. Relat. Mater.* **6** (1997) 390–393.

188. Y. Yang, S.F. Franzen and C.L. Olin, *Cells Mater.* **6**(4) (1996) 339–354.

189. A. Alanazi, C. Nojiri, T. Noguchi *et al.*, *ASAIO J.* **46**(4) (2000) 440–443.

190. A. Alanazi, C. Nojiri, T. Noguchi, Y. Ohgoe, T. Matsuda, K. Hirakuri, A. Funakubo, K. Sakai and Y. Fukui, *Artificial Organs* **24**(8) (2000) 624–627.

191. N.A. Morrison, I.C. Drummond, C. Garth, P. John, D.K. Milne, G.P. Smith, M.G. Jubber and J.I.B. Wilson, *Diam. Relat. Mater.* **5**(10) (1996) 1118–1126.

192. R.S. Butter and A.H. Lettington, DLC for biomedical applications (Reviews), *J. Chem. Vapour Depo.* **3** (1995) 182–192.

193. Veli-Matti Tiainen, *Diam. Relat. Mater.* **10** (2001) 153–160.

194. S.P.J. Higson and Pankaj M. Vadgama, *Analytica Chemica Acta* **300** (1995) 77–83.

195. S.P.J. Higson and Pankaj M. Vadgama, *Biosensors Bioelectron* **10**(5) (1995) viii.

196. C. Du, X.W. Su, F.Z. Cui and X.D. Zhu, *Biomaterials* **19** (1998) 651–658.

197. F.Z. Cui and D.J. Li, *Surf. Coat. Technol.* **131** (2000) 481–487.

198. V.I. Ivanov-Omskii, L.K. Panina and S.G. Yastrebov, *Carbon* **38** (2000) 495–499.

199. G.A. Dyuzhev, V.I. Ivanov-Omskii, E.K. Kuznetsova and V.D. Rumyantsev *et al.*, *Mol. Mat.* **8** (1996) 103–106.

200. Z. Bengali and L.D. Shea, *MRS Bulletin* **30**(9) (2005) 659.

201. D.E. Packham (ed.), *Handbook of Adhesion* (Longman Sci. and Technol., 1992).

202. T. Young, *Trans. Roy. Soc. London* **95** (1805) 66.

203. C.J. Van Oss, *Ann. Rev. Microbiol.* **32** (1978) 19–39.

204. J. Tsibouklis, M. Stone, A.A. Thorpe, P. Graham, V. Peters, R. Heerlien, J.R. Smith, K.L. Green and T.G Nevell, *Biomaterials* **20** (1999) 1229–1235.

205. J. McLaughlin, B. Meenan, P. Maguire and N. Jamieson, *Diam. Relat. Mater.* **8** (1996) 486–491.

206. L.J. Yu, X. Wang, X.H. Wang and X.H. Liu, *Surf. Coat. Technol.* **128–129** (2000) 484.

207. R.S. Butter, D.R. Waterman, A.H. Lettington, R.T. Ramos and E.J. Fordham, *Thin Solid Films* **311** (1997) 107–113.

208. Y. Mitsuyu, H. Zhang and S. Ishida, *Trans. ASME* **123** (2001) 188.

209. P.B. Leezenberg, W.H. Johnston and G.W Tyndall, *J. Appl. Phys.* **89** (2001) 3498.

210. M. Grischke, K. Trojan and H. Dimigen, *Proc. Joint 4th Int. Symp. Trends and New Applications in Thin Films 11th Conf. On High Vacuum, Interfaces and Thin Films* (1994), p. 433.

211. R. Memming, *Thin Solid Films* **143** (1986) 279.

212. N.F. Mott and E.A. Davis, *Electronic Processes in Non-Crystalline Materials*, 2nd edn. (Oxford Uni. Press, 1979).

213. A. Ilie, J. Robertson, N. Conway, B. Kleinsorge and W.I Milne, *Diam. Relat. Mater.* **8** (1999) 549–553.

214. B. Meyerson and F.W. Smith, *J. Non-Cryst. Solids* **35–36** (1980) 435–440.
215. J.C. Angus *et al.*, in Plasma Deposited Thin Films, eds. J. Mort and Jansen F (CRC Press, FL, 1986), p. 89.
216. B. Meyerson and F. Smith, *Solid State Commun.* **34** (1980) 531.
217. H. Dimigen, H. Hubsch and R. Memming, *Appl. Phys. Lett.* **50** (1987) 1056.
218. C.P. Klages and R. Memmin, *Mater. Sci. Forum* **52–53** (1989) 609.
219. M. Wang, K. Schmidt, K. Reichelt, X. Jiang, H. Hubsch and H. Dimigen, *J. Mater. Res.* **7** (1992) 1465.
220. M. Vogel, O. Stenzel, W. Grunewald and A. Barna, Electrical and optical properties of amorphous carbon layers: Limits of the isotropic layer model, *Thin Solid Films*, **209**(2) (1992) 195–206.
221. H. Tsai *Mat. Sci. Forum* **52–53** (1989) 71.
222. M.H. Brodsky (ed.) *Amorphous Semiconductors* (Springer-Verlag, 1979).
223. A. Bozhko, A. Ivanov, M. Berrettoni, S. Chudinov, S. Stizza, V. Dorfman and B. Pypkin, *Diam. Relat. Mater.* **4** (1995) 488–491.
224. A. Bozhko, S.M. Chudinov, M. Evangelisti, S. Stizza and V.F. Dorfman, *Mater. Sci. Eng.* **C5** (1998) 265–269.
225. A. Bozhko, M. Shupegin and T. Takagi, *Diam. Relat. Mater.* **11** (2002) 1753–1759.
226. L.W. Swanson and P.R. Davis, *Methods of Experimental* Vol. 22, eds. R.L. Park and M.G. Legally (Acad. Press London, 1995), p. 2.
227. I.D. Baikie, http://www.rgu.ac.uk/subj/skpg/kpintro.htm (1996).
228. S.M. Sze, *Physics of Semiconductor Devices*, 2nd edn. (J. Wiley & Sons, Inc., 1981).
229. E.H. Rhoderick and R.H. Williams, *Metals-Semiconductor contacts*, 2nd edn. (Clarendon Press, Oxford, 1988), p. 10.
230. A. Schroeder, Gilbert Francz, Arend Bruinink, Roland Hauert, Joerg Mayer and Erich Wintermantel, *Biomaterials* **21** (2000) 449–456.
231. D.P. Magill, A.A. Ogwu, J.A. McLaughlin and P.D. Maguire, *J. Vac. Sci. Technol.* **A19**(5) (2001) 2456–2462.
232. H. Luth, *Surfaces and Interfaces in Solid Materials*, 3rd edn. (Springer, Germany, 1995), p. 438.
233. P.M. Gundry and F.C. Tompkins, *Experimental Methods in Catalytic Research*, ed. R.B. Anderson (Acad. Press, NY, 1968), p. 116.
234. M.C. Desjonqueres and D. Spanjaard, *Concepts in Surface Physics*, 2nd edn. (Springer, Germany, 1996), p. 394.
235. M. Pfeiffer, K. Leo and N. Karl, Fermi level determination in organic thin films by the Kelvin probe method, *J. Appl. Phys.* **80**(12) (1996) 6880–6883.
236. W.A. Zisman, *Rev. Sci. Instrum.* **3** (1932) 367.
237. I.D. Baikie, S. Mackenzie, P.J.Z. Estrup and J.A. Meyer, *Rev. Sci. Instr.* **62**(5) (1991) 1326.
238. I.D. Baikie, K.O. van der Werf, H. Oerbekke, J. Broeze and A. van Silfhout, *Rev. Sci. Instr.* **60**(5) (1989) 930.
239. I.D. Baikie and G.H. Bruggink, *Mat. Res. Soc. Symp. Proc.* **309** (1993) 35.
240. S.L. Goodman, S.L. Cooper and R.M. Albrecht, *J. Biomater. Sci. Polymer Edn.* **2**(2) (1991) 147–159.

241. J.L. Brash, *Macromol. Chem. Suppl.* **9** (1985) 69.
242. L.K. Lambrecht, B.R. Young, R.E. Stafford, K. Park, R.M. Albrecht, D.F. Mosher and S.L Cooper, *Thrombosis Res.* **41** (1986) 99.
243. O. Wildner, T. Lipkow and J. Knop, Increased expression of ICAM-1, E-selectin and V-CAM-1 by cultured endothelial cells upon exposure to haptens, *Exp. Dermatol.* **1** (1992) 191–198.
244. S. Moncada and J.R. Vane, *Pathobiology of Endothelial Cells*, eds. H.L. Nossel and H.J. Vogel (New York Academic Press, 1982), pp. 253–285.
245. M.A. Gimbrone, Jr., in *Vascular Endothelium in Hemostasis and Thrombosis*, ed. M.A. Gimbrone Jr. (Churchill Livingstone, Edinburgh, 1986), pp. 1–13.
246. M.A. Gimbrone, Jr., *Ann. N. Y. Acad. Sci.* **516** (1987) 5–11.
247. T.K. Chan and V. Chan, Antithrombin III, the major modulator of intravascular 326 coagulation is synthesised by human endothelial cells, *Thrombosis and Haemostasis* **46** (1981) 504–506.
248. C. Busch, C. Ljungman, C.-M. Heldin, E. Waskson and B. Obrink, Surface properties of cultured endothelial cells. *Haemostasis* **8** (1979) 142–148.
249. E.A. Jaffe, Synthesis of factor VIII by endothelial cells, *Ann. N. Y. Acad. Sci.* **401** (1982) 163–170.
250. D.F. Mosher, M.J. Doyle and E.A. Jaffe, Secretion and synthesis of thrombospondin by cultured human endothelial cells, *J. Cell Biol.* **93** (1982) 343–348.
251. E.W. Ades, F.J. Candal, R.A. Swerlick, V.G. George, S. Summer, D.C. Bosse and T.J. Lawley, *Invest. Dermatol.* **99** (1992) 683–690.

CHAPTER 11

NANOCOATINGS FOR ORTHOPAEDIC AND DENTAL APPLICATION

Weiqi Yan

Bone and Joint Research Institute and
Department of Orthopaedic Surgery
The Second Affiliated Hospital, Medical College, Zhejiang University
Hangzhou, China 310009
wyan@zju.edu.cn

1. Introduction

1.1. *Clinical Background*

Clinical applications of biomaterial implants have led to a remarkable increase in the quality of life for millions of patients each year. A well-known field with biomaterial implants is that of orthopedics and dentistry in which implants made from metals, ceramics, glass or polymers have had a critical role in the reconstruction of total hip and knee joints, spine, teeth systems, and in the repair of large bony defects of fracture and tumors resection. Worldwide, the number of patients requiring and receiving biomedical implants in the treatment of skeletal problems is rapidly increasing. This increasing demand arises from an aging population with higher quality of life expectations. In the United States and in Europe, more than 800,000 hip and knee arthroplasties and over 500,000 dental implants are being performed annually [1–3]. Orthopaedic and dental applications represent approximately 55% of the total biomaterials market, with increasing tendency. To date, tens of millions of people have received medical implants. Most frequently, however, the results were mixed and confounding both in success and in failure. Implant loosening, post-surgical infection, fracture nonunion, and unpredictable periodontal regeneration are still issues

573

of concern. As the average life expectancy increases, there is a growing demand to minimize implant failure rates and to extend implant lifetime [4, 5]. Thus, the development of a new generation of biomaterials to aid in the regeneration and healing of bone and soft tissues is an important goal of clinical research.

1.2. *Biomimetic Nanoscale Biomaterials*

Research on biomaterial surface characteristics is the key for continued success in the application of biomaterials. Orthopedic and dental biomaterials can be broadly classified into ceramics, metals, polymers and composite. The materials have to withstand the effects of a most hostile environment. The success of these implants depends on acquiring stable fixation of the device at the bony site. Thus, current clinical biomaterials research is to design newly improved implant materials and techniques to not only induce controlled and guided growth of bone cells, but also to assist in rapid healing. In addition, these implants should result in formation of a characteristic interfacial layer that has adequate biomechanical properties to promote bone-implant osseointegration.

With the advancement of nanotechnology and the knowledge that human bone is a nanophase living material, it is believed that the use of nanoscale biomaterials to the area of surgical implants will exert a much more positive cellular reaction at the site of application than is currently observed in traditional macroscale implants. To design successful bone implants, bonelike nanophase structure of successfully proven biomaterials will be very promising, because of that cells and tissue in the body are accustomed to interacting with nanostructured surfaces on a daily basis. The mechanical properties of nanophase ceramics are closer to bonelike toughness because of reactions at the grain boundaries which make these materials very attractive for bone implant applications. Biomimetic processes have attracted much attention in recent years due to their significant applications in orthopaedic and dental areas. Recent trends towards the convergence of biomedicine and nanotechnology have accelerated the development of nanocomposite and nanocoatings for medical implants and devices [6, 7]. In orthopaedics and dentistry, the popularity of coated implants appears to be increasing and their applications continue to expand. Surfaces that contain micro- and nanoscale features in a well-controlled manner significantly affect cellular and subcellular function. Although the optimal micro/nanostructure for desired osseointegration is still a subject of debate, these studies show the importance of developing more active interfaces with micro- and nanoscale architecture that can incorporate osteotactic biomolecules to enhance apposition of bone from existing bone surfaces and stimulate new bone formation [6, 8–10].

The advances in nanotechnology point to new ways for developing inexpensive and effective medical devices. Implant surface with nanocoatings hold promise to solve several such problems present in conventional coatings. The novel implants and prostheses can help to provide useful functions and long-term stability for the human body. The ability to engineering biomaterials at the nanoscale and nanocomposite coatings will greatly impact biomedical devices, therapeutics, and strategies for health care. Undoubtedly, biomaterials have had a major impact on the practice of contemporary medicine and patient care in improving the quality of lives of humans. Modern biomaterial practice still takes advantage of developments in the traditional and nanoscale material field but is also aware of the biocompatibility and biofunctionality of medical implants.

2. Properties of Bone Implants

When selecting materials used as bone implants, there are several specific properties that must be considered, such as compressive, tensile, and shear strengths, stiffness, ductility, fatigue endurance and the various expansion coefficients. Most importantly these materials must be biocompatible. In order for a material to be biocompatible it must not corrode when exposed to bodily fluids, be non-toxic to the body, and promote cellular adhesion.

2.1. *Concept of Biocompatibility*

Currently, cellular, molecular, and genetic level research is producing exciting new information that has improved understanding of the biocompatible concept. Applications of this knowledge could lead to development of new generations of biomaterial implants as well as to improved strategies for their evaluation and utility. Therefore, whether a material is biocompatible must account for the interactions between tissues and biomaterials, in which not only can biomaterials affect biological responses, but also that the milieu of the body can affect materials.

Biocompatibility is the basic prerequisite for any material implanted in a body and refers to the ability of a material to perform with an appropriate host response in a specific application. It can be defined as state of mutual existence between the biomaterial and the physiologic atmosphere, in such a way that neither one produce undesirable effects to the other part.

2.2. *Classification of Biomaterial Implants*

The biomaterials can be classified as synthetic materials with biomedically artificial origin (metals, ceramics, polymers) and biologically natural origin (collagen, chitin, elastin). When a synthetic material is placed within the bone, bony tissue reacts towards the implant in a variety of ways depending on the material type and the extent of their biocompatibility. In general, there are three terms in which a biomaterial implant may be described in representing the tissues responses: biotolerant, bioinert, and bioactive materials. *Biotolerant materials* release substances but in non-toxic concentrations that may lead to only benign tissue reactions such as formation of a fibrous connective tissue capsule or weak immune reactions that cause formation of giant cells or phagocyts. These implants are characterized by the fact that a fibrous tissue layer will always develop around the materials [11–14]. These implant materials include stainless steels or bone cement consisting of polymethylmethacrylate (PMMA). *Bioinert materials* exhibit minimal chemical interaction with adjacent tissue but also do not show positive interaction with living tissue [15]. As a response of the body to these materials, usually a fibrous capsule might form around bioinert implants. Through the interface only compressive forces will be transmitted. Examples of these are titanium and its alloys, ceramics such as alumina, zirconia and titania, and some polymers. *Bioactive materials* are those that elicit positive bone responses which could ultimately result in bone growth or regeneration at the *in vivo* site of application (see Table 2.1) [16]. They can bond to bone through a bonelike apatite layer on implant surface. In contrast to bioinert materials, there is chemical bonding along the bone-implant interface. In addition to compressive forces, to some degree tensile and shear forces can also be transmitted through the interface. Prime examples of these materials are synthetic hydroxyapatite (HA), calcium phosphates and bioglasses [16–20]. It is generally considered that bioactivity of these materials is associated with the formation of a bonelike carbonate apatite layer on the implant surface that is chemically and crystallographically equivalent to the mineral phase in nature bone [21–23]. Hence, these bioactive materials can act as a scaffolding for bone cells migration and proliferation when implanted in bone, leading to new bone growth.

Table 2.1.   Application and biological behavior of biomaterial implants [16].

Material	Application	Biological behavior
Stainless (austenitic) steel	Osteosynthesis (bone screws)	biotolerant
Bone cement (PMMA)	Fixation of implants	biotolerant
cp-titanium	Acetabular cups	bioinert
Ti6Al4V alloy	Shafts for hip implants, tibia	bioinert
CoCrMo alloy	Femoral balls and shafts, knee implants	bioinert
Alumina	Femoral balls, inserts of acetabular cups	bioinert
Zirconia (Y-TZP)	Femoral balls	bioinert
HD-polyethylene	Articulation components	bioinert
Carbon (graphite)	Heart valve components	bioinert
CFRP	Inserts of acetabular cups	bioinert
Hydroxyapatite	Bone cavity fillings, coatings, ear implants, vertebrae replacement	bioactive
Tricalcium phosphate	Bone replacement	bioactive
Tetracalcium phosphate	Dental cement	bioactive
Bioglass	Bone redacement	bioactive

2.3. *Osteogenesis Around Bone Implants*

After implantation of a biomaterial in the human bone, not only does the healing bone approach the biomaterial, but bone formation (osteogenesis) can also begin on the implant surface and extend outward to the surrounding bone. Upon implantation, a series of events are initiated around the implant surface. Hence the surface of bone implant plays an essential role in the final bone formation.

It has been accepted that there are three modes for osteogenic response according to bone growth around an implant. They are called distance osteogenesis, contact osteogenesis and bonding osteogenesis. *Distance osteogenesis* is a gradual process that bone healing begins at a distance and secondarily reaches the implant surface. Bone growth does not occur at the implant surface. There will always be an intervening connective tissue between the bone and implant. Biotolerant implants and some bioinert implants with inapposite fit in bone bed often result in distance osteogenesis with a fibrous layer separating the implant from bone. *Contact osteogenesis* occurs when bone cells migrate to the implant surface and begin new bone apposition onto the implant. Bone is formed directly on the implant surface. Press-fit bioinert implants like titanium can give rise to contact osteogenesis with apposition of newly bone to the implant surface. *Bonding osteogenesis* is a mode of active bone growth process that direct mineral apposition and proliferatoin of bone cells occur at the bone–implant interface. The new bone grows primarily on and bonds directly to the implant surface. The bioactive implants, like hydroxyapatite ceramic, are capable of creating a substantial continuity and bonding osteogenesis on their surfaces. From both a mechanical and integrative standpoint, bonding osteogenesis is considered to be a clinically superior mode of bone healing [24–28].

2.4. *Materials for Orthopaedic and Dental Use*

Bone implants have traditionally been made of metals and alloys. Metallic biomaterials consisting of various steel formulations resulted in detrimental tissue reactions and turned out to be a failure, during the early years of the twentieth century. A small number of people are hypersensitive to nickel and chromium. The high load applications required for many orthopedic and dental implants restrict the selection of materials. The ideal materials for orthopaedic and dental application should fulfil the following criteria: (1) excellent biocompatibility, (2) strength (such as sufficient tensile, compressive and torsional strength, stiffness and fatigue resistance) for function recovery, (3) corrosion free, (4) workability (easy for manufacture and use), (5) optimal esthetics, (6) inexpensive, and (7) minimal influence on subsequent radiological imaging.

Clinically, a number of different metals, ceramics, polymers and composite systems have been used for bone prosthesis [29–33]. The most widely used materials in orthopaedic and dental surgery are titanium and its alloys. The advantages of titanium are: (1) high corrosion resistance, (2) greater modulus of elasticity, less stress shielding with titanium implants and lightness (it weighs 40% less than stainless steel), (3) highly biocompatible, forming a direct interface with the bone, (4) the ability to be facbricated into structure surfaces, allowing optimization of the morphology and porosity to fit the living bone, and (5) less interference with MRI or CT or radiotherapy. The disadvantages of titanium are: (1) expense, (2) sensitivity to stress risers, and (3) poor bone bonding ability (easily get encapsulated by fibrous tissue between the bone and implant).

In orthopaedics, such materials are used not only for prosthetic devices in load-bearing condition (Fig. 2.1) but also for internal fixation during fracture healing. Currently, most artificial joints consist of a metallic component such as titanium alloy or Co–Cr alloys articulating against a polymer such as ultrahigh molecular weight polyethylene (UHMWPE). In dentisty,

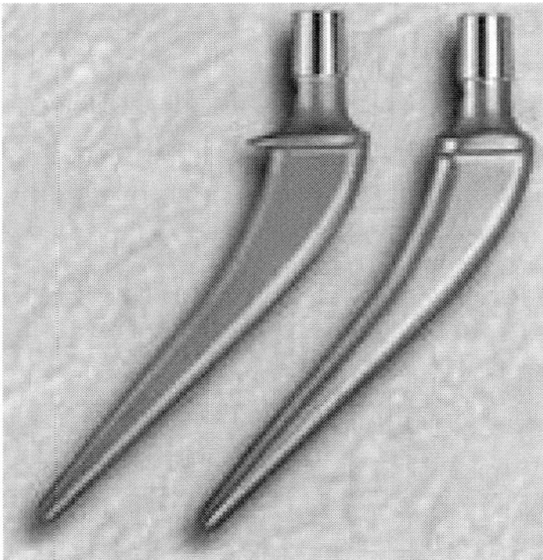

Fig. 2.1.   Titanium alloy stems used for total hip replacement.

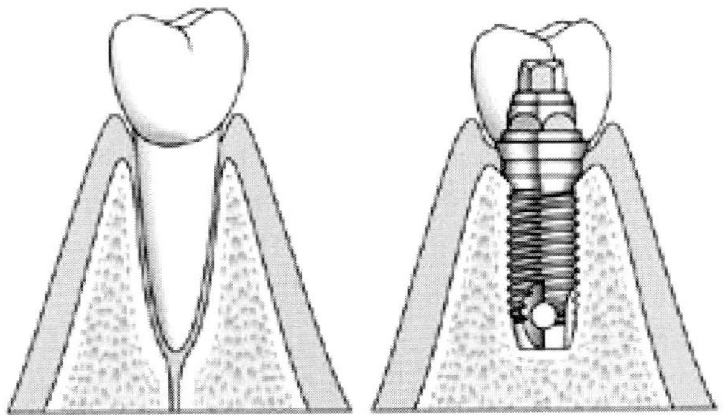

Fig. 2.2.   Nature tooth and dental implant.

commercial titanium for dental implants is the material around which the theory of osseointegration developed [34–36]. The term was coined by Brånemark in 1977 to describe intimate bone to implant contact on a histological level [37]. A loaded implant that has at least 50% average surface area contact with bone and, at its passage through cortical bone (Fig. 2.2), 90% surface area contact, may be considered as osseointegrated. Osseointegration is the process in which clinically asymptomatic rigid fixation of alloplastic materials is achieved and maintained in bone during functional loading [27, 38]. Implants of c.p. Ti showed better osseointegration than implants of Co–Cr, stainless steel and even Ti–6Al–4V [36, 37, 39, 40]. This may be because of the high dielectric constant of TiO_2 which makes up the protective surface layer. A high dielectric constant is supposed to inhibit the movement of cells to the implant surface and so promote a positive biological response [41–43].

Ceramics, glasses and glass ceramics are also widely used in clinic as bone substitutes. Many clinical applications of calcium phosphate biomaterials such as hydroxyapatite (HA) and tricalcium phosphate (TCP) have been made in the repair of bone defects, bone augmentation and coatings for metal implants. The main characteristics of ceramics are shown in Fig. 2.3.

The ceramic materials used in bone reconstructive surgery can be bioinert, bioactive or bioresorbable. The main drawbacks for the reconstructive applications of ceramic materials are their rigidity and fragility. At present the applications of calcium phosphate ceramics are mainly focused on bone defect filling, both in dental and orthopaedic surgery. Hydroxyapatite is also widely used as coatings in clinical application to improve osteoconductive properties of prostheses. It appears to be optimized in implants that mimic not only bone structure, but also bone chemistry [26, 28, 44–47]. Osteoconduction refers a process where the bone grows for apposition, starting from a bone tissue already existent, through structures that serve as structure to the migration of osteoblastic cells. Osteoconductive properties of a graft or implant depend significantly on its architectural structure. Clinically, porous structure such as tricalcium phosphate porous implants result in more ingrowth with new bone formation [48], because bone implants with

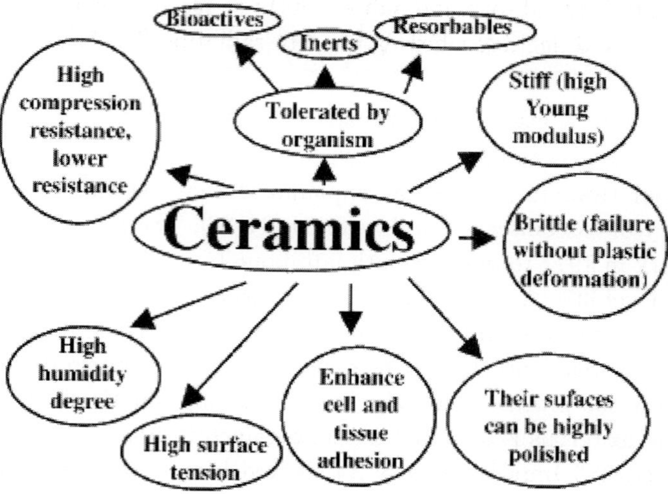

Fig. 2.3.   Main features of ceramic materials.

greater porosity and surface area allow osteoconduction to proceed more quickly and completely.

A common characteristic of osteoconductive implants is a time-dependent, kinetic bioactivity of the surface. The surface forms a biologically bonelike apatite layer that provides the bonding interface with living tissues. More recent applications are the production of calcium phosphate based cements, preparation of biphasic mixtures in an attempt to obtain a mineral component of bone tissue as similar as possible to biological apatites, and manufacturing cellular carrying substrates and biochemical factors for tissue engineering. Calcium phosphate cements which can be resorbed and injected are being commercialized by various international corporations, with slight differences in their compositions and preparation. Calcium phosphate has a variety of origins and has been investigated as potential biocement. However, there are problems related to their mechanical toughness, the curing time, and application technique on the osseous defect [49–53]. Research on improving these new cements are still underway.

Because the practice of medicine and surgery requires sterile products, decisions regarding choice of biomaterials for a specific application should include consideration of sterilization of the final products. Moist heat and high pressure (such as steam autoclaves), ethylene oxide gas, and gamma radiation are procedures commonly used in sterilizing biomedical materials and devices [54, 55]. Special care should be taken with polymers that do not tolerate heat, absorb and subsequently release ethylene oxide (a toxic substance), and degrade when exposed to radiation [56, 57].

3. Bone Structure and Formation

There are two types of bone: mature or *lamellar bone*, and immature or *woven bone*. Both types consist of cells, and a partly organic, partly mineral extracellular matrix of collagen fibres embedded in a ground substance providing strength and stiffness. Periosteum is the dense layer of fibroconnective tissue covering bone. It is thicker in children. Bone provides skeletal support and protection for the soft tissues, mechanical advantage for muscles and a center for mineral metabolism. The majority of bone is formed from mesenchymal tissues by mineralization of a cartilage model, endochondral ossification, but some is formed by direct mineralization of mesenchymal layers without a cartilage model, intramembranous ossification. Unmineralized bone is known as osteoid.

3.1. *Bone and Cells*

Lamellar bone is an organized, regular, stress-orientated structure which may be cortical or cancellous. Woven bone has a high cell turnover and content, a less organized structure and higher water content, making it weaker and more deformable than lamellar bone. Woven bone is formed first in the embryonic skeleton, and following a fracture, before being replaced by organized lamellar bone. Bone remodels and changes shape according to load and stress — Wolff's law.

1. Cortical or compact bone has tightly packed haversian systems, or osteons, with a central haversian canal for nerves and vessels. The slow turnover of cortical bone occurs at the cement line, where osteoblasts and osteoclasts are active, which is the margin of the osteon. (Fig. 3.1.)

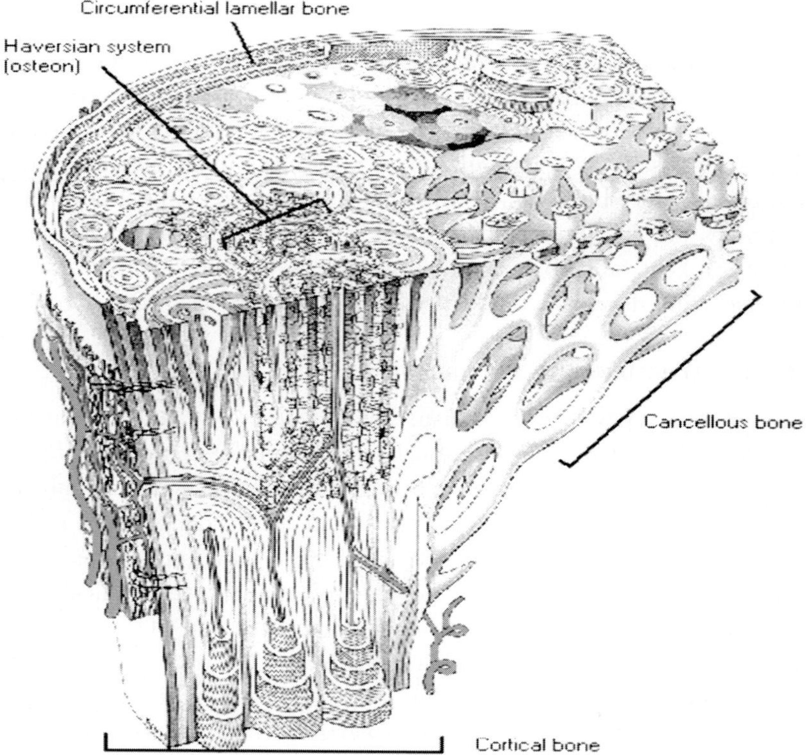

Fig. 3.1.   Microstructure of bone.

2. Cancellous or trabecular bone undergoes faster turnover and is less dense. By weight, only 20% of the skeleton is cancellous bone, which is more elastic than cortical bone with a smaller Young's modulus. (Fig. 3.1.)

There are three cell types in bone:

1. Osteoblasts, derived from undifferentiated mesenchymal cells or stromal cells, synthesize and secrete bone matrix. Osteoblasts have similar characteristics to fibroblasts.
2. Osteocytes, the most common cell line, control the microenvironment, and the concentration of calcium and phosphate in bone matrix. Osteocytes are derived from osteoblasts trapped in the matrix, communicating with each other by long cytoplasmic processes, called canaliculi. Osteocytes are stimulated by calcitonin and inhibited by parathyroid hormone (PTH).
3. Osteoclasts, derived from circulating monocyte progenitors and therefore stem cells, are multinucleate giant cells responsible for resorbing bone under stimulation by PTH, vitamin D, prostaglandins, thyroid hormone and glucocorticoids. They are inhibited by calcitonin. Resorption occurs at the cell surface by the enzymatic activity of tartrate-resistant acid phosphatase and carbonic anhydrase. Osteoprogenitor cells line the Haversian canals and periosteum, before being stimulated to differentiate into osteoblasts and other cells.

In bone, organic and mineral components form non-cellular bone, called the matrix. The organic component consists of type I collagen for tensile strength, proteoglycans for compressive strength, glycoproteins and phospholipids. The collagen is a triple helix of tropocollagen. Tensile strength is increased by cross-linking. Forty percent of the matrix is made up by the organic component, which itself is 90% type I collagen. The inorganic component is mineral, mostly poorly crystalline calcium hydroxyapatite, which mineralizes bone and provides compressive strength. A small amount of the inorganic component is osteocalcium phosphate.

At the nanoscale lever, bone is composed of nanostructured collagen and hydroxyapatite. Thus bone is a nanophase living material and cells in the body are accustomed to interacting with nanostructured surfaces on a daily basis. Figure 3.2 shows the relationship between microstructure and nanostructure of bones [58, 59].

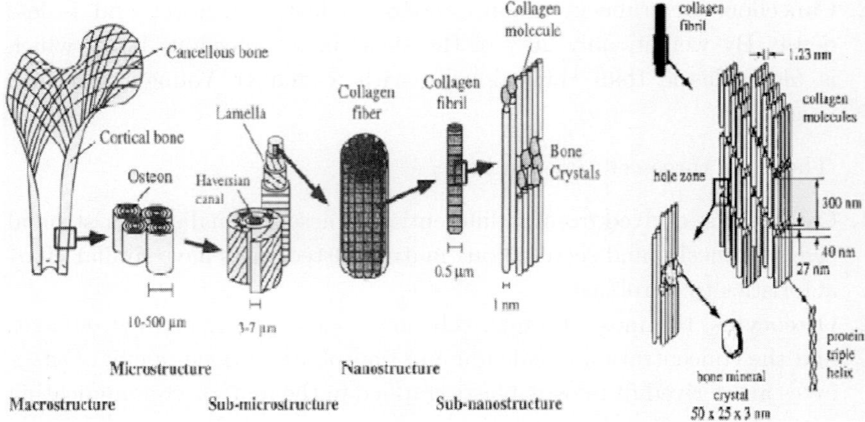

Fig. 3.2.   Schematic representation of macrostructure to nanostructure in human bone.

3.2. Bone Formation

The formation of bone is mostly by mineralization of a cartilage model,
endochondral ossification and sometimes directly by intramembranous
ossification.

3.2.1. Endochondral Ossification

Endochondral ossification is the principal system of bone formation and
growth. Initially a cartilaginous model, derived from mesenchymal tissue
condensation, is formed. Vascular invasion, with osteoprogenitor cells dif-
ferentiating into osteoblasts, leads to the formation of a primary center of
ossification. The epiphysis is a secondary center at the bone ends, leaving
an area between for longitudinal growth the physis. The physis consists of
three zones for chondrocyte growth and transformation:

1. The reserve zone is where resting cells store products required for subsequent maturation. This zone is on the epiphyseal side of the physis.
   Some disease affects this zone.
2. The proliferative zone is an area of cell proliferation and matrix
   formation, with longitudinal growth as the chondrocytes become organized or stacked in an area of high oxygen tension and high proteoglycan
   concentration. Achondroplasia is a defect in this zone.

3. The hypertrophic zone has three areas of maturation, degeneration and provisional calcification. In this zone, cells first accumulate calcium as they increase in size and then die, releasing calcium for mineralization. Osteoblasts are active in this zone, with calcification of the cartilage, promoted by low oxygen tension and low proteoglycan. Mucopolysaccharide diseases affect the zone, with chondrocyte degeneration, and in rickets there is reduced provisional calcification. Physeal fractures occur through the zone of provisional calcification.

3.2.2. *Intramembranous Ossification*

Intramembranous ossification is a process of bone formation without a cartilage precursor during embryonic life, and in some areas during fracture healing. Mesenchymal cells aggregate into layers or membranes, followed by osteoblast invasion and bone formation.

The micro-environment of bone formation is under hormonal control, and knowledge of this control is expanding rapidly. This hormonal control is especially relevant to fracture healing. The principal agent is bone morphological protein (BMP), a member of the TGF-b superfamily. There are several types of BMP responsible for formation of bone, cartilage and connective tissue, the hallmark being the induction of endochondral bone formation, by the differentiation of cells into cartilage and bone phenotypes. This also involves other factors including TGF-b, IGF and FGF. In the clinical setting, additional BMP may help stimulate new bone formation to bridge defects, which integrates well with host bone [60–65].

3.3. *Bone Properties*

Bone is formed by a series of complex events involving mineralization of ECM proteins rigidly orchestrated by specific cells with specific functions of maintaining the integrity of the bone. Bone is an intimate composite of an organic phase (collagen and noncollagenous proteins) and an inorganic or mineral phase. Bone consists of the mineral calcium hydroxyapatite $[Ca_{10}(PO4)_6(OH)_2]$, collagen and glycosaminoglycans which bind collagen. Hydroxyapatite is strong in compression, whereas collagen gives bone flexibility and tensile strength. The weakest area is the cement line between osteons. The inorganic to organic ratio is proximately 75 to 25 by weight and 65 to 35 by volume [66].

When a bone is stressed beyond its ultimate strength bone fractures occur. The compressive and tensile strength of bone is proportional to the density. The porosity of cortical bone is 3–30%. Cancellous bone can have a porosity of 90%. Cancellous bone has a lower ultimate strength, is 10% as stiff, but 500% as ductile as cortical bone. Bone is more ductile in the young, leading to the characteristic greenstick and plastic deformation types of failure. With increasing age, as well as osteoporosis, bone becomes more brittle. The modulus of elasticity reduces by 1.5% per annum. In the elderly, bone may be osteoporotic — cortical bone is resorbed endosteally, widening the intramedullary canal. The bony trabeculae of cancellous bone become thinner. The resultant reduction in bone density weakens the bone and the ultimate strength reduces by 5–7% per decade [67, 68].

Bone geometry are: (1) the longer a bone, the greater the potential bending moment, (2) the larger the cross-sectional area, the greater the stress to failure, (3) the area moment of inertia measures the amount of bone in a cross-section and the distance from the neutral axis. The more bone further from the axis, the stronger and stiffer the bone. Wolff's law (1892) states that bone is laid down in response to stress and resorbed where stress is absent. Consequently there can be localized areas of reduced density (after plate removal due to stress shielding) or systemic reduction after prolonged recumbency.

Bone has a considerable margin of safety between the forces experienced in daily activity and those required to cause a fracture. The greater the area under the stress/strain curve, the more energy will be released at the time of fracture; thus osteoporotic fractures are not associated with as much soft tissue damage as fractures of denser bone. Bone exhibits viscoelasticity. The ultimate strength increases when loaded at a faster rate. Thus with a fast rate of loading, bone stores more energy before failure and this excess energy is released at failure. Bone does not have the same strength if loaded in different directions, a property known as anisotropicity. Bone is less strong and less stiff when stressed from side to side, than from end to end. In general, bone is strongest and stiffest in the direction in which it is loaded *in vivo* (longitudinally for a long bone). Even when tested in the same direction, bone is strongest in compression, weakest in tension, with shear being intermediate. In common with most materials, bone fails at lower stresses when repeatedly stressed. A fracture typically occurs after a few repetitions of a high load, or many repetitions of a lower load. The ability of bone to repair microfractures protects it from this phenomenon. Quantitative differences of the mechanical properties between natural bone and bioinert ceramics are shown in Table 3.1.

Table 3.1. Comparison of mechanical properties of alumina, zirconia (Y-TZP) and bone [66].

Properties	Aluminia	Y-TZP	Bone[1]
Density [g/cm^3]	3.98	6.08	1.7–2.0
E-modulus [GPa]	380–420	210	3–30
Compressive strength [MPa]	4000–5000	2000	130–160
Tensile strength [MPa]	350	650	60–160
Flexural strength [MPa]	400–560	900	100
Fracture toughness [MNm$^{-3/2}$]	4–6	>9	2–12

[1]The lower values refer to trabecular (spongy) bone, the higher values to cortical (dense) bone.

3.4. *Bone Remodeling*

Bone is a dynamic tissue, in constant resorption and formation, permitting the maintenance of bone tissue, the repair of damaged tissue and the homeostasis of the phosphocalcic metabolism. This process is called bone remodeling. There are two phases in this process. *Resorption:* cells called osteoclasts dissolve some tissue on the bone's surface, creating a small cavity. This process usually takes place over a few days. *Formation:* cells called osteoblasts fill the cavities with new bone. This process takes place over a few months.

Approximately 30% of bone mass is remodeled in a year. This is necessary for normal skeletal maintenance. Bone matrix is produced and mineralized by osteoblasts, as described in the previous chapter. Bone resorption is done by bone-resorbing cells, osteoclasts. Osteoclasts originate from the hematopoetic-macrophage lineage. Their mononuclear precursors use vascular routes to enter skeletal sites, where they fuse to become active, multinucleated osteoclasts. By a mechanism still unknown, osteoclasts are guided to appropriate sites to be resorbed. There are theories postulating that osteocytes, osteoblasts and bone-lining cells regulate the ionic flow between the syncytium and the extracellular fluid of bone [69–71]. This communication network seems to influence two bone cell activities: strain-related adaptive remodeling and mineral exchange.

Osteoclasts go through several steps during bone resorption, including attachment to the bone surface, polarization of the cell surface into three distinctive membrane compartments, formation of a sealing zone, resorption, final detachment and eventual cell death. Resorption of bone leads to a release of the growth factors buried in the bone matrix, such as TGF-β, BMPs and other factors that activate and recruit osteoblasts to form new

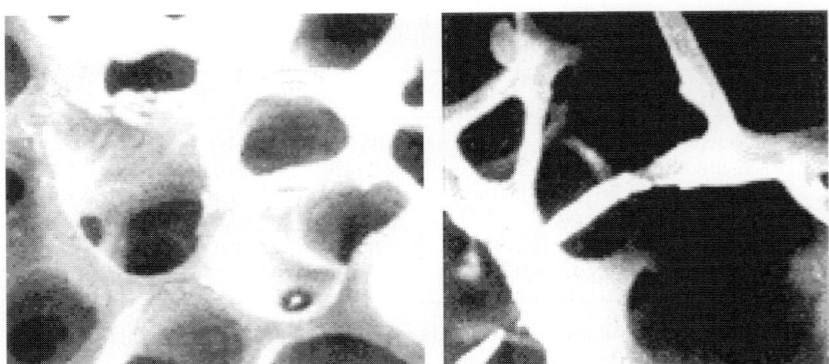

Fig. 3.3.   Contact microradiography of normal bone (left) and porous bone (right).

bone at the resorption site. Osteoblasts, in turn, produce growth factors, such as BMPs and PDGF, which are embedded in the newly synthesized bone matrix, and cytokines, which modulate osteoclast activity, including IL-11, IL-6, IL-8 and TNF-β [59, 67, 69, 72].

Bones are living, growing tissue. During our lifetime, bone is constantly being renewed. The old bone is removed and the new bone is laid down. Through this remodeling process, about 5% of cortical bone and 20% of trabecular bone is renewed. Although cortical bone makes up 75% of the total volume, the metabolic rate is 10 times higher in trabecular bone since the surface area to volume ratio is much greater (trabecular bone surface representing 60% of the total). Bone remodeling occurs throughout life, but only up to the third decade is the balance positive. It is precisely in the third decade when the bone mass is at its maximum, this is maintained with small variations until the age of 50. From then on resorption predominates and the bone mass begins to decrease (see Fig. 3.3).

4. Bone Healing Around Implants

The interactions in the bone implant interface are initiated from the time of implant insertion. The complex physiologic processes, comparable to those of fracture healing, are regulated by different factors and participation of several cell types (see Fig. 4.1). The reactions at the interface between the implant and bone tissue are of outmost importance for the final bone healing, prognosis and implant success. Responses to the implant may occur both locally and systemically. The physiological environment is aggressive and the body is well equipped to defend itself against foreign objects. The

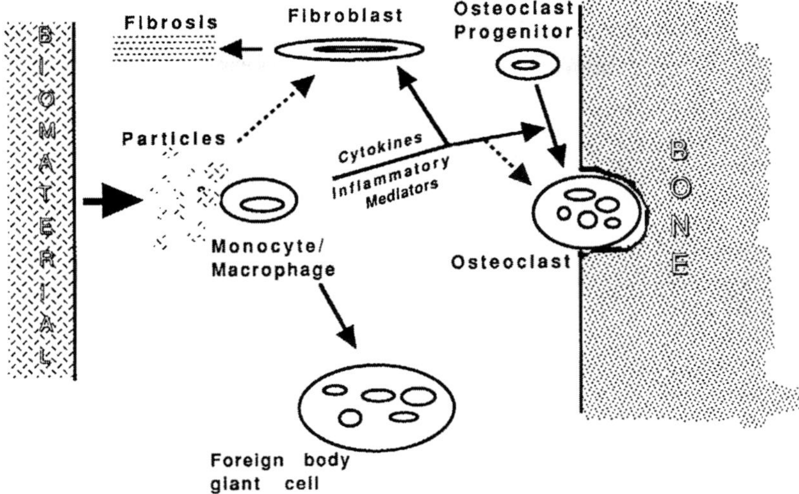

Fig. 4.1. Biological response to biomaterial surface after implantation.

local host response to an implant can be considered a modification to the
normal wound healing process, which consists of two overlapping phases.
The first phase is inflammation, in which cells are attracted to the site of
the injury, damaged tissue and dead cells are removed and foreign bodies
such as microbes are attacked by the immune system. Repair, the second
phase, begins with the re-vascularization of the wound area, then continuity
is re-established by the laying down of new connective tissue. An implant
within the wound site will prolong inflammation and, in the worst case, will
act as a persistent source of irritation. With a perfectly inert material the
inflammation would resolve and the implant would become encapsulated
by a thin fibrous layer (see Fig. 4.2).

The bone repair to the injury of implant insertion is similar to its
response to fracture healing. Bone healing is a specialized form of tissue
healing and similar to other histopathological repair processes. An understanding of normal bone formation and structure is essential. Bone consists
of mesenchymal cells embedded in a mineralized extracellular matrix for
strength and stiffness, with type I collagen predominating. The periosteum,
with an inner cambium or vascular layer and outer fibrous layer, plays a vital
role in bone healing. The process of fracture healing is similar to the process
occurring at an active growth plate. Woven (immature) bone formed within

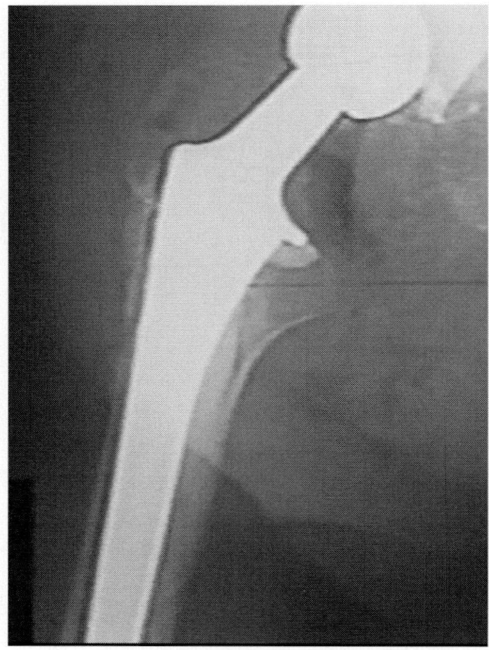

Fig. 4.2.   Implant encapsulated by fibrous tissue becomes loosened after THR surgery.

callus after a fracture, with rapid turnover and irregular mineralization giving it a blurred radiographic appearance, is replaced by lamellar (mature) bone. Bone healing consists of three distinct phases: inflammation, repair and remodeling. Trauma from the implantation procedure results in hemorrhage. There will typically be a gap between the implant surface and the surgically prepared bone site that fills with blood, fibrin clot and callus. The formation of callus, in response to fracture movement, is a vital part of healing called indirect healing and is followed by a process of remodeling. Rigid internal fixation interferes with this natural process, leading to direct or primary healing without abundant callus formation. Through the complex and dynamic process of bone healing and bone remodeling, the functional stability of bone implant is established and maintained. In order to acquire the ideal conditions for healing, the installed implant must obtain primary stability.

After implantation of a material, the bone response event starts with organization of the haematoma into granulation tissue and a change of environment from slightly acidic, to neutral and subsequently slightly alkaline, promoting alkaline phosphatase activity. Periosteal vessels are stimulated by angiogenic fibroblast growth factors to produce capillary buds. Osteoclasts, derived from monocytes and monocytic progenitors, resorb non-viable bone, possibly under stimulus from prostaglandins. Proliferation of the mesenchymal stem cells (MSCs) leads to formation of a matrix of fibrous tissue and cartilage [73]. After mineralization this forms woven bone, described as fracture callus. Hard callus is formed at the periphery by intramembranous bone formation, and soft callus is formed in the central region, with a higher cartilage content, by endochondral ossification. Eventually a mass of callus bridges the fracture margins and clinical union has occurred. Initially the matrix consists of types I, III and V collagen, glucosaminoglycans, proteoglycans and other non-collagenous proteins acting as enzymes, growth factors and mediators. Some of this collagen is converted to types II and IX collagen and then type I collagen predominates during osteogenesis, mineralization and remodeling [74–77].

There are several physical signs of bone healing or growth and a few of the more important identifiers follow. The first cellular indications of bone healing is the presence of fibroblasts which are responsible for new collagen fiber deposition during bone healing; however, these particles are not desirable in the case of implantation because they form a fibrous capsule around the implant reducing the ability of the bone–implant fixation. The next two cellular responses which are desirable in all bone healing and fixation cases are the presence of osteoblasts and osteoclasts; osteoblasts, or bone forming cells, are found aligned to the surface of both original and developing bone while osteoclasts, or bone resorbing cells are important for the use of bioresorbable materials but in excess may result in osteoporosis [74, 77]. Finally, hydroxyapatite (HA) fibers (a calcium phosphate based ceramic naturally found in bone), of dimensions 2–5 nm wide and 50 nm long, have been found to align in natural bone structure. In general, the type of protein absorption to the bone surface determines whether or not bone cells will adhere and if the healing site is healthy; also, the calcium deposition is an index of the mineralization of the bone matrix, another indication of positive bone response. This starts during the repair phase and sees the conversion of woven bone to lamellar bone, with restoration of the normal matrix. The process may take several months or years, and is demonstrated radiographically by the cortical and trabecular patterns returning to normal. The bone returns to its normal shape by responding to loading and stress (Wolff's Law).

5. Implant Surface Modifications and Coatings

Implant biomaterials interact with the body through their surfaces. Consequently, the properties of the surfaces of a material are critically important in determining biocompatibility, biological responses, bone–implants interface reaction and, ultimately, implant integration and performance success. Changes in surface properties can then lead to altered biological responses. In order to obtain desirable tissue-implant interactions and improve osteointegration, many attempts aimed at modifying implant surface composition, surface energy, surface topography and morphology have been made during last decades. Both *in vivo* and *in vitro* studies have shown that modification of material through changes in surface properties while maintaining the bulk properties of the implant can optimize functionality of implant materials. The surface modification approaches according to the surface properties being altered can be classified into physicochemical, morphologic, or biological modifications.

The physicochemical characteristics, including surface energy, surface charge, and surface composition, have been altered in different treatments with the aim of changing both material and biological responses. The morphologic modifications of surface topography by roughness, groove and porous coating have been shown to encourage tissue ingrowth, which would increase fixation of implants in the bone owing to biomechanical interlocking, known as biological fixation (Fig. 5.1). The biological surface modification is used to control cell and tissue responses to an implant by immobilizing biomolecules, such as fibronectin, RGD-containing peptides growth factors, and heparin/heparan sulfate-binding peptides, on biomaterials in an attempt to give implants the ability to induce cell growth, activity, and/or differentiation [5, 78–82].

5.1. *Bioactive Material Coatings*

The most important improvement in the surface osteoconductive and bioactive properties resulted from surface bioactive coatings. Calcium phosphate ceramic coatings on orthopaedic and dental metal (Ti or Ti alloy) implants have been extensively investigated because of their chemical similarity to bone mineral. The coated implants have been shown to be capable of conducting bone formation and forming a chemical bond to bone, which are drastically different from those of the uncoated materials both *in vitro* and *in vivo* [25, 28, 83–85].

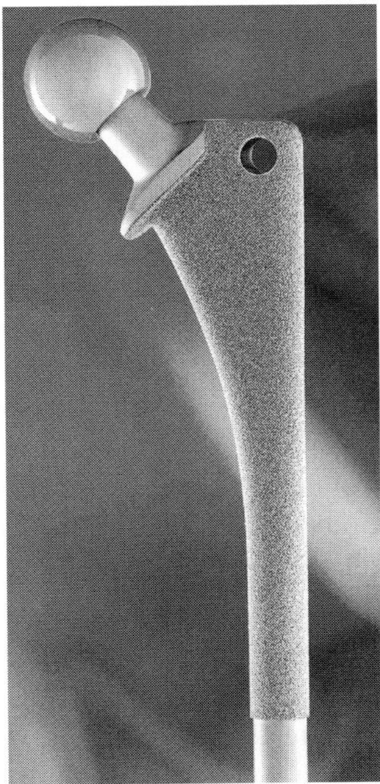

Fig. 5.1.   Porous coating on total hip femoral stem.

In orthopaedic and dental implant application, surface coatings have been used to impart bioactivity to the metal substrates to enhance and accelerate skeletal fixation. The technique whereby such coatings are produced has recently undergone a revolutionary change, which has had profound consequences for their potential to serve as carrier systems for drugs and osteoinductive agents. Osteoinductive factors are able to stimulate bone formation extraskeletally by recruitment of mesenchymal cells without the presence of osteogenic or osteoconductive substances. With the appropriate geometry, bioactive coatings have been recently investigated in combination with osteoconductive factors such as bone morphogenetic proteins (BMPs), transforming growth factor beta (TGF-β) as well as mesenchymal stem cells (MSCs) that allow regeneration of hard tissues using tissue engineering [86–88].

5.2. *Hydroxyapatite-Coated Implants*

Many approaches have been applied for depositing the coatings with different techniques. Plasma spray technique is the most extensively examined and widely used technique for coating of bone implant. HA coatings by plasma spraying have been demonstrated to enhance the bone–implant bonding strength and bony ingrowth compared to uncoated implants in various situations. There are mainly two interfaces that contribute to implant fixation: the *implant–coating interface*, and the *coating–bone interface*. The properties of the coating and the bonding strength to the metal substrate are important for biological performance. The bonding strength of the implant–coating interface is reduced *in vivo*, most likely due to disintegration of the coating because of resorption of part of the coating. Rapidly resorbable coatings such as TCP have shown to exhibit poor fixation compared with HA-coated implants although the implants had equal bone ingrowth [44, 89]. The coating–bone interface is supposed to be bonded chemically. Bone itself might be weaker than the coating–bone interface. Analyzes of implants subjected to tensile test and push-out test have shown bone fragments cohesive on the implant surface [46, 90–92]. Failure observed at the bone–Ti interface in the case of uncoated Ti implant, and at the coating–Ti interface in the case of hydroxyapatite-coated implant, is attributable to the stronger bond between the bone and the bioactive coating compared with that between the Ti and the coating.

Hydroxyapatite (HA)-coated orthopedic and dental implants have been used clinically for nearly 20 years. Many types of HA-coated implants are available commercially, and a number of clinical studies evaluating the effect of HA coating have been reported [57, 93, 94]. Although many investigations showed high success rates with HA coating, the long-term stability of the fixation is still in controversy. Concerns regarding the coatings on microbiological susceptibility, resorption, fatigue, and delamination in long-term application have been pointed out [45, 50, 95–98]. It has some intrinsic drawbacks related to the extremely high processing temperature, making coating of intricate shapes and incorporation of growth factors impossible. This process cannot provide uniform coatings on porous metal surfaces and irregular surface of the implants [89, 91, 98–101]. Another main concern in this clinical application is the possibility that HA coatings increase polyethylene wear or third-body wear and osteolysis [45, 46, 51, 96].

There exist advantages and limitations of these conventional coating techniques, including plasma spraying, high-velocity oxygen-fuel spraying, electrophoretic deposition, sol–gel deposition, hot isostatic pressing, frit enamelling, ion-assisted deposition, pulsed laser deposition, electrochemical deposition, sputter coating and chemical vapor deposited diamond-like carbon (DLC) coating, etc. It has been considered that the bioactive ceramic coatings generated by these existing methods, being composed of large, partially molten hydroxyapatite particles, were prone to delamination and degradation in a biological milieu.

5.3. *Biomimetic Coatings on Titanium-Based Implants*

The development of new coating technology is one of the major goals in improving function and longer lifespan of total joint replacements and endosseous dental roots. All the bioactive materials can form a bonelike apatite layer on their surfaces in the living body, and bond to bone through the surface apatite. Nowadays, the formation of nano-apatite can be induced on implant surface by functional groups, such as Si–OH, Ti–OH, Zr–OH, Nb–OH, Ta–OH, –COOH, and PO_4H_2. These groups, having specific structures with negatively charge, induce apatite formation via formations of an amorphous calcium compound as shown Fig 5.2. Recent research has shown that it is possible to deposit apatite layers on bone implants under physiological conditions of temperature and pH by the so-called biomimetic process, during which bioactive agents can be co-precipitated [73, 102–105].

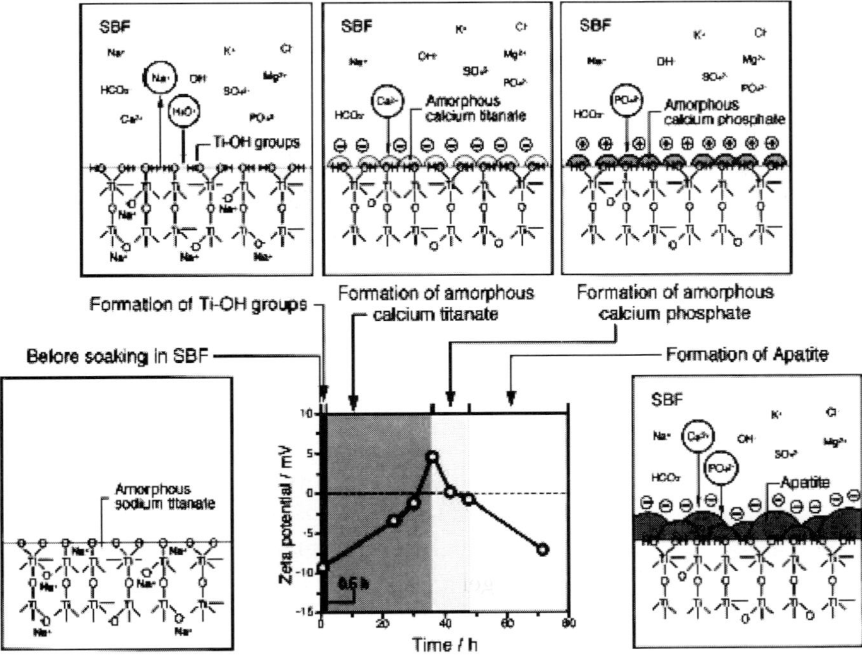

Fig. 5.2.   The process of nano-apatite formation on titanium surface in a simulated body fluid [112].

Since these molecules are integrated into the inorganic latticework, they are released gradually *in vivo* as the layer undergoes degradation. This feature enhances the capacity of these coatings to act as a carrier system for osteogenic agents. By this means, titanium-alloy implants can be rendered osteoinductive as well as osteoconductive by incorporating BMP into the crystal latticework of coatings [75, 106–108]. Using animal ectopic ossification models, titanium-alloy implants bearing a coprecipitated layer of BMP-2 and calcium phosphate has been shown to induce bone formation in the peri-implant region, resulting in the sustainment of osteogenic activity that is essential for the osseointegration of implants [109].

The biomimetic process is one of the most promising techniques for producing a bioactive coating on metal implants. This allows deposition of the desired coating of homogeneous bone-like composition (Fig. 5.2). Nano-apatite coatings have been obtained using this method, by which nano-apatite coating can grow directly on bone implants from simulated physiological solutions at ambient conditions (Fig. 5.3) [110, 111]. The biomimetic approach is considered to have more advantage in coating

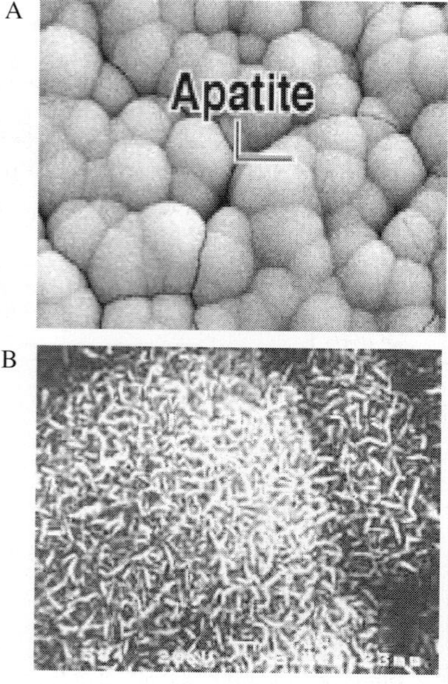

Fig. 5.3.   SEM photographs of nano-apatite on the surface of titanium metal after alkali- and heat-treatment subsequent soaking in an SBF for 3 days. (A) magnification ×5,000, (B) ×10,000.

temperature-sensitive polymeric scaffolds and implants with complex geometry than the traditional line-of-sight and high temperature plasma spray coating technique. The nanocrystalline, nanoporous microstructure shows adsorption of serum proteins, making the apatite coating a potential carrier for therapeutic agents. The nano-apatite coatings have also been found to be capable of promoting bone growth and affecting the behavior of human osteoblasts (Fig. 5.4).

The biomimetic approach involves the nucleation and growth of bone-like crystals upon a pretreated substrate by immersing this in a super-saturated solution of calcium phosphate under physiological conditions of temperature (37°C) and pH (7.4). The method is simple to perform, is cost-effective and may be applied even to heat-sensitive, nonconductive and porous materials of large dimensions and with complex surface geometries. This biomimetic process also consists of a chemical treatment in an alkaline solution, followed by a heat treatment and ending with an immersion in a simulated body fluid. The apatite formation mechanism on pure Ti and its alloys is thought that once the apatite nuclei are formed, crystals spontaneously start to grow by consuming calcium and phosphorous ions from the surrounding solution. The crystal growth is controlled by the ion transfer through the interface between the substrate and the fluid as shown Fig. 5.2 [112].

The biological effects of nano-apatite coating could trigger the cells in the bone tissue to an intrinsic production of growth factors, thus leading to a rapid new bone formation. This effect will then be achieved without any external addition of growth factors. In the case of the implants used

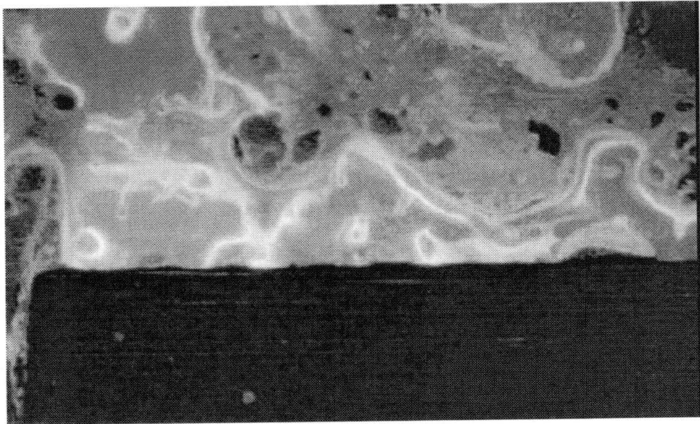

Fig. 5.4.   Fluorescence labels of new bone formation on the nano-apatite coated implant during remodeling of the deposited bone matrix.

for orthopedics or dental applications, such as hip and maxillae facial prostheses, this process allows the deposition of apatite layer on the surface of the implant. Such methods are used to improve the bioactivity of materials such as titanium alloys, tantalum, alumina, and biodegradable polymer composites [112–115]. This method has the following advantages in comparison with the traditional methods: (1) it is a low temperature process that can be applied to any temperature sensitive substrate, (2) it forms apatite crystals, similar to those of bone, showing good bioactivity and good reabsorption characteristics, (3) it can be deposited even in porous substrates or implants of complex geometries, and (4) it can incorporate bone growth factors.

Biomimetic bone implants may be the next development in the orthopeadic and dental fields. At present, no biomimetic implant system is available commercially, primarily because of the need to ensure the absence of undesired host–tissue reactions. Combining the concepts of biomimetic nanoapatite and bone implants may result in a nanophase osteogenesis on orthopaedic and dental implants. Patients with challenging situations, such as poor bone quality and quantity, will benefit from an improved, predictable treatment modality, shorter initial healing times and better long-term performance of the orthopeadic and dental implant. Understanding implant geometry, chemistry and bioactivity and the interactions between these factors is the key to future improvements in implant design and to ensuring progress in this exciting and rewarding field of orthopeadics and dentistry.

5.4. *Hybrid Coatings with Nanomaterials*

Nanophase biomaterials investigated so far include ceramics, metals, polymers, composites, carbon nanofibers and nanoparticles. Each type of material has distinct properties that can be advantageous for specific bone regeneration applications. In searching for new generation of bone substitutes that are bioactive and biodegradable, nanophase composites made of nano calcium phosphate (usually nano-HA) minerals and organic agents have received much attention recently because they mimic the basic compositions of bone. The nanoscale modification of bone implant surfaces by incorporating functional biological agents into the existing nanostructure materials could enhance the regeneration and healing of bone and soft tissues. The unique nanomaterials with nanoporous microstructure allows strong selective adsorption of molecular architectures, such as serum proteins and collagen, making hybrid coating as a potential carrier for therapeutic agents [67, 116–120].

However, the drawback of the synthetic nanocomposite is the low mechanical strength that limits the composite using for load-bearing applications. An alternative way is to apply a thin film of nano-HA/collagen hybrid coating to the surface of metallic implants. Many artificial hip parts, especially the acetabular components, are fixed to bone directly, and a bioactive surface could accelerate bone growth and implant fixation. Biomimetic approach may also grow a bone mineral-like nanoapatite coating on joint implants from simulated physiological solutions. Bioactive coating on metallic implants also established a model of studying the mineralization process of collagen-based biological tissues. The biomimetic approach showed to be advantageous in coating temperature-sensitive polymeric tissue engineering scaffolds and implants with porous surface or complex geometry than the traditional line-of-sight high-temperature plasma spray coating technique [67, 111, 115]. The hybrid nanoapatite coating by biomimetic approach also proved capable of promoting bone ingrowth [119, 121–123]. To apply a bonelike nanostructure coating on metallic implants, a novel coating technique needs to be developed since many common coating methods such as plasma spray and sol-gel are not applicable. Implantable nanophase devices and nanoscaffolds for use to further minimize implant failure rates and to extend device lifetimes are goals of nanophase biomaterial researchers. In the foreseeable future, the most important clinical application of nanomaterials will probably be in nanophase modification or hybrid coatings of implant surface. These applications take advantage of the unique properties of nanomaterials as carrier for drugs or therapeutic agents or for new strategies to controlled release, cell targeting, and regeneration stimulation.

6. Conclusion and Future Work

Clinical application of implanted devices is a well-acknowledged practice in the field of orthopaedic and dental surgery. Scientific research and clinical experience suggest that the successful exploitation of these devices mainly depends on initial implant fixation with strong bone bonding (the bonding of the implant to surrounding bone), considered as both anatomical congruency and load-bearing capacity. The osseointegration process is influenced by a wide range of factors: anatomical location, implant size and design, surgical procedure, loading effects, biological fluids, age and sex, and, in particular, surface characteristics. For this reason, numerous attempts have been aimed at modifying implant surface composition and morphology to optimise implant-to-bone contact and improve integration. Preliminary interactions between implanted materials and biological environment are governed by the surface properties; they control the amount and quality of cell adhesion on the surface, consequently cell/tissue growth and regeneration. Thus, surface properties govern new bone tissue formation and implant osseointegration. The osteoconductivity of titaniuim-alloy implants used in dental and orthopaedic surgery was first improved by modifying their surfaces either chemically or physically. Later, further improvement in osteoconductivity was achieved by coating these implants with a layer of calcium phosphate. For many years, these inorganic coatings were prepared under highly unphysiological conditions, which precluded the incorporation of proteinaceous osteogenic agents. Hence, they could be rendered osteoinductive only after their formation, by the superficial adsorption of osteogenic growth factors upon their surfaces. With the advent of the biomimetic coating process, bone growth factors could be coprecipitated with the inorganic elements. A slow growth of nanostructured apatite film on the implant surface could be formed from the simulated body fluid, resulting in the formation of a strongly adherent, uniform nanoporous layer. The layer was found to possess a stable nanoporous structure and bioactivity. Nanophase coating could, therefore, promote osseointegration, a crucial development for the clinical success of orthopaedic/dental implants. In addition to improving the efficacy of orthopaedic/dental implants, these novel biomaterial formulations could be used to explore alternative strategies for bone-related tissue-engineering applications.

The biomimetic approach that appears to offer new promise is the use of biologically-inspired nanophase materials or composites comprised of nanoscale structures. Nanophase materials investigated so far include ceramics, metals, polymers, composites, and carbon nanofibers/nanotubes. The rapidly escalating costs of health care characterized by the shift from acute to chronic conditions, aging of the population, persistence of health disparities, and new public health challenges call for the acceleration of basic understanding of the complexity of biological systems and increase the need for more effective strategies for orthopaedic/dental implants. The hybrid coatings with nanophase biomaterials provide a new method to design implant surfaces for next generation prostheses. The specific role of nanophase modified materials will have future potential of biologically-inspired nanomaterials for various prosthetic applications.

References

1. http://www.aaos.org/wordhtml/press/hip_knee.htm.
2. http://www.azcentral.com/health/0617newhips17/html.
3. M. Parker and K. Gurusamy, *Cochrane Database Syst. Rev.* 2006 Jul. 19; 3:CD001706.
4. A.M. Minino and B.L. Smith, *Natl. Vital. Stat. Rep.* **49** (2001) 12.
5. F.C.M. Driessens, J.A. Planell and F.J. Gil, Marcel Dekker Inc., New York, 1995, p. 855.
6. J. Ma, H. Wong, L.B. Kong and K.W. Peng, *Nanotechnology* **14** (2003) 619–623.
7. A. de la Isla, W. Brostow, B. Bujard, M. Estevez, J.R. Rodriguez, S. Vagas and V.M. Castano, *Mat. Resr. Innovat.* **7** (2003) 110–114.
8. T.J. Webster, C. Ergun, R.H. Doremus, R.W. Siegel and R. Bizios, *Biomaterials* **21** (2000) 1803–1810.
9. X. Huang, W. Yan, D. Yang, J. Feng and Y. Feng, *Key. Eng. Mater.* **311** (2006) 953–956.
10. C. Du, F.Z. Cui, Q.L. Feng, X.D. Zhu and K. de Groot, *J. Biomed. Mater. Res.* **42** (1998) 540–548.
11. A.P. Praemer, S. Furner and D.P. Rice, American Academy of Orthopaedic Surgeons, Rosemont, IL, 1999, pp. 34–39.
12. G. Daculsi, *Biomaterials* **19** (1998) 1473–1478.
13. M. Vallet-Regí, *Dalton Trans.* **28** (2006) 5211–5220.
14. P. Ducheyne and Q. Qiu, *Biomaterials* **20**(23–24) (1999) 2287–2303.
15. D. Buser, J. Ruskin, F. Higginbottom, R. Hartdwick, C. Dahlin and R. Schenk, *Int. J. Oral. Maxillofacial Implants* **10** (1995) 666–681.
16. D. Williams (ed.), Pergamon Press and The MIT Press, 1990, pp. xvii–xx.
17. C.E. Cambell and A.F. Von Recum, *J. Invest. Surg.* **2** (1989) 51–74.
18. M.V. Thomas, D.A. Puleo and M. Al-Sabbagh, *J. Long. Term Eff. Med. Implants* **15**(6) (2005) 585–597.
19. H. Plenk Jr., *J. Biomed. Mater. Res.* **43**(4) (1998) 350–355.
20. C. Knabe, C.R. Howlett, F. Klar and H. Zreiqat, *J. Biomed. Mater. Res. A* **71**(1) (2004) 98–107.
21. J.L. Ong, D.L. Carnes and K. Bessho, *Biomaterials* **25**(19) (2004) 4601–4606.
22. F. Barrere, C.M. van der Valk, R.A. Dalmeijer, G. Meijer, C.A. van Blitterswijk and K. de Groot *et al.*, *J. Biomed. Mater. Res.* **66A**(4) (2003) 779–788.
23. W. David, Thomas, (ed.), London: Dordrecht (Kluwer Academic Publishers, 2004).
24. K.C. Dee and D.A. Puleo (eds.), Rena BiziosHoboken, N.J. Wiley-Liss, 2003.
25. K. de Groot, R. Geesink and C.P. Klein, *J. Biomed. Mater. Res.* **21** (1987) 375–381.
26. R. Geesink, K. de Groot, CPAT. Klein, *J. Bone Joint Surg. [Br]* **70-B** (1988) 17–22.

27. K. Soballe, E.S. Hansen, H. Brockstedt-Rasmussen *et al.*, *J. Arthroplasty* **6** (1991) 307–316.

28. R.G.T. Geesink and M.T. Manley (eds.) (New York: Raven Press, 1993).

29. X.F. Walboomers, H.J.E. Croes, L.A. Ginsel and J.A. Jansen. *Biomaterials* **19** (1998) 1861–1868.

30. S.L. Goodman, P.A. Sims and R.M. Albrecht, *Biomaterials* **17** (1996) 2087–2095.

31. M.M. Mulder, R.W. Hitchock and P.A. Tresco, *J. Biomat. Sci. Polymer Edn.* **9** (1998) 731–748.

32. J.E. Sanders, C.E. Stiles and C.L. Hayes, *J. Biomed. Mat. Res.* **52** (2000) 231–237.

33. J. Cappello, J.W. Crissman, M. Srissman *et al.*, *J. Control. Rel.* **53** (1998) 105–117.

34. T.W. King and C.W. Patrick Jr., *J. Biomed. Mat. Sci.* **51** (2000) 383–390.

35. T. Uehara, K. Takaoka and K. Ito, *Clin. Oral Implants Res.* Oct **15**(5) (2004) 540–545.

36. T.J. Griffin and W.S. Cheung, *J. Prosthet. Dent.* Aug **92**(2) (2004) 139–144.

37. P.I. Branemark, B.O. Hansson, R. Adell, U. Breine, J. Lindstrom, O. Hallen *et al.* Scand, *J. Plast. Reconstr. Surg. Suppl.* **16** (1977) 1–132.

38. T. Arektsson and L. Sennerby, *J. Periodontol.* **71**(6) (2000) 1054–1055.

39. I. Abrahamsson, T. Berglundh, J. Wennstrom and J. Lindhe, *Clin. Oral Implants Res.* **7** (1996) 212–219.

40. R. Adell, U. Lekholm, B. Rockler, P.I. Branemark, J. Lindhe, B. Eriksson *et al.*, *Int. J. Oral Maxillofac. Surg.* **15** (1986) 39–52.

41. D. Brocard, P. Barthet, E. Baysse, J.F. Duffort, P. Eller, P. Justumus *et al.*, *Int. J. Oral Maxillofac. Implants* **15** (2000) 691–700.

42. T. Berglundh and J. Lindhe, *Clin. Oral Implants Res.* **8** (1997) 117–124.

43. I. Abrahamsson, T. Berglundh and J. Lindhe, *J. Clin. Periodontol.* **24** (1997) 568–572.

44. S.D. Cook, K.A. Thomas, J.E. Dalton, T.K. Volkman, S. Thomas, I. Whitecloud and J.F. Kay, *J. Biomed. Mater. Res.* **26** (1992) 989–1001.

45. W.N. Capello, J.A. D'Antonio *et al.*, *Clin. Orthopaed. Relat. Res.* **355** (1998) 200.

46. R.J. Furlong and J.F. Osborn, *J. Bone Joint Surg.* **73B** (1991) 741.

47. D.C.R. Hardy, R.l. Frayssinet *et al.*, *J. Bone Joint Surg.* **73B** (1991) 732.

48. C.L. Tisdel, V.M. Goldberg *et al.*, *J. Bone Joint Surg.* **76A** (1994) 159.

49. R.G.T. Geesink, *Clin. Ortho. Relat. Res.* **261** (1990) 39.

50. R.G.T. Geesink, K. De Groot and C.T. Klein, *Clin. Ortho. Relat. Res.* **225** (1987) 147.

51. J.E. Lemons, Bioceramics, *Clin. Ortho. Relat. Res.* **261** (1990) 153.

52. W.R. Lacefield, Hydroxyapatite coatings, *Ann. NY Acad. Sci.* **523** (1988) 72.

53. C.A. Engh, *Clin. Orthopaed. Relat. Res.* **176** (1983) 52.

54. A. Piattelli, M. Degidi, M. Paolantonio, C. Mangano and A. Scarano, *Biomaterials* **24**(22) (2003) 4081–9.

55. F. Reno, F. Lombardi and M. Cannas, *Biomaterials* **24**(17) (2003) 2895–2900.

56. C.J. Sychterz, K.F. Orishimo and C.A. Engh, *J. Bone Joint Surg. Am.* **86-A**(5) (2004) 1017–1022.

57. R.H. Hopper Jr., C.A. Engh Jr., L.B. Fowlkes and C.A. Engh, *Clin. Orthop. Relat. Res.* Dec. **429** (2004) 54–62.

58. S.C. Cowin *et al.*, *Handbook of Bioengineering*, R. Skalak and S. Chien (eds.) (McGraw Hill, New York, 1987).

59. I. Malsch, *The Industrial Physicist* (2002).

60. I. Sivin, *Drug Saf.* **26**(5) (2003) 303–335.

61. C. Du, F.Z. Cui, X.D. Zhu and K. deGroot, Three-dimensional nano-HAp/collagen matrix loading with osteogenic cells in organ culture, *J. Biomed. Mat. Res.* **44** (1999) 407–415.

62. J. Liu, A.Y. Kim, L.Q. Wang *et al.*, Self-assembly in the synthesis of ceramic materials and composites, *Advan. Coll. Interface Sci.* **69** (1996) 131–180.

63. J. Voldman, M.L. Gray and M.A. Schmidt, *Ann. Rev. Biomed. Eng.* **1** (1999) 401–525.

64. J. Farhadi, C. Jaquiery, A. Barbero, M. Jakob, S. Schaeren, G. Pierer, M. Heberer and I. Martin, *Plast. Reconstr. Surg.* **116**(5) (2005) 1379–1386.

65. R.E. Schreiber, K. Blease, A. Ambrosio, E. Amburn, B. Sosnowski and T.K. Sampath, *J. Bone Joint Surg. Am.* **87**(5) (2005) 1059–1068.

66. J.D. Currey, *Bones: Structure and Mechanics* (Princeton University Press, Princeton, NJ, 2002).

67. C.M. Langton and C.F. Njeh (eds.), Bristol and Philadelphia: Institute of Physics Publishing, 2003.

68. J. Homminga, B.R. McCreadie, T.E. Ciarelli, H. Weinans, S.A. Goldstein and R. Huiskes, *Bone* **30** (2002) 759–764.

69. A. Rubinacci, I. Villa, F. Dondi Benelli, E. Borgo *et al.*, *Calcif. Tissue Int.* **63**(4) (1998) 331–339.

70. S.J. Brown, P. Pollintine, D.E. Powell, M.W. Davie and C.A. Sharp, *Calcif. Tissue Int.* **71** (2002) 227–234.

71. M. Marenzana, A.M. Shipley, P. Squitiero, J.G. Kunkel and A. Rubinacci, *Bone* **37**(4) (2005) 545–554.

72. R.A. Hipskind and G. Bilbe, *Front Biosci.* **3** (1998) 804–816.

73. B.S. Yoon and K.M. Lyons, *J. Cell. Biochem.* **93**(1) (2004) 93–103.

74. P.J. Marie, F. Debiais and E. Hay, *Histol. Histopathol.* **17**(3) (2002) 877–885.

75. N. Loveridge, J. Power, J. Reeve and A. Boyde, *Bone* **35**(4) (2004) 929–941.

76. Y. Xiao, H. Haase, W.G. Young and P.M. Bartold, *Cell Transplant.* **13**(1) (2004) 15–25.

77. J.J. Stepan, F. Alenfeld, G. Boivin, J.H. Feyen and P. Lakatos, *Endocr. Regul.* Dec. **37**(4) (2003) 225–238.

78. A.S. Blawas and W.M. Reichert, Protein patterning, *Biomaterials* **19** (1998) 595–609.

79. Y. Ito, Surface micropatterning to regulate cell functions, *Biomaterials* **20** (1999) 2333–2342.

80. Y. Ito, L.-S. Li, T. Takahashi *et al.*, *J. Biochem.* **121** (1997) 514–520.

81. T. Takezawa, M. Yamazaki, Y. Mori *et al.*, *J. Cell Sci.* **101** (1992) 495–501.

82. J.-M. Laval, J. Chopineau and D. Thomas, *Trends Biotechnol.* **13** (1995) 474–481.

83. W.J.A. Dhert, C.T. Klein *et al.*, *J. Biomed. Mater. Res.* **25** (1991) 1183.

84. R.D. Bloebaum, L. Zou *et al.*, Analysis of particles in acetabular components from patients with osteolysis, *Clin. Orthopaed. Relat. Res.* **338** (1997) 109.

85. R.D. Bloebaum, D. Beeks *et al.*, *Clin. Orthopaed. Relat. Res.* **298** (1994) 19.

86. A.K. Shah, J. Lazatin, R.K. Sinha, T. Lennox, N.J. Hickok and R.S. Tuan, *Biol. Cell.* Mar. **91**(2) (1999) 131–142.

87. A.T. Raiche and D.A. Puleo, *J Biomed. Mater. Res. A.* **69**(2) (2004) 342–350.

88. M. Lind, *Act. Orthop. Scand. Suppl.* **283** (1998) 2–37.

89. K. Soballe, E.S. Hansen, H. Brockstedt-Rasmussen, C.M. Pedersen and C. Bunger, *Acta. Orthop. Scand.* **61** (1990) 299–306.

90. S.D. Cook, K.A. Thomas, J.E. Dalton and J.F. Kay, *Semin. Arthroplasty* **2**(4) (1991) 268–279.

91. S.D. Cook, K.A. Thomas, J.E. Dalton, T.K. Volkman, S. Thomas, I. Whitecloud and J.F. Kay, *J. Biomed. Mater. Res.* **26** (1992) 989–1001.

92. H. Lin, H. Xu, X. Zhang and K. de Groot, *J. Biomed. Mater Res.* **43**(2) (1998) 113–122.

93. F.F. Buechel, R.A. Rosa and M.J. Pappas, *Clin. Orthop. Relat. Res.* **248** (1989) 34–49.

94. K. Soballe and S.J. Overgaard, *Bone Joint Surg.* **78B** (1987) 66–71; 689–691.

95. W.L. Jaffe and D.F. Scott, *J. Bone Joint Surg.* **78A** (1996) 1918–1935.

96. R.G.T. Geesink, K. de Groot and C.P.A.T. Klein, *J. Bone Joint Surg.* **70B** (1988) 17–22.

97. H.F. Morris, S. Ochi, J.R. Spray and J.W. Olson, *Ann. Periodontol.* **5**(1) (2000) 56–67.

98. S.D. Cook, K.A. Thomas, J.F. Kay and M. Jarcho, *Clin. Orthop.* **230** (1988) 303–312.

99. S. Joschek, B. Nies, R. Krotz and Goferich, *Biomaterials* **21** (2000) 1645–1658.

100. M. George, M. Mueller and J.A. Shepperd, in: J-A. Epinette and MT. Manley (eds.) (France: Springer-Verlag, 2004) pp. 211–216.

101. P.A. Butler Manuel, S.E. James and J.A. Shepperd, *Bone Joint Surg. [Br]* **74-B** (1992) 74–77.

102. A. Jokstad, U. Braegger, A. Carr *et al.*, *Int. Dent. J.* **53** (2003) 409–443.

103. F. Barrere, P. Layrolle, C.A. van Blitterswijk and K. de Groot, *Bone* **25** (Suppl.) (1999) 107S–111S.

104. M. Shirkhanzadeh and G.Q. Liu, *Mater. Lett.* **21** (1994) 115–118.

105. H.B. Wen, J.R. de Wijn, C.A. van Blitterswijk and K. de Groot, *J. Biomed. Mater. Res.* **46** (1999) 245–252.

106. Z. Schwartz and B.D. Boyan, *J. Cell. Biochem.* **56** (1994) 340–347.

107. M. Tanahashi, T. Yao, T. Kokubo, M. Minoda, T. Miyamoto, T. Nakamura and T. Yamamuro, *J. Am. Ceram. Soc.* **77** (1994) 2805–2808.

108. J.J. Hwang, K. Jaeger, J. Hancock and S.I. Stupp, Organoapatite growth on an orthopaedic alloy surface, *J. Biomed. Mater. Res.* **47** (1999) 504–515.

109. F. Barrere, C.M. van der Valk, G. Meijer, R.A. Dalmeijer, K. de Groot and P. Layrolle, *J. Biomed. Mater. Res.* **67B**(1) (2003) 655–665.

110. W.Q. Yan, T. Nakamura, K. Kawanabe, S. Nishigochi, M. Oka and T. Kokubo, *Biomaterials* **18**(17) (1997) 1185–1190.

111. P. Li, *J. Biomed. Mater. Res. A.* **66**(1) (2003) 79–85.

112. T. Kokubo, H.-M. Kim and M. Kawashita, *Biomaterials* **24** (2003) 2161–2175.

113. J. Ma, W. Huifen, L.B. Kong and K.W. Peng, *Nanotechnology* **14** (2003) 619.

114. N. Costa and P.M. Maquis, *Med. Eng. Phys.* **20** (1998) 602.

115. F.A. Muller, L. Jonasova, A. Helebrant, J. Strnad and P. Greil, *Biomaterials* **25** (2004) 1187.

116. X. Zhu, O. Eibl, L. Scheideler and J. Geis-Gerstorfer, *J. Biomed. Mater. Res.* **79A** (2006) 114–127.

117. S.S. Liao, F.Z. Cui, W. Zhang and Q.L. Feng, *J. Biomed. Mater. Res. B Appl.* **69**(2) (2004) 158–165.

118. J.L. Ong, D.L. Carnes and A. Sogal, *Int. J. Oral. Maxillofac. Impl.* **14** (1999) 217–225.

119. A. Krout, H. Wen, E. Hippensteel and P. Li, *J. Biomed. Mater. Res.* **73A** (2005) 377–387.

120. Y. Liu, K. de Groot and E.B. Hunziker, *Ann. Biomed. Eng.* **32**(3) (2004) 398–406.

121. L. Jonasova, F.A. Muller, A. Helebrant, J. Strnad and P. Greil, *Biomaterials.* **25**(7–8) (2004) 1187–1194.

122. A. Chun, J. Moralez, H. Fenniri and T.J. Webster, *Nanotechnology* **15** (2004) S234.

123. T.J. Webster and J.U. Ejiofor, *Biomaterials* **25** (2004) 4731.

INDEX